Countercurrent Chromatography

CHROMATOGRAPHIC SCIENCE

A Series of Monographs

Editor: JACK CAZES
Sanki Laboratories, Inc.
Sharon Hill, Pennsylvania

Volume 1: Dynamics of Chromatography
J. Calvin Giddings

Volume 2: Gas Chromatographic Analysis of Drugs and Pesticides
Benjamin J. Gudzinowicz

Volume 3: Principles of Adsorption Chromatography: The Separation of Nonionic Organic Compounds (out of print)
Lloyd R. Snyder

Volume 4: Multicomponent Chromatography: Theory of Interference (out of print)
Friedrich Helfferich and Gerhard Klein

Volume 5: Quantitative Analysis by Gas Chromatography
Joseph Novák

Volume 6: High-Speed Liquid Chromatography
Peter M. Rajcsanyi and Elisabeth Rajcsanyi

Volume 7: Fundamentals of Integrated GC-MS (in three parts)
Benjamin J. Gudzinowicz, Michael J. Gudzinowicz, and Horace F. Martin

Volume 8: Liquid Chromatography of Polymers and Related Materials
Jack Cazes

Volume 9: GLC and HPLC Determination of Therapeutic Agents (in three parts)
Part 1 edited by Kiyoshi Tsuji and Walter Morozowich
Parts 2 and 3 edited by Kiyoshi Tsuji

Volume 10: Biological/Biomedical Applications of Liquid Chromatography
Edited by Gerald L. Hawk

Volume 11: Chromatography in Petroleum Analysis
Edited by Klaus H. Altgelt and T. H. Gouw

Volume 12: Biological/Biomedical Applications of Liquid Chromatography II
Edited by Gerald L. Hawk

Volume 13: Liquid Chromatography of Polymers and Related Materials II
Edited by Jack Cazes and Xavier Delamare

Volume 14: Introduction to Analytical Gas Chromatography: History, Principles, and Practice
John A. Perry

Volume 15: Applications of Glass Capillary Gas Chromatography
Edited by Walter G. Jennings

Volume 16: Steroid Analysis by HPLC: Recent Applications
Edited by Marie P. Kautsky

Volume 17: Thin-Layer Chromatography: Techniques and Applications
Bernard Fried and Joseph Sherma

Volume 18: Biological/Biomedical Applications of Liquid Chromatography III
Edited by Gerald L. Hawk

Volume 19: Liquid Chromatography of Polymers and Related Materials III
Edited by Jack Cazes

Volume 20: Biological/Biomedical Applications of Liquid Chromatography IV
Edited by Gerald L. Hawk

Volume 21: Chromatographic Separation and Extraction with Foamed Plastics and Rubbers
G. J. Moody and J. D. R. Thomas

Volume 22: Analytical Pyrolysis. A Comprehensive Guide
William J. Irwin

Volume 23: Liquid Chromatography Detectors
Edited by Thomas M. Vickrey

Volume 24: High-Performance Liquid Chromatography in Forensic Chemistry
Edited by Ira S. Lurie and John D. Wittwer, Jr.

Volume 25: Steric Exclusion Liquid Chromatography of Polymers
Edited by Josef Janča

Volume 26: HPLC Analysis of Biological Compounds: A Laboratory Guide
William S. Hancock and James T. Sparrow

Volume 27: Affinity Chromatography: Template Chromatography of Nucleic Acids and Proteins
Herbert Schott

Volume 28: HPLC in Nucleic Acid Research: Methods and Applications
Edited by Phyllis R. Brown

Volume 29: Pyrolysis and GC in Polymer Analysis
Edited by S. A. Liebman and E. J. Levy

Volume 30: Modern Chromatographic Analysis of the Vitamins
Edited by Andre P. De Leenheer, Willy E. Lambert, and Marcel G. M. De Ruyter

Volume 31: Ion-Pair Chromatography
Edited by Milton T. W. Hearn

Volume 32: Therapeutic Drug Monitoring and Toxicology by Liquid Chromatography
Edited by Steven H. Y. Wong

Volume 33: Affinity Chromotography: Practical and Theoretical Aspects
Peter Mohr and Klaus Pommerening

Volume 34: Reaction Detection in Liquid Chromatography
Edited by Ira S. Krull

Volume 35: Thin-Layer Chromatography: Techniques and Applications, Second Edition, Revised and Expanded
Bernard Fried and Joseph Sherma

Volume 36: Quantitative Thin-Layer Chromatography and Its Industrial Applications
Edited by Laszlo R. Treiber

Volume 37: Ion Chromatography
Edited by James G. Tarter

Volume 38: Chromatographic Theory and Basic Principles
Edited by Jan Åke Jönsson

Volume 39: Field-Flow Fractionation: Analysis of Macromolecules and Particles
Josef Janča

Volume 40: Chromatographic Chiral Separations
Edited by Morris Zief and Laura J. Crane

Volume 41: Quantitative Analysis by Gas Chromatography, Second Edition, Revised and Expanded
Josef Novak

Volume 42: Flow Perturbation Gas Chromatography
N. A. Katsanos

Volume 43: Ion Exchange Chromatography of Proteins
Shuichi Yamamoto, Kazuhiro Nakanishi, and Ryuichi Matsuno

Volume 44: Countercurrent Chromatography: Theory and Practice
Edited by N. Bhushan Mandava and Yoichiro Ito

Countercurrent Chromatography

THEORY AND PRACTICE

edited by

N. Bhushan Mandava
TODHUNTER, MANDAVA, & ASSOCIATES
WASHINGTON, D.C.

Yoichiro Ito
NATIONAL HEART, LUNG, AND BLOOD INSTITUTE
NATIONAL INSTITUTES OF HEALTH
BETHESDA, MARYLAND

MARCEL DEKKER, INC. **New York and Basel**

Library of Congress Cataloging-in-Publication Data

Countercurrent chromatography : theory and practice / edited by
N. Bhushan Mandava, Yoichiro Ito.
p. cm.
Includes index.
ISBN 0-8247-7815-4
1. Countercurrent chromatography. I. Mandava, N. Bhushan
II. Ito, Yoichiro
QP519.9.C68C68 1988
543'.0894—dc19

MARCEL DEKKER, INC.
270 Madison Avenue, New York, New York 10016

Current printing (last digit):
10 9 8 7 6 5 4 3 2 1

PRINTED IN THE UNITED STATES OF AMERICA

Preface

Historically, the separation of similar molecules by their differential solubility between two immiscible phases has provided the basis for several highly selective separatory techniques, such as paper chromatography, countercurrent distribution (CCD), and liquid-liquid chromatography. The great selectivity of these techniques arises from the fact that the partition coefficient is the reflection of the intermolecular forces which can be strongly modified by conditions of temperature and phase composition. Thus, by a judicious choice of conditions, highly specific interactions can often by exploited that permit the resolution of selected components from very complex mixtures.

In separation science, several chromatographic methods, such as paper chromatography, high-resolution column chromatography, and high-performance liquid chromatography (HPLC), were developed on the basis of liquid-liquid partition principles, but these methods require a solid support that sometimes presents such problems as loss of samples by adsorption and chemical degradation of compounds. To date, more limited recognition has been attained by alternative separation systems, also based on solvent-solvent partitioning but not requiring the use of a solid support matrix. These include the classic separatory funnel, cascade extraction train, partition chromatography, Craig's CCD apparatus, cascade percolator train, and Ronor columns, to name a few older partition methods. Except for partition chromatography and CCD, they did not gain wide acceptance in laboratory scale separations despite the fact that some did show very good resolving power. The fragility and inconvenience of the apparatus, and its unadaptability to separations of varying scale, may account for some of these failures.

The technology of countercurrent chromatography (CCC) evolved from keen observation that two immiscible liquids flowing countercurrently in a helical tube rotating in an acceleration field become uniformly segmented in the coils of the helix. Separation of both soluble and particulate samples was demonstrated. Although tedious and limited to microgram samples, the value of the technique was immediately acknowledged by the commercial development of an instrument called the coil planet centrifuge by Sanki Engineering, Ltd. (Japan).

In the late 1960s, a systematic program to design and evaluate an extensive series of separation devices was initiated. With the exception of droplet countercurrent chromatography and locular countercurrent chromatography, the instruments are based on a helical coil which is acted on by either a gravitational or centrifugally induced acceleration field. Aside from their mechanical features, the devices may be classified on the basis of the cyclic variation in the direction and intensity of this field relative to the coil. The more recently designed chromatographic devices incorporate an ingenious flow-through system which permits high-speed rotation of the coil without kinking of the flexible influent and effluent lines, thereby avoiding the need for an often troublesome rotating fluid seal. With the introduction of a new family of high-performance countercurrent chromatographic devices, this field of separation science is now comparable in many respects to other chromatographic systems, especially to HPLC, in speed, resolution, efficiency, and sample size as well as in separation capabilities.

The process of countercurrent chromatography is essentially liquid-liquid chromatography in which the stationary liquid bed is retained in the column by an acceleration field rather than by a solid supporting matrix. Adsorption effects are thereby eliminated. The technique is particularly advantageous in the preparative (mg to g) range for polar and labile organic compounds and bioparticulate materials such as cells and cell fragments. Virtually any two-phase system, either aqueous or nonaqueous, may be employed, and some CCC instruments are particularly useful for the two-phase aqueous polymer systems.

This book consists of essentially two parts. An introductory chapter covers the basic chromatographic principles, which are generally applicable to any chromatographic science. Part I, dealing with the theory and instrumentation, is divided into three chapters on historical developments that have led to what is now known as countercurrent chromatography, instrumentation that describes the state of the art in the design of various CCC devices, and theory of countercurrent chromatography. Part II deals with applications of countercurrent chromatography. This section discusses the application of droplet countercurrent chromatography, rotation locular countercurrent chromatography, and centrifugal countercurrent chromatography to natural products isolation. Other applications of countercurrent chromatography include agrochemicals, drugs, peptides, and biopolymers, each of which is described in a separate chapter. Chapter 12, at the end of the book, was introduced to show how promising CCC is for the future with tandem methods such as CCC/MS as powerful techniques of separation science. A glossary of terms commonly used in chromatography and a list of CCC instruments now commercially available are included in Part III.

The contributors, who are specialists in their respective fields, deserve special credit for illustrating the power of CCC in their specialties. We express our sincere thanks to all contributors for their ready cooperation and for presenting their work very lucidly and candidly. Finally, we thank our spouses for their patience and understanding while we completed the tasks of writing and editing this book. The publisher's support and the series editor's enthusiasm for this book are greatly appreciated.

N. Bhushan Mandava
Yoichiro Ito

Contents

Contributors

Reimar C. Bruening* Department of Chemistry, Columbia University, New York, New York

Walter D. Conway Department of Pharmaceutics and Medicinal Chemistry, School of Pharmacy, State University of New York at Buffalo, Amherst, New York

Fadila Derguini Department of Chemistry, Columbia University, New York, New York

Frederick J. Hanke Division of Entomology and Parasitology, College of Natural Resources, University of California at Berkeley, Berkeley, California

Kurt Hostettmann Institute of Pharmacognosy and Phytochemistry, School of Pharmacy, University of Lausanne, Lausanne, Switzerland

Yoichiro Ito Laboratory of Technical Development, National Heart, Lung, and Blood Institute, National Institutes of Health, Bethesda, Maryland

Current affiliation: Department of Chemistry, University of Hawaii at Manoa, Honolulu, Hawaii

Martha Knight Peptide Technologies Corporation, Trinity College, Washington, D.C.

Isao Kubo Division of Entomology and Parasitology, College of Natural Resources, University of California at Berkeley, Berkeley, California

N. Bhushan Mandava* Office of Pesticides and Toxic Substances, U.S. Environmental Protection Agency, Washington, D.C.

Gary T. Marshall Division of Entomology and Parasitology, College of Natural Resources, University of California at Berkeley, Berkeley, California

Andrew Marston Institute of Pharmacognosy and Phytochemistry, School of Pharmacy, University of Lausanne, Lausanne, Switzerland

David G. Martin Cancer and Viral Diseases Research, The Upjohn Company, Kalamazoo, Michigan

Koji Nakanishi Department of Chemistry, Columbia University, New York, New York

John M. Ruth Pesticide Degredation Laboratory, Agricultural Envirnomental Quality Institute, Beltsville Agricultural Research Center, ARS, USDA, Beltsville, Maryland

Ian A. Sutherland Engineering Department, National Institute for Medical Research, Mill Hill, London, England

Current affiliation: Senior Partner, Todhunter, Mandava, & Associates, Washington, D.C.

chapter one

Introduction to Chromatography

N. Bhushan Mandava*
Office of Pesticides and Toxic Substances
U.S. Environmental Protection Agency
Washington, D.C.

John M. Ruth
Agricultural Environmental Quality Institute
Beltsville Agricultural Research Center, ARS, USDA
Beltsville, Maryland

In preparation for a description of countercurrent chromatography (CCC), it is convenient to summarize briefly the general principles which underlie all forms of chromatography. The theoretical foundations of countercurrent chromatography resemble most closely those of liquid chromatography and countercurrent distribution, and for that reason the theory of elution chromatography will receive attention in preference to other techniques. Frontal and displacement chromatography, treated in terms of solute fronts, are less applicable to countercurrent chromatography and will be mentioned only briefly.

The function of the theoretical treatment is to describe the behavior of solutes, which in elution chromatography involves solute zone profiles and positions. As discussed by Morris and Morris, this requires two kinds of information: (a) the properties of the solutes and (b) the properties of the chromatographic system. Although the specific properties of solutes are of less interest in the chromatography of macromolecules, polar molecules, and ionic molecules than the properties of the chromatographic systems, the solute properties are important in explaining the behavior of small molecules in nonpolar solvent systems [28. Topics covered include:

1. General chromatographic principles, theory, and fundamental differences in chromatographic systems as they relate to chromatographic retention and ideal linear chromatography.

**Current affiliation*: Todhunter, Mandava, & Associates, Washington, D.C.

2. Liquid-liquid partition chromatography. This part will be treated in more detail because the same phenomena are found in coutercurrent chromatography. Included are solvent systems, support materials, some description of the operation of chromatography, and a few examples of applications.
3. Countercurrent distribution. The basic principles and processes will be discussed, but the details of application will not be included.

The treatment of other chromatographic systems—liquid-solid chromatography based on adsorption principles, gel permeation (size exclusion) chromatography, ion-exchange chromatography, and affinity chromatography—will be avoided because they have little application in countercurrent chromatography.

I. HISTORICAL DEVELOPMENTS

The origins of the chromatography, discussed by Novak and Janak [29], go back to Runge's experiments [1] on capillary analysis. Just before the turn of the century, Davy [2] observed changes in the composition of crude petroleum when it came in contact with rocks displaying adsorptive activity. These reports can be considered as part of the development of chromatography. It was only about 80 years ago that Tswett [3] devised methods for the separation of petroleum extracts on columns of inorganic adsorbents and coined the word "chromatography." He demonstrated the separation of the petroleum ether extract of the coloring substances of leaves into a series of green and yellow bands. The apparatus was a column of calcium carbonate in a glass tube, through which the light petroleum solvent was made to percolate continuously. The importance of Tswett's work was not recognized until the beginning of the 1930s, when Kuhn et al. [4] applied chromatographic principles to the separation of natural substances.

Another milestone in the development of chromatography was reached in 1941 when Martin and Synge [5] reported their discovery of liquid-liquid partition chromatography. In their work one liquid was used as a sorbent and another liquid was allowed to percolate through the former, thus making the technique a chromatographic process. Their work established the foundations for chromatography and set a precedent for the development of other forms of chromatography and other separation techniques, including:

1. Paper chromatography
2. Countercurrent chromatography
3. Ion-exchange chromatography
4. Electrophoresis

5. Thin-layer chromatography
6. Gel permeation chromatography
7. Gas-liquid chromatography
8. High-performance liquid chromatography

It is said that when one of the great discoveries in chromatographic science was submitted for publication, the journal of a renowned chemical society rejected it. The importance of the work was eventually emphasized by a Nobel Prize. Such instances are not uncommon in the history of scientific society publications. The paper on brassinolide [25] was rejected by the journals of two learned societies. After it was published in *Nature* and its impact was realized, the journals of those societies carried news stories.

Stahl [6] notes the comment of a well-known critic: "Only amateurs consider that an idea must be brand new in order to be good; important in reality is not he who first had the idea but he who expressed it better." This comment applies to the work of Twsett, who not only coined the term "chromatography" but also defined his method as a general means of separating substances, including colorless substances.

It was in 1906 that Goppelsroeder [7] published his book entitled *Suggestions for the Study of Capillary Analysis, Based on Capillarity and Adsorption Phenomena.* Subsequently (38 years later) Consden et al. [8] developed a practical technique which is known as paper chromatography. The chronological developments in separation techniques after Tswett's discovery of chromatography, based first on adsorption principles and then on partition phenomena, are shown in Table 1.

II. CHROMATOGRAPHIC PRINCIPLES

Martin [24] defines chromatography as "the uniform percolation of a fluid through a column of more or less finely divided substance, which selectively retards, by whatever means, certain components of the fluid." As Morris and Morris noted, this definition includes mixtures of gases and liquids, and avoids any description of the mechanisms responsible for selective retardation. The term "column" refers to various experimental arrangements, from the most commonly used cylindrical tube to the thin sheets of porous material used in paper chromatography [28]. The words "finely divided substance" denote a packing material responsible for the separation process in adsorption chromatography or refer to a support for another active phase in partition chromatography. As Morris and Morris point out, the features listed are common to chromatographic separations: (a) they are cascade processes and (b) separation results from the differential migration of the components of the mixture through the two-phase system.

Table 1 Chronological Developments in Separation Techniques

Separation system	Authors	Year	Ref.
Partition chromatography	Martin and Synge	1941	[5]
Paper chromatography (PC)	Consden, Gordon, and Martin	1944	[8]
Countercurrent distribution (CCD)	Craig	1944	[9]
Ion-exchange chromatography (IEC)	Mayer and Thompkins	1947	[10]
	Samuelson	1963	[11]
Electrophoresis	Haugaard and Kroner	1948	[12]
Thin-layer chromatography (TLC)	Ismailov and Shraiber	1938	[13]
	Kirchner, Miller, and Keller	1951	[14]
Gel permeation chromatography (GPC)	Barrer	1945	[15]
	Porath and Flodin	1959	[16]
Gas chromatography (GC)	Claesson	1946	[17]
	James and Martin	1952	[18]
High-performance liquid chromatography (HPLC)			[19]

In words of Novák and Janák [29], "Chromatography can be characterized as a separation method based on the differential migration of solutes through a system of two phases, one of which is mobile." Chromatographic systems can therefore be classified [29] according to:

1. The state of aggregation of the phases
2. The physical arrangement of the phases, and
3. The mechanism underlying the distribution equilibrium

The four possible systems derived from solid, liquid, and gaseous phases are liquid-liquid, liquid-solid, gas-liquid, and gas-solid chromatography. Of the two systems, liquid-liquid and liquid-solid, that may be called "liquid chromatography," it is liquid-liquid chromatography whose principles are directly applicable to countercurrent chromatography. The physical arrangements of the phases in chromatography can be classified as columnar or planar systems. Depending on the geometry of the chromatographic column support, the columnar systems can be further divided into such types as packed columns, capillary columns, and so forth (Table 2).

In liquid-liquid systems of all types, they [29] note that an important feature is the way in which the stationary liquid phase is formed. There are various ways of achieving a stationary liquid phase. As shown in the following sections, it is produced in countercurrent chromatography by surface tension or by centrifugal force, as in droplet countercurrent chromatography (DCCC) or in rotation locular countercurrent chromatography (RLCCC), respectively.

Chromatographic systems may be divided into two categories, two-dimensional and three-dimensional, in which the solute-solvent interactions take place on the surface or in the interior of the phase. These two geometries characterize "adsorption chromatography" and "partition chromatography," respectively [29].

A chromatographic system is characterized by a distribution isotherm, which is a curve showing the equilibrium relationship of concentration in the stationary phase to concentration in the mobile phase, at a constant temperature. When the distribution constant is independent of the solute concentration, the distribution isotherm (see Section IV.D) is linear. The term "linear chromatography" is applied to this case. When the distribution constant varies with solute concentration, a curved (nonlinear) isotherm is observed, giving "nonlinear chromatography."

III. CHROMATOGRAPHIC TECHNIQUES

The chromatographic system allows us to perform the following three types of operation:

1. Elution chromatography
2. Frontal chromatography, and
3. Displacement chromatography (with a solid adsorbent)

Table 2 Examples of the Geometrical Types of Chromatographic Systems

Columnar	Planar
Packed columns	Paper chromatography
Capillary columns	Thin-layer chromatography
Liquid-liquid columns	

In elution technique, the mixture of solutes is introduced at the column inlet onto the stationary phase (solid support in liquid chromatography or stationary liquid phase in CCC). The solute bands begin to migrate as the mobile phase starts passing through the stationary phase, and begin to separate, depending on the migration velocities of the individual components. The mobile phase, being essentially nonsorbed on the stationary phase, does not interfere with that mechanism. The migration velocities are inversely proportional to the corresponding distribution coefficients. In the chromatogram (Fig. 1a), the distances of the peak maxima from the

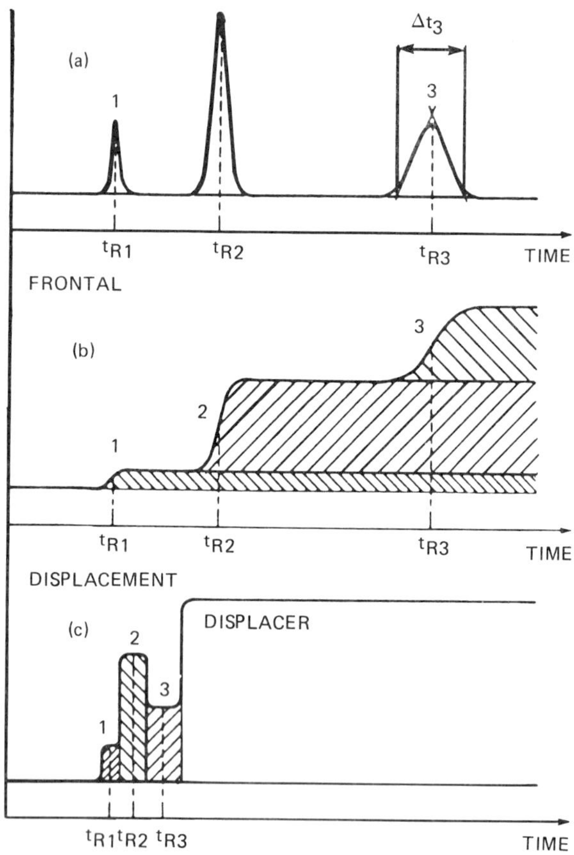

Fig. 1 Chromatographic records: (a) elution chromatography; (b) frontal chromatography; (c) displacement chromatography. R = response. (From J. Novák and J. Janák, in *Liquid Column Chromatography*, (Z. Deyl, K. Macek, and J. Janák, eds.), Elsevier, Amsterdam, 1975, pp. 3–10.)

starting point depend on the solute properties, and the peak areas are proportional to the total amounts of the components.

The status of frontal chromatography (Fig. 1b) appears to be well defined by Morris and Morris [28]: "Although this method of chromatographic analysis is not strictly a practical separation method, the simple theoretical treatment forms a useful introduction to the more complex methods of operation." The mobile phase is the sample mixture itself, flowing continuously. The column is eventually saturated with each solute. The first solute to reach saturation in the column appears pure in the effluent. Other solutes, successively achieving saturation of the column, then mix in the output, which finally returns to the input mixture composition.

In displacement technique (Fig. 1c), the mobile phase (in contrast to that in elution chromatography) is adsorbed more strongly than any of the solutes in the sample mixture, so that each solute follows those less strongly adsorbed. Each emerging zone is internally pure but touches others at its boundaries. In elution chromatography each zone not only contains, in the ideal case, a single component but also is separated by distance along the column axis and by time of elution from its neighboring zones.

In elution and displacement methods, components may be recovered either by elution from the column or by stopping the run before zones reach the exit and extracting portions of the column packing. Displacement and frontal methods have a place in the historical development, but modern practice in liquid chromatography is essentially restricted to the elution techniques of open columns, high performance liquid chromatography (HPLC), and CCC.

IV. DEFINITIONS AND EQUATIONS

A. Capacity Ratio D_C and Distribution Constant K_C

During the progress of the chromatographic elution process, the effluent composition leaving the column outlet is described mathematically as either a differential (Fig. 2a) or an integral (Fig. 2b) form. In this chapter and the following one, the subscript C will be used to mark quantities defined for liquid-liquid partition chromatography. The subscript D will indicate those defined for countercurrent distribution. The reason for that is that the distribution coefficient is defined differently in liquid-liquid partition chromatography and in countercurrent distribution. In chromatography, it is usually expressed as the concentration in the stationary phase divided by the concentration in the mobile phase. In countercurrent distribution, it is usually defined as the concentration in the upper phase (usually, but not always, mobile) divided by the concentration in the lower phase, so that the two are most often, but not always, reciprocals of each other.

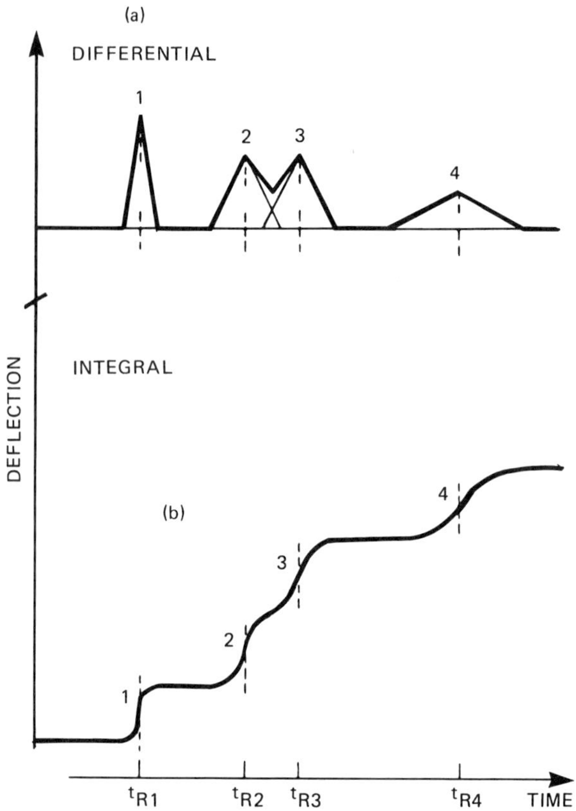

Fig. 2 Schematic representation of a differential (elution) chromatogram (a) and the corresponding integral chromatogram (b). (From J. Novák and J. Janák, in *Liquid Column Chromatography*, (Z. Deyl, K. Macek, and J. Janák, eds.), Elsevier, Amsterdam, 1975, pp. 3–10.)

Let us define t_R to be the total time taken by a solute molecule to pass through the column (retention time) and t_m and t_s to be the times spent by the molecule in the mobile and stationary phases, respectively [28]. Then

$$t_R = t_m + t_s \tag{1}$$

The ratio of the net retention time ($t_R - t_m$) of a solute component to the time t_m spent in the mobile phase is usually called the "capacity ratio," which is denoted by D_C.

$$D_C = \frac{t_R - t_m}{t_m} \tag{2}$$

The quantity t_R is the total retention time of a component of the mixture chromatographed. To express the same thing in a different way, the value of D_C is the ratio of the total amount of the component in the stationary phase to the total amount in the mobile phase of the chromatographic system. That is,

$$D_C = \frac{\text{mass } M_s \text{ of solute in the stationary phase}}{\text{mass } M_m \text{ of solute in the mobile phase}} \tag{3}$$

By Eq. (3), which defined it, $D_C = M_s/M_m$. Any particular solute molecule, in traversing the column, contributes to M_s and M_m in proportion to the time t_s and t_m, respectively, that it spends in the stationary and mobile phases, and all molecules behave alike in that respect. Therefore,

$$D_C = \frac{t_s}{t_m} \tag{4}$$

Solving Eqs. (1) and (4) for t_m and t_s, we obtain

$$t_m = \frac{t_R}{1 + D_C} \tag{5}$$

$$t_s = t_R \frac{D_C}{1 + D_C} \tag{6}$$

and

$$t_R = t_m(1 + D_C) \tag{7}$$

Equation (7) is convenient for gas chromatography. It can be converted into the equivalent volumetric form for liquid chromatography by means of the identities [28]

$$V_R = Ft_R \quad \text{and} \quad V_m = Ft_m \tag{8}$$

in which V_R is the elution volume, V_m is the total mobile phase volume or column void volume, and F is the flow rate, to give

$$V_R = V_m(1 + D_C) \tag{9}$$

Solving for D_C,

$$D_C = \frac{V_R - V_m}{V_m} \tag{10}$$

The distribution coefficient K_C is defined to be

$$K_C = \frac{C_s}{C_m} \tag{11}$$

in which C_s and C_m are the equilibrium concentrations of the solute component in the stationary and mobile phases, respectively. Therefore, the value of K_C is dimensionless. Then it is seen from Eqs. (3) and (11) that

$$D_C = K_C \frac{V_s}{V_m} = K_C \frac{\phi_s}{\phi_m} \tag{12}$$

where K_C is the distribution coefficient; V_s and V_m are the volumes of the stationary and mobile phases, respectively, in the column; and ϕ_s and ϕ_m are the cross-sections of the respective phases. Morris and Morris call Eqs. (7) and (9) the most important ones in the theory of chromatography. They can be rewritten in various forms, suited to the needs or conventional notations of different types of chromtatography. For liquid-liquid partition chromatography, our main interest in this book, Eq. (9) may be combined with Eq. (12), $D_C V_m = K_C V_s$, to give

$$V_R = V_m + K_C V_s \tag{13}$$

In a similar way, corresponding relationships may be derived for adsorption or ion-exchange chromatography. For column chromatography, Eq. (13) is perhaps the most useful form of the basic equation because [28]:

1. It relates the retention volume V_R and the void volume (mobile phase volume in the column) V_m directly to the volume V_s of the stationary phase through a dimensionless parameter K_C, a thermodynamic property of the solute in the two-phase system, and
2. The void volume V_m can always be measured directly in any chromatographic system by means of a marker solute for which K_C is equal to 0 in the system. Measurement of V_s is usually difficult, although methods are now available.

If R_C, called the relative rate parameter or retention coefficient, is defined by the equation $R_C = V_m/V_R$, and the value of V_R from Eq. (9) is substituted into the definition, the basic equation can be expressed in another useful form, in terms of the relative rates of movement of solute and solvent molecules through the chromatographic system:

$$R_C = \frac{V_m}{V_R} = \frac{V_m}{V_m(1 + D_C)} \tag{14}$$

$$R_C = \frac{1}{1 + D_C} \tag{15}$$

Eq. (15) is used in paper chromatography and thin-layer chromatography. Rearranging Eq. (15) gives

$$D_C = \frac{1 - R_C}{R_C} \tag{16}$$

In Eq. (16), $(1 - R_C)$ is the fraction of the solute in the stationary phase and R_C is the fraction in the mobile phase.

With Eqs. (1)–(16), we can deal with any type of chromatography. For liquid chromatography, Eq. (13) is probably the most directly applicable in its theoretical treatment. The derivation of Eq. (9) involved the following assumptions [28]:

1. Constancy of the mass distribution ratio (D_C) throughout the column, which in turn requires assuming both a linear distribution isotherm and a constant phase ratio throughout the column
2. Establishment of instantaneous equilibrium between the phases
3. Negligible longitudinal transport of solutes by diffusion in both phases, and
4. Identical behavior of all the molecules in the solute zone

Chromatography performed under these conditions can be referred to as "ideal chromatography" or as "ideal linear chromatography."

B. Adsorption Coefficeint

Let M_S be the total mass of solute retained in all of the stationary phase as above and let M_{SP} be the total mass of the stationary phase itself. Then, if M_C is defined as the mass of solute per unit mass of the stationary phase,

$$M_c = \frac{M_s}{M_{sp}} \tag{17}$$

Let $C_m = M_m/V_m$ be the concentration of solute in the mobile phase as defined above. The adsorption coefficeint a may be defined as the ratio of M_c to C_m.

$$a = \frac{M_c}{C_m} \tag{18}$$

The constant a has the dimensions of a reciprocal concentration.

To obtain a dimensionless constant, Snyder adopted "the concept of a discrete surface volume V_a composed of the molecules of the adsorbed solvent and solute on the surface of the adsorbent [28]." The volume V_a can be expressed as ml/g of adsorbent, and an adsorption coefficient K_A, analogous to the partition coefficeint K_C, can then be defined such that

$$a = V_a K_A \tag{19}$$

C. Retardation Factor and R_F

Equation (6) can be rearranged to read as follows [5,19]:

$$\frac{V_m}{V_R} = \frac{V_m}{V_m + K_C V_s} = R_C \tag{20}$$

where R_C is the so-called retardation factor. Theoretically R_C is identical to the quantity R_F employed in planar techniques [20,29].

$$R_F = \frac{\text{distance of the solute zone from the origin}}{\text{distance of the solvent front from the origin}} \tag{21}$$

For all practical purposes, Eq. (22) holds true [21]:

$$R_C = \frac{R_F}{f} \tag{22}$$

where f is a factor whose value ranges from 0.8 to 0.9.

D. Distribution Isotherms

The distribution of solute between two liquid phases or between adsorbent and liquid phase will, in general, depend on the equilibrium concentrations in the liquid or adsorbent phases involved. That distribution was described above by the coefficient defined in each case as a ratio of concentrations existing in the two phases at a fixed temperature. It must be kept in mind that the coefficient does not necessarily remain fixed unless the concentrations also remain constant. That is, the coefficient is at least potentially a function of three variables, the temperature and the two phase concentrations.

The relationship between the solute concentrations in two liquid phases, at a constant temperature, is known as the *partition isotherm*. That between the amount of solute adsorbed per unit weight of adsorbent and the corresponding equilibrium concentration of the solute in the liquid phase at a constant temperature is called the *adsorption isotherm*. The course of the operation in any chromatographic system is mainly determined by the nature of the isotherm [28].

The isotherms most frequently found in chromatography are of three types (Fig. 3). Curve A is a typical linear isotherm in which the distribution ratio is independent of the concentrations. Curve B indicates another type which is frequently encountered in chromatography. In this case, the distribution ratio decreases with increasing concentration, and therefore a smaller fraction of the total amount of solute present is held in the stationary phase at higher concentrations. Isotherms of type B are common in adsorption chromatography. Curve C, in which the distribution ratio increases with concentration (a larger percentage of the solute is in the stationary phase), is less frequently encountered. A fourth type of distribution isotherm occasionally found (not shown in Fig. 1) is sigmoid in shape [28].

In the great majority of cases, the distribution isotherm has the form of either curve A or curve B. At sufficiently low concentra-

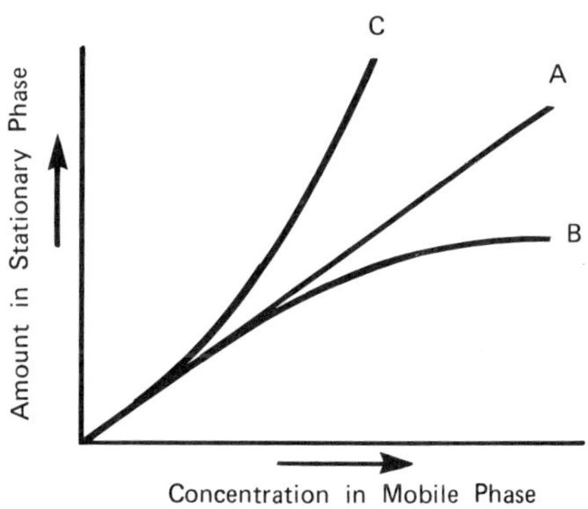

Fig. 3 Distribution isotherms. (From C. J. O. R. Morris and P. Morris, *Separation Methods in Biochemistry*, 2nd. ed., Halsted Press, John Wiley, New York, 1976.)

tions nonlinear isotherms approach a linear form. It is desirable to work in the concentration range in which the isotherm is linear [28].

In the absence of association and dissociation phenomena, partition isotherms are usually linear over a wider concentration range. Because of the constancy of the partitition coefficient over a wide range of concentrations, which is the basis for the Nernst partition law, this property is ideal for zonal separation. In situations where the linearity is affected by intermolecular associations or by electrolytic dissociations, which are temperature-dependent, the range of linearity may be extended by working at elevated temperatures. It should be noted, however, that a nonlinear isotherm is not always disadvantageous [28].

E. Resolution

The extent of separation of two solutes by the chromatographic process is described numerically by an empirical measure, R_s, called resolution, which can be calculated by means of Eq. (23),

$$R_s = \frac{2(t_{R_2} - t_{R_1})}{(\Delta t_2 + \Delta t_1)} \tag{23}$$

in which t_{R_2} and t_{R_1} are the retention times of the two components, and Δt_2 and Δt_1 are the time-widths of the respective zones [29].

F. Theoretical Plates

In 1941 Martin and Synge [5] adapted from distillation theory the concept of the equivalent theoretical plate as a measure of column performance. According to the theoretical plate model, the column can be divided into hypothetical layers, all equal in thickness, which are perpendicular to its geometrical axis. The thickness (length in the axial direction) of a plate is such that the concentrations in the two phases are at equilibrium within the plate. Each plate is analogous to a simple separatory funnel, and the combined efficiency of the series of plates is that of an equal number of such simple operations repeated. The axial length (thickness) of this column layer is called the height equivalent to a theoretical plate (HETP or H). To summarize the details of the model:

1. The column is divided into a number of transverse sections, called "plates."
2. All of the plates are equal in length.
3. The cross-sectional area of each plate is equal to that of the column, which is constant throughout the length of the column.

4. The ratio of the stationary phase to the mobile phase within each plate is constant, and equal to the ratio for the column as a whole.
5. The boundary between the phases within a plate is freely permeable to all solutes.
6. The length of the plate is the value required to make the mobile phase leaving the plate to be in equilibrium with the stationary phase within the plate.

The chromatographic resolution is dependent on the separation efficiency, expressed by the number N of theoretical plates:

$$N = 16\left(\frac{t_R}{\Delta t}\right)^2 \tag{24}$$

where Δt is again the time-width of the peak. The number of theoretical plates is determined by the column length L and the height equivalent to a theoretical plate H as expressed in Eq. (25):

$$H = \frac{\sigma 2}{L} = \frac{L}{N} \tag{25}$$

The parameter H is dependent on the nature of the chromatographic system and on the operating conditions employed, particularly on the movement of the mobile phase. In Eq. (25), σ is the standard deviation of the length of the chromatographic zone.

$$H \approx \frac{(\Delta z)^2}{16L} \tag{26}$$

or

$$\Delta z \approx 4(HL)^{1/2} \tag{27}$$

where Δz is the approximate width of the chromatographic zone as measured at the column outlet [27–29].

G. Estimation of N by Tangent Method

Of several methods for the estimation of N, the tangent method is widely used because of its practical convenience. In Fig. 4, tangents are drawn at the points of inflection of the Gaussian curve to intersect the baseline at points A and B, which are separated by a distance W_t. The value of N is calculated from Eq. (28):

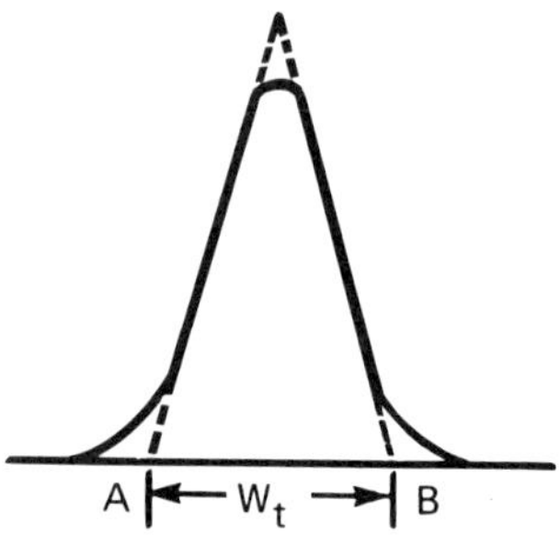

Fig. 4 The measurement of equivalent theoretical plates: tangent method. (From C. J. O. R. Morris and P. Morris, *Separation Methods in Biochemistry*, 2nd. ed., Halsted Press, John Wiley, New York, 1976.)

$$N = 16 \frac{V_R^2}{W_t^2} \tag{28}$$

An equivalent definition of N is given by Eq. (29):

$$N = \frac{X}{H} = \frac{X^2}{\sigma^2} \tag{29}$$

where X is the linear distance of migration of the zone apex and σ is the standard deviation [28].

H. Zone Spreading or Zone Broadening

The increase in the axial length of a solute zone as it travels along the column toward the exit is called zone spreading or broadening. It acts to degrade or reverse the separation for which the chromatographic process was designed. Spreading is mainly caused [27] by imperfections in the flow rate of the mobile phase (including any remixing in connecting tubes or dead volumes), by longitudinal diffusion of the solute in either phase, and by the failure to achieve sorption equilibrium. The potential problem with diffusion is complex, depending on several factors, as discussed more fully in the reference cited. Any chemical reaction changing the nature of the sample material (for example, hydrolysis, oxidation, or polymerization) may cause large changes in its physical properties. This may result in broadening of the mixed zone of alteration product (formed over a time interval in the passage through the column) and any remainder of the original solute, or even in the absence of any eluted zone if the product is permanently adsorbed in the column. Flow rate problems may originate in volume effects in sample injection, in input or output lines connected to the column, or in the detector. More detailed discussions may be found in the references given [22, 27,28].

I. Gaussian Distribution of Solute Through the Elution Zone

In Fig. 5, V and C represent mobile phase volume and solute concentration [28]. V_R is the retention volume of the solute, $\bar{C}$ is the mean value of C, and σ is the standard deviation of the solute distribution as expressed by

$$C = \bar{C} \exp -\left[\frac{t^2}{2}\right] = \frac{1}{\sigma\sqrt{2\pi}} \exp -\left[\frac{N'(V_R - V)^2}{2VV_R}\right] \tag{30}$$

For large N, N' ≈ N. The integration parameter t, found from Eq. (30), is

$$t = (V_R - V)\sqrt{\frac{N}{V_R V}} \tag{31}$$

When V_R is large, t can be approximated by Eq. (32):

$$t \approx \frac{(V_R - V)}{V_R}\sqrt{N} \tag{32}$$

For $0.00 \leqslant t \leqslant 4.00$, reference [26] gives values of y (our C). The amount A_t of solute in any part of an elution zone, corresponding to the shaded area in Fig. 5, may be found there [26], or from Eq. (33):

$$A_t = \frac{1}{\sqrt{2\pi}}\int_0^t e^{-t^2/2}\,dt \tag{33}$$

VI. CHROMATOGRAPHIC RETENTION PROCESS

Chromatography is a very complicated process because the basic mechanism is controlled by several factors, including:

1. Flow of the mobile phase
2. Diffusion of liquids used for mobile and stationary phases
3. Diffusion of solute within the phases
4. Equilibration of solute between the phases, and

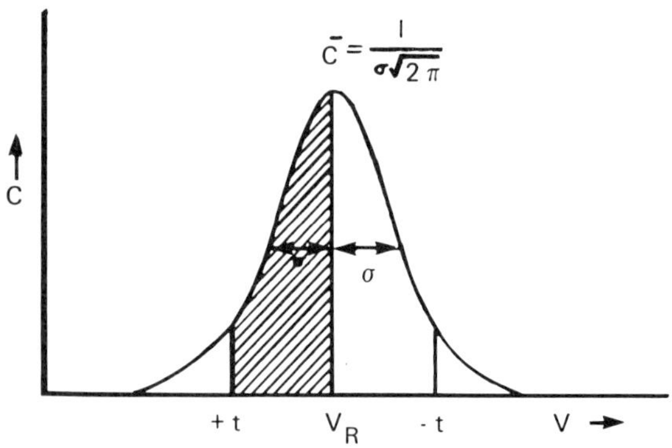

Fig. 5 Properties of the Gaussian error function. (From C. J. O. R. Morris and P. Morris, *Separation Methods in Biochemistry*, 2nd. ed., Halsted Press, John Wiley, New York, 1976.)

5. Solute mass balance during its transport (by mobile phase flow and molecular diffusion) in the chromatographic column

The separation of solute molecules in a mixture is achieved by varying retention of the solute molecules on the column, either by partition or by adsorption, depending on the type of chromatography under consideration. In order to comprehend the process of chromatographic retention, the concept of ideal linear chromatography was developed. This simplified model incorporates the following assumptions, as listed by Novák et al. [27]:

1. Infinitely rapid equilibration of solute between the phases
2. Negligibly slow longitudinal solute diffusion
3. Constant partition coefficient, independent of the concentration of solute in the phases, and
4. Piston flow of the mobile phase

In Fig. 6, the z-axis represents the column axis. Above it is a diagram of an axial section of the column, indicating its division into theoretical plates. The superimposed broken curve shows the concentration profile along the axis, through a solute zone. The simplified model with a linear relationship between C_m and C_s (constant partition coefficient) is quite useful, because (1) a linear system is more easily susceptible to an approximate mathematical treatment, and (2) real systems approach linearity of behavior over some concentration range (longer or shorter, depending on the individual system). In the treatment of Novák et al. [27], the column is given an overall cross-sectional area ϕ. The fractions of that cross-sectional area occupied by the mobile phase and by the stationary phase are ϕ_m and

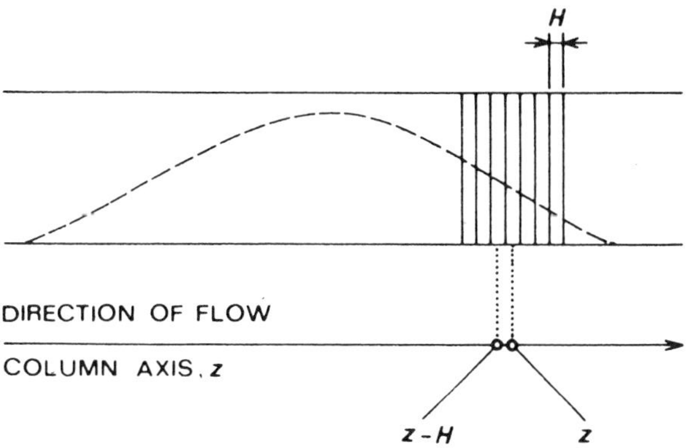

Fig. 6 Schematic representation of the plate model. (From J. Novák, J. Janák, and S. Wičar, in *Liquid Column Chromatography* (Z. Deyl, K. Macek, and J. Janák, eds.), Elsevier, Amsterdam, 1975, pp. 25–43.)

ϕ_s, respectively. A migrating solute zone in the column may be described schematically as in Fig. 6, in which the dashed curve is a superimposed graph of the solute concentration $C = C_m + C_s$ in the column. If we use that same figure again, we may for the present purpose consider the thickness H of the slab shown there to be now replaced by a differential dz. We wish to consider the change $\partial n/\partial t$ of the total number n of moles of solute (in both phases) with time t within the slab (volume element ϕdz). The quantity $\partial n/\partial t$ (see Section IV.I above), the difference between the numbers of moles of solute brought into and carried out of the volume element, is

$$\frac{\partial n}{\partial t} = - U \frac{\partial n_m}{\partial z} \tag{34}$$

Equation (34) assumes that (a) the column axis is oriented in the direction of coordinate z and (b) the gradient of solute concentration in the zone is positive with respect to the mobile phase velocity U. In Eq. (34), dn_m/dz represents the instantaneous gradient of the number of moles of solutes in the mobile phase within the volume element.

At any time and in any volume element of the zone, the total number n of moles of solute is equal to the sum of the numbers in the stationary and mobile phases in that element, which are always in equilibrium. The equilibrium constant is the distribution constant K_C defined in Eq. (11). That is, Eq. (35)

$$n = n_m + n_s \tag{35}$$

holds, where

$$n_s = K_C n_m \frac{V_s}{V_m} \tag{36}$$

In Eqs. (35) and (36), n_m and n_s denote the numbers of moles of solute present in the mobile and the stationary phases, respectively. The distribution constant K_C was defined in Eq. (11) as the equilibrium ratio C_s/C_m of the concentrations of the solute in the stationary and mobile phases, respectively. That is,

$$K_C = \frac{n_s/V_m}{n_m/V_s} = \frac{n_s V_m}{n_m V_s} \tag{37}$$

where V_s and V_m are the volumes of the stationary and mobile phases.

Substituting n_s from Eq. (36) into Eq. (35), we obtain

$$n = n_m + K_C n_m \left(\frac{V_s}{V_m}\right) = n_m \left[1 + K_C \left(\frac{V_s}{V_m}\right)\right] \tag{38}$$

Further, substituting n from Eq. (38) into Eq. (34) and rearranging, we get

$$\frac{\partial n_m}{\partial t} \left[1 + \left(\frac{K_C V_s}{V_m}\right)\right] = - U \frac{\partial n_m}{\partial z} \tag{39}$$

The quotient $K_C V_s/V_m$ is the capacity ratio D_C of Eq. (12):

$$(1 + D_C) \frac{\partial n_m}{\partial t} = -U \frac{\partial n_m}{\partial z} \tag{40}$$

Division of both sides of Eq. (40) by the volume $\phi_m dz$ of the mobile phase that is present together with stationary phase in the total volume element converts the number n_m of moles of solutes in the mobile phase to the concentration C_m of the solute in the mobile phase of the volume element. This yields Eq. (41), which can be used as a definition of ideal linear chromatography:

$$(1 + D_C) \frac{\partial C_m}{\partial t} = -U \frac{\partial C_m}{\partial z} \tag{41}$$

The solution of this partial differential equation gives Eq. (42), which is the basic chromatographic retention equation:

$$t_R = \frac{L(1 + D_C)}{U} \tag{42}$$

In this equation, t_R is the retention time and L is the length of the chromatographic column. Replacing L in Eq. (42) by use of the expression for zone migration velocity, $U_i = L/t_R$, and comparing with Eq. (15) gives

$$\frac{U_i}{U} = R = \frac{1}{1 + D_C} \tag{43}$$

Equation (43) merely restates the definition of the retardation factor R_C, which was seen to approximate the R_f value of thin-layer chromatography. In terms of the mobile phase hold-up (dead retention) time $t_m = L/U$, Eq. (42) can be rewritten

$$t_R = t_m(1 + D) \tag{44}$$

In Eq. (44), the retention time t_R is replaced by the retention volume V_R if both sides are multiplied by the colume flow rate of the mobile phase:

$$V_R = V_m(1 + D) \tag{45}$$

The volume V_m is the void volume (mobile phase hold-up or dead retention volume) of the column. With the help of Eq. (12), in which V_s is the total volume of the stationary phase present in the column, Eq. (45) is converted into Eq. (46)

$$V_R = V_m + K_C V_s \tag{46}$$

which is Eq. (13) above.

Some of the concluding remarks in the discussion of ideal linear chromatography by Novák et al. [27] are especially applicable to the case of countercurrent chromatography. In CCC, each phase of essentially every solvent system may be expected to be a mixture of components. A preequilibrated system of two liquid phases is always required, with a minimum of two components. Since some degree of mutual solubility can hardly be avoided, it is reassuring that mixed solvents offer no special challenge. The reference of Novák et al. [27] to adsorption effects exerted by the solid support in a packed column serves as a reminder of the advantages arising from the absence of any solid support in CCC, in which the mechanism of the retention process remains simple.

Although a simple ideal linear model is a very useful tool for the description of some basic general features, it of course can not yield information on any behavior whose existence was already precluded in the construction of the model. Novák et al. [27] point out that in the present case the ideal linear system says nothing about zone broadening, because no such assumptions were put into the model. They suggest that the simplest yet most realistic model is that of nonideal but linear chromatography: "In this case, the zone spreading due to non-ideal effects, such as finite rate of solute equilibration and appreciable longitudinal diffusion of solute, can be assumed to be symmetrical with respect to the center of the zone. Hence the centre of the zone can be taken as a region very near to sorption equilibrium, and the velocity of migration of the centre of the zone should approach the velocity expected in ideal linear chromatography."

VI. INSTRUMENTATION FOR CHROMATOGRAPHY

The essential component of the columnar chromatographic system is the column in which the stationary phase is retained. The stationary phase can be either a solid or a liquid material, which is purged with a mobile phase consisting of either a gaseous substance (as in gas chromatography) or a liquid (as in liquid chromatography). The solute mixture to be chromatographed is introduced into the inlet of the column through an injection port and is passed through the bed of the chromatographic column, in which the separation of the zones by differential migration takes place, as the solute mixture passes through the stationary phase in dynamic equilibrium with the mobile phase. The resulting zones (bands) from the solute mixture elute from the outlet of the column. In practice, the separated components are collected into different fractions and are identified by further analysis. In modern chromatographic instruments, the eluate is monitored with a detector. The recorded elution profile presents a series of peaks, each peak representing the separated solute component as well as the amount present in the solute mixture being chromatographed. As a part of the progress in the automation of chromatographic instruments, it has become common practice nowadays to analyze the separated components quantitatively, by integrating the measurements of the separated solute components or the chromatographic profile peaks representing them. There are several detectors available, such as ultraviolet (UV, standard and variable wavelength) monitors and refractive index (RI) monitors. The reader is advised to consult the standard reference books on chromatography, which describe the detectors, integrators, and computer peripherals for the complete automation of chromatography [6,23,28].

To sum up. the three essential functional regions of a modern liquid chromatographic instrument [30] are:

1. A solvent reservoir and a pump to provide flow of the mobile phase. Provision for gradient formation, a filter, degassers, and other appratus may be present.

2. A chromatographic column with some means for sample injection and a suitable oven or compartment with instrumentation for heating or cooling and temperature regulation.
3. The output section of the chromatograph, consisting of detectors and fraction collectors, with means for the measurement of peak areas (amounts of solutes) available in more complete systems.

In recent years, for detection purposes, tandem methods in which the chromatograph is coupled to spectroscopic instruments such as infrared (IR) and mass spectrometers (MS) have come into increasing use. Such combinations as GC/MS, LC/MS, GC/FTIR, LC/FTIR, and CCC/FTIR, interfaced to automatic data acquisition and storage systems, are the most up-to-date tools of the chromatographer. The latest one, CCC/MS, is the subject of another chapter in this book.

REFERENCES

1. F. F. Runge, *Farbenchemie*, Vol. 3, Mittler, Berlin, 1850.
2. D. T. Davy, *Proc. Amer. Phil. Soc.*, *36*, 112 (1897).
3. M. S. Tswett, *Proc. Warsaw Soc. Nat. Sci., Biol. Soc.*, *14*, No. 6 (1903); *Ber. Deut. Bot. Ges.*, *24*, 234, 316, 384 (1906).
4. R. Khun, A. Winterstein, and E. Lederer, *Hoppe-Seyler's Z. Physical Chem.*, *197*, 141 (1931).
5. A. J. P. Martin and R. L. M. Synge, *Biochem. J.*, *35*, 1358 (1941).
6. E. Stahl, *Thin-Layer Chromatography. A Laboratory Handbook*, Springer-Verlag, New York, 1969.
7. F. Goppelsroeder, *Anregungen Zum Studium der auf Capillaritats und Adsorptionserscheinungen beruhenden Capillaranalyse*, Heidelberg and Lichtenhan, Basel, 1906.
8. R. Consden, A. H. Gordon, and A. J. P. Martin, *Biochem. J.*, *38*, 224 (1944).
9. L. C. Craig, *J. Biol. Chem.*, *155*, 519 (1944).
10. S. W. Mayer and E. R. Tompkins, *J. Am. Chem. Soc.*, *69*, 2866 (1947).
11. O. Samuelson, *Ion Exchange in Analytical Chemistry*, John Wiley, New York, 1963.
12. G. Haugaard and T. D. Kroner, *J. Am. Chem. Soc.*, *70*, 2135 (1948).
13. N. A. Ismailov and M. S. Shraiber, *Farmatzija* (Moscow), *3* (1938); *cf. Chem. Abstr.*, *34*, 855 (1940).
14. I. G. Kirchner, J. M. Miller, and G. I. Keller, *Anal. Chem.*, *23*, 420 (1951).
15. R. M. Barrer, *J. Soc. Chem. Ind.* (London), *64*, 130 (1945).

16. J. Porath and P. Flodin, *Nature 183*, 1657 (1959).
17. S. Claesson, *Ark Kemi Mineral. Geol., A23,* No. 1 (1946).
18. A. T. James and A. J. P. Martin, *Biochem. J., 50*, 679 (1952).
19. C. S. G. Phillips, *Disc. Faraday Soc., 7*, 241 (1949).
20. A. L. Le Rosen, *J. Am. Chem. Soc., 67*, 1683 (1945).
21. J. C. Giddings, G. H. Steward, and A. L. Ruoff, *J. Chromatogr., 3*, 239 (1960).
22. G. J. Giddings, *Dynamics of Chromatography*, Marcel Dekker, New York, 1965.
23. Z. Deyl, K. Macek, and J. Janák, eds., *Liquid Column Chromatography: A Survey of Modern Techniques and Applications, "J. Chromatography Library,"* Vol. 3, Elsevier, Amsterdam, 1975.
24. A. J. P. Martin, *Ann. Rev. Biochem., 19*, 517 (1950).
25. M. D. Grove, G. F. Spencer, W. K. Rohwedder, N. B. Mandava, J. F. Worley, J. D. Warthen, Jr., G. L. Steffens, J. L. Flippen-Anderson, and J. C. Cook, Jr., *Nature, 281*, 216 (1979).
26. W. H. Beyer, ed., *CRC Handbook of Tables for Probability and Statistics*, 2nd ed., The Chemical Rubber Co., Cleveland, Ohio, 1968, pp. 127–134.
27. J. Novák, J. Janák, and S. Wičar, in *Liquid Column Chromatography* (Z. Deyl, K. Macek, and J. Janák, eds.), Elsevier, Amsterdam, 1975, pp. 25-40.
28. C. J. O. R. Morris and P. Morris, *Separation Methods in Biochemistry*, 2nd, ed., Halsted Press, John Wiley, New York, 1976.
29. J. Novák and J. Janák, in *Liquid Column Chromatography*, (Z. Deyl, K. Macek, and J. Janák, eds.), Elsevier, Amsterdam, 1975, pp. 3–10.
30. M. Krejčí, Z. Pechan, and Z. Deyl, in *Liquid Column Chromatography*, (Z. Deyl, K. Macek, and J. Janák, eds.), Elsevier, Amsterdam, 1975, pp. 101–168.

Part 1

THEORY AND INSTRUMENTATION

chapter two

The Origins of Countercurrent Chromatography

N. Bhushan Mandava*
Office of Pesticides and Toxic Substances
U.S. Environmental Protection Agency
Washington, D.C.

John. M. Ruth
Agricultural Environmental Quality Institute
Beltsville Agricultural Research Center, ARS, USDA
Beltsville, Maryland

I. COUNTERCURRENT DISTRIBUTION

The operating principles of extraction are based on the concept of heterogeneous equilibrium in multiphase liquid-liquid partition systems. Extraction processes are primarily concerned with the fraction of a given solute in a given phase at equilibrium. The fraction in a single phase is the portion of the solute molecules that will be removed or transferred at a given step, when that particular phase is separated. This underlying principle in the extraction process is based on the distribution law which was first recognized by Nernst [1] in 1891. The distribution law states that a solute dissolved in one phase in equilibrium with another, immiscible phase will distribute itself between the two phases, so that the ratio of the concentrations in the two phases is constant at a fixed temperature. This can be represented by Eq. (1):

$$K = \frac{C_1}{C_2} \tag{1}$$

where C_1 and C_2 are the concentrations of the solute molecule in the first and second phases, respectively, and K is usually called the "distribution constant" or the "partition constant" [4].

Current affiliation: Todhunter, Mandava, & Associates, Washington, D.C.

However, this simple law may not hold true in all experiments because of (a) molecular interactions (e.g., association and dissociation) between the solute and the solvent molecules, and (b) the effects of concentration of the solute molecules. Experimentally, the validity of the distribution law for a given solute over a wide concentration range in a selected solvent system can be determined from partition isotherms. These isotherms (Craig and Craig [4]) are the graphical representation of the solute concentrations in two phases. Therefore, the behavior of the partition coefficients can be determined from the linearity or nonlinearity of the isotherms (see Chap. 1 on chromatography for further details).

According to the distribution law, the separation of the components depends on distribution between two immiscible liquids. This distribution law was applied to extraction processes involving partition of the components of the solute mixture between two immiscible liquid phases. However, liquid-liquid partition is not limited to two-phase systems, but can be extended to multiphase systems [13].

Partition in multiple liquid-liquid systems may be defined as a process consisting of several successive liquid-liquid partitions or distributions. However, because of experimental difficulties, both in the selection of suitable solvents for the preparation of immiscible multiphase systems and in the development of convenient phase separation devices, advances in multiple liquid-liquid partition systems have not been made beyond three-phase systems. In this book, we limit our discussion to two-phase systems, although we have recognized, on the flow chart in Fig. 1, the existence of the three-phase system. Two-phase systems may be classified as stepwise or as continuous phase transfer systems, each of which is further subdivided according to feeding modes which are either batchwise or continuous. Stepwise phase transfer, which can be operated either in batchwise or in continuous transfer modes, constitutes countercurrent distribution, a technique thoroughly investigated as a separation method. Lyman Craig, a pioneer in CCD, studied CCD extensively and devised several CCD instruments. He proved that CCD was not only superior to other preparative methods of his time, including other chromatographic methods, but was also applicable to the separation of a wide range of material types, including monomers, polymers, cells, and particles. Craig's CCD is a discontinuous process, simple in theory and operation, and is analogous to the extractions that we usually carry out in separatory funnels in the laboratory. The subject matter for this section is CCD in the meaning of Craig and his associates. The underlying principles of CCD, as well as representative examples of applications to the separation of compounds, will be discussed in order to get an understanding of the distribution process before taking up CCC, which is based on CCD and liquid chromatography [4–6, 8–9].

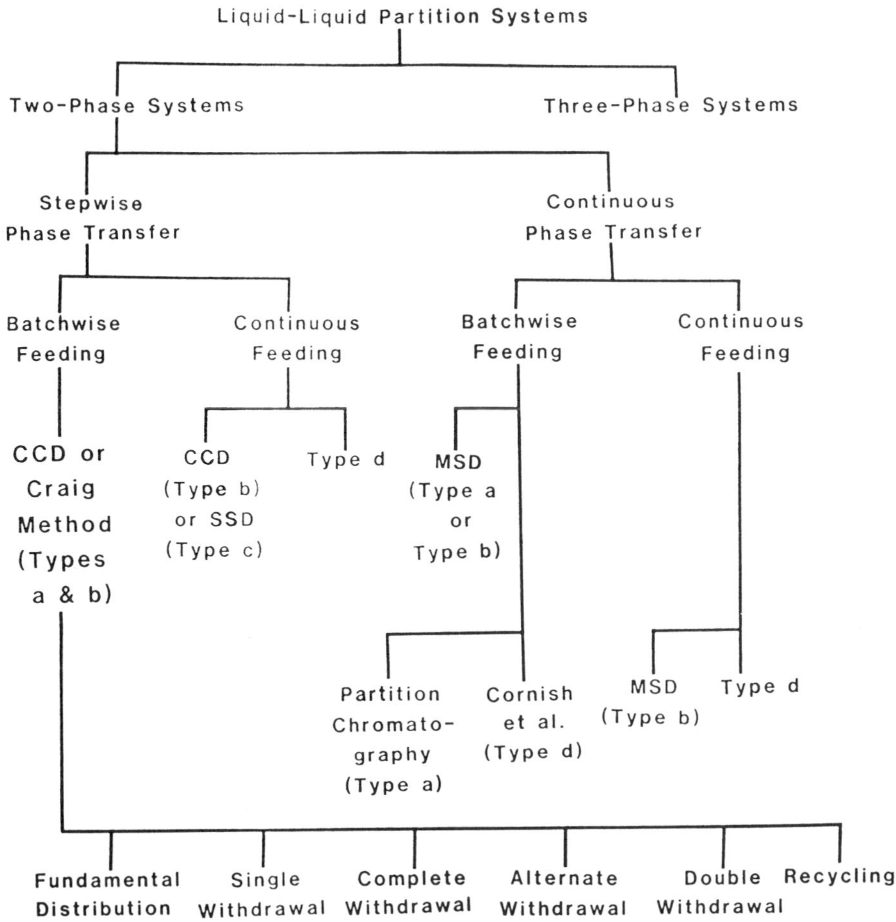

Fig. 1 Classification of multiple liquid-liquid partition systems. Definitions. CCD: countercurrent distribution; SSD: steady state distribution; MSD: Martin-Synge type distribution; Type a: phase m moves stepwise while phase s is quiescent; Type b: phase m moves stepwise while phase s is quiescent, and the countercurrent is effected by changing the position of the partition units; Type c: phases m and s move stepwise and alternately, in opposite directions; Type d: phases m and s move stepwise and simultaneously, in opposite directions: Type e: The lowest phase is stationary, and the upper two phases move stepwise.

It is obvious that liquid-liquid partition systems can process large quantities of material (Table 1). In chromatographic systems the samples size is limited to gram quantities, while liquid-liquid partition systems can handle up to kilogram (even ton) quantities.

A. Mechanism of the Distribution Process

Separation of solute mixtures can be accomplished by extraction, which was extensively studied by Jantzen [2] and Cornish et al. [3] in the 1930s. They investigated extractions either in multiple contact countercurrent manner or in the truly continuous, differential manner. The theory that was developed was completely tested by experiment. It is very adaptable to experimental study, and the partition ratio can be adjusted over a broad range with a variety of solvent pairs. It should be remembered that separation by extractions is always a countercurrent process, even though it is performed in discrete stages.

Separation procedure basically depends on the partition, transfer, and recombination of various fractions. The distribution of solutes between phases is assumed to follow a linear partition (distribution) isotherm. Here the partition coefficient K_D is defined as the ratio of the solute concentrations between the upper and the lower phases:

Table 1 Comparsion of Chromatographic and Multiple Liquid-Liquid Partition Methods for Separation of Quantities of Solutes

Quantity	Method
ng–g	Gas chromatography, thin-layer chromatography, and paper chromatography
μg–mg	Column chromatography and high-performance liquid chromatography
mg–g	CCD and MSD
g–kg	Preparative Martin-Synge distribution (MSD) [10,29–30], countercurrent distribution (CCD) [4–9], steady-state distribution (SSD) [34–35], and counter double current distribution [28], methods of Fischer [26], O'Keeffe [24], and Watanabe-Morikawa [31]
kg–Tonne	The methods of V. Dijck [32] and Jantzen [2]

Source: From Ref. 13.

$$K_D = \frac{\text{concentration of solute in upper phase}}{\text{concentration of solute in lower phase}} = \frac{C_U}{C_L} \quad (2)$$

whereas in liquid-liquid partition chromatography, the partition coefficient becomes:

$$K_C = \frac{\text{concentration of solute in stationary phase}}{\text{concentration of solute in mobile phase}} = \frac{C_s}{C_m} \quad (3)$$

Usually, separations by partition chromatography are accomplished by keeping the upper phase mobile. In that case (upper phase mobile), the partition coefficeint K_D for separation by the distribution process is the reciprocal of the partition coefficient K_C of the same solute in the same solvent system as defined for liquid-liquid partition chromatography (Morris and Morris [9]):

$$K_D = \frac{1}{K_C} \quad (4a)$$

However, if the lower phase in partition chromatography is mobile, then

$$K_D = K_C \quad (4b)$$

B. Theoretical Considerations

For a simple description of the theory of separation by distribution, let us assume that

1. Each solute behaves independently during an extraction involving two solutes.
2. The partition ratios of the solutes remain constant over the entire concentration range, provided that we employ sufficiently dilute solutions.
3. The two phases exist in equal volumes.
4. The partition ratio for the selected solute in the two-phase system is set at 1.

Now, if we arrange the separatory funnels in a series numbered 0, 1, 2, . . . , r from left to right (Fig. 2), and if a unit quantity (1 g) of the solute is dissolved in the lower phase (L_0) and the upper phase (U_0) is moved over and added to it, the extraction process is in the 0 transfer stage. The two immiscible phases in separatory funnel 0 attain equilibrium and separate into two phases

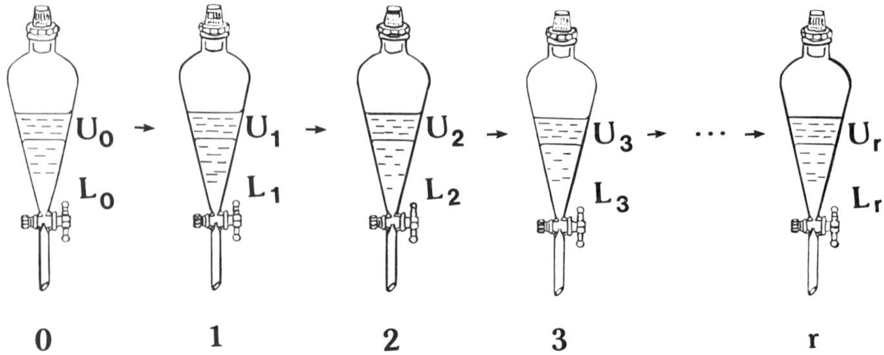

Fig. 2 Countercurrent extraction in separatory funnels, keeping the lower phase stationary.

(upper and lower). Then we shift the upper phase U_0 from funnel 0 over the lower phase L_1 in funnel 1, and simultaneously the upper phase U_1 (fresh solvent) over the lower phase L_0 in funnel 0. Now we have completed transfer 1. Since we began with a unit partition coefficient and equal volumes of two phases, one-half of the solute will have migrated to funnel 1 in the upper layer. Therefore, the total, in both layers combined, in each of funnels 0 and 1 will be 0.50 g. These amounts are shown in a tabular form in Fig. 3 on the appropriate line for the number of transfers.

Now let us consider transfer 2, which involves the equilibration of funnels 0 and 1, and the shifting of the upper layers such that U_0 is over L_2, U_1 is over L_1, and U_2 (fresh solvent) is over L_0. After transfer 2 has been carried out, Fig. 3 shows the total fractions of the solute (both phases still present) in each of funnels 0, 1, and 2 as 0.25, 0.5, and 0.25, respectively. The process can be continued indefinitely as shown in Figs. 2 and 3.

Let us now examine the table (Fig. 3) which represents an expansion corresponding to

$$(x + y)^n = 1 \tag{5}$$

where x represents the fraction remaining in the lower phase and y represents the fraction being transferred, and n corresponds to the number of transfers carried out (Fig. 3). Expanding the binominal expression, we obtain:

$$(x + Y)^n = x^n + nx^{n-1}y + \frac{n(n-1)}{2!}x^{n-2}y^2 + \cdots + y^n \tag{6}$$

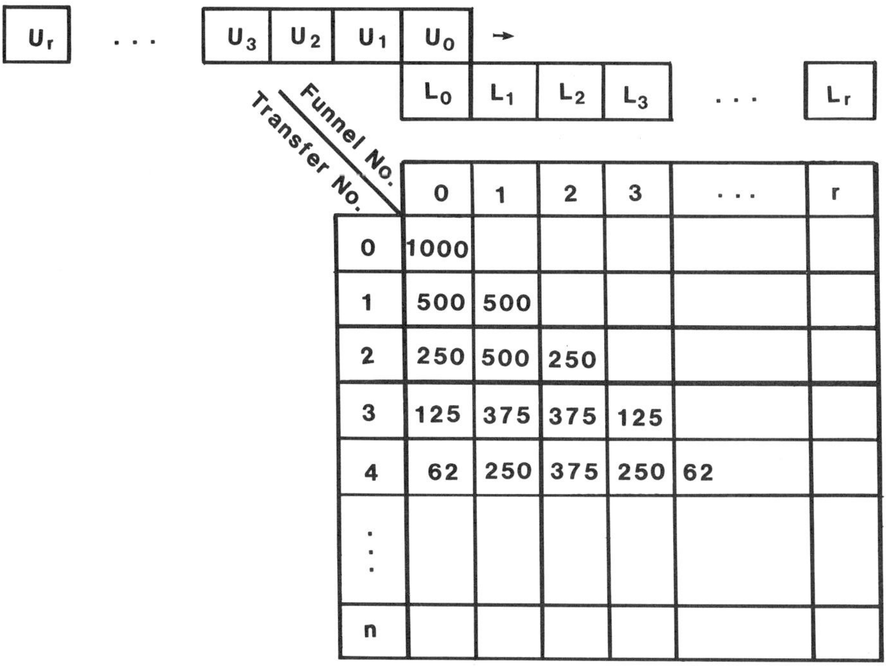

Fig. 3 Countercurrent distribution in the separatory funnel experiment. (Adapted from Ref. 4, p. 172.)

$$(x + y)^n = T_0 + T_1 + T_2 + \cdots + T_n \qquad (7)$$

where T_0, T_1, . . . , T_n represent the amount in each funnel (Figs. 2 and 3).

At any stage, the distribution coefficient $K_D = y/x$ by definition. Also, $x + y = 1$, so that we have a simple pair of simultaneous equations. Solution for x and y gives

$$x = \frac{1}{1 + K_D} \qquad (8)$$

and

$$y = \frac{K_D}{1 + K_D} \qquad (9)$$

When these values for x and y are substituted in to Eq. (6), the binomial expression becomes

$$\left(\frac{1}{1 + K_D} + \frac{K_D}{1 + K_D}\right)^n = 1 \tag{10}$$

when equal volumes of two phases are used. This expression applies just as well for unequal volumes, provided that K_D is multiplied by the ratio of the two volumes. That is,

$$D_D = \frac{Q_U}{Q_L} = \frac{C_U V_U}{C_L V_L} = K_D \frac{V_U}{V_L} = K_D R_D \tag{11}$$

where D_D is the mass distribution ratio, Q_U and Q_L are the quantities of solute in the upper and lower phases, V_U and V_L are the volume of the upper and lower phases, and R_D is the phase volume ratio V_U/V_L.

Note that when $K_D = 1/K_C$, the mass distribution coefficient D_D is the reciprocal of the chromatographic capacity factor D_C:

$$D_D = \frac{1}{D_C} \tag{12}$$

We can also establish relationships between the lower and upper layer fractions x and y, the solute quantities Q_U and Q_L, and the mass distribution coefficient D_D.

$$x = \frac{Q_L}{Q_U + Q_L} = \frac{1}{1 + D_D} \tag{13a}$$

and

$$y = \frac{Q_U}{Q_U + Q_L} = \frac{D_D}{1 + D_D} \tag{13b}$$

In the special case in which $V_U = V_L$, Eqs. (13a) and (13b) reduce to (14) and (15), respectively:

$$x = \frac{1}{1 + K_D} \tag{14}$$

$$y = \frac{K_D}{1 + K_D} \tag{15}$$

Equations (8) and (14) are the same. Equations (9) and (15) are also identical. As has already been discussed, the relationships between x and K_D in Eq. (8) or Eq. (14), or between y and K_D in Eq. (9) or Eq. (15), hold true only where phase volumes are equal during the distribution process. When unequal phase volumes are employed, we must use Eqs. (11) and (13) to derive any of the parameters.

In Fig. 3, when the funnels are numbered 0, 1, 2, 3, . . . , r (see top line), the funnel numbered r corresponds to the rth term of the binomial expansion. In the example above, we assumed that a unit quantity (1 g = 1000 mg) of solute was equally distributed in equal volumes of upper and lower phases. That means that x = y, x = 0.5 g or 500 mg, and y = 0.5 g or 500 mg. Substituting the values of x and y in the binomial expression (2) for four transfers (see Fig. 3):

$$(x + y)^4 = x^4 + 4x^3y + 6x^2y^2 + 4xy^3 + y^4$$

$$(0.5 + 0.5)^4 = 0.0625 + 0.250 + 0.375 + 0.250 + 0.0625$$

$$= T_0 + T_1 + T_2 + T_3 + T_4 \qquad (16)$$

As shown in Figs. 2 and 3, the separation procedure can be expanded to any number of transfers. We can calculate the amount of solute distributed into each funnel from the binomial expansion, even though the operations become somewhat complex [4,156].

To demonstrate the theory of separation by this distribution process in a simple manner, we have used a series of separatory funnels. The fractions resulting from the transfers are tabulated in Fig. 3. Perhaps 10 or 12 such transfers might be carried out with individual contacting units (separatory funnels), but it becomes less practical for larger numbers of transfers, unless a train of contacting units is used which enables many simultaneous extractions to be made. To accompish this objective, Craig and coworkers devised several countercurrent distribution units which vary in size and number of transfers, as described in [4–6]. These devices (CCD trains) consist of a series of glass tubes in which fractionation takes place as a result of transfers between the tubes. Table 2 gives an example for the distribution of fractional parts of a solute with a partition coefficient of 1 when we perform CCD involving eight transfers. The amounts represented in each tube (Table 2) from a unit quantity (1 g = 1000 mg) can be calculated directly from the binomial expression, since the fraction in the rth tube (Fig. 3) is the rth term of the binomial expansion [Eq. (6)].

Instead of using Eqs. (6) or (10), we can employe Eqs. (17) and (18) as general formulas for calculating the fraction $T_{n,r}$ of a solute present in the rth tube for n transfers.

Table 2 Distribution of Fractional Parts for a Solute with $K_D = 1$

Transfer no. (n) \ Tube no. (r)	0	1	2	3	4	5	6	7	8
	Amount in tube (mg)								
0	1000								
1	500	500							
2	250	500	250						
3	125	375	375	125					
4	62	250	375	250	62				
5	31	156	313	313	156	31			
6	16	93	234	313	234	93	16		
7	8	54	164	274	274	164	54	8	
8	4	31	109	219	274	219	109	31	8

Source: Adapted from Ref. 4.

$$T_{n,r} = \frac{n!}{(n - r)!r!} x^{n-r} y^r \tag{17}$$

or

$$T_{n,r} = \frac{n!}{(n - r)!r!} \left(\frac{1}{1 + K_D}\right)^n K_D^r \tag{18}$$

$T_{n,r}$ represents the rth term of the binomial expansion $(x + y)^n = 1$. The sum of all $T_{n,r}$ terms is 1 since no solute is withdrawn during a distribution. K_D is equal to the mass distribution ratio D_D only when equal volumes of two immiscible phases are used. If unequal volumes of the phases are used, K_D should be multiplied by the ratio of the volumes of the two phases, as indicated in Eq. (11). It must also be remembered that K_D is used for the case in which the upper phase is the one transferred. If the lower phase is transferred, $1/K_D$ is substituted for K_D in Eq. (14).

Curve 1 ($K_D = 1$) in Fig. 4 represents the data from Table 2, with the fraction of solute present in a tube plotted against the serial number of the tube. Curve 2 ($K_D = 0.333$) and curve 3 ($K_D = 3$) are also shown in Fig. 4. As the number of transfers is increased, the curve appraoches the normal error curve. Curve 1 is already symmetrical because $K_D = 1$.

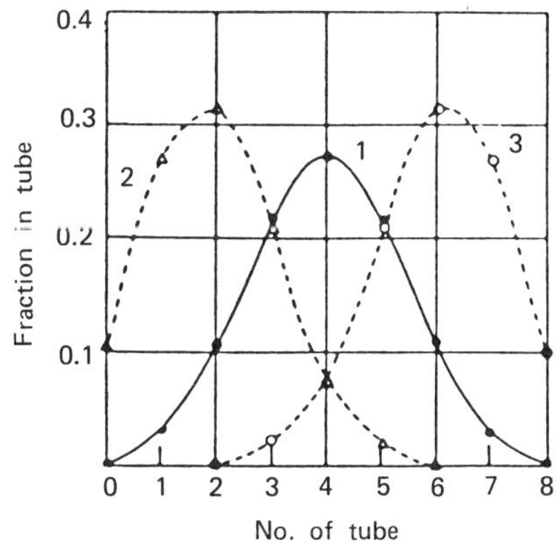

Fig. 4 Countercurrent distribution curves. (Reprinted with permission from L. C. Craig and D. Craig, in *Technique of Organic Chemistry*, Volume III, Part I, (A. Weissberger, ed.) Interscience, New York, 1956. Copyright 1956 John Wiley & Sons, Inc.)

If a solute mixture consists of two components, each will tend to distribute itself independently. Because the partition ratios for two components will be different, two bands, overlapping or otherwise, will be obtained. Such a situation is shown in Fig. 5, which represents the separation of propionic and butyric acids by CCD. Curve 1 represents the extraction by eight transfers, while curve 2 (butyric acid, K_D = 0.485) and curve 3 (propionic acid, K_D = 2.06) represent each compound under the same CCD conditions. The total solute in each tube (curve 1) is the sum of curves 2 and 3 (individual acids). Using the binomial equations, we can also calculate the degree of separation in each tube and the percentage recovery for any given purity.

The resolving power of solute separation depends on the number of transfers in CCD. In other words, the effect of passing to higher and higher numbers of transfers is seen in progressive dilution of solution and, as a result, the solute molecules scatter in more and more tubes as the distribution proceeds. An example representing 10–1000 transfers for a solute with K_D of 1 is shown in Fig. 6. Although more sample is required where higher numbers of transfers are carried out, the clear advantage of large numbers of transfers is that the peak that results from such a separation is narrow and well resolved. It is then possible to resolve other peaks, if the sample to be separated contains two or more compounds, provided that they differ considerably in their partition coefficients. In Fig. 6, (10–1000 transfers), the amount of solvent remains the same in all the three operations. It will be seen that curve A spreads through all fractions and therefore cannot show any other band, even though the second band might have a partition ratio considerably different

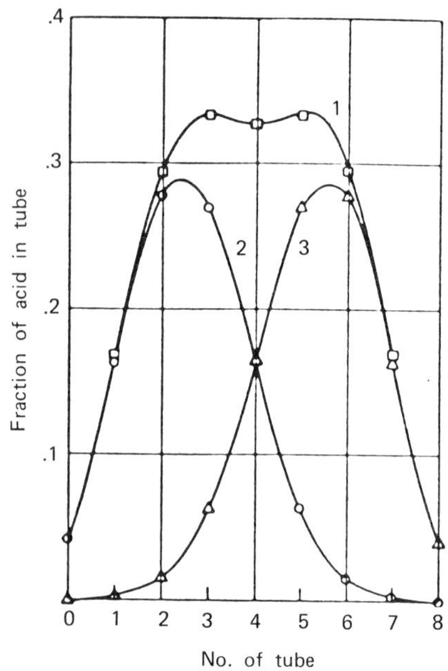

Fig. 5 Countercurrent distribution curves. (Reprinted with permission from L. C. Craig and D. Craig, in *Technique of Organic Chemistry*, Volume III, Part I, (A. Weissberger, ed.) Interscience, New York, 1956. Copyright 1956 John Wiley & Sons, Inc.)

from that of the first. With larger numbers of transfers (curves B and C), there is sufficient space on either side of the peak to permit observing other peaks, if their partition ratios are sufficiently different.

In separation science, the isolation of individual components from a mixture is of paramount importance. A high degree of purity can be achieved by increasing the number of transfers. It must be remembered that purity is never 100%, and that the separation and recovery of one component from the others is likewise never 100% effective. Craig and Craig [4] gave an example of the CCD separation of a binary mixture containing solutes A and B, whose partition ratios were 0.707 and 1.414. Figure 7 shows their separation after 100 transfers. Recovery of each solute was 96% with 95% purity. For smaller numbers of transfers, the recovery rate of each component in a given state of purity is lower. For example, for 24 transfers, only 15.5% of each compound could be recovered with 98% purity, as contrasted to 93% recovery with 100 transfers.

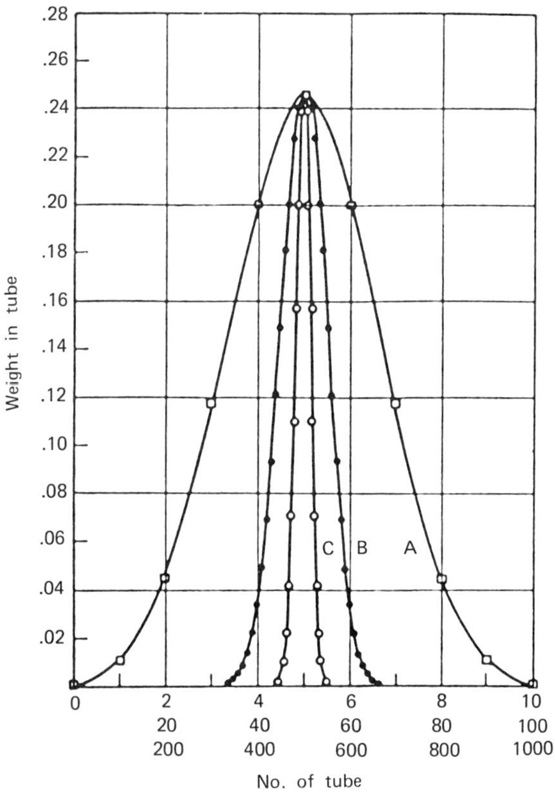

Fig. 6 Comparative distributions with increasing numbers of transfers. (Reprinted with permission from L. C. Craig and D. Craig, in *Technique of Organic Chemistry*, Volume III, Part I, (A. Weissberger, ed.) Interscience, New York, 1956. Copyright 1956 John Wiley & Sons, Inc.)

C. Calculation of Theoretical Curves

Craig et al. [4–6,8,17], Hecker [11], and others [13] studied CCD with discontinuous transfer extensively and developed useful formulas to predict the shape of the curves during liquid-liquid distribution. As already mentioned, Eqs. (17) and (18) are general expressions for the fractional quantity in any stage, corresponding to a certain term in the binomial expansion, Eq. (6).

The ratio of the fractional quantities of solute present in any two adjacent stages can be calculated [9] from

$$\frac{T_{n,r}}{T_{n,r-1}} = \left(\frac{n - r + 1}{r}\right) D_D \qquad (19)$$

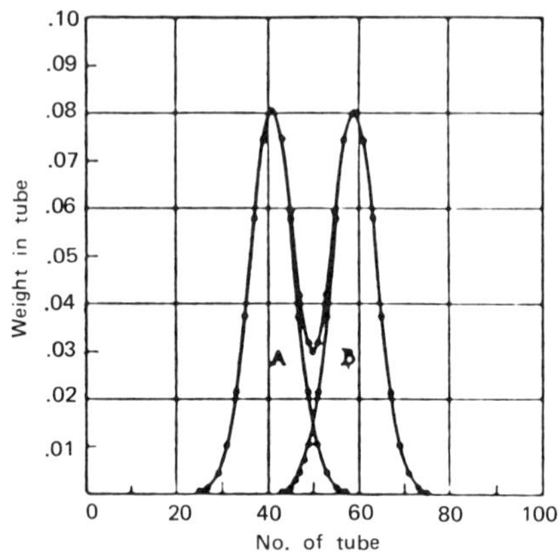

Fig. 7 100-transfer distribution of a binary mixture. (Reprinted with permission from L. C. Craig and D. Craig, in *Technique of Organic Chemistry*, Volume III, Part I, (A. Weissberger, ed.) Interscience, New York, 1956. Copyright 1956 John Wiley & Sons, Inc.)

and

$$\frac{T_{n,r}}{T_{n,r+1}} = \left(\frac{r+1}{n-r}\right)\frac{1}{D_D} \tag{20}$$

and the quantity in stage zero can be expressed as

$$T_0 = \frac{1}{(1 + D_D)^n} \tag{21}$$

where $D_D = (V_U/V_L)K_D$ as in Eq. (11). Since we can evaluate n and r for any particular distribution and can determine the quantities in adjacent stages of an experimental distribution, the value of D_D can be calculated. Evaluating the experimental mass distribution D_D over the entire range is one way to test the curve for homogeneity. As has already been noted, for a peak containing only one solute, K_D will remain constant throughout the region of the curve. Any deviation will indicate the contamination of the solute under investigation.

Williamson and Craig [18] emphasized the usefulness of Eqs. (19) and (20) for plotting the theoretical shape of a peak in a countercurrent distribution. They pointed out that by using the experimentally measured peak as a starting point, the values in adjacent stages can be derived. Then each successive stage may be calculated by means of a sequence of simple ratios.

Another equation applicable to the determination of theoretical distribution curves is the following:

$$r_{max} = ny = \frac{nD_D}{1 + D_D} \tag{22}$$

This approximate Eq. (22) can be used for estimating the distribution coefficient. In this equation, r_{max} refers to the number of the stage in which the maximum is found. Craig observed this relationship empirically. On the basis of probability considerations, Brock [19] derived Eq. (22), which is exact only for $D_D = 1$ or for $n = \infty$.

Equation (22) can be expressed as

$$K_D = \frac{r_{max}}{R_D(n - r_{max})} \tag{23}$$

where $R_D = V_U/V_L$, and $D_D = R_D K_D$. With Eq. (23), the K_D for a solute can be calculated from its peak position and the number of transfers performed.

For distributions involving large numbers of transfers ($n \geqslant 20$), the shape of the binomial distribution can be closely approximated by the Gaussian error curve [5,6,19]. Then the CCD curve may be expressed as

$$T_{r,z} = T_{r,r_{max}} \exp\left(-\frac{z^2(1 + R_D K_D)^2}{2nR_D K_D}\right) \tag{24}$$

where

$$T_{r,r_{max}} = \frac{1 + R_D K_D}{\sqrt{2\pi n R_D K_D}} \tag{25}$$

The peak maximum is found where $z = 0$. $T_{r,z}$ represents the peak height at distances $\pm z$ stages on either side of the maximum. Since $R_D = V_U/V_L$ and $D_D = R_D K_D$, D_D can be substituted for $R_D K_D$ in Eqs. (24) and (25).

The standard deviation of the peak location can be determined from

$$\sigma = \sqrt{nxy} = \frac{\sqrt{nR_D K_D}}{R_D K_D + 1}$$

Note that 99.7% of the solute will be contained in a region extending for 3σ on either side of the zone maximum. For all practical purposes, the peak width W can be derived from

$$W = \frac{6\sqrt{nR_D K_D}}{R_D K_D + 1} \tag{27}$$

Equation (27) is valid only for symmetrical zone profile curves. As n increases, W also increases, but the fraction of the total number of tubes occupied by W decreases. As in chromatography, W is proportional to $\sqrt{n}$ while the distance between the maxima of two adjacent solute zones is proportional to n. Therefore, any increase in n will result in improved resolution of the adjacent solute zones. In a 50-stage train, 2–4 components can be fully resolved, while in a 100-stage train, 10–20 components can be completely separated.

For additional information concerning the number of stages required for separation of solutes with different D_D values, the reader is referred to the original papers [21–22]. The best resolution of two components A and B is obtained when

$$D_{D,A} D_{D,B} = 1 \tag{28}$$

where $D_{D,A}$ and $D_{D,B}$ are the mass distributions for A and B. They may be adjusted to conform with Eq. (28) by varying the phase volume ratio R_D. Under restriction (28)

$$R_D = \frac{1}{\sqrt{K_{D,A} K_{D,B}}} \tag{29}$$

where $K_{D,A}$ and $K_{D,B}$ are the respective partition coefficients and are fixed by the choice of solvent system. Additional details of the theoretical treatment can be found in extensive reviews [4–6,8–12].

Countercurrent distribution theory is closely related [9] to the theoretical plate model of chromatography, which was in turn taken over from its earlier development in the study of fractional distillation. In CCD, the plates are clearly visible as the tubes of the instrument. Morris and Morris [9] note the correspondence of the boundary between chromatographic stationary and mobile phases with the interface between the separated phases in countercurrent distribution. In CCD, the achievement of solute equilibrium between phases is accomplished by shaking, which produces a dispersion of fine droplets of one phase in the other, increasing the interfacial surface area and reducing the distance traversed by molecules between the phase interior and the interface. After sufficient mixing, the shaking is stopped, and the phases separate. The mobile phase is then transferred quantitatively to the next tuve, where it meets a fresh portion of the stationary phase. At the same time, a fresh portion of the mobilw phase is added to the first tube, just as would be done in a manual procedure using a sequence of separatory funnels. Shaking is then started again to produce equilibration in both tubes. Alternate equilibrations and transfers, with the addition of fresh mobile phase at each transfer, continue until the initial portion of the mobile phase has reached the last tube. This is the process that Craig and Post [20] called the fundamental distribution process. The main difference between CCD and chromatography is that in CCD equilibrations and transfers are done alternately, in stepwise fashion, while in chromatography all of the operations take place simultaneously and continuously [9].

D. Operating Procedures

Among several methods of operation of a countercurrent distribution train, Craig lists four methods: (a) fundamental distribution, (b) single withdrawal, (c) completion of squares, and (d) recycling.

Fundamental Distribution

This is the first example shown in Figs. 1 and 2. In this type of distribution, the number of transfers n is either equal to or less than the number of stages ($n \leqslant r$).

Single Withdrawal

In this procedure, fractionation is continued in the same manner as in the fundamental distribution. After every transfer, an aliquot of the upper phase is brought into the first tube and the effluent emerging from the last tube ($n = r^*$) is collected as fraction 0, 1, 2, ... This method is analogous to column elution in chromatography. The curve from the withdrawn series of aliquots can be obtained from Eq. (30):

$$T_{W,n} = \left(\frac{(n-1)!}{r^*!(n-r^*-1)!}\right) \frac{D_D{}^{r^*+1}}{(D_D+1)^n} \tag{30}$$

where r^* is the number of the higher stage in the CCD train and n is the transfer number [22]. The fundamental procedure would have been completed with the r^*th transfer. The terms of the withdrawn series commence with $r^* + 1$ and continue as $r^* + 2$, $r^* + 3$, ... Other such parameters in the theoretical distribution curve, such as the transfer number n* and position of the peak maximum, can be calculated in a manner similar to that used in the fundamental distribution [156].

Completion of Squares

This is similar to single withdrawal, but after the solvent front reaches the end of the CCD train, n = r, no additional upper phase solvent is introduced. In other words, after the desired number of transfers in fundamental distribution [23], the process is continued without further addition of the upper phase. All the upper phase solvent leaves the distribution train and forms one effluent series, as in the single withdrawal method. The distribution train contains only the lower phase solvent, which will contain solutes of lower partition coefficients (for further details on this method, consult the paper by Bush and Densen [23]).

Recycling

The aforementioned procedures are very useful for relatively short distribution trains containing 50–60 tubes. When large trains are employed, the recycling and partial withdrawal procedures have been found very practical. To resolve components of low K_D values, the last tube of the train may be connected to the zeroth tube and the operation repeated. In this manner, we can conserve the solvent and can increase the number of stages in the distribution train.

An example shown in Fig. 8 gives an idea of how these procedures can be employed for a separation of solute mixtures. In the case of a hypothetical ternary mixture, a partial resolution of three components with K_D values of 0.25, 0.60, and 2.0 is obtained after a 50-transfer fundamental distribution (Fig. 8A). Continuation of the distribution to 90-transfer yields the transfer of the solute with K_D of 2 out of the train with the emerging upper phase (Fig. 8B), which forms an effluent series in elution chromatography. Here each emerging fraction is indicated by the transfer number, and this process is called the single-withdrawal method. The remaining two solutes can be completely resolved by the recycling method in which we set the train to recycle for 20 more transfers by bringing the emerging upper phase back to tube zero (Fig. 8C). This recycling method has been reported to be most useful for separation of mixtures with low separation factor (which is the ratio of partition coefficients).

There are other operating procedures, such as double and partial withdrawal methods, whose details can be found in the extensive treatises [4–6,8–12].

E. Other Countercurrent Distribution Schemes

Thus far we have described the discontinuous countercurrent distribution system which was the most successful method investigated by

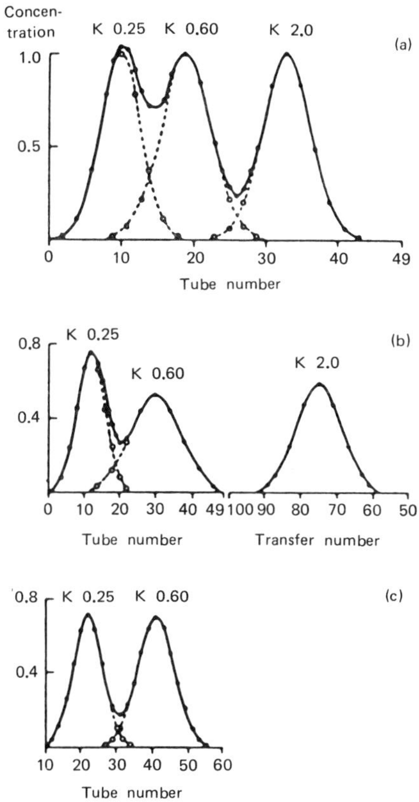

Fig. 8 Distribution curves of a mixture of solute with K_D values of 0.25, 0.60, and 2.0. (a) Fundamental distribution after 50 transfers. (b) Completion of squares distribution after 100 transfers. (c) Recycling of the two solutes with K_D values of 0.25 and 0.60 after 110 transfers, following removal of the solute of K_D of 2.0 by the single-withdrawal procedure. (Reprinted with permission from T. P. King and L. C. Craig, in *Methods of Biochemical Analysis* (D. Glick, ed.), Vol. 10, Wiley–Interscience, New York, 1962, pp 201-228. Copyright 1962 John Wiley & Sons, Inc.)

Craig and his associates for the separation of any type of compounds on a laboratory scale. Another scheme that has been widely used on industrial scale separations is the continuous countercurrent distribution method. This subject was extensively reviewed by Scheibel [10]. Hynninen [13–15] used the term "Martin-Synge distribution" (MSD) to differentiate it from Craig's discontinuous stepwise phase process of countercurrent distribution (CCD). In continuous flow partition methods, the partition units are connected to each other, and the phase transfer is continuous or semicontinuous (pulsative) As compared to Craig's CCD, continuous phase transfer causes an excessive broadening of the concentration zones in the longitudinal direction of the device. In continuous feeding, it is possible that (a) the equilibrium condition is seldom attained because the mass transfer process between the phases is slow, (b) the dynamics of liquids cause longitudinal eddy diffusion (turbulence), and (c) molecular diffusion is possible between the partitioning units. In spite of these disadvantages, several continuous-feeding instruments were devised which were designed to minimize longitudinal mixing of the solutes, to maintain constant equilibrium between the phases, and to achieve instant mass transfer.

In order to attain continuous feeding, the countercurrent operation with movement and withdrawal of the solvent must be employed, so that the input can be divided into two fractions. A few units operating in this manner were devised by O'Keeffe et al. [24], Heitala [25], Fischer et al. [26], and Signer and Arm [27]. Details of the operation of the continuous feeding devices may be found in the literature [9,10,13,16].

Another distribution train described by Post and Craig [28] allows a procedure termed "counter double current distribution" (CDCD), which is basically a steady-state procedure. In this system, the sample is continuously introduced at the center of the train while upper and lower phases are both transferred, but in opposite directions. The procedure can be adjusted in such a manner that steady streams of two separated components emerge from the opposite ends of the train. Therefore, it may be possible in CDCD to separate those solutes for which $K_D > 1$ by moving them in one direction while those with $K_D < 1$ will travel in the opposite direction. The solute with $K_D = 1$, tending to accumulate at the center, will eventually move to one end of the train.

At least two more CCD devices based on steady-state distribution (SSD) have been reported. The reader may consult the works of O'Keeffe [24] and Alderweireldt [34–35], as well as the other references on continuous feed distribution systems [24–35], many of which were extensively used in industrial scale separations. The theoretical treatment for continuous feed distribution systems was also worked out [13,33]. As will be seen in the following chapters, the basic ideas of continuous feed distribution methods were adapted in countercurrent chromatography (CCC). For example, the "helix countercurrent chromatograph" [11,37] may be considered a modification of the U-tube apparatus of Jantzen [2], and the "droplet countercurrent chromatograph" (DCCC) [38] may be regarded as a modification of the apparatus of Kies and Davis [36].

Hynninen [13] compared the Craig discontinuous CCD method with the continuous feed transfer method, also known as the Martin-

Synge distribution (MSD), on the basis of (a) real separation powers, (b) reproducibility, and (c) adaptability for analytical or preparative purposes. He also considered suitability for nonideal situations, automation, and the main structural features of some important CCD and MSD units [24–36]. When one looks into the developments in the instrumentation for liquid-liquid partitions, it is apparent that CCC has evolved from CCD and MSD. Therefore it can be considered as an extention of the distribution of solute molecules in liquid-liquid partition, utilizing the principles of CCD and liquid-liquid partition chromatography.

F. Applications

Craig's countercurrent distribution method was extensively used in organic chemistry and biochemistry. It was the method of choice for the separation of any type of molecule, from small molecules to macromolecules, as well as for the separation of particles and cells. In recent years, CCD has been supplanted by advances in chromatography, especially high-performance liquid chromatography (HPLC), because of the ease of operation, suitability for analytical scale separations, and, above all, short separation times. However, CCD is very useful for cleaning up crude extracts of natural products before the active principles in trace quantities are identified by other assay methods. Craig, his associates, and others [4–12] have investigated CCD extensively, developing suitable solvent systems and applying CCD to carbohydrates, fats and lipids, steroids, amino acids, peptides, proteins, and nucleic acids and their derivatives. A good account of CCD applications may be found in [4–12]. Albertsson has summarized the CCD applications in biopolymers, particles, and cells. Because of its excellent resolving power, milder conditions, and absence of molecular size limitations, CCD has been used for the separation of proteins, bacteria, and viruses. Further details of separations, instrumentation, and solvent selection, which are outside the scope of this book, may be found in the references cited above.

II. LIQUID-LIQUID PARTITION CHROMATOGRAPHY

The term "liquid chromatography" (LC) is broadly used to cover the processes both of liquid-liquid chromatography (LLC) and of liquid-solid chromatography (LSC). In LLC, separation of solute molecules is achieved on the basis of their partition differences between the stationary liquid phase which is fixed to a solid support and the mobile phase serving as a chromatographic eluting solvent. This method is also referred to as partition chromatography, a name first used by Martin and Synge [30].

Liquid-solid chromatographic (LSC) operation generally involves the principles of one or more of the following:

1. Adsorption
2. Gel permeation or size exclusion
3. Affinity
4. Ion exchange

Depending on the mechanism of separation involved, LSC may then be called (a) adsorption chromatography, (b) gel permeation, or gel exclusion, or gel chromatography, (c) affinity chromatography, or (d) ion-exchange chromatography. However, LSC does not completely exclude partition chromatography either.

In this book, descriptions of the chromatographic systems based on liquid-solid chromatography are not included, but modern instruments for high-performance liquid chromatography (HPLC) are designed to work on the operating principles 1–4 above (LSC), as well as on partition principles, depending on the packed columns employed for HPLC work. Details concerning LSC may be found in standard treatises on the subject.

Some discussion of liquid phases, sorbents, and chromatographic retention in LLC is included in the following sections because CCC depends on the principles of partition chromatography and countercurrent distribution. Tswett's discovery in 1903 of column chromatography is considered the first liquid chromatographic technique [40]. This simplest method of LC, a form of liquid-solid chromatography, consists of a column packed with a suitable adsorbent such as silica gel or alumina which is wetted with an eluting solvent. A solution of the mixture to be resolved is added to the top of the column, and then the mixture is eluted with a continuous flow of solvent. Separation is based on the differential partition of the solute components in the mixture between the solid support and the liquid phase. Detection is made either by visual inspection of the bands on the column, for colored materials (hence the name chromatography), or by collection and analysis of fractions. Other chromatographic methods include (a) those in which the solid packing material of the column is coated with a liquid film (liquid-liquid chromatography) and (b) those in which the solid support is present as a sheet rather than as a column (thin-layer and paper chromatography).

The original technique of liquid-solid chromatography (LSC) is still widely used, especially as a preparative method, but the major emphasis today is on the use of high-performance liquid chromatography (HPLC) for both analytical and preparative separations.

Table 3 summarizes the differences between classical and high-performance liquid chromatography. HPLC can be used in both the

Table 3 Differences between Classical and High-Performance Liquid Chromatography

Characteristics	Classical liquid chromatography	High-performance liquid-chromatography
1. Efficiency (HETP in mm)	~0.5	~0.1
2. Output (number of plates/sec)	0.2	5
3. Pressure	Fraction to ~5 kg/cm^2	>5, usually 20–300 kg/cm^2
4. Flow rate (mm/min)	5–50	600
5. Time for each run	Hours to days	Minutes to hours
6. Equipment	Simple column and accessories	Chromatograph
7. Orientation	Predominately preparative	Predominantly analytical
8. Detection of separation	Usually analytical, in fractions; sometimes continuous, with a detector	Always with a detector
9. Scale	Microgram to kilogram (even tons)	Nanogram to microgram

Source: From Ref. [122], p. 102.

normal phase mode (liquid-solid column chromatography) and the bonded phase mode (analogous to liquid-liquid partition chromatography), as well as in such other modes as ion-exchange and size exclusion (gel permeation) chromatography. Therefore, HPLC has been considered to be the most versatile fo the chromatographic methods.

The major advantage of the liquid chromatographic methods is that they are applicable, in principle, to all compounds that can be dissolved in some solvent and are not restricted to compounds that can be volatilized. In addition, column liquid chromatography can readily be scaled up to perform preparative separations.

A. Terminology

Martin and Synge [30] coined the term "liquid-liquid chromatography" for their chromatographic system, which involves the partition of the solute molecules between the mobile liquid phase and the stationary liquid phase affixed to a solid support (matrix). Lester-Smith [41] later suggested the term "partition chromatography" for this type of chromatographic separation. Since all chromatographic systems involve partition between the mobile and the stationary phases and since the term "gas-liquid partition chromatography" has been widely used, Morris and Morris [9] preferred the term "liquid-liquid partition chromatography" to distinguish it from other chromatographic processes. Other terms, "liquid-liquid chromatography" and "liquid-liquid column chromatography," have also been used to describe the partition chromatographic process of Martin and Synge. All these terms are used synonymously, and no preferred usage was chosen for this book. In discussing countercurrent chromatography, when we use the term "liquid chromatography," we obviously mean "liquid-liquid partition chromatography."

Martin and Synge regarded the distribution process in liquid-liquid chromatography as solely due to the partition phenomenon. However, later work showed that adsorption effects sometimes play a dominant role in the separation of solute moelcules because (a) the calculated values of partition coefficients differed greatly from experimental results and (b) the variation between experimental and calculated partition coefficients varied for different supporting materials. In order to suppress or minimize adsorption effects, several approaches involving the use of more polar solvents, different solid phases, or a combination of both were found useful in maintaining the distribution process by partition. Essentially, liquid-liquid partition chromatography involves a two-phase solvent system and a supporting solid matrix. They will be described in the following section.

B. Solvent Systems

Well-chosen solvent systems are important to the attainment of suitable distribution ratios of partition coefficients for the type of chromatographic separation to be performed. The following parameters must be taken into account for optimum resolution of solute molecules in liquid-liquid partition chromatography.

1. Distribution ratios in the range 1–100 (while CCC and paper chromatography require distribution ratios of 0.5–9.0) for optimum resolution.
2. Relative volumes of stationary and mobile phases because of the relationship:

$$D_C = K_C \frac{V_s}{V_m} \tag{31}$$

in which D_C is the mass distribution ratio, K_C is the partition coefficient, and V_m and V_s are the volumes of mobile and stationary phases. A wide variation of the magnitude of the term D_C is generally not possible because the ratio V_s/V_m is usually determined by the supporting solid phase and by the method of column packing.

As pointed out by Morris and Morris [9], the solvent system for liquid-liquid partition chromatography must fulfill the following conditions:

1. If supporting phases such as silica gel or Celite are to be employed, the solvent system must form two immiscible phases.
2. The components of the solvent system must not react irreversibly with the solutes.
3. The mobile phase in the solvent system must permit the recovery of the solutes from it (use of easily volatile solvent is encouraged).
4. The partition isotherms of the solutes in the solvent system should be linear over a considerable concentration range (essential for a satisfactory chromatographic zone structure).
5. The separation factor (a ratio of the partition coefficients for similar solutes) should be as large as possible (very difficult to satisfy).

In liquid-liquid partition chromatography, the most important practical problems encountered are (a) the selection of suitable solvents for mobile and stationary liquid phases, (b) the choice of the supporting solid phase, (c) the selection of packing conditions and column operating conditions so that the stationary phase remains immobilized on the solid support matrix. The reader may refer to the publication by F. A. von Metzch [42], who compiled information on solvent systems and their miscibilities for countercurrent distribution and liquid-liquid partition chromatography. Some of the more commonly used solvents are listed in Table 4. It should be kept in mind that the terms "stationary" and "mobile" phases are for "normal" operating conditions, under which the more polar solvent is customarily used as a stationary phase and the eluting "mobile" phase solvent is of nonpolar type. Under "reversed phase" conditions, the less polar solvent can be used as the "stationary" phase.

Table 4 Some Solvents Useful in the Preparation of Stationary and Mobile Phases for Liquid-Liquid Chromatography

Stationary phase	Mobile phase	Water-miscible, for mobile phase
(A) Normal Phase	1-Butanol	1-Propanol
	2-Butanol	2-Propanol
Water	1-Pentanol	Acetone
Aqueous acids and bases	Benzyl alcohol	Tetrahydrofuran
Buffers	Phenol	Pyridine
Salt solutions	Isobutyric acid	Lutidine
Aqueous methanol and ethanol	Methyl ethyl ketone	Collidine
	Ethyl acetate	
Alcohols (methyl, ethyl, and furfuryl)	Amyl acetate	
	Benzene	
Glycols (ethylene and propylene)	Toluene	
	Chloroform	
Ethylene glycol monoethyl ether	Carbon tetrachloride	
	Cyclohexane	
Formamide	Hydrocarbon mixtures	
Nitromethane	Ethers (ethyl and isopropyl)	
(B) Reversed Phase		
Butanol		
Octanol		
Chloroform		
Hydrocarbons		
Hydrocarbon-ester mixtures		

Source: Ref. 19.

C. Sorbents and Supports

In liquid-liquid partition chromatography, the term "sorbent" is defined as a liquid with a sorption activity fixed onto a solid support which may either be inert or may affect the properties of the fixed liquid. There are three general types of liquid-liquid systems [123]:

1. Systems in which the amount of the stationary liquid exceeds that necessary for coating the carrier with a monomolecular layer of the liquid.
2. Systems in which the amount of the stationary liquid is less than what would correspond to a monomolecular layer.
3. Systems in which the active sorbing phase is bound chemically to the support surface.

From the foregoing description, it is evident that there is no strict distinction between liquid-liquid partition systems. The support surface may exert its sorbing property even when the amount of stationary phase considerably exceeds the amount necessary for covering the carrier with a monomolecular layer.

A large amount of stationary phase on the support surface permits a higher capacity ratio to be obtained on shorter columns or on columns with a smaller weight of the support. In practice, 15–20% (w/w) of the stationary liquid is fixed on the support, and then the sorbing effect in most cases is not noticed.

If supporting solid phases such as silica gel or Celite are to be employed, the solvent system must form two immiscible liquid phases (polyacrylamide gel supporting systems are an exception to these conditions because only one phase is present). The essential requirements of the supporting solid phases are that they should [9,123]:

1. Possess a high capacity for the stationary phase
2. Retain the stationary phase firmly
3. Not interact with the solute molecules, and
4. Be finely divided for maximum retention of the stationary phase

The other two features of the solid support that need to be taken into consideration are

1. Interaction with the components of the solvent system, and
2. Complexing with the stationary phase, which is sometimes favorable for separation

The following paragraphs summarize data on some generally used supporting phases [9]:

1. *Kieselguhr* (Celite), composed of the siliceous skeletons of diatoms (diatomaceous earth), has high porosity and thus has a high retention capacity for liquids, which are held in the pores by surface forces. Several chromatographic grades of kieselguhr (such as Chromosorb A, G, P, and W, Celite 545, Hyflo Super-cel) are available for use either in normal phase or as silanized derivatives in reversed phase chromatography.

2. *Cellulose,* although less used as a supporting phase for liquid-liquid chromatography, is available as purified cellulose powders from various suppliers, including Whatman and Nagel.

3. *Sephadex,* a crosslinked dextran, has many properties which make it an excellent supporting matrix for liquid-liquid chromatography. Three forms of Sephadex—G-25, G-200, and LH-20—have been found to be suitable supporting phases for the separation of various mixtures.

4. *Silica*, in the gel form, was the first supporting matrix used in this form of chromatography. Several forms of silica, such as spherical microporous beads (Porasil, Sperosil) and its surface-bonded materials (Durapak, Chromosord, and Carbowax), are now available as excellent supporting phases for liquid-liquid partition chromatography.

5. *Synthetic polymers,* originally produced for use in gas-liquid chromatography, have been found useful as support materials for liquid-liquid chromatography. They include several forms of Chromosorbs (Johns-Manville), Poragels and Porapaks (Waters), and Corvics (ICI) as well as the halogenated hydrocarbons such as PTFE, Kel-F, and Teflons, all of which have excellent retention capacity for the stationary phases and make excellent support media in liquid-liquid partition chromatography.

Supports can be classified into three categories:

1. Supports with a large specific surface area (adosrbents such as silica gel)
2. Supports (same as those used in gas chromatography) with (a) a small specific surface area, (b) a low sorption activity of the surface, and (c) a large volume of internal pores of relatively large pore diameter (0.1–1 μm) in the support
3. Supports of controlled surface porosity with a small specific surface area (usually several m^2/g)

In liquid-liquid partition chromatography, the stationary liquid phase is applied to the support in such a way that the first portions of the applied liquid cover the support surface with a monomolecular layer, subsequent portions may contribute to the addition of more molecular layers, and the final portions of the liquid collect in the carrier pores. Care should be exercised to prevent the formation of an excessively thick liquid layer. With an increasing amount of liquid, the interface between

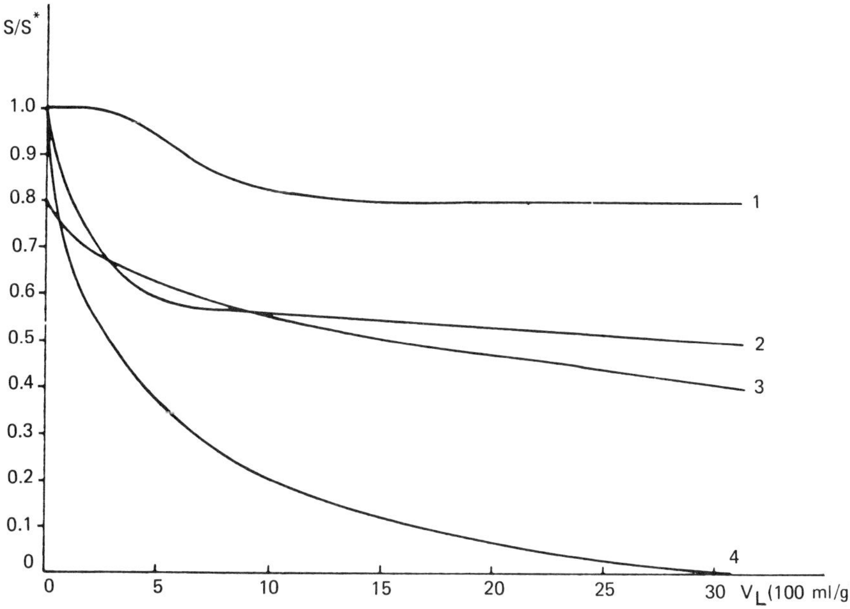

Fig. 9 Decrease in the adsorbent specific surface area as a consequence of introducing a liquid. S = surface area of wet adsorbent; S* = surface area of dry adsorbent; V' = volume of coated liquid. 1, Silica gel with a chemically modified surface; 2, silica gel; 3, Chromosorb W; 4, Rysorb. (Reprinted with permission from J. Janák, J. Čoupek, M. Krejčí, O. Mikeš, and J. Turková, in *Liquid Column Chromatography*, (Z. Deyl, K. Macek, and and J. Janák, eds.), Elsevier, Amsterdam, 1975. Copyright 1975 Elsevier Scientific Publishing Company.)

the mobile phase and the stationary phase decreases as shown in Fig. 9. The decrease of the surface area is characteristic and is a function of the pore volume and the specific surface area. Relative values listed in Fig. 9 are expressed for unit surface area of the original dry carrier. Specific surface areas of some supports are given in Table 5.

D. Mobile Phases

Equilibrium Between the Mobile and Stationary Phases

Since the fundamental process in liquid-liquid partition chromatography is distribution, the technique can be considered as a repeated distribution of solutes between two immiscible liquids. A high efficiency of separation may be accomplished by a frequent repetition of the elementary distribution process, which is effected by the

Table 5 Specific Surface Areas of Some Supports

Support material	Specific surface area (m^2/g)
1. Charcoal	400–1000
2. Silica gel	5–400
3. Alumina	100–200
4. Diatomaceous earth	1–7
5. Corasil	7
6. Zipak	1
7. Surface-etched beads	0.4

Source: From Ref. [123], p. 184.

movement of at least one phase. From these basic characteristics, the following conclusions can be drawn [14].

1. The choice of stationary and mobile phases should be made in such a manner that they are very poorly soluble but are sufficiently good solvents for all the solute components of the sample.
2. The distribution isotherm for the selected phases should be either a straight line or at least close to a straight line, in order to obtain symmetrical elution zones.
3. It is generally desirable to suppress maximally the adsorption of the sample on the carrier used, a situation analogous to that in gas-liquid chromatography.
4. If the sample and the mobile phase are poorly soluble in the chosen stationary phase, the controlling mechanism may include the process of adsorption of the sample on the surface of the stationary liquid.
5. Because two completely immiscible and mutually insoluble liquids do not exist, the experimental conditions should be so chosen that the passage of the mobile phase through the column does not result in wasting out of the stationary phase and changing of the column properties.

Selection of Liquid-Liquid Partition Systems

The choice of the partition system in liquid-liquid partition chromatography is based on its selectivity, which is a property characteristic of a given system of two phases with respect to the solute components under investigation. We can try discussing the effect of a single phase on the system selectivity and use each component of the chosen liquid pair, especially when both are of low molecular weight, as either the stationary or the mobile phase. Since the calculations for selectivity are based on the relative retention of the separated components, it is irrelevant which of the liquids was chosen for the stationary phase or the mobile phase.

As has already been mentioned, an important criterion for selectivity in liquid-liquid partition chromatography is the immiscibility of the chromatographic phases. The liquid phases are chosen either empirically (a common practice) or with the aid of the theory of solutions. The main reasons for selecting them empirically are (a) the inadequacy of the theory of solutions for giving an effective and sufficiently precise prediction of the partition coefficients in any given system; (b) a lack of such essential data as molar volumes, critical temperatures, and solubility parameters; and (c) to a certain extent, also the difficult calculations involved. The approach based on the theory of solutions involves the determination of partition coefficients from the characteristic parameters of the sample and both liquid phases [43]. For this approach, any one of the following procedures can be used [14]:

1. The theory of regular solutions as proposed by Hildebrand and Scott [44,45]
2. The method of structural increments, as reported by several workers [46—52]
3. The sum of the partition coefficients, which may sometimes be useful when choosing a system of phases

Although other procedures are equally important and are used by others, only the first method, which is based on the phenomenon of polarity, will be discussed here. In general, it can be said that one of the conditions necessary for achieving adequate selectivity in a liquid-liquid partition chromatographic system consists of those differences in the properties of the phases which are usually characterized as differences in polarity (Tables 6 and 7). The polarity of a liquid phase is the result of intermolecular reactions in the liquid, which are divided into nonspecific (caused by dispersion forces) and specific (caused by other interactions) types. The different types and intensities of these interactions are evident not only from the different behavior of liquids as solvents, but also from their boiling points and from their latent heats of vaporization. The stronger are the intermolecular interactions, the more effective the given liquid usually as a solvent for polar solutes, and the higher its heat of vaporization. Its magnitude, relative to unit volume, was used by Hildebrand and Scott [44,45] for the quantitative characterization of the magnitude of the intermolecular forces.

Table 6 Selected Analytical Properties of Some Common Mobile Phases

Mobile phase	Refractive index, n	Dielectric constant, ε	UV cutoff (nm)	B.P. (°C)	Viscosity at 20°C (cp)
n-Pentane	1.358	1.844	210	36	0.23
n-Decane	1.412	1.991	210	174	0.92
Cyclohexane	1.427	2.023	210	81	1.00
Cyclopentane	1.406	1.965	210	49.5	0.47
1-Pentane	1.371	2.100	—	29.2	—
Carbon disulfide	1.626	2.641	380	45	0.37
Xylene	1.50	2.270-2.568	290	136-145	0.62-0.81
Diisopropyl ether	1.368	3.880	220	69	0.37
Toluene	1.496	2.379	285	110.8	0.59
Chlorobenzene	1.525	5.708	—	132	0.80
Benzene	1.501	2.284	280	80.2	0.65
Diethyl ether	1.354	4.005	220	34.5	0.23
Chloroform	1.489	4.006	245	64.2	0.57
Dichloromethane	1.424	9.008	245	40	0.44
Methyl ethy ketone	1.381	18.5	330	79.6	—
Acetone	1.359	20.7	330	56	0.32
Dioxane	1.422	2.209	220	101.4	1.54
Ethyl acetate	1.372	6.02	260	77.2	0.45
Methyl acetate	1.362	6.68	260	57.5	0.37
Amyl alcohol	1.410	13.9	210	138.3	4.10
Aniline	1.586	6.89	330	184.4	4.40
Pyridine	1.510	12.3	305	115.4	0.90
Isopropanol	1.378	18.3	210	82.2	2.81
n-Propanol	1.385	20.1	210	87.2	2.21
Ethanol	1.361	24.40	210	78.3	1.21
Methanol	1.331	32.63	210	64.7	0.60

Table 6 (Continued)

Mobile phase	Refractive index, n	Dielectric constant, ε	UV cutoff (nm)	B.P. (°C)	Viscosity at 20°C (cp)
Ethylene glycol	1.432	37.7	210	197.2	19.1
Acetic acid	1.372	6.15	—	118.1	1.21
Water	1.380	80.37	—	100	1.00

Source: From Ref. [122], p. 154.

Table 7 Eluotropic Series of Solvents

Solvent[a]	Boiling point	Dielectric constant ε, 20°C	Interfacial tension with water in dynes/cm (20°C)	Viscosity in cp (20°C)
n-Hexane	68.7	1.890	51.4	0.326
Heptane	98.4	1.924	50.4	0.408
Cyclohexane	81.4	2.023	—	1.02
Carbon tetrachloride	76.8	2.238	45.1	0.969
Benzene	80.1	2.284	35.0	0.652
Chloroform	61.3	4.806	27.7	0.580
Diethyl ether	34.6	4.34	9.7	0.233
Ethyl acetate	77.1	6.02	6.3	0.455
Pyridine	115.3	12.3	b	0.974
Acetone	53.5	20.7	b	0.316
Ethanol	78.5	24.30	b	1.2
Methanol	64.6	33.62	b	0.597
Water	100	80.37	b	1.005

[a]Additional solvents and their properties can be found in *Handbook of Chemistry and Physics*, CRC Press, Boca Raton, (59th ed., 1972–1973).
[b]Miscible in water.
Source: Ref. 17.

Hildebrand and Scott [44,45] employed the solubility parameter δ as a measure of polarity. The solubility parameter is expressed as

$$\delta = \left(\frac{\Delta E^V}{V}\right)^{1/2} \tag{32}$$

where ΔE^V is the molar heat of evaporation and V is the molar volume. The solubility parameters for liquids used as mobile phases range between 6 for perfluoroalkanes and 21 for water, as shown in Table 8. When liquids are arranged in order of increasing δ, the series obtained is ordered very similarly to some eluotropic series. Hence, the solubility parameter can also be considered to be a measure of the elution power of a solvent in liquid-liquid partition systems. It is important to note that there is no direct relationship between ε^0, characterizing the elution power in liquid-solid systems, and the solubility parameter. Although there is a close relationship between the solubility parameter and actual elution behavior, appreciable discrepancies are found in some instances, probably because of the effect of specific interactions. Therefore, the magnitude of δ may not always be a sufficiently reliable guide, but it is a good indicator for solvent selection in liquid-liquid partition chromatography [14].

E. Chromatographic Column Operations

Generally, the method for packing columns for liquid-liquid partition chromatography differs from those for adsorption and ion-exchange columns because the stationary liquid and supporting solid phases must be mixed before introduction into the column. Two types of column-packing operations are adapted. For supporting materials involving silica, kieselguhr, and synthetic polymers, a slurry is prepared and equilibrated by mixing before introduction into the column. The column thus packed is homogeneous and as hard as possible. Thereafter, a solution of the solute sample, preferably dissolved in the mobile phase as solvent, is added at the top of the column. After that, the column is eluted with the mobile phase. The rest of the chromatographic process is similar to liquid-solid adsorption chromatography (Morris and Morris [9]).

Packing with cellulose or crosslinked dextrans (Sephadex) requires an entirely different technique. The solid support is pre-equilibrated with the appropriate phase of the two-phase mixture for 24 hr at constant temperature. It is then packed into the column by conventional methods. These operations include washing the column with the stationary phase. For example, a well-packed kieselguhr column should have efficiency corresponding to 15–20 equivalent

Table 8 Eluotropic and Solubility Parameters for Selected Solvents

Solvent	ε^0 (Al_2O_3)[a]	δ[b]
Fluoralkane	−0.25	6.0
$CFCl_2$-CF_2Cl	—	6.2
Isooctane	0.01	7.0
Diisopropyl ether	0.28	7.0
n-Pentane	0.00	7.1
n-Hexane	0.01	7.3
n-Heptane	0.01	7.4
Diethyl ether	0.38	7.4
n-Decane	0.04	7.8
Cyclopentane	0.05	8.1
Cyclohexane	0.04	8.2
Tetrachloromethane	0.18	8.6
Ethyl acetate	0.58	8.6
Ethyl bromide	0.37	8.8
m-Xylene	0.26	8.8
Toluene	0.29	8.9
Chloroform	0.40	9.1
Tetrahydrofuran	0.45	9.1
Benzene	0.32	9.2
Methyl ethyl ketone	0.51	9.3
Acetone	0.56	9.4
Dichloromethane	0.42	9.6
Chlorobenzene	0.30	9.6
1,2-Dichloroethane	0.49	9.7
Dioxane	0.56	9.8
Carbon disulfide	0.15	10.0
n-Propanol	0.82	10.2

Table 8 (Continued)

Solvent	ε^0 $(Al_2O_3)^a$	δ^b
Ethanol	0.88	11.2
Acetonitrile	0.65	11.8
Acetic acid	1.00	12.4
Methanol	0.95	12.9
Water	—	21.0

[a]Elution strength in liquid-solid systems for elution from alumina.
[b]Solubility parameters calculated from boiling points.
Source: Ref. 14.

theoretical plates per cm. A new equilibrium is attained when the column is eluted with the mobile phase prior to introduction of the solute sample solution. The other requirements during the operation of elution chromatography are (a) maintenance of constant temperatures, (b) optimal flow rates of 2–4 ml/cm^2 for analytical separations, and (c) preferred flow rates of 15–20 ml/cm^2 for preparative purposes.

F. Applications of Liquid-Liquid Partition Chromatography

Since the introduction of this chromatographic technique by Martin and Synge, it has been widely used for the separation of small and macromolecules. Because the literature on the application of this method has been extensively covered in the form of several review articles and books, we will not discuss it in detail. To illustrate the importance of the Martin and Synge technique as a powerful separation method, we will cite here a few examples from the published literature.

Because of the convenience, ease of operation, and resolving power of other methods (e.g., gas chromatography and liquid-solid adsorption chromatography), applications of liquid-liquid partition chromatography to simple aliphatic and aromatic hydrocarbons, alcohols, phenols, aldelydes, ketones, and heterocyclic compounds have not come into wide use. The method has, however, been used for the separation of steroids, lipids, organic acids, peptides, nucleic drates, as well as the separation of amino acids, peptides, nucleic acids and their derivatives, and proteins. Table 9 lists a few examples, giving the solid support but not the more complex details of the stationary and mobile phases. References are provided in this table, so that the reader may consult the original articles. For further reading in the general area of HPLC, a number of references [71–81] are available.

III. COUNTERCURRENT CHROMATOGRAPHY

A. Historical Development of Liquid-Liquid Partition Systems

It has been a well-documented phenomenon that a mixture consisting of similar molecules can be separated into individual components on the basis of their differences in solubility between two immiscible phases. This phenomenon subsequently led to the discovery of several highly selective separatory techniques such as paper chromatography, countercurrent distribution, and liquid-liquid chromatography. The high selectivity of these techniques arises form the fact that the partition coefficient is a fundamental reflection of the intermolecular forces, which can be greatly influenced by conditions of temperature and phase composition. Thus, by a judicious choice of conditions, highly specific interactions can often be exploited that permit the resolution of selected components from very complex mixtures.

In separation science, chromatography is the principal tool for resolving complex mixtures into individual components. Chromatographic techniques such as paper chromatography, high-resolution column chromatography, and high-performance liquid chromatography were developed on the basis of liquid-liquid partition principles. All of these methods require a solid support that sometimes causes problems, such as loss of samples by adsorption, and chemical decomposition depending on the nature and reactivity of the components in the mixture. To date, more limited recognition has been attained by alternate separation systems which are also based on solvent-solvent partition principles but use no solid supporting matrix. The chronological development of some of these systems is summarized in Table 10.

Extraction employing the separatory funnel and similar devices is a very interesting chemical art. The origin of this classical liquid-liquid partition device is difficult to trace. Perhaps it originated with the alchemists. When two immiscible solvents containing solutes are shaken in a separatory funnel, the separated solutes are partitioned between the phases. The ratio K_C of the solute concentrations in the phases is called the partition coefficient. If the partition coefficients of two different solutes are considerably different, a one-step extraction in a separatory funnel may suffice for their separation. As the nature of solutes becomes more similar, the

Table 9 Examples of the Application of Liquid-Liquid Partition Chromatography

Compound classes	Stationary phase (solvent system)[a]	Mobile phase	Solid support	Ref.
(1) Steroids	Aq. methanol, acetic acid, formamide	Mixtures of various hydrocarbons	Kieselguhr (Celite or Hyflo Super-cel	[54]
	Several solvents	Several solvents	Sephadex derivatives	[55]
(2) Carbohydrates				
Monosaccharides	(Ethylacetate-1-propanol-water (4:2:1)]		Celite 545	[56]
	[Salt-ethanol]		Modified Sephadex and Dowex	[57]
Polysaccharides (mucopolysaccharides)	3% Laurylamine in n-butanol	Aq. NaCl and n-butanal mixture	Celite 545	[58]
(3) Amino acids and their derivatives	[Heptane-pentane-2-butanone]		Silica	[30,59]
Peptides	[Several solvents including 1-butanol-aq. aq. acetic acid-benzene-pyridine]		Sephadex	[60] [60]
(4) Lipids	[Chloroform-methanol-water-acetic acid]		Sephadex G-25	[61]
	[Benzene-methanol]		Silica	[62]

(5) Organic acids, including fatty acids	[Methanol-isooctane]		Silica	[63]
	Heptane	Acetonitrile-methanol	Silanized Celite 545 (reversed phase)	[64]
(6) Nucleic acids and	[Ethyl acetate-2-ethoxyethanol-water (4:1:2)]		Celite 545, Microcel E	[65]
	[2-Ethoxyethanol-2-butoxyethanol-1.25 M phosphate buffer (ph 6.88)-triethylamine (2:1:6: 0.1—0.5)]		Sephadex G-25	[66]
(7) Porphyrins	Tri-n-butyl phosphate	0.01 N HCl or 0.066 M phosphate buffer	Polytrifluorochloroethylene (Hostaflon C2, Hoechst) (reversed phase)	[67,68]
(8) Proteins	[Aq. $(NH_4)_2SO_4$-ethyl Cellosolve]		Celite	[69]
(Ribonuclease)	[Ethylene glycol alkyl ethers-aqueous salt solutions]		Silanized Celite	[70]

[a]The alternative heading "[Solvent system]" refers to the cases in which the separated phases are not listed. It is understood that the solvent system will split into two phases.
Source: Compiled from text of Ref. 19.

Table 10 Liquid-Liquid Partition Systems

Apparatus	Inventor(s)	Year of invention	Ref.
1. Separatory funnel	—	—	—
2. U-tube cascade unit	Jantzen	1932	[2]
3. Discontinuous feed distribu-bution unit	Watanabe & Morikawa	1933	[31]
4. Packed/perforated plate column	Van Dijek	1935	[82]
5. Continuous multislope extractor	Van Dijek et al.	1937	[83]
6. Spinning rotor column	Schutze et al.	1938	[84]
7. Cascade extraction train	Martin and Synge	1941	[29]
8. Partition chromatography	Martin and Synge	1941	[30]
9. Discontinuous CCD	Craig	1944	[17]
10. Column with mix and settle zones	Scheibel	1948	[85]
11. Discontinuous feed unit	O'Keeffe et al.	1949	[24]
12. Cascade percolator train	Kies and Davis	1951	[36]
13. Large glass apparatus—CCD	Craig et al.	1951	[86]
14. Rotating disk apparatus (Ronor column)	Signer and Arm	1952	[27]
15. Spinning rotor columns	Spence and Streeton	1952	[90]
	Short	1953	[91]
16. Compartmental perculator column	Keppes	1954	[92]
17. Continuous feeding unit	Fischer et al.	1954	[26]
18. Microdiscontinuous CCD	Bell et al.	1956	[93]
19. Automated 3-phase CCD	Meltzer et al.	1958	[94,95]
20. More compact CCD	Raymond	1958	[96]
21. Alternate withdrawal decanta-tion unit	Alderweireldt	1958	[97]
22. High-efficiency continuous mix-settle	Hietala	1960	[25,98]

Table 10 (Continued)

Apparatus	Inventor(s)	Year of invention	Ref.
23. Counter double current distribution system	Post and Craig	1963	[28]
24. Countercurrent centrifuge	Tavel and Bollinger	1968	[99]

Source: Ref. 15a.

difference in their partition coefficients decreases, requiring a multistep extraction for separation.

The successful development of modern laboratory scale countercurrent extraction began in the 1930s, although mixer-settler units, as well as the countercurrent extraction units, were employed in industrial processes prior to that time [2,31,82] (based on an original concept that was developed by Jantzen [2]). In 1941 Martin and Synge [29] described an extraction apparatus of the cascade type. During the design of this apparatus, it occurred to them to support one liquid by adsorption into a porous solid matrix held in a tube and to percolate a second immiscible liquid phase through it. These studies led to the discovery of partition chromatography [30]. Martin and his coworkers abandoned further development of countercurrent devices and proceeded to develop paper chromatographic processes in which one phase is supported on a matrix that is ideally inert.

A few years later, Craig [17], following along the lines of Martin and Synge, invented a discontinuous countercurrent distribution (CCD) apparatus which is a true liquid-liquid distribution train. Subsequently, Craig and his coworkers [86] in 1951 reported the development of a larger, all glass, automatic unit. The Craig apparatus performs a discontinuous, stepwise process in which the phases are shaken until they attain equilibrium. The phases are then allowed to separate by settling, after which all portions of one phase are simultaneously removed from their respective tubes and transferred as discrete batches to the next higher tubes without any mixing of the individual portions. Although the Craig distribution is commonly referred to as a countercurrent process, only one phase is moved, and in a true sense it is only a differential migration method. It is quite analogous to partition chromatography, except for the stepwise movement of the mobile phase.

The original Craig distribution apparatus has undergone some modifications which adapt the technique for handling both large and small samples. In these instruments [27,93,96] as many as 1000 distribution steps can be efficiently accomplished. Following these developments, Post and Craig [28] proceeded to describe a distribution train which allows a procedure called counter double current distribution to be performed. This is a steady-state procedure in which the sample is continuously introduced in the center of the train while the upper and the lower phases are both transferred, but in opposite directions. The procedure can be adjusted in such a way that the steady streams of two separated components emerge from opposite ends of the train. Subsequent developments based on the steady-state procedure include an apparatus which permits transfer of either the upper or lower phase [34,35]. It also permits the ratio of upper to lower phase solvent volume passing through the train to be varied by changing the ratio of the number of upper and lower phase transfers.

While the original CCD apparatus was undergoing several modifications as described above, other partition systems were reported. For example, Scheibel [85] described countercurrent extraction columns with mixing and settling zones in 1948. These and other, similar columns are widely used industrially. In 1952, Signer [27, 87–89] described a rotating disk column which was later marketed by Brinkmann as the Ronor column. This apparatus contained up to 400 stages and could easily hand a 10-g sample. Other devices include a spinning rotor column by Spence and Streeton [90] (originally investigated by Schutze et al. [84]), a highly efficient, relatively large-capacity continuous mixer-settler by Kietala [25], a successful cascade percolator train by Kies and Davies [36], the compartmental percolator column by Keppes [92], and other units that are listed in Table 10. Although these devices have not all achieved commercial success, they contributed to the development of countercurrent chromatographic instruments. The cascade percolator train appears to have influenced the development of the droplet countercurrent chromatograph by Tanimura et al. [38] in 1970, and the spiral countercurrent apparatus of Jantzen [2] led to the design of the helix CCC [103], described elsewhere in this book.

The countercurrent distribution (CCD) method has gained wide acceptance in the fields of organic chemistry and biochemistry since its discovery in 1944. Although little used today for analytical separations, apparatus based on the Craig design still has its advantages for the separation of such polar compounds as polypeptides, other macromolecules, and other natural products. It is particularly adaptable for large-scale separations. A few disadvantages of CCD are the following: (a) the complexity of the apparatus, (b) its fragility, (c) the formation of emulsions that are sometimes hard to break, (d) the fact that the volume of solvent required is fixed by the size of the apparatus and is often excessive, and (e) the long separation times.

Many of the analytical applications of CCD have been superseded by recent advances in HPLC, which is claimed to be much faster, easier to operate, and much more economical of sample size. However, in instances where adsorption or sample denaturation occurs on columns, CCD has had advantages over HPLC. The very high sample capacity of CCD is still unequaled. For example, the CCD apparatus manufactured by Labortec, Münchenstein, Basel contains over 200 stages of 2 L each and is capable of processing samples of kilogram size.

B. Countercurrent Chromatography

Ito and his coworkers, first in Japan and later in the National Institutes of Health in Bethesda, Maryland, noticed that immiscible liquids in coiled, narrow-bore tubes could be segmented and made to undergo countercurrent flow. Phase separation will occur in the gravitational field with tubing of about 2 mm internal diameter, but it is enhanced by rotation of the column in a centrifugal field, particularly in tubes of very small (0.2 mm) internal diameter. The process is continuous. A sample is introduced at the column inlet, and the separated constituents are monitored in the column effluent as in chromatorgraphy. In other words, Ito and his coworkers [100] combined centrifugal and, later gravitational force with the principle of the Archimedes screw to cause two liquids to counterflow past the sample placed initially at the interface of the liquids in a helical coil of plastic tubing. Although there were at least two prior studies [101,102] employing helical devices, they did not introduce centrifugal force and were not employed for chromatography. Ito's instrument [100] therefore introduced new principles into countercurrent distribution. There is no solid support, thereby eliminating adsorption effects and enabling the separation, which is achieved entirely by liquid-liquid partitioning, to be predicted from a knowledge of the partition coefficients and the column efficiency. Although Ito's first centrifugal chromatograph [100], called the coil planet centrifuge, could accept samples of only a few μL (μg—mg samples) and was fairly tedious to load, it was reported to be commercially successful and was marketed in Japan as the CPC Analyzer. This was only the first of a series of unique countercurrent chromatographic instruments developed by Ito and his coworkers, whose inventions are summarized chronologically in Table 11.

The devices designed by Ito and his collaborators use either gravitational or centrifugal force, and all utilize a form of countercurrent chromatography (CCC). Droplet countercurrent chromatography (DCCC) is based not on a coiled tube but on the passage of small droplets of one phase through a series of vertical tubes that hold the immiscible stationary liquid. DCCC is widely used, especially in Japan and Europe, because of the availability of commercial instru-

Table 11 Countercurrent Chromatographic Instruments

Instrument	Authors	Year reported	Ref.
1. Coil planet centrifuge (CPC)	Ito et al.	1966	[100]
2. Helix CCC	Ito and Bowman	1970	[103]
3. Droplet CCC	Tanimura et al.	1970	[38]
4. Locular CCC—rotation and gyration	Ito and Bowman	1970	[104]
5. Flow-through coil planet centrifuge—pulley drive	Ito and Bowman	1971	[37,105]
6. Elution centrifuge—planetary pulley drive	Ito et al.	1972	[106,107]
7. Angle rotor CCC	Ito and Bowman	1975	[108]
8. Preparative CCC—slow rotating helix	Ito and Bowman	1977	[109,110]
9. Horizontal flow-through coil planet centrifuge	Ito and Bowman	1978	[111-113]
10. Nonsynchronous coil planet centrifuge	Ito et al.	1979	[114,115]
11. Toroidal coil planet centrifuge	Ito	1980	[116,117]
12. Preparative CCC—rotating coil assembly	Ito and Bhatnagar	1981	[118]
13. High-speed CCC	Ito	1981	[119,120]

Source: *Source*: Ref. 15b.

ments. A great many difficult separations have been achieved by DCCC. In the locular CCC aparatus, the stationary liquid is segmented in small individual cells separated by very narrow passages.

The main feature of the CCC instruments introduced by Ito and his collaborators after 1971 is the introduction of either a pulley drive unit or a gear drive unit which does not require rotating seals in the influent and effluent lines. This was a major advance in CCC because it provides for continuous inlet and outlet lines without the effects of frictional heating and problems of wear and leakage associated with rotating seals. Additionally, elimination of rotating seals, which are prone to contamination, permits the interior of the instrument to be sterilized for certain biological applications.

The versatile CCC instruments developed by Ito and collaborators are efficient forms of countercurrent chromatographs, permitting separations of samples ranging from a few milligrams to grams. The earlier designs of CCC took long separation times. That problem has been overcome by the latest multilayer instruments [119,120] that handle samples on both the analytical and the preparative scales. The greatest advantage of the multilayer instruments is their reduction of the separation time, making CCC comparable to HPLC both in speed and in resolving power. They make possible the application of CCC to the separation of virtually all types of compounds, as well as particles such as blood cells and tissue cells.

As mentioned earlier, countercurrent distribution played an important role in separation science for about 35 years, until most of its analytical applications were supplanted by HPLC because of the speed, resolving power, and ease of operation of HPLC. However, CCD retains certain special advantages over HPLC because of its superiority for the separation fo polar biological materials. CCC is in many ways superior to CCD. It is fast and convenient, and can employ a wide variety of solvent systems. The capacity of the apparatus can be varied by changing the column size to accommodate samples of different size, making solvent usage more efficient. A comparison of CCD and CCC for separation of bovine insulin is shown in Fig. 10. Prior to Craig's work, the impurity, a deaminated insulin, could not be resolved either by electrophoresis or by ultracentrifugation. The separation of bovine insulin by CCC was reported to be superior and is achieved in less than 6 hr [121].

C. Terminology Used in CCD and CCC*

In laboratory practice, the terms countercurrent extraction and countercurrent distribution are used synonymously. They mean the process whereby a stream of one phase of a two-phase liquid-liquid system is caused to flow past the second phase, which is held stationary in the apparatus as either a continuous or a segmented bed. Only one effluent stream is produced.

In industrial practice, the term countercurrent extraction signifies a process in which two influent streams enter from opposite ends of a contactor and flow in opposite directions, resulting in two effluent streams. One variant of discontinuous countercurrent distribution designed by Post and Craig [24] is called counter double current distribution, to denote that both phases move in opposite directions.

In centrifugal countercurrent chromatography, analogous to CCD, one phase remains stationary while the second phase is passed

*Preferred terminology for CCC.

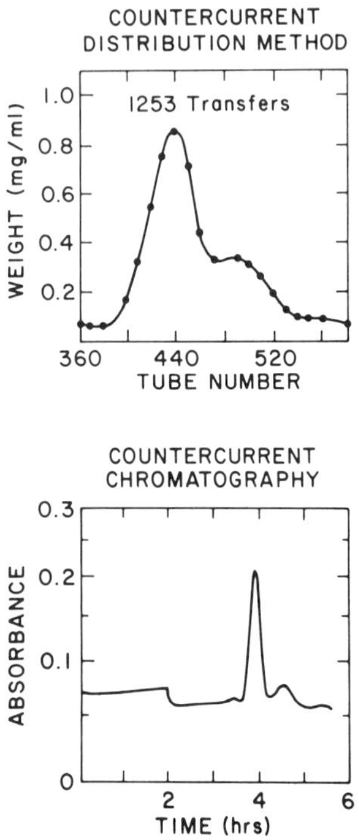

Fig. 10 Comparison of CCD and CCC for separation of bovine insulin. Countercurrent chromatography using flat-coiled helix column wound from 90 m of 0.55-mm i.d. PTFE tubing. Phase system: 2-butanol/dichloroacetic acid/water (1:0.03:1). Sample: 0.5 ml of 1% (w/v) solution of bovine insulin in lower phase. Speed: 700 rpm, 30.7-cm radius. Flow-rate: 6 ml/hr. (From Y. Ito, R. E. Hurst, R. L. Bowman, and E. K. Achter, in *Separation and Purification Methods* (E. S. Perry, C. J. van Oss, and E. Grishka, eds.), Volume 3, Marcel Dekker, Inc., New York, 1974, 1975. Copyright 1974, 1975 Marcel Dekker, Inc.)

through the segments of stationary phase. A true counterflow of liquids is produced in the coil planet centrifuge [100]. Although it is feasible, continuous counterflow of both liquids is yet to be demonstrated in a centrigugal chromatograph of the horizontal flow-through type.

Since countercurrent chromatography can be operated by using either phase as the stationary phase, the following terminology (denoted by symbols K_E and K_R) is suggested in this book.

$$K_E = \frac{\text{concentration of the solute in mobile phase}}{\text{concentration of the solute in stationary phase}}$$

$$K_R = \frac{\text{concentration of the solute in stationary phase}}{\text{concentration of the solute in mobile phase}}$$

Therefore,

$$K_E = \frac{1}{K_R}$$

Similar relationship holds true for K_D and K_C which are used in countercurrent distribution and partition chromatography, respectively, except as discussed with Eqs. (4a) and (4b).

As indicated by their subscripts (E for elution and R for retention), solute with the highest K_E (or lowest K_R) is eluted first and solute with the highest K_R (or lowest K_E) is retained longest in the column. These new terms for partition coefficients also provide a convenient notation for the choice of the mobile phase by inserting it in parentheses. For example, K_E(u/1) indicates the upper phase mobile and the lower phase stationary and K_R(aq/nonaq), the aqueous phase stationary and the nonaqueous phase mobile.

REFERENCES

1. W. Nernst, *Z. Physik. Chem.*, *8*, 110 (1891).
2. E. Jantzen, *Das fraktionierte Distillieren and das fraktionierte Verteilen*, Dechema Monograph, Vol. 5, No. 48, Verlag Chemie, Weinheim, 1932, p. 81.
3. R. E. Cornish, R. C. Archibald, E. A. Murphy, and H. M. Evans, *Ind. Eng. Chem.*, *26*, 397 (1934).
4. L. C. Craig and D. Craig, *Laboratory Extraction and Countercurrent Distribution* in *Technique of Organic Chemistry*, Vol. 3, Part I, A. Weissberger (ed.), Interscience, New York, 1956, pp. 149–332.
5. L. C. Craig, in *Comprehensive Biochemistry* (M. Florkin and E. H. Stotz, eds.), Vol. 4, Elsevier, Amsterdam, 1962, pp. 1–31.
6. T. P. King and L. C. Craig, *Countercurrent Distribution* in *Methods of Biochemical Analysis* (D. Glick, ed.), Vol. 10, Interscience, New York, 1962, pp. 201–228.
7. E. Stahl, in *Thin-Layer Chromatography. A Laboratory Handbook* (E. Stahl, ed.; M. R. F. Ashworth, trans.), Springer-Verlag, New York, Inc., New York, 1969, p. 202.
8. L. C. Craig and T. P. King, *Encycl. Ind. Chem. Anal.*, *1*, 506 (1966).
9. C. J. O. R. Morris and P. Morris, *Separation Methods in Biochemistry*, 2nd ed., John Wiley, New York, 1976.

10. E. G. Scheibel, in *Technique of Organic Chemistry*, 2nd ed. (A. Weissberger, ed.), Vol. 3, Part I, Interscience, New York, 1956, pp. 322–393.
11. E. Hecker, *Verteilensverfahren im Laboratorium*, Verlag Chemie, Weinheim, 1955.
12. J. R. Weisiger, in *Organic Analysis* (J. Mitchell, M. Kolthoff, E. R. Proskauer, and A. Weissberger, ed.), Vol. 2, Interscience, New York, 1954, p. 277.
13. P. H. Hynninen, Multiple Liquid-Liquid Partition. I, II, and III. *Anal. Biochem.*, *74*, 376–391, 392–404, 405–418 (1976).
14. O. Mikeš and R. Vespalec, in *Liquid Column Chromatography*, (Z. Deyl, K. Macek, and J. Janák, eds.), Elsevier, Amsterdam, 1975, pp. 233–280.
15a. N. B. Mandava, Y. Ito, and W. D. Conway, *American Laboratory*, *14* (10), 62 (1982).
15b. W. D. Conway, in *CRC Handbook of Natural Pesticides: Methods*, *2*, CRC Press, Boca Raton, Florida, 1985, pp. 103--146.
16. R. E. Treybal, *Liquid Extraction*, McGraw-Hill, New York, 1963.
17. L. C. Craig, *J. Biol. Chem.*, *155*, 519 (1944).
18. B. Williamson and L. C. Craig, *J. Biol. Chem.*, *168*, 687 (1947).
19. R. M. Bock, *J. Am. Chem. Soc.*, *72*, 4269 (1950).
20. L. C. Craig and O. Post, *Anal. Chem.*, *21*, 500 (1949).
21. J. D. Gregory and L. C. Craig, *Ann. N.Y. Acad. Sci.*, *53*, 1015 (1951).
22. P. L. Nichols, *Anal. Chem.*, *22*, 915 (1950).
23. M. T. Bush and P. M. Densen, *Anal. Chem.*, *20*, 121 (1947).
24. A. E. O'Keeffe, M. A. Dolliver and G. T. Stiller, *J. Am. Chem. Soc.*, *71*, 2452 (1949).
25. P. Hietala, *Ann. Acad. Sci. Fennicae, AII*, 100 (1960).
26. W. Fischer, G. Braune, W. Dietz, and O. Jubermann, *Angew. Chem.*, *66*, 317 (1954).
27. R. Signer and H. Arm, *Helv. Chem. Acta*, *50*, 46 (1967).
28. O. Post and L. C. Craig, *Anal. Chem.*, *35*, 641 (1963).
29. A. J. P. Martin and R. L. M. Synge, *Biochem. J.*, *35*, 91 (1941).
30. A. J. P. Martin and R. L. M. Synge, *Biochem. J.*, *35*, 1358 (1941).
31. S. Watanabe and K. Morikawa, *J. Soc. Chem. Ind. Japan*, *36*, 585B (1933).
32. W. J. D. Van Dijek and A. Schaafama, U.S. Patent 2,245,945, 1941.
33. S. Stene, *Arkiv Kemie Mineral., Geol.*, *18*H, No. 18 (1944).
34. Alderweireldt and M. Verzele, *Bull. Soc. Chem. Belg.*, *70*, 703 (1961).
35. F. Alderweireldt, *Anal. Chem.*, *33*, 1920 (1961).
36. M. W. Kies and P. L. Davis, *J. Biol. Chem.*, *189*, 637 (1951).
37. Y. Ito and R. L. Bowman, *Anal. Chem.*, *43*, 70A (1971).

38. T. Tanimura, J. J. Pisano, Y. Ito, and R. L. Bowman, *Science, 169,* 54 (1970).
39. P. A. Albertsson, *Partition of Cell Particles and Macromolecules,* 2nd ed., Interscience, New York, 1971.
40. M. S. Tswett, *Proc. Warsaw Soc. Natl. Sci., Biol. Sec., 14,* No. 6 (1903); *Ber. Deut. Bot. Ges., 24,* 234, 316, 384 (1906).
41. See p. 361 of Ref. 9 above.
42. F. A. Von Metzch, *Angew. Chem., 65,* 568 (1953); *68,* 323 (1956).
43. J. F. K. Huber, C. A. M. Meijers, and J. A. R. Hulsman, *Anal. Chem., 44,* 111 (1972).
44. J. H. Hildebrand and R. L. Scott, *Regular Solutions,* Prentice-Hall, Englewood Cliffs, N.J., 1962.
45. J. H. Hildebrand and R. L. Scott, *Solubility of Non-Electrolytes,* Dover, New York, 1964.
46. C. H. Deal, E. R. Derr, and M. N. Papadoupoulos, *Ind. Eng. Chem., Fund. 1,* 17 (1962).
47. E. R. Derr, C. H. Deal, and G. J. Pierotti, *Am. Soc. Test. Mater. (ASTM), Spec. Tech. Publ. No. 244,* 111 (1957).
48. M. L. Huggins, *Ann. N.Y. Acad. Sci., 43,* 1 (1942).
49. D. E. Martire, in *Gas Chromatography 1966* (A. B. Littlewood, ed.), Elsevier, Amsterdam, 1967, p. 21.
50. G. J. Pierotti, C. H. Deal, E. L. Derr, and P. E. Porter, *J. Am. Chem. Soc. 78,* 2989 (1956).
51. G. J. Pierotti, C. H. Deal, and E. L. Derr, *Ind. Eng. Chem., 51,* 95 (1959).
52. A. P. Kudchadker, G. H. Alani, and B. J. Zwolinski, *Chem. Rev., 68,* 659 (1968).
53. J. Novák, J. Janák, and S. Wičar, in *Liquid Column Chromatography: A Survey of Modern Techniques and Applications* (Z. Deyl, K. Macek, and J. Janák, eds.), "J. Chromatography Library," Vol. 3, Elsevier, Amsterdam, 1975, pp. 25–43.
54. L. L. Engel, C. B. Cameron, A. Stoffyu, J. A. Alexander, O. Klein, and N. D. Troflomov, *Anal. Biochem., 2,* 114 (1961).
55. J. Ellingboe, E. Nystrom and J. Sjovall, *Methods Enzymol., 14,* 317 (1969).
56. R. H. Hall, *Anal. Biochem., 4,* 395 (1962).
57. O. Samuelson, in *Ion Exchange* (J. A. Marinsky, ed.), Vol. 2, Marcel Dekker, New York, 1969.
58. G. S. Bernson, R. Roseman, and A. Dorfman, *Biochem. Biophys. Acta., 17,* 75 (1955).
59. L. Kesner, E. Muntwyler, G. E. Griffin, and T. Abrams, *Anal. Chem., 35,* 83 (1963).
60. D. Yamashiro, D. Gillesen, and V. du Vigneaud, *J. Am. Chem. Soc., 88,* 1310 (1966).
61. A. N. Siakotos and G. Rouser, *J. Am. Oil Chem. Soc., 42,* 913 (1965).

62. D. G. McConnell, R. L. Hoffmann, G. J. Elman, and C. D. Evans, *J. Am. Oil Chem., Soc.*, *44*, 641 (1967).
63. L. L. Ramsey and W. I. Patterson, *J. Assoc. Off. Agric. Chemists.*, *31*, 144, 441 (1948).
64. O. S. Privett and E. C. Nickell, *J. Am. Oil Chem. Soc.*, *40*, 189 (1963).
65. R. H. Hall, *Biochemistry*, *4*, 661 (1965).
66. K. H. Muench and P. Berg, *Biochemistry*, *5*, 970 (1966).
67. H. J. Mundschenk, *J. Chromatogr.*, *37*, 431 (1968).
68. H. J. Mundschenk, *J. Chromatogr.*, *40*, 393 (1969).
69. A. J. P. Martin and R. R. Porter, *Biochem. J.*, *49*, 215 (1951).
70. R. R. Porter, *Biochem. J.*, *53*, 320 (1953).
71. L. R. Snyder and J. J. Kirkland, *Introduction to Modern Liquid Chromatography*, 2nd ed., John Wiley, New York, 1979.
72. M. J. E. Golay, *Anal. Chem.*, *29*, 928 (1957). A. I. M. Keulemans, *Gas Chromatography*, Reinhold, New York, 1957.
73. L. S. Ettre, *J. Chromatogr.*, *65*, 235 (1979).
74. B. L. Karger, L. R. Snyder, and C. Hovrath, *An Introduction to Separation Science*, John Wiley, New York, 1975.
75. L. S. Ettre and C. Hovrath, *Anal. Chem.*, *47*, 422A (1975).
76. R. W. Yost, L. S. Ettre, and R. D. Conlon, *Practical Liquid Chromatography: An Introduction*, Perkin-Elmer, Norwalk, Conn., 1980.
77. E. K. Johnson and R. Stevenson, *Basic Liquid Chromatography*, Varian Associates, Pato Alto, Cal., 1978.
78. P. E. Bristow, *Liquid Chromatography in Practice*, Help, Winslow, England, 1976.
79. R. J. Hamilton and P. A. Sewell, *Introduction to High Performance Liquid Chromatography*, Halstead Press, New York, 1977.
80. J. N. Done, J. H. Knox, and J. Loheac, *Application of High-Speed Liquid Chromatography*, Interscience, New York, 1974.
81. A. Pryde and M. T. Gilbert, *Applications of High Performance Liquid Chromatography*, Halstead Press, New York, 1979.
82. W. J. D. Van Dijck, U.S. Patent 2,011,186, 1935.
83. W. J. D. Van Dijck and V. S. Ruys, *Perfumery Essential Oil Records*, *28*, 91 (1937).
84. H. G. Schutze, W. A. Quebedaux, and H. L. Lochte, *Ind. Eng. Chem. Analyt. Ed.*, *10*, 675 (1938).
85. E. G. Sheibel, *Chem. Eng. Progr.* *44*, 681, 771 (1948).
86. L. C. Craig, W. Hausmann, E. H. Ahrens, Jr., and E. J. Harbenist, *Anal. Chem.*, *23*, 1236 (1951).
87. R. Signer, *Chimia*, *6*, 243 (1952).
88. R. Signer, *Chimia*, *10*, 95 (1956).
89. R. Signer, *Chimia*, *12*, 83 (1958).
90. R. Spence and R. J. W. Streeton, *Analyst*, *77*, 578 (1952).
91. J. F. Short, *J. Chem. Soc.*, *1*, 1278 (1953).

92. A. Keppes, *Bull. Soc. Chim. Biol.*, *48*, 1243 (1954).
93. P. H. Bell, K. S. Howard, R. G. Shepherd, B. M. Finn, and J. H. Meisenhelder, *J. Am. Chem. Soc.*, *78*, 5059 (1956).
94. H. L. Meltzer, *J. Biol. Chem.*, *233*, 1327 (1958).
95. H. L. Meltzer, J. Buchler, and Z. Frank, *Anal. Chem.*, *37*, 721 (1965).
96. S. Raymond, *Anal. Chem.*, *30*, 1214 (1958).
97. F. C. Alderweireldt, *Bull. Soc. Chim. Belg.*, *67*, 225 (1958).
98. P. Hietala, *Acta Chem. Scand.*, *14*, 212 (1960).
99. P. V. Tavel and W. Bollinger, *Helv. Chim. Acta*, *51*, 278 (1968).
100. Y. Ito, M. Weinstein, I. Aoki, R. Harada, E. Kimura, and K. Nunogaki, *Nature*, *212*, 985 (1966).
101. E. Jantzen, *Das fraktionierte Distillieren und das fraktionierte Verteilen*, Verlag Chemie, Weinheim, 1932.
102. G. Michot and H. F. Wiegandt, *Ind. Eng. Chem. Process Res. Dev.*, *10*, 586 (1974).
103. Y. Ito and R. L. Bowman, *Science*, *167*, 281 (1970).
104. Y. Ito and R. L. Bowman, *J. Chromatogr. Sci.*, *8*, 315 (1970).
105. Y. Ito and R. L. Bowman, *Science*, *173*, 420 (1971).
106. Y. Ito, R. L. Bowman, and F. Noble, *Anal. Biochem.*, *49*, 1 (1972).
107. Y. Ito and R. L. Bowman, *Science*, *182*, 391 (1973).
108. Y. Ito and R. L. Bowman, *Anal. Biochem.*, *65*, 310 (1975).
109. Y. Ito and R. L. Bowman, *J. Chromatogr.*, *136*, 189 (1977).
110. Y. Ito and R. L. Bowman, *Anal. Biochem.*, *78*, 506 (1977).
111. Y. Ito, *Anal. Biochem.*, *100*, 271 (1979).
112. Y. Ito, *J. Chromatogr.*, *188*, 33 (1980).
113. Y. Ito, *J. Chromatogr.*, *188*, 43 (1980).
114. Y. Ito, P. Carmeci, and I. A. Sutherland, *Anal. Biochem.*, *94*, 249 (1979).
115. I. A. Sutherland and Y. Ito, in *Cell populations*, "Methodological Surveys in Biochemistry Series." (E. Ried, ed.), Vol. 9, 1979, p. 217.
116. Y. Ito, *J. Chromatogr.*, *192*, 75 (1980).
117. Y. Ito, *Anal. Biochem.*, *102*, 150 (1980).
118. Y. Ito and R. Bhatnagar, *J. Chromatogr.*, *207*, 171 (1981).
119. Y. Ito, *J. Chromatogr.*, *214*, 122 (1981).
120. Y. Ito, J. Sandlin, and W. G. Bowers, *J. Chromatogr.*, *244*, 247 (1982).
121. Y. Ito, R. E. Hurst, R. L. Bowman, and E. K. Achter, *Separation and Purification Methods*, *3*, 133 (1974).
122. M. Krejčí, Z. Pechan, and Z. Deyl, in *Liquid Column Chromatography* (Z. Deyl, K. Macek, and J. Janák, eds.), Elsevier, Amsterdam, 1975, pp. 101–168.
123. J. Janák, J. Čoupek, M. Krejčí, O. Mikeš, and J. Turková, in *Liquid Column Chromatography* (Z. Deyl, K. Macek, J. Janák, eds.), Elsevier, Amsterdam, 1975, pp. 169–231.

chapter three

Principles and Instrumentation of Countercurrent Chromatography

Yoichiro Ito
National Heart, Lung, and Blood Institute
National Institutes of Health
Bethesda, Maryland

I. THE COIL PLANET CENTRIFUGE [1,2]

A. Introduction

The coil planet centrifuge system is the original and memorable model which opened the door to the whole development of countercurrent chromatography. As was introduced elsewhere [1] this apparatus was originally constructed for separation of cell particles while other capabilities such as osmotic fragility measurement and solute partitioning were later accidentally discovered in the course of preliminary studies on general behavior of the liquids and solid particles in a coiled tube treated with the apparatus.

The original coil planet centrifuge differs from all other members of the coil planet centrifuge in that it has no flow-through mechanism and therefore solute partitioning takes place in an end-closed coiled tube. Consequently, samples are loaded beforehand in the middle portion of the coil at the interface of the two immiscible solvents which undergo countercurrent flow as the coil is slowly rotated in a strong centrifugal force field. In spite of this primitive feature of the apparatus, the system had been extremely useful for fundamental studies of partition process in the rotating coiled tube and results of these studies have served as an important guidance in the course of developing a variety of flow-through coil planet centrifuges. In virtue of this great contribution, the apparatus may be deserved for the detailed description of these preliminary results. Below, the description of the apparatus is initiated by the analysis of the motion of a particle in a slowly rotating coiled tube (by Dr. Marvin A. Weinstein), which may give useful information not only for the separation of solid particles with a single-solvent system but also for the behavior of a small droplet suspended in the immiscible liquid in the rotating coiled tube.

B. Analysis of the Motion of a Particle in the Rotating Helical Tube [1]

We consider a tube filled with a fluid of density ρ_0 and coiled into a helix of radius R, with its axis horizontal. We suppose that a spherical particle of radius a, density ρ, is placed in the tube and presume that it is required to study the motion of the particle when the helix undergoes uniform rotation about its axis with angular velocity ω.

As a first approximation we neglect the lateral motion in the tube and assume that in a given turn of the coil the particle moves in a vertical circle of radius R. The position of the particle can then be specified by the angle θ, as indicated in Fig. 1.

With this approximation, the particle is acted on by only two forces, the Stokes drag:

$$F_s = -\ 6\pi a \eta\ R \left(\frac{d\theta}{dt} - \omega\right)$$

where η is the viscosity of the fluid, and the net gravitational force tangent to the circular path:

$$F_g = -\left(\frac{4\pi}{3}\right) a^3(\rho - \rho_0)g \sin\theta$$

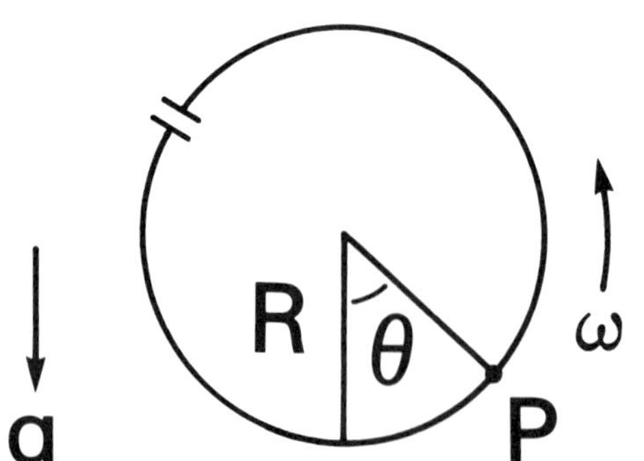

Fig. 1 Particle in a rotating helical tube.

The equation of motion is therefore

$$\frac{4\pi}{3} a^3 \rho R \frac{d^2\theta}{dt^2} = F_s + F_g$$

And on introducing the total angle of rotation of the helix $x = \omega t$, this can be written in the convenient form:

$$\frac{d^2\theta}{dx^2} + \frac{1}{\omega\tau}\left\{\frac{d\theta}{dx} - 1 + \frac{\omega_e}{\omega}\sin\theta\right\} = 0 \qquad (1)$$

where τ is the relaxation time:

$$\tau = \frac{2}{9}\left(\frac{\rho a^2}{\eta}\right) \qquad (2)$$

and ω_e is the critical angular velocity:

$$\omega_e = \frac{2}{9}\,\frac{(\rho - \rho_o)ga^2}{\eta R} = \frac{\overline{V}_e}{R} \qquad (3)$$

where $\overline{V}_e$ is the equilibrium Stokes velocity. We consider separately the motion for

$$\frac{\omega_e}{\omega} > 1 \qquad \text{and for} \qquad \frac{\omega_e}{\omega} < 1$$

$\omega_e/\omega > 1$

In this case Eq. (1) has a singular point at $(\omega_e/\omega)\sin\theta = 1$, that is, at $\theta = \theta_e = \sin^{-1}(\omega/\omega_e)$, $0 \leqslant \theta \leqslant \pi/2$, and at $\theta = \pi - \theta_e$. θ_e and $\pi - \theta_e$ are the angles at which the drag of the liquid is just balanced by the gravitational force so that the particle is in equilibrium when at rest. It is readily seen from physical considerations that the equilibrium at θ_e is stable while at $\pi - \theta_e$ is unstable. More precisely, an analysis in the neighborhood of the singular points shows that $\pi - \theta_e$ is an unstable saddle point while θ_e is a stable node if $\omega\tau \leqslant (1/4)\tan\theta_e$ and a stable spiral point if $\omega\tau > (1/4)\tan\theta_e$. Consequently, after a time of the order of τ, the particle will remain fixed at θ_e. Its angular velocity relative to the tube will then be

$$\omega_{rel} = \omega\left\{\frac{d\theta}{dx} - 1\right\} = -\,\omega \qquad (4)$$

that is, it will spiral down the helix at a rate independent of its size and density.

$\omega_e/\omega < 1$, $\omega\tau << 1$

When $\omega_e/\omega < 1$, Eq. (1) has no singular points. The character of the motion, however, depends to some extent on the size of $\omega\tau$ and we shall consider in detail only the physically interesting case $\omega\tau << 1$. It can readily be shown that within this limit, after a time of the order of τ, the angular velocity of the particle adjusts itself in such a way that the inertial term $\omega\tau(d^2\theta/dx^2)$ becomes negligible, so that Eq. (1) reduces to

$$\frac{d\theta}{dx} = 1 - \frac{\omega_e}{\omega} \sin\theta \tag{5}$$

The particle then always rotates in the same direction as the tube, but more slowly than the tube when $0 < \theta < \pi$ and more rapidly when $\pi < \theta < 2\pi$. The two effects do not quite cancel out, and the net effect is that, again, the particle spirals down the tube in a direction opposite to that of the tube rotation.

To determine the mean angular velocity of this spiralling motion, we first calculate the time required for the particle to traverse one turn of the coil. We have from Eq. (5):

$$x(2\pi) - x(0) = \int_0^{2\pi} \frac{d\theta}{1 - \dfrac{\omega_e}{\omega}\sin\theta} = \frac{2\pi}{\sqrt{1 - \left(\dfrac{\omega_e}{\omega}\right)^2}}$$

The mean angular velocity of the particle relative to the tube is therefore

$$\bar{\omega}_{rel} = \omega\left\{\frac{2\pi}{x(2\pi) - x(0)} - 1\right\} = -\omega\left\{1 - \sqrt{1 - \left(\frac{\omega_e}{\omega}\right)^2}\right\} \tag{6}$$

which joins continuously on the value at $\omega_e/\omega = 1$ [Eq. (4)].

When ω is close to ω_e, $\bar{\omega}_{rel}$ is rather insensitive to the size and density of the particle. When $\omega_e/\omega << 1$, however, Eq. (6) becomes:

$$\bar{\omega}_{rel} = \frac{1}{2}\left(\frac{\omega_e}{\omega}\right)^2 \tag{7}$$

so that, in agreement with experiment, the rate of motion down the tube is proportional to a^4 and $(\rho - \rho_0)^2$. Thus, when $\omega e \ll \omega \ll 1/\tau$, the method should be quite effective in segregating particles of different size and density. In order to apply the results of the above analysis to the coil planet centrifuge, the value of g should be replaced by that of the centrifugal acceleration acting on the axis of the rotating helical tube.

C. Apparatus

The results of the above analysis have encouraged us to build a novel type of the centrifuge which produces a planetary motion of the coil holder—a slow rotation about its own axis and fast revolution around the central axis of the centrifuge.

One of our prototype coil planet centrifuges is illustrated in Fig. 2. The main body of the apparatus consists of three parts, each capable of rotation as a unit:

Part I: a pair of coil holder (No. 6) and interchangeable planetary gears (No. 7) with shaft
Part II: frame consisting of a pair of arms (No. 4) and disks (No. 3) bridged with links (No. 5)
Part III: central shaft (No. 1) and central sun gear (No. 2) which interlocks with the planetary gears (No. 7) of Part I

Simultaneous rotation of parts II and III at a different angular velocity results in a revolution and a rotation of part I as a planet. The relative rotation of part I is determined by the difference in angular velocity between parts II and III and also by the tooth ratio between the central (No. 2) and peripheral (No. 7) gears. In this apparatus, the rotation revolution ratio is adjustable into four steps, i.e., one rotation of part I per 1000, 2000, 3000, and 5000 revolutions of part II. The rate of revolution of part II may be regulated up to 1600 rpm, which is equivalent to 350g on the axis of the coil holder. Each coil holder, measuring 24 cm in length and 3 cm in diameter, has six grooves in its periphery into which six coiled tubes can be loaded—a thick strong vinyl cylinder is then slipped over the holder to hold the coiled tubes in place. Another coil planet centrifuge, developed for the separation of particles and erythrocyte osmotic fragility test, has a similar capacity except that the rotation revolution ratio can be adjusted to 1:100, 1:200, 1:300, and 1:500. The coiled separation column was prepared from a narrow-bore flexible plastic tube, typically 1 mm i.d. or less, by winding it tightly onto a glass rod of 24 cm in length and 6 mm in diameter, making as many turns as possible. After the test samples were introduced into the coil, the coil was sealed by making knots in each end.

Fig. 2 The coil planet centrifuge. Simultaneous rotation of the frame (3, 4, and 5) and the central shaft (1) at a different angular velocity results in a revolution and a rotation of the coil holders (6) as a planet.

A series of preliminary experiments has been performed to explore the potential capability of the apparatus. The results are summarized in Fig. 3 where the coils are schematically drawn uncoiled to illustrate the distribution of their column contents before and after centrifugation.

Single Medium

When the tube is filled with a single medium and a particle mixture is introduced at one end, centrifugation separates the particles ac-

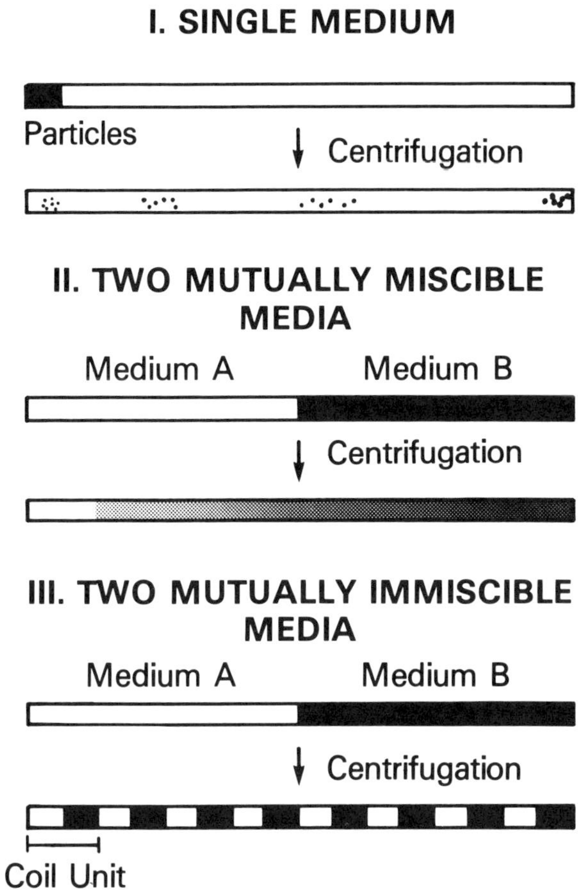

Fig. 3 Three basic applications of the coil planet centrifuge. Tubes are shown uncoiled before and after centrifugation. (I) When the coil is filled with a single medium and the particle mixture is introduced at the tail of the coil, centrifugation separates the particles according to their size and relative density. (II) When the coil is filled with two mutually miscible media, one on the head side and the other on the tail side as illustrated, centrifugation produces a stable gradient between the two media. This gradient can be used for determination of osmotic fragility of erythrocytes. (III) When the coil is similarly filled with two immiscible media, centrifugation produces countercurrent movement of the two media, which are finally distributed through the coil as alternating segments. Samples introduced beforehand at the interface between the two media are distributed through the coil according to their partition coefficients.

cording to the difference in size and relative density. This can be applied to the separation of cells or intracellular particles.

Two Mutually Miscible Media (Gradient Method)

When the tube is filled with two mutually miscible media, the heavier in one half and the lighter in the other half, centrifugation produces a density gradient. After centrifugation for some time, however, the gradient reaches a fairly stable state which is determined by various factors, such as the inner diameter of the tubing, the density difference and viscosity of the media used, the centrifugal force applied, etc. Such a gentle gradient is considered to be stable and in practice is introduced into the tube by means of a gradient maker in a few minutes. A stable linear gradient made between physiological saline solution and distilled water can be used for the osmotic fragility test of blood cells. Red cells pipetted into the saline side of the tube are forced to travel through the gradient down to the point where hemolysis occurs. The distribution of hemoglobin thus formed indicating the osmotic fragility of the sample. The most remarkable characteristic of this method is that each cell in the tube is exposed to a gradually increasing hypotonic environment for hemolysis within a desired period. Because of the highly reproducible results with a minute amount of blood sample (2–3 μL), this method is most suitable for medical investigations on neonatal jaundice and other hematological disorders and for screening tests of blood at blood banks.

Besides these examples, the gradient method has broad applications in scientific research including particle separation by density gradient, salting out of chemicals by salt gradient, and elution with pH gradient.

Two Mutually Immiscible Media (Countercurrent Chromatography)

When two mutually immiscible media are used similarly, centrifugation forces these two media to interchange and in the final stage each unit of the coil is occupied by the two media nearly half and half as illustrated in Fig. 3, III. Consequently, a small amount of a sample injected beforehand at the interface of the two media, i.e., the middle portion of the tube, is distributed inside the tube according to the partition coefficient. This new countercurrent distribution method which may be called countercurrent chromatography is applicable to microgram amounts of chemicals with a high efficiency that is expected to be equivalent to some thousand units of a conventional apparatus. When polymer phase systems are used, the centrifugal force can be adjusted in such a way that it overcomes the viscosity of the media and separation is completed in a shorter time.

D. Mechanism of Countercurrent Process

The mechanism of the above countercurrent phenomenon in the rotating coil was further studied with a simple rotary device shown in Fig. 4. In this model slow rotation of the coil produces an Archimedean screw force which tends to drive all objects in the coil toward one end of the coil. This end is called the head and the other end, the tail. When the coil contains two immiscible solvent phases, one phase on the head side and the other on the tail side, rotation of the coil results in a countercurrent flow of the two solvent phases across the interface. The process continues until the hydrodynamic equilibrium is established between the two solvent phases throughout the coil where the two phases are evenly distributed as alternating segments from the head toward the tail, any excess of either phase remaining on the tail side of the coil. This counter current process is schematically illustrated in Fig. 5.

The coiled tube at the top is filled with the upper phase (clear) at the head end and the lower phase (shaded) in the tail end forming an interface at the bottom of the coiled tube. When the tube is

Fig. 4 Simple rotating coil assembly. The rotational speed of the coil is adjusted by changing the distance between the rim of the coil holder and the center of the turntable.

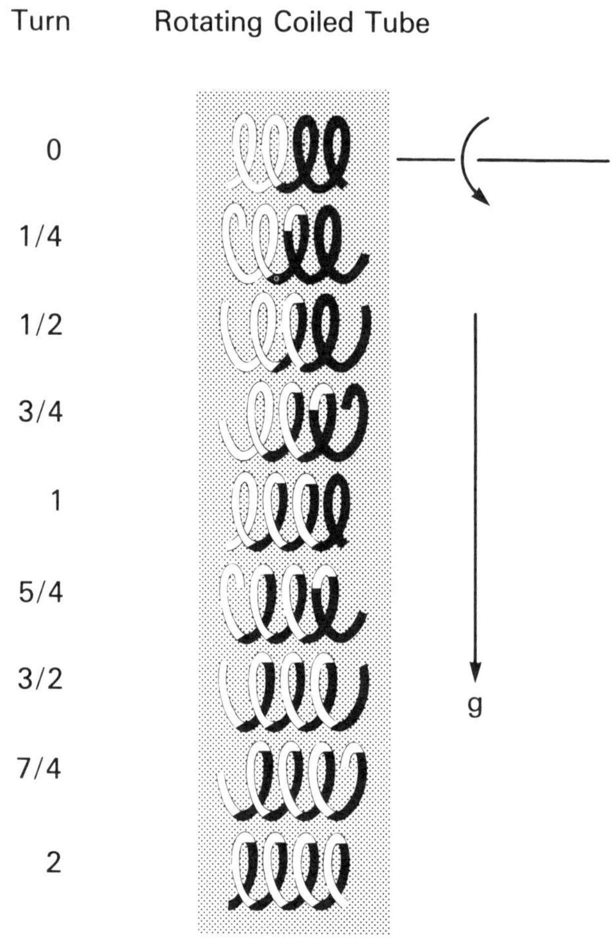

Fig. 5 Mechanism of countercurrent process of two immiscible media in a rotating helical tube.

subjected to a slow rotation under a gravitational field as indicated, the two phases begin to interchange through the interface. In each half rotational cycle of the coil, a new interface is formed at the bottom of the coil through which the countercurrent process continues. The process occurs four times during two complete rotational cycles when the two phases finally establish the hydrodynamic equilibrium state. At this stage each coil unit is occupied by approximately equal volumes of the two phases as illustrated, and the whole length of the tube contains alternating segments, each about half a coil unit in length, of the two phases. Further rotation of the coil results in an alternating countercurrent process in which the pattern of segments remains the same, but in which the segments of one of the two phases pass alternately backward and forward through segments of the other phase. If initially the direction of rotation had been in the opposite sense, the original head becomes the tail and the countercurrent process proceeds through the interface appearing at the top of the coiled tube.

Under a slow rotation of the coiled tube as illustrated in Fig. 5, the two phases interchange to form symmetrical segments, each occupying about 50% of the space in the coil unit throughout the coil, regardless of the head–tail relationship of the coil. Under a higher rotational speed of the coil, the segmented pattern of the two phases becomes asymmetrical. The lower phase advancing toward the head end of the coil fails to occupy 50% of the space in the coil unit so that, at the end of the second rotational cycle, the coil shows alternating segments consisting of over 50% upper phase and below 50% lower phase, while any excess of the lower phase totally remains at the tail end of the coil. The continued rotation of the coil, however, results in a gradual adjustment of the volume ratio of the two phases in the coil units starting from the head end of the coil, and finally the two phases reach the hydrodynamic equilibrium state where alternating segments of the two phases are distributed more symmetrically throughout the coil as shown in Fig. 5 (bottom).

As mentioned earlier, this unique countercurrent process can be applied to countercurrent chromatography if the sample solution is introduced beforehand at the interface of the two phases. Under a slow rotation of the coil, the solutes are subjected to an efficient continuous partition process and eventually distributed throughout the length of alternating segments of the two phases in the coil according to their relative partition coefficients. It is apparent that the ratio of the area of the interface to the volume of the two phases plays an important role in the efficiency of partitioning, and therefore we expect to obtain better results with a fine tube; in practice, good resolution has been obtained with a tube measuring less than 1 mm in internal diameter. To enable the countercurrent process to take place inside such a tube, however, necessitates enhancement of the gravitational acceleration field, this being achieved by the use of the coil planet centrifuge described earlier.

E. Separation of Various Test Samples

Potential capability of the coil planet centrifuge has been successfully demonstrated on separations on various kinds of colored samples with a broad spectrum ranging from small molecules (basic dyes) to macromolecules (algal proteins) and cell particles (erythrocytes). The use of the colored samples facilitated the observation of sample distribution within the coil through the transparent column wall.

Separation of Basic Dyes

A mixture of four basic dyes was partitioned with a conventional two-phase solvent system composed of isoamyl alcohol/ethanol/acetic acid/water (4:2:1:5). The solvent mixture was thoroughly equilibrated in a separatory funnel at room temperature and separated before use. The sample solution was prepared by dissolving the sample mixture in the lower phase to make the concentration of each component 0.03% for methylene blue, 0.05% for neutral red, 0.08% for basic fuchsin, and 0.17% for methyl green. The sample loading was initiated by injecting the lower phase into the head end of the coiled tube to just beyond the half-way point. Then, the sample solution was injected from the head end, after admitting an air bubble to 2–5 mm long between the lower phase and the sample solution to prevent their mixing. Typically, the sample volume was 30 μL; three or four coil units were filled with the sample and the actual amount depended on the tubing size used. Next, the upper phase was slowly injected through the head end, introducing the similar air bubble at 1-cm intervals in the first 10 cm, until the sample solution reached a position a few turns before the desired starting location and the coiled tube at the tail end was sealed with two knots. Then, pressure was gradually applied by injecting more medium until the sample reached the desired portion of the tube while the air bubble were compressed and finally absorbed. The end of the tube on the head side was clamped and tied. The tube was then inserted into a groove of the coil holder of the coil planet centrifuge, balanced, and centrifuged under the selected conditions of rotation and revolution for desired periods of time.

Figure 6 shows the typical chromatogram of methyl green (M.G.), methylene blue (M.B.), neutral red (N.R.), and basic fuchsin (F.B.). The first coil displays the separation of the mixture and the rest of the coils show the chromatogram of each dye to demonstrate the reproducibility of the method. This separation was performed with 6 m of 0.35-mm i.d. tubing centrifuged at 300g, with a relative coil rotation of 0.25 rpm, for 10 hr. The total column capacity was 0.6 ml, and the samples contained 2 μg of methylene blue, 3 μg of neutral red, 3 μg of basic fuchsin, and 20 μg of methyl green. Appearance of the sample bands markedly varies with the sample size and may be broadened by a heterogeneity of the sample: F.B. is a mixture of three components and M.G. has a minor component. The asymmetrical distribution of N.R. may be due to the aberration from the linear isotherm at high concentration.

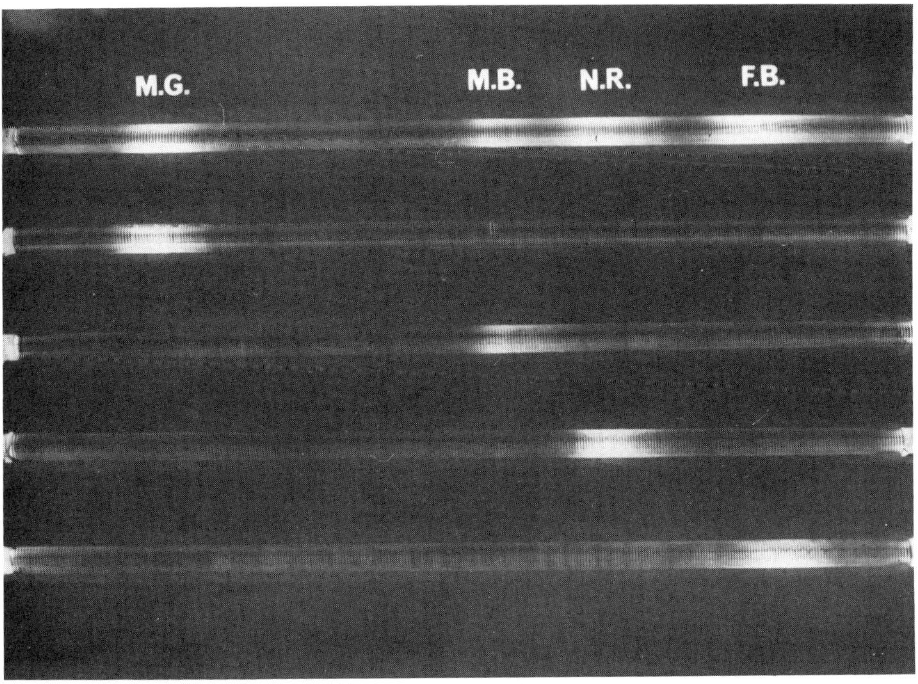

Fig. 6 Separation of basic dyes with a conventional two-phase solvent system. M.G.: methyl green; M.B.: methylene blue; N.R.: neutral red; F.B.: basic fuchsin. Solvent system: isoamyl alcohol/ethanol/acetic acid/water (4:2:1:5).

Separation of Algal Proteins

Separation of algal proteins, phycoerythrin and phycocyanin, was performed with a polymer phase system introduced by Albertsson [3]. The composition of the applied phase system is listed in Table 1. The phase mixture was allowed to equilibrate at room temperature and separated before use. Both phycoerythrin and phycocyanin were extracted from dried Asakusa-nori (*Porphyra tenera*) with an aqueous solution and purified through a Sephadex column according to the method described elsewhere. The final mixture was dissolved in 0.5 ml of the upper phase used for separation. The sample loading and coil planet centrifugation were performed according to the procedure used for separation of the basic dyes previously described.

Figure 7 illustrates the chromatogram obtained from about 100 μg of a mixture of phycoerythrin and phycocyanin with 0.4-mm i.d. coil after 6 hr of centrifugation. By measuring the peak width, the partition efficiency was roughly estimated as 100 theoretical plates.

Table 1 Polymer Phase System for Separation of Algal Proteins

20% (w/w) dextran 500	35.0 g
30% (w/w) PEG 6000	14.7 g
0.05M KH_2PO_4	10.0 ml
0.05M K_2HPO_4	10.0 ml
0.22M KCl	10.0 ml
H_2O	20.3 ml

Separation of Erythrocytes

Capability of the present method has been further extended to partition of cells with polymer phase systems listed in Table 2. A different approach was required to separate the human and rabbit erythrocytes. When the countercurrent separation was performed in the usual manner (i.e., loading one phase, then the sample, and then

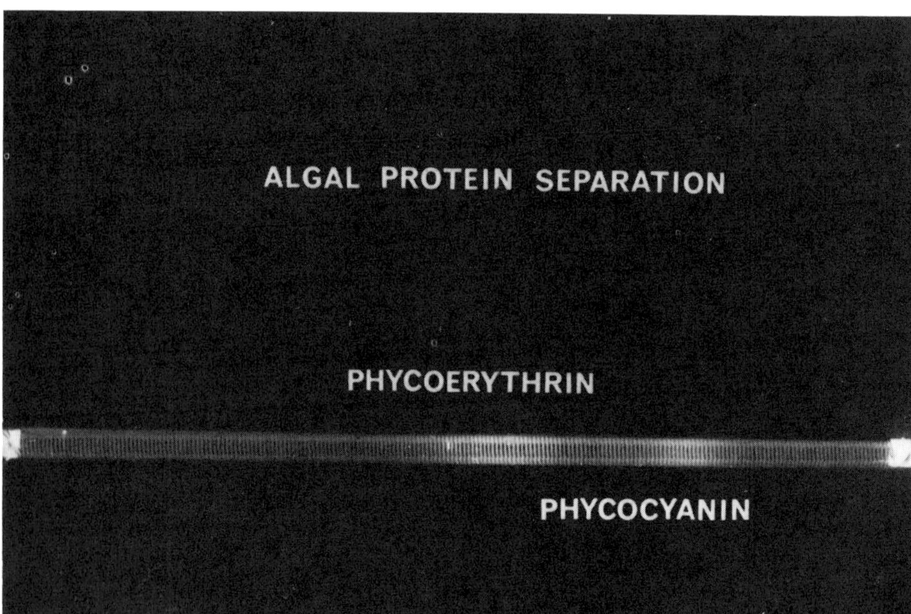

Fig. 7 Separation of algal proteins with a polymer phase system. Composition of the solvent system is listed in Table 1.

Table 2 Polymer Phase Systems for Separation of Erythrocytes

System A		System B	
20% (w/w) dextran 500	25.0 g	20% (w/w) dextran 500	25.0 g
30% (w/w) PEG 6000	13.3 g	30% (w/w) PEG 6000	13.3 g
0.55 M NaH_2PO_4	5.0 ml	0.05 M NaH_2PO_4	5.0 ml
0.55 M Na_2PO_4	5.0 ml	0.05 M Na_2HPO_4	5.0 ml
25% human albumin solution	1.0 ml	25% human albumin solution	1.0 ml
H_2O	50.7 ml	1.5 M NaCl	10.0 ml
		H_2O	40.7 ml

the second phase followed by centrifugation), erythrocytes showed a tendency to move with a dextran-rich lower phase regardless of the partition coefficient. We found, however, the hydrodynamic equilibrium state following the completion of the countercurrent process forced the cells to move toward either the head or the tail according to the partition coefficient, centrifugal force, and rotational speed. In other words, with both centrifugal force and rotational speed properly fixed, the cells move through the coiled tube toward a direction determined by the partition coefficient. This finding led us to carry out a gradient countercurrent separation on cell particles utilizing the hydrodynamic equilibrium state of the two phases. We chose the pair of phase systems A and B shown in Table 2—centrifugation under the hydrodynamic equilibrium state with system A alone directed cells toward the tail, and that with system B alone directed cells toward the head. Consequently, when the tube was filled with the media, the upper phase of system A in the head end half and lower phase of system B in the tail end half, the countercurrent process produced a gradient between these two media and the further treatment under the hydrodynamic equilibrium state distributed the cells as expected, at a given portion of the tube according to the partition coefficients.

The separation was actually performed as follows: The upper phase of A and the lower phase of B (Table 2) were used as the separating media. The lower phase of A and the upper phase of B were also used with similar results. The upper phase was injected first, followed by the lower phase. Without sample loading the coils were loaded into the coil planet centrifuge and centrifuged until the two phases distributed themselves along the helix. We used a 1/100 relative rotation of 1500 rpm for 30 min. This produced a gradient along the length of the coil, and the coils were ready for the partition process under the hydrodynamic equilibrium state. About 5 μl of the human and rabbit erythrocyte mixture suspended in the upper phase of A was injected into a point of about 10 coil units from midpoint of the coil toward the head end with a syringe and fine needle; the hole was sealed with heat by locally applying a heated iron. Centrifugation was resumed until the described separation was achieved which was usually one hour.

Figure 8 shows the separation of 5 μl of packed human and rabbit erythrocyte mixture by the gradient method. The tube at the top shows separation of human and rabbit cells, whereas in the middle and the bottom tubes each sample is separately distributed to demonstrate the reproducibility of the method.

F. Quantitative Studies on Various Physical Factors

Having established that the method can separate a variety of materials, we studied the sum of the factors affecting the efficiency of

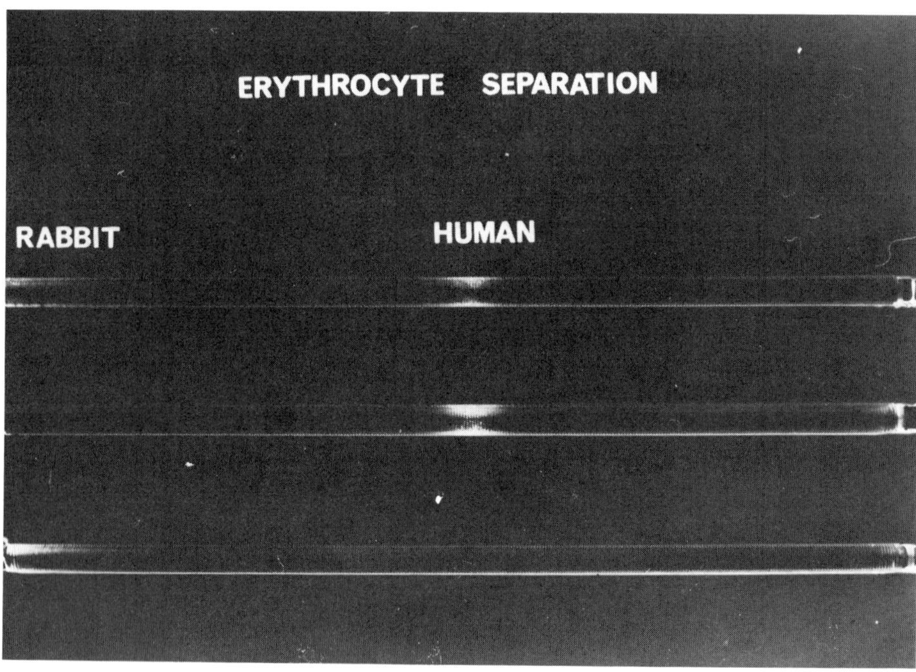

Fig. 8 Separation of human and rabbit erythrocytes with a pair of polymer phase systems using a gradient method. Compositions of the polymer phase systems are listed in Table 2.

the separation process. Among the various factors involved in this method, the following are considered to influence efficiency:

1. Number of coil units in which the countercurrent process takes place
2. Internal diameter of the tube used
3. Rotational speed of the coil or "contact time" (the time required for the media to pass through a given number of the coil units)
4. Centrifugal force applied
5. Helical diameter of coiling
6. Tubing material
7. Sample dose
8. Molecular weight and shape of the solute
9. Interactions between two phases and inner surface of the tube
10. Viscosity of the media
11. Vibration applied to the coiled tube during partitioning

We studied the effect of some of these factors with the methylene blue on the alcohol-acid system described previously.

Number of Coil Units

Figure 9 shows how the efficiency rises as countercurrent process proceeds in relation to the internal diameter of the tube used. The

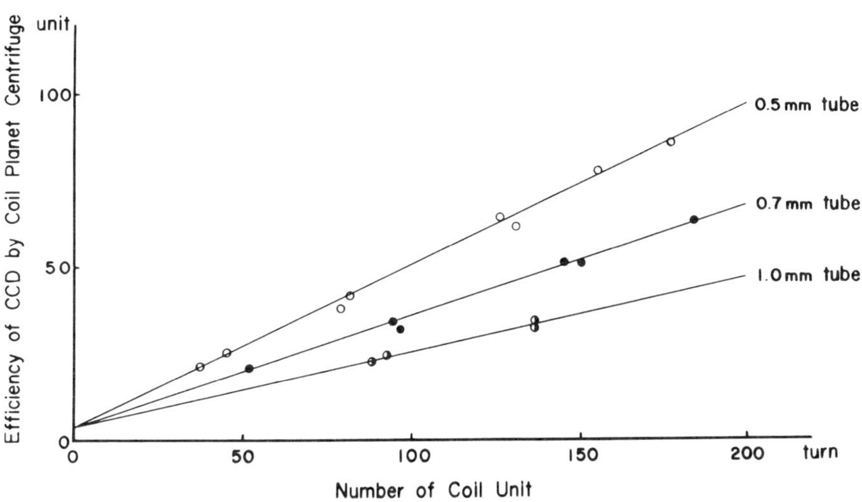

Fig. 9 Effects of number of helical turns and internal diameter of the tube on partition efficiency.

vertical axis represents the efficiency expressed by CCD plate number and the number of coil units are plotted on the horizontal axis. Circles, dots, and black and white mosaic circles represent the results obtained from the tube measuring 0.5, 0.7, and 1.0 mm in internal diameter, respectively. Three straight lines drawn through each group clearly indicate that the efficiency rises in proportion to the number of coil units through which countercurrent separation was performed. Each point represents one experiment.

Internal Diameter of the Tube

Figure 9 also shows the efficiency of these three groups at 150 turns as 74 units for the 0.5-mm tube, 52.5 units for the 0.7-mm tube, and 36.5 units for the 1.0-mm tube, the ratio being close to (1/0.5):(1/0.7):(1/1.0). This indicates that the efficiency alo rises in inverse proportion to the internal diameter of the tube used.

Contact Time of the Media

The influence of the contact time of the two phases on efficiency is shown in Fig. 10. In the figure, the partition efficiency, expressed at the CCD plate number, is plotted on a logarithmic scale against the time taken for the media to partition through 100 coil units in a 0.7-mm tube. Though the rise of the efficiency is exponential and tends toward an asymptote, it may be concluded that within a certain range the efficiency rises nearly in proportion to the square root of the contact time for a given number of turns of the tube. Contact time is related to the relative rotational speed of the coil unit as mentioned above.

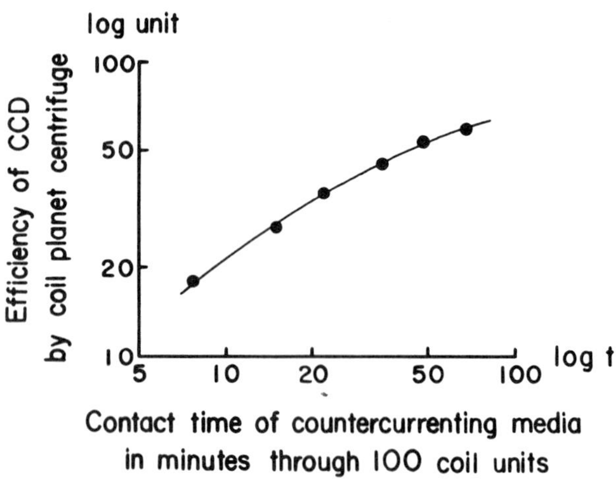

Fig. 10 Effects of contact time (coil rotation) on partition efficiency.

Centrifugal Force

The effect of the centrifugal force upon the efficiency was studied by comparing similar tubes treated under 50g and 350g with the same rotational speed or contact time. The results indicated that strong centrifugal force slightly improves the partition efficiency within the applied range.

As stated previously, the centrifugal force influences the volume ratio between the upper and lower phases in a coil unit. Under a given centrifugal force and rotational speed, the volume of the advancing phase in a coil unit decreases as the tube becomes smaller and/or the viscosity of the media increases. Inasmuch as polymer phase systems (which are extremely useful for separation of biologically unstable components) are usually of high viscosity, enhancement of the centrifugal force is indispensable for practical use of the apparatus for these systems.

Helix Diameter

Influence of the helix diameter on the partition efficiency was examined by winding the tube onto the cylindrical vinyl protector measuring 3.7 cm in outer diameter. The data showed that the efficiency resulted from the coiled tube with helical diameter of 6 mm and 3.7 cm was quite similar. Where this relationship holds, increase of the helical diameter has the advantage of providing a larger capacity so that larger samples may be run. It should be noted that the length of the tube also increases with the helix diameter. Therefore, for a given length of the tube, the partition efficiency decreases with the increase of the helix diameter.

Tubing Material

The hardness of the tubing was found to have considerable influence on the partition efficiency; the harder the tube, the better result is expected. Some organic solvents adsorb to the wall and some can easily pass through the wall of vinyl, polyethylene, or polypropylene tube. For certain solvents, therefore, the use of PTFE tubing is preferred.

Sample Size

The influence of the size of the sample on the partition efficiency was studied in experiments in which the volume and concentration of the sample were varied. The results were, however, confused to some extent by the deformation and shift of the distribution curve of methylene blue, probably due to nonlinear isotherm at high concentrations. Nevertheless, it may be stated that the sample size can be increased up to nearly 5% of the total volume of the media (as in the case of the Craig's apparatus) without considerably altering the partition efficiency, provided that the linear isotherm is held.

Other Factors

The data so far available do not provide clear evidence on the influence of the rest of the factors except that the use of viscous media and macromolecular samples considerably lowers the partition efficiency. The interactions between the two phases and the inner surface of the tube are complex and require further investigation. In most experiments with methylene blue, the lower phase was used as the advancing phase (which moves in the tail—head direction). However, it is also possible to carry out countercurrent experiments by advancing the upper phase. When the results obtained from these two methods were compared, the partition efficiency obtained by advancing the lower phase always exceeded that by advancing the upper phase. This can be accounted for by difference in viscosity between these two phases, inasmuch as the mass transfer resistance is, as illustrated in Fig. 5, mainly located in the advancing phase. However, it may also be explained on the basis of the difference between the ways the two phases exchange. When the lower phase is advancing as shown in Fig. 5, phase interchange takes place in such a way that the upper phase floats up through the lower phase along the inner wall surface of the tube. Because the upper phase used has an affinity to polyethylene, it will form a thin layer over the inner surface of the tube and the solute partition will proceed smoothly through its broad interface. Also in this situation the lower phase advancing toward the head of the coil can occupy nearly 50% of the space in each coil unit. On the other hand, when the upper phase is advancing (as represented by Fig. 5 turned upside down, provided that the clear phase is the lower phase and the shaded phase is the upper phase), the manner of the phase interchange is such that the lower phase sinks down into the upper phase from the top of the coiled tube. Because of a lack of the affinity to the wall surface, the sinking lower phase will fail to form a layer over the inner surface of the tube and instead breaks into droplets which quickly reach the bottom.

In this situation, one may argue that the size of these droplets could be reduced so as to improve efficiency. To examine this hypothesis, we prepared the alcohol-phosphoric acid phase system. The influence of emulsification of the phase system on the partition efficiency was tested under the maximal centrifugal force of 350g and relative rotation of 1/5000 for 4 hr. It turned out that the efficiency of methylene blue distribution rose with the reduction of phosphoric acid concentration until the concentration became 10^{-4} M and then began to decline with a marked tailing of methylene blue when the media showed intensive emulsification. Although no conclusion can be drawn until the effects on various phase systems under a broader range of centrifugal force have been studied, the above experiment suggests a possibility that the size of droplets may be an important factor in this method. Vibration of the apparatus may cause similar effects by producing small droplets which contribute to the partition process.

G. Summary

The microscale liquid-liquid partition method (countercurrent chromatography) we have described is similar to Craig's countercurrent distribution system in that the two immiscible phases pass through each other. However, our process is continuous and is not susceptible to analysis based on equilibrium distribution. Our method is related to chromatography, but we offer the advantage that neither phase is restrained by a carrier or support. Furthermore, this method, being characterized by a continuous countercurrent under a centrifugal force, provides the remarkable advantage of enabling viscous polymer phase systems to be used for purification and analysis of biologically unstable components including proteins, nucleic acids, viruses, cells, etc. Although it is possible to scale up our method, we feel it will have greatest applications where microsamples must be separated. Thus, the separation with the present method covers a range of substances from particles as large as erythrocytes to small molecules such as basic dyes.

II. PRINCIPLE OF COUNTERCURRENT CHROMATOGRAPHY

A. Introduction

Countercurrent chromatography (CCC) uses an intriguing behavior of two immiscible solvents in a coiled tube. There are two basic systems of CCC [4,5] to be described here. The first system uses a stationary coiled tube to produce a hydrostatic equilibrium of the two phases and therefore was named the hydrostatic equilibrium system (HSES). The second system uses a rotating coiled tube to produce a hydrodynamic equilibrium of the two phases and likewise was named the hydrodynamic equilibrium system (HDES). Each basic CCC system branches out to produce a variety of CCC schemes each having its specific advantages. Actually, all CCC schemes developed in the past have stemmed from these two basic CCC systems. Recent discovery of unilateral HDES [6,8] has further revolutionized the CCC technology to open a rich domain of applications including continuous extraction [7,8], high-speed CCC [7,9], and dual CCC [10].

B. Hydrostatic Equilibrium System (HSES)

Model System [4,5]

The HSES uses a stationary coil which is placed in the horizontal position. A model of this basic CCC system shown in Fig. 11 consists of a coiled tube with five coil units or helical turns. In Fig. 11a the coil is first filled with the lower (heavier) phase of an equilibrated two-phase solvent system and the upper (lighter) phase is slowly introduced through one end of the coil. The upper phase then pushes the lower phase downward in the first coil unit until the interface reaches the bottom of the coil where the upper phase starts to percolate upward through the lower phase affected by the gravity. This retains the lower phase in one half of the first coil unit. This process is repeated in each coil unit to produce alternating segments of the upper and the lower phases throughout the coil. Once this so-called hydrostatic equilibrium state is established, continued flow of the upper mobile phase results in the replacement of the upper phase only, leaving the lower phase stationary in each coil unit.

Figure 11b illustrates the similar process when the lower phase is used as the mobile phase. The coil is first filled with the upper phase and the lower phase is slowly introduced at one end of the coil. In this case, the lower phase immediately starts to percolate downward through the upper phase by the effect of gravity and quickly reaches the bottom of the first coil unit. The lower phase then pushes the upper phase upward until the interface reaches the top of the coil. This process is repeated in each coil unit to produce alternating segments of the two phases throughout the coil. After this hydrostatic equilibrium state is established, the mobile lower phase only displaces the lower phase while retaining the stationary upper phase in each coil unit.

These demonstrations clearly illustrate a characteristic feature of this basic CCC system. No matter which phase is moved, a similar hydrostatic equilibrium state is established where the two phases are distributed through the entire length of the coil in a pattern of alternating segments. Figure 11 also illustrates the effects of liquid-wall interaction which manifests itself in the flowing pattern of the mobile phase through each stationary phase segment. When the mobile phase has a strong wall surface affinity, it flows smoothly along the wall of the tube to form a continuous stream (Fig. 11a). When the mobile phase lacks the wall surface affinity, it forms multiple droplets into the stationary phase segments (Fig. 11b).

Hydrostatic equilibrium states of the two solvent phases illustrated in Fig. 11 introduce all essential physical factors for performing CCC. The coil is divided into five partition units to retain a large amount of the stationary phase. In each partition unit, the two solvent phases form characteristic interfaces while each phase is continuously mixed by the effects of the flowing mobile phase. Consequently, a basic CCC system is established. Solutes locally introduced at the inlet of the coil are subjected to a partition process between the flowing mobile phase and the retained stationary phase in each partition unit and finally eluted out with the mobile phase according to the order determined by their relative partition coefficients. If the partition process in each partition unit is highly efficient, this model is expected to yield a partition efficiency close to five theoretical plates.

Experiments to determine actual partition efficiency of this basic model system have been performed with a similar model consisting of more coil units. However, the results usually indicated that the number of theoretical plates yielded by a large-bore coil such as 5 mm i.d. or greater is much smaller than the number of the coil units in the applied column. Apparently, both interfacial area and degree of the phase mixing are not sufficient in such a large-bore coil. On the other hand, the use of a small-bore coil such as 2 mm i.d. or less tends to produce a plug flow (the mobile phase pushes the stationary phase segment as a plug) resulting in carryover of the stationary phase. Thus, the present model is impractical for the separation unless the performance is substantially improved. The modification to improve the performance of this basic model system has been successfully made in two different directions, one for preparative scale separations and the other for analytical scale separations.

Development of CCC Schemes Based on HSES [5]

Figure 12 illustrates various CCC schemes developed from the basic HSES. In the series of preparative schemes (top), the coil geometry of the original model is greatly modified to improve the partition efficiency while keeping the bore size of the column relatively large. In all these preparative schemes one coil unit of the basic model system is modified in the following manner. One half of the coil unit which is entirely filled with the mobile phase and therefore considered as an inefficient column space is reduced to a fine connection tube, while the other half of the coil unit which contains both mobile and stationary phases, and thus provides the effective column space, is made into a straight column. This column space is further divided into multiple partition units each providing broad interface area and efficient mixing of the phases. All the schemes thus created have completely lost their original helical appearance; hence they are also named the nonhelix CCC.

The first scheme in the preparative series is called droplet countercurrent chromatography (DCCC) [11–13]. In this scheme the mobile phase is so selected that it has less wall surface affinity than the stationary phase. When such mobile phase is introduced

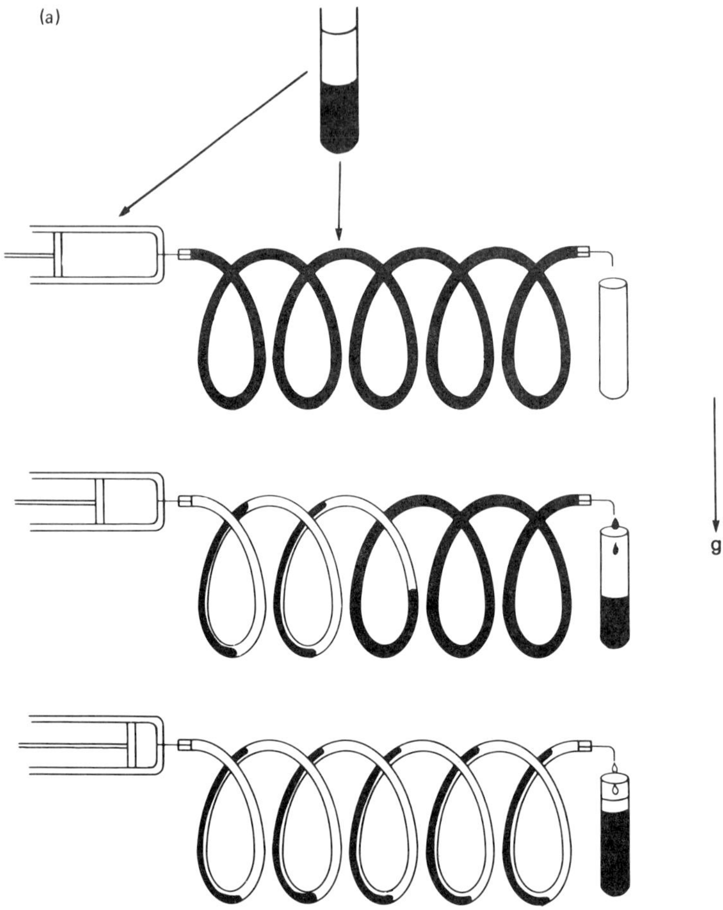

Fig. 11 Principle of the hydrostatic equilibrium system (HSES). The coil is first filled with the stationary phase, either the lower phase (a) or the upper phase (b), and the mobile phase is introduced at one end of the coil (top). Due to the effect of gravity, the mobile phase percolates through the stationary phase on one side of the coil (middle). When the mobile phase has a strong wall surface affinity, it flows smoothly along the wall of the tube (a). When the mobile phase lacks the wall surface affinity, it forms multiple droplets in the stationary phase segments (b). This process continues until the mobile phase reaches the outlet of the coil. Thereafter, the continued elution displaces only the mobile phase, hence a large amount of the stationary phase is permanently retained in each helical turn (bottom). Consequently, solutes introduced locally at the inlet of the coil are subjected to a partition process between the two phases in each helical turn and finally eluted out through the outlet of the coil in the order of their partition coefficients.

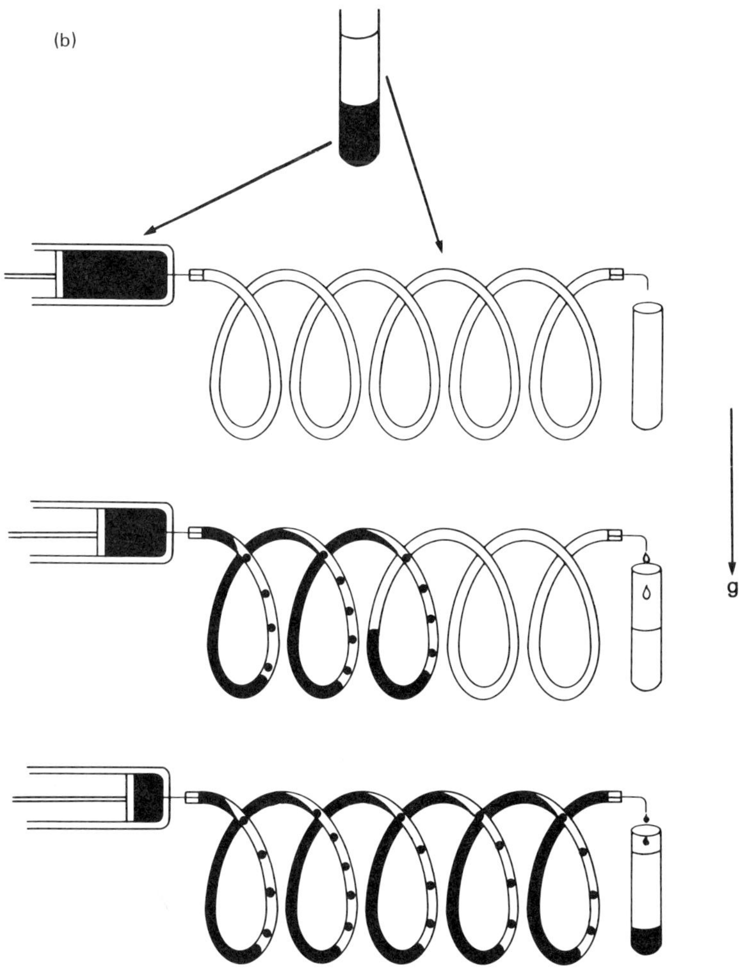

Fig. 11 (Continued)

into the vertical column previously filled with the stationary phase, it forms discrete droplets which extend across the diameter of the column. These droplets tend to form at regular intervals and move through the stationary phase at a constant rate so that the entire length of the column is again divided into multiple partition units corresponding to the number of droplets observed in the column. In each of these newly formed partition units, the efficient partition process takes place due to vigorous stirring of both the stationary and the mobile phases by the effect of the moving droplet.

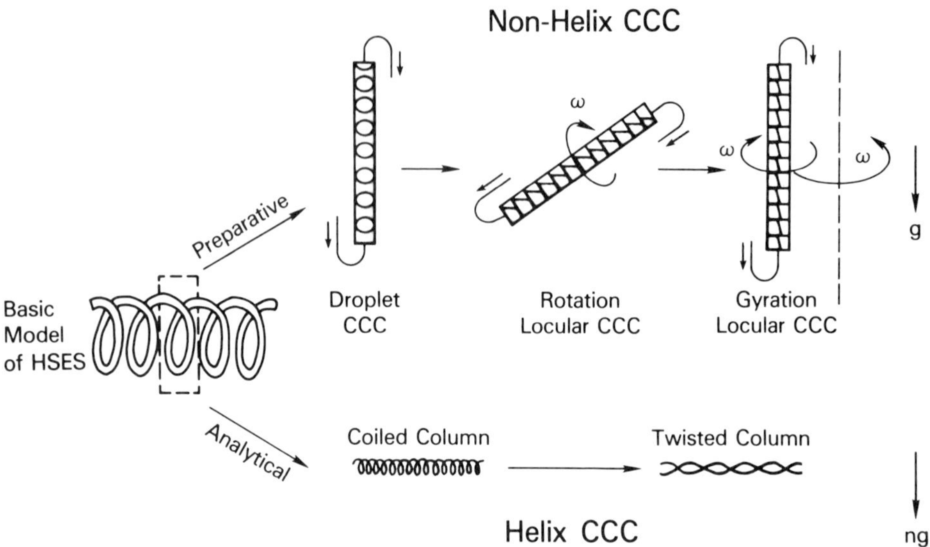

Fig. 12 Development of efficient CCC schemes from the basic model of HSES. The preparative CCC schemes (top) are developed by reducing one side (inefficient space) of the coil to a fine connection tube while the other side (efficient space) of the coil is transformed into a straight column. As a result, these preparative columns have completely lost their original helical appearance, hence these schemes are called nonhelix CCC. In the analytical CCC schemes (bottom), both the tube and helical diameters are reduced while the original helical conformation is preserved. In order to reduce the pressure necessary to elute the column under a centrifugal force field, the coil is further modified to make a twisted form with a ropelike appearance. These schemes may be called the helix CCC as opposed to the nonhelix CCC above.

In the second preparative scheme, called rotation locular countercurrent chromatography (RLCCC) [11,13], the column space is divided into multiple small compartments or locules by placing centrally perforated disks at regular intervals. Each locule becomes an efficient partition unit when the column is titled and rotated around its own axis. In each locule the mobile upper phase displaces the lower stationary phase above the level of the perforation open to the next locule while forming an interface across the locule. Both retention of the stationary phase and the interface area of the two phases are optimized by selecting a column angle at around 30° from the horizontal plane while the effective stirring within each phase is maintained by gentle rotation of the column.

The third preparative scheme, called gyration locular countercurrent chromatography (GLCCC) [11,13], uses a vertical locular column. In this scheme mixing of the two phases is carried out more efficiently by a gyrating motion of the column. Gyration introduces a planetary motion to the column which revolves around the central axis of the apparatus and simultaneously counterrotates about its own axis at the same angular velocity. This produces a rotating centrifugal force field to circulate the two phases and their interface within each locule in such a fashion that the interface is always oriented tangentially to the vector formed by the sum of the gravitational and the rotating centrifugal force fields.

A quite different approach has been taken to develop the analytical scheme called the helix countercurrent chromatography (HCCC) [11,13,14]. Figure 12 illustrates two types of columns used for this scheme (bottom). In the first type both tube and helix diameters are greatly reduced while the overall helical configuration remains the same. To solve the problem of plug flow formation which tends to occur in a small-bore coil, the column is subjected to a centrifugal force field to hold the stationary phase segment in each coil unit while the mobile phase is continuously eluted through the coil. Reduction of the tube diameter renders a number of advantages. A length of the tube to make one coil unit in the original model system can form an enormous number of coil units or partition units by reducing the helical diameter. It also increases the interfacial area of the two solvent phases in relation to their volumes and provides more efficient mixing of the phases in each coil unit. Thus this scheme can yield an extremely high partition efficiency which is comparable to that in a refined gas chromatography.

While the use of a longer column increases the yield of theoretical plates, it also builds up hydrostatic pressure within the column, which would eventually result in rupture of the tube. The minimum pressure at the inlet of the column necessary to sustain flow linearly increases with the acting centrifugal force, density differences between the two phases, the number of coil units, the helical diameter, etc. Among these factors the helical diameter, or more precisely the distance between the pair of interfaces across the diameter of the tube at the top and the bottom of each coil unit, can be minimized by eliminating the central core of the coil. This second type called the twisted column is prepared by folding the tube in two and twisting along its length to make a ropelike structure. Although this column configuration reduces mixing effects of the two phases and decreases the partition efficiency per coil unit, it permits greater retention of the stationary phase and allows the use of a much longer column to yield a maximum number of theoretical plates at a given operational condition. Accumulation of the hydrostatic pressure in the column also necessitates the development of leak-free flow-through centrifuge devices. The use of rotating seals in the conventional centrifuge systems would limit the maximum applicable pressure to 50 psi or less. As described in the following section, a series of seal-free flow-through centrifuge schemes has been introduced. These centrifuge devices enable continuous elution of the mobile phase, either aqueous or nonaqueous phase, through the rotating column against a high pressure of several hundred pounds per square inch.

C. Hydrodynamic Equilibrium System (HDES) [4,5]

Model System

The model of this basic CCC system consists of a horizontal coiled tube slowly rotating around its own axis. Clearly this model differs from that of the HSES only with respect to this slow rotation applied to the coil. Nevertheless, this simple motion introduces a new feature to the system which involves complex hydrodynamic interactions of the two solvent phases in the coil.

Demonstration to show the basic phenomena of this two-phase interaction is initiated by filling the coil with water, closing both ends, and rotating the coil slowly so that the centrifugal force effects may be ignored. The water confined in the coil then simply follows the tubing around. The superimposed gravitational field has no effect because the water is of homogeneous density. Next, some air bubbles and glass beads are introduced (Fig. 13a). Now the water still follows the tubing around, but both the air bubbles and glass beads are observed to migrate in the same direction toward one end of the coil which will be called the head and the other end, the tail. The motion of the suspended objects results from interplay of buoyancy (which tends to keep them in the top or bottom of the coil) and hydrodynamic drag (which tends to move them with the water). In fact, any suspended object, whether heavier or lighter than the water, migrates toward the head of the coil. In the extreme case where the buoyant forces overpower the hydrodynamic drag, e.g., large, heavy beads, the beads remain at the bottom while the coil undergoes helical screw motion past the bead. Hence, for every revolution, the bead advances one turn of the coil toward the head. In the more general case, the suspended object tries to follow the water around but is consecutively retarded and accelerated relative to the water by gravity. This time, a more gradual migration toward the head is observed with a number of revolutions being required to advance the object one turn of the coil.

The second stage is to repeat similar experiments (Fig. 13b) using two immiscible solvents such as chloroform and water previously equilibrated in a separatory funnel. In order to facilitate observation of the two-phase behavior in the coil, a dye may be used to color one

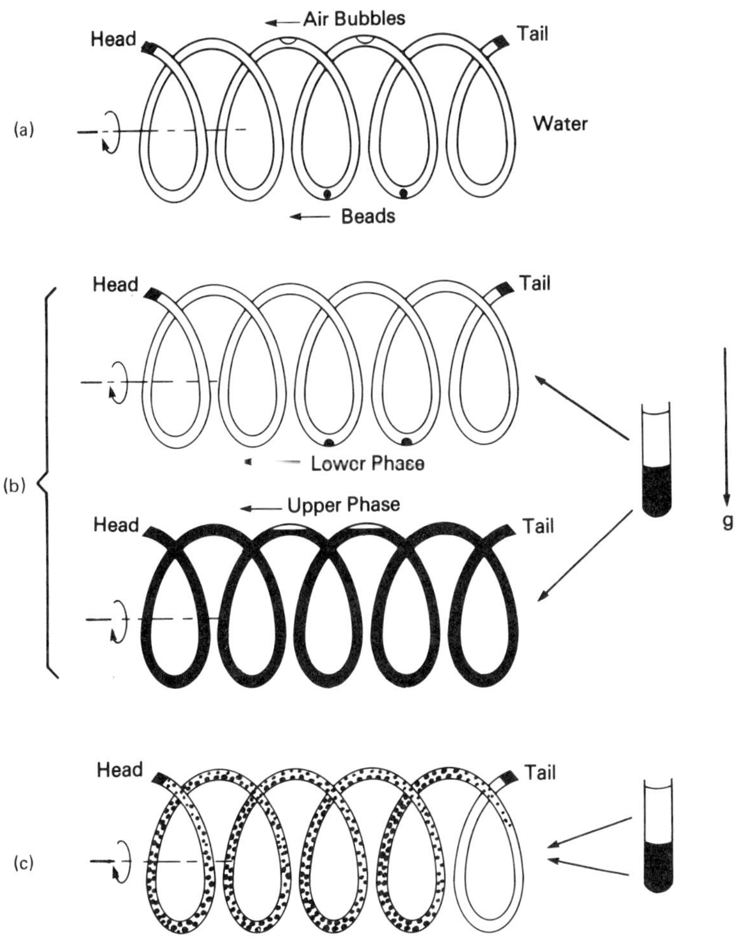

Fig. 13 Motion of various objects in a rotating coil. (a) A rotating coil exhibits "Archimedean screw" effect. Any object, either lighter (air bubbles) or heavier (glass beads) than the water, moves toward one end of the coil. This end is called the head and the other end, the tail of the coil. (b) A similar phenomenon is observed with the two immiscible solvent phases. A small amount of the lower (heavier) phase suspended in the upper (lighter) phase travels toward the head (top) while a small amount of the upper phase suspended in the lower phase also travels toward the head of the coil (bottom). These experiments suggest that the relative volumes of the two phases govern their behavior in the rotating coil. (c) Two-phase solvent system introduced in a slowly rotating coil soon establishes a hydrodynamic equilibrium state whereby each solvent phase occupies nearly equal space in each helical turn and any excess of either phase remains at the tail of the coil.

of the phases for contrast. Sudan III exclusively colors the non-aqueous phase and acid fuchsin generally favors the aqueous phase. The results of the experiments will show that either phase can migrate toward the head of the coil if suspended as a droplet while the same phase can act as a suspending medium if used to fill the coil. This clearly indicates that these behaviors of the two phases are governed by the relative volumes of the two phases occupying the free space in the coil.

Finally, the coil is initially filled with nearly equal amounts of the two phases (Fig. 13c). Under this condition, the principle that suspended droplets migrate toward the head of the coil still applies but the question of which phase acts as suspending medium depends on which phase is in excess within various portions of the coil. After a few minutes of rotation, the two phases usually reach a hydrodynamic equilibrium state where, starting from the head end, each helical turn is occupied by nearly equal volumes of the two phases. Any excess in a given coil unit will simply spill over into the next coil unit and subsequently accumulate at the tail end of the coil. Once this hydrodynamic equilibrium is established, continued rotation of the coil results in local mixing of the two phases while the overall distribution of the two phases remains the same throughout the coil.

Now let us observe more closely the behavior of the two phases in the rotating coil in this hydrodynamic equilibrium state. We first notice that the portion of the coil containing the two phases gives different profiles when viewed from each side. Careful observation will disclose that one of the phases with less wall surface affinity is usually found in droplet state. These droplets are populated more densely on one side of the coil and less densely on the other. The features of asymmetrical droplet distribution are schematically illustrated in Fig. 14 for both large-bore (a) and small-bore (b) coils. Obviously this uneven distribution of droplets is produced from the interplay of the gravity and the rotation of the coil. The fact that this overall distribution pattern of the droplets remains unchanged with respect to the observer, while the coil is rotating, indicates that these droplets are repeating an oscillatory motion along the coil synchronously with the rotation. More precisely, when the droplets are present at the right half of the coil unit in Fig. 14, they move toward the head, and when carried over to the left side of the coil unit by the rotation of the coil, they move toward the tail. While the rates of their motion are similar on both sides of the coil unit, the time for the droplets to stay at the right half of the coil unit will be longer than at the left half of the coil unit due to the rotation of the coil. Thus, at any given moment more droplets are observed at the right half of the coil unit. The size, shape, and motion of these droplets are greatly affected by the internal diameter of the coil as well as the rotational speed. In a large-bore coil (a) the

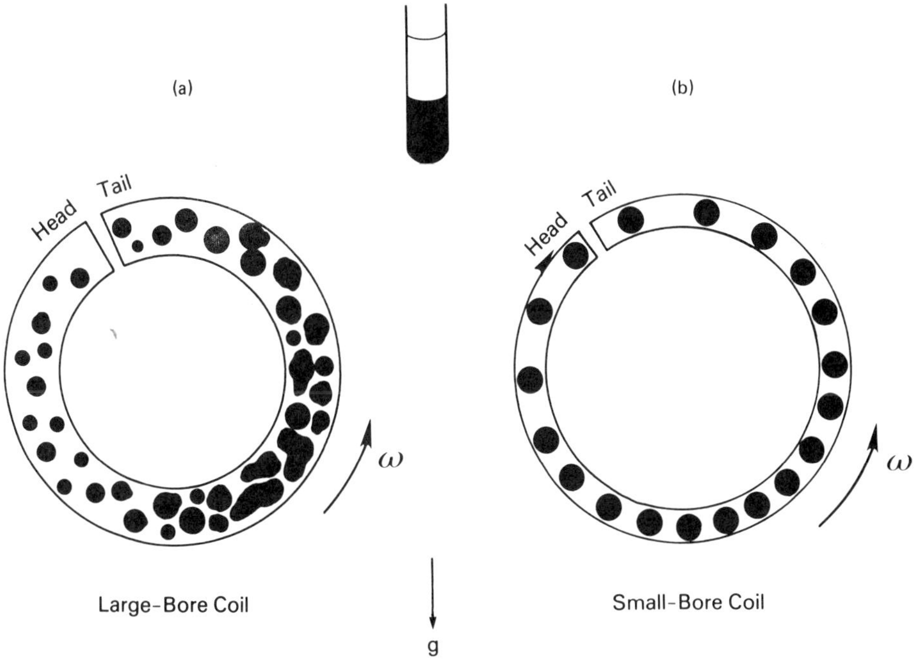

Fig. 14 Motion of two-phase solvent system in a rotating coil. One phase with less affinity to the wall surface tends to form multiple droplets in the other phase. Due to the combined effects of rotation and gravity, these droplets populate more densely on one side of the coil (a). When the tube diameter is small, each droplet can occupy the space across the diameter of the tube (b).

droplets tend to coalesce at the right half of the coil unit to form irregular-shaped droplets which are constantly broken down into smaller droplets after being carried over to the left side of the coil unit. In a small-bore coil (b) where the individual droplets can occupy the space across the diameter of the coil, the droplets become discrete and uniform in size oscillating along the tube without coalescence.

The experiments with this basic model may be extended by adding a flow-through mechanism to the system simply by means of a pair of rotary seals (Fig. 15). The coil is first completely filled with one of the phases, either the lower (a) or the upper (b) phase, and the other phase is pumped through the head end of the rotating coil. As soon as the mobile phase reaches the first coil unit, the two phases interact to establish the hydrodynamic equilibrium

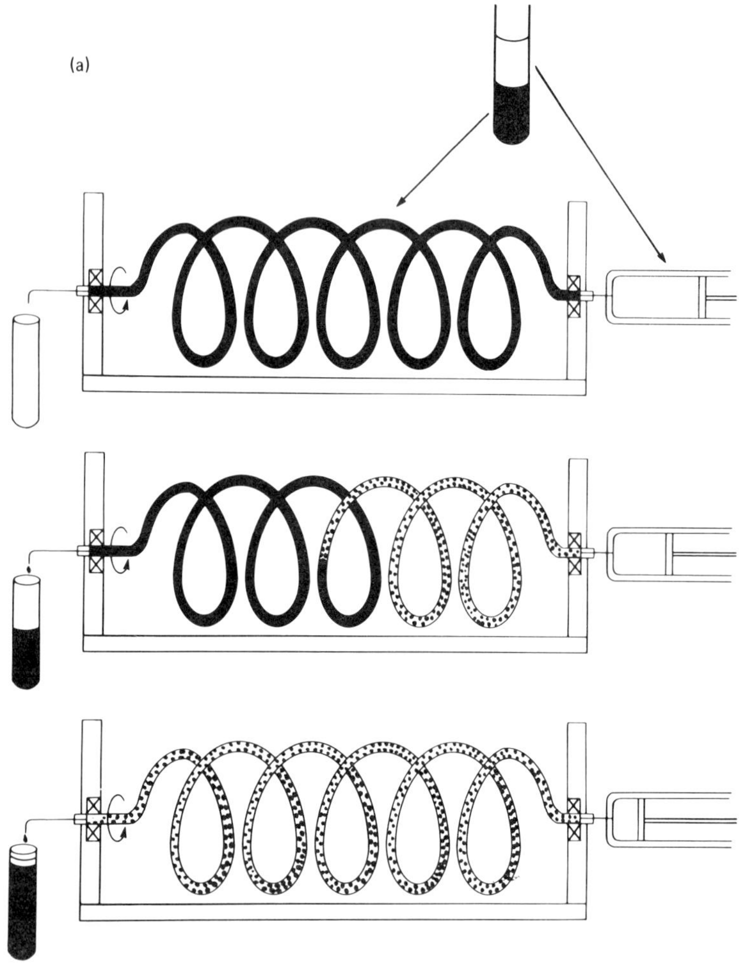

Fig. 15 Principle of the hydrodynamic equilibrium system (HDES). The coil is first filled with the stationary phase, either the lower phase (a) or the upper phase (b), and the mobile phase is introduced at the head of the coil while the coil is slowly rotated around its own axis (top). As soon as the mobile phase meets the stationary phase in the coil, a hydrodynamic equilibrium is established between the two phases (middle). This process continues until the mobile phase reaches the tail of the coil. Thereafter, the continued elution results in the displacement of the mobile phase only and a large amount of the stationary phase is permanently retained in each helical turn of the coil (bottom). Consequently, solutes introduced locally at the head of the coil are subjected to an efficient partition process between the mobile and the stationary phases and separated according to their partition coefficients.

(b)

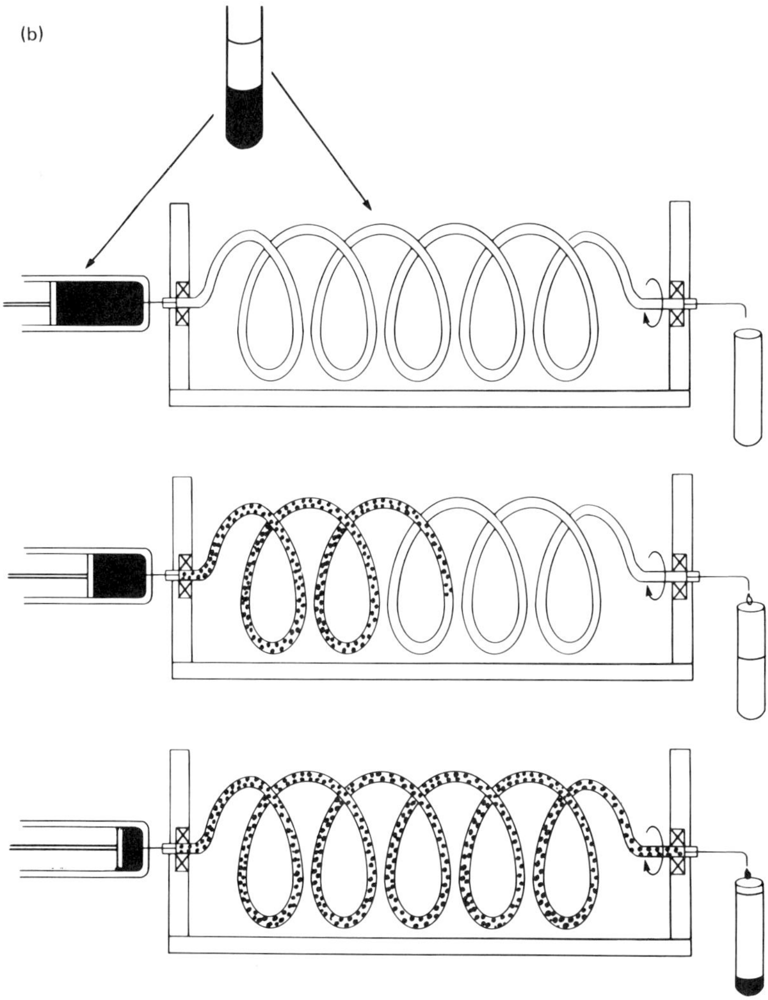

Fig. 15 (Continued)

where one phase with less wall surface affinity is split into multiple droplets which oscillate synchronously with the coil rotation as described earlier. While continuous pumping of the mobile phase keeps breaking this equilibrium state at the head end of the coil, the two phases can quickly react to restore the equilibrium by readjusting their relative volumes in each coil unit. Thus the stationary phase pushes newly introduced excess amounts of the mobile phase toward the tail end of the coil. As the elution proceeds, the length of coil under the hydrodynamic equilibrium steadily increases while the stationary phase is collected from the tail end of the coil. Finally the entire length of the coil becomes occupied with the two phases in the hydrodynamic equilibrium state when the mobile phase starts to elute through the tail end of the coil. Thereafter, continued pumping of the mobile phase only displaces the same phase while retaining a large amount of the stationary phase in each coil unit. This hydrodynamic equilibrium established under the external introduction of flow is different from that in the end-closed coil in that it consists of more mobile phase and less stationary phase, the difference increasing with the flow rate of the mobile phase.

The HDES demonstrated by this basic model again produces all physical factors essential for efficient partition process. The coil is capable of retaining either phase as the stationary phase and of providing multiple partition units which could outnumber the coil units. In every portion of the coil, broad interface is formed by droplets which violently but locally oscillate in the suspending phase to reduce mass transfer resistance and at the same time prevent laminar flow spreading of sample bands. Consequently, solutes locally introduced at the head end of the coil are subjected to an efficient partition process between the two phases at every portion of the coil and separated according to their relative partition coefficients in a manner analogous to that in liquid chromatography but in the absence of the solid support.

Clearly the present model system has various advantages over the hydrostatic equilibrium model system previously described. As illustrated in Fig. 15, the rotating coil has no inefficient free space which is completely occupied by the mobile phase and therefore every portion of the coil is considered to be efficient column space. At any portion of the coil, one of the phases forms droplets regardless of the choice of the mobile phase. Each droplet becomes an efficient stirrer of the other phase by its violent local oscillation. By virtue of these ideal features for partition, this simple model system itself can become an efficient countercurrent chromatographic scheme [15–17].

Unilateral HDES [6–8]

This particular form of HDES was accidentally discovered during the hydrodynamic studies with a rotating coil assembly. The characteristic features of the unilateral HDES are schematically illustrated in Fig. 16. The rotating coil at the top contains equal volumes of two immiscible solvent phases while the rest of the coils are shown uncoiled to indicate overall distribution of the two solvent phases along the length of the coil.

In the basic HDES previously described (left), slow rotation of the coil evenly distributes the two solvent phases from the head

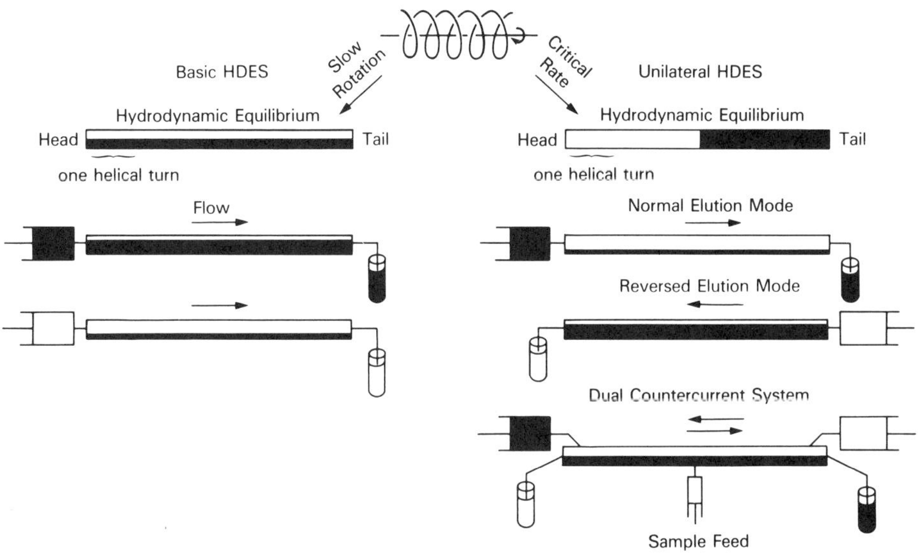

Fig. 16 Principle of the unilateral hydrodynamic equilibrium system (unilateral HDES). The rotating coil (top) contains two immiscible solvents in equal amounts. The rest of the coils are schematically drawn uncoiled to illustrate relative volume distribution of the two solvent phases along the length of the coil. In the basic HDES (left), the two solvent phases are evenly distributed from the head toward the tail of the coil. Consequently, the elution of either phase through the head of the coil produces retention of the other phase within the coil where the amount of the mobile phase always dominates that of the stationary phase. In the unilateral HDES (right), the two solvent phases are completely separated in the coil, one phase entirely occupying the head side and the other phase the tail side of the coil. This hydrodynamic equilibrium condition produces high stationary phase retention under the proper elution mode. The system also allows dual-countercurrent chromatography by simultaneously introducing the two solvent phases through the respective end of the coil which is equipped with five flow channels (bottom).

toward the tail and any excess of either phase remains at the tail of the coil. Consequently, elution with either phase from the head toward the tail results in the retention of the other phase in the coil. However, maximum attainable retention of the stationary phase may be about 50% of the total column capacity, which decreases gradually with increased flow of the mobile phase. Thus the depletion of the stationary phase in the coil eventually limits the application of high flow rates in this system. This problem can be solved by the unilateral HDES.

As the rotational speed of the coil is increased, the distribution pattern of the two solvent phases becomes altered in such a way that one phase dominantly occupies the head while excess of the other phase occupies the tail. When the rotational speed of the coil eventually reaches the critical range, the two solvent phases are completely separated along the length of the coil, one phase (head phase) entirely occupying the head side and the other phase (tail phase) the tail side of the coil as illustrated in Fig. 16 (right). This unilateral distribution of the two solvent phases clearly indicates that the head phase, if introduced at the tail, would move through the tail phase toward the head and the tail phase, if introduced at the head, would move through the head phase toward the tail. Consequently, this unilateral hydrodynamic equilibrium condition can be effectively utilized for performing CCC in the following two different ways: The coil is first entirely filled with the tail phase and the head phase is pumped through the tail of the coil. Alternatively, the coil is entirely filled with the head phase and the tail phase is pumped through the head of the coil. In both cases, the system can maintain a high retention of the stationary phase under a high flow rate of the mobile phase, resulting in excellent peak resolution of solutes in a short period of time.

The system also allows simultaneous introduction of the two solvent phases through the respective ends of the coil. This dual-countercurrent system requires an additional flow channel at each end of the coil to collect the effluent and a sample feed line at the middle portion of the coil. Recently, this system has been successfully applied to foam separation as described later in detail [10].

Development of CCC Schemes Based on HDES

Although the simple model systems described above are quite practical for preparative scale separations, the performance of the system is greatly improved in terms of versatility and partition efficiency by enhancing the acceleration field by means of centrifugation. Application of the centrifugal force enables the use of a small-bore column for analytical scale separations by eliminating plug flow phenomena. In preparative scale separation, centrifugation greatly reduces the separation time by allowing retention of the stationary phase under a higher flow rate of the mobile phase. Under the centrifugal force field the rotational speed of the coil can be increased to attain more efficient mixing of the phases within the limit where emulsification of the phases begins to take place.

Because of these advantages, efforts have been made to develop various types of centrifuge devices which can produce a rotating centrifugal force field with respect to the coiled column. All these devices are called the coil planet centrifuge (CPC) because they induce a planetary motion to the coiled column. Development of these centrifuge devices has been directed to achieve several major objectives as follows:

1. Elimination of the use of rotating seals which would cause a number of complications and limit the practical application of the device
2. Application of a particular mode of planetary motion to provide a centrifugal force field ideal for performing the desired type of CCC
3. Optimization of the acting angle of the acceleration field with respect to the coil to stabilize retention of the stationary phase
4. Freely adjustable rotational speed of the coiled column under any given revolutional speed to optimize phase retention and mixing

Various types of the flow-through CPC devices have been successfully developed to achieve these objectives as described later in this chapter.

D. Summary

In order to elucidate the general principle of countercurrent chromatography (CCC), the two basic CCC systems are described. One is the hydrostatic equilibrium system (HSES) which uses a hydrostatic equilibrium of the two immiscible solvent phases in a stationary coil and the other is the hydrodynamic equilibrium system (HDES) which uses a hydrodynamic equilibrium of the two phases established in a slowly rotating coil.

In the model of each basic CCC system (Figs. 11 and 15), the coil is filled with one phase of an equilibrated two-phase solvent system followed by the elution with the other phase. Both basic CCC systems allow retention of the stationary phase in each coil unit while the two phases maintain the respective type of equilibrium state. Consequently, the solutes locally introduced at the inlet of the coil are subjected to a partition process between the two phases in each coil unit and eluted out through the outlet of the coil in the order of their partition coefficients as in liquid chromatography but in the absence of the solid support.

Each basic system has its own specific merits in performing CCC. The HSES permits a stable retention of the stationary phase with a minimum risk of emulsification of the solvents. The HDES provides broad interface and efficient mixing of the two phases to yield an efficient partition process. Various characteristic features inherent in these two basic CCC systems are summarized in Table 3.

Table 3 Comparison Between Two Basic CCC Systems

	Hydrostatic equilibrium system (HSES)	Hydrodynamic equilibrium system (HDES)
Motion of coil	Stationary	Rotation around its own axis
Functional symmetry	Symmetrical	Rotation creats asymmetry to the coil. The suspended object moves toward the head determined by the handedness of the coil and direction of the rotation.
Direction of elution	The mobile phase can be introduced from either end of the coil with identical results.	The mobile phase should be introduced from the head end. Elution from the tail end will result in loss of the stationary phase.
Partition efficiency	Low	High

Retention of the stationary phase	Stable and close to 50%. However, no return of the stationary phase once carried over to the next coil unit.	Sensitive to rotational speed and interfacial tension of the two phases. Poor retention when solvents are emulsified. Carried over stationary phase can be returned to the original coil unit by the screwing force.
Number of partition units per coil unit	Usually one	Can be more than one
Interfacial area of the two phases	Relatively small especially when the mobile phase has wall surface affinity.	Large regardless of the choice of the mobile phase.
Mixing of the phases	Mild	Excellent
Efficient column space	50%	100%

A variety of useful CCC schemes have been developed from those basic systems. While a series of nonhelix preparative CCC schemes have been developed from the HSES, most of the CCC schemes have retained their original helical column configuration and instead great efforts have been made to introduce various types of seal-free flow-through centrifuge devices to establish highly efficient CCC schemes for both preparative and analytical separations. Recent discovery of the unilateral HDES has led to the development of high-speed CCC which accomplished an epoch-making advance in the CCC technology.

III. FLOW-THROUGH CENTRIFUGE SCHEMES FREE OF ROTATING SEALS*

In the past, various types of flow-through centrifuge devices have been developed for performing CCC. In many of these devices the use of rotating seals is entirely eliminated [5]. Although the rotating seal is commonly used in the conventional flow-through centrifuge devices, it becomes a potential source of various complications such as leakage, corrosion, contamination, and sample denaturation. The occurrence of any of these complications would largely limit the versatility of the device. Thus, the elimination of the rotating seals provides a number of advantages as follows:

1. The system is leak-free under a high pressure of several hundred psi.
2. It readily facilitates the use of multiple flow channels.
3. Samples are free from mechanical damage at the site of the seal.
4. It eliminates the heat dissipation problem at the seal.
5. It elimiantes the contamination problem.
6. It permits the use of corrosive solvents.
7. It minimizes the dead space at the seal which would cause unnecessary sample band broadening.

---

*Most recent development of new coil planet centrifuge schemes has necessitated an innovation of nomenclature which conveniently provides simple descriptive terms for all types of rotary-seal-free flow-through centrifuge schemes. In addition, our successful efforts in derivation of the general formulas for computation of acceleration generated by various modes of synchronous planetary motions have yielded diagrammatic presentation of the centrifugal force field characteristic of each scheme in the same format, thus greatly facilitating the comparative studies. All these recent advances in the present subject are supplemented in Appendixes 1–3 at the end of this chapter.

Because of these advantages, these seal-free CCC schemes permit a broad application for the separation and purification of various biological materials.

In the following, three classes of the seal-free flow-through centrifuge schemes are described and simple mathematical analyses applied on the typical planetary motions are reviewed.

A. Three Major Schemes for Seal-Free Centrifuge

General Description

Figure 17 illustrates the principle of various types of seal-free flow-through centrifuge schemes together with their mutual relationship. Each diagram schematically illustrates the orientation and motion of the cylindrical coil holder equiped with a bundle of flow tubes, the other end of which is tightly supported at the point marked with a black circle on the central axis of the centrifuge.

These seal-free flow-through centrifuge schemes developed for CCC are classified into three categories according to their mode of planetary motion. They are called synchronous, nonplanetary, and nonsynchronous schemes as indicated at the top of the figure. In the nonplanetary scheme, the coil holder simply rotates around the central axis of the centrifuge as in the conventional centrifuge system. On the other hand, both synchronous and nonsynchronous schemes produce planetary motions of the coil holder; hence they are called the coil planet centrifuges (CPC). In the synchronous schemes the rotation and revolution of the holder are synchronized to give a rotation/revolution ratio of 1:1, while in the nonsynchronous schemes the rotational rate of the holder is independently adjustable with respect to the revolution.

All these schemes are also numbered I through VII in the order of their complexity. Scheme I in the synchronous scheme is the simplest form which gradually evolves toward the most complex forms of nonsynchronous schemes VI and VII as indicated by arrows. Following this evolution pathway may facilitate elucidating the seal-free mechanism involved in all these flow-through centrifuge schemes.

Synchronous Schemes (Schemes I—IV)

In scheme I, the holder undergoes the simplest mode of synchronous planetary motion by keeping its axis always parallel to and at a given distance from the central axis of the centrifuge. In order to avoid the flow tubes being twisted, the holder synchronously counterrotates about its own axis while revolving around the central axis of the centrifuge. As a result, the motion of the holder becomes similar to the circular motion of a beaker observed when a chemist gently mixes its contents with his hand. This simple principle can be applied

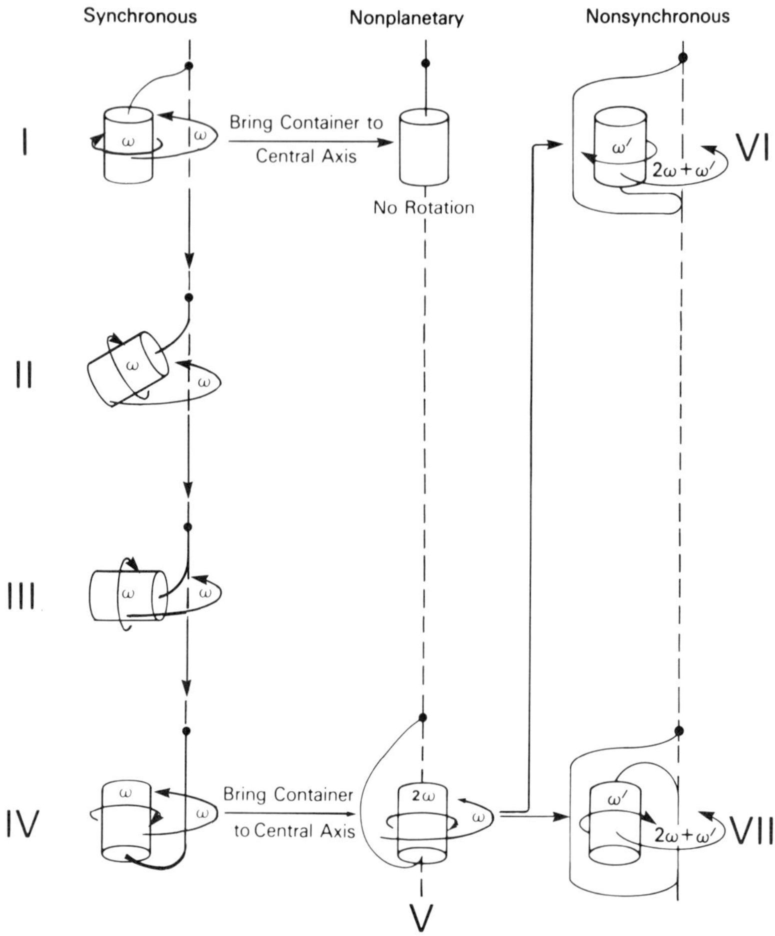

Fig. 17 A series of seal-free flow-through centrifuge schemes and their mutual relationship. These schemes are numbered I–VII in the order of complexity. They are also classified into three categories according to the mode of the planetary motion. Schemes I–IV on the left are called the synchronous schemes because of the synchronous planetary motion of the holder; the holder rotates about its own axis once during one revolution around the central axis of the centrifuge. Scheme V at the middle is called the nonplanetary scheme because it shows a simple rotation of the holder around the central axis of the centrifuge. Schemes VI and VII on the right are called the nonsynchronous schemes because of the nonsynchronous planetary motion where the rotational rate of the holder is freely adjustable at a given revolution rate. For new descriptive nomenclature for these centrifuge schemes, see Chap. 3, Appendix Fig. 162.

equally well to other types of synchronous schemes with tilted (scheme II), horizontal (scheme III), and even inverted (scheme IV) orientations of the holder. Each scheme displays its own characteristic centrifugal force field which can be efficiently utilized for performing CCC.

The synchronous planetary motion of scheme I produces a homogeneous centrifugal force field which rotates in a plane perpendicular to the axis of the holder. Thus, at any given moment every point on the holder is subjected to an identical force regardless of its location on the holder. When the coil is mounted on the holder parallel to the axis of the holder, the system becomes identical to the basic model for the HDES except that the gravitational field acting on the basic model is now replaced by the centrifugal force field. The CCC apparatus based on this type of planetary motion is called the flow-through CPC (scheme I) and successfully used for both analytical and preparative separations [18,19].

In scheme II with a tilted orientation of the holder, the centrifugal force field displays an extremely complex pattern which involves three-dimensional fluctuation of individual force vectors. Every point on the holder is subjected to a different centrifugal force field at any given moment. In addition, the centrifugal force vector acting on each point on the holder changes both magnitude and direction during one revolutional cycle of the holder. To ease understanding of the effects of these forces, the centrifugal force vector may be divided into two components in such a way that the first component always acts along the axis of the holder and the second component rotates in the plane perpendicular to the axis of the holder. In order to make use of this centrifugal force field for performing CCC, the coiled column is made by winding a tube on a fine flexible core which is again coiled around the holder to form a configuration of the coiled helix of what is called a *toroidal coil*. Under this column geometry each coil unit experiences two effective forces: the first component acts to trap the stationary phase as the gravity in the basic HSES model and the second component acts to produce efficient mixing of the two solvent phases as in the basic HDES model. Relative strength of these two force components is determined by the inclination of the holder. Thus, this type is considered to be a hybrid between the two basic systems. The apparatus is called the angle rotor flow-through CPC or the flow-through CPC (scheme II) and has been most successfully used for analytical scale separations [20].

Scheme III with a perpendicular orientation of the holder to the centrifuge axis also produces a complex centrifugal force field which is similarly divided into two components. The first component steadily acts along the axis of the holder whereas the second component rotates around the axis of the holder in a peculiar manner. As shown in the following subsection, the relative magnitude of the component is determined largely by the location of the point on the holder. When the effect of this second component is minimized by the choice of proper location and orientation for the column, the apparatus can produce a rather stable radial centrifugal force field as in the conventional centrifuge system. This force can then act on the coiled helix column on the holder as does the gravity in the basic HSES model. This particular member of the coil planet centrifuge is named the elution centrifuge or the flow-through CPC (scheme III) and mainly used for analytical scale separations [21].

In scheme IV with the inverted orientation of the holder, the holder rotates about its own axis and revolves around the central axis of the centrifuge at the same angular velocity in the same direction. This synchronous planetary motion is similar to that in scheme I except that the rotation of the holder about its own axis is in the reversed direction. Despite this similarity, scheme IV produces a quite different pattern of the centrifugal force field. While all force vectors are confined within a plane perpendicular to the holder axis as in scheme I, they display either rotating or oscillating pattern depending on the location of the point on the holder. The point located close to the holder axis is subjected to a rotating force field somewhat similar to that in scheme I. The point located remote from the holder axis is subjected to a centrifugal force field which is always directed outwardly from the holder. These vectors also display the fluctuation of their magnitude accompanied with the angular oscillation during each revolutional cycle of the holder.

As described later in detail, this centrifugal force field renders an extremely versatile feature to the present CCC scheme. The coiled column mounted onto the holder is subjected to either rotating or oscillating centrifugal force field depending on the location and orientation of the column on the holder, each force field being effectively used for performing CCC. The rotating centrifugal force field, available in the vicinity of the holder axis, enables an efficient mixing of the two phases within a small-bore coil as in scheme I previously described and is utilized for performing analytical scale CCC based on the HDES [2,23]. On the other hand, the oscillating force field, available at some distance away from the holder axis, provides a more stable retention of the stationary phase with the efficient local mixing of the two phases which acts to reduce mass transfer resistance. Consequently, this scheme is considered to be a modified HSES and capable of performing both analytical [24,25] and preparative [26,27] separations at high partition efficiencies. The capability of the present scheme is further extended by mounting a tube directly around the holder to make a simple coil with a coaxial orientation to the holder. Under this column orientation, the two-phase solvent system in the coil establishes a peculiar form of a hydrodynamic equilibrium where one of the phases, usually the upper phase, dominates on the head side of the coil and in some cases two solvent phases are completely separated along the length of the coil with the upper phase on the head side and the lower phase on the tail side. As mentioned earlier, this unilateral hydrodynamic equilibrium of the two solvent phases can yield high retention of the stationary phase for both high-speed CCC [28] and continuous extraction [8,29]. The system further permits the dual countercurrent operation where the two phases literally flow countercurrently to each other through the coiled column [6,8]. In addition, another unique feature of the present scheme is created when the axis of the centrifuge is placed in the horizontal position. The holder, then, rotates with respect to the gravitational field as in the basic HDES model. This feature gives a remarkable advantage in performing a large-scale preparative CCC with a slowly rotating coiled column [26,27,30]; hence the apparatus was named the horizontal flow-through CPC or the flow-through CPC (scheme IV).

Nonplanetary Scheme (Scheme V)

Scheme IV in the synchronous series is also a transitional form to scheme V called the nonplanetary scheme. When the holder of scheme I is brought to the central axis, the counterrotation of the holder cancels out the revolution resulting in zero rotation of the holder as illustrated in Fig. 17. However, a similar treatment applied to the scheme IV holder gives an amazing result. Because the rotation and the revolution of the holder in scheme IV are both in the same direction, transference of the holder to the central axis of the centrifuge results in the summation of these two motions. Consequently, the centrally located holder rotates at the doubled angular velocity of 2ω. This produces a stable, radially acting centrifugal force field as in the conventional centrifuge system. The central shift of the holder also causes the formation of a large loop of the tube bundle which revolves around the central axis of the centrifuge at angular velocity ω.

Since the system eliminates all complications arising at the site of the rotating seal, the nonplanetary centrifuge is ideal for on-line plasmapheresis and blood cell separations [31,32]. For the CCC application of this nonplanetary scheme, the coil holder is modified into a shallow bowl to accommodate a long, fine-coiled tube around its periphery; hence the apparatus is called the toroidal coil centrifuge. Under this arrangement, every helical turn of the column is subjected to a stable centrifugal force field to retain the stationary phase segment in a manner similar to that in the basic HSES model. This toroidal coil centrifuge has a great capacity for retaining the stationary phase in a small-bore column and is useful for analytical scale separations, especially for partition of cells and macromolecules with polymer phase systems [33–36].

Nonsynchronous Schemes (Schemes VI and VII)

The nonplanetary scheme of scheme V provides a base for the nonsynchronous schemes. In order to form scheme VI (Fig. 17, right top), the holder of the nonplanetary scheme is again shifted away from the central axis of the centrifuge while keeping a portion of the tube bundle fixed at the rotating base. Then, the holder is subjected to a planetary motion on the rotating base in such a way that the holder revolves around the central axis of the apparatus and counterrotates about its own axis at the same angular velocity ω'. When viewed from the earth, the holder revolves around the central axis at $2\omega + \omega'$ and simultaneously counterrotates about its own axis at ω', which gives a rotation/revolution ratio of $-\omega'/(2\omega + \omega')$. Similarly, scheme VII (Fig. 17, right bottom) is derived from the nonplanetary scheme. In this case the holder of the nonplanetary scheme is again inverted and shifted away from the central axis while keeping a portion of the tube bundle fixed at the center of the rotating base. Under this orientation an additional planetary motion can be applied to the holder in such a way that it revolves around the central axis of the centrifuge and rotates about its own axis at the same angular velocity ω' in the same direction. Consequently, with respect to the earth, the holder revolves around the central axis of the centrifuge at $2\omega + \omega'$ and rotates about its own axis at ω', which gives a rotation/revolution ratio of $\omega'/(2\omega + \omega')$. In both types of the nonsynchronous schemes described above, any combination between the rates of the rotation and revolution is available by selecting suitable values for ω and ω'.

These nonsynchronous schemes have an advantage over other centrifuge schemes in that the rotational rate of the coiled column is freely adjustable under a given centrifugal force field provided by revolution. Application of a reduced rotational rate of the coiled column relative to the revolutional speed becomes essential when the polymer phase systems [3] are to be employed for partition of cells or macromolecules. Because of the aqueous-aqueous phase composition, the polymer phase systems have an extremely low interfacial tension which gives a high tendency of emulsification. Thus, CCC with the polymer phase systems necessitates gentle mixing of the two phases in the coiled column by applying a slow coil rotation while applying a strong centrifugal force field to maintain the satisfactory level of stationary phase retention in the coiled column. The nonsynchronous flow-through CPCs (scheme VI) have been constructed and successfully used for performing partition of cells and macromolecules with the polymer phase systems [37]. The nonsynchronous schemes are also useful for cell elutriation with a single-phase physiological solution [37,38]. Although these nonsynchronous schemes require highly elaborate mechanical designs, they are most versatile among all seal-free flow-through centrifuge schemes.

B. Mathematical Analysis of Acceleration Produced by Synchronous Planetary Motion

The planetary centrifuge schemes produce various patterns of the rotating acceleration field. Among the synchronous planetary motions illustrated in Fig. 17, schemes I and IV produce the simplest forms of orbits which are always confined in a plane perpendicular to the holder axis. Consequently, the mathematical analysis on these two schemes is conveniently performed by the aid of a two-dimensional coordinate system. On the other hand, in schemes II and III, where each acceleration vector displays a three-dimensional motion, the analysis of acceleration becomes much more complex and necessitates the use of a three-dimensional coordinate system. Below, reviewing the simple mathematical analyses on three representative synchronous planetary motions (schemes I, III, and IV) will elucidate the centrifugal force field characteristic of each scheme.

Acceleration Produced by Planetary Motion Scheme I [5,19,39]

The planetary motion to be analyzed is illustrated in Fig. 18A (left), where the axis of the centrifuge is in the horizontal position. As described earlier, the cylindrical coil holder revolves around the central axis of the centrifuge (center of revolution) and simultaneously counterrotates about its own axis (center of rotation), both at the same angular velocity as indicated by the arrows. During this synchronous planetary motion, the center of rotation maintains its orientation always parallel to and at a distance R from the center of revolution.

The motion of an arbitrary point P located at distance r from the center of rotation on the holder can then be analyzed by the aid of a a coordinate system shown in Fig. 18B (left). In order to simplify the analysis, the coordinate system is selected in such a way that the center of revolution is fixed at the center of the coordinate system and the center of rotation is initially located on the x-axis as shown in the diagram. The arbitrary point on the holder initially locates at point P_0 and forms an angle θ_0 with the x-axis. After a lapse of time t, the arbitrary point moves to point P (x, y). Due to the revolution and the counterrotation of the holder, the angle formed between the rotational radius (a line drawn from the arbitrary point to the center of rotation) and the x-axis is given by $\theta_0 + \omega t - \omega t = \theta_0$ and therefore remains unchanged. The location of the arbitrary point P (x, y) is then expressed by

$$x = R \cos \theta + r \cos \theta_0 \tag{8}$$

$$y = R \sin \theta + r \sin \theta_0 \tag{9}$$

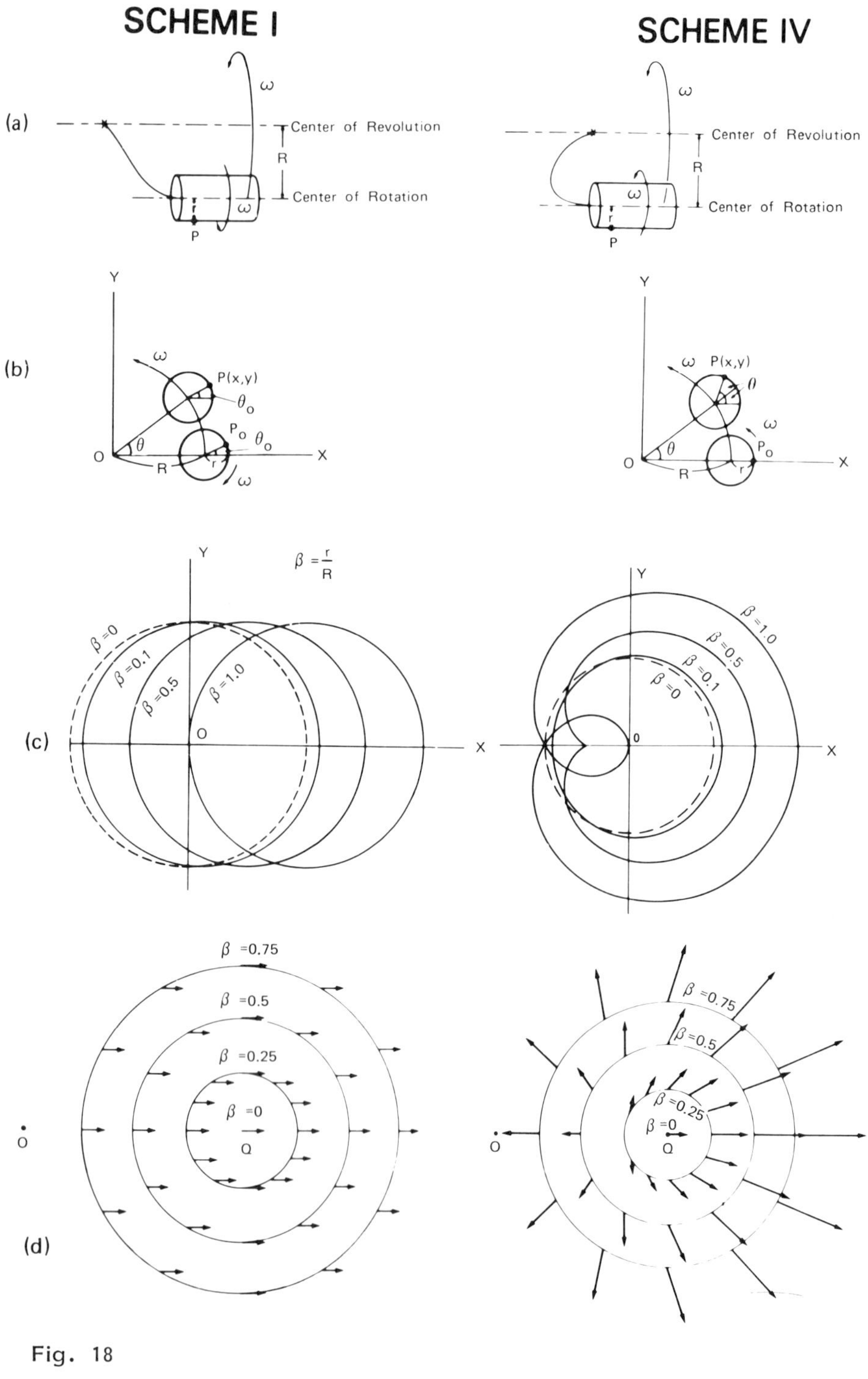
SCHEME I
SCHEME IV
(a)
ω
Center of Revolution
R
Center of Rotation
r
P
(b)
Y
ω
P(x,y)
θ_o
P_o
O
θ
R
r
X
(c)
$\beta = \frac{r}{R}$
β =0
β =0.1
β =0.5
β =1.0
O
X
(d)
β =0.75
β =0.5
β =0.25
β =0
O
Q

Fig. 18

From these equations, the orbit of the arbitrary point is easily derived by eliminating the variable θ and

$$(x - r\cos\theta_0)^2 + (y - r\sin\theta_0)^2 = R^2 \quad (10)$$

is obtained. This represents a circle with radius R centered at point $(r\cos\theta_0, r\sin\theta_0)$. The fact that the radius of the circle R is independent of both θ_0 and r indicates that every point located on the holder has a circular orbit with radius R. Such orbits are drawn in Fig. 18C (left) where the dotted circle ($\beta = 0$) indicates the orbit of the center of rotation and three solid circles, each for the point located at a distance of 0.1R ($\beta = 0.1$), 0.5R ($\beta = 0.5$), and 1R ($\beta = 1.0$) away from the center of rotation and $\theta_0 = 0$.

The net magnitude of acceleration α and its angle γ formed from the x-axis at the arbitrary point on the holder are further computed from the second derivatives of Eqs. (8) and (9), and

$$\alpha = \{(d^2x/dt^2)^2 + (d^2y/dt^2)^2\}^{1/2} = R\omega^2 \quad (11)$$

$$\gamma - \pi = \tan^{-1}\{(d^2y/dt^2)/(d^2x/dt^2)\} = \omega t = \theta \quad (12)$$

This shows that the point is subjected to a constant magnitude $R\omega^2$ of acceleration which rotates around the point at a uniform rate of ω. The acceleration vector, expressed by magnitude $R\omega^2$ forming angle $(\theta + \pi)$ from the x-axis, is totally independent of both r and θ_0, which determine the location of the point on the holder. This indicates that at any given moment every point on the holder is subjected to the identical acceleration field acting in a plane perpendicular to the holder axis. Figure 18D (left) shows the distribution of the centrifugal force vectors (same magnitude as the acceleration but acting in the opposite direction or at angle $\gamma - \pi = \theta$) on the holder when the center of rotation is crossing the x-axis on its positive side.

The results of the above analysis clearly indicate that no matter where the coiled column is mounted on the holder, all portions of the coil are subjected to the identical centrifugal force field. This fact permits the freedom to mount multiple coiled columns around the holder at any location to produce the exactly same effects, provided

---

Fig. 18 Mathematical analysis of synchronous planetary motion of scheme I (left) and scheme IV (right). (a) Synchronous planetary motion of the holder. (b) Coordinate system for analysis. (c) Orbits of the arbitrary points on the holder. (d) Distribution of centrifugal force vectors.

that the axis of the coil is parallel to the axis of the holder. Consequently, scheme I is considered to be an ideal HDES, where the gravity in the basic HDES model is replaced by a strong centrifugal force field. In the basic model the coil rotates in the gravitational field, whereas in shceme I the centrifugal force rotates around the coil to produce the equivalent effect. By the use of this enhanced acceleration field, scheme I enables efficient CCC with a small-bore coiled column.

Acceleration Produced by Planetary Motion

Scheme IV [5,39]

The planetary motion of scheme IV is illustrated in Fig. 18A (right), where the axis of the centrifuge is placed in the horizontal position. The coil holder revolves around the central axis of the centrifuge (center of rotation) at the same angular velocity ω and in the same direction as indicated by the arrows. As in the synchronous planetary motion of scheme I, the holder always maintains its axis parallel to and at a distance R from the center of revolution.

The motion of an arbitrary point located on the holder at distance r from the center of rotation is analyzed by the aid of a coordinate system shown in Fig. 18B (right). In order to simplify the analysis, the coordinate system is selected so that the center of revolution is located at the center of the coordinate system whereas both the center of rotation and the arbitrary point are initially located on the x-axis as illustrated. After the lapse of time t, the center of rotation circles around the central point o by $\theta = \omega t$ and the location of the arbitrary point P (x, y) is given by

$$x = R \cos \theta + r \cos 2\theta \qquad (13)$$

$$y = R \sin \theta + r \sin 2\theta \qquad (14)$$

From these equations, the orbit of the arbitrary point is computed by eliminating the variable θ. As illustrated in Fig. 18C (right), the orbit thus obtained displays a great variety of shape according to the locations of the point on the holder. As indicated in the figure, the location of the point on the holder is conveniently expressed by $\beta = r/R$, i.e., the ratio between the radius of rotation (r) and the radius of revolution (R). When $\beta \leqslant 0.25$, the orbit is a single circular loop. As the β-value increases, it becomes heart-shaped ($\beta = 0.5$) and then forms a double loop ($\beta = 1.0$) which gradually approaches a double circle with greater β-values. These results suggest that the acceleration field not only changes its pattern according to the location of the point on the holder but also fluctuates periodically during each revolutional cycle of the holder.

The acceleration acting on the arbitrary point is further calculated from the second derivatives of Eqs. (13) and (14) as

$$\frac{d^2x}{dt^2} = -\ R\omega^2(\cos\ \theta + 4\beta\ \cos\ 2\theta) \tag{15}$$

$$\frac{d^2y}{dt^2} = -\ R\omega^2(\sin\ \theta + 4\beta\ \sin\ 2\theta) \tag{16}$$

which gives the absolute net magnitude

$$\alpha = [(d^2x/dt^2)^2 + (d^2y/dt^2)^2]^{1/2} = R\omega^2(1 + 16\beta^2 + 8\beta\ \cos\ \theta)^{1/2} \tag{17}$$

acting at the angle relative to the x-axis:

$$\gamma_x = \pi + \tan^{-1}[(d^2y/dt^2)/(d^2x/dt^2)]$$

$$= \pi + \tan^{-1}[(\sin\ \theta + 4\beta\ \sin\ 2\theta)/(\cos\ \theta + 4\beta\ \cos\ 2\theta)] \tag{18}$$

provided $R \neq 0$ and $\beta = r/R$. (If $R = 0$, $\alpha = 4r\omega^2$ and $\gamma_x = 2\theta + \pi$ which give the acceleration produced by the nonplanetary scheme, scheme V in Fig. 17.) In order to visualize the effect of the acceleration on the behavior of the two solvent phases in the coiled column, it is more convenient to express the acting angle of the acceleration vector with respect to the rotating holder. The angle formed between the acceleration vector and the radius of rotation, i.e., the line drawn from the arbitrary point P to the center of the holder, is given by

$$\gamma = \gamma_x - 2\theta - \pi = \tan^{-1}[-\ \sin\ \theta/(4\beta + \cos\ \theta)] \tag{19}$$

The changes of magnitude α and the angle γ during one revolutional cycle of the holder are respectively illustrated in Figs. 19 and 20 where several curves are drawn according to the β-values. In Fig. 19 the α-values for $\beta > 0$ show patterns undulating in such a way that it becomes greatest at $\theta = 0°$ (location of the point most distant from the axis of revolution) and smallest at $\theta = 180°$ (location of the point nearest to the axis of revolution). At $\theta = 180°$, α decreases as the β-value increases from 0 to 0.25 where α becomes minimized.

Figure 21 illustrates undulating patterns of angle γ around the radius of rotation or the $\gamma = 0$ line for various β-values. The γ-values on the ordinate are chosen so that the bottom line or $\gamma = -180°$ continually joins the top line or $\gamma = 180°$ to form a complete circle of 360°. Thus, curves crossing over these lines indicate angles rotating

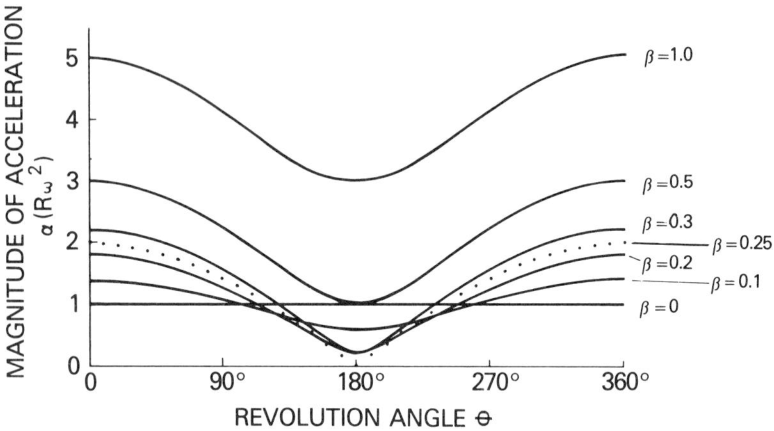

Fig. 19 Undulation of the net magnitude of acceleration for various β-values during one revolutional cycle.

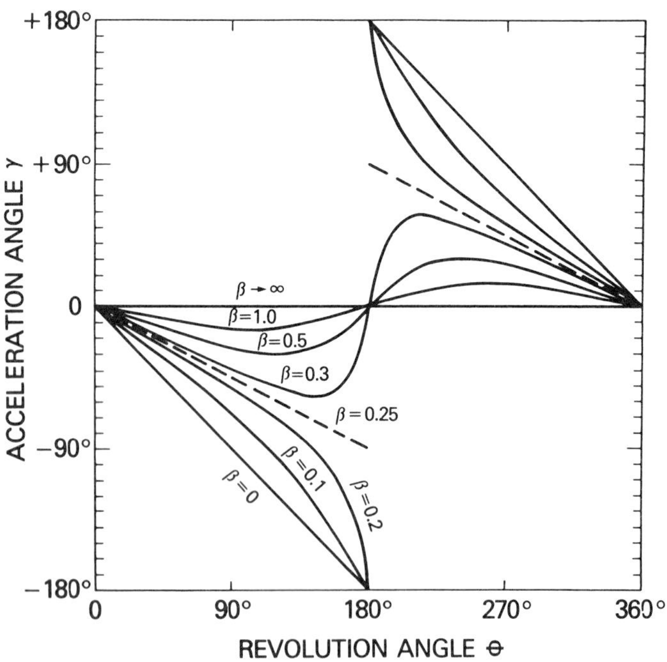

Fig. 20 Changing pattern of the acting angle of acceleration for various β-values during one revolutional cycle.

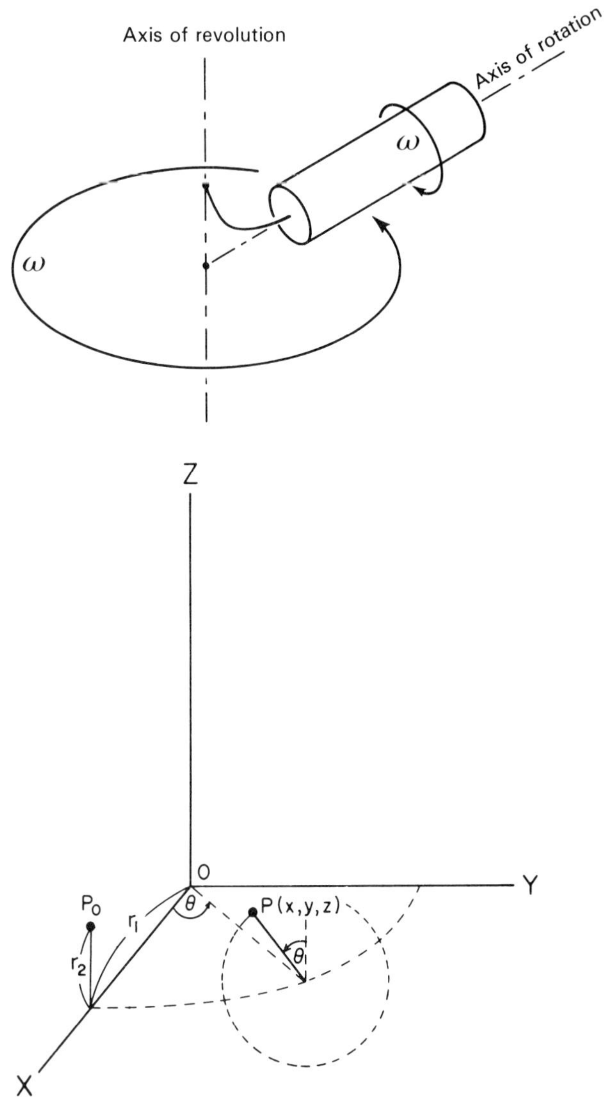

Fig. 21 Mathematical analysis of synchronous planetary motion for scheme III. (a) Synchronous planetary motion of the holder. (b) Coordinate system for analysis.

around the arbitrary point while other curves show angles swinging around the $\gamma = 0$ line. At $\beta = 0$, or the location of the point on the axis of rotation, γ forms a straight line across the ±180° lines indicating that the acceleration field rotates uniformly around the point as in the holder of scheme I but in the opposite direction. When $0 < \beta < 0.25$, the acceleration field still rotates around the point but the rate of rotation is not uniform in such a way that the field changes its direction rather slowly around $\theta = 0°$ and more quickly around $\theta = 180°$. When β becomes greater than 0.25, the rotational motion of the field changes into a swinging motion where the angle moves back and forth around the $\gamma = 0$ line with its amplitude of angular oscillation decreasing with greater β-values. Here again the acceleration changes its direction rather slowly around $\theta = 0°$ and more quickly around $\theta = 180°$, especially with β-values close to 0.25. When β-values become infinite, the γ-line approaches a straight line of $\gamma = 0$, which indicates a stable acceleration field as observed in the conventional centrifuge system.

From Eqs. (17) and (19), the relative centrifugal force vectors acting at various locations on the holder are obtained and illustrated in Fig. 18D (right). Three circles centered at Q correspond to β-values of 0.25, 0.5, and 0.75 as labeled while point Q, the center of rotation, is at $\beta = 0$. A set of arrows around each circle indicates the distribution pattern of the centrifugal force field at a given moment. It is important to note that the arbitrary point on the holder experiences these vectors in sequence during each revolutional cycle. As clearly shown in the figure, when $\beta > 0.25$, the vector is always directed outwardly from the circle. The vector also undulates in its relative magnitude and direction according to the location on the holder at given β-values. As the β-value increases, the magnitude of the relative centrifugal force vector becomes greater whereas the amplitude of angular oscillation around the axis of rotation is reduced. This distribution pattern of the force vector provides a great advantage when the scheme is used for performing CCC.

When a coiled column is mounted at the periphery of the holder at a location $\beta > 0.25$, two immiscible solvents in the coil are separated by the outwardly directed centrifugal force in such a way that the heavier phase occupies the outer portion and the lighter phase the inner portion of each turn of the coil as observed in the basic HSES model. However, fluctuation of the centrifugal force field produces vigorous mixing of the two phases at their interface to promote the partition process. Consequently, with this column geometry, the present scheme becomes an efficient HSES which can be applied to both preparative [22,40] and analytical [23,24] CCC.

The capability of the present scheme is further expended by applying a coiled column coaxially to the coil holder. The separation column is simply made by winding a piece of tubing directly onto the holder making a desired number of helical turns. With this column configuration, the centrifugal force separates the two solvent phases into two layers, the lighter phase in the inner portion and the heavier phase in the peripheral portion of each helical turn, while the flucutating force field enables passing one of the phases toward the head of the coil. This scheme becomes a modified version of the HDES where one of the phases dominates at the head of the coil and in some cases the two phases are completely separated along the length of the coil with one phase on the head side and the other phase on the tail side. Recent studies have shown that this particular scheme is capable of retaining a large amount of the stationary phase against a high flow rate of the mobile phase and, therefore, becomes an ideal scheme for continuous countercurrent extraction [8] as well as high-speed preparative CCC [9,28].

Acceleration Produced by Planetary Motion

Scheme III [21]

The scheme to be analyzed is illustrated in Fig. 21A where a cylindrical column holder revolves around the central axis of the apparatus and simultaneously rotates about its horizontal axis at the same angular velocity in the indicated direction. During this synchronous planetary motion, the axis of the holder is always confined in the horizontal plane and maintains the radial orientation with respect to the central axis of the centrifuge in such a way that any given point on the holder axis always remains at the same distance from the central axis of the centrifuge. A mathematical analysis on the motion of an arbitrary point on the holder is performed by the aid of a three-dimensional coordinate system. In Fig. 21B an arbitrary point is located at r_1 from the axis of revolution and at r_2 from the axis of rotation at the starting point P_o. The axis of rotation then revolves around the z-axis at angular velocity ω while the arbitrary point simultaneously rotates around the axis of the rotation at the same angular velocity in the indicated direction. Consequently, the arbitrary point travels always on a spherical surface centered at point o with a radius of $(r_1{}^2 + r_2{}^2)^{1/2}$ and after the lapse of time t the location of the point, P (x, y, z), is expressed by

$$x = r_1 \cos \theta + r_2 \sin^2 \theta \tag{20}$$

$$y = r_1 \sin \theta - r_2 \sin \theta \cos \theta \tag{21}$$

$$z = r_2 \cos \theta \tag{22}$$

where $\theta = \omega t$. Since the net centrifugal acceleration field:

$$\alpha = [(d^2x/dt^2)^2 + (d^2y/dt^2)^2 + (d^2z/dt^2)^2]^{1/2} \tag{23}$$

gives a rather complicated picture which exhibits the three-dimensional change of its acting direction during one revolution cycle, it is convenient to express it in terms of two components, α_1 (centrifugal acceleration parallel to the axis of the holder) and α_2 (centrifugal acceleration in a plane perpendicular to the axis of the holder).

The first component is given by the equation:

$$\alpha_1 = -(d^2x/dt^2) \cos \theta - (d^2y/dt^2) \sin \theta \qquad (24)$$

which reduces to

$$\alpha_1 = r_1\omega^2 (1 - 2\beta \cos \theta) \qquad (25)$$

where $\beta = r_2/r_1$ and $r_1 \neq 0$. [If $r_1 = 0$, $\alpha_1 = -2r_2\omega^2 \cos \theta$ which is apparent if we expand Eq. (25) before inserting $r_1 = 0$.]

In Fig. 22, the value of α_1 expressed in terms of $r_1\omega^2$ is plotted against angle θ during one revolutional cycle. Four lines are drawn according to the β-values of 1, 1/2, 1/4, and 0 as indicated. It

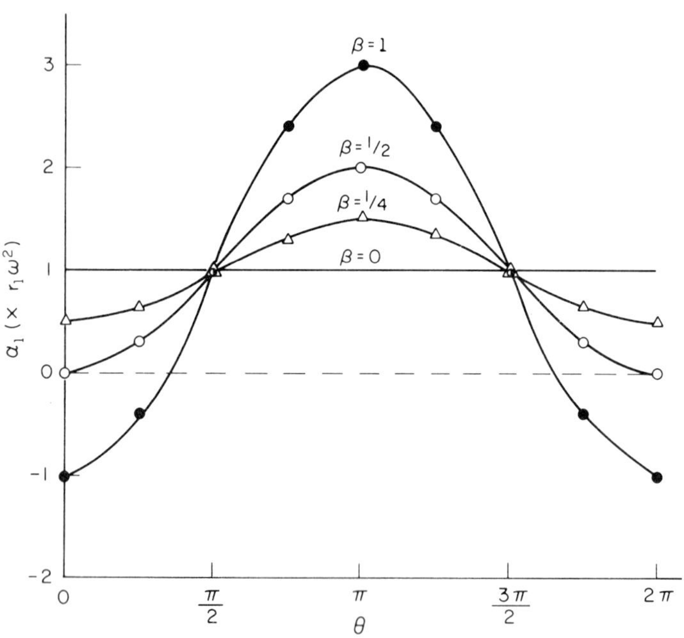

Fig. 22 Fluctuation of α_1 expressed in terms of $r_1\omega^2$ during one revolution cycle.

illustrates that α_1 oscillates around the $\beta = 0$ line once in each revolution with an amplitude dependent on the β-value. When β exceeds 1/2, α_1 crosses below the broken 0 line, indicating that the acceleration acts momentarily in the opposite direction to cause the liquids to move toward the center of revolution. As the β-value decreases, the amplitude of the oscillation becomes smaller and finally reduces to zero at $\beta = 0$ providing a stable acceleration field similar to that in the conventional centrifuge system.

The second component, which acts in the plane perpendicular to the axis of the holder, is given by

$$\alpha_2^2 = [(d^2x/dt^2) \sin \theta - (d^2y/dt^2) \cos \theta]^2 + (d^2z/dt^2)^2 \qquad (26)$$

or

$$\alpha_2 = r_2\omega^2 (1 + 3 \sin^2 \theta)^{1/2} = r_2\omega^2 [(5 - 3 \cos 2\theta)/2]^{1/2} \qquad (27)$$

Figure 23 shows the relationship between α_2 (in terms of $r_2\omega^2$) and θ during one revolution. Note that α_2 oscillates twice in each revolution between the values of $r_2\omega^2$ and $2r_2\omega^2$, and that the wave form is not accurately sinusoidal. The value of α_2 is independent of r_1 and reduces to 0 at $r_2 = 0$. The acting direction (γ) of α_2 in the plane perpendicular to the holder axis is calculated by

$$\tan \gamma = [(d^2x/dt^2) \sin \theta - (d^2y/dt^2) \cos \theta]/(-d^2z/dt^2) \qquad (28)$$

which becomes

$$\tan \gamma = 2 \tan \theta \qquad (29)$$

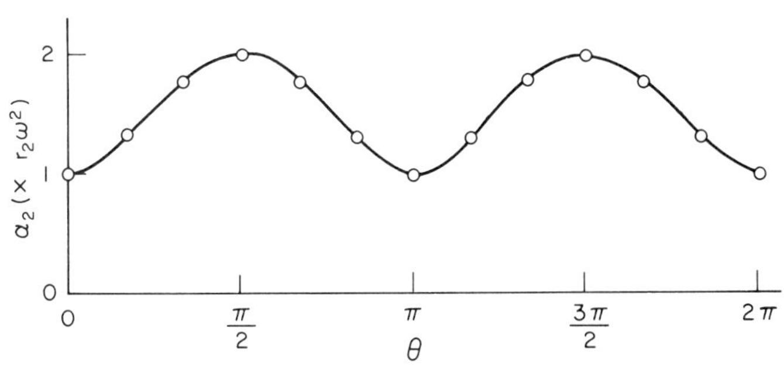

Fig. 23 Fluctuation of α_2 expressed in terms of $r_2\omega^2$ during one revolution cycle.

indicating that the direction of α_2 undulates about the θ line twice during each revolutional cycle (Fig. 24).

The results of the above analysis show that the present synchronous planetary motion causes an oscillating centrifugal force field. The relative amplitude of the oscillation becomes greater as the distance (r_1) from the axis of revolution decreases or the distance (r_2) from the axis of the rotation increases. Thus, the most stable centrifugal force field can be obtained at or near the axis of the holder with a great radius (r_1) of revolution. On the other hand, the mixing or vibration, if desired, can be attained at the location remote from the axis of the holder with a small radius (r_1) of revolution. In order to perform CCC, the coiled column is prepared by winding a small-bore tube onto a flexible core which is in turn coiled around the holder to make a coiled helix configuration. Under this column orientation, the first component, α_1, acts to retain the stationary phase, either the upper or the lower phase, in each helical turn of the column as in the basic HSES model while the second component, α_2, produces mixing of the two phases at their interfaces to promote the partition process.

C. Summary

A series of seal-free flow-through centrifuge schemes is described (Fig. 17). The synchronous schemes characterized by the synchronous planetary motion produce either a rotating or an oscillating centrifugal force field which is efficiently used for performing CCC. Schemes I, II, and III are mainly used for small-scale analytical separations. Scheme IV is most versatile and covers various applications such as both preparative and analytical CCC, continuous

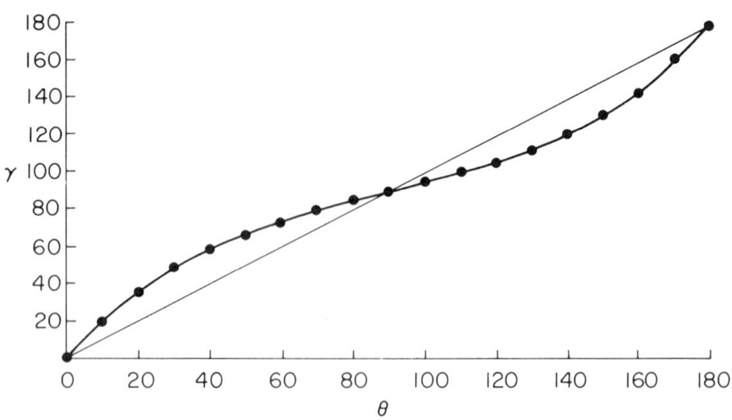

Fig. 24 Undulation of γ (direction of α_2) during half revolution.

countercurrent extraction, foam CCC, etc. The nonplanetary scheme V produces a stable radially acting centrifugal force field as in the conventional centrifuge systems and is useful for analytical separation. The nonsynchronous schemes, including schemes VI and VII, provide freely adjustable rotational rate of the coiled column at a given revolutional rate. These schemes have a great potential in separations of cells and macromolecules with polymer phase systems and also in cell elutriation with a physiological solution.

Mathematical analyses have been performed on typical synchronous planetary motions which include schemes I, III, and IV. The centrifugal force field characteristic of each planetary motion provides useful information for understanding for motion of two solvent phases in the rotating coiled column.

IV. NONHELIX COUNTERCURRENT CHROMATOGRAPHY

As stated earlier, this group of preparative CCC methods has been developed from the basic HSES model which consists of a stationary coiled tube (Fig. 12). Due to a radical modification of the column configuration, the system totally lost its original helical appearance; hence it acquired the term nonhelix CCC. There are three CCC schemes included in this group, i.e., droplet CCC (DCCC) [11–13], rotation locular CCC (RLCCC) [11,13], and gyration locular CCC (GLCCC) [11,13]. All these schemes use a column assembly consisting of multiple straight tubes which are interconnected in series with transfer tubes. They are most useful for preparative scale separations with a variety of conventional two-phase solvent systems used in CCD. These CCC schemes in general share common advantages over CCD in terms of partition efficiency, separation times, and simplicity of operation.

A. Droplet CCC (DCCC) [11–13]

This simple preparative CCC scheme has stemmed from the basic model of HSES by an accidental discovery of droplet flow in a modified coiled tube. In the course of developing a preparative CCC scheme, a flat coil was prepared by winding a tube around a piece of a long flat aluminum plate. The coil was then placed on the floor in such a way that the axis of the coil is in the horizontal orientation and the flat sides of the coil in the vertical orientation. When the flat coil was eluted with a two-phase solvent system, we have observed that the upper aqueous phase with low wall surface affinity formed discrete droplets which rose quickly through the lower phase on one side of the coil. Close observation of the individual droplets disclosed an active interfacial motion which suggested efficient mass transfer capability. Under ideal conditions each droplet of the mobile phase could become a plate if kept more or less discrete throughout the system [11]. The efforts have been made along this thought to develop an efficient preparative CCC schemes which can yield partition efficiencies of up to 1000 theoretical plates (TP). Because of the beautiful arrangement of droplets in the column, the scheme was named droplet countercurrent chromatography (DCCC), although droplet formation is commonly observed in many other CCC schemes.

DCCC has a great advantage over other CCC schemes in its simplicity that the apparatus can be easily and economically constructed by the individual scientist in an ordinary chemical laboratory. The apparatus also became commercially available through Tokyo Rikakikai Co. Ltd., Tokyo, and Büchi Laboratory-Techniques Ltd., Flawil, Switzerland. Because of its simplicity and availablity of the apparatus, the method has been widely used throughout the world. A number of applications reported by the users of DCCC will be reviewed in Part II by Hostettmann and Marston.

Principle of DCCC

The principle of the present scheme is illustrated in Fig. 25 where three straight tubular columns are placed in the vertical position. These columns are interconnected in series with transfer tubes which run from the top to the bottom between the neighboring columns. In this illustration, the syringe feeds the solvent through the bottom of the first column while the effluent is collected through the top of the third column. However, if necessary, the whole column can be inverted so that the solvent is introduced from the top of the first column and collected through the bottom of the third column. In order to demonstrate the principle of this system, the entire column space, including the free space in the connection tubes, is first filled with one phase of an equilibrated two-phase solvent system. This becomes the stationary phase and the other the mobile phase. Although the stationary phase can be the upper or the lower phase, it should have a stronger affinity to the internal surface of the column than the mobile phase. Then the mobile phase is introduced into the column in such a way that it enters the column from the top if the mobile phase is the lower phase and it enters the column from the bottom if the mobile phase is the upper phase, the latter being the case in Fig. 25. Because of the low wall surface affinity, the mobile phase immediately forms a discrete droplet in the stationary phase. Under the ideal conditions the droplet can occupy the column space across the diameter of the tube. Then, due to the effect of the gravity, the droplet soon starts to move either downward or upward through the stationary phase at a uniform rate. Under a constant pumping rate of the mobile phase, these droplets are formed one after another at regular intervals so that the entire

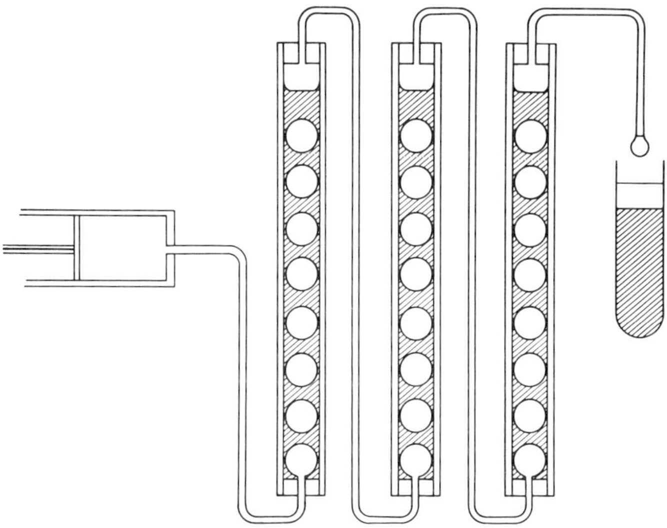

Fig. 25 Principle of DCCC. Three column units interconnected with transfer tubes are first filled with the stationary phase (shaded). The mobile phase with less affinity to the wall surface (clear) introduced at the bottom of each column unit forms multiple droplets in the stationary phase at regular intervals to divide the column space into a number of partition units. Under this steady equilibrium condition, solutes introduced at the inlet of the column are partitioned between the mobile droplet phase and the stationary phase and eluted out in the order of their partition coefficients.

column space is soon divided into the alternating segments of the moving droplets and the suspending stationary phase as illustrated. In this steady state equilibrium, the stationary phase is constantly forming a thin layer between the internal wall surface and the advancing droplet while each droplet is also subjected to violent internal stirring due to the drag force produced by its movement through the column. Consequently, solutes introduced locally at the inlet of the column are subjected to an efficient partition process between the moving droplet phase and the stationary phase and eluted out from the outlet of the column in the order of their partition coefficients.

The Original DCCC Apparatus

Figure 26 shows a column assembly of the original DCCC model constructed in our laboratory in collaboration with T. Tanimura, a visiting scientist from the University of Tokyo. It consists of 300

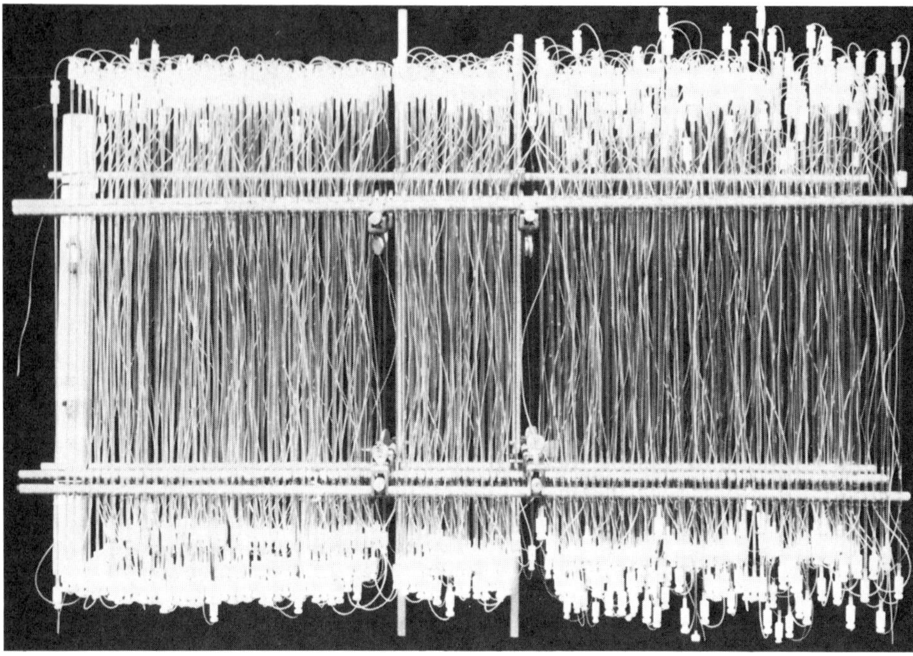

Fig. 26 Apparatus for DCCC. The column assembly consists of 300 glass column units (each 60 cm long and 1.8 mm i.d.) interconnected with transfer tubes. It has a total capacity of 540 ml. The sample chamber (a section of Mohr pipette) is mounted at the left of the assembly.

silylated Pyrex glass tubes, each 60 cm long with 1.8 mm i.d. and 0.6 mm wall thickness, which are interconnected with 0.55 mm i.d. standard wall PTFE (polytetrafluoroethylene) tubing. The column has a total capacity of about 540 ml including about 80 ml of the free space in the connection tube. Because the maximum column pressure would reach over 10 atmospheres during separation, particular efforts were made to perfect leak-proof junction throughout the column. Each junction should also provide a geometry suitable for droplet formation. In this prototype the junctions were made as follows: The interconnection tubes were passed through a Delrin fitting which with suitable tubing adapters secured it to the glass tubing. The base of the fitting was fitted over the glass tubing which was then flanged. A sleeve of polyethylene tubing, 2.92 mm i.d., 3.73 mm o.d., and approximately 1 cm long, was placed over the glass to adapt the Delrin to the flange. Inserted into the glass tubing was a 1-cm length of AWG No. 18 standard wall PTFE tubing containing a polyethylene tip with 0.76 mm i.d. and 1.22 mm o.d. for the production of droplets of suitable size. the PTFE transfer tubing was passed through the male threaded head of the Delrin fitting and into the PTFE adaptor in the glass tubing. The transfer tubing has a tendency to slip out under high pressure. It was held in place with a short sleeve of shrinkable PTFE 1203 Penntube WTF. When shrunk, the sleeve bound to the transfer tubing and provided an anchor to the male Delrin fitting [12].

The capability of the apparatus was tested on separation of DNP amino acids with a two-phase solvent system composed of chloroform, acetic acid, and 0.1 N HCl at a 2:2:1 volume ratio equilibrated at room temperature. Sample solution was prepared by dissolving seven DNP amino acids (Sigma Chemical Co.), 2–10 mg each, in 3 ml of a 1:1 mixture of the upper and the lower phases. The column was first entirely filled with the stationary lower phase and the above sample solution was totally placed in a 5-ml sample chamber (made from a modified 5-ml Mohr pipette with a suitable high-pressure fitting) located on the flow path between the pump and the column. The upper aqueous phase was then pumped through the top of the sample chamber at a flow rate of 16 ml/hr. A Beckman Accu-Flo pump, equipped with a side arm reservoir for damping pulsations, was used to provide a constant flow. The above operational condition yielded ideal droplet formation as shown in Fig. 27. These droplets were uniformly spaced at about 0.5-cm intervals throughout the entire length (60 cm) of the column. Fractions, each 3 ml, were collected to determine the absorbance at 350 nm.

Figure 28 shows a typical chromatogram obtained by the original DCCC apparatus. Symmetrical, completely resolved peaks were obtained for each component in the mixture. The time required for the separation was 70 hr. Partition efficiency of the separation can be calculated from the conventional gas chromatographic formula [41]:

$$N = \left(\frac{4R}{w}\right)^2 \qquad (30)$$

where N is the number of the theoretical plates (TP); R, the retention time referred to the peak maximum; and w, the peak width. The efficiency of the separation calculated from the peak of DNP-L-alanine (partition coefficient 0.56) was 900 TP which in turn gives an HETP (height equivalent to a theoretical plate) value of 20 cm. The analysis time of 70 hr could not be shortened with the present design without decreasing the number of columns, as the mobile phase was pumped at a near maximum rate. Although 30 mg of mixture was used in this separation, 300 mg has been separated without significant loss of peak resolution.

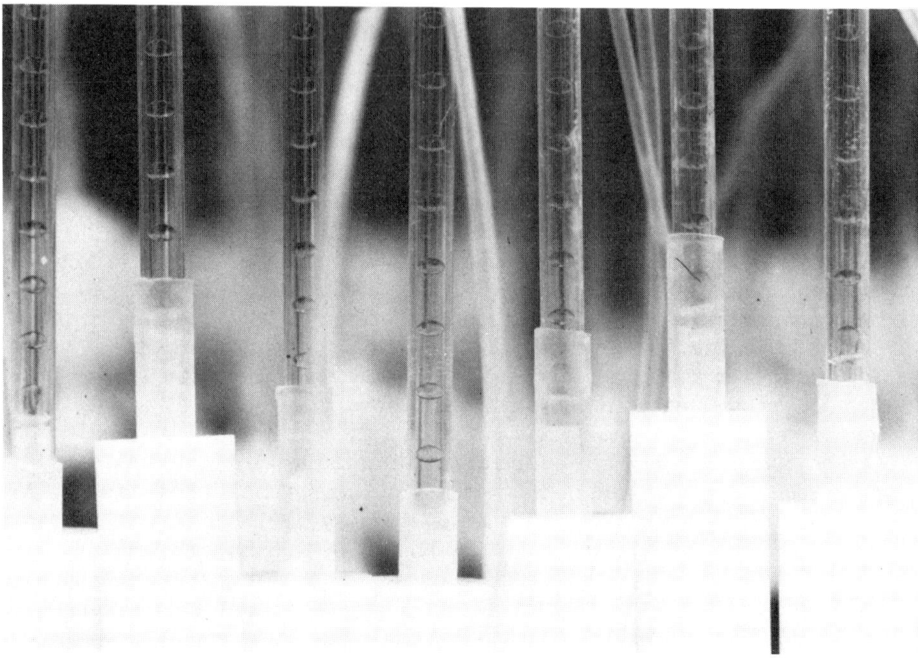

Fig. 27 Close-up view of the apparatus showing droplets in columns of the stationary phase liquid.

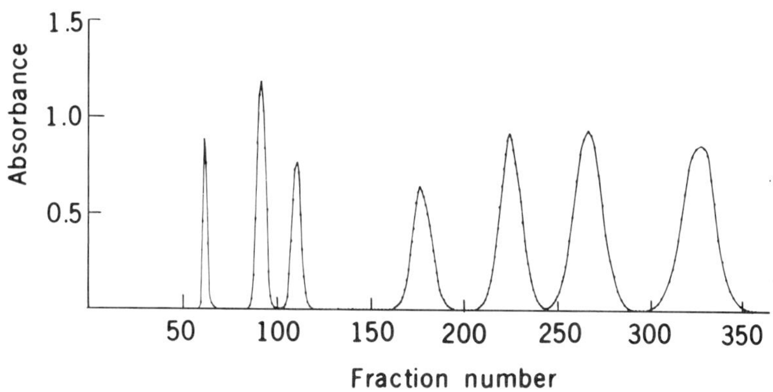

Fig. 28 Separation of DNP (dinitrophenyl) amino acids with DCCC. Peaks (partition coefficients are shown in parentheses) in order of elution are: DNP-L-ornithine (>100), DNP-L-serine (3.8), DNP-L-threonine (2.4), N,N'-diDNP-cystine (0.94), DNP-β-alanine (0.71), DNP-L-alanine (0.56), and DNP-L-proline (0.45).

Column Fabrication

The column assembly for DCCC is easily fabricated in the research laboratory. Among various types of tubing materials PTFE and glass tubes are almost exclusively used in the past because of their favorable properties such as chemical inertness, high rigidity, and good transparency. With the PTFE column, the choice of the mobile phase is limited to the aqueous phase. The glass column may permit the use of either aqueous or nonaqueous phase as a mobile phase without treatment. However, a silylated glass column will give substantially better results when the aqueous phase is used as the mobile phase.

The optimum internal diameter of the column depends highly on the interfacial tension of the two-phase solvent system to be applied. In general, the greater the interfacial tension, the larger the preferred diameter. In practice, the most useful solvent systems have a moderate interfacial tension of around 10 dyne/cm and can generate desirable size droplets in a 2-mm i.d. column.

In order to make a column, tubes are cut in a uniform length (20–60 cm) to obtain the desired number of column units, which in turn are mounted vertically onto the support at a regular interval. Interconnections between the neighboring units (the upper terminal of one unit to the lower terminal of the next unit) is made with a piece of PTFE tubing of the suitable diameter. In a short column such as 50- to 100-unit assembly where the column pressure remains relatively low, interconnection is made by simply inserting the transfer tube into each end of the unit. For a longer column where the column pressure is expected to be high during separation, suitable connectors, now available in the market, should be applied at each junction to prevent leakage of the solvents.

Choice of Solvent System

The unique feature of DCCC is that the efficiency of separation totally depends on the formation of the droplets. To produce the best results, the size of the droplets should be optimized so that the individual droplets barely occupy the column space across the diameter of the tube (Fig. 25). The formation of too large droplets which tend to advance at a slug pace would result in carryover of the stationary phase from the column. If the size of the droplets is too small, the droples quickly reach the terminal of the column or coalesce to form a large droplet at the middle portion of the column. In either case they fail to form a regular arrangement of the droplets through the length of the column.

The generation of droplets having suitable size and mobility are governed by factors such as the difference in specific gravities of the two solvent phases, viscosity of the solvents, the surface tension, the flow rate of the mobile phase, the diameter of the inlet tip, and the diameter of the column. Among these factors, the interfacial tension between the two solvent phases is considered to be the most dominant factor. Solvent systems with a high interfacial tension tend to generate large droplets and produce a plug flow phenomenon in a regular diameter column. Such solvent systems include many binary systems (hexane/water, chloroform/water, etc.) and some ternary systems (hexane/methanol/water, etc.). Solvent systems with an extremely low interfacial tension often fail to form stable droplets even in a small-diameter column. Such solvent systems include non-aqueous-nonaqueous phase systems (chloroform/ethylene glycol, etc.) and aqueous-aqueous polymer phase systems (dextran/polyethylene glycol, etc.). Except for these particular cases above described, the solvent systems used for CCD and other conventional extraction systems can be adapted to DCCC by optimizing other factors mentioned earlier. Various two-phase solvent systems and their applications to DCCC will be fully described by Hostettmann and Marston in Part II.

Recently, Murayama et al. [42] reported a new CCC apparatus which can carry out DCCC under a centrifugal force field. The use of a strong centrifugal force effectively prevents the plug flow phenomenon of two solvent phases in a narrow space of the column and produces efficient chromatographic separations at high flow rates, The method permits universal application of two-phase solvent systems including aqueous-aqueous polymer phase systems. The apparatus is now commercially available through Sanki Engineering, Ltd., Kyoto, Japan. The design of the apparatus and its applications will be described in detail by Bruening and Nakanishi in Part II.

B. Rotation Lacular CCC (RLCCC) [11,13]

Although DCCC described earlier is simple and capable of separating a relatively large quantity of materials, it possesses inherent limitations arising from the fact that the efficiency of the method depends entirely on droplet formation. Some solvent systems fail to form suitable droplets in the column and therefore cannot be utilized for DCCC. The partition efficiency of DCCC in DNP amino acid separation with the chloroform/acetic acid/0.1 N HCl (2:2:1) system measured 20 cm in HETP, that is, approximately 40 pairs of droplets and stationary phase segments are needed to produce one plate. This lack of efficeincy in DCCC may be largely attributed to the mass transfer resistance of the moving droplets and/or stationary phase segments which would inevitably increase by the use of more viscous solvent systems.

In order to improve the versatility and the partition efficiency of DCCC, a new type of column called the locular column has been devised. The locular column is prepared by placing multiple centrally perforated partitions in a straight tubular column, thus dividing the column space into numerous partition units called locules. Efficient solute partitioning takes place through the interface formed between the two phases in each locule. Gentle but effective stirring of both phases is achieved in each locular space by rotating the tilted column in the gravitational field while the undesirable longitudinal spreading of the sample bands toward the neighboring locules is minimized by small perforations channeling between the locules. Consequently, the present scheme yields a higher partition efficiency and allows more universal application of the solvent systems. The scheme also permits either phase to be used as the mobile phase in a given column assembly. Recently, RLCCC apparatus became commercially available through Tokyo Rikakikai Co. Ltd., Tokyo. The design of the commercial model and its application to purification of various natural products are described in detail by Kubo et al. in Part II.

Principle of RLCCC

The mechanism of RLCCC is schematically illustrated in Fig. 29 where a locular column is held at angle α from the horizontal plane. The column is first filled entirely with the lower phase and the upper phase is pumped through the first rotating seal into the column at the lower terminal, while the column is rotated around its own axis at a moderate rate. Due to the effect of the gravity, the upper phase partially displaces the lower phase in the first locule down to the level of the exit hole which leads to the second locule, this process being continued in each locule until the mobile phase reaches the outlet of the column. Once this hydrostatic equilibrium state is established throughout the column as shown in the figure, continuous elution with the mobile upper phase results in the displacement of the upper phase only, leaving the stationary lower phase within each locule of the column. When the lower phase is to be the mobile phase, the column is first filled with the stationary upper phase and the mobile lower phase is introduced through the upper terminal of the column. Then, the countercurrent process will similarly proceed, leaving the stationary upper phase within each locule of the column. In either case, a large amount of the stationary phase is retained in each locular space while the mobile phase is continuously eluted through the column. Consequently, solutes introduced at the inlet of the column are subjected to a continuous multistep partition process and finally eluted out through the outlet of the column in the order of their partition coefficients. During separation, the gravity contributes to the phase separation and retention of the stationary phase while the rotation of the column induces efficient stirring of each phase to reduce the mass transfer resistance. Under the ideal condition, each locule should become close to a plate which is equivalent to the HETP value of several millimeters.

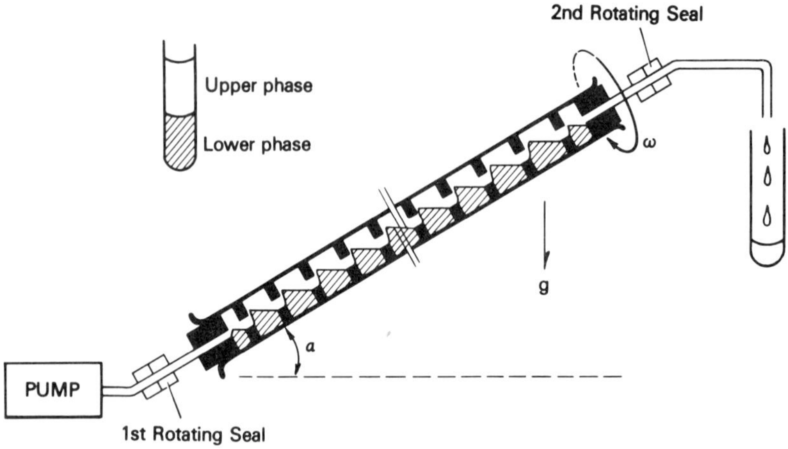

Fig. 29 Principle of RLCCC. A locular column is held at angle α and rotated about its own axis at angular velocity ω. The column is initially filled with the stationary lower phase (shaded) followed by elution with the mobile upper phase (clear) through the first rotating seal. Due to the effects of gravity, the mobile phase displaces the stationary phase in each locular space down to the level of the exit hole to the next locule. Once this equilibrium state is established throughout the column as shown in the figure, continued elution of the mobile phase displaces the upper phase only, leaving the lower phase stationary in each locular space. Consequently, solutes introduced locally at the inlet of the column are subjected to an efficient partition process between the two phases in each locular space and eluted out through the outlet of the column in the order of their partition coefficients.

In order to make a long locular column assembly, multiple column units are interconnected in series with transfer tubes and mounted in parallel around the rotary shaft. The column assembly thus arranged would produce an excellent result if each column unit is placed close to the rotary shaft where the centrifugal force produced by rotation is negligible.

The Original RLCCC Apparatus [11,13]

A simple rotary device was made to rotate the locular column assembly at various rates with desired inclination. It consists of a rotary shaft or column holder (50 cm long and 1.25 cm diameter) equipped with a rotating seal at each end. The rotational speed of the rotary shaft is made adjustable in several steps up to 300 rpm with a gear assembly connected to a synchronous motor. The whole apparatus can be tilted at the desired angle up to 45° from the horizontal plane by lifting either end which is then supported by a metal block placed at the bottom.

Partition disks for the locular column were punched out from a PTFE sheet with a borer, the central hole being drilled with a syringe needle of suitable diameter. The types of column units were prepared by spacing these partition disks (2.8 mm o.d. and 1 mm thick with a 0.8-mm-diameter hole or 4.8 mm o.d. and 1.2 mm thick with a 1.2-mm-diameter hole) into the respective size PTFE tubing (3.5 mm o.d. and 2.6 mm i.d. or 5.1 mm o.d. and 4.6 mm i.d., each 47 cm long) at every 3-mm interval to make over 100 locules. In each column, the desired number of the column units (10 for the short column and 50 for the long column) were mounted symmetrically around the rotary shaft. Connections between the column units and the rotating seals and between the neighboring column units were made of small-bore PTFE tubing (1 mm o.d. and 0.6 mm i.d.). The solvents were pumped by either a syringe driver (Harvard Apparatus Co., Inc.) or a Beckman Accu-Flo pump at flow rates ranging from 2 to 40 ml/hr. The eluate was continuously monitored through an LKB Uvicord II and then fractionated with an LKB fraction collector.

To examine the capability of the scheme, we used DNP amino acids (Sigma Chemical Co.) as test samples and a two-phase solvent system composed of chloroform/acetic acid/0.1 N HCl at a 2:2:1 volume ratio. DNP amino acids with suitable partition coefficients (see Figs. 30–32) were dissolved in a portion of the upper aqueous phase to make the concentration of each component at 0.5–1.0 g% where the solubility permitted. Before each separation the column was entirely filled with the stationary lower phase and 10–40 μl of the sample solution was introduced into the entrance of the column by gentle suction. In preliminary studies, separations were performed with the short column under the sets of conditions of various rotational speeds, column angles, and flow rates to find the optimal operational conditions. Under the optimized conditions determined by the above preliminary studies, nine DNP amino acids were separated on the long locular column assembly consisting of 50 units. For quantitative analysis, 32 drops of the eluate were collected in each tube and mixed with 3 ml of 95% ethanol for determination of the optical density at 350 nm with a Beckman DU spectrophotometer.

Figure 30 shows a typical chromatogram of nine DNP amino acids on the long column consisting of over 5000 locules each 2.6 mm i.d. and 3 mm long with a total capacity of about 100 ml which includes a free space in the connection tubes estimated at about 5%. The column was rotated at 180 rpm at an angle of 30° from the horizontal plane. The upper phase was pumped at a flow rate of 5 ml/hr with a Beckman Accu-Flow pump. The maximum column pressure during the operation measured 20 psi. The efficiency estimated according to the equation used for gas chromatography, Eq. (30), is about 3000 TP,

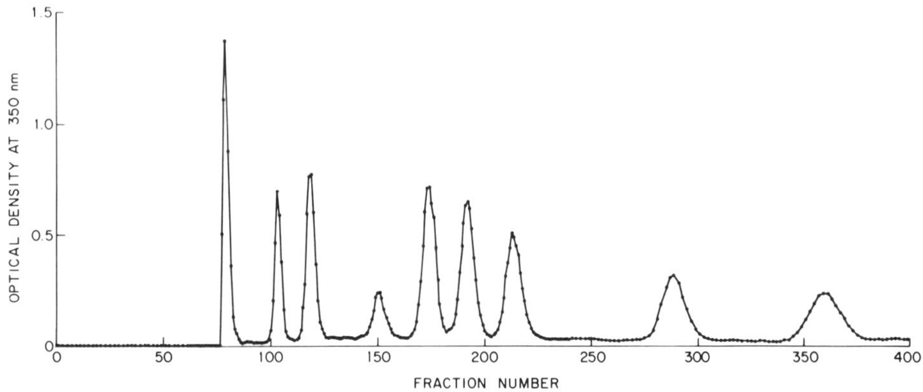

Fig. 30 Chromatogram of DNP amino acids in RLCCC. Peaks and the partition coefficients are, from left to right, N-DNP-L-arginine (>50), N-DNP-L-aspartic acid (3.8), N-DNP-D,L-glutamic acid (1.9), N,N'-diDNP-L-cystine (0.94), N-DNP-β-alanine (0.71), N-DNP-L-alanine (0.56), N-DNP-L-proline (0.45), N-DNP-L-valine (0.26), N-DNP-L-leucine (0.18), in order of elution. Column efficiency is about 3000 TP and the total elution time 70 hr.

Fig. 31 Effects of the rotational speed upon the separation of DNP amino acids in RLCCC. Separations are improved with the increase of the rotational speed up to 180 rpm (d). Loss of peak resolution at 240 rpm (e) is probably due to partial cancellation of the gravity by the centrifugal force produced by rotation. The experimental details are as follows: The column used consists of 1100 locules, each 4.6 mm in diameter and 3 mm long, with total capacity of 60 ml including 1.5% free space in connection tubes. Sample DNP amino acids and their partition coefficeints are N-DNP-L-arginine (>50), N-DNP-L-aspartic acid (3.8), N-DNP-D,L-glutamic acid (1.9), and N,S-diDNP-L-cysteine (0.62). The phase system is composed of chloroform, glacial acetic acid, and 0.1 N HCl (2:2:1). In each run, 10 μl of the sample solution (each component 0.5–1.0% in upper phase) was used. The upper phase was pumped with a Harvard syringe driver to maintain a uniform flow and 32 drops of eluate was collected in each tube to measure absorbance at 350 nm with a Beckman DU spectrophotometer.

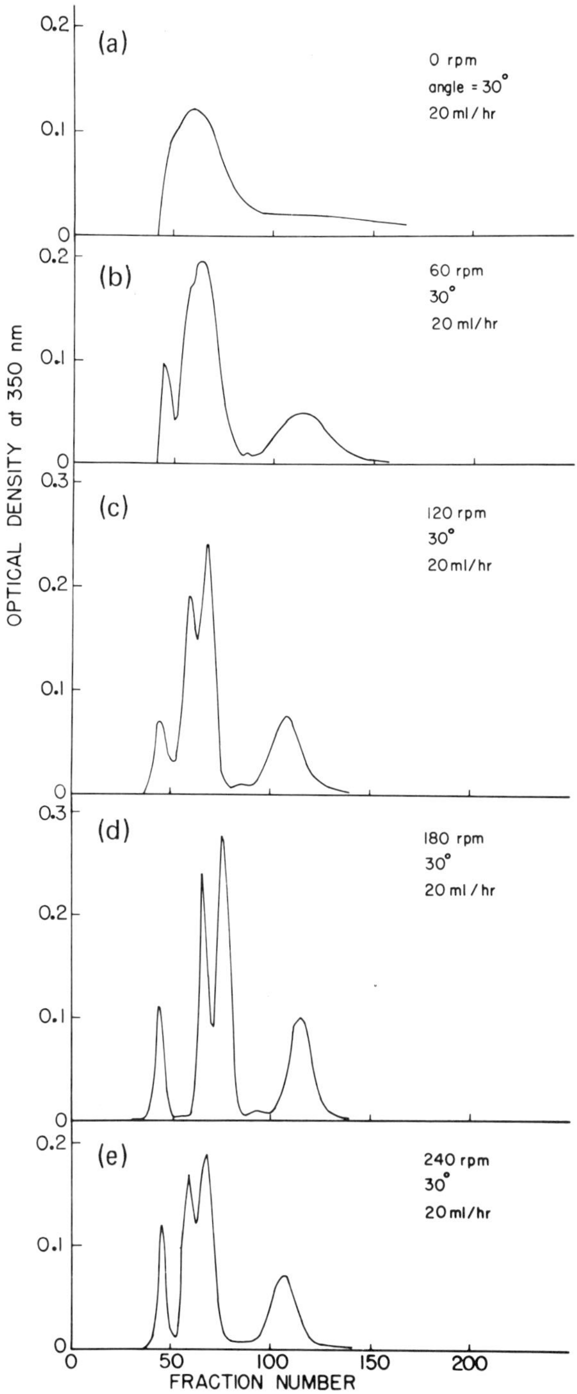

OPTICAL DENSITY at 350 nm
FRACTION NUMBER
(a)
0 rpm
angle = 30°
20 ml/hr
(b)
60 rpm
30°
20 ml/hr
(c)
120 rpm
30°
20 ml/hr
(d)
180 rpm
30°
20 ml/hr
(e)
240 rpm
30°
20 ml/hr
0
0.1
0.2
0.3
50
100
150
200

Fig. 31

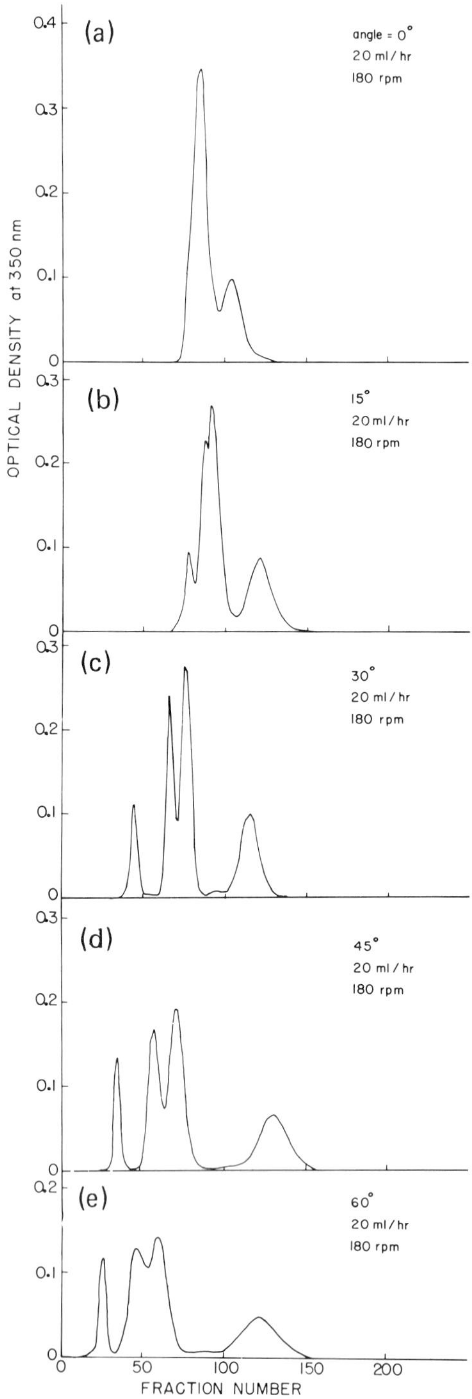

(a)
angle = 0°
20 ml/hr
180 rpm
(b)
15°
20 ml/hr
180 rpm
(c)
30°
20 ml/hr
180 rpm
(d)
45°
20 ml/hr
180 rpm
(e)
60°
20 ml/hr
180 rpm
OPTICAL DENSITY at 350 nm
FRACTION NUMBER

Fig. 32

that is, equivalent to over 50% partition efficiency per locule. Time required for elution of the first peak was 11 hr and that of the final peak 60 hr. The present column has a sample loading capacity of 1.0 ml without appreciable loss of resolution provided that the linear distribution isotherm holds. The column capacity and the sample size can be further increased without loss of efficiency by using a column of wider diameter. The method seems to work for most of the two-phase solvent systems used for conventional extraction systems including the Craig countercurrent distribution apparatus. The rotating seals, which is the potential source of leakage and contamination, may be eliminated from the present scheme by adapting the design for the seal-free flow-through centrifuge scheme (scheme V) illustrated in Fig. 17.

Effects of Various Factors on Separation

Preliminary studies made on short columns indicated that in general the partition efficiency rises with increase of column length, decrease of the locular length, and slow flow rate. With a given column and flow rate, the efficiency increases with the rotational speed of the column up to 180 rpm as illustrated in Fig. 31. Further increase of the rotational speed, however, results in loss of peak resolution probably due to the partial cancellation of the gravitational force by the centrifugal force produced by rotation of the column. The favorable effect of the rotation upon the partition process may be viewed on the basis of two different phenomena, i.e., (a) the efficient stirring of each phase induced by viscous drag of the solvent and (b) formation of a thin film of either phase over the internal surface of the locule. These phenomena are enhanced by increasing the diameter of the locules, and it was found that at a given linear flow rate of the mobile phase, the column efficiency was quite insensitive to the locular diameter. Column angle is another important factor which determines both volume ratio and interfacial area between the two phases in each locular space. In Fig. 29 it is visualized that the volume of the upper phase in each locule would increase with the decrease of the α value from $+90°$ to $-90°$. Since the mobile phase usually covers the exit opening of the locules, $\alpha = 0°$ gives the

---

Fig. 32 Effects of the column angle upon the separation of DNP amino acids in RLCCC. Results clearly indicate that the best peak resolution is given at around 30° (c). The column angle determines the retained stationary phase volume in each locular space: the greater the angle, the higher the retention level. Change of the volume ratio between the two phases in the locular column is reflected in remarkable shifts of the early peaks. For experimental conditions see Fig. 31.

volume ratio of the mobile phase to the stationary phase being greater than 1. Under this horizontal orientation of the column, the interfacial area of the two phases is also below the maximum attainable value. Therefore, the optimal column angle, where each locule is occupied by a nearly equal amount of the two phases with a near maximum interfacial area, should be found at $90° > \alpha > 0°$ if the mobile phase is the upper phase and at $0° > \alpha > -90°$ if the mobile phase is the lower phase. The experiments to find the optimal column angle were performed with the short column under a flow rate of 20 ml/hr and a rotational speed of 180 rpm. The results of the experiments clearly indicated that the optimal angle for the present column is around 30° from the horizontal plane as illustrated in Fig. 32.

C. Gyration Locular CCC (GLCCC) [11,13]

The present scheme has been developed to further improve the partition efficeincy of the locular column used in the previous scheme. RLCCC has an inherent limitation in that a high rotational speed of the column produces a centrifugal force field which would disturb the integrity of the shape of the interface and the motion of the two phases in the locular space. Under an ideal condition shown in Fig. 29, the liquid interface in the locular space is held horizontal by the gravitational force while the rotation of the column introduces efficient mixing in each phase. However, the centrifugal force field generated by the column rotation tends to alter the shape of the liquid interface from the horizontal to a circular plane coaxial to the locular column where the upper phase is surrounded by the lower phase occupying in the periphery. Although this transformation increases the liquid interfacial area, the two phases rotate with the column resulting in a loss of the mixing effect. In practice separations are performed with multiple column units each eccentrically mounted around the rotary shaft. In this case, the centrifugal force produced by the rotation modifies the shape of the liquid interface in such a way that it becomes a portion of a circular plane across the locular space. Here again, the two phases and their interface rotate with the column resulting in loss of the mixing effect. At a given rotational speed this detrimental effect is more pronounced on the column units mounted remotely from the axis of rotation. As a result, the scheme fails to apply a high rotational speed effectively or limits the number of column units mounted around the rotary shaft.

The new scheme called gyration locular CCC (GLCCC) has been introduced to solve all these problems. GLCCC also eliminates the necessity for the use of rotating seals by adapting the design used in the flow-through CPC previously described (see scheme I in Fig. 17).

Principle of GLCCC

The principle of GLCCC is illustrated in Fig. 33a where a locular column is placed in the vertical position at a distance r away from the central axis of the apparatus. Flow tubes of the column are tightly supported at the top stationary member of the apparatus on the central axis. Both column and flow tubes are rotated around the central axis of the apparatus at the angular velocity ω and simultaneously counterrotated about its own axis at the same angular velocity, the motion being identical to the synchronous planetary motion of scheme I in Fig. 17. As explained earlier, the motion prevents twisting of the flow tubes and therefore eliminates the need for the rotating seal. Under this planetary motion, a homogeneous acceleration field $r\omega^2$ rotates around the column at angular velocity ω while the column always maintains its fixed orientation with respect to the earth as described in the previous section.

To demonstrate the mechanism of GLCCC, the column is first filled with the stationary phase (lower phase) and the mobile phase (upper phase) is introduced into the column at the bottom end. Under the combined force field formed by the gravity and the rotation of the column, the mobile phase displaces the stationary phase in the first locule up to the level where the interface formed between the two phases meets the exit hole to the second locule. This process is repeated in each locule until all free space in the column is occupied by nearly equal amounts of the two phases as illustrated in Fig. 33a. Once this hydrostatic equilibrium is established, continued flow of the mobile phase displaces the mobile phase only, leaving the stationary phase retained in the column. Thus, solutes introduced locally at the entrance of the column are subjected to an efficient partition process in each locular space and separated according to their relative partition coefficients as in RLCCC.

The motion of the two phases in the locular space is easily visualized on the cross-section through the middle portion of the locule. Figure 33b shows successive positions of the locule during one complete gyration cycle. The upper phase (clear) and the lower phase (shaded) are seen separated by the centrifugal force forming an interface perpendicularly to the vector of the force. With the steadily changing direction of the centrifugal force, both liquids and their interface rotate with respect to "x" marked on the column wall. Thus, the viscous drag force and the wall surface affinity of the liquid induce efficient mixing within each phase to accelerate the partition process. In the present scheme, applicable rotational speed, hence the degree of phase mixing, becomes much greater than that in RLCCC because the centrifugal force field which produces ill effect to RLCCC is now efficiently utilized to separate the two phases to form a more stable interface in the locular space.

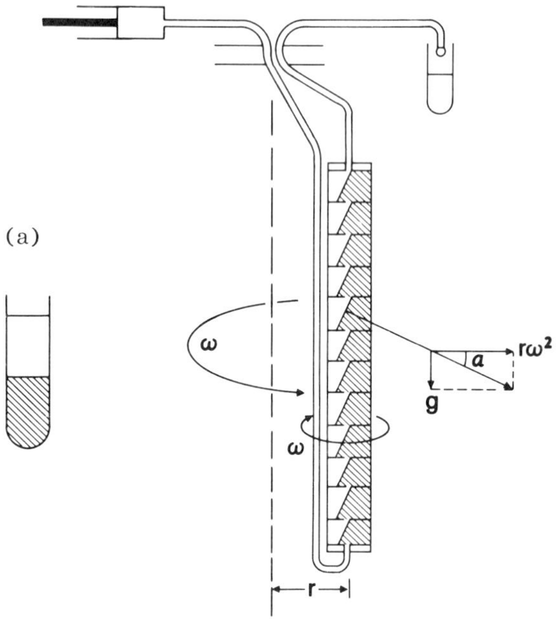

Fig. 33 Principle of GLCCC. (a) Longitudinal section through the axis of the locular column. The locular column held in the vertical position is subjected to a gyrational motion (revolution around the central axis of the apparatus and simultaneous counterrotation about its own axis at the same angular velocity) which generates a rotating centrifugal force field with a relative magnitude of $r\omega^2$. The column initially filled with the stationary phase (shaded) is eluted with the mobile phase (clear) through the bottom end of the column. The mobile phase then displaces the stationary phase in each locular space up to a level where the liquid interface meets the exit hole to the next locule. The gravity and the rotational speed of the column determine the inclination ($90° - \alpha$) of the liquid interface and the volume ratio between the two phases in the column. Eventually all locular spaces in the column are equilibrated with the two phases as shown in the figure and the continued flow results in the mere displacement of the mobile phase while leaving the stationary phase within each locule. Consequently, solutes introduced locally at the inlet of the column are subjected to an efficient partition process in each locular space and separated according to their partition coefficients. (b) Successive positions of the locular column during one gyrational cycle. Cross-section through the middle portion of the locule shows the liquid interface perpendicularly formed to the centrifugal force vector. With the gyrational motion of the column, both the centrifugal force vector and the liquid interface rotate with respect to the locular wall having the mark "X."

(b)

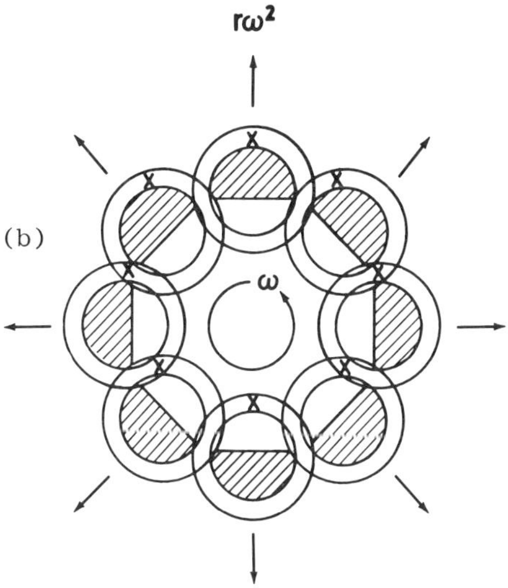

Fig. 33 (Continued)

As illustrated in Fig. 33a, the interface of the two phases on the vertical section through the locule is a straight line perpendicular to the summation of the centrifugal ($r\omega^2$) and gravitational (g) force vectors. Angle α here again determines both volume ratio and interfacial area of the two phases in the locular space. Since the mobile phase usually covers the entire area of the exit hole in the locule, the maximum interfacial area with near 50% retention of the stationary phase is obtained by eluting the column with either the upper phase upward or the lower phase downward.

The Original GLCCC Apparatus

Figure 34 shows the design of the apparatus. In Fig. 34a, a pair of large wheels rotates synchronously at angular velocity ω while a plate eccentrically bridging these wheels conveys the motion to the column secured on the plates. Radius r of the column gyration is equal to the distance on the wheel between the center and the eccentric connection.

The gyrator based on the above design is shown in Fig. 34b. A pair of identical rotary shafts is mounted in the vertical position on the rectangular support. The motor mounted on the back of the support synchronously drives these shafts by means of a pair of identical toothed pulleys and toothed belts. A cylindrical metal block equipped with an eccentrically positioned shaft is mounted on each end of the rotary shafts to support a pair of bridges horizontally at

(a)
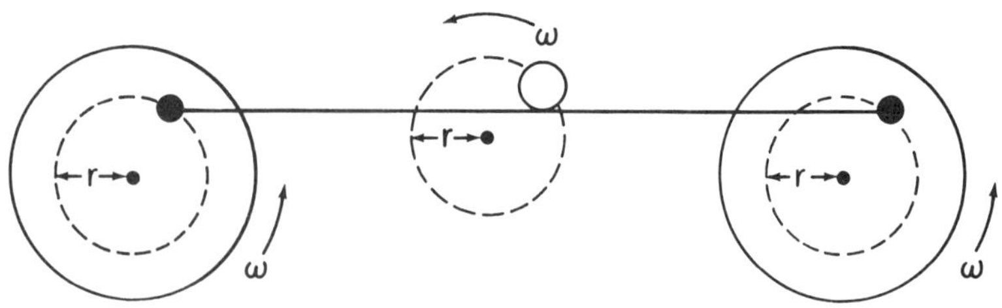

Fig. 34 Design of the apparatus for GLCCC. (a) Design principle. A pair of large wheels synchronously rotate at angular velocity ω, while a plate eccentrically bridging these wheels conveys the motion to the column secured on the plate. (b) Photography of the apparatus. A pair of identical rotary shaft vertically mounted on the rectangular support is synchronously driven by the motor located at the back of the support. At the top and the bottom of each rotary shaft mounted is a cylindrical metal block equipped with an eccentrically positioned shaft to support a pair of horizontal bridges through ball bearings. A perforated metal plate vertically mounted on the bridges holds a bundle of locular column units connected in series. This gyrator has a fixed gyrational radius of 2.5 cm.

the top and the bottom of the rotary shafts through bearings. A perforated plate mounted on the bridges holds a bundle of locular columns connected in series. The counterbalance is provided by a pair of brass blocks eccentrically mounted on the rotary shafts. In this gyrator, r has a fixed value of 2.5 cm and the gyration speed is continuously adjustable up to 1500 rpm.

Effects of Various Factors on Separation

Preliminary experiments to evaluate the potential capability of GLCCC have been carried out with a set of DNP amino acid samples and two-phase solvent system composed of chloroform/acetic acid/0.1 N HCl (2:2:1). To facilitate the comparative studies, separations were performed with a short locular column used in the previous studies for RLCCC (Figs. 30 and 31).

Figure 35 a–d shows the effects of gyration speed on the separation at a uniform flow rate of 20 ml/hr maintained by a Harvard syringe driver. Efficiency rises with increase of gyration speed up to 500 rpm (c) where the column efficiency exceeds 900 TP or over 80% partition efficiency per locule. Further increase of gyration speed at 800 rpm, however, results in a lower resolution which is clearly observed in the elevated trough between the second and third peaks (d). The loss of peak resolution, which occurs at the increased gyration speed at 800 rpm, may be caused by the enhanced

(b)

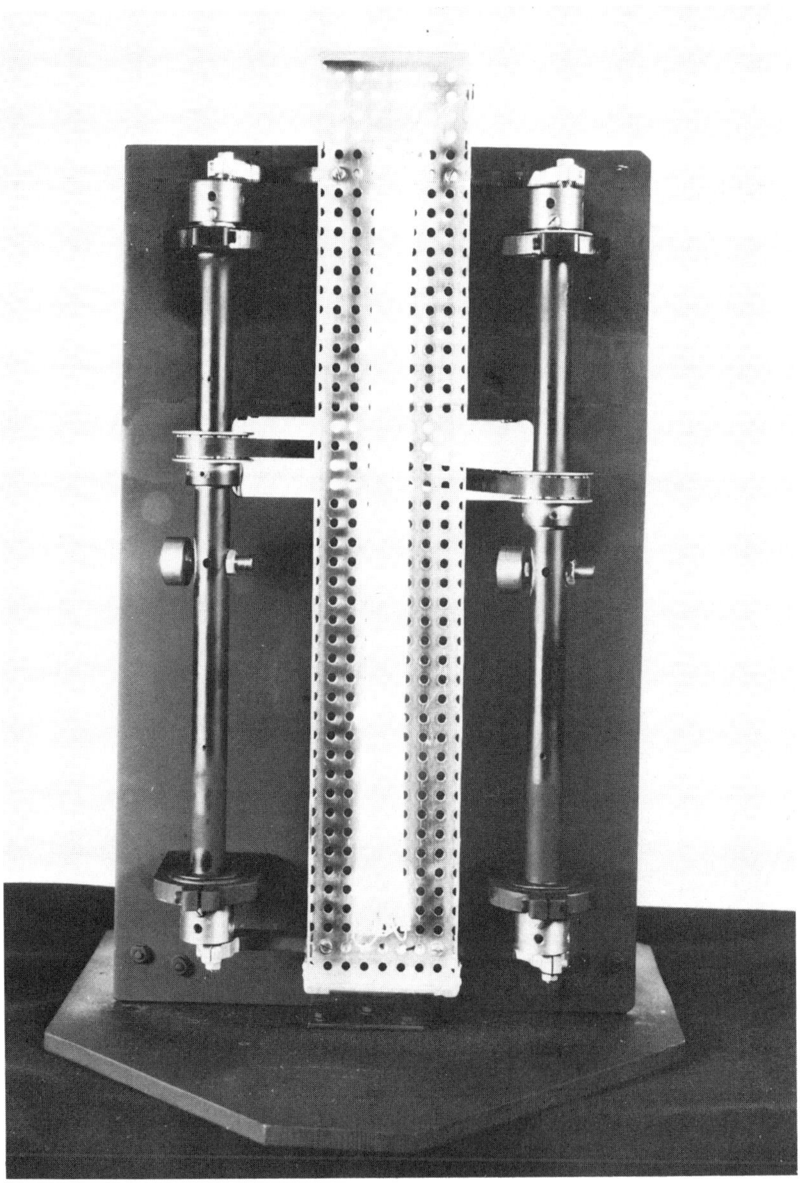

Fig. 34 (Continued)

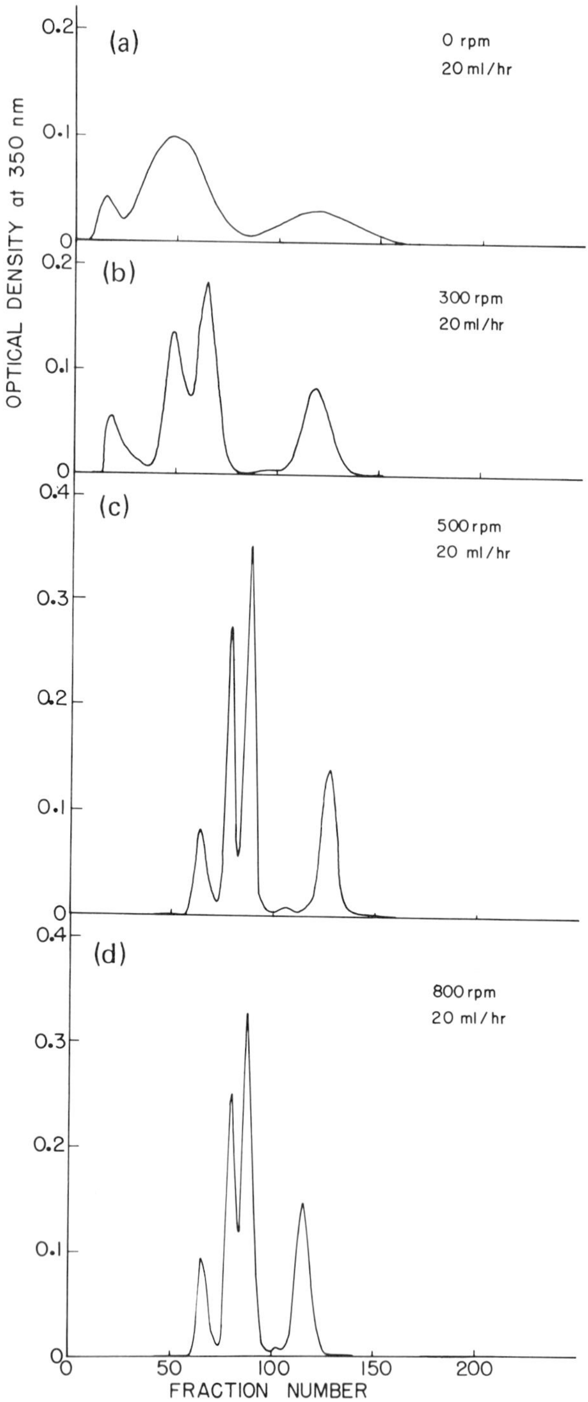

OPTICAL DENSITY at 350 nm
(a)
0 rpm
20 ml/hr
(b)
300 rpm
20 ml/hr
(c)
500 rpm
20 ml/hr
(d)
800 rpm
20 ml/hr
0.4
0.3
0.2
0.1
0
0
50
100
150
200
FRACTION NUMBER

Fig. 35

vibration of the column which led to the loss of the stationary phase and the decreased interfacial area in the locular space. The increased centrifugal force field at 800 rpm gives an inadequately small α value of 7° (Fig. 33a) which may have also contributed to the loss of peak resolution. The decreased retention level of the stationary phase in the column at 800 rpm gyration speed is evidenced by the reduced retention volume of the first peak and the shortened distance between the peaks.

The effects of the flow rate upon the separation are shown in Fig. 36a–e. At a given gyration speed of 500 rpm the best result appears at 20 ml/hr (b). Either increase or decrease of the flow rate lowers the peak resolution. This relatively high optimal flow rate together with the extremely high partition efficiency (over 80% in each locule) may suggest that a solute gradient might be formed and partially preserved even within an individual locule in the present scheme which provides a symmetrical circular motion of the interface in each locular space. (In RLCCC, the nearly horizontal liquid interface in the locular space tends to swing back and forth during each rotation due to the asymmetrical action of the vector derived from the summation of the gravitational and centrifugal acceleration, the resultant longitudinal mixing of the liquids interfering with formation of the solute gradient within the locule.) The increase of the flow rate above 20 ml/hr resulted in a gradual loss in peak resolution. An experiment has been performed to improve the peak resolution at a 40 ml/hr flow by applying a higher gyration speed of 800 rpm (Fig. 36e). The results, however, showed additional loss of the peak resolution mainly due to the depletion of the stationary phase from the column.

Carryover of the stationary phase, which tends to increase with vibration of the apparatus at a high gyration speed, seems to be the limiting factor of this method. Two devices have been successful in eliminating this complication. The use of a column with threaded partition holes prevents the carryover, if the direction of gyration is chosen to screw the carried over droplets back into the original locule. Also, we have found that the surface nature of the partition hole greatly related to this phenomenon. When the exit hole of each

---

Fig. 35 Effects of gyrational speed upon the separation of DNP amino acids in GLCCC. Column efficiency increases with the gyrational speed up to 500 rpm (c). Further increase of the gyrational speed at 800 rpm (d) results in loss of peak resolution probably due to the depletion of the retained stationary phase and decrease of the liquid interface area in the column. For the comparative studies, the locular column, sample and solvent system used here are the same as those used in the previous experiments in RLCCC (see caption under Figs. 31 and 32).

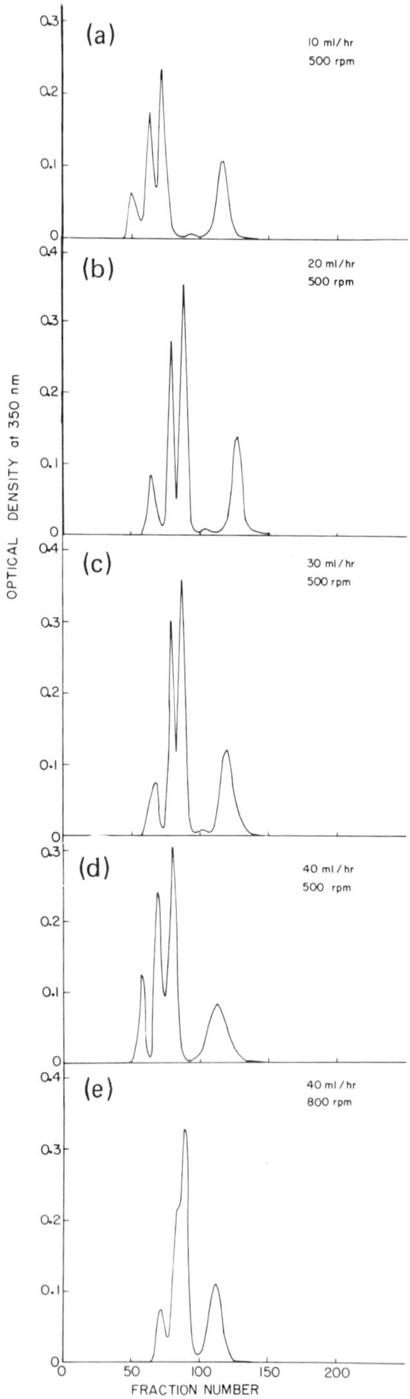

(a)
10 ml/hr
500 rpm
(b)
20 ml/hr
500 rpm
(c)
30 ml/hr
500 rpm
(d)
40 ml/hr
500 rpm
(e)
40 ml/hr
800 rpm
OPTICAL DENSITY at 350 nm
FRACTION NUMBER
0
0.1
0.2
0.3
0.4
50
100
150
200

Fig. 36

locule is made of the material more wettable by the mobile phase, the exit hole becomes less accessible for the stationary phase and, therefore, less tendency for carryover is expected. In order to make such a locular column more convenient for the choice of the mobile phase (either the aqueous or nonaqueous phase may be used), we embedded a small glass bead (1.2 mm o.d., 0.5 mm i.d., and 0.7 mm long) partially into each PTFE partition hole to make one side wettable to the aqueous phase and the other side wettable to the nonaqueous phase. The use of such partitions completely eliminated carryover of the stationary phase at 800 rpm and yielded a high partition efficiency close to one plate per locule.

The results of these preliminary experiments furnish useful guidance for the development of this method. In order to shorten the time required for the separation at a high efficiency, faster mixing and therefore higher gyration speed is necessary. Once the gyration speed is determined, the optimal condition for both phase volume ratio and interfacial area can be satisfied by angle α for a given column. Thus, the major requirements of the design of the apparatus will be finely adjustable radius of gyration, well-balanced system, and uniform speed of gyration to minimize carryover of the stationary phase. Though a high speed of gyration should be applied, the resultant centrifugal force is as small as several g value by selecting the proper radius r for the optimal angle α.

In spite of these promising results in the preliminary experiments, further development of LCCC had been discouraged by tedious works required for preparation of the locular column and finally interrupted by the discovery of a new scheme called the flow-through coil planet centrifuge (CPC), which utilizes a simple coil as the separation column. (Ironically, the flow-through CPC was originally designed and constructed in the author's laboratory to investigate the effects of the gyration radius on the partition efficiency of the locular column.) However, after years of intensive development of various CCC schemes using helical columns, it is felt that the locular column could be a better choice for large-scale industrial applications where the use of the helical column necessitates a bulky core and suffers from a loss of partition efficiency due to a large internal diameter of

---

Fig. 36 Effects of flow rate on the separation of DNP amino acids in GLCCC. At 500 rpm, the optimal flow rate is found at 20 ml/hr (b). Marked broadening of the peaks occurs at 40 ml/hr (d). Increase of rpm at this flow rate results in a higher efficiency in terms of the theoretical plates but poor resolution between the peaks due to depletion of the stationary phase volume in the column (e). For the comparative studies, the locular column, sample, and solvent system used here are the same as those used in the previous experiments in RLCCC.

the tube. Although the production of the locular column is substantially more expensive compared with the coiled column, the locular column is more amenable to incorporate various additional devices such as mixers, baffles, etc., within the individual locules to improve partition efficiency.

D. Monolithic Integrated Flow Circuit (MIFC): Novel Column Design for Nonhelix CCC

Fabrication of MIFC

Recently, Kolobow et al. [43] developed a new column design which can be ideal for nonhelix CCC. Using a molding technique, complex repetitive flow channels were formed into sheets of polypropylene. Two of these sheets were fused to form a closed separation chamber with a complete set of locules and integral transfer tubes. The process of making MIFC locular column is schematically illustrated in Fig. 37.

In Fig. 37a a mold of one half of the flow channels and mirror images of the same are machined into a 1/2 in. thick aluminum jig plate using a numerically controlled milling machine. The polypropylene sheets (0.015 in.) are dried at 220°F for 2--6 hr to remove any moisture. Two such sheets are aligned in the mold. The purpose of the 1/8 in. thick silicone rubber pad shown in Fig. 37b is to force each polypropylene sheet in a hydraulic press into the machined cavity after a suitable temperature has been reached (280°F). Following this and while in the press the temperature is raised to 310°F to relieve stress in the formed sheet. After cooling, the rubber pad and Teflon sheets are removed and the molds containing the polypropylene sheets realigned. The sheets are then fused under pressure at 330°F (Fig. 37c). When cooled, the completed plastic molding is removed and polypropylene tubing fused to the inlet and outlet ports. After testing for leaks, the unit is ready for use.

Two types of locular MIFC columns were produced for performing CCC. Each plate was made of two 0.015 in. thick polypropylene sheets measuring 26 cm × 40 cm. One plate (spherical locular column) consisted of 1174 spherical locules, each 6 mm in diameter, with a total capacity of 134 ml including the volume of the transfer channels. The second type (cylindrical locular column) consisted of 642 cylindrical locules, each measuring 6 mm in diameter and 10 mm long, with a total capacity of 175 ml.

Preliminary Results with MIFC Locular Column

Using the standard set of DNP amino acid samples and two-phase solvent system composed of chloroform/acetic acid/0.1 N HCl (2:2:1), the performance of the two types of MIFC locular columns was evalua-

(a)

(b)

(c)

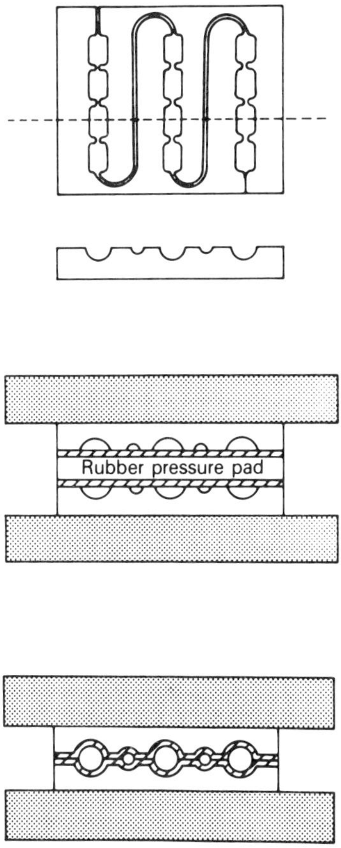

Fig. 37 MIFC-forming process. (a) Aluminum mold for MIFC; (b) forming process; (c) fusing process.

ted by measuring partition efficiencies in terms of theoretical plates (TP) and peak resolution, separation speed, stationary phase retention, etc. In addition to RLCCC and GLCCC previously described, oscillation locular CCC (OLCCC) was also tested. The OLCCC apparatus consisted of a column holder plate connected to a reciprocating piston device and produces vertical oscillatory motion to the locular column mounted onto the holder plate.

Figure 38 shows chromatograms obtained from the above three types of MIFC CCC devices. Overall results indicated that among all devices GLCCC yields the most efficient separations with over 90% partition efficiency in each locule. OLCCC also yields a high partition efficiency while the retention of the stationary phase seems

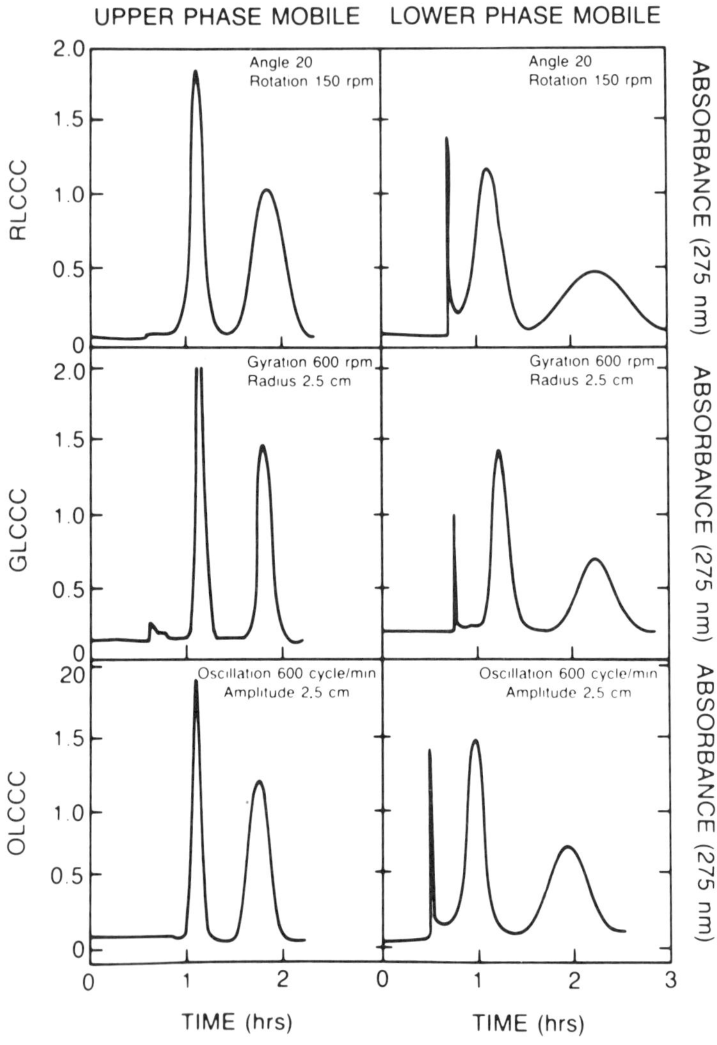

Fig. 38 Typical chromatograms obtained from three different LCCC schemes: RLCCC; GLCCC; and OLCCC. In all separations the experimental conditions are as follows: Column: MIFC with 642 cylindrical locules and a total capacity of 175 ml. Solvent system: chloroform/acetic acid/0.1 N HCl (2:2:1). Sample: DNP-glu and DNP-ala mixture 2 ml each component at 0.5 g% in the upper aqueous phase. Flow rate: 120 ml/hr.

less stable and the system tends to produce a steady carryover of the stationary phase.

The MIFC columns used in the preliminary studies were made of polypropylene sheets which are permeable to some organic solvents such as chloroform, hexane, etc. In order to maintain the original volume and phase composition of the stationary phase in the column during the run, the columns were carefully wrapped with aluminum foil to prevent leakage of the fumes. This problem may be largely eliminated in future by using more suitable materials such as PTFE, FEP (fluorinated ethylene propylene), etc. Although the MIFC column requires further improvement to effect the performance, it has a great potential to become an ideal locular column for CCC. The advantages are: The column is easily fabricated to a desired conformation. The column has no internal connections or junctions except for the inlet and outlet lines and, therefore, the risk of leakage of solvent is minimum. In addition, the column is light, compact, nonfragile, and relatively inexpensive and is suitable for mass production.

E. Summary

Three types of preparative CCC schemes are described. Droplet CCC (DCCC) uses a vertical tubular column. The mobile phase with less wall surface affinity forms multiple droplets in the stationary phase to divide the column space into a number of partition units. Consequently, the choice of the solvent systems to form stable droplets of suitable diameter becomes the core of this method. The apparatus for DCCC is inexpensive and the separation column is easily fabricated in the research laboratory.

Locular CCC (LCCC) uses a locular column which is prepared by spacing centrally perforated disks into a tubular column at regular intervals to divide the column space into multiple partition units called locules. In rotation locular CCC (RLCCC) the locular column is mounted lengthwise onto an inclined rotary shaft. Due to the effect of the gravity, two solvent phases form a horizontal interface in each locular space while the rotation efficiently mixes each phase to promote partition process. In this scheme, efficiency increases with the rotational speed of the column until the centrifugal force field generated by the rotation cancels out the mixing effect. Gyration locular CCC (GLCCC) uses the same locular column vertically mounted onto the column holder of the gyrator. Gyration, which is a nonrotational revolution or a synchronous planetary motion identical to scheme I in Fig. 17, produces a rotating centrifugal force field with respect to the column so that the two phases and their interface rotate in the locular space to accelerate the partition process. The scheme allows application of a high gyration speed of the locular column to yield over 80% partition efficiency in each locule.

Both RLCCC and GLCCC permit the universal application of the conventional two-phase solvent systems where either phase can be used as the mobile phase. The time-consuming procedure of fabrication of the locular columns may be replaced by a simple and inexpensive molding method of MIFC [43] recently developed in our laboratory.

V. HELIX COUNTERCURRENT CHROMATOGRAPHY WITH NONPLANETARY CENTRIFUGES

The nonplanetary centrifuge is an ordinary type of centrifuge in which the rotor simply rotates around the central axis of the apparatus to produce a stable centrifugal force field. This conventional style centrifuge system has been utilized for performing CCC by arranging a coiled column around the periphery of the centrifuge bowl to form what is called the "toroidal coil."

A. Toroidal Coil Centrifuge with a Rotating Seal [11,13,14]

In the original coil planet centrifuge [1,2] described earlier (Sec. I), the separation is performed with an end-closed coiled tube, thus limiting the versatility of the method. Introduction of the flow-through mechanism to the helix CCC has first demonstrated a useful capability of CCC that permits continuous elution, on-line monitoring, and fractionation as applied in liquid chromatography. However, this early method is rather primitive and somewhat impractical in that it employs a rotating syringe to pump the mobile phase and a rotating seal to collect the effluent from the rotating column. Nevertheless, it has first shown a great potential of countercurrent chromatography that could yield a high partition efficiency comparable to that of a refined gas chromatography.

Principle of Toroidal Coil Centrifuge

Figure 39 shows a schematic design of the present scheme. The centrifuge bowl holds a toroidal coil around its periphery and a syringe on the central axis of the centrifuge. The solvent is introduced from the rotating syringe into the toroidal coil by pushing the syring plunger with a syringe drive through a rotary joint, while the effluent from the column is collected through a single-channel rotating seal located at the center of the rotary joint.

The principle of the present system is considered to be identical to the basic HSES model system previously illustrated in Fig. 11, except that the gravity in the basic model system is now replaced by a strong centrifugal force field which prevents a plug flow phenomenon tending to occur in the capillary column. The toroidal coil

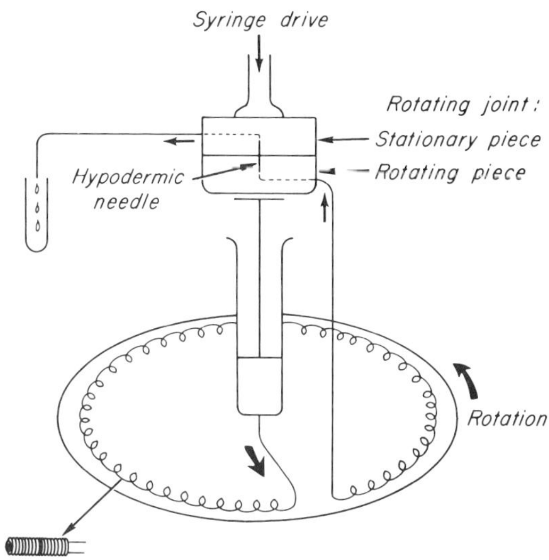

Fig. 39 Schematic design of the original apparatus. The coiled separation column is circumferentially accommodated inside the centrifuge rotor. The solvent is introduced into the column from the syringe mounted at the center of the rotor while the effluent from the column is collected through a rotating seal located at the center of the rotary joint above the syringe plunger.

is first entirely filled with one phase (stationary phase) of an equilibrated two-phase solvent system. Then the other phase (mobile phase) is introduced into the column from the syringe while the apparatus is rotated at a given revolutional speed. Under this orientation of the column, each helical turn of the column is subjected to a stable centrifugal force field which acts perpendicularly to the axis of the toroidal coil. Aided by this stable centrifugal force field, the mobile phase partially displaces the stationary phase in each helical turn of the toroidal coil and eventually the entire length of the column is divided into alternating segments of the mobile and the stationary phases. After this hydrostatic equilibrium state is established throughout the column, continued flow of the mobile phase displaces the mobile phase only, the stationary phase being permanently retained in each helical turn of the column. Solutes introduced beforehand at the inlet of the column are subjected to an efficient partition process between the two phases in each stationary phase segment and eluted out from the outlet of the column in the order of their partition coefficients in a manner analogous to that of liquid chromatography but in the absence of solid supports.

Design of the Original Apparatus

Figure 40 illustrates the essential features of the apparatus. A Model PR-2 International centrifuge fitted with a modified Helix tractor head (International Equipment Co.) provides the centrifugal field and holds the coiled column, syringe, and a rotating seal. The toroidal coil was accommodated at the periphery of the Helix tractor head and a special high-pressure syringe of 10-ml capacity was mounted at the center of the rotor to pump the mobile phase into the column. Pressure of 16–20 atmospheres was used in these experiments so that it was easier to push a rotating syringe than to make a high-pressure dual rotating seal. The plunger of the syringe was pushed through a thrust bearing by a multispeed syringe driver (Harvard Apparatus Co.). Effluent from the column at approximately normal atmospheric pressure was conducted from the rotating system through a rotating seal located at the center of the thrust bearing above the syringe plunger. A stable centrifugal force field was maintained by running the centrifuge on DC power from a Lambda LH 122FM power supply that was regulated constant to 0.01%.

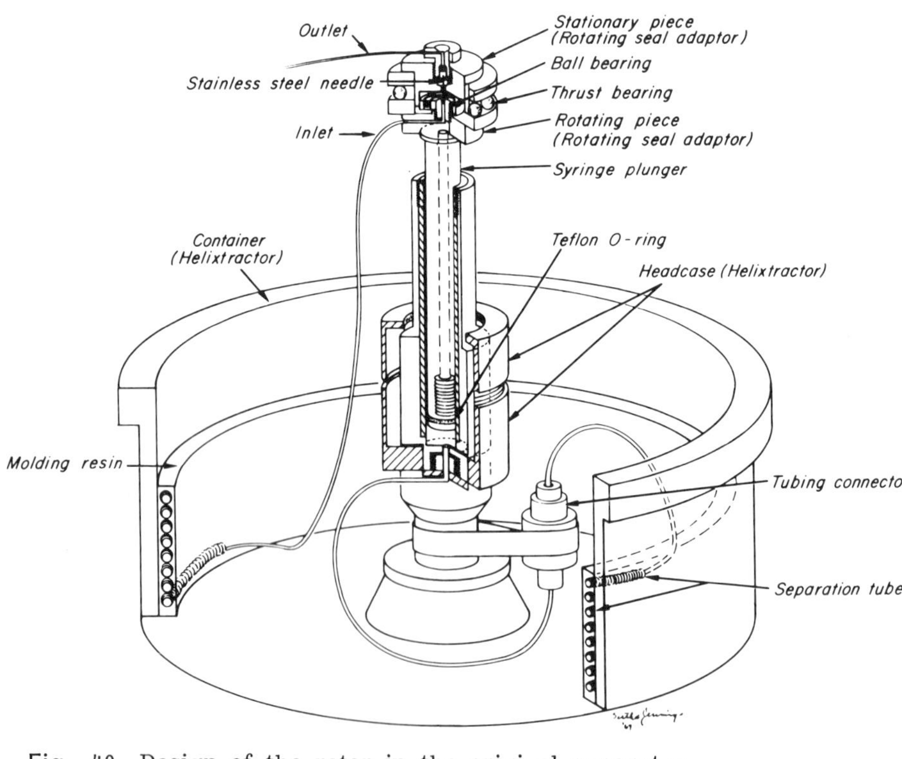

Fig. 40 Design of the rotor in the original apparatus.

The separation column was prepared in the following manner. A small-bore PTFE tubing of 0.2 mm i.d. was produced from 0.28-mm i.d. tubing (No. 32, Zeus Industrial Products, Raritan, NJ) by pulling it through a plastic die. Two types of the toroidal coil were prepared from the tubing as prepared above. In the first type, the tubing was wound tightly on a flexible rod support which was subsequently coiled in a number of turns around the inside of the centrifuge head (Fig. 40). Alternatively, the tubing was folded in two and twisted along its length to provide a ropelike structure in which the individual strands had the appearance of a stretched helix of small diameter (the twisted column in Fig. 12). The coil was then wound around a drum support which fitted into the centrifuge head. If the centrifugal field (and hence the operating pressure) was high, both coil forms were encased in a transparent resin for additional support. An LKB Ultrorac fraction collector was used for fractionation of the eluate and a Beckman DU spectrophotometer for measurement of the absorbance of obtained fractions.

Performance of the Apparatus

The capability of the apparatus was examined by the standard set of samples and solvent system which are extensively used throughout the development of the various CCC schemes. The two-phase solvent system was composed of chloroform, acetic acid, and 0.1 N HCl at a 2:2:1 volume ratio equilibrated at room temperature. Dinitrophenyl (DNP) amino acids (Mann Research Lab., NY) with suitable partition coefficients (see Table 4) were chosen as samples because they are conveniently analyzed by measuring the absorbance at a single wavelength. These samples were mixed and dissolved in a portion of the lower phase at concentrations of 1 g% where the solubility permitted. The sample solution was kept at 4°C in darkness until used.

The separation column was first filled with the stationary lower phase and the sample solution (5 μl) was drawn into the sample port located at the proximal end of the column by gentle suction. Usually 2–3 μl of chloroform was also added to compensate for evaporation losses. The syringe was then filled with the mobile upper phase and the necessary connections were made. Centrifugation was performed at 900–950 rpm with flow rates of 125, 200, 320, 510, and 820 μl/hr at room temperature. Each fraction consisting of 8 drops of the eluate was mixed with 95% ethanol to make a final volume of 3 ml for determination of absorbance at 350 nm with a 1-cm light path quartz cell.

Chromatograms of DNP amino acids obtained under the optimum conditions were illustrated in Fig. 41. Figure 41a shows the chromatogram obtained for nine DNP amino acids on a coiled column of 8000 turns with a helical diameter of 0.85 mm and prepared from a 40-m length of the 0.2-mm i.d. PTFE tubing. Figure 41b shows the

Table 4 Summary of DNP Amino Acid Separation

			Data for Fig. 41A		Data for Fig. 41B	
Sample[a]	Conc. (%)	P.C.[b]	R[c] (hr)	N[d] (TP)	R[c] (hr)	N[d] (TP)
N-DNP-Δ-L-ornithine	0.5	>100	6.8	—	10.3	—
N-DNP-L-aspartic acid	1	3.8	8.1	5200	12.5	8500
N-DNP-DL-glutamic acid	1	1.9	9.1	4500	14.3	6300
N,N'-diDNP-L-cystine	0.2	0.94	11.5	4500	18.1	8200
N-DNP-β-alanine	1	0.71	13.3	4900	21.0	8600
N-DNP-L-alanine	1	0.56	15.1	4800	23.3	8500
N-DNP-L-proline	1	0.45	17.5	4600	27.4	8600
N-DNP-L-valine	1	0.26	27.4	3000	—	—
N-DNP-L-leucine	1	0.18	39.6	2500	—	—

[a]In order of elution.
[b]Partition coefficient.
[c]Retention time referred to the peak maximum.
[d]Efficiency in terms of theoretical plates.

chromatogram of seven DNP amino acids obtained on a twisted column consisting of 17,000 turns and prepared from an 80-m, 0.2 mm i.d. PTFE tubing. In both experiments, the upper phase was used as the mobile phase at a flow rate of 125 μl/hr. The results of the experiments are summarized in Table 4. The partition efficiencies were expressed in terms of the theoretical plates according to Eq. (30). The values obtained for the twisted column are relatively uniform and extremely high, whereas those obtained for the coiled column tend to decrease with increased retention time.

The results obtained under various operational conditions indicated that the efficiency increases with tube length, smaller bore, and slower flow. Also, with a given length of tubing, the efficiency is improved with increasing the number of helical turns by decreasing the helix diameter. The significance of the first three factors is obvious by analogy with liquid chromatography. However, the unique factor of the number of helical turns may merit further discussion. As stated earlier, each helical turn of the coil under a hydrostatic equilibrium state shows an asymmetrical distribution of the two phases. One half of the helical turn is an efficient column space and contains both the mobile and stationary phases while the other half is an inefficient free space entirely occupied by the mobile phase (Fig. 11). At the boundary of these two spaces segmentary interface is present at the top and the bottom of each helical turn. The length of the efficient column space is approximately half that of the uncoiled tube and therefore independent of the number of turns. However, the number of segmentary interfaces is equal to twice the number of helical turns. For a given length and bore of tubing, therefore, the total area of interface, and hence the partition efficiency, will increase with the number of turns. Also, since the inefficient free space in each helical turn decreases in length with the helix diameter, laminar flow spreading of the solute band within this space and broadening of the resultant solute peaks will be reduced.

The relationship between phase viscosity and column efficiency is of interest. Although either phase may be used as the mobile phase, the more viscous phase is preferred. Under these circumstances undesirable longitudinal diffusion within the free space is slower and the enhanced transfer of solutes from within the stationary phase to the phase interface promotes rapid attainment of equilibrium conditions.

Theoretically, the efficiency of CCC for a given phase system may be increased by using a greater length of more tightly wound narrower tubing. The practical limit, however, is governed by the centrifugal force field and the injection pressure necessary to promote interfacial flow. The minimum pressure of the syringe plunger, necessary to sustain flow, may be calculated from the equation:

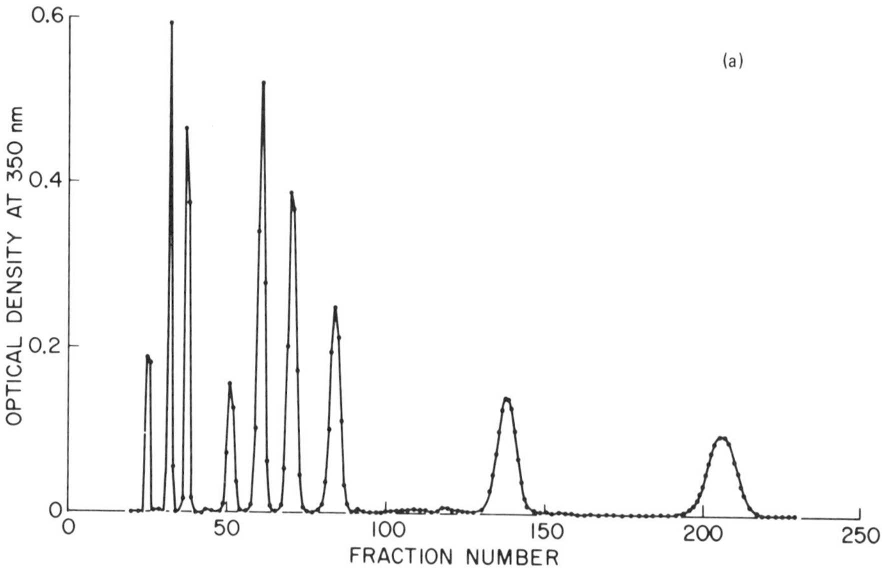

Fig. 41 Separations of DNP amino acids with the original toroidal coil centrifuge. (a) Chromatogram of 9 DNP amino acids with a coiled column. The column was prepared from a 40-m long, 0.2-mm i.d. PTFE tube coiled onto a 0.85-mm i.d. core to make 8000 helical turns. (b) Chromatogram of 7 DNP amino acids with a twisted column. The column was prepared from a 40-m long, 0.2-mm i.d. PTFE tube which was folded in two and then twisted to make a total of 17,000 turns. See Table 4 for more details.

$$P = R\omega^2(\rho_1 - \rho_u)nd \tag{31}$$

where R denotes the distance from the center of rotation to the axis of the coil; ω, the angular velocity of rotation; n, the number of helical turns; d, the helical diameter; ρ_1 and ρ_u, the density of the lower and the upper phases, respectively. Although some additional factors, such as liquid drag force, initiation of flow at the interface against interfacial tension, etc., may also contribute to the column pressure, these factors are relatively insignificant compared with the hydrostatic pressure described above.

By the aid of a manometric device attached to the present apparatus, the validity of the above formula was confirmed. An end-closed capillary tube was coiled around the syringe barrel and the open end was connected to the hole made through the center of the syringe plunger. A drop of mercury was introduced in the capillary before

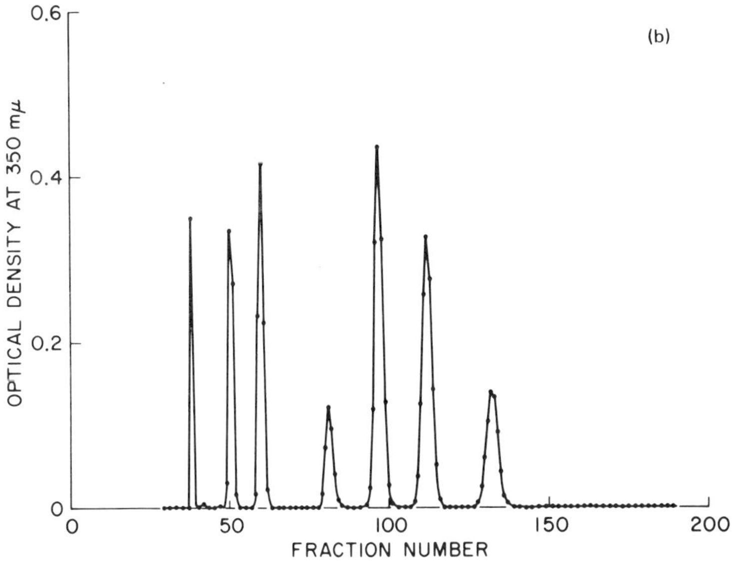

Fig. 41 (Continued)

starting the centrifuge run. As the elution proceeded, the internal pressure of the syringe rose to compress the air in the capillary tube and push the mercury toward the closed end of the capillary. From the position of the mercury observed under the stroboscopic illumination, the pressure was computed. The result of the experiment is shown in Fig. 42 where the internal syringe pressure is plotted against the length of the coil containing the two phases in the hydrostatic equilibrium state. The experimental data closely coincide with the theoretical figure expressed by the dotted line. These data also indicate that at a low flow rate the pressure produced by viscous resistance of the solvent is insignificant in comparison to the opposing hydrostatic pressure created by the density difference of the two phases under the applied centrifugal force field. The maximum column pressure measured during the separations was 16 atmospheres for the twisted column and 20 atmospheres for the coiled column. Since 20 atmospheres is close to the maximum capacity of the syringe pump to deliver the solvent and also the critical level to rupture the tube, the twisted column has an advantage in reduction of the hydrostatic pressure by minimizing the helical diameter.

Although the present scheme has successfully demonstrated a potential capability of CCC to achieve an extremely high partition efficiency, the practical application of the method is largely limited by the use of the rotating syringe for elution of the mobile phase.

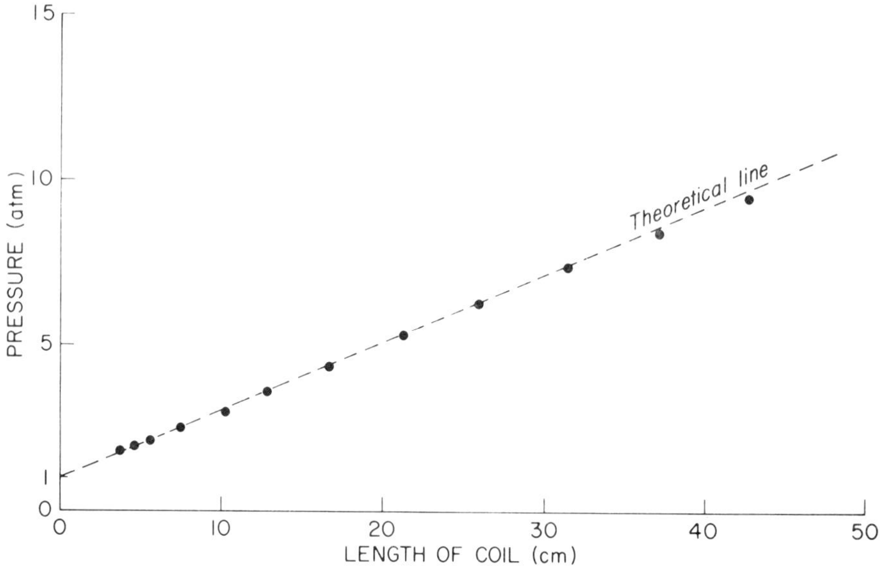

Fig. 42 Measurement of column pressure in the toroidal coil centrifuge. The column pressure measured with a manometric device is plotted against the length of the coiled column equilibrated with the two solvent phases. The data coinsides with the theoretical value expressed by the dotted line.

When a slow elution rate of less than 1 ml/hr is to be applied under a high column pressure, the syringe fails to maintain a uniform flow and instead produces an intermittent flow at various intervals. The leakage of the solvent at the seal of the syringe plunger was found to be a common complication which may be improved by choosing a syringe with a small diameter at the sacrifice of the total capacity of the mobile phase eluted. The system also lacks versatility of applying either stepwise or gradient elution mode which is extremely useful in both liquid chromatography and CCC. In the recent model described in the following, the performance of the apparatus has been greatly improved by incorporating a seal-free flow-through device (Fig. 17) whereby the continuous elution through the rotating column is possible with an ordinary elution pump commonly used in liquid chromatography.

B. Toroidal Coil Centrifuge Without Rotating Seal [33,34]

The original toroidal coil centrifuge previously described has disclosed a great potential to yield a high partition efficiency comparable to that in refined gas chromatography. The undesirable features of the previous scheme, i.e., the use of rotating syringe and a rotating seal, have been entirely eliminated in the new flow-through centrifuge device which allows the solvent to move in and out through the rotating column without rotating seals. The principle of this nontwisting system appears to have been vaguely known for a number of decades among the inventors of various spinning machines without useful application. As far as the author's present knowledge is concerned, the concept of this twist-free rotating scheme was first disclosed by Wrench in 1967 [44] and several years later Adams applied the system to various optical and/or electrical devices [45]. Application of this principle to high-speed centrifugation has been conducted in the author's laboratory to achieve an on-line plasmapheresis without complications caused by the use of rotating seals [31,32,46]. The same apparatus has been successfully applied to CCC by using a toroidal coil as a separation column. The capability of the scheme was first demonstrated by a standard test system, i.e., on separation of DNP amino acids with a two-phase solvent system composed of chloroform/acetic acid/0.1 N HCl at a 2:2:1 volume ratio [33]. Because of the stable centrifugal force field produced by the present scheme, low interfacial tension two-phase solvent systems such as polymer phase systems can also be applied for separations of cell particles and macromolecules [34].

Principle and Design of the Toroidal Coil Centrifuge Free of Rotating Seals

The principle of the present scheme is illustrated in Fig. 43. A bundle of flexible tubes connected to a bowl is supported stationary at the other end, forming a loop as shown in the figure. When the bowl rotates at an angular velocity of 2ω around the vertical axis and the loop simultaneously revolves around the same axis at ω, the tube bundle remains free from twisting. In so doing, the tube bundle counterrotates around its own axis at ω. Consequently, the system allows flow in and out through the rotating bowl without the use of rotating seals. The toroidal coil coaxially accommodated in the periphery of the bowl is subjected to a radial centrifugal force field as in the original system. The principle of this scheme is somewhat more logically described as the nonplanetary scheme in Fig. 17 (scheme V).

Overall design of the centrifuge rotor is illustrated in Fig. 44a. The motor directly drives the rotary frame which consists of three parallel horizontal plates rigidly linked together. The frame holds three rotary elements: a column container (center), a countershaft (right), and a tube holder (left). A stationary pulley rigidly mounted on the motor housing coaxially to the motor shaft is coupled through a toothed belt to an identical pulley mounted at the bottom

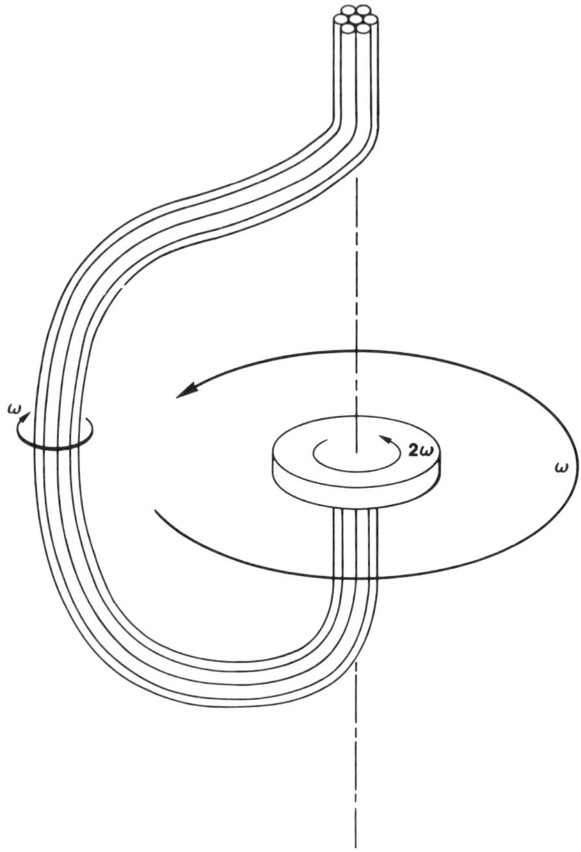

Fig. 43 Principle of seal-free flow-through centrifuge scheme. The central bowl rotates at 2ω and the tube bundle revolves at ω, both in the same direction around the central axis of the diagram. This motion allows the central bowl to rotate without twisting the tube bundle. The system is identical to scheme V of the nonplanetary scheme illustrated in Fig. 17.

of the countershaft. This coupling produces a counterrotation of the countershaft with respect to the rotating frame. This motion is further conveyed to the central container by 1:1 ratio gear coupling between the countershaft and the central container. Under this arrangement, the central container rotates at ω, with respect to the frame while the latter also rotates at ω in the same direction. Consequently, the central container rotates at 2ω with respect to the outside observer. To support a bundle of the flow tubes under the centrifugal force field, the tube holder is actively counterrotated by means of coupling a pair of identical toothed pulleys, one mounted around the tube holder and the other on the countershaft. A pair of flow tubes connected to the toroidal coil in the central container is first led through the central hole of the container, then passed through the tube holder, and finally supported tightly at the top stationary member of the centrifuge as illustrated. These tubes are thoroughly lubricated with grease and protected by a piece of flexible vinyl tubing at each supported place. When this procedure is carefully followed, these tubes can maintain their function for many months of use. Figure 44b schematically shows a toroidal coil circumferentially arranged inside the central container. In practice a longer column can be accommodated in the container by making as many turns as necessary around the periphery of the container. Our test system permits rotation of the central container up to 2000 rpm which produces about 750g at the periphery.

Performance of the Apparatus

The capability of this CCC scheme has been demonstrated on separation of a set of DNP amino acids (Sigma Chemical Co.) on a two-phase system composed of chloroform/acetic acid/0.1 N HCl at a 2:2:1 volume ratio. The column was prepared by winding PTFE tubing (Zeus Industrial Products, Raritan, NJ), 0.38 mm i.d. and 24 m long, onto a nylon tube 1.5 mm o.d. and 300 cm long, to make approximately 3000 helical turns with a total capacity of about 3 ml. The column was first filled with the stationary lower phase followed by sample loading (20 μl containing 0.25–0.5 g% of each component). Then the mobile upper phase was pumped with a syringe drive (Harvard Apparatus Co.) at a rate of 0.3 ml/hr while the rotor was spun at 750 rpm (100g). The eluate was continuously monitored through an LKB Uvicord III to measure the absorbance at 280 nm.

Figure 45 shows a typical chromatogram obtained with the present method. Nine DNP amino acids are separated at efficiencies ranging between 4000 and 1250 TP as calculated according to the formula used for gas chromatography [Eq. (30)]. In consideration of the diameter and length of the column used in separation, the above result is comparable to those obtained by the previous scheme with a much smaller diameter column.

One particular advantage of the present CCC scheme is that the system allows an excellent retention of the stationary phase for the polymer phase systems [3] which are useful in partitioning cell particles and macromolecules. Potential capability of the present scheme to utilize the polymer phase systems has been demonstrated in separation of bacterial cells. Figure 46 shows partition behavior of two *E. coli* strains in a polymer phase system composed of 5% (w/w)

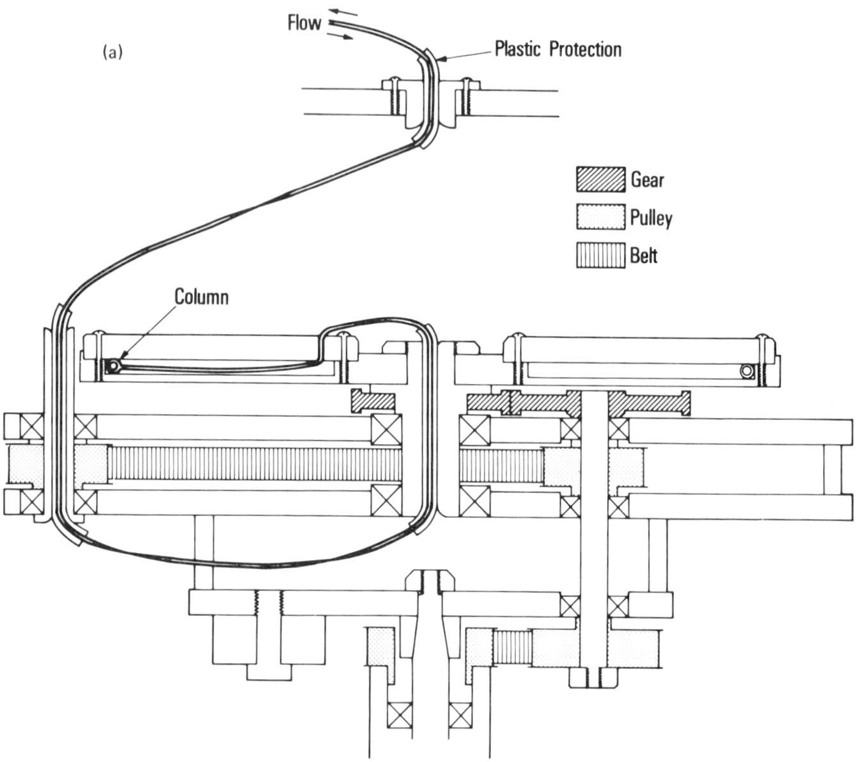

Fig. 44 Design of the seal-free toroidal coil centrifuge. (a) Cross-sectional view of the rotor. The rotary frame consisting of three parallel plates bridged together with links holds three rotary elements: the central column container, the countershaft (right), and the central column container, the countershaft (right), and the hollow shaft or tube support (left) all mounted through ball bearings. The stationary toothed pulley rigidly mounted on the motor housing coaxially around the motor shaft is coupled with a toothed belt to an identical pulley on the countershaft. This coupling causes the countershaft to rotate on the rotary frame. The motion of the countershaft is then conveyed to the central container by a 1:1 ratio gear coupling. Consequently, the central container rotates at the doubled speed with respect to the outside observer. The hollow shaft is also actively rotated by coupling a pair of identical pulleys between the countershaft and the hollow shaft. These mechanical arrangements produce a motion required for the seal-free centrifuge scheme illustrated in Fig. 43. (b) Schematic view of the torodial coil. The coiled column is circumferentially arranged inside the column container. Longer column can be accommodated by making multiple turns around the periphery of the container.

(b)

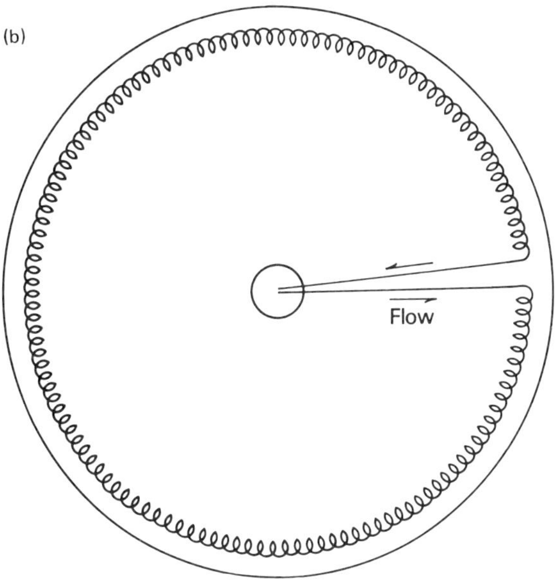

Fig. 44 (Continued)

dextran 500 and 4% (w/w) polyethylene glycol 6000 with various concentration of NaCl. In each diagram, percentages of cells partitioned in each phase and at the interface are plotted against the NaCl concentration in the polymer phase system. Although both types of cells show similar tendency to distribute toward the lower phase as the NaCl concentration increases, the partition patterns of these two strains are quite different. *E. coli* ATCC 11303 distributes almost entirely in the upper phase at low NaCl concentrations whereas *E. coli* ATCC 8739 distributes mostly at the interface at the NaCl concentrations over 0.01 M. Under these circumstances a gradient elution in the decreasing NaCl concentration would elute *E. coli* ATCC 11303 earlier than *E. coli* ATCC 8739, since cells distributed at the interface tend to remain within the column as if they were distributed in the lower phase.

Because of the high viscosity of the polymer phases, it was necessary to use a short column with a relatively large internal diameter. Separations of *E. coli* strains were performed with a toroidal coil consisting of 500 helical turns (6.5 mm o.d. core) of 1-mm i.d. tubing with a 12-ml capacity. For the gradient elution, a pair of polymer phase systems were prepared; the starting medium has a 0.02 M NaCl concentration and the ending medium, zero NaCl concentration. The column was first filled with the lower phase of

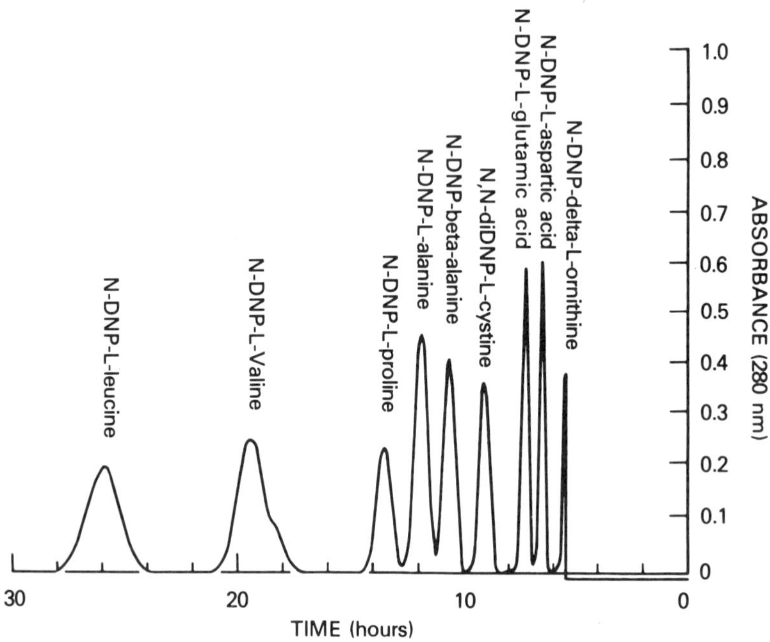

Fig. 45 Separation of DNP amino acids with the seal-free toroidal coil centrifuge. Chromatogram was obtained with a column consisting of 3000 helical turns of a 24-m long, 0.38-mm i.d. PTFE tube with a total capacity of about 3 ml.

the starting medium and a mixture of the two *E. coli* strains suspended in the starting medium was injected at the inlet of the column. Then the apparatus was rotated at 700 rpm while the column was eluted with the upper mobile phase of the ending medium via a 5-ml capacity mixing chamber initially filled with the upper phase of the starting medium to produce an exponential NaCl gradient as shown in Fig. 47a.

Figure 47a shows the obtained data where cell population in the yielded fractions are plotted against the elution time (which also corresponds to the NaCl concentration shown in Fig. 47a above). Three curves drawn in the chart indicate each separate experimental run. In the first experiment, cells were separated into two peaks labeled A and B. Cells from each peak were collected and rerun separately under the identical operational conditions to demonstrate reproducibility of the method. As shown in the figure, the two *E. coli* strains were separated each into a discrete peak which was reproduced by the rerun of the harvested cells. The rate of the

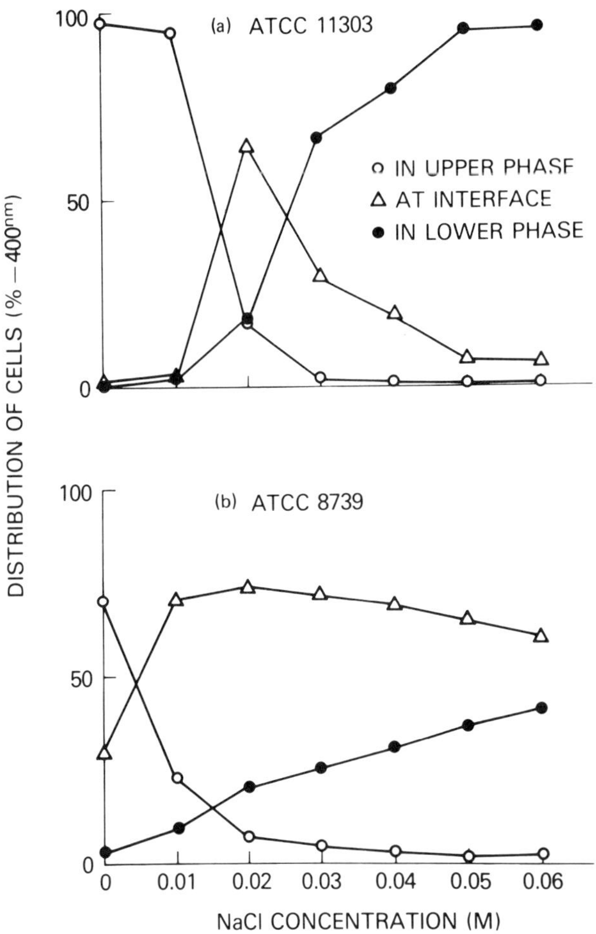

Fig. 46 Partition behavior of two *E. coli* strains in a polymer phase system composed of 5% (w/w) dextran 500, 4% (w/w) polyethylene glycol 6000, and various NaCl concentrations. Both strains show a similar tendency to partition toward the interface and the lower phase as the NaCl concentration is increased. However, at the low NaCl concentration strain ATCC 11303 (top) populates more densely in the upper phase than strain ATCC 8739 (bottom). This difference in partition behavior becomes a base for separation of these two strains with NaCl gradient elution.

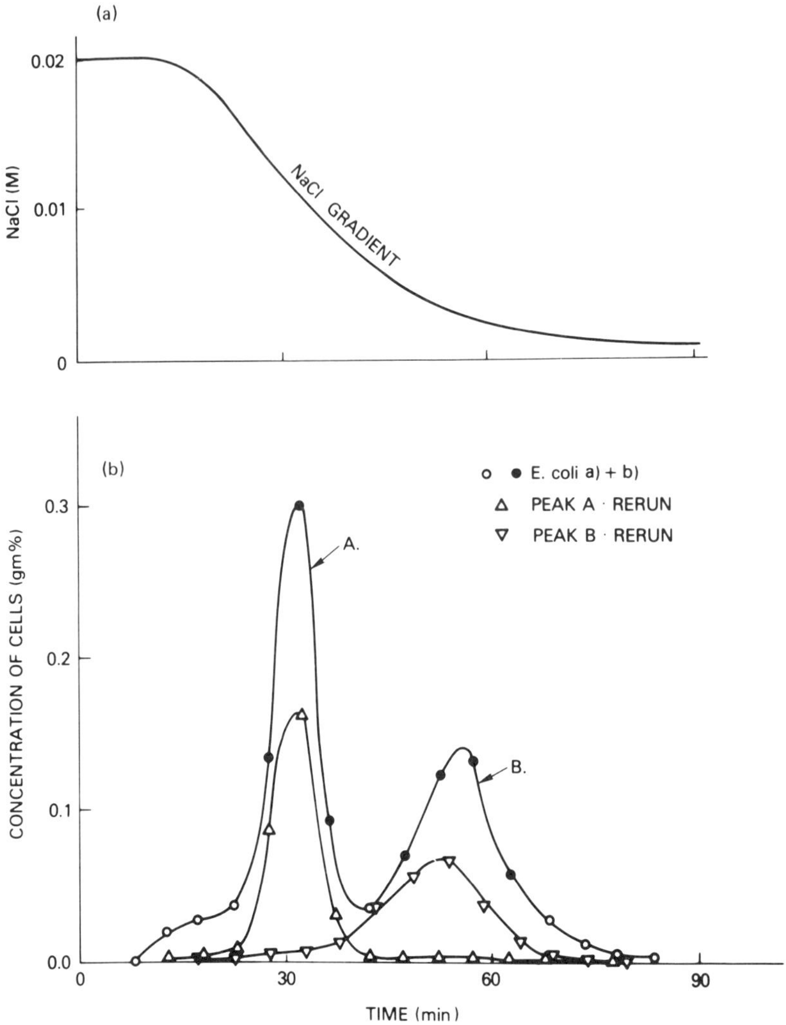

Fig. 47 Separation of two *E. coli* strains with the polymer phase system by the seal-free toroidal coil centrifuge. (a) An exponential NaCl gradient in the polymer phase system used for separation of two *E. coli* strains shown in Fig. 46. (b) Chromatograms obtained from three separate experiments. In the first experiment, a mixture of the two strains was separated into two peaks (A and B). Then, fractions for each peak were collected and rerun separately under the identical operational conditions to demonstrate the reproducibility of the method.

reproducibility was approximately 85% in each peak. Further experiments on partition of cells and macromolecules will be described by Sutherland in Chap. 11.

The present scheme has been bound to work well for small cells and cell organelles. However, for larger particles such as blood cells, the stable centrifugal force field promotes sedimentation of particles at the peripheral portion of the coil. This complication can be eliminated by introducing a slow rotation to the coiled column which is possible with the use of nonsynchronous flow-through coil planet centrifuge described later in this chapter.

Future Development

The toroidal coil centrifuge is a reliable CCC scheme which permits universal application of two-phase solvent systems. The capability of the toroidal coil centrifuge may be further expanded toward various directions with the use of modified coiled columns. A few examples of such applications are discussed below.

1. *Microanalytical column.* As described earlier, analytical scale separations have been successfully performed with the toroidal coil centrifuge equipped with a long, small-diameter, coil column (0.38 mm i.d.) under a slow flow rate of 0.3 ml/hr. The problem in using such analytical columns was found to be a high hydrostatic pressure accumulated in the coil column which can eventually rupture the tube or overcome the capacity of the pump used for the elution of the mobile phase. This limitation is, however, greatly lessened by the recent development in HPLC technology which has introduced a thick-wall PTFE tubing and a high-pressure metering pump for elution. The combined use of the thick-wall capillary coiled column (0.1 mm i.d.) and a high-pressure pump in the toroidal coil centrifuge technique will permit elution under a pressure of 1000 psi. The present seal-free toroidal coil centrifuge is capable of producing a centrifugal force field of about 700g (2000 rpm) which is seven times that previously used for the DNP amino acid separations. Application of this strong centrifugal force field would prevent the plug flow phenomenon and produce a satisfactory retention of the stationary phase in a narrow capillary column provided that the optimum flow rate is applied for elution of the mobile phase. Unfortunately, the elution pumps now available for HPLC are generally designed for flow rates ranging from 6 to 600 ml/hr or greater, which are much higher than the rates needed for this CCC scheme. Development of a high-pressure elution pump which can produce a reliable flow rate in a range between 0.01 and 1 ml/hr is considered to be essential for performing the microanalytical CCC.

2. *Flat toroidal coil.* The configuration of the toroidal coil may be altered into a flat shape to perform various modes of separations.

The core support of the column used in our preliminary studies was a doughnut-shaped plate made of either a cardboard or a plastic material and had a typical dimension of 30 cm o.d., 15 cm i.d., and 0.2 cm in thickness to fit into the rotor of the toroidal coil centrifuge. The flat toroidal coil is made by winding the PTFE tubing around the core support as shown in Fig. 48. In analogy to the preparative CCC schemes developed from the HSES (Fig. 11), one side of the coil which provides effective column space for separation is made of a large-diameter tube while the other side of the coil is made of a small-diameter tube which serves as a transfer tube. The large diameter portion of the tube is arranged in the radial position to promote the separation process. Although such a flat toroidal coil can be made by connecting multiple pieces of tubing with two different diameters, it is more conveniently prepared from a single piece of heat-shrinkable PTFE tubing by applying heat at regular intervals to make alternating segments of the column units and transfer tubes each of a desired length. This nonjunction column is leakproof and can be used safely under a high pressure with minimum risk of failure.

The flat torodial coil can be useful not only for performing CCC but also in many other experiments. For the CCC application, the column is first filled with the stationary phase and then eluted with the mobile phase in a proper direction so that a large volume of the

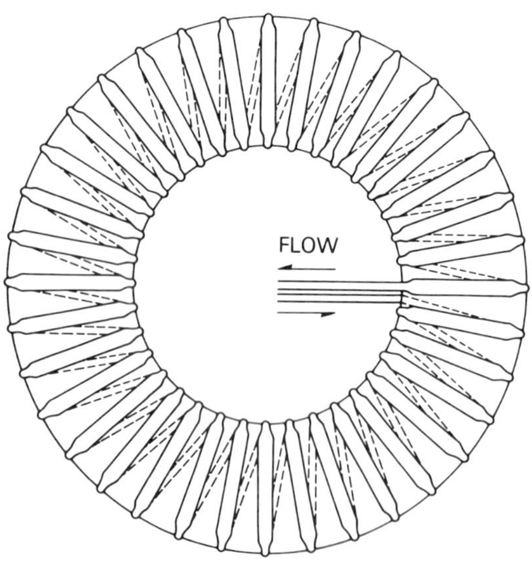

Fig. 48 The flat toroidal coil. The column uses a doughnut-shaped plate to support the flat coil which consists of alternating segments of the thick column units and the thin transfer tubes.

stationary phase is retained in each column unit. In order to effectively utilize the radially acting centrifugal force field, the heavier phase should be introduced from the proximal end of the column units and the lighter phase from the peripheral end. The method is somewhat similar to DCCC but this new scheme will permit application of a much faster flow and produce more efficient mixing of the solvents by the vigorous motion of the droplets accelerated under a strong centrifugal force field. A similar but more elaborated centrifugal CCC apparatus has been successfully developed by Sanki Engineering, Ltd., Kyoto. As described later by Bruening and Nakanishi in Part II, the apparatus is equipped with a pair of highly refined rotary seals which can function under 60 atmospheric pressure [42].

The flat-coiled column is also capable of retaining small particles in the column units under continuous elution with a single-phase medium such as saline solution. Since these retained particles in each column unit are constantly agitated by the counterflowing medium, the scheme may be efficiently applied for elutriation of cells and cell organelles. The scheme also permits the use of live cells such as erythrocytes as a solid support matrix to perform an affinity chromatography. A loosely packed erythrocyte column for this cell membrane affinity chromatography is easily obtained by introducing an adequate volume of the whole blood into the column and then eluting with an isotonic saline solution at a proper rate. Subjected to a steady flow of the saline solution through the peripheral end of the column unit, the erythrocytes are well washed and retained in each column unit for many hours without visible evidence of hemolysis. In each separation, a sample mixture dissolved in a physiological solution is injected locally into the flowing stream at the inlet of the column. Under these circumstances, the solutes which have an affinity to the cell membrane (crystal violet, brilliant cresyl blue, etc.) are retained longer in each column unit while the solutes having no affinity to the membranes (Evans blue, L-tyrosylglycylglycine, etc.) are quickly eluted out from the column [47]. The toxic substance which produces hemolysis may also be quantitatively detected by monitoring the concentration of hemoglobin released from the lysed cells. The substances which produce cell aggregation such as lectins cause instantaneous sedimentation of cells starting at the first column unit which can be observed under the stroboscopic illumination. This process continues toward the neighboring column units until all lectins are adsorbed to the membranes. Preliminary experiments showed that small amounts of lectins can be quantitatively measured by the number of column units containing dark-colored, short, hard segments of aggregated erythrocytes [47].

In addition to the various applications with the toroidal coil, the seal-free flow-through centrifuge provides a rich field of application in continuous blood separation by means of a separation chamber equipped with a circular or a spiral channel [31,32]. Recently developed MIFC molding technique [42] can be used to produce a variety of complex repetitive flow channels in an annular plate which is in turn embedded into a proper plastic resin in a shallow centrifuge bowl. A series of interconnected conical locules radially arranged around the periphery of the bowl (with the tips directed toward the periphery) has been used for multistage elutriation of various cells with isotonic solution in our laboratory [48].

C. Summary

The nonplanetary flow-through centrifuge which provides a simple rotation of the rotor can be utilized for performing CCC. A coiled column is accommodated circumferentially inside the rotor to form a toroidal coil. A stable radially directed centrifugal force field traps either lighter or heavier stationary phase in each helical turn as the other phase continuously elutes through the column. Consequently, the solutes introduced locally at the inlet of the column are subjected to a continuous partition process through the column and eluted out in the order of their partition coefficients. The potential capability of this scheme has been demonstrated by DNP amino acid separations with a conventional two-phase solvent system and the partition of *E. coli* strains with aqueous-aqueous polymer phase systems. The present scheme also offers a variety of other useful applications by either modifying the column configuration or applying different types of separation columns.

VI. SYNCHRONOUS COIL PLANET CENTRIFUGES: EARLY INSTRUMENTATION (SCHEMES I, II, AND III)

Among various flow-through centrifuge schemes developed for performing CCC, the synchronous CPCs are simplest in mechanical design yet they give an extremely versatile performance. As illustrated in Fig. 17 (schemes I–IV), all schemes share a common feature of synchronous planetary motion: the coil holder revolves around the central axis of the centrifuge and simumlaneously rotates about its own axis at the same speed. This planetary motion allows continuous elution through the rotating column without the use of rotary seals.

A. Flow-Through Coil Planet Centrifuge (Scheme I)

Principle and Design of the Apparatus (13,18,19)

The principle of the present scheme is illustrated in Fig. 49. The separation column is a helix (Fig. 49a). Flexible feed and return

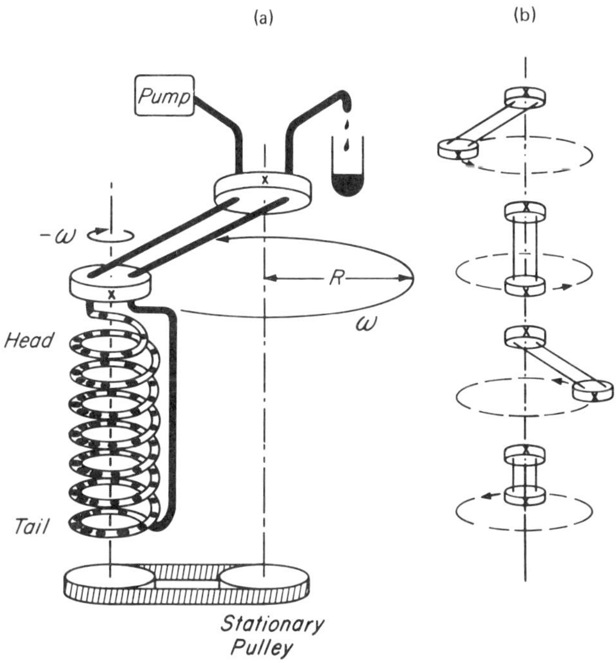

Fig. 49 Principle of the flow-through coil planet centrifuge. (a) Helical separation column with flexible feed and return tubes supported by the moving and stationary disks. (b) Successive positions of the stationary and moving disks.

flow tubes are supported by the moving disk at the top of the helix and the stationary disk fixed to the center of the upper frame of the centrifuge. The helix and moving disk revolve around the axis of the centrifuge but are not permitted to rotate with respect to the stationary disk. The fixed orientation of the helix is maintained by coupling a pulley on the helix holder to a stationary pulley of equal diameter on the axis of the centrifuge drive with a toothed belt. This coupling causes a counterrotation ($-\omega$) of the helix, canceling out the rotation (ω) of the helix induced by its revolution. The feed and return tubes do not twist because the moving disk does not rotate with respect to the stationary disk, as indicated by the "x" marks (Fig. 49b) showing in successive positions of the stationary and moving disks. Because the helical column maintains a fixed orientation while it revolves, the radially directed centrifugal force rotates with respect to the column as previously described (Fig. 18).

Figure 50 shows the original flow-through CPC constructed by modifying a conventional floor-type centrifuge (International Equipment

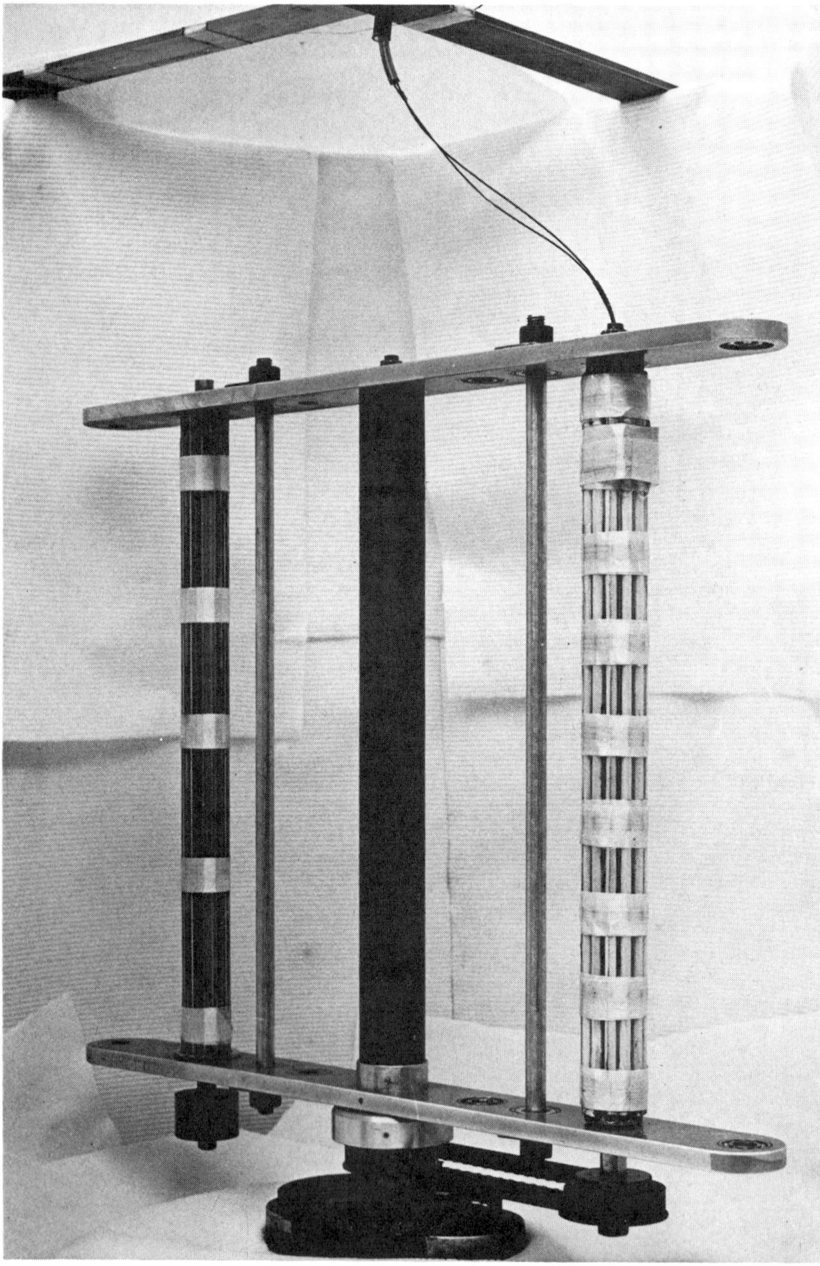

Fig. 50 The flow-through coil planet centrifuge.

Company, Model II). The column holder (right) and the counterweight holder (left) are vertically supported by a pair of rotary arms with bearings. The desired planetary motion of the column holder is established by applying a toothed belt between a pair of toothed pulleys of the same size, one fixed to the bottom of the column holder and the other to the stationary tube mounted on the motor housing at the axis of the centrifuge drive. The radius of revolution is adjustable to 30.7, 20.2, 13,6, and 8.6 cm. The revolutional speed is continuously adjustable up to 700 rpm at a 30.7-cm radius (160g).

The column is prepared by winding the PTFE tubing onto a plastic core, 5—10 mm diameter and 45 cm long, to make a column unit. Two to twelve column units are interconnected in series by tail-head connection and tightly mounted in parallel into the peripheral grooves of the holder. Both feed and return tubes are passed through the center hole at the top of the holder and then supported at a height of about 25 cm above the center of the apparatus. These tubes, protected with a piece of greased silicone rubber tubing at the holder, have not failed or stretched appreciably even after having been used many times.

In each separation the column is filled with the stationary phase and a sample solution is introduced through the feed tube. After the desired revolutional speed is established, the mobile phase is pumped with either a metering pump or a syringe drive, while the effluent from the outlet of the column is monitored with an LKB UV monitor at 280 nm.

Studies on Retention of the Stationary Phase [19]

Although the present technique is simple and versatile in separation, successful separation depends on satisfying two requirement for phase behavior in the equilibrated coil. The first requirement is to retain an adequate amount of the stationary phase in the column, and the second requirement is to provide a rich segmentation of the two phases to reduce mass transfer resistance. For a given solvent system, these requirements should be satisfied by choosing a combination of instrumental variables such as the tube bore, revolutional speed and radius, flow rate, etc.; hence, the phase distribution diagram has been devised to indicate the steady-state relative volumes occupied by the two phases as a function of revolutional speed for a given column mounted at a given radius, using a given flow rate.

A set of short PTFE columns measuring 1.2, 0.85, 0.55, and 0.38 mm i.d. was used for this study. Each column was prepared by winding the tube onto a pair of plastic core, 45 cm long and 5 mm diameter (for 1.2-mm bore tube, 7 mm core was used to prevent excessive flattening of the tube), and mounted onto the holder in such a way that the fluids were introduced at the top of the first column unit and collected at the top of the second through the junction situated at the bottom to cancel out the effect of gravity. The capacity of each column and the dead space in the flow tubes were measured before use.

In each experiment, the column is filled with the stationary phase, and the mobile phase is introduced at a constant linear flow rate of 1 cm/sec (that is, 24 ml/hr for 1.2-mm, 12 ml/hr for 0.85-mm, 6 ml/hr for 0.55-mm, and 2.4 ml/hr for 0.38-mm i.d. tubes) while the column is rotated at the desired speed until the mobile phase begins to elute and the steady state is established. The volume percentage of the aqueous phase occupying the column space is then calculated from the column capacity, the dead space volume, and the eluted stationary phase volume and plotted against the applied revolutional speed in rpm to obtain the phase distribution diagram.

Figure 51 shows a phase distribution diagram of hexane/water system in the 1.2-mm bore column. The ordinate indicates percentage of the aqueous phase volume occupying the column space and the abscissa the applied revolution speed in rpm. Two points are measured for each speed; open circles for the aqueous phase as the mobile phase and closed circles for the aqueous phase as the stationary. As the flow rate is decreased, these two points move closer and will finally reduce to one point which represents the true equilibrium composition under no-flow conditions. A pair of curves drawn through each point shows some common features among various phase systems examined. The upper curve starts to move down earlier and displays a sharp spike before it becomes nearly horizontal, whereas the lower curve starts to rise later and smoothly approaches the upper curve.

The phase distribution diagram may be conveniently divided into three regions as shown in Fig. 51. The plug flow region located at the low-speed range produces plug flow regardless of the choice of the mobile phase. At the spike region, the plug flow is induced only if the nonaqueous phase is used as the mobile phase, while the mobile aqueous phase produces a poor phase segmentation. The equilibrium region, following the spike region, provides an ideal condition because either phase can be used as the mobile phase without plug flow to form rich segmentation of the two phases.

Adequate separations are usually obtained only if the amount of stationary phase for the aqueous phase is indicated by the distance suitable conditions can easily be determined. The percentage of the stationary phase for the aqeuous phase is indicated by the distance between the lower curve and the 0% line, and that for the nonaqueous phase is read by the distance between the upper curve and the 100% line. Although the best separation is expected from the flat equilibrium regions, the spike regions can also be applied by pumping the aqueous phase is an adequate equilibrium region is not available.

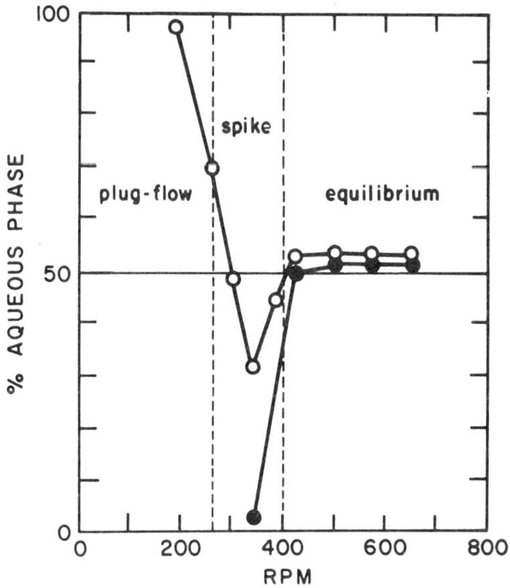

Fig. 51 Phase distribution diagram of hexane/water system. The ordinate indicates percentage of the aqueous phase volume occupying the column space and the abscissa the applied revolutional speed in rpm. On the basis of the relative motion of the two phases, the phase diagram is divided into three regions. The plug flow region produces a plug flow regardless of the choice of the mobile phase. At the spike region, the plug occurs only if the nonaqueous phase is used as the mobile phase, while the mobile aqueous phase produces a poor phase segmentation. The equilibrium region gives a rich phase segmentation regardless of the choice of the mobile phase.

The phase distribution pattern varies with the phase parameters and instrumental variables. Figure 52 shows the phase distribution diagrams for nine phase systems in relation to various column i.d. at a 30.7-cm revolutional radius. Among the physical properties of the phase system, the interfacial tension plays the most important role. A high interfacial tension tends to shift the equilibrium region toward the high speed as seen in hexane/water and ethyl acetate/water. On the other hand, a low interfacial tension causes emulsification in the equilibrium region with loss of the retained stationary phase, indicated by a rise of the upper curve and a fall of the lower curve as seen in n-BuOH/CH_3COOH/H_2O and sec-BuOH/H_2O. The interfacial tension is a variable open to some experimental manipulation by addition of some intermediate component(s) to the phase sys-

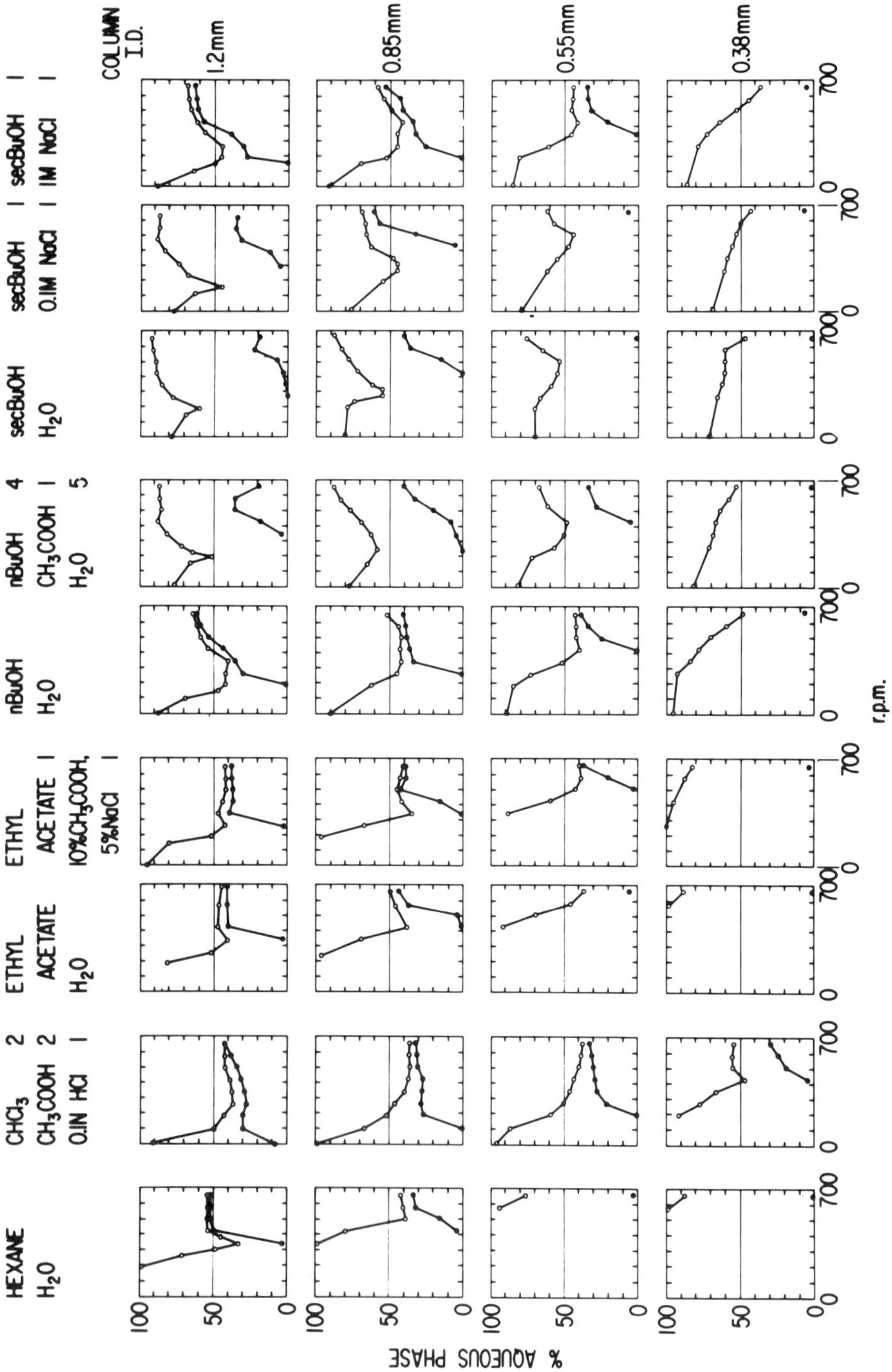

HEXANE
H2O
CHCl3 2
CH3COOH 2
0.1N HCl 1
ETHYL
ACETATE
H2O
ETHYL
ACETATE 1
10% CH3COOH,
5% NaCl 1
nBuOH
H2O
nBuOH 4
CH3COOH 1
H2O 5
secBuOH
H2O
secBuOH 1
0.1M NaCl 1
secBuOH 1
1M NaCl 1
COLUMN
I.D.
1.2mm
0.85mm
0.55mm
0.38mm
% AQUEOUS PHASE
100
50
0
0
700
r.p.m.

Fig. 52 Phase distribution diagrams of nine phase systems in relation to the internal diameter of the column. In each diagram, flat equilibrium region represents the ideal condition. Hexane/water and ethyl acetate systems give plug flow in a small-bore column due to high interfacial tension. The chloroform system shows excellent phase distribution throughout because a large amount of acetic acid reduces interfacial tension. BuOH systems have relatively low interfacial tension. Satisfactory phase distribution of n-BuOH/H_2O system is impaired by adding acetic acid, where the lowered interfacial tension results in emulsification of the phases. Unsatisfactory phase distribution of sec-BuOH system is improved by adding a salt which increases interfacial tension as well as the density difference between the two phases.

tem. Substances which are more or less evenly distributed in both phases tend to reduce the interfacial tension, whereas substances which are soluble in one phase only tend to increase it. In practice, the high interfacial tension is adjusted by adding methanol or acetic acid ($CHCl_3/CH_3COOH/0.1$ N HCl), and the low interfacial tension by salts (sec-BuOH/1 M NaCl). Also, the interfacial tension should be adjusted to meet the tube diameter. The large bore (1.2 mm) favors a relatively high interfacial tension, whereas the small bore (0.38 mm) requires a low interfacial tension.

The density difference and the viscosity of the phases are less significant, and it is not easy to isolate the effect of these variables. Since the net screwing force acting on the phase segments depends highly on the density difference of the two phases, a minimum requirement for the density difference must exist for a given set of other variables. The value of the density difference for the nine listed phase systems range from 0.337 g/cm^3 (hexane/water) to 0.049 g/cm^3 (n-BuOH/CH_3COOH/H_2O). Since ethyl acetate/water shows an excellent phase distribution with a 1.2-mm bore column in spite of the small density difference (0.065 g/cm^3), the unsatisfactory phase distribution of n-BuOH/CH_3COOH/H_2O (0.049 g/cm^3) and sec-BuOH/H_2O (0.103 g/cm^3) is not likely to be solely due to the small density difference. However, a relatively large density difference of the two phases may yield a profitable effect for low interfacial tension phase systems as suggested in $CHCl_3/CH_3COOH/0.1$ N HCl (0.228 g/cm^3) and sec-BuOH/1 M NaCl (0.188 g/cm^3). As to the effect of the viscosity on the phase distribution, it is interesting to note that the true equilibrium composition of the viscous BuOH systems tend to shift toward the 100% line at the equilibrium region for a 1.2-mm bore column.

The internal diameter of the column is an important instrumental variable in that decreasing of the tube bore causes a remarkable shift of the equilibrium region toward higher revolutional speeds in such a way that the location of the spike region appears to be inversely proportional to the column diameter. With very small-bore columns, our test system fails to provide suitable conditions especially for high interfacial tension phase systems because of the limited revolutional speed available. Thus, for microanalytical purposes, the apparatus should produce at least 1000 rpm at 30.7 cm of radial position.

The effect of the helix diameter on the phase distribution has not been carefully studied but the preliminary studies suggest that a small helix diameter increases the amount of the stationary phase. The revolutional radius determines the centrifugal force field for a given revolutional speed. The comparison of the phase distribution obtained at between 8.6- and 30.7-cm radii under the equivalent centrifugal force field reveals an important fact that the location of the spike region is determined by the applied force field. The greater radius also increases the stationary phase volume especially in the low interfacial tension phase systems. However, high interfacial tension phase systems yield a suitable phase distribution in 1.2-mm bore column even at an 8.6-cm radius. Obviously, a high revolutional speed applied at a short radius promotes emulsification for the low interfacial tension solvent systems. The flow rate mainly affects the retained stationary phase volume. The distance between the upper and the lower curves becomes shorter as the flow rate decreases, and the unsatisfactory phase distribution of the low interfacial tension groups is greatly improved at their spike and early equilibrium regions by decreasing the flow rate.

The above studies revealed two main problems. One is the plug flow of the high interfacial tension phase systems in a small-bore column, and the other is the emulsification of the low interfacial tension phase systems in a large-bore column. The first problem may not be so serious because extremely high interfacial tension phase systems such as hexane/water are not commonly used for separation, and the problem may be solved simply by increasing the revolutional speed and/or radius by refining the apparatus. However, the second problem of emulsification is important because the low interfacial tension phase systems are widely used for separation of various biological substances such as polypeptides, nucleic acids, polysaccharides, etc.

Further studies have shown that the coiled helix configuration of the column largely prevents the emulsification even at a high revolutional speed. The preliminary studies have been carried out by winding a PTFE tube onto a flexible core, either a round pipe or a flat strip, which is in turn mounted around the column holder at various angles. When the angle of the helix column is 60°, the stationary phase retention of low interfacial tension butanol phase systems was highly stabilized even at the maximum revolutional speed of 650 rpm at 30.7 cm radius as shown in Fig. 53.

Application of the Apparatus [4,13,18,19,49]

The potential capability of the flow-through CPC has been demonstrated on separations of various compounds ranging from small molecules to macromolecules with conventional two-phase solvent systems.

1. *DNP Amino Acids [4,13,18]*. A set of nine dinitrophenyl (DNP) amino acids were separated using a two-phase solvent system composed of chloroform/acetic acid/0.1 N HCl (2:2:1). The partition coefficients of the various DNP amino acids in this phase system varied from 100 to 0.18. Analytical and preparative chromatograms are shown in Fig. 54a and b, respectively. In each case, the nine DNP amino acids were well resolved in order of decreasing partition

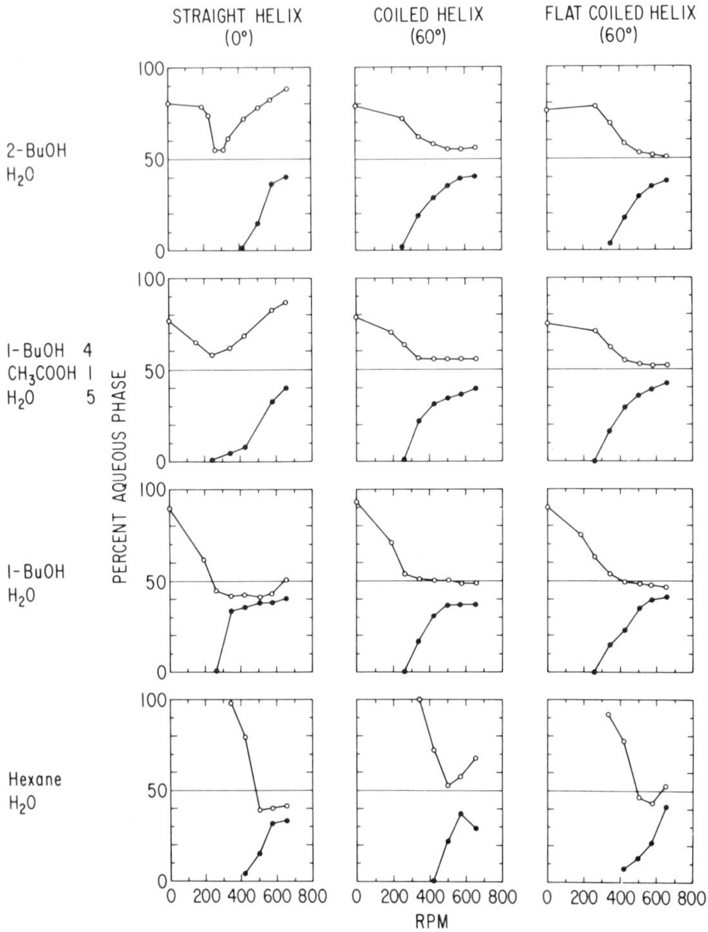

Fig. 53 The effect of column configuration on phase retention. Note the improvement in phase retentions shown for the top two phases in the coiled helix and flat coiled helix columns.

coefficient. When the efficiency of separation was estimated using Eq. (30), the analytical column yielded 10,000 to 3000 TP for the various peaks, while the preparative column yielded 2000 to 500 TP. The time required per transfer was 0.7–2.4 sec for the analytical column and 3–12 sec for the preparative column, compared to several minutes per transfer required for the countercurrent distribution method. To assess the relative performance of countercurrent chromatography and countercurrent distribution, the following rough comparison can be made. If two solutes are barely resolved in a countercurrent chromatography at efficiency of 100 TP, equivalent separation by countercurrent distribution would require migration of the solute peaks through 100 tubes using the same solvent system.

2. *Catecholamine Metabolites [4,49].* The analysis of catecholamine metabolites from urine is shown in Fig. 55. The metabolites, 3-methoxy-4-hydroxyphenylhydroxyacetic acid (VMA) and 3-methoxy-4-hydroxyphenylacetic acid (HVA), were well separated from each other. HVA was completely separated from other UV-absorbing components in urine. VMA elutes close to several unknown compounds which, however, do not interfere with colorimetric assays for VMA. Characteristic profiles are obtained for various neurosecretory tumors, and the technique does not require preliminary concentration or purification of the urine sample.

3. *Dipeptides [4,19].* A mixture of dipeptides was separated using a n-butanol/aqueous dichloroacetic acid solvent system as shown in Fig. 56. Note the resolution of sequential isomers. As previously mentioned, it is possible to change the composition of the mobile phase during elution. A gradient in dichloroacetic acid concentration (0–2%) was employed to achieve the separation of the same dipeptides with excellent results as shown in Fig. 57. Two features should be noted. First, the analysis time has been reduced almost by a factor of 3. Second, and of equal importance, the efficiency of the separation of the last peak has actually been increased by roughly the same factor. As a consequence, there is much less dilution of the later peaks. In fact, this phenomenon can be exploited to concentrate solutes if the solute has a high solubility in the initial stationary phase. This is illustrated in Fig. 58 for the same dipeptides. Note that although the resolution of the early peaks is degraded, the resolution of the later peaks is unaffected by a 33-fold increase in sample size.

4. *Insulin [4].* Figure 59 compares the partition of bovine insulin in the sec-butanol/aqueous dichloroacetic acid system by countercurrent distribution and by countercurrent chromatography. The countercurrent distribution data reported by Harfenist and Craig revealed a second component, identified as a deamidated insulin, which was not separable by either electrophoresis or ultracentrifugation at that time. Countercurrent chromatography shows a similar result. Because of the low interfacial tension of this phase system, it was necessary to use a flat coiled helix column rather than the usual straight helix. The column was a flat helix column wound from 90 m of 0.55-mm i.d. PTFE tubing, using the phase system: sec-butanol/dichloroacetic acid/water (100:3:100). A sample of 0.5 ml of 1% solution of bovine insulin in lower phase was separated at a speed of 700 rpm at a 30.7-cm radius and a flow rate of 6 ml/hr.

5. *Antibiotics.* In addition to the above application with the original apparatus, separations of various types of antibiotics have

(a)

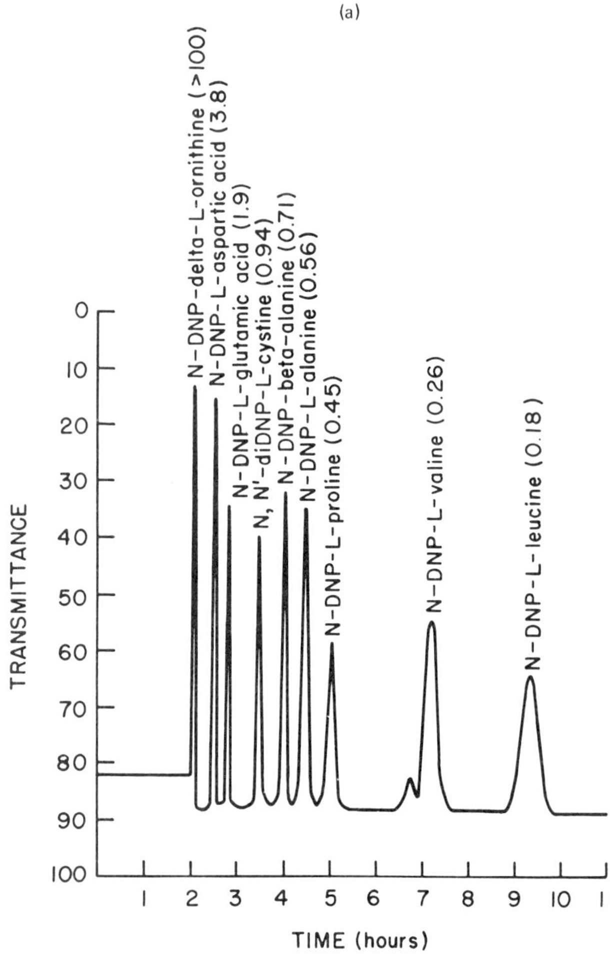

Fig. 54 Separation of DNP amino acids. (a) Analytical column: straight helix wound from 100 m of 0.30-mm i.d. PTFE tubing, helix diameter 5 mm, total capacity 8 ml. 10-μl sample contained about 1% (W/V) concentration of each component (where solubility permitted), dissolved in upper phase of solvent system. Speed 550 rpm, radius 30.7 cm, elution with upper phase at 2.4 ml/hr. Phase system was chloroform/acetic acid/0.1 N HCl (aqueous) (2:2:1). (b) Preparative column: straight helix wound from 100 m of 1.4-mm i.d. PTFE tubing, helix diameter 1 cm, total capacity 140 ml. Sample: 100 μl of sample solution used in (a) above, diluted with upper phase to 2 ml (sample size was adjusted to facilitate UV monitoring), speed 520 rpm, radius 8.6 cm, flow rate 60 ml/hr.

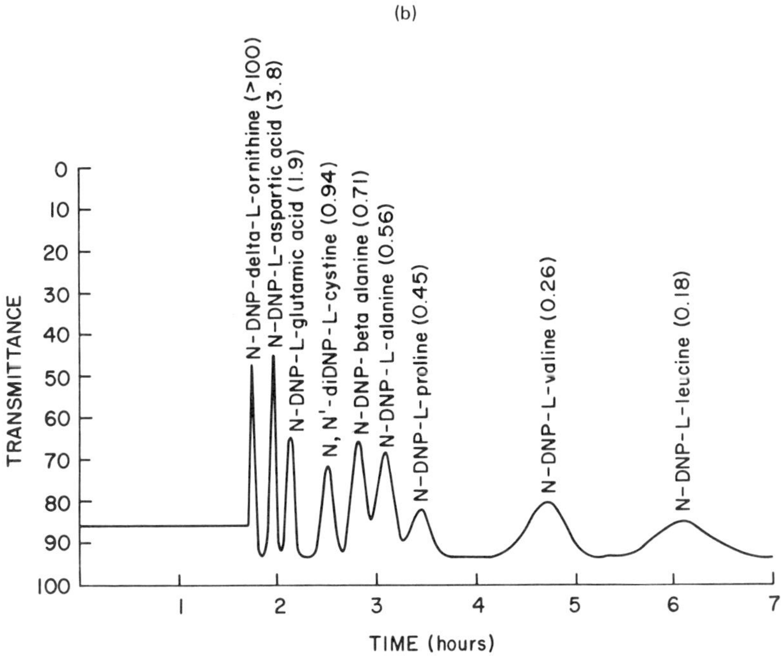

Fig. 54 (Continued)

been successfully performed by a scientific group at the National Institute for Medical Research, Mill Hill, London, using a flow-through CPC equipped with a self-balancing system. Candicidine and related polyene antibiotics including levorin, trichomycin, and hemycin were separated with a solvent system composed of chloroform/methanol/borate buffer (pH 8.2) (4:4:3) [50]. Separations of quinozaline antibiotics, including quinomycin C, echinomycin, and triostin A, were performed with two different solvent systems—trichloroethane/methanol/water (7:3:1) and n-heptane/ethyl acetate/acetone/water (1:1:3:1) [51].

6. *Miscellaneous*. The flow-through CPC has been found to be versatile and useful for other types of separations such as foam separation with gas-liquid systems [52] and elutriation of particles with single-phase solvents [53]. The foam separation was demonstrated on enrichment and stripping of rhodamine B present in a large volume of sample solution and also on a chromatographic separation of rhodamine B and Evans blue. In these experiments, sodium dodecyl sulfate was employed as a foam-producing agent to collect rhodamine B. Recently, the method has been remarkably improved

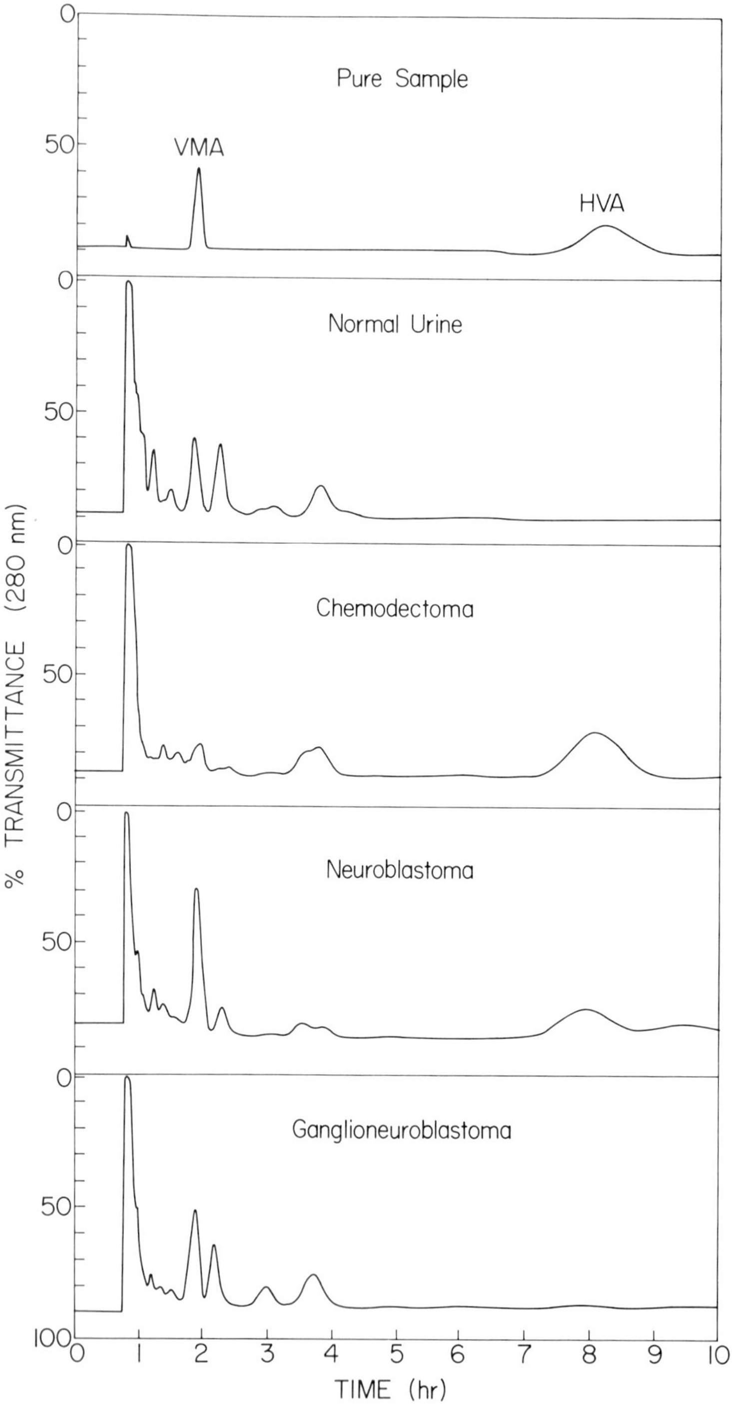

Pure Sample
VMA
HVA
Normal Urine
Chemodectoma
Neuroblastoma
Ganglioneuroblastoma
% TRANSMITTANCE (280 nm)
TIME (hr)
0
50
100
0
1
2
3
4
5
6
7
8
9
10

Fig. 55

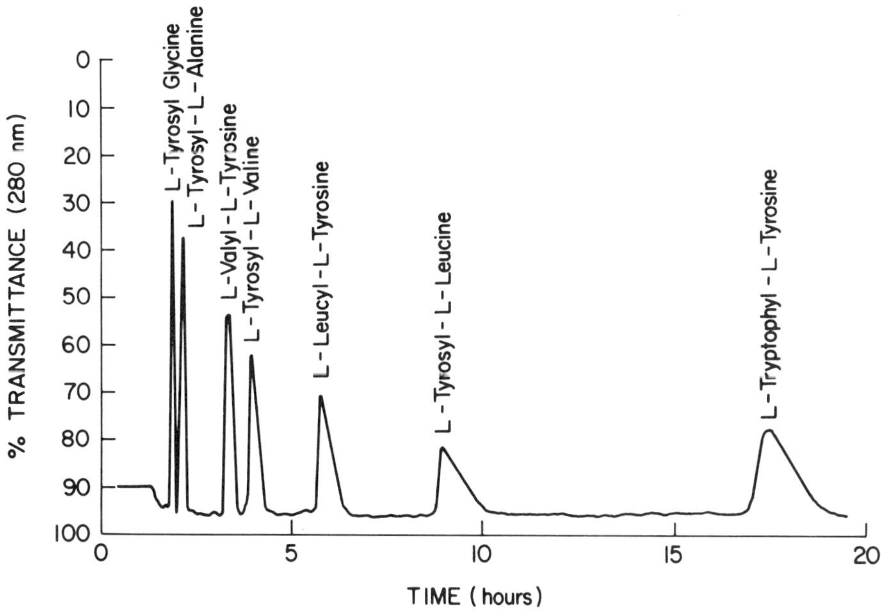

Fig. 56 Separation of dipeptides by isocratic elution. Solvent system: 1-butanol/1% aqueous dichloroacetic acid (1:1): Straight helix column wound from 140 m of 0.55-mm bore tubing. Speed 720 rpm, radius 20.2 cm, flow rate 12 ml/hr.

by utilizing the unilateral hydrodynamic equilibrium (see Fig. 16) which allows dual gas-liquid countercurrent flow through a coiled column as described later in detail. In particle elutriation, Sephadex particles were separated according to their size difference by eluting with acetone through the rotating coil.

B. The Elution Centrifuge [4,21,54]

The elution centrifuge is a member of the coil planet centrifuge in that the column holder undergoes a planetary motion, i.e., rotation around its own horizontal axis and revolution about the vertical axis

---

Fig. 55 Analysis of catecholamine metabolites from urine. Phase system: ethyl acetate/5% w/v NaCl, 10% v/v acetic acid in water (1:1). Lower phase used as mobile phase. Straight helix column wound from 60 m of 0.85-mm i.d. PTFE tubing. Sample: 0.9 ml of urine with 50 mg of NaCl and 0.1 ml of glacial acetic acid added. Speed 720 rpm, radius 20.2 cm, flow rate 24 ml/hr.

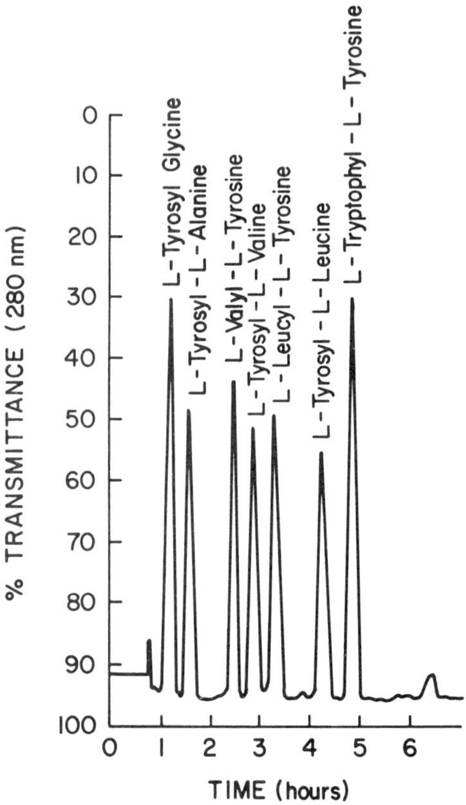

Fig. 57 Separation of dipeptides by gradient elution. Gradient elution: solvent system composed of 1-butanol/0.1 M aqueous ammonium formate with gradient in dichloroacetic acid concentration from 1% (v/v) to 0%, with lower phase used as mobile phase. Straight helix column wound from 70 m of 0.55-mm i.d. tubing. Speed 720 rpm, radius 20.2 cm, sample 0.3 ml, flow rate 6 ml/hr.

of the centrifuge to permit continuous elution without the use of rotating seals. The acceleration field produced by this planetary motion, however, is quite unique among other coil planet centrifuges as described previously (see Figs. 22–24). Under a proper design for the column holder and the column geometry, the scheme could yield a strong radial centrifugal force field somewhat similar to that in the conventional centrifuge system which can be utilized for helix countercurrent chromatography on both conventional and polymer phase systems. Capability of this analytical scheme has been demonstrated in separation of DNP amino acids on chloroform/

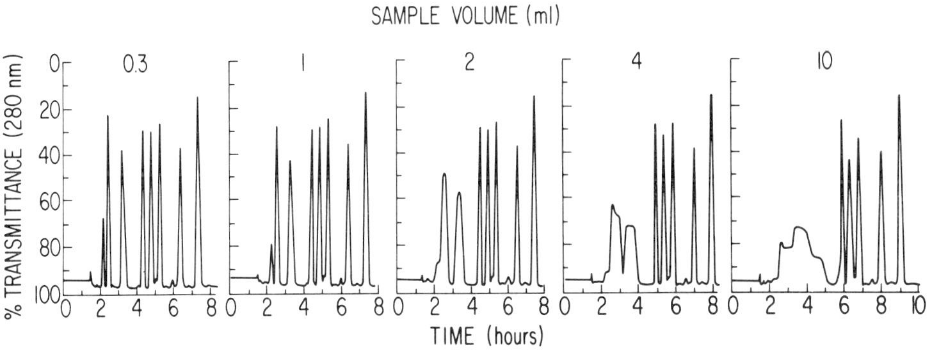

Fig. 58 Effect of sample volume on gradient elution separation of dipeptides. To demonstrate the solute-concentrating capability of the gradient elution technique, the sample was diluted with mobile phase before loading. The earlier peaks become affected at 2 ml. However, efficiency of separation of later peaks is not degraded even at 10 ml, where the solute concentration at the peak maximum is estimated to be 10 times as high as that in the applied sample solution.

acetic acid/0.1 N HCl (2:2:1) phase system and also in partition of polynucleotides on polymer phase systems.

Principle

The mechanism of the elution centrifuge is previously illustrated in Fig. 21a. A cylindrical holder containing the separation column is held in a horizontal position. Both feed and return tubes are led through the center of the holder and then supported tightly at the center of the apparatus by a guide tube fixed to the top frame of the centrifuge. The holder revolves around the central axis of the apparatus in the horizontal plane and simultaneously rotates about its own axis at the same angular velocity in the indicated direction. This synchronous rotation of the holder unwinds the twist of the inflow and outflow tubes caused by revolution, eliminating the need for rotating seals.

As described earlier, this synchronous rotation adds a peculiar effect to the centrifugal force field at the column holder. Simple mathematical analysis previously described (see Figs. 22–24) shows that every point on the column holder except for those located on the axis of rotation is subjected to a periodic fluctuation of the centrifugal force. Under some conditions, the acceleration vector can even change its sign to point toward the axis of the centrifuge. This rather startling result is produced when $\beta > 0.5$ or the radius of the

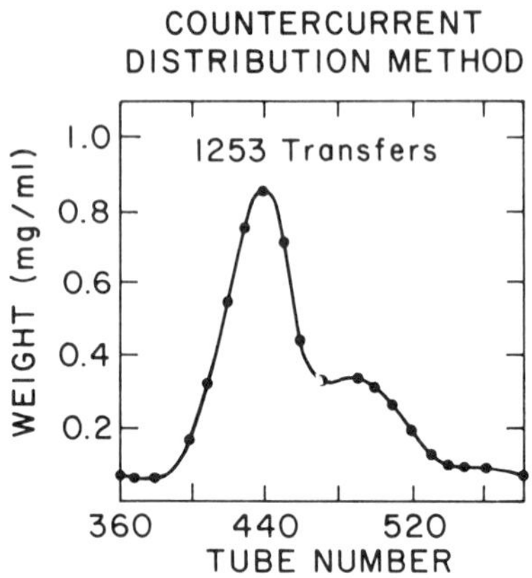

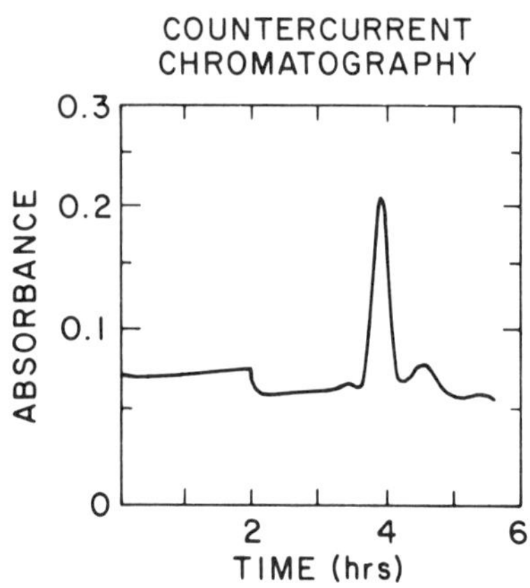

Fig. 59 Partition of bovine insulin. Comparison between countercurrent distribution method and countercurrent chromatography. Countercurrent chromatography using flat coiled helix column wound from 90 m of 0.55-mm i.d. PTFE tubing. Phase system: 2-butanol/dichloroacetic acid/water (1:0.03:1 v/v/v). Sample: 0.6 ml of 1% (w/v) solution of bovine insulin in lower phase, speed 700 rpm, radius 30.7 cm, flow rate 6 ml/hr.

coil holder is greater than or equal to one-half the radius of the centrifuge arm. For the smaller β, the centrifugal force vector gradually reduces its magnitude and oscillatory motion. Because of this centrifugal force field, the radius of the coil holder in our prototype was limited to values less than one-tenth the centrifuge arm or $\beta < 0.1$ to stabilize the retention of the stationary phase in the coil.

Apparatus

A test system was constructed by modifying a conventional centrifuge. The centrifuge head (Fig. 60A) was newly designed as a rectangular aluminum box (30 × 12.5 × 11.5 cm) with a center hole through the bottom plate to fit the motor shaft. A vertical septum was placed on each side parallel to and 10 cm from the end-plate. One the column side a thrust bearing (a) and an ordinary ball bearing (b) were embedded in the septum and the end-plate, respectively, to hold the column holder (3.6 cm i.d. and 25 cm long) (c) horizontally. A counterweight (d) made of an aluminum cylinder and equipped with an adjusting screw (e) at the periphery was mounted on the opposite side. The planetary motion of the holder was maintained by applying a pair of toothed pulleys (f) of the same size, one fixed to the holder and the other to the motor housing at the center of the apparatus. These pulleys were coupled by a toothed belt (g), which passed through a bottom hole of the centrifuge head and over a pair of toothed idler pulleys (h) mounted on the bottom plate of the centrifuge head. These idler pulleys function to change the direction of the running belt by 90° (Fig. 60B). The coiled column was mounted around the cylindrical support (shown in Fig. 60C with a twisted column used for DNP amino acid separation) which fitted tightly into the column holder (c). Both feed and return flow tubes (i) were passed through the center hole of the support and then led through the stationary guide tube (j) vertically projected down from the center of the top frame of the centrifugw. These tubes, if lubricated with grease and protected by a piece of silicone rubber tubing throughout the freely moving portion, are well preserved for many weeks of use. Revolutional speed of the apparatus is continuously adjustable up to 2000 rpm.

Three types of the columns (Fig. 61) were prepared and tested to determine the optimal column configurations for both conventional and polymer phase systems [4,54]. The straight helix column was prepared by winding a tube onto a rigid rod and mounting it parallel to the axis of the support. The twisted column was prepared by folding a tube in two and twisting along its length to make a ropelike configuration. The twisted tube was then wound around the column support. The coiled helix column was prepared by winding a tube onto a flexible tube which was coiled tightly around the support to make a double-helix configuration or a toroidal coil. Preliminary studies have indicated that the twisted column gives the highest efficiency for the conventional phase systems while the coiled helix column is most suitable for the polymer phase systems.

Applications of the Original Apparatus

1. *DNP amino acid separation on chloroform/acetic acid/0.1 N HCl (2:2:1) [21].* In order to demonstrate the capability of the

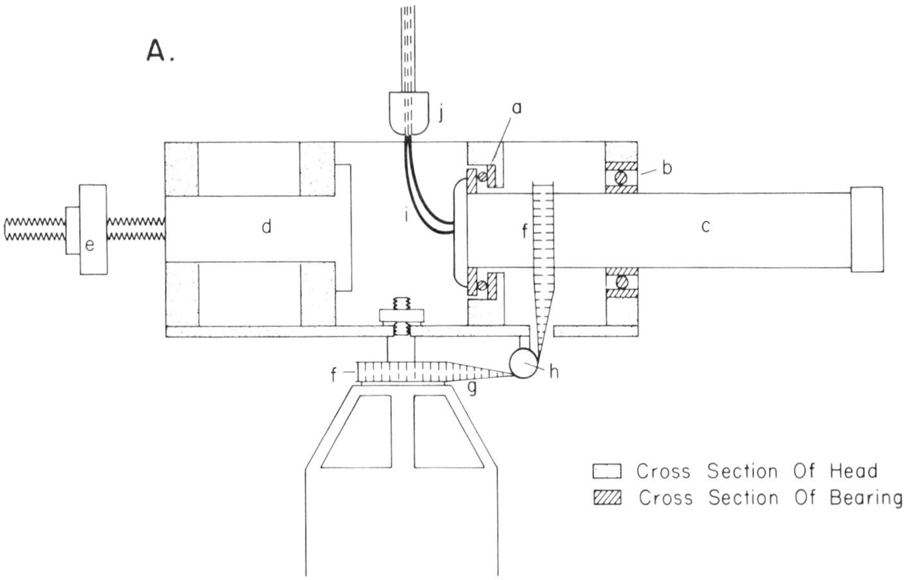

Fig. 60 Design of elution centrifuge: (A) column holder (c) horizontally supported by thrust bearing (a) and ordinary ball bearing (b) is counterbalanced by aluminum block (d) equipped with adjusting screw (e). A pair of toothed pulleys (f), one fixed to column holder and the other to motor housing at center, is coupled with toothed belt (g) through a pair of idler pulleys (h). Flow tubes (i) are supported by stationary guide pipe (j) projected down from top frame of centrifuge. (B) Belt scheme. (C) Column.

method, nine DNP amino acids were separated on a two-phase solvent system composed of chloroform, acetic acid, and 0.1 N hydrochloric acid (2:2:1) using a twisted column. The column was made from a 0.2-mm i.d., 60-m-long PTFE tubing by folding it in two and twisting together to make approximately 10,000 turns or a total of 20,000 turns. The twisted pair of tube was subsequently wound around a cylindrical support (20 cm long and 2.5 cm in diameter) to make three layers covering a length of 9 cm at the peripheral portion of the support (Fig. 60C) so that the β-value becomes less than 0.1 at all locations. The column was filled with the stationary lower phase and 3 μl of sample solution, containing each component at about 1% where solubility permitted, was introduced through the feed tube. The upper mobile phase was then pumped with a syringe driver at a rate of 120 μl/hr while the apparatus was spun at 700 rpm in which

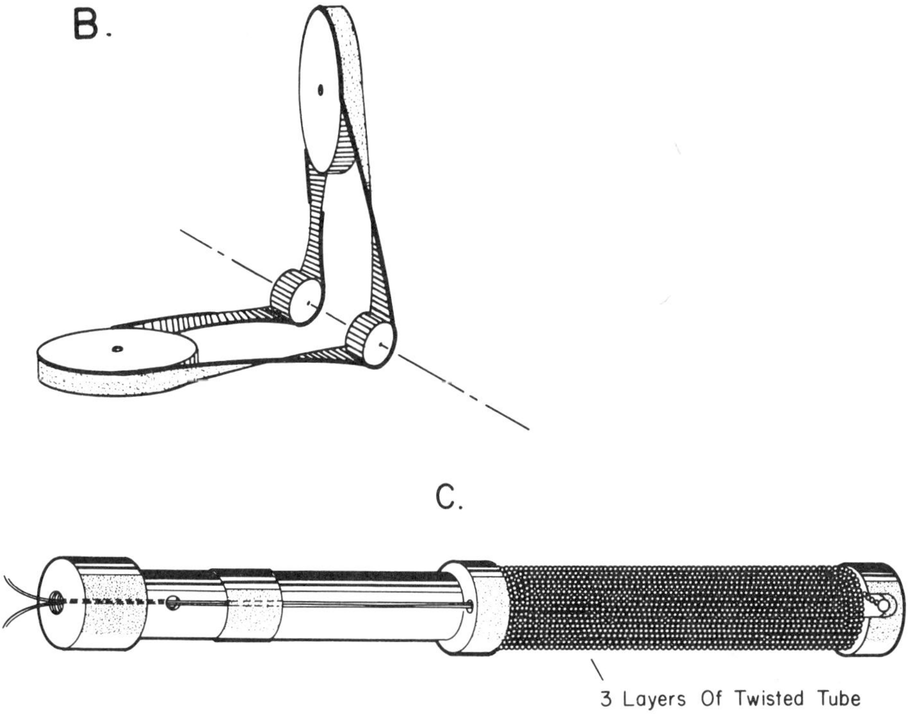

Fig. 60 (Continued)

the mean magnitude of the radial centrifugal force field ranged approximately from 80 to 125g. The eluate was monitored with an LKB Uvicord II at 280 nm. The separation was performed at room temperature.

Figure 62 shows the result of DNP amino acid separation. Nine DNP amino acids were eluted out within 54 hr at an efficiency ranging between 10,000 and 6500 TP estimated according to the gas chromatographic formula [Eq. (30)]. The efficiency increases with tube length, number of coil units in a given length of the tube, and slower flow rate as in the case of the toroidal coil centrifuges (see Sec. V). The applicable revolutional speed ranges from 600 to 1000 rpm. Either faster flow or slower revolutional speed tends to reduce the amount of the stationary phase retained in the column, while high revolutional speed produces an excessive hydrostatic pressure in the column. The fact that the present scheme yields a resolving power equivalent to that in other helix countercurrent chromatographic schemes clearly indicates that the fluctuation of the centrifugal force field described earlier has no significant effect on the separation, at least under the condition chosen for the present study ($\beta < 0.1$).

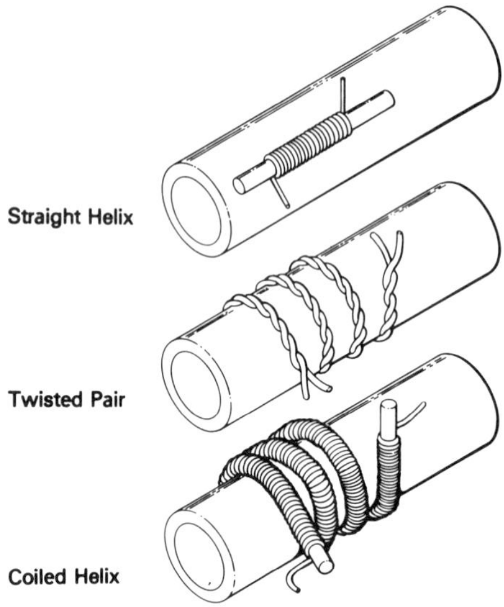

Fig. 61 Three types of column configuration used with the elution centrifuge.

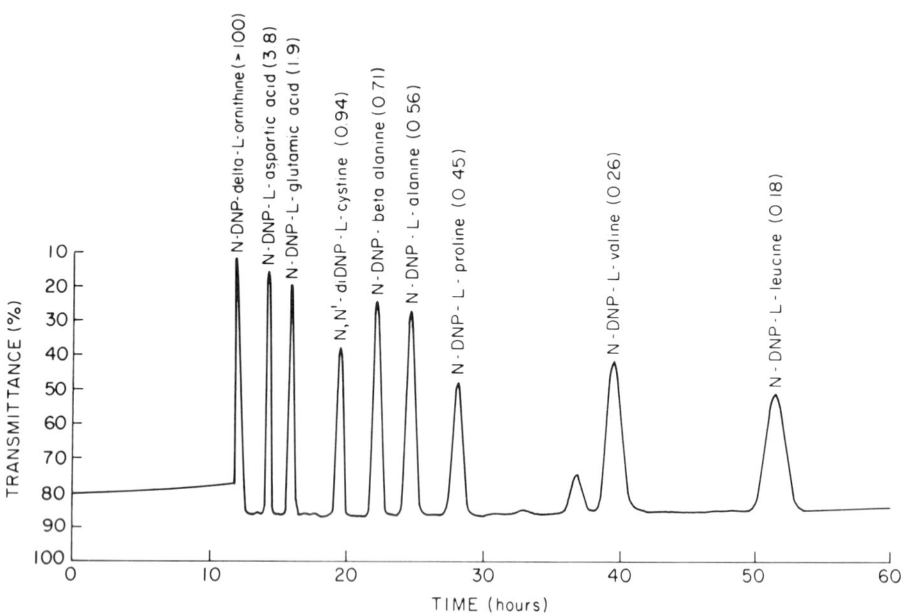

Fig. 62 Separation of DNP amino acids using the elution centrifuge. Column: twisted pair of 20,000 turns of 0.2-mm i.d., 60-m long. Phase system: chloroform/acetic acid/0.1 N HCl (2:2:1).

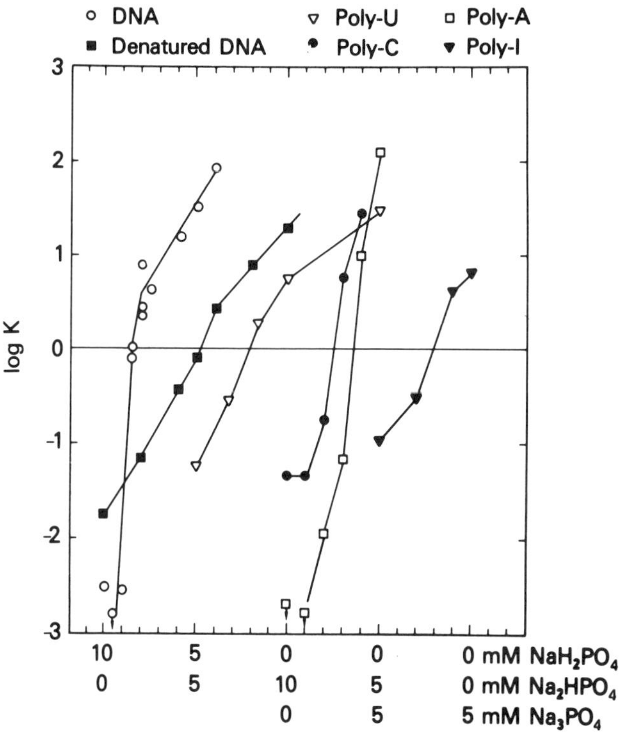

Fig. 63 Partition behavior of various polynucleotides in 4% (w/w) PEG 6000/5% (w/w) dextran 500 polymer phase systems with different sodium phosphate compositions. K: Partition coefficient.

2. *Partition of Polynucleotides on Polymer Phase System [4,54].* Partition behaviors of various polynucleotides on polymer phase systems have been reported by Albertsson [3] as illsutrated in Fig. 63, where the partition coefficients of various polynucleotides in a dextran/polyethylene glycol system are plotted as a function of different mixtures of sodium phosphate. We have utilized the above system to demonstrate the general capability of the elution centrifuge for performing CCC with polymer phase systems.

The polymer phase system used in our study consists of 5% w/w dextran T 500 (Pharmacia), 4% w/w polyethylene glycol 6000 (Union Carbide Corporation), and 10 mM sodium phosphate of various compositions. The phase mixture was gently degassed by suction and equilibrated at 18°C. The same temperature was maintained during the elution by cooling the apparatus. Four polynucleotides (Sigma)—polyuridylic acid (poly U), polycytidylic acid (poly C), polyadenylic acid (poly A), and polyinosinic acid (poly I)—were selected as samples. Figure 61 shows three column configurations examined. Each column was made of PTFE tubing 5 m long and 0.55 mm i.d. The straight helix column was prepared by winding a single piece of tubing onto four rigid cores (each 12 cm long and 3 mm in diameter) in series and symmetrically mounted around the column support parallel to the axis of the support. The twisted column was prepared by folding a tube in two and twisting it along its length to make a ropelike configuration. The twisted tube was then wound around the column support. The coiled helix column was made by winding a tube onto a flexible core, 50 cm long and 3 mm in diameter, which was in turn coiled around the column support. Each column was placed into the column holder at its periphery to minimize the effect of the Coriolis force previously analyzed.

Using these three types of columns retention of the polymer phase system (10 mM NaH_2PO_4) was measured under various revolutional speeds at a given flow rate of 2.8 ml/hr. The results are shown in Fig. 64 (top), where the percentage of the retained stationary phase volume relative to the total column capacity (open

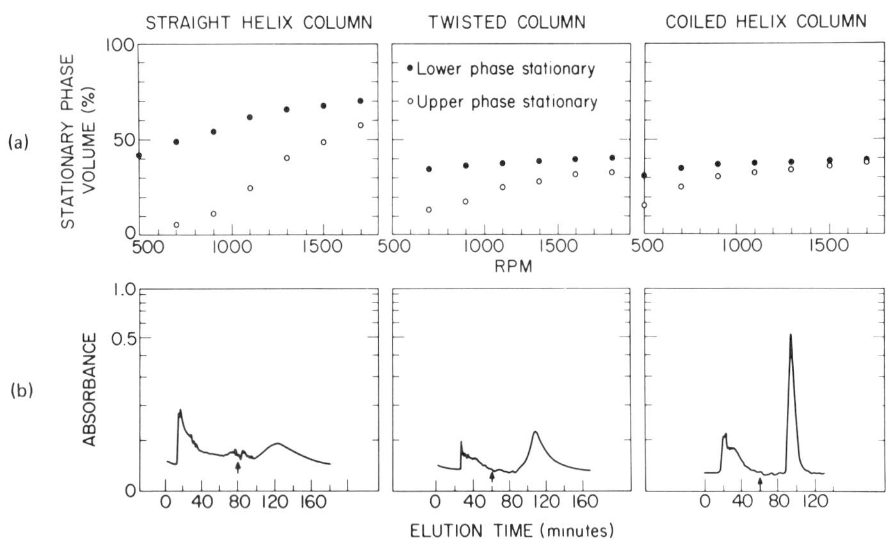

Fig. 64 Performance of three types of separation column (a) Retention of the stationary phase. All three columns show sufficient retention of more than 30% above 1000 rpm, if the lower phase is stationary. (b) Column efficiency examined on stepwise elution. The first elution eliminates impurities while the second elution, indicated by arrow, elutes the poly U sample. Sharpness of the peak obtained by the coiled helix column indicates the highest efficiency.

circles for the upper phase and solid circles for the lower phase) is plotted against the applied revolutional speed. All columns show a sufficient amount of retention (30% or more) of the lower phase at the revolutional speeds of over 1000 rpm (250g). Among three column configurations tested, the coiled helix column yielded the most stable retention for both phases while the highest retention was achieved from the straight helix column.

The partition efficiency of each column was evaluated on stepwise elution of poly U sample. The column was first filled with the stationary lower phase (10 mM NaH_2PO_4) and the sample was eluted with the upper phase (10 mM NaH_2PO_4) at 1500 rpm for 1 hr followed by the second elution with the upper phase (10 mM NaH_2PO_4:10 mM Na_2HPO_4 = 1:3) at a flow rate of 2.8 ml/hr. The eluate was continuously monitored with an LKB Uvicord II at 254 nm. Figure 64 (bottom) shows the results of the experiments. The first elution eliminates most of the impurities originally present in the sample, while the second elution (indicated by arrows) elutes poly U as a single peak. The sharpness of the elution peak obtained from the coiled helix column clearly indicates the highest efficiency of the column types tested.

To demonstrate the general capability of the method, the partition behavior of polynucleotides was studied with a coiled helix column (6 m long, 0.55 mm i.d., and 1.5 mm helix diameter) using both stepwise and gradient elution techniques. For the gradient elution, the use of 2-ml capacity mixing chamber provided an exponential gradient between the starting and ending phases. Each polynucleotide sample was dissolved in a phase mixture (10 mM NaH_2PO_4) at a concentration of 0.12 g%, and 0.1 ml (for stepwise elution) or 0.15 ml (gradient elution) was charged in each run. In order to eliminate impurities present in the sample solution, the elution was carried out in two steps: The sample was first isocratically eluted with the upper phase (10 mM NaH_2PO_4) for about 1 hr, eliminating most of the impurities while retaining the polynucleotide in the column. This was followed by the second elution with the upper phase either with given sodium phosphate compositions or in an exponential gradient mode between 10 mM NaH_2PO_4 and Na_3PO_4.

Figure 65 shows chromatograms of poly U, poly C, poly A, and poly I obtained by the second elution with the indicated phase compositions. In both stepwise and gradient elutions, the peaks from different polynucleotides are clearly resolved and eluted in the order of the partition coefficients determined from Fig. 63. The minor peaks observed in the gradient elutions presumably represent high molecular weight impurities.

These results indicate the feasibility of sharp separation of polynucleotides with polymer phase systems by means of the elution centrifuge method. With further improvements in apparatus and the use

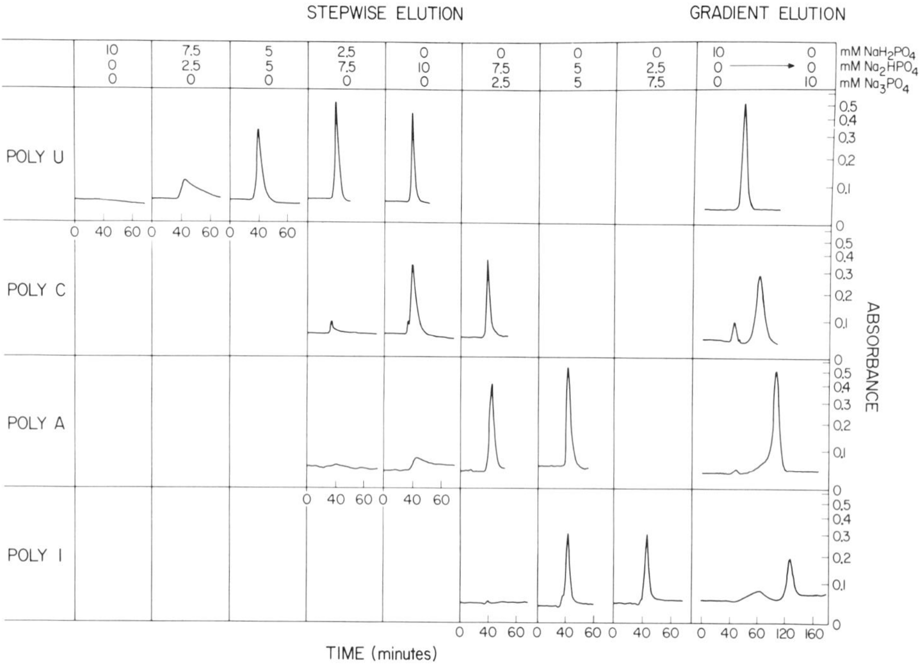

Fig. 65 Countercurrent chromatograms of polynucleotides with stepwise (left) and gradient (right) elutions, using the phase systems composed of 5% (w/w) dextran T-500, 4% (w/w) polyethylene glycol 6000, and 10 mM sodium phosphate of the indicated compositions.

of shallower gradient, separation of biological macromolecules having only small differences in partition coefficient may be feasible.

C. Angle Rotor Coil Planet Centrifuge (Scheme II) [20]

This apparatus is considered to be a hybrid between the previously described two types of CPCs, the flow-through CPC and the elution centrifuge, hence retaining the advantages of both parent schemes. The flow-through CPC with a parallel holder orientation (0°) to the centrifuge axis can yield a high partition efficiency in a short elution time by vigorous mixing of the two phases while it fails to retain low interfacial tension polymer phase systems which have a high tendency of emulsification. The elution centrifuge with a perpendicular holder orientation (90°) to the centrifuge axis provides a stable centrifugal force field for retention of polymer phase systems while the separation generally requires a long elution time at a low flow rate to overcome the increased mass transfer resistance. Since these contrasted features of the two schemes are apparently derived from the different orientation of the holder, the optimum holder angle must lie between 0 and 90°, which would provide universal application of the two-phase solvent systems with a high partition efficiency.

Search for the optimum holder angle has been successfully performed with two types of angle rotors, the first rotor with a fixed holder angle (45°) and the second rotor with an adjustable angle holder (30, 20, and 10°). The optimum holder angle was found at 30° from the centrifuge axis with the second rotor.

Design of the Adjustable Angle Rotor (30, 20, and 10°)

Figure 66 shows the design of the adjustable angle rotor fixed at the 30° position. A pair of bearing blocks (a), holding the aluminum column holder (b) (32 mm i.d. and 25 cm long) on each side of the rotor, is sandwiched between a pair of rectangular aluminum plates (c) each equipped at the middle with a pin (d) that fits into the pinhole (d) made in the side plate (e) of the rotor. This arrangement allows the column holder blocks to be rotated around the pins without disturbing the bearing alignment and then bolted to the side plates at a selected angle. A toothed pulley (f) mounted on the column holder is coupled to a stationary pulley (g) of equal diameter on the axis of the centrifuge through a toothed belt (h) by the aid of a pair of idler pulleys (i) mounted on the bottom plate (j) of the rotor. This coupling establishes counterrotation of the column holder to prevent twisting of flow tubes (o). The positions of the pulley (f) on the column holder and idler pulleys (i) are also adjusted for each angle to provide an adequate tension to the running belt (h).

In the light of test results obtained by the 45° angle rotor, several design items were improved. Properly sealed ball bearings (k) were employed, which can hold lubricant much longer and permit easy lubrication with a syringe needle inserted between the inner ring of the bearing and the plastic seal. Between the column holder and the inner race of each bearing, a piece of bakelite tube (l) was tightly placed to minimize heat conduction and also prevent rotation of the column holder against the bearings. In order to cool the column, several side holes (m) were made at the top and bottom of the column holder to allow air flow inside the holder during the run. A cylindrical column support (n) made of a delrin rod was tightly secured at the top and bottom of the holder to prevent vibration during operation.

Two types of column configurations were employed. The straight helix column was prepared by winding PTFE tubing onto a rigid plastic rod to make a column unit. Multiple column units

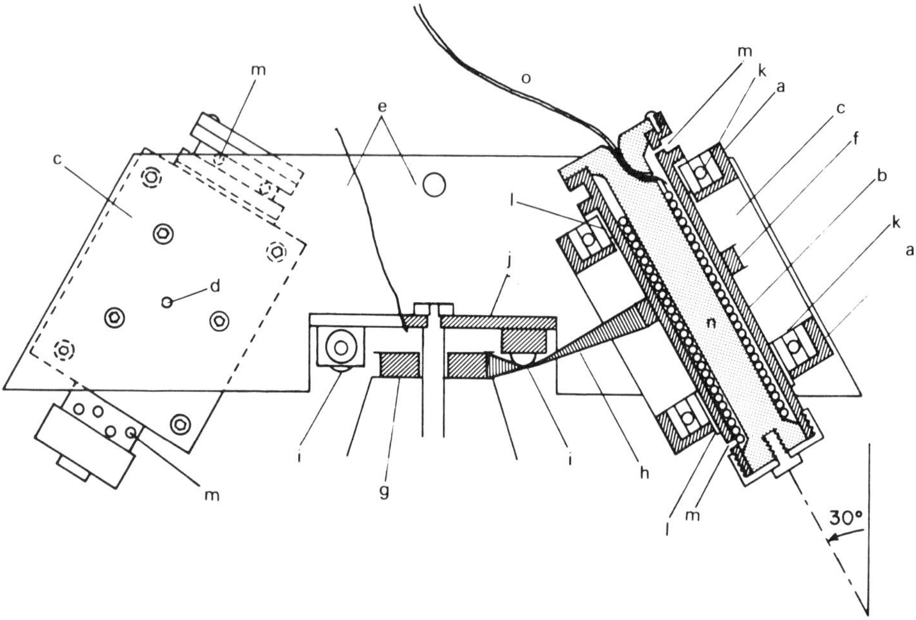

Fig. 66 Diagram of the adjustable angle rotor fixed at the 30° position. (a) bearing block, (b) column holder, (c) side plate of column holder block, (d) pin and pin hole in side plate, (e) toothed belt, (i) idler pulley, (j) bottom plate of rotor, (k) ball bearing, (l) bakelite pipe, (m) side hole in column holder, (n) column support carrying a coiled helix column, (o) feed and return tubes. Rotor angle is adjusted to 30, 20, or 10° from the vertical by rotating each column holder block around the pin hole (d) and bolting to the side plates (e). Counterrotation of the column holder is established by coupling pulleys (f) and (g) with a toothed belt (h) to prevent twisting of the flow tubes (o). Idler pulleys (i) change direction of the running belt (h).

were interconnected in series and tightly mounted in parallel onto the column support. The coiled helix column (shown in Fig. 66) was prepared by winding PTFE tubing onto a flexible core which in turn was tightly coiled around the column support. In order to minimize any adverse effects of the Coriolis force [21] previously described, these columns were placed within 1.3 cm from the central axis of the column holder with a mean revolutional radius of 21 cm. Both feed and return tubes (o) were passed through the center hole of the column support and then held tightly at the center of the upper member of the centrifuge. The flow tubes served several months of operation, when protected with a piece of greased silicone rubber tube at each supported end.

Performance of the apparatus has been evaluated at revolutional speeds ranging between 500 and 1300 rpm during a period of 6 months. Leakage of lubricant from the bearing seal was much reduced and weekly lubrication seems satisfactory. The column temperature became close to the controlled ambient temperature with a maximum rise of no more than a few degrees in centigrade after several hours of operation.

Performance of the Apparatus with Various Solvent Systems

Capability of the angle rotor (30°) has been evaluated on separations of biological materials with a variety of two-phase solvent systems of different physical properties. In order to demonstrate analytical potential of the method, a fine coiled helix column, 0.3 mm i.d. and 40 m long with a helix diameter of 1.2 mm, was used throughout unless otherwise specified. In each separation, the column was filled with the stationary phase and a sample solution was introduced through the feed tube followed by the elution with the mobile phase using a syringe drive (Harvard Apparatus) at the indicated flow rate while the rotor was run at the indicated revolutional speed. Separations were performed at 25°C unless otherwise indicated. Eluate was monitored through a UV monitor (LKB Uvicord II) at either 276 or 254 nm except for the polymer phase system. In separation of the colony-stimulating factor with the polymer phase system, eluted fractions were added to human bone marrow cell culture for colony count, while the total protein contents were estimated by turbidity measurement at 300 nm in 1 M perchloric acid solution.

1. *Chloroform/acetic acid/0.1 N HCl (2:2:1) [55].* This phase system has a moderately low interfacial tension, low viscosity, and a great density difference between two phases, which is ideal for countercurrent chromatography. Figure 67a shows a chromatogram of nine dinitrophenyl (DNP) amino acids at a flow rate of 0.5 ml/hr at 500 rpm, using the upper aqueous phase as a mobile phase. Sample volume was 10 μl containing each component in the lower phase at about 1% where solubility permits. All components were eluted out within 17 hr. Figure 67b shows a similar chromatogram obtained at a higher flow rate of 1.5 ml/hr at 1000 rpm. The separation was completed within 6 hr without serious loss of the resolution.

2. *Ethyl acetate/10% acetic acid, 5% NaCl (1:1) [49].* This phase system has a high interfacial tension and low viscosity and tends to produce a plug flow in a fine-bore column in the vertical rotor [19]. However, the present method overcomes the problem of

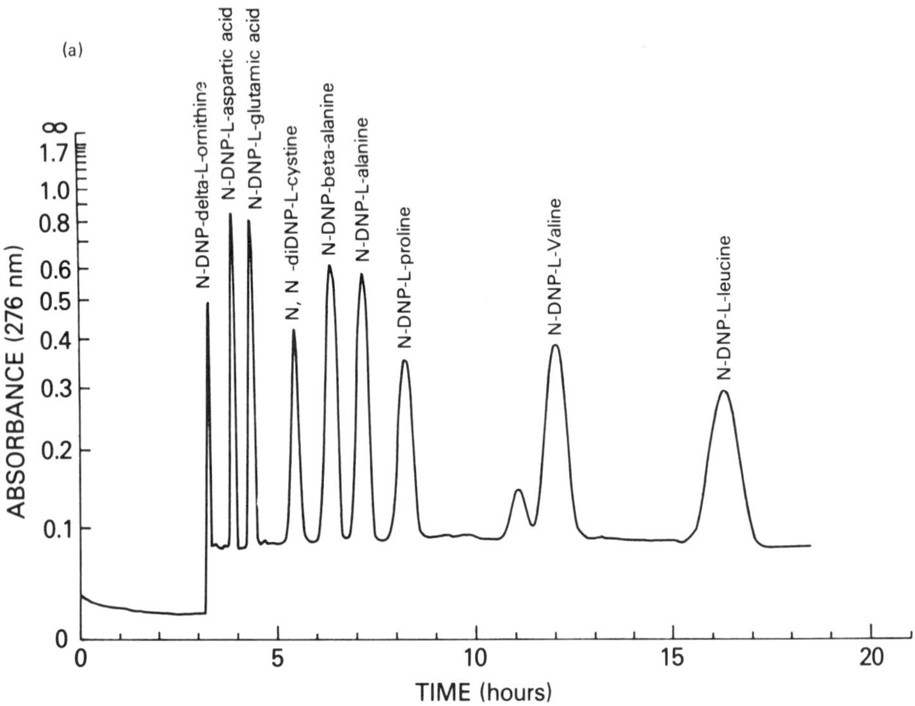

Fig. 67 Chromatograms of DNP amino acids on chloroform/acetic acid/0.1 N HCl (2:2:1) at different flow rates. (a) Elution at 0.5 ml/hr. (b) Elution at 1.5 ml/hr.

plug flow and five catecholamine metabolites were well separated and eluted out in 13 hr (Fig. 68a). Applied sample volume was 20 μl, containing each component at 0.5 g/100 ml in the lower phase. The lower phase was used as the mobile phase at a flow rate of 1 ml/hr at 1200 rpm. The same solvent system was applied to detect urinary catecholamine metabolites from patients (Fig. 68b). Each urine sample was prepared by adding NaCl (5%) and acetic acid (10%) according to the phase composition and 100 μl loaded, while the pure sample was dissolved in the lower phase at 0.5 g/100 ml for each component and 20 μl charged. Abnormal levels of VMA (4-hydroxy-3-methoxymandelic acid) and HVA (4-hydroxy-3-methoxyphenylacetic acid) were detected from 100 μl of urine. VMA was chromatographed with several other compounds but none of them interfered with any of the colorimetric assays for VMA. HVA was completely separated from other compounds. These results are quite comparable with

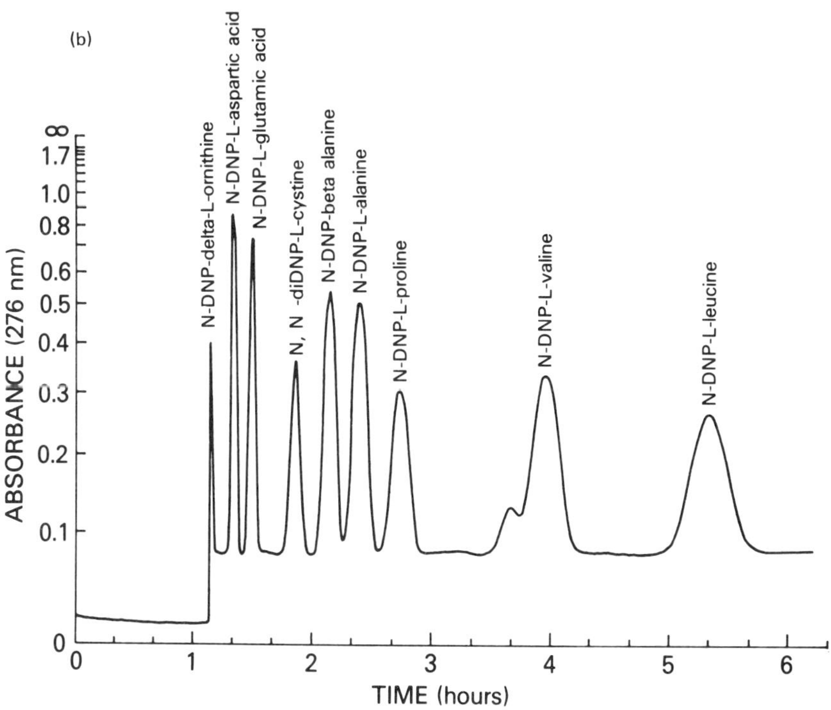

Fig. 67 (Continued)

those obtained on a ninefold sample size with the flow-through CPC illustrated in Fig. 55 [4].

3. *N-BuOH/aqueous systems.* These two-phase systems have moderately low interfacial tension and relatively high viscosity, and are extremely useful for partition of various biological compounds. Figure 69a shows a chromatogram of purines and pyrimidines on a two-phase system composed of n-BuOH/1 M potassium phosphate (pH 6.5) at 1:1 volume ratio [56], obtained at a flow rate of 0.5 ml/hr at 1200 rpm. Sample was prepared by dissolving each component in the upper phase at 0.1 g/100 ml and 10 μl charged. Since partition coefficients of these compounds favor the lower phase, the upper organic phase was used as a mobile phase for better resolution of peaks. Thus, the present method permits a choice of the mobile phase according to partition coefficients of the solutes of interest.

Figure 69b shows a chromatogram of a set of dipeptides using a gradient elution technique. A linear gradient was applied between 1% and 0% dichloroacetic acid on the base of n-BuOH and 0.1 M ammonium

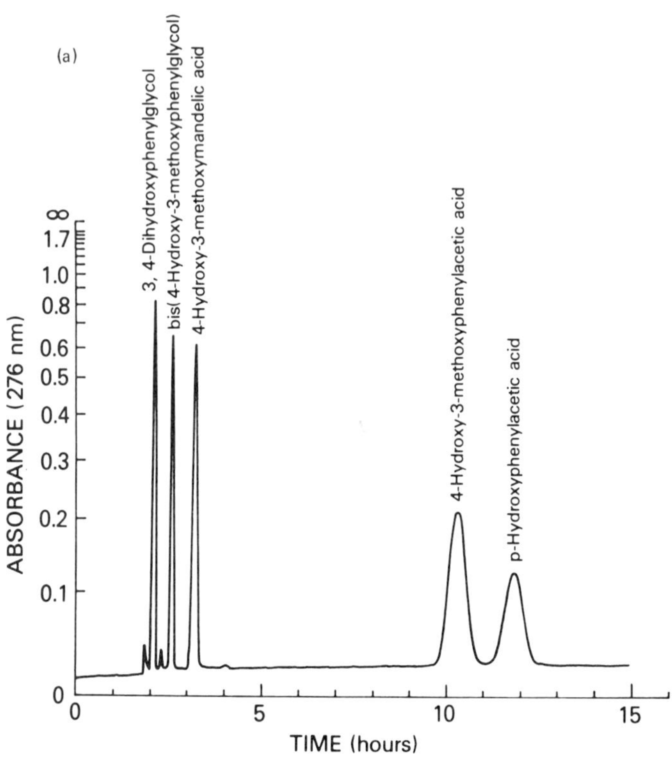

Fig. 68 Chromatogram of catecholamine metabolites on ethyl acetate/ 10% acetic acid, 5% NaCl (1:1). (a) Mixture of five samples. (b) Urine samples: abnormal levels of VMA (4-hydroxy-3-methoxymandelic acid) and HVA (4-hydroxy-3-methoxyphenylacetic acid) were detected from 100 μl of urine sample.

formate at 1:1 volume ratio [19] using the lower phase as a mobile phase. Sample size was 100 μl containing each component at 0.2 g/100 ml in the lower phase. All components were separated and eluted out in 6 hr at a flow rate of 1 ml/hr at 1200 rpm. Two pairs of isomers, L-valyl-L-tyrosine vs. L-tyrosyl-L-valine and L-leucyl-L-tyrosine vs. L-tyrosyl-L-leucine, were completely separated in this method.

4. *Sec-BuOH/3% aqueous dichloroacetic acid (1:1) [4].* This two-phase system is characterized by low interfacial tension and high viscosity with a high tendency of emulsification. Figure 70 shows a countercurrent chromatogram of bovine insulin compared with the partition data by countercurrent distribution method on a similar

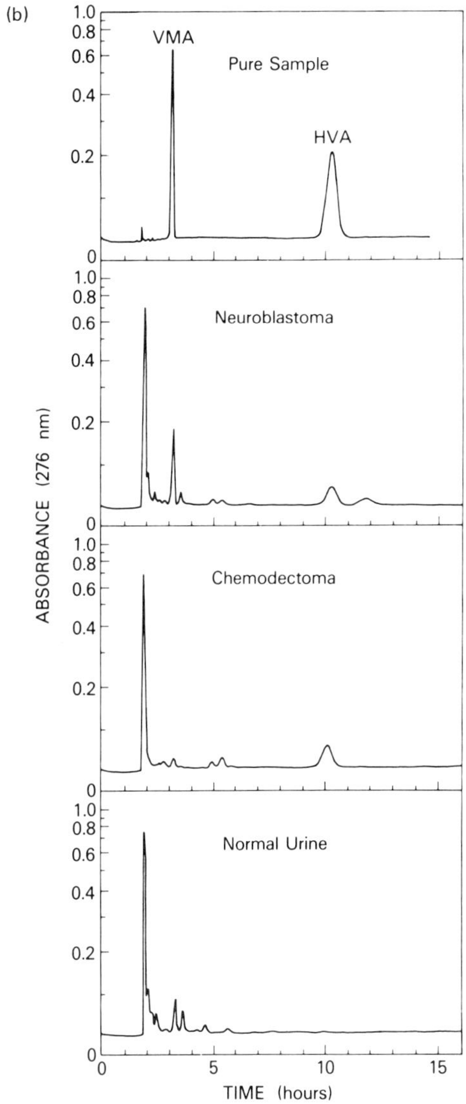

Fig. 68 (Continued)

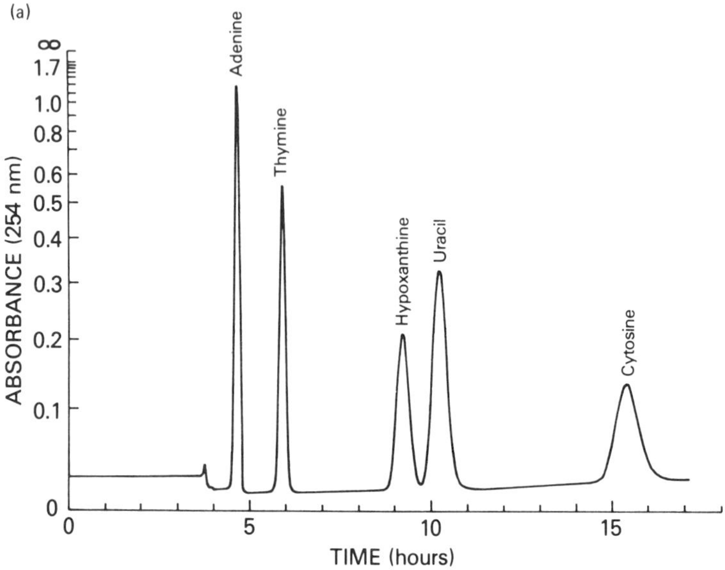

Fig. 69 (a) Chromatogram of purines and pyrimidines on n-BuOH/1 M potassium phosphate (1:1). The upper organic phase was used as the mobile phase. (b) Chromatogram of a set of dipeptides using a gradient elution. A linear gradient of 1–0% dichloroacetic acid was applied on the base of n-BuOH/0.1 M ammonium formate using the lower phase as the mobile phase.

phase system reported by Harfenist and Craig [57]. The CCD revealed a second component, identified as a deamidated form of insulin which was not separable by either electrophoresis or ultracentrifugation. Countercurrent chromatography completely separated the second component from the main peak. Applied sample size was 1 mg dissolved in 50 μl of the lower phase and the separation was performed at a flow rate of 1 ml/hr at 1200 rpm. Chromatogram of bovine insulin obtained with the angle rotor CPC shows substantially improved peak resolution compared with the result obtained with the flow-through CPC (see Fig. 59).

5. *Polymer phase system [3, 58]*. In general, polymer phase systems have extremely low interfacial tension, high viscosity, and small density difference between the phases and require a considerable time for phase separation. Because of the aqueous/aqueous phase composition, application to the conventional chromatography becomes difficult and therefore the partition relies on a time-consuming countercurrent distribution method. In order to demonstrate appli-

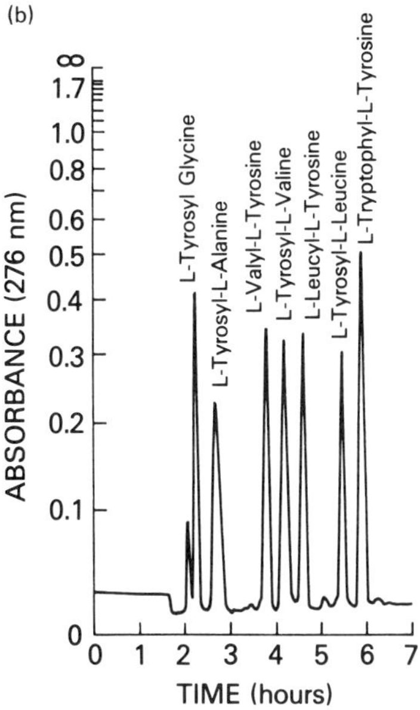

Fig. 69 (Continued)

cability of polymer phase systems to the present method, preliminary studies were performed on separation of colony-stimulating factor (CSF) [59] from human embryonic kidney supernatant using a charged poly(ethylene glycol) phase system [12] composed of 6% trimethylaminopolyethylene glycol (supplied by courtesy of Dr. G. Johansson), 6% dextran T 500, and 0.05 M sodium phosphate.

The pH range of the polymer phase system suitable for partition of CSF was determined on test tube experiments and the toxicity of the upper and lower phases was ruled out on bone marrow cell culture. Subsequently, the separation was performed by applying an exponential gradient elution between the starting (pH 4.4) and ending (pH 6.3) media, using the upper phase as a mobile phase. Two-milliliter human embryonic kidney supernatant (Flow Laboratories) was lyophilized and dissolved in 2 ml of the upper phase of the starting medium. A coiled helix separation column, 0.55 mm i.d., 18 m long, and 4 ml capacity, was first filled with the lower phase of the starting medium and then 1.2 ml of the above sample solution was introduced at a rate of 0.6 ml/hr at 1300 rpm, followed by gradient elution for 15 hr at 18°C. Thirty fractions, each 0.3 ml, were obtained during the run and then 10 fractions were collected

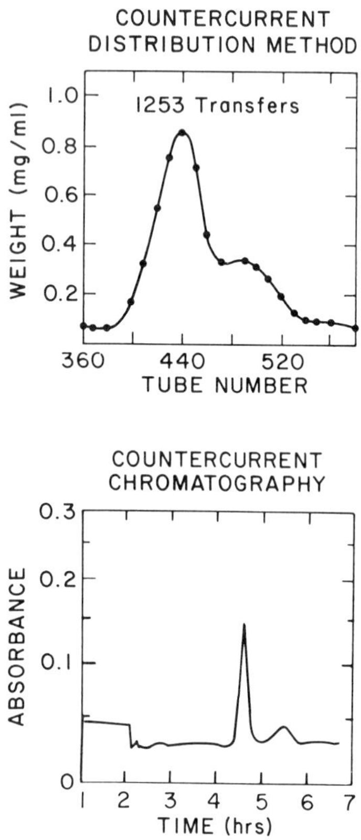

Fig. 70 Partition of bovine insulin by countercurrent distribution method and countercurrent chromatography. Countercurrent distribution (top) demonstrates a second component which is hardly separable by other separation methods. Countercurrent chromatography (bottom) completely separates the second component from the main peak.

from the column. These fractions were added to human bone marrow cell culture for evaluation of CSF activity by colony count, while the total protein was measured by turbidity in 1 M perchloric acid solution at 300 nm with a Beckman spectrophotometer.

Figure 71 shows the results of separation. CSF was fairly well separated from the rest of the proteins. The CSF activity at the peak was more than twice that of the original sample solution.

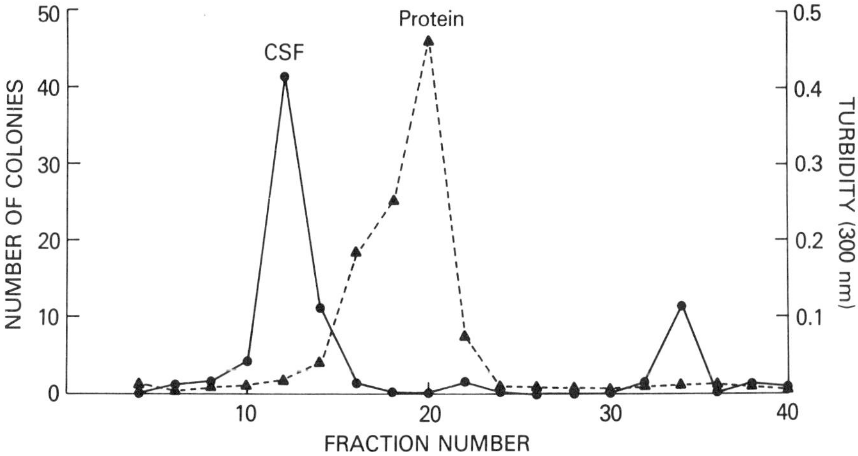

Fig. 71 Chromatogram of CSF (colony-stimulating factor) from human embryonic kidney supernatant using a charged polyethylene glycol polymer phase system. CSF (solid line) is fairly well separated from the total protein (dotted line).

VII. SYNCHRONOUS COIL PLANET CENTRIFUGE: RECENT INSTRUMENTATION (SCHEMES IV, IV + I)

Since the late 1970s the CCC technology has gained a remarkable advance by introducing a variety of highly efficient CCC schemes. All these schemes have been derived from a particular type of the coil planet centrifuge which provides an epicyclic mode of synchronous planetary motion (scheme IV) to produce a complex centrifugal force field as illustrated in Figs. 18–20. Series of hydrodynamic studies have shown that the system can yield extremely high partition efficiency in both analytical and preparative separations provided that the proper choice of column configuration is made with suitable orientation and location of the column on the holder. Consequently, the horizontal flow-through CPC has become one of the most useful CCC apparatuses intensively used in research laboratories.

Recent discovery of the unilateral hydrodynamic equilibrium system (Fig. 16) has led to the development of high-speed CCC that enables high peak resolution and speedy separation comparable to that of preparative HPLC. Further development of this system has introduced the dual countercurrent system in which the two phases literally undergo countercurrent flow through a coiled tube. The method has been successfully applied to foam separation in our

(a)

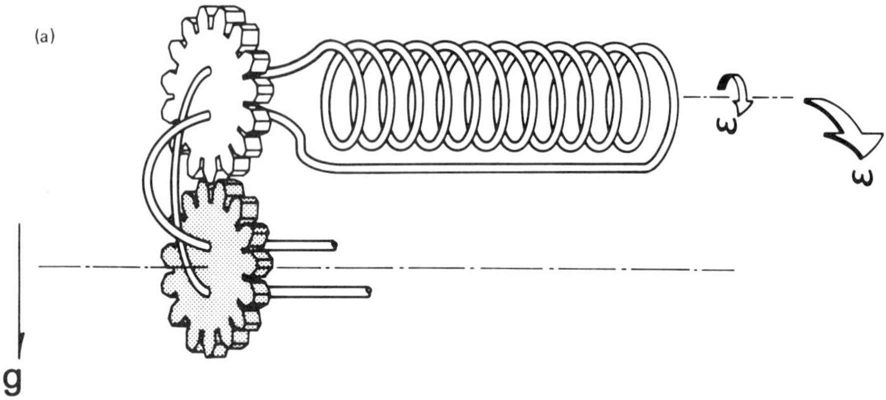

Fig. 72 Principle of the horizontal flow-through coil planet centrifuge. (a) The helical column is horizontally supported by a mobile gear that rolls around an identical stationary gear (shaded) mounted at the central axis of the centrifuge. Rolling motion of the mobile gear around the stationary gear produces a planetary motion of the helical column, a revolution about the central axis of the apparatus, and rotation about its own axis at the same angular velocity as indicated by a pair of arrows. (b) Successive positions of the black and white flow tubes during one complete revolution of the mobile gear. This planetary motion prevents twisting the flow tubes. Note that both tubes pass each other around 45° and 225°. A black dot on the mobile gear helps to visualize the motion of the helical column which rotates twice with respect to the gravitational field and once with respect to the centrifugal force field produced by revolution.

laboratory. Because of the distinctive performance, these most advanced forms of CCC based on the unilateral hydrodynamic equilibrium system are discretely dealt with under the heading "High-Performance CCC" (Sec. X) later in this chapter.

A. The Horizontal Flow-Through Coil Planet Centrifuge [26,27]

Principle and Design of the Original Apparatus

The principle of the horizontal flow-through CPC is illustrated in Fig. 72a. The separation column is a helix supported horizontally and fixed to a gear that rolls around a stationary gear (shaded) mounted on the central axis of the centrifuge. The feed and return flow tubes of the helix are led through the mobile gear and then tightly

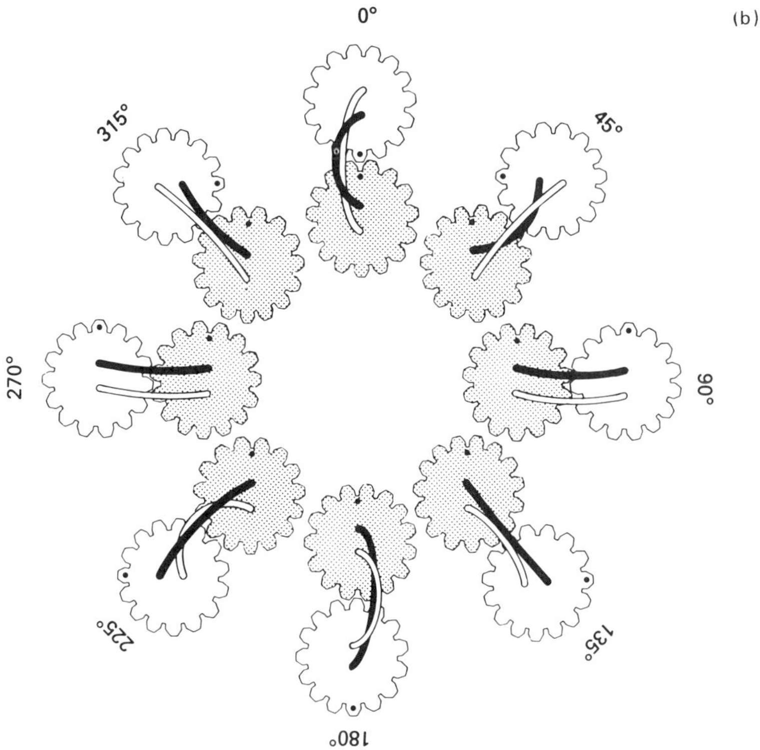

Fig. 72 (Continued)

supported by the stationary gear as illustrated. Rotation of the mobile gear around the stationary gear produces a planetary motion of the helix, i.e., revolution around the central axis of the apparatus and rotation about its own axis at the same angular velocity. Thus, during one complete revolution of the mobile gear, the helix rotates twice with respect to the gravity and once with respect to the centrifugal force field introduced by revolution.

This planetary motion prevents twisting of the flow tubes as they rotate with the mobile gear. Figure 72b illustrates motion of the flow tubes (white and black) during one revolution of the mobile gear around the stationary gear (shaded). Note that the black tube is pulled under the white tube between 0 and 45° while the white tube similarly goes under the black tube between 180 and 225°. A dot on the mobile gear helps to visualize the planetary motion of the helix connected to the gear.

The unique capability of the present scheme derives from the fact that the helical column rotates with respect to both gravitational and centrifugal force fields. When a large-bore preparative column is eluted with an extremely low interfacial tension phase system, a slow rotation of the column in the gravitational field avoids emulsification of the phases which would result in carryover of the stationary phase. For a fine-bore analytical column, a high revolutional speed can create a strong centrifugal force field that allows the two phases to counterflow through a narrow opening of the column without plug flow as in the flow-through coil planet centrifuge previously described.

The design of the apparatus is illustrated in Fig. 73. The rotary frame of the centrifuge consists of a pair of aluminum plates rigidly linked and driven by the motor (Electro-Craft Co.) around the hollow stationary shaft fixed at the central axis of the centrifuge. The frame holds a pair of rotary shafts, 54 cm long and 1.9 cm in diameter, symmetriclaly spaced at a distance of 15 cm from the center of the apparatus. Each rotary shaft is equipped with a mobile gear which engages an identical stationary gear fixed on the stationary shaft. In order to secure the remote end (right) of the stationary shaft, a short coupling pipe is extended from the aluminum plate of the rotary frame and supported through a bearing by a standing

Fig. 73 Photograph of the original horizontal flow-through CPC.

support. The helical column can be mounted on either side of the rotary shaft and a counterweight is placed on the other side to obtain a proper balancing of the centrifuge. The revolutional speed of the test system is continuously adjustable from 0 to 300 rpm.

The coiled separation column was prepared from 2.6-mm i.d. PTFE tubing with 0.4- to 0.5-mm wall thickness. The tubing was wound onto an aluminum pipe, 48 cm long and 1.25 cm o.d. with 1-mm wall thickness, to make a short column or column unit. One column unit has approximately 100 helical turns with a total capacity of about 25 ml. A long column for preparative scale separations was prepared by connecting 10 column units in series with PTFE tubes 0.45 mm i.d. and 15 cm in length. Application of these small tubes appeared to be necessary to prevent pulsative surging flow of the two-phase solvent system under a strong centrifugal force field which would cause unnecessary broadening of sample bands. The column was mounted as close as possible by using fuse holders symmetrically arranged around the rotary shaft to clamp the ends of each column unit. Then the column was secured in place by wrapping fiberglass tape in several places (Fig. 73). The flow tubes (0.85 mm i.d. PTFE) of the helical column were first passed through the center hole of the rotary shaft and then through the side hole of the coupling pipe to lead into the opening of the stationary shaft. The moving portion of the flow tubes was lubricated with grease and protected with a piece of silicone rubber tube at each supporting point. The tubes, if properly protected, can maintain their integrity over several months of use.

Performance of the Apparatus

Partition efficiency of the apparatus was studied with a single-column unit under various operational conditions; the results are summarized in Fig. 74a.

Figure 74a shows separation of DNP-L-glutamic acid and DNP-L-alanine on the two-phase system composed of chloroform/acetic acid/ 0.1 N HCl (2:2:1). In each separation the partition efficiency is easily estimated by measuring the height of the trough between the two peaks. These diagrams clearly demonstrate the presence of two

---

Fig. 74 Effects of revolutional speed and flow rate on separation of two selected samples on typical two-phase solvent systems. (a) Separation of DNP-L-glutamic acid in DNP-L-alanine on chloroform/ acetic acid/0.1 N HCl (2:2:1) phase system. (b) Separation of L-valyl-L-tyrosine and L-tryptophyl-L-tyrosine on n-butanol/acetic acid/water (4:1:5) phase system. In both (a) and (b), two optimum revolutional speeds are present, one around 10–25 rpm and the other 300 rpm.

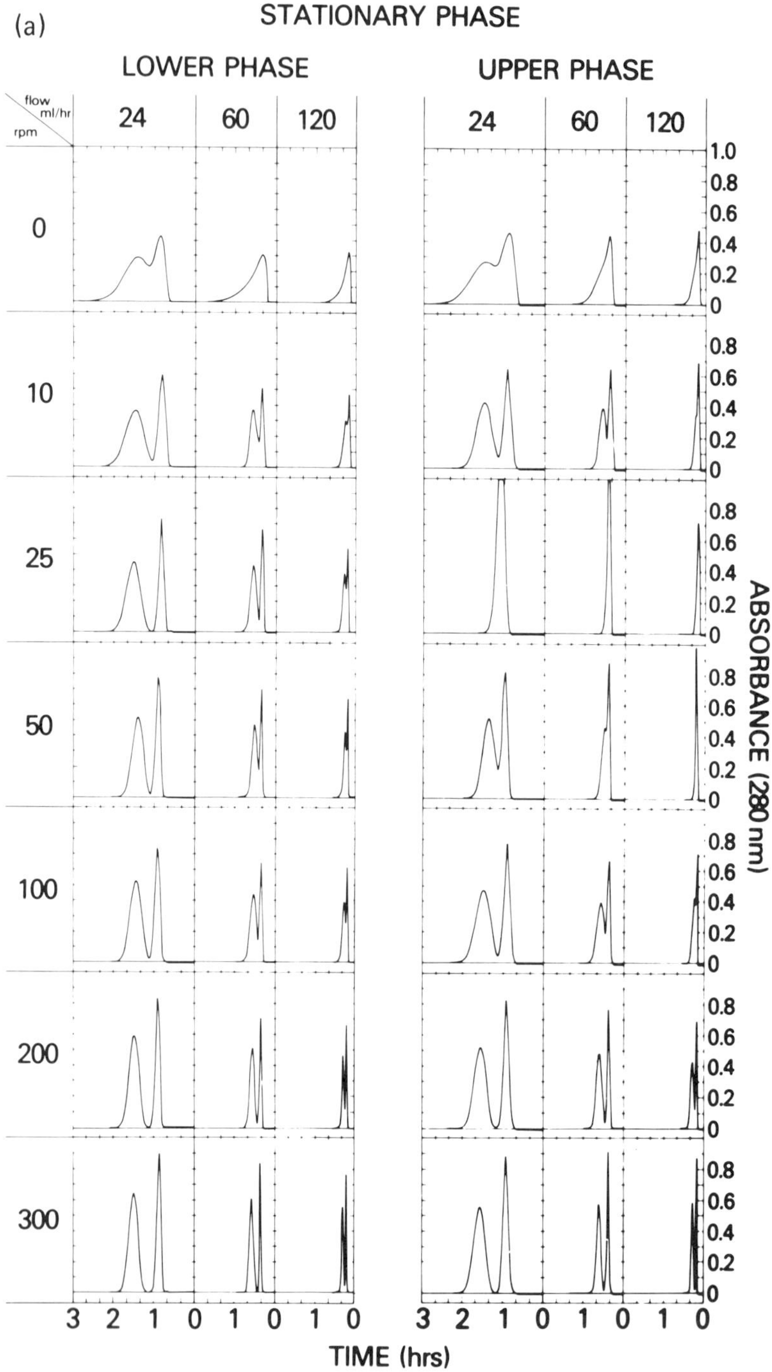

(a)
STATIONARY PHASE
LOWER PHASE
UPPER PHASE
flow ml/hr
rpm
24
60
120
24
60
120
0
10
25
50
100
200
300
1.0
0.8
0.6
0.4
0.2
0
ABSORBANCE (280 nm)
3 2 1 0 1 0 1 0
3 2 1 0 1 0 1 0
TIME (hrs)

Fig. 74

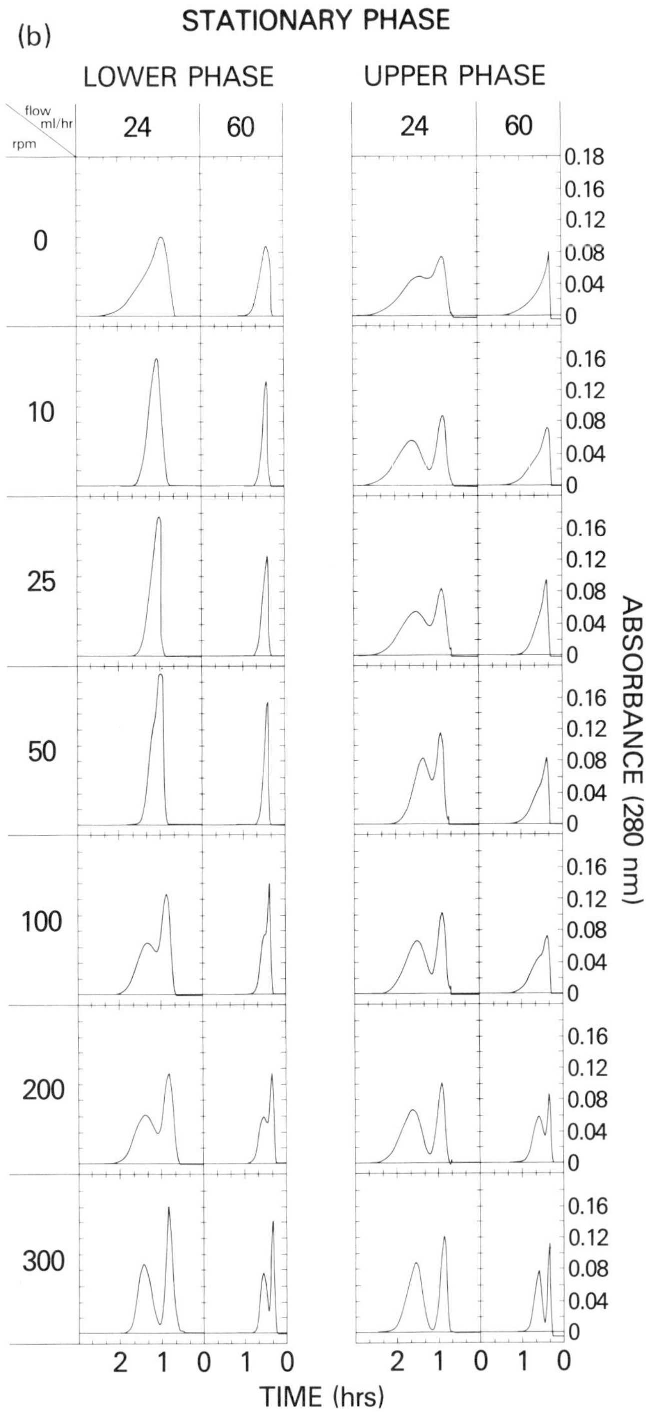

Fig. 74 (Continued)

optimum revolutional speeds. The efficiency sharply rises with revolution from 0 to 10 and 25 rpm where gravity plays a major role in partition as the centrifugal force produced by revolution is negligibly small. When the revolutional speed reaches a range of 25–50 rpm, the gravity fails to retain a satisfactory amount of the stationary phase against the flow and yet centrifugal force is not strong enough to compensate gravity. As a result, efficiency falls with the depletion of the retained stationary phase manifested by a decreased retention time of the second peak and a shortened time lapse between the two peaks at a given flow rate. Further increase of revolutional speed up to 300 rpm results again in a sharp rise of efficiency where the centrifugal force plays a major role in partition.

Figure 74b shows a diagram similarly obtained on separation of L-valyl-L-tyrosine and L-tryptophyl-L-tyrosine with the n-butanol/ acetic acid/water (4:1:5) phase system. In spite of contrasting physical properties of this n-butanol phase system with those of the former chloroform phase system, presence of the two optimum revolutional speeds is evident—one around 10 rpm and the other at 300 rpm.

The unique capability of the present scheme among other countercurrent chromatographic systems derives from the fact that it has two optimum operational conditions, one at a slow revolutional speed in the gravitational field and the other at a high revolutional speed utilizing the centrifugal force field. When a large-bore preparative column is eluted with an extremely low interfacial tension phase system, a slow revolutional speed prevents emulsification of the phases which would cause carryover of the stationary phase. For a small-bore analytical column a high revolutional speed can create a strong centrifugal force field that allows the two phases to counterflow through a narrow opening of the column without plug flow. Thus, the present scheme is capable of using a variety of two-phase solvent systems for both preparative and analytical scale separations.

Using a long column consisting of 10 column units, preparative scale separations were performed under the optimum operational conditions predetermined by preliminary experiments. Figure 75a shows separations of a set of DNP amino acids on the chloroform phase system using a 10-ml sample size in each separation. As indicated in Fig. 75, two chromatograms on the right were obtained by a slow revolutional speed using gravity and the other two chromatograms on the left by a high revolutional speed using the centrifugal force field. When efficiencies of separation are calculated according to the gas chromatographic formula, all separations show over 1000 theoretical plates in the early peaks which decrease to several hundred theoretical plates in the later peaks. Although a good resolution is achieved under the slow revolutional speed, similar results can be obtained under the high revolutional speed with a remarkably shorter elution time. Figure 75b shows separation of peptides on the n-butanol phase system using a 10-ml sample size in each separation. The efficiencies range from 1000 to 100 theoretical plates in these separations. Here again application of high revolutional speed greatly decreases separation time without significantly affecting resolution.

With commonly used two-phase solvent systems, versatility of the horizontal flow-through coil planet centrifuge was demonstrated on preparative scale separations. The sample size may be increased to grams by using a larger bore helical column by utilizing a slow revolutional speed to prevent emulsification of the solvent. For analytical separation, a high revolutional speed enables the use of a small-bore column to prevent plug flow as in the flow-through coil planet centrifuge previously described.

Recently, Tian You Zhang at New Technology Applications Research Institute of Beijin, Beijin, China constructed a modified model of the horizontal flow-through CPC and successfully applied it to separations of various biological samples such as antibiotics, agricultural chemicals, and various active principles extracted from medicinal herbs [60].

B. The Combined Horizontal Flow-Through CPC (Schemes I + IV) [22—24,39,61]

The original horizontal flow-through CPC above described has a pair of holders, one carrying a coiled separation column and the other a counterweight for balancing the centrifuge. In this new horizontal flow-through CPC, each holder provides a different mode of synchronous planetary motion. One holder (scheme I planetary motion) produces efficient mixing of two solvent phases in a narrow-bore coil to yield analytical scale separations while the other holder (scheme IV planetary motion) provides stable retention of the stationary phase in a large-bore coil for preparative scale separations. Combinations of these two column holders was accomplished to allow continuous elution through both columns without complications arising from the use of the rotary seal device. Consequently, the combined horizontal flow-through CPC is capable of performing both small- and large-scale separations with excellent results. Because of this added versatility and higher revolutional speed provided by the improved design of the planetary system, the present apparatus became the first CPC intensively used in research laboratories.

Principle

Figure 76 illustrates the design principle of the combined horizontal flow-through CPC. Each diagram labeled A—D corresponds to schemes I—IV in Fig. 17 and shows the orientation and motion of a cylindrical column holder equipped with a bundle of flow tubes, the end of which is tightly supported by a stationary member marked X

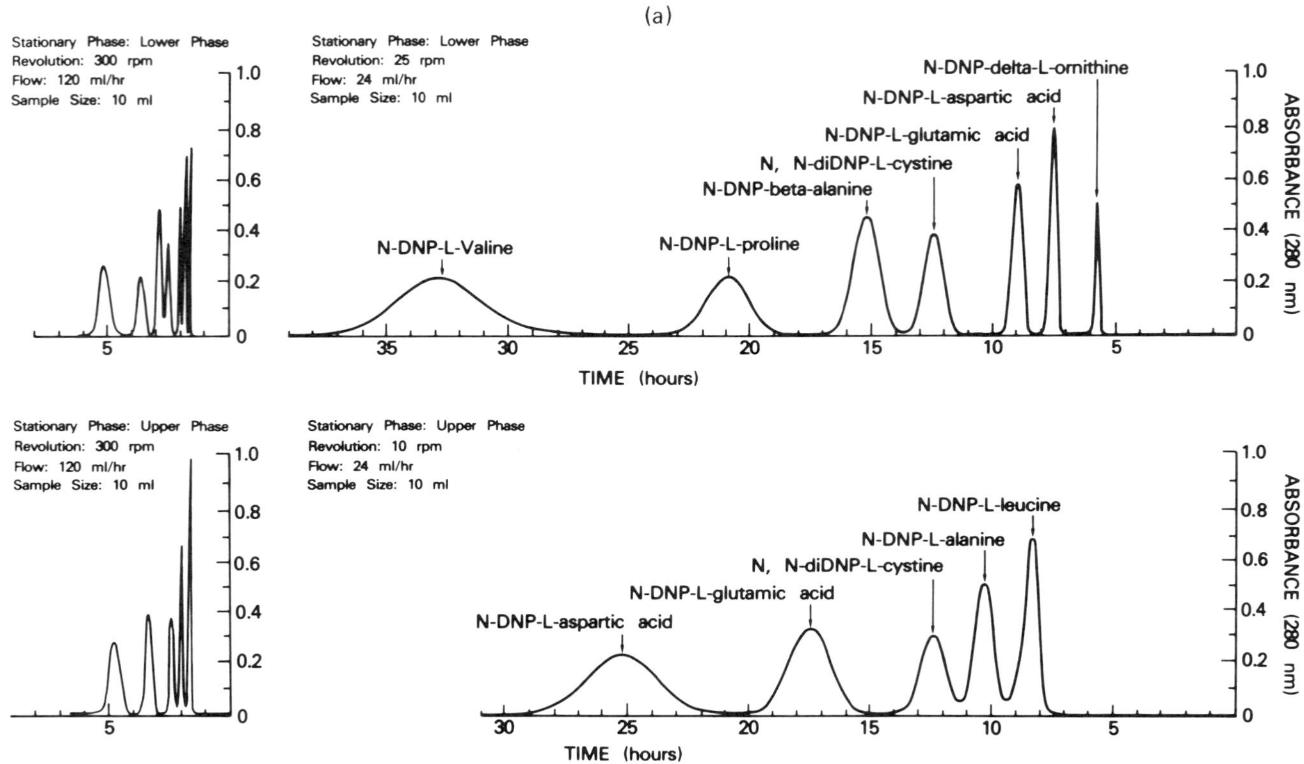

Fig. 75 Preparative scale separations under the two optimum revolutional speeds. (a) Separation of a set of DNP amino acids on chloroform/acetic acid/0.1 N HCl (2:2:1) phase system. (b) Separation of a set of peptides on n-butanol/acetic acid/water (4:1:5) phase system. Application of a higher revolutional speed greatly decreases elution time without significantly affecting resolution.

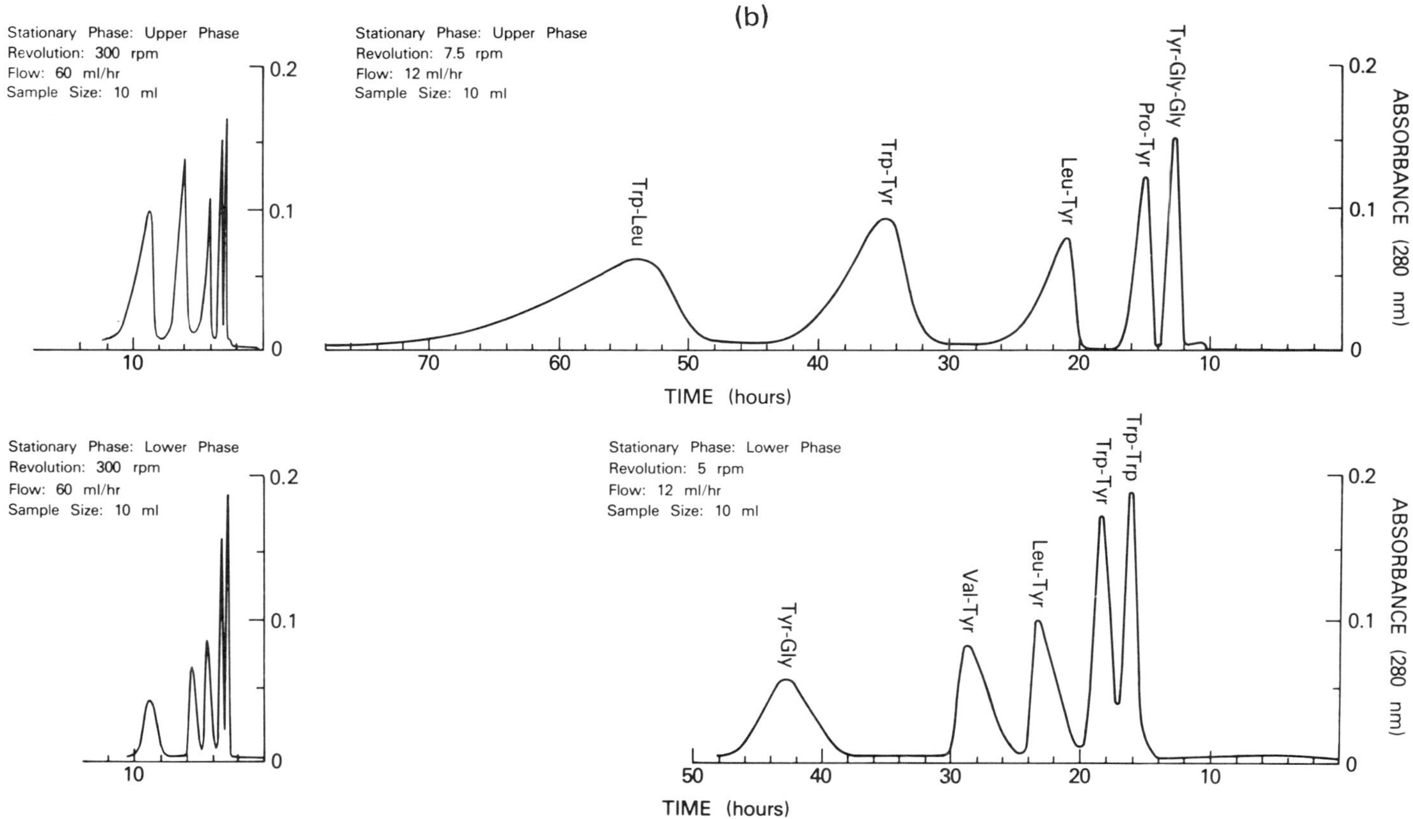

Fig. 75 (Continued)

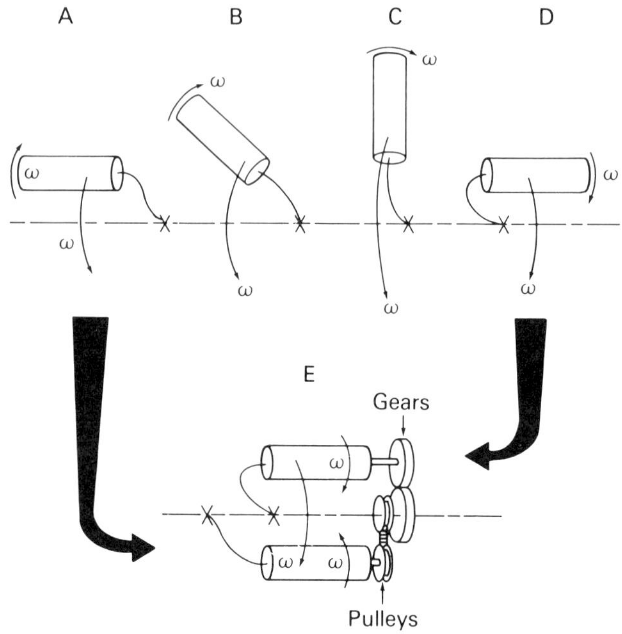

Fig. 76 Principle of the seal-free design of the combined horizontal flow-through coil planet centrifuge. A–D show various types of planetary motion of the column holder which prevent twisting of the flow tubes. Systems A and D can be combined in one apparatus E without interfering with each pathway of the flow tubes.

on the central axis of the centrifuge. In A the holder maintains a parallel orientaiton of its axis with respect to the axis of the centrifuge. The synchronous counterrotation of the holder prevents twisting of flow tubes as in the flow-through CPC described earlier. The orientation of the holder, however, can be altered without twisting the flow tubes by lifting the axis of the holder by 45° (B), 90° (C), and even 180° (D). In the latter case of D, the system becomes identical to that in the horizontal flow-through CPC where the holder synchronously rotates in the same direction as the revolution. Because of the different geometry of the flow tubes together with the symmetrical orientation of the holder, systems A and D are conveniently paired on one apparatus as shown in E where a pair of identical gears and pulleys provides a respective planetary motion to each column holder. Consequently, separations can be carried out simultaneously in both columns without the use of rotating seals.

In order to perform countercurrent chromatography, a coiled column is mounted onto each holder with the coil axis parallel to the holder axis. Under this orientation of the column, the centrifugal force field produced by either mode of the planetary motions enables continuous mixing of the two phases while retaining one of the phases stationary in each turn of the coil as the other phase is eluted through the column. Consequently, solutes locally introduced at the inlet of the column are subjected to an efficient partition process between the retained stationary phase and the flowing mobile phase and separated according to their relative partition coefficients as in the conventional liquid chromatography but without solid supports.

Because of the different modes of the planetary motion, each column holder of the present apparatus provides its own specific application for countercurrent chromatography. The gear side holder gives stable retention of the stationary phase in a large-bore preparative column whereas the pulley side holder provides efficient mixing of the phases in a narrow-bore tube to yield microscale separations with extremely high efficiency. Obviously this unique feature of the present apparatus has a great advantage over the previous models in that both large and small-scale separations are possible in one apparatus.

Design of the Apparatus

Figure 77 illustrates the design of the prototype on a cross-sectional view through the central axis of the apparatus. The motor (Electro-Craft Model E650) drives the rotary frame through a pair of toothed

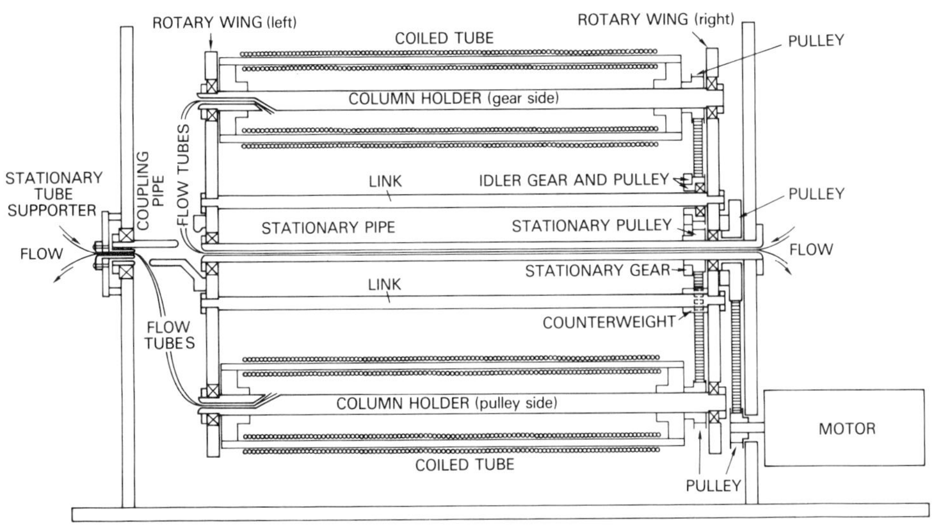

Fig. 77 Design of the combined horizontal flow-through CPC in cross-sectional view through the central axis.

pulleys and a toothed belt around the stationary pipe mounted on the central axis of the centrifuge. The rotary frame consists of a pair of rotary wings rigidly bridged with links and holds a pair of column holders symmetrically located at 15 cm from the central axis of the apparatus. In order to introduce the desired planetary motion to each column holder, a set consisting of the stationary pulley and gear is coaxially mounted around the stationary pipe. The stationary pulley is directly coupled with a toothed belt to an identical pulley mounted on the pulley side column holder (bottom) to produce a counterrotation. The stationary gear is first engaged to an identical idler gear mounted around the link through a bearing. The idler gear is coaxially connected to an idler toothed pulley which in turn is coupled with a toothed belt to an identical pulley mounted on the gear side column holder. To stabilize the centrifuge system, a short coupling pipe is mounted coaxially on the left end of the rotary frame and supported by the stationary member of the apparatus with a ball bearing.

The separation column is prepared by winding PTFE tubing onto a metal pipe to make a short column or column unit. The long large-bore column is made by connecting the desired number of column units, usually 10 units, in series. The long small-bore column for microscale separations can be prepared from one piece of tubing without interconnection between the column units. Each column holder is equipped with two different levels to mount columns, one at 3.5 cm from the axis of the holder for large-bore columns and the other (not shown in the figure) at 2.0 cm for small-bore columns. The columns are symmetrically arranged around each column holder and tightly fixed with screws through a hole made at each end of the column unit core. When two different columns are mounted on the holders, the counterweight should be added on the holder carrying the lighter column.

The flow tubes from the column mounted on the pulley side holder are first passed through the center hole of the holder and then led through a side hole of the coupling pipe to reach the stationary tube supporter extending from the stationary member of the centrifuge. On the gear side holder, flow tubes are similarly passed through the center hole of the holder and led through another side hole of the coupling pipe to enter the opening of the stationary pipe on the central axis of the centrifuge. These flow tubes are lubricated with silicone grease and protected by a piece of flexible plastic tubing at each supported portion to prevent direct contact against metal parts. If this protection is carefully followed, these tubes can maintain their function almost permanently.

Figure 78 shows the combined flow-through CPC equipped with both analytical (top) and preparative (bottom) columns. The revolutional speed of the apparatus is continuously adjustable from 0 to 600 rpm with a speed control unit shown on the right.

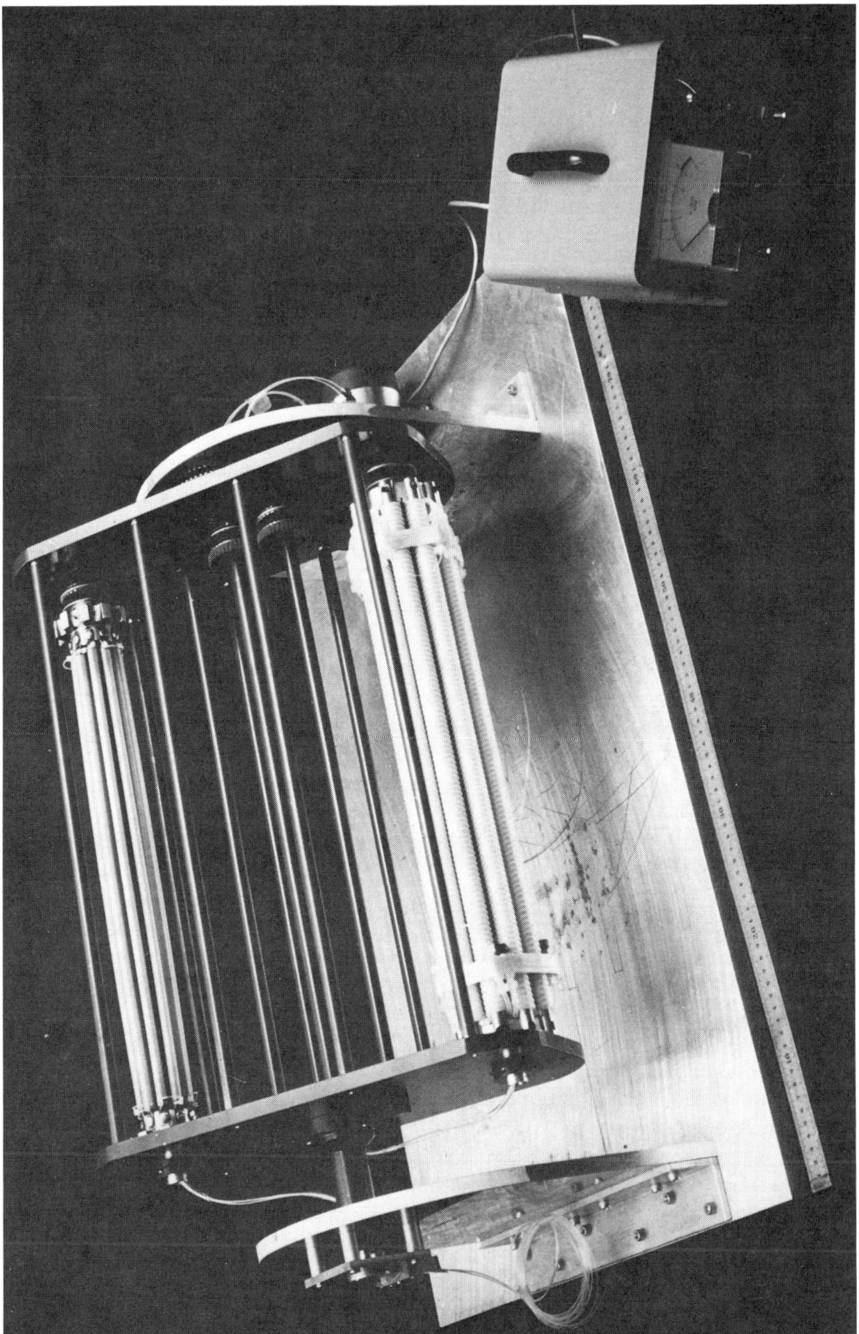

Fig. 78 Overall view of the combined horizontal flow-through coil planet centrifuge.

Studies on Retention of the Stationary Phase

A series of preliminary studies has been performed on the retention of the stationary phase using commonly employed two-phase solvent systems with a broad spectrum of interfacial tension and hydrophobicity. Overall results of the experiments obtained with 2.6-mm i.d. coil are summarized in Fig. 79.

In each diagram retention percentage of the stationary phase is plotted against the revolutional speed in rpm. Several curves drawn in each diagram indicate the effect of flow rates applied. The elution of the mobile phase is usually directed from the head to the tail end of the column but for the gear side column the elution is also reversely applied from the tail to the head end of the coil. In general, retention of over 30% is considered to be satisfactory and that near 50% is ideal. In addition to the net amount of retention, the inclination of the curves should also be considered. When an operational conditions is chosen at a portion of the curve showing a steep inclination, a slight shift of the revolutional speed will produce a considerable change in retention percentage. Therefore, for the reproducible results, it is desirable to select the operational conditions at or near horizontal portion of the curve.

Retention curves obtained from the pulley side column display a great variety in shape according to interfacial tension and viscosity of the two-phase solvent systems. For a low-interfacial-tension, high-viscosity n-butanol system, retention declined sharply with the revolutional speed where stroboscopic observation revealed intensive emulsification of the phases in the coiled column. For other phase systems, retention curves follow a characteristic pattern with the rate of revolution. At a low revolutional speed around 200 rpm, the curves of nonaqueous stationary phases exhibit a sharp rise which becomes more pronounced in the solvent systems with higher interfacial tension. After this critical range of rpm, the retention becomes stable but the retention levels for the nonaqueous phase always exceed those for the aqueous phase in the same solvent system. These complex patterns of retention may be related to the various physical factors of the solvent system such as solvent—wall interaction, strong molecular cohesive forces, and higher viscosity of the aqueous phase, etc.

On the other hand, the retention curves for the gear side column follow a quite different pattern. In all solvent systems the curves smoothly approach the stable levels at a relatively low rpm and then display a long plateau up to the maximum revolutional speed of 600 rpm. These retention levels at the plateau are mostly within an ideal range for both nonaqueous and aqueous stationary phases. Furthermore, the reversed elution mode of the tail-to-head elution also gives similar retention levels where the same procedure for the pulley side column results in complete loss of the stationary phase

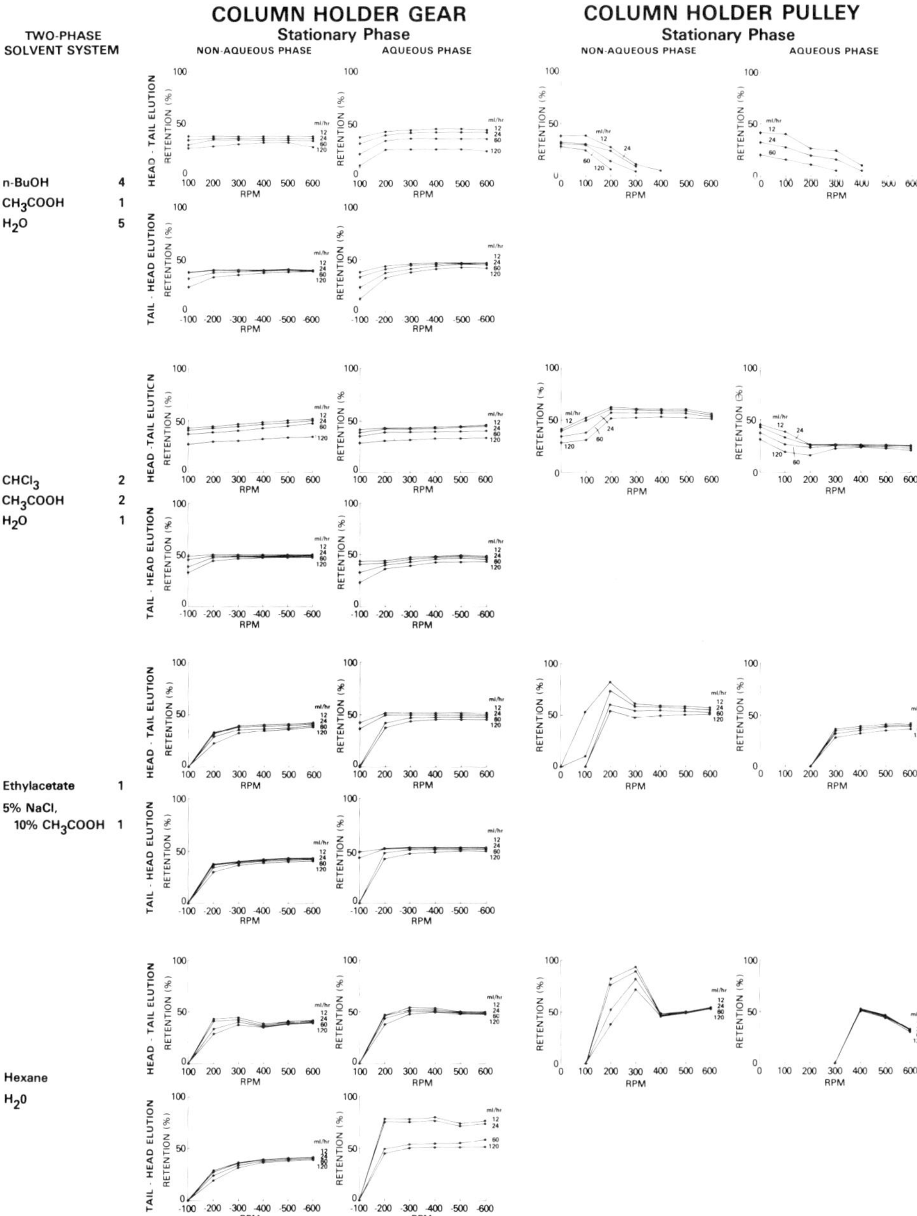

Fig. 79 Effects of revolutional speed and flow rate on retention of the stationary phase in a large-bore column. Column: 2.6 mm i.d. and helical diameter 1.25 cm (β = 0.25).

above a relatively low rpm. Stroboscopic observation of the gear side column reveals complete separation of the two phases in each turn of the coil with undulating interfaces swinging back and forth during each cycle of revolution. This finding is consistent with the results of analysis on the acceleration field described earlier. Overall results indicate that for the use of a large-bore column the gear side holder has greater advantages over the pulley side holder in that satisfactory retention is available for a broad spectrum of solvent systems under a wide range of revolutional speeds and flow rates.

The experimental results obtained from the 0.55-mm i.d. coil is similarly illustrated in Fig. 80. Two curves drawn in each diagram indicate the data obtained under the flow rates of 6 ml/hr and 2.4 ml/hr. The overall results reveal characteristic features of the gear side and pulley side columns observed in the previous studies using the large-bore column. Compared with the previous data, the retention of the aqueous phases is much less than that of the nonaqueous phases, especially in pulley side column. This result suggests that in the small-bore column the effects of solvent–wall interaction on retention becomes more significant and the retention strongly favors the nonaqueous phase which has an affinity to the PTFE tube. In the n-butanol phase group, the pulley side provides a satisfactory retention for the nonaqueous phases whereas retention for the aqueous phase requires a greater centrifugal force field to reach a suitable level.

Studies on Partition Efficiency

When the coiled column enables retention of the stationary phase, elution of the mobile phase results in separation of the solutes according to their relative partition coefficients. However, the resolution of the solute peaks is affected by various factors such as geometry of the column, physical properties of the two phases, acting centrifugal force field, and applied flow rate. In order to achieve an efficient separation, the operational condition should provide broad interfacial area and/or efficient mixing of the two phases to minimize mass transfer resistance, especially for viscous phase systems. On the other hand, excessive mixing of less viscous phase systems may cause undesirable sample band spreading along the length of the coil to lower the peak resolution. Under a set of other operational conditions, revolutional speed determines the degree of mixing and therefore can be used as a parameter to optimize the operational conditions.

In the present studies, two typical phase systems were selected for separation of samples with suitable partition coefficients, i.e., chloroform/acetic acid/0.1 N hydrochloric acid (2:2:1) for separation of 2,4-dinitrophenyl (DNP)-D, L-glutamic acid and DNP-L-alanine, and n-butanol/acetic acid/water (4:1:5) for separation of L-valyl-L-

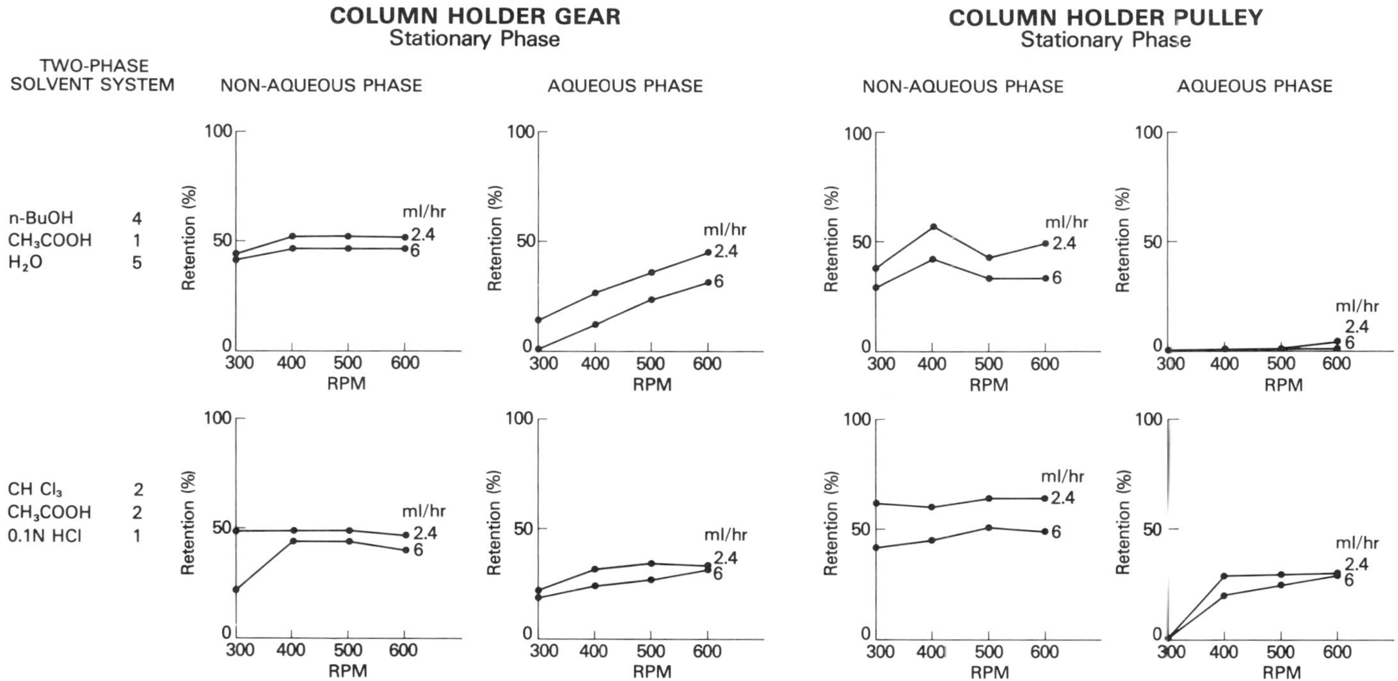

Fig. 80 Effects of revolutional speed and flow rate on retention of the stationary phase in a small-bore column. Column: 0.55 mm i.d. and 0.68 cm in helical diameter (β = 0.13).

tyrosine and L-tryptophyl-L-tyrosine. The DNP-amino acid sample mixture was prepared by dissolving each component at 0.5 g% in the stationary phase, 0.2 ml of the solution being applied for the large-bore column and 0.01 ml for the small-bore column. The peptide mixture was similarly prepared to obtain final concentrations of 1.0 g% for L-valyl-L-tyrosine and 0.3 g% for L-tryptophyl-L-tyrosine, 0.2 ml of this solution being applied for the large-bore column and 0.05 ml for the small-bore column.

In each separation, the column was first filled with the stationary phase followed by sample injection through the sample port. Then the mobile phase was eluted at a given flow rate while the apparatus was run at a desired revolutional speed. The eluate was continuously monitored through an LKB Uvicord III at 280 nm.

1. *Partition Efficiency of Large-Bore Column.* The column employed was identical to the large-bore column used for retention studies and consisted of 100 helical turns of 2.6-mm i.d. PTFE tubing coiled onto a 1.25-cm o.d. pipe with a total capacity of about 26 ml. The column was mounted on each column holder 3.5 cm from its axis or $\beta = 0.23$, if not otherwise indicated.

Figure 81a shows the results of DNP amino acid separation on the chloroform/acetic acid/0.1 N hydrochloric acid (2:2:1) phase system obtained from both gear side and pulley side columns at a fixed flow rate of 60 ml/hr under various revolutional speeds ranging from 0 to 600 rpm. Each diagram indicates one experiment where the partition efficiency is estimated from resolution of the two peaks by measuring the relative height of the trough between the peaks.

The efficiency obtained from the gear side column sharply increases with the revolutional speed up to 400 rpm. The peak resolution obtained by the head–tail elution mode is always better than that by the tail–head elution mode at a given rpm. The pulley side column gives quite different results. The efficiency becomes maximum at relatively slow revolutional speeds of around 100–200 rpm where further increase of the revolutional speed results in a sharp decrease of the peak resolution. Also, in the pulley side column, the results obtained by the stationary nonaqueous phase are much better than those obtained by the stationary aqueous phase, while the choice of the stationary phase makes little difference in resolution for the gear side column. These results clearly indicate that the gear side column is much superior to the pulley side column when a large-bore coiled tube is used.

Figure 81b similarly shows effects of flow rate and revolutional speed on DNP amino acid separation with the same solvent system using the gear side column. The data show that the best peak resolution is given by combination of revolutional speed and flow rate. At a slow flow rate of 24 ml/hr, the highest resolution is found at 400 rpm for the nonaqueous stationary phase and at 200 rpm for the aqueous stationary phase. When the flow rate of 120 ml/hr is applied, the best results are given at around 500 rpm regardless of the choice of the stationary phase. The partition efficiency may also be expressed in terms of theoretical plates using Eq. (30). In this case the highest efficiency is given under the slow flow rate of 24 ml/hr while the time required to yield one theoretical plate is minimized at the highest flow rate of 120 ml/hr.

Figure 81c shows the results of peptide separation with the n-butanol/acetic acid/water (4:1:5) phase system using the gear side column. Overall results are found to resemble those obtained by DNP amino acid separation with the chloroform phase system. The partition efficiency increases with the revolutional speed up to 500–600 rpm in all groups. The inferior peak resolution at the highest flow rate of 120 ml/hr may be explained on the basis of high viscosity of the nonaqueous phase which increases mass transfer resistance.

Further experiments were performed to examine the effects of β-values or the location of the column on the holder on separation of the same peptides with the n-butanol phase system. For this test the same column was mounted on the gear side column holder 6 cm away from its axis or at $\beta = 0.40$ by applying a pair of extension blocks. The results of the experiments obtained by this modification are summarized in Fig. 81d where the separations obtained at $\beta = 0.40$ under 60 ml/hr flow rate are compared with those previously obtained at $\beta = 0.23$ under the otherwise identical conditions. These data clearly indicate that the column located on the holder at $\beta = 0.23$ yields a higher partition efficiency than that at $\beta = 0.40$ throughout all experimental conditions applied. This finding agrees with the results previously obtained in analysis of the acceleration field in that mixing of the two phases in the coiled column becomes less efficient on the gear side holder as the β-value increases. However, the advantages of applying large β-values for the gear side column should not be neglected for the application of polymer phase systems which have a great potential in partition of cells and macromoelcules under mild environments. Because of the aqueous-aqueous phase composition the polymer phase systems possess an extremely low interfacial tension; they tend to be emulsified under violently undulating centrifugal force fields. With the use of large β-values, satisfactory retention of the polymer phase systems may be feasible.

Here it should be noted that the large-bore column mounted on the gear side holder can retain a satisfactory amount of stationary

---

Fig. 81 Partition efficiency of large-bore column. (a) DNP amino acid separation with the gear side and pulley side columns. Effects of elution mode and revolutional speed. (b) DNP amino acid separation with the gear side column. Effects of revolutional speed and flow rate. (c) Oligopeptide separation with the gear side column. Effects of revolutional speed and flow rate. (d) Oligopeptide separation with the gear side column. Effects of β-values.

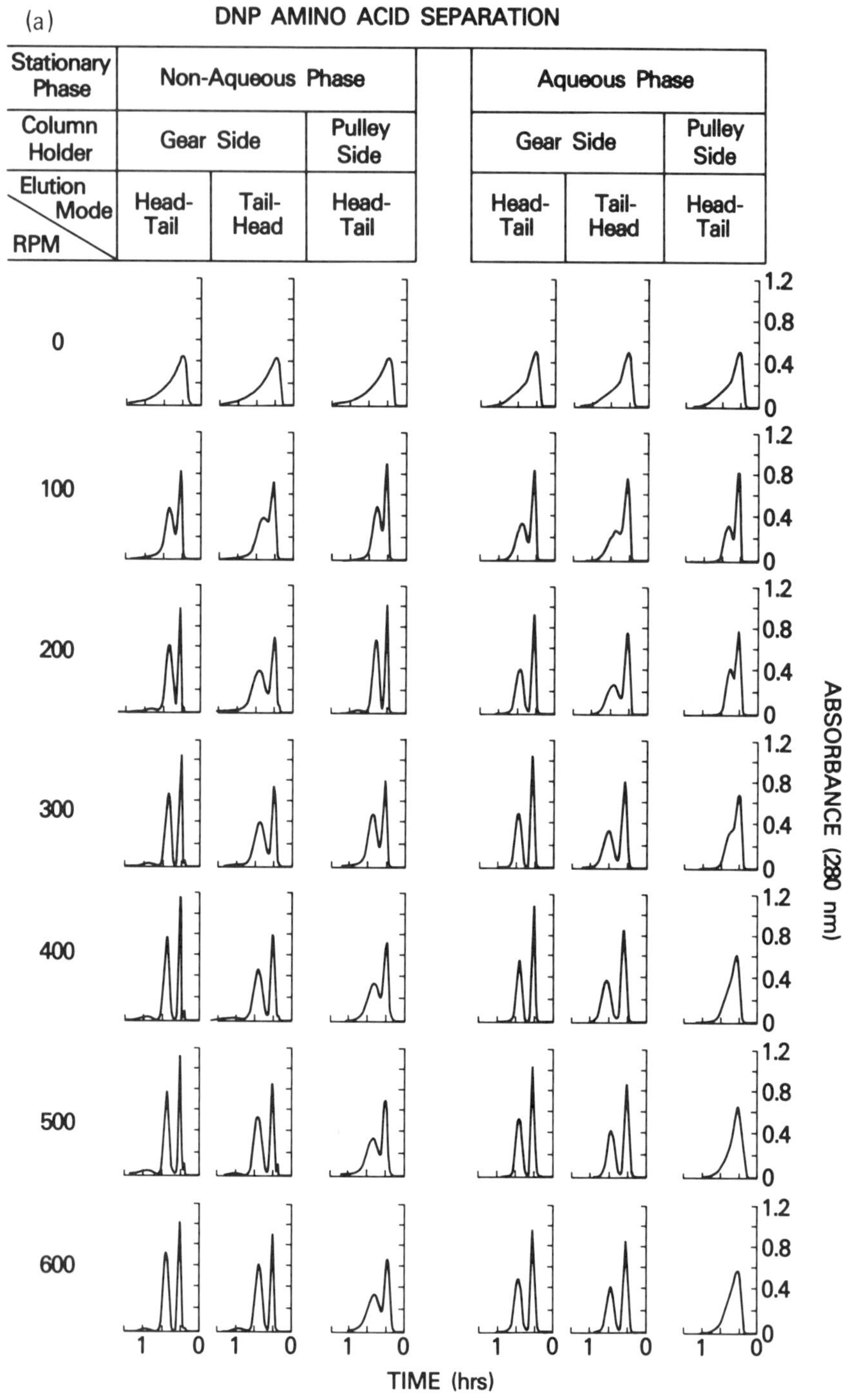
(a)
DNP AMINO ACID SEPARATION
Stationary Phase
Non-Aqueous Phase
Aqueous Phase
Column Holder
Gear Side
Pulley Side
Gear Side
Pulley Side
Elution Mode
RPM
Head-Tail
Tail-Head
Head-Tail
Head-Tail
Tail-Head
Head-Tail
0
100
200
300
400
500
600
1.2
0.8
0.4
0
ABSORBANCE (280 nm)
1
0
TIME (hrs)

Fig. 81

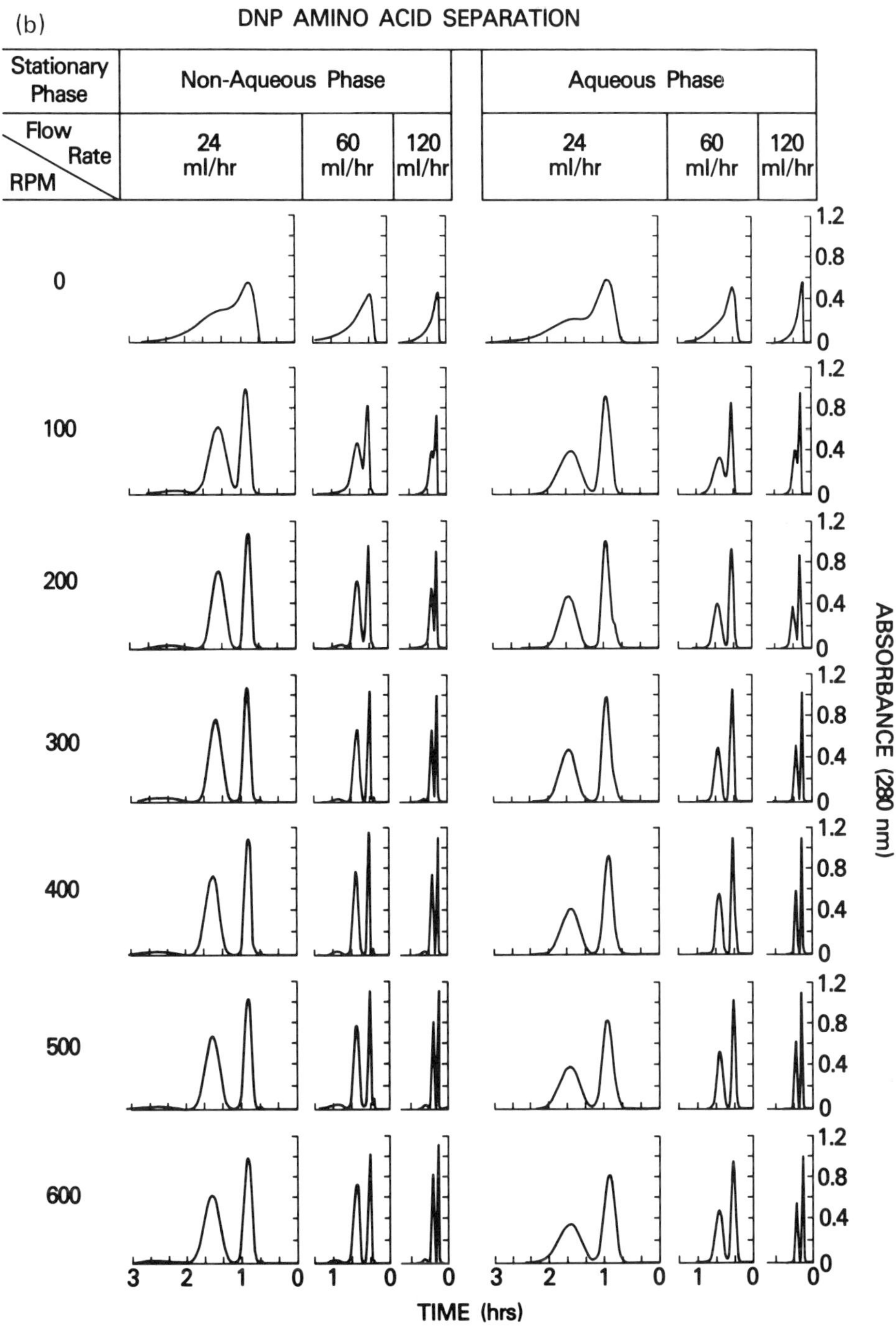

Fig. 81 (Continued)

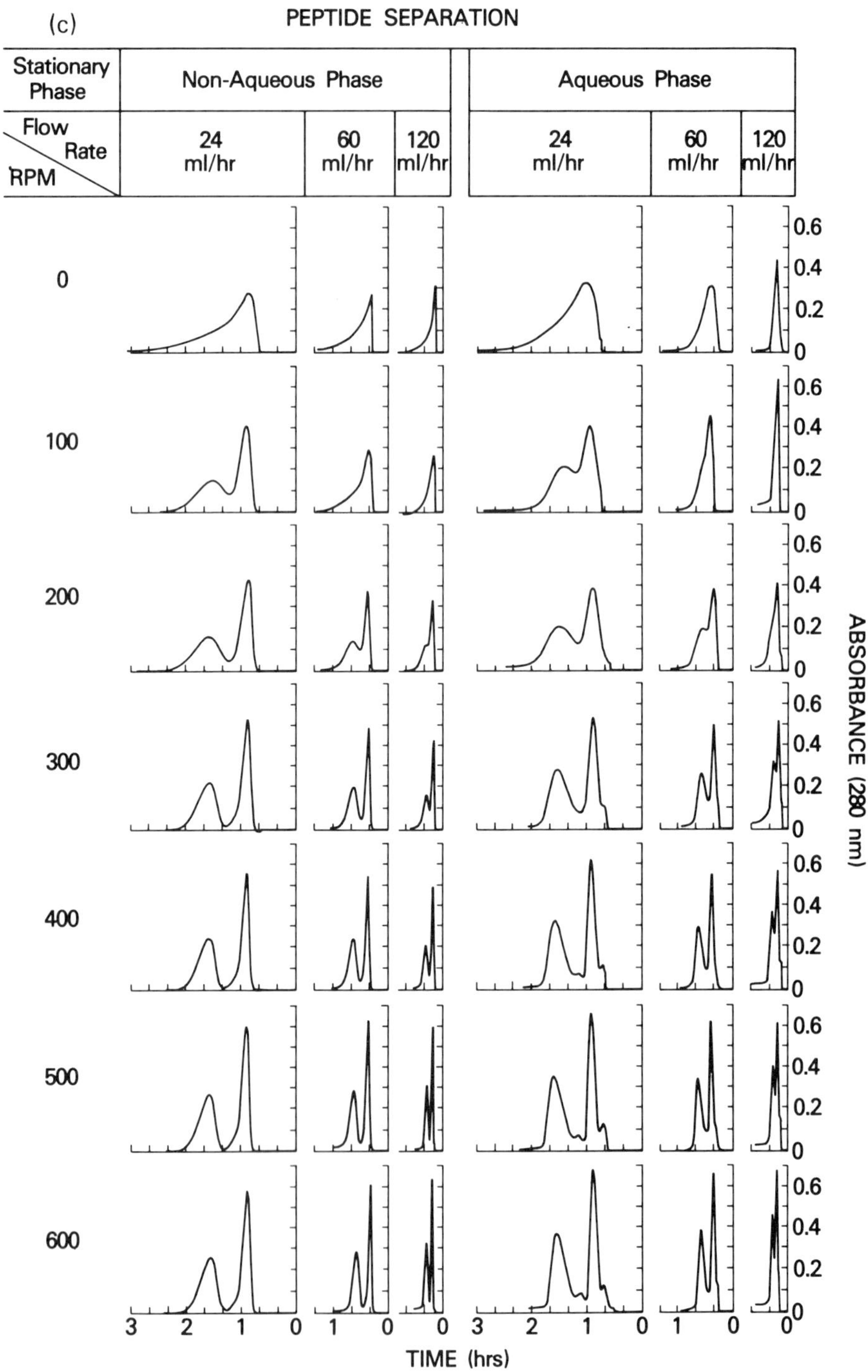

Fig. 81 (Continued)

(d)

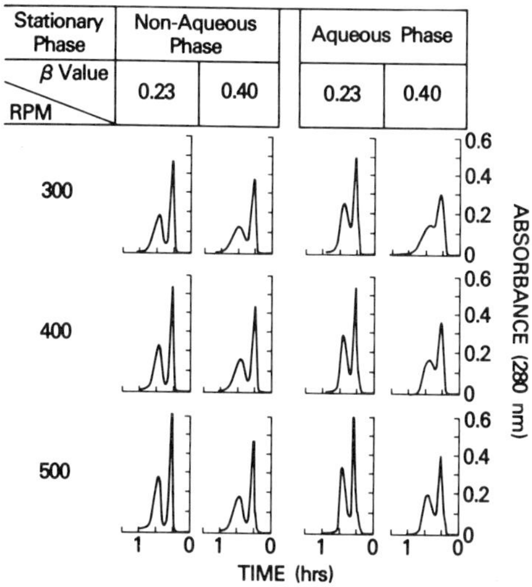

Fig. 81 (Continued)

phase when the apparatus is rotated slowly. In this case the gravitational acceleration field establishes the hydrodynamic equilibrium state of the two phases in the coiled tube. Previous experiments have shown that the optimum revolutional speeds for such gravitational separations are 25 rpm for DNP amino acids and 7.5 rpm for peptides using the respective two-phase solvent systems applied in the present studies. Although peak resolutions obtained under the slow revolutional speeds are quite satisfactory, similar or even superior results can be attained under faster revolutional speeds within much shorter elution times.

2. *Partition Efficiency of Small-Bore Column.* The small-bore column used in these studies is identical to that used in the retention studies. The column consisted of about 340 helical turns of a 0.55-mm i.d. PTFE tube coiled on 0.68-cm o.d. stainless-steel tubing with a total capacity of about 2.4 ml. The column was mounted on each column holder at a location of 2 cm from its axis or at β = 0.13.

Figure 82a summarizes the results obtained by separation of the two DNP amino acids on the chloroform phase system using combinations of various revolutional speeds and flow rates. Overall results indicate that the pulley side column yields more efficient separation, especially when the nonaqueous phase is used as the stationary phase. Figure 82b shows similar data obtained on separation of the two peptides on the n-butanol phase system. As expected from the results

(a)

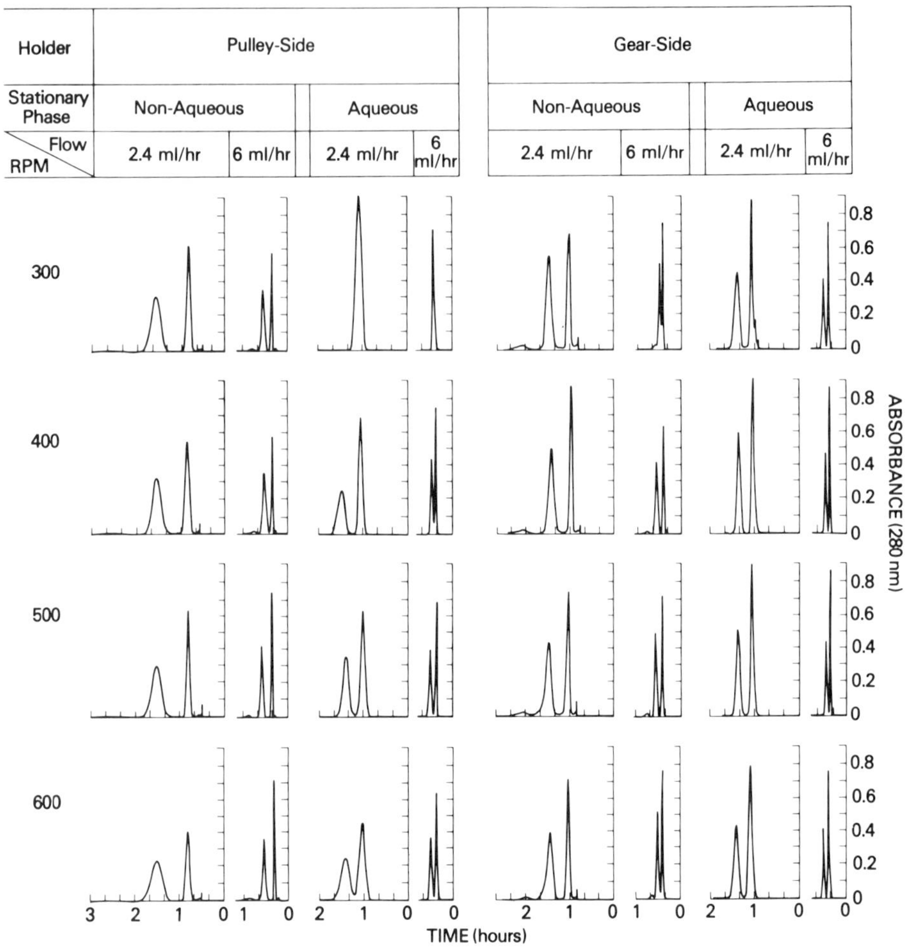

Fig. 82 Partition efficiency of small-bore column (0.55 mm i.d.). (a) DNP amino acid separation. Effects of revolutional speed and flow rate. (b) Oligopeptide separation. Effects of revolutional speed and flow rate.

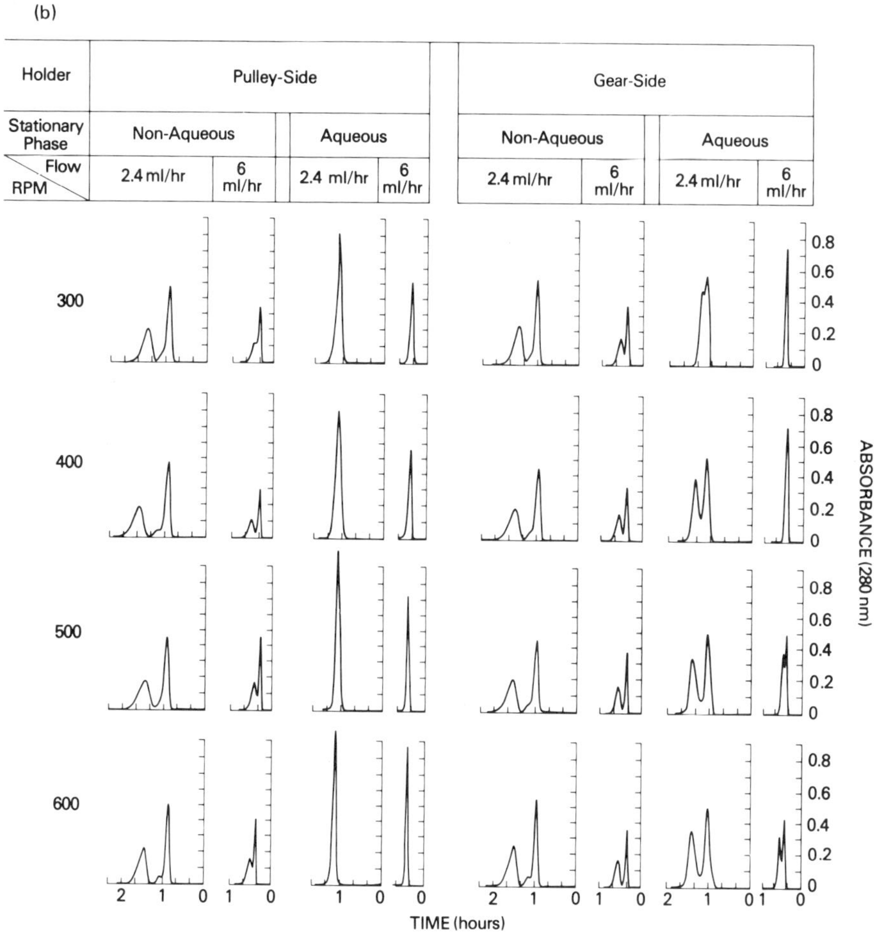

Fig. 82 (Continued)

of the retention studies, no separation was observed in the pulley side column when the aqueous phase was used as the stationary phase. Although the pulley side yields slightly higher resolution at 2.4 ml/hr of flow rate with the stationary nonaqueous phase, the gear side column gives more uniform results and permits the choice of the stationary phase.

Studies on Performance of Long Columns

The foregoing studies on retention and partition efficiency with a single-column unit provided optimum operational conditions which can be directly applied to a separation in a long column consisting of a desired number of the identical column units connected in series. In order to obtain successful separation, however, some precautions should be taken to avoid unnecessary sample band broadening which tends to occur in a long preparative column.

Theoretically, the partition efficiency should increase linearly with the length of the coiled column. However, our experiments have shown that the partition efficiency obtained with a long column is sometimes significantly lower than that estimated from the results obtained with the short column, especially for a large-bore column. The loss of efficiency may be largely attributed to high-pressure buildup in the column which causes considerable expansion of the column and occasional surging of the solvent upon a slight shift of revolutional speed. This detrimental effect, however, can be minimized by applying a narrow-bore tubing at each junction between the column units. A small-bore column has a relatively strong resistance against expansion and a very long column can be prepared from one piece of tubing without risk of losing peak resolution.

The choice of the solvent phases for preparation of the sample solution is important when a large sample volume is to be applied. If the sample contains solutes with high partition coefficients which favor the partition to the mobile phase, the sample should be dissolved in the stationary phase. In this way the mobile phase eluted through the column enables concentration of the solutes into sharp sample bands. Although other solutes with lower partition coefficients present in the sample solution may initially form broader sample bands, they are subjected to a longer partition process or greater dilution in the column, and therefore the initial band width becomes less significant for the peak resolution.

1. *Performance of Large-Bore Column.* Taking the above precautions, preparative scale separations were performed with a large-bore column which is equivalent to 10 column units in series with a piece of 10-cm long, 0.4-mm i.d. PTFE tubing at each junction between the column units. The column was arranged symmetrically around the gear side column holder 3.5 cm away from its axis or $\beta = 0.23$. The two-phase solvent systems and samples were those used in the foregoing studies on partition efficiency. In each experiment the column was first filled with the stationary phase followed by sample injection through the sample port. Then the mobile phase was eluted at the indicated flow rate while the apparatus was run at the optimum revolutional speed determined by the previous experiments. The eluate from the column was continuously monitored with an LKB Uvicord III at 280 nm through a 1.8-mm light path flow cell.

Figure 83a shows chromatograms obtained with a long, large-bore column on the gear side holder using the samples and two-phase systems indicated on the chart. In both DNP amino acid and peptide separations, impurities originally present in the sample solutions were well separated out from the major peaks. The partition efficiencies calculated from these chromatograms according to Eq. (30) range between 950 and 1950 theoretical plates, showing an efficiency loss of 10–20% compared with the figures estimated from the results obtained with the single-column unit. In these separations, the sample volume can be increased by diluting with the stationary phase up to 10 ml without significant loss of peak resolutions.

Figure 83b shows chromatograms of a set of DNP amino acids obtained with the large-bore column mounted on the gear side holder under the optimum operational conditions. The upper pair of chromatograms were obtained by using the upper aqueous phase as the mobile phase, and the lower pair of chromatograms were obtained by using the lower nonaqueous phase as the mobile phase. In all of these chromatograms the DNP amino acids were well resolved and eluted as symmetrical peaks with partition efficiencies ranging from 550 to 2500 TP. It is apparent from these chromatograms that a flow rate of 60 ml/hr gave better peak resolution than a flow rate of 120 ml/hr, but with the faster flow rate separations were completed in a shorter period of time. These results allow a choice of flow rates for the separation of biological materials depending on their stability and their relative partition coefficients.

The chromatograms in Fig. 83b can also serve as a guide to the choice of mobile phase. For example, DNP-L-aspartic acid and DNP-L-glutamic acid have partition coefficients which favor partition to the upper aqueous phase. When these compounds are eluted with the aqueous phase as shown in the upper chromatograms, they appear shortly after the solvent front and form very sharp peaks. When the same samples are eluted with the lower nonaqueous phase as shown in the lower chromatograms, they are retained in the column for much longer periods of time and produce broader but more resolved peaks. In separations of several components having similar partition coefficients which favor a particular phase, higher resolution should be obtained by using that phase as the stationary phase and eluting with the other phase. In this manner, the solutes are subjected to the partitioning process for longer periods of time and yield fractions of higher purity.

Figure 83c shows chromatograms of a set of oligopeptides obtained with the large-bore column mounted on the gear side holder. Separations were performed at a revolutional speed of 500 rpm using flow rates of 24 ml/hr and 60 ml/hr with both phases as the mobile phase. In all chromatograms the five peptides are well resolved with partition efficiencies ranging between 200 and 2000 TP. In contrast to the runs with the DNP amino acids, the peaks are quite asymmetrical. This skewing resulted from nonlinear distribution isotherms of the samples in this particular solvent system employed and could be significantly decreased by lowering the concentration of peptides in the sample solution.

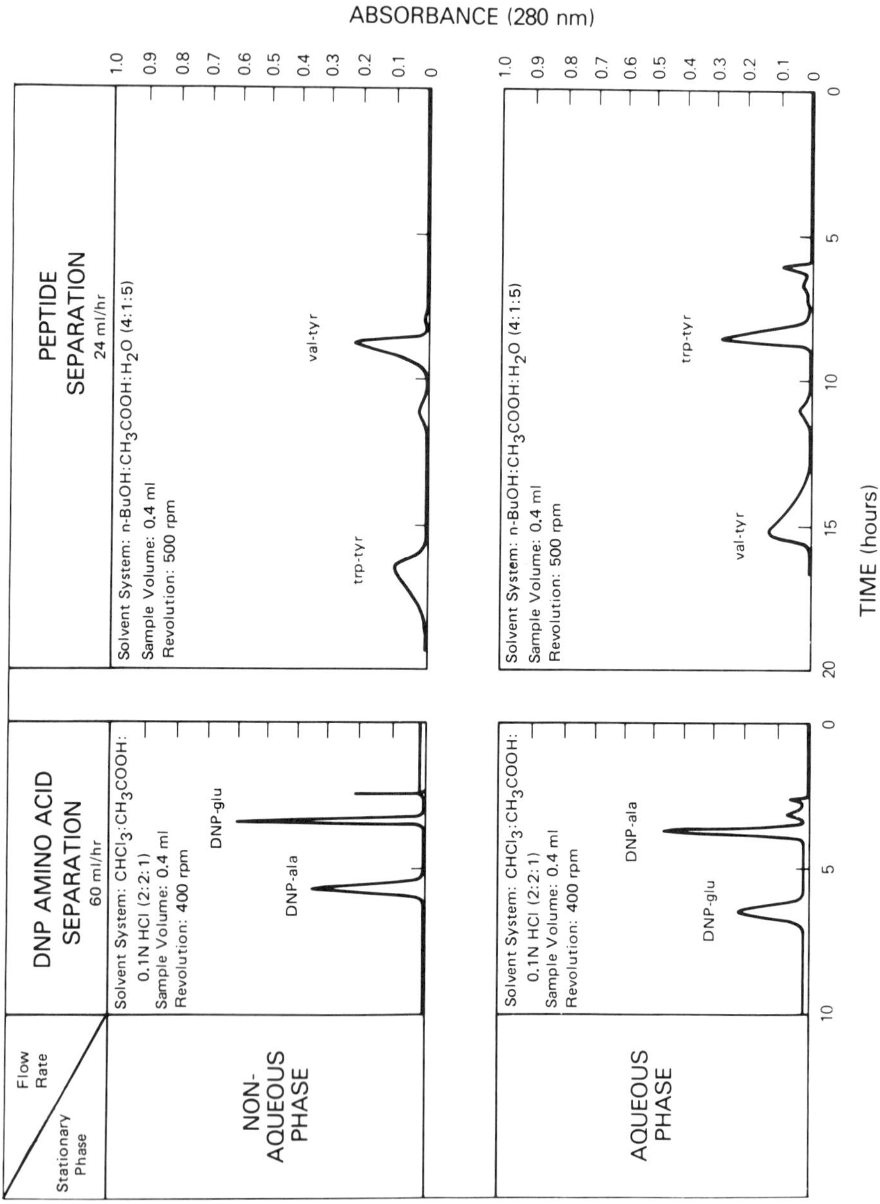

(a)
Stationary Phase
Flow Rate
DNP AMINO ACID SEPARATION
60 ml/hr
PEPTIDE SEPARATION
24 ml/hr
NON-AQUEOUS PHASE
AQUEOUS PHASE
Solvent System: CHCl$_3$:CH$_3$COOH: 0.1N HCl (2:2:1)
Sample Volume: 0.4 ml
Revolution: 400 rpm
Solvent System: n-BuOH:CH$_3$COOH:H$_2$O (4:1:5)
Sample Volume: 0.4 ml
Revolution: 500 rpm
DNP-glu
DNP-ala
val-tyr
trp-tyr
ABSORBANCE (280 nm)
1.0
0.9
0.8
0.7
0.6
0.5
0.4
0.3
0.2
0.1
0
TIME (hours)
0
5
10
15
20

Fig. 83

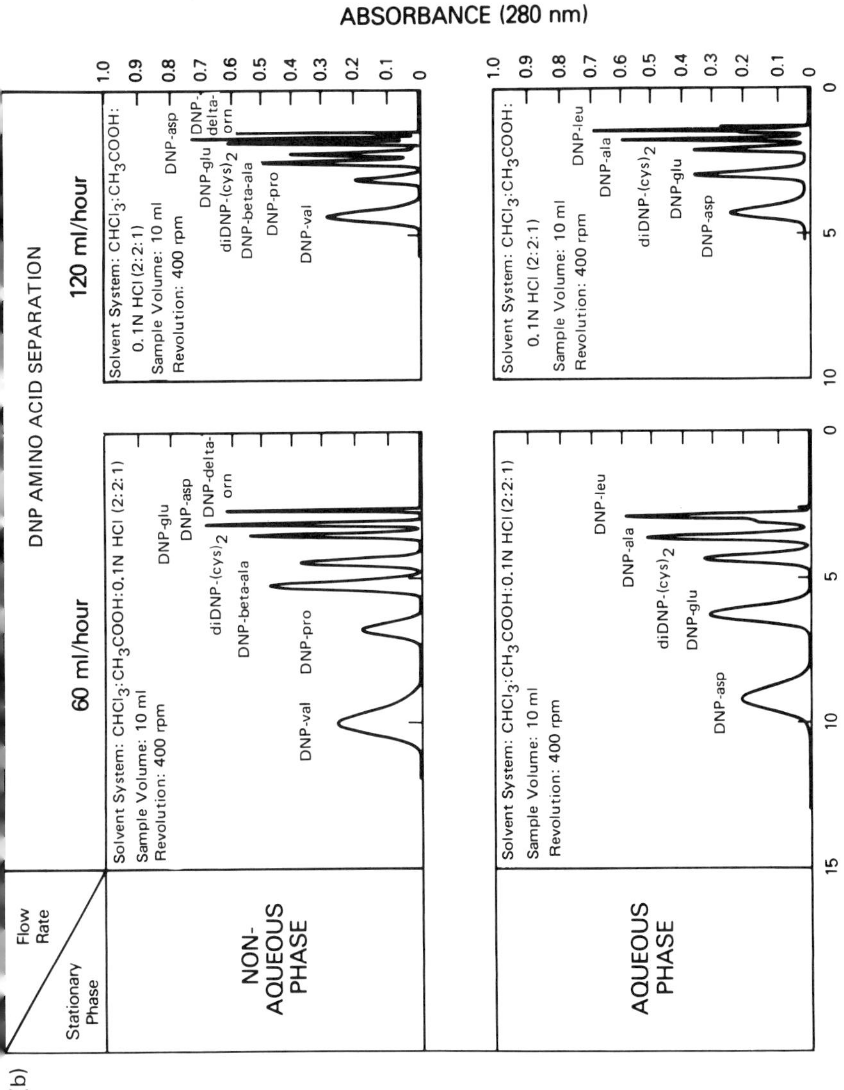

Fig. 83 (Continued)

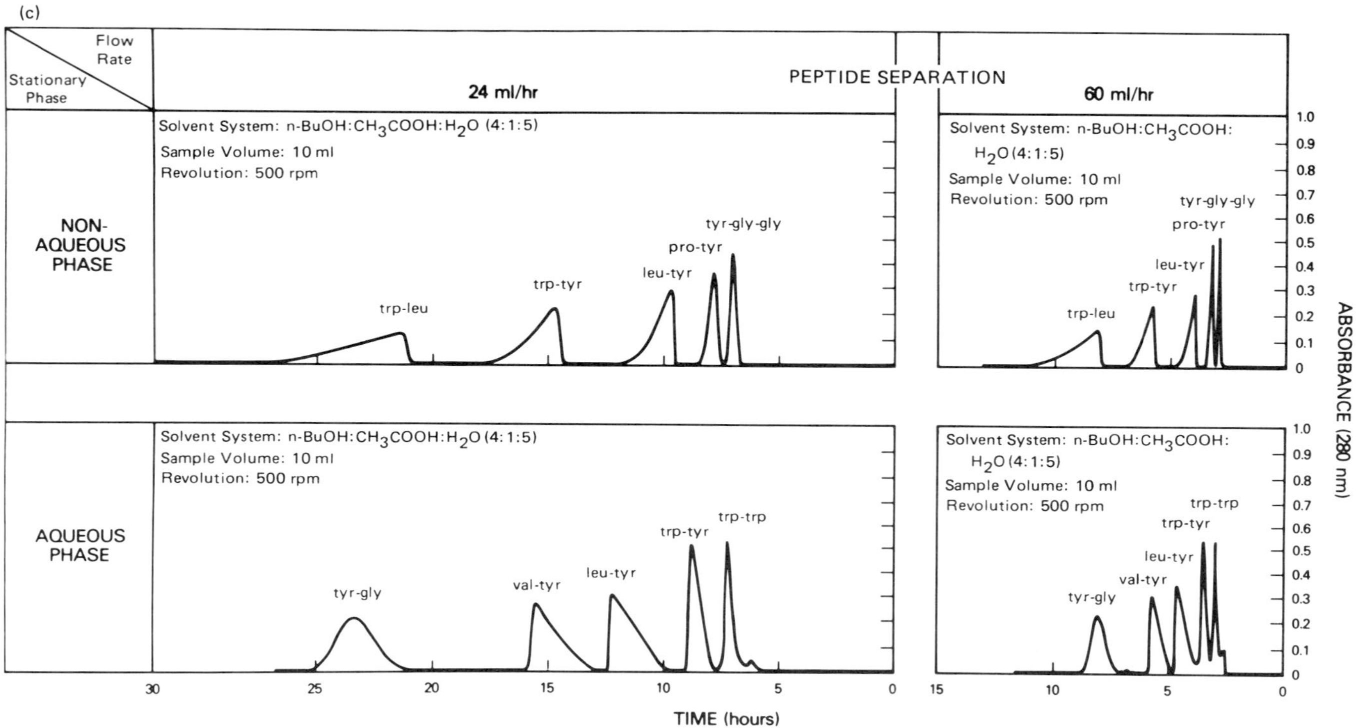

Fig. 83 Chromatograms obtained with a long preparative column on the gear side holder. (a) Separations of the test samples of DNP amino acids and oligopeptides. (b) Separations of a set of DNP amino acids. (c) Separations of a set of oligopeptides.

2. *Performance of Small-Bore Column.* Figures 84a and b show chromatograms of a set of DNP amino acids obtained with the small-bore column mounted on the pulley side holder under the optimum conditions of flow rate and revolutional speed indicated in the figures. All components were separated with much greater resolution as compared to the separations obtained with the large-bore column. The partition efficiencies calculated from these chromatograms range from 1800 to 400 TP. Comparison between two chromatograms (a) and (b) clearly indicates that the elution with the aqueous phase gives higher peak resolution and shorter retention time for the solvent front. This indicates the effects of solvent–wall interaction in the narrow-bore column on the retention of the stationary phase and the partition efficiency. When the stationary phase has an affinity to the tubing wall, a greater amount of the stationary phase is retained in the column resulting in higher peak resolution. Therefore, for the analytical column, it is recommended that the nonaqueous phase be used as the stationary phase. Figure 84c shows a chromatogram of a set of oligopeptides obtained with the small-bore column mounted on the pulley side holder. The separation was performed by eluting with the lower aqueous phase at 2.4 ml/hr under 500 rpm of the revolutional speed. Different from the results obtained in DNP amino acid separations, the peak resolution obtained with the small-bore column is very similar to that with the large-bore column mounted on the gear side holder (see Fig. 83c top left).

Application of the Apparatus

1. *Prostaglandins and Their Metabolites [62,63].* Separations of radiolabeled prostaglandins E_2, $F_{2\alpha}$, A_2, and B_2 and thromboxane B_2 were successfully performed with a solvent system composed of chloroform, acetic acid, and water at a 2:2:1 volume ratio. Figure 85a shows a chromatogram of prostaglandins obtained with a small-bore column (0.55 mm i.d. and 24 ml capacity) by eluting with the upper aqueous phase at a flow rate of 2.4 ml/hr and at a revolutional speed of 500 rpm. Figure 85b shows a chromatogram of prostaglandins B_2, F_2, and $F_{2\alpha}$ with a large-bore column (2.6 mm i.d. and 260 ml capacity) by eluting with the lower nonaqueous phase at 24 ml/hr. The present method offers an advantage over thin-layer chromatography in that the separations are performed within a coiled column which protects sensitive prostaglandins from exposure to oxygen. Because the method allows quantitative recovery of the sample loaded, it is ideal for separations of radiolabeled compounds.

2. *Antitumor Drugs [64,83].* Figure 86 shows a chromatogram of daunorubicin (first peak) and its reduction products, 7-deoxydaunorubicinol aglycone (second peak) and 7-deoxydaunorubicinol aglycone (third peak). The separation was performed with a solvent system composed of chloroform, ethylene chloride, hexane, methanol,

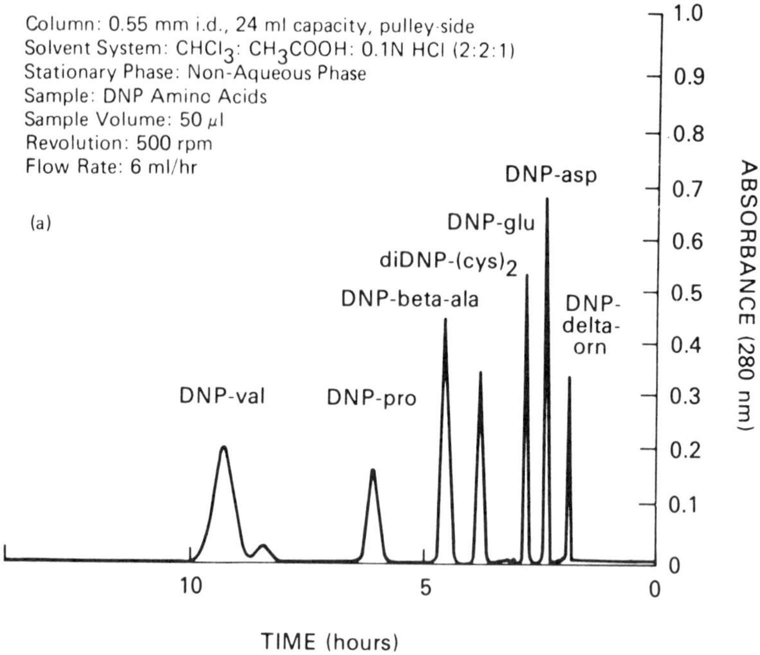
Column: 0.55 mm i.d., 24 ml capacity, pulley-side
Solvent System: CHCl3: CH3COOH: 0.1N HCl (2:2:1)
Stationary Phase: Non-Aqueous Phase
Sample: DNP Amino Acids
Sample Volume: 50 μl
Revolution: 500 rpm
Flow Rate: 6 ml/hr
(a)
DNP-val
DNP-pro
DNP-beta-ala
diDNP-(cys)2
DNP-glu
DNP-asp
DNP-delta-orn
ABSORBANCE (280 nm)
1.0
0.9
0.8
0.7
0.6
0.5
0.4
0.3
0.2
0.1
0
10
5
0
TIME (hours)

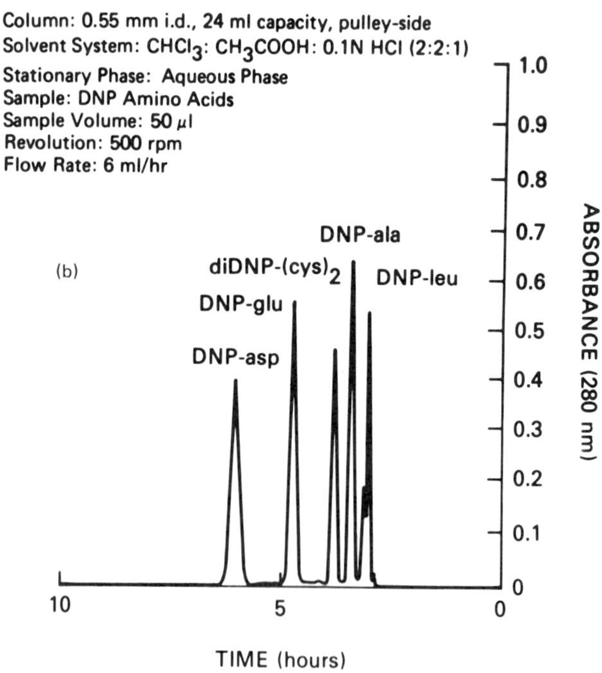
Column: 0.55 mm i.d., 24 ml capacity, pulley-side
Solvent System: CHCl3: CH3COOH: 0.1N HCl (2:2:1)
Stationary Phase: Aqueous Phase
Sample: DNP Amino Acids
Sample Volume: 50 μl
Revolution: 500 rpm
Flow Rate: 6 ml/hr
(b)
DNP-asp
DNP-glu
diDNP-(cys)2
DNP-ala
DNP-leu
ABSORBANCE (280 nm)
1.0
0.9
0.8
0.7
0.6
0.5
0.4
0.3
0.2
0.1
0
10
5
0
TIME (hours)

Fig. 84

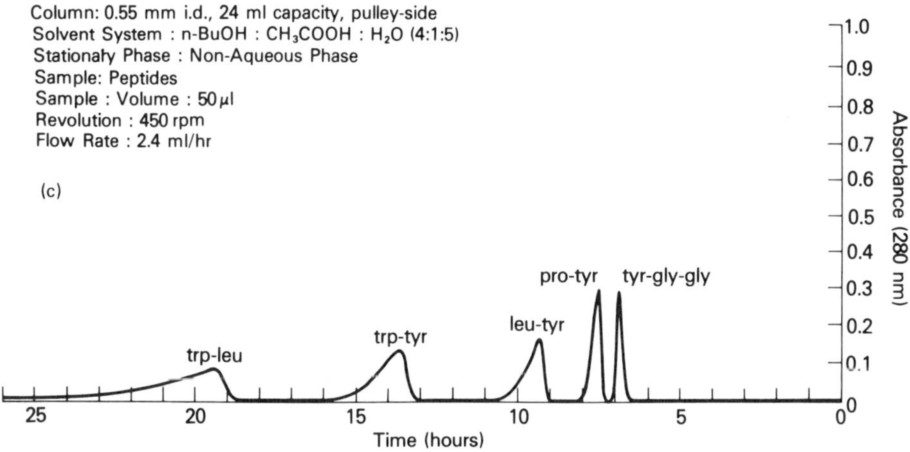

Fig. 84 Chromatograms obtained with a long analytical column on the pulley side holder. (a) Separations of a set of DNP amino acids with the upper phase mobile. (b) Separations of a set of DNP amino acids with the lower phase mobile. (c) Separations of a set of oligopeptides with the lower phase mobile.

and water (2:2:2:7:2) in a large-bore column. The application of the same sample to an HPLC column has produced contamination with the solid support material which interfered with the subsequent mass spectrometric analysis [83]. Diaziquone and its reaction products were separated with a solvent system composed of dichloroethane, methanol, and 10 mM sodium phosphate of maleate (pH 6.1) (3:3:1) using the lower phase as the mobile phase [64].

3. *p-Nitrophenol Polar Conjugates [65].* In conjunction with developing a new method of partition coefficient measurement, three polar conjugates of p-nitrophenol (PNP) (PNP-glucose, PNP-sulfuric, acid ester, and PNP-glucuronide) were separated on a two-phase solvent system of ethyl acetate/KH_2PO_4 (pH 4.5) (1:1) with a short preparative column (2.6 mm i.d. and 25 ml capacity). Studies showed that the molarity of KH_2PO_4 plays the major role in determination of partition coefficients. Complete resolution of these three compounds was obtained with 0.5 M KH_2PO_4.

4. *Gramicidin [61, 63].* A gramicidin mixture was efficiently separated with a two-phase solvent system of chloroform/benzene/methanol/water (15:15:23:7) using both large- and small-bore columns. Figure 87a shows a chromatogram of gramicidins with a large-bore column under the operational conditions indicated in the figure. All major components were well resolved while each component is further

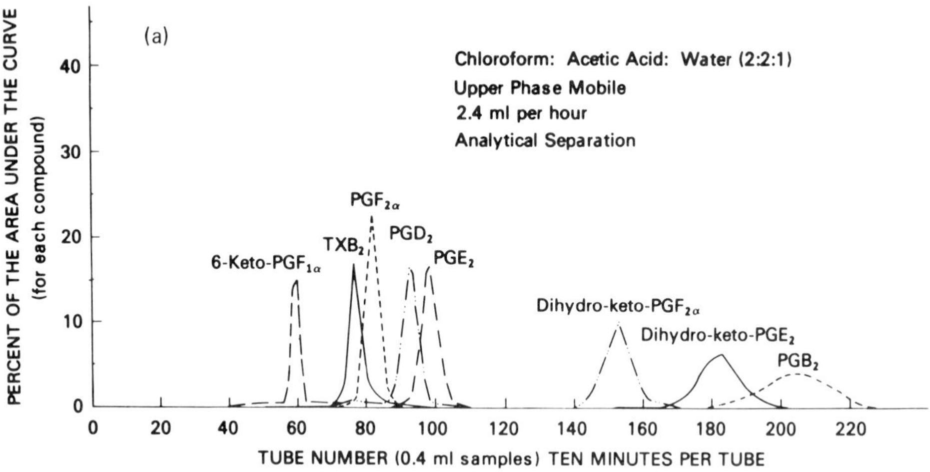

Fig. 85 Separation of radiolabeled prostaglandins on chloroform/acetic acid/water (2:2:1). (a) Chromatogram obtained with the analytical column mounted on the pulley side holder. (b) Chromatogram obtained with the preparative column mounted on the gear side holder.

partially resolved into valine and isoleucine analogs. Complete resolution of each analog was accomplished with a small-bore column at a reduced flow rate of 2.4 ml/hr as shown in Fig. 87b.

5. *Synthetic Peptides.* Separations of synthetic peptides have been most intensively carried out by the combined horizontal flow-through CPC with a large-bore column which yields reliable retention of the butanol solvent systems. Applied samples include 5, 6, 9, and 15 amino acid peptides [40,66,67], des-enkephalin-γ-endorphin [68] and a number of cholecystokinin fragments, sulfated or unsulfated [69–71], etc. The detail description of peptide separations is made in Chap. 10 by M. Knight.

C. Toroidal Coil Planet Centrifuge [24,25]

Principle and Design of the Apparatus

The toroidal coil planet centrifuge is a new variety of the horizontal flow-through CPC which is equiped with a long, fine coiled tube would again on a large cylindrical coil holder. The holder undergoes a planetary motion around the central axis of the apparatus in such a way that the holder revolves around the central axis of the apparatus while it synchronously rotates about its own axis in

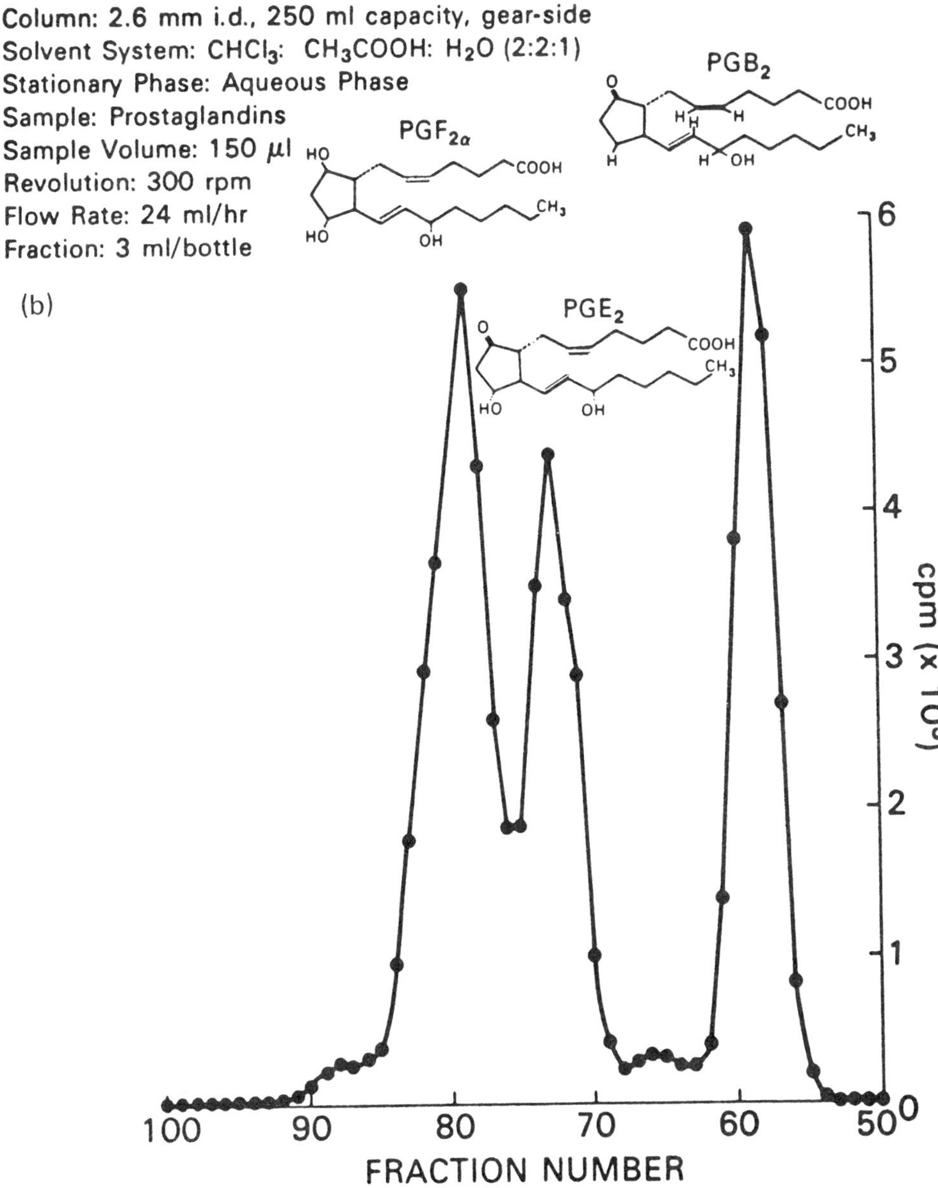

Fig. 85 (Continued)

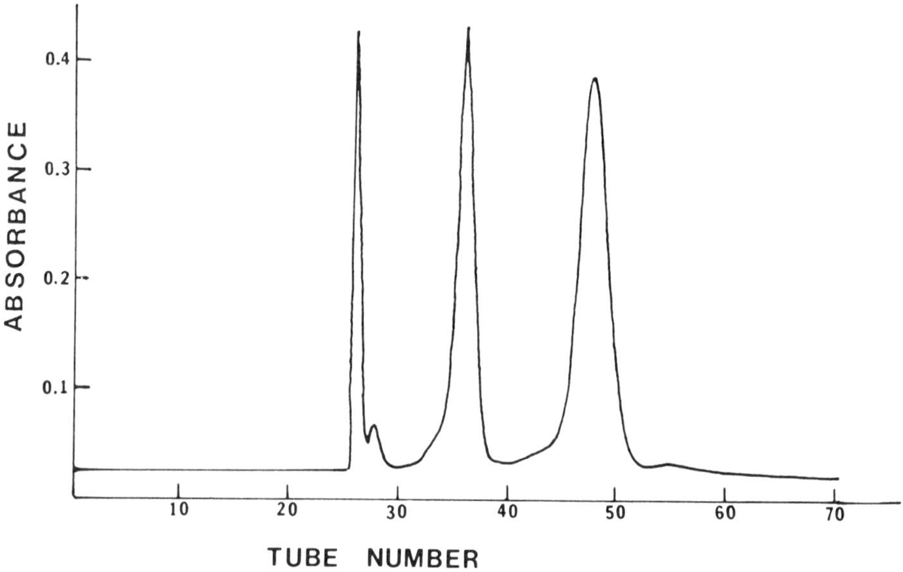

Fig. 86 Separation of daunorubicin reduction products on chloroform/ethylene chloride/hexane/methanol/water (2:2:2:7:2). Peaks are, from left to right, daunorubicin, 7-deoxydaunorubicinol aglycone, and 7-deoxydaunorubicin aglycone.

the same directions. This synchronous planetary motion of the holder not only allows continuous elution without using rotating seals but also provides an ideal centrifugal force field for performing countercurrent chromatography. As described earlier, the centrifugal force vector produced by this planetary motion is predominantly directed radially from the central axis of the holder, periodically changing both magnitude and angle during each revolutional cycle. When a two-phase solvent system is introduced into the toroidal coil, the centrifugal force field separates the two phases in each turn of the coil so that the heavier phase occupies the outer portion and the lighter phase the inner portion of the toroidal coil. This produces numerous discrete segments of the two phases alternatingly arranged along the length of the column, each pair of the segments providing a partition unit similar to the Craig countercurrent distribution apparatus, but in an extremely small scale. When either the heavier or lighter phase is continuously eluted through the coil, it will percolate through the segments of the other phase which is held by the centrifugal force in each coil unit to perform an efficient solute partitioning. This partition process is aided by the fluctuating force field which efficiently mixes the two phases to reduce mass transfer

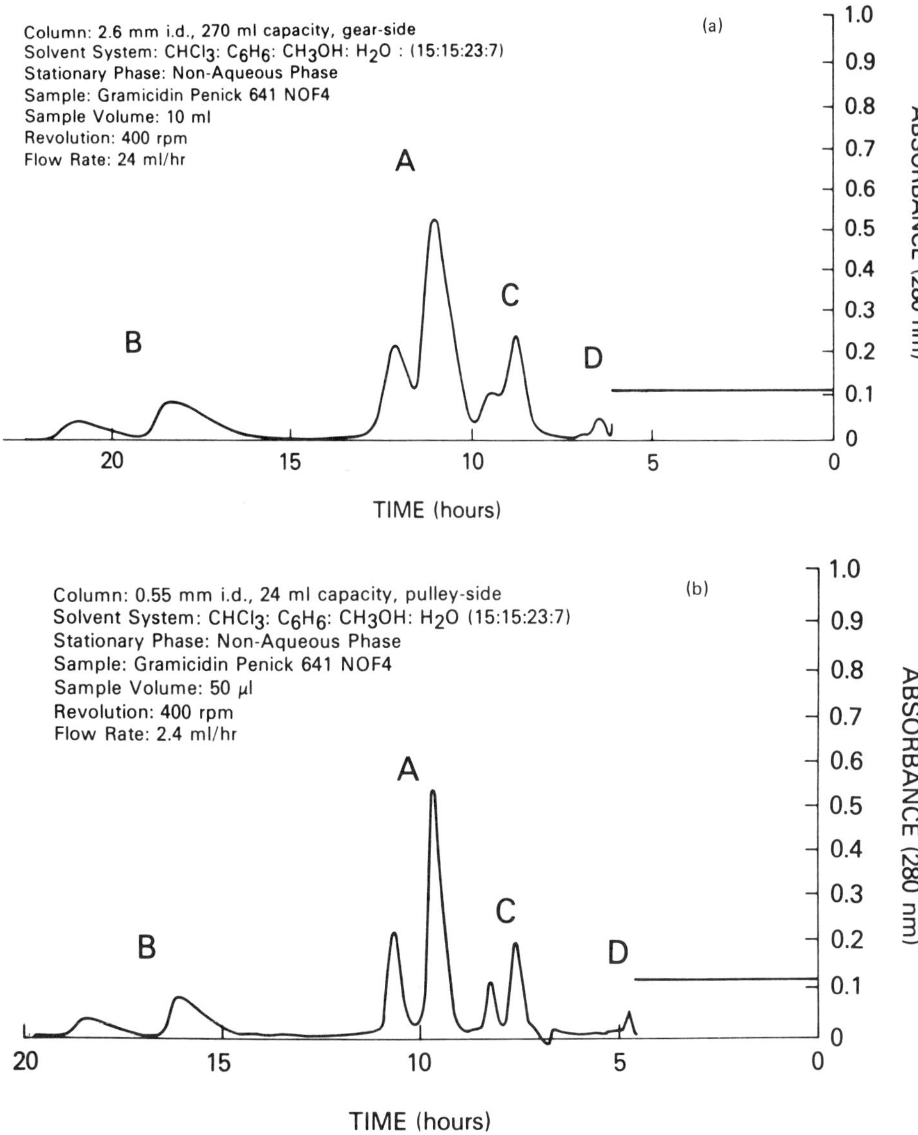

Fig. 87 Separations of gramicidins on chloroform/benzene/methanol/water (15:15:23;7). (a) Chromatogram obtained with the preparative column mounted on the gear side holder. (b) Chromatogram obtained with the analytical column mounted on the pulley side holder.

Fig. 88 Photograph of the toroidal coil planet centrifuge. (a) Overall view of the apparatus. (b) The toroidal coil mounted on the holder.

resistance. Consequently, solutes introduced into the coil are subjected to an efficient partition process in each turn of the coil and separated according to their relative partition coefficients as in the conventional liquid chromatography but in the absence of solid supports.

Figure 88a shows a photograph of our prototype. The rotary frame consists of a pair of circular aluminum plates rigidly bridged with multiple aluminum links and is driven by a motor (Electro-Craft)

Fig. 88 (Continued)

around the stationary pipe mounted on the central axis of the centrifuge. The rotary frame holds a pair of cylindrical holders symmetrically spaced at a distance of 10 cm from the central axis of the apparatus. The holder having the coiled column is 15 cm in diameter (β = 0.75) while the other holder with a counterweight, 10 cm in diameter (β = 0.5). Each coil holder is equipped with a plastic gear (Winfred M. Berg, Inc.) which is engaged to an identical stationary gear mounted on the central stationary pipe. With this gear arrangement, each holder undergoes the desired planetary motion, i.e., revolution around the central axis of the apparatus and rotation about its own axis at the same angular velocity in the same direction. In order to mechanically stabilize the centrifuge system, a short coupling pipe is coaxially mounted on the free end (right side) of the rotary frame while the other end of the coupling pipe is supported by the wall of the outside enclosure of the apparatus through a ball bearing.

The coiled separation column is prepared by winding PTFE tubing onto a flexible core which is again coiled around the holder to form a coiled helix configuration as shown in Fig. 88b. A counterweight is applied to the other holder to balance the centrifuge. The flow tubes from the separation column are first passed through the center hole of the column holder shaft and then led into the coupling pipe through a side hole to enter the opening of the central stationary pipe. The flow tubes pass through the central stationary pipe and emerge at the left side of the centrifuge enclosure. These flow tubes are thoroughly lubricated with silicone grease and also protected with a piece of plastic tubing at each supported portion to prevent direct contact against metal parts. When this protection is provided, the flow tubes can maintain their function almost permanently. The revolutional speed of the prototype is continuously adjustable up to 1000 rpm (450g) with a Motomatic speed control unit (Electro-Craft).

A Chromatronix Cheminert metering pump was used for elution of the solvent and an LKB Uvicord III for monitoring the absorbance at 280 nm.

Studies on Performance of the Apparatus

In order to evaluate the partition capability of the apparatus, a series of experiments was performed to study the degree of stationary phase retention and partition efficeincy in short, coiled columns under various revolutional speeds and flow rates.

Two types of coiled columns were prepared each from a 7-m long, 0.55-mm i.d. PTFE tube by winding it onto a flexible nylon core of two different diameters. One column had a core diameter of 5 mm and the other column, 1.5 mm. In both columns, the total capacity measured was approximately 2 ml. These columns were mounted on the column holder having β value of 0.75. Two typical phase systems were selected, chloroform/acetic acid/0.1 N hydrochloric acid (2:2:1) for partition of DNP amino acids and n-butanol/acetic acid/water (4:1:5) for partition of oligopeptides. For each phase system a pair of samples with suitable partition coefficients was selected: N-DNP-L-glutamic acid and N-DNP-L-alanine for the chloroform phase system; L-valyl-L-tyrosine and L-tryptophyl-L-tyrosine for the n-butanol phase system. The DNP amino acid sample mixture was prepared by dissolving each component in the aqueous phase at 0.5 g% and 10 μl of this solution was charged in each separation. The dipeptide sample mixture was similarly prepared to make final concentration of 1 g% for L-valyl-L-tyrosine and 0.3 g% for L-tryptophyl-L-tyrosine, and 20 μl was applied to each run.

In each separation the column was first filled with the stationary phase followed by sample injection through the sample port. Then the column was eluted with the mobile phase while the apparatus was run at the desired revolutional speed. The eluate was continuously monitored through an LKB Uvicord III at 280 nm and then collected into a graduated cylinder to measure the volume of the stationary phase eluted from the column.

Figure 89 summarizes the data for retention of the stationary phase under various operational conditions. In each diagram, the retention is plotted against the applied revolutional speed at two different flow rates. The retention of near 50% is ideal but that over 30% is considered to be satisfactory if the inclination of the curve is relatively small. Operation at the near horizontal portion of the curve ensures minimum carryover of the stationary phase even if there are shifts of revolutional speed. The overall results indicate that the satisfactory level of retention is obtained over a wide range of revolutional speeds and flow rates for both solvent systems. The 5-mm core column gives a slightly higher level of retention than the 1.5-mm core column. Because of the solvent-wall interaction in a narrow-bore tube, the nonaqueous phase having an affinity to the PTFE tube gives substantially higher level of retention than the aqueous phase under a given set of operational conditions. The retention levels of the n-butanol phase system (Fig. 89b) is generally lower than those of chloroform phase system (Fig. 89a) due to its higher viscosity and the smaller difference in density between the two phases. Satisfactory retention levels obtained from these typical phase systems with contrasting physical properties suggest that the present method permits universal application of the two-phase solvent systems.

Figure 90a summarizes the results of DNP amino acid separations with both 5- and 1.5-mm core columns performed at revolutional speeds ranging between 400 and 1000 rpm. Flow rates of 6 and 2.4 ml/hr were applied using both nonaqueous and aqueous phases as the stationary phase. In each chromatogram, partition efficiency can

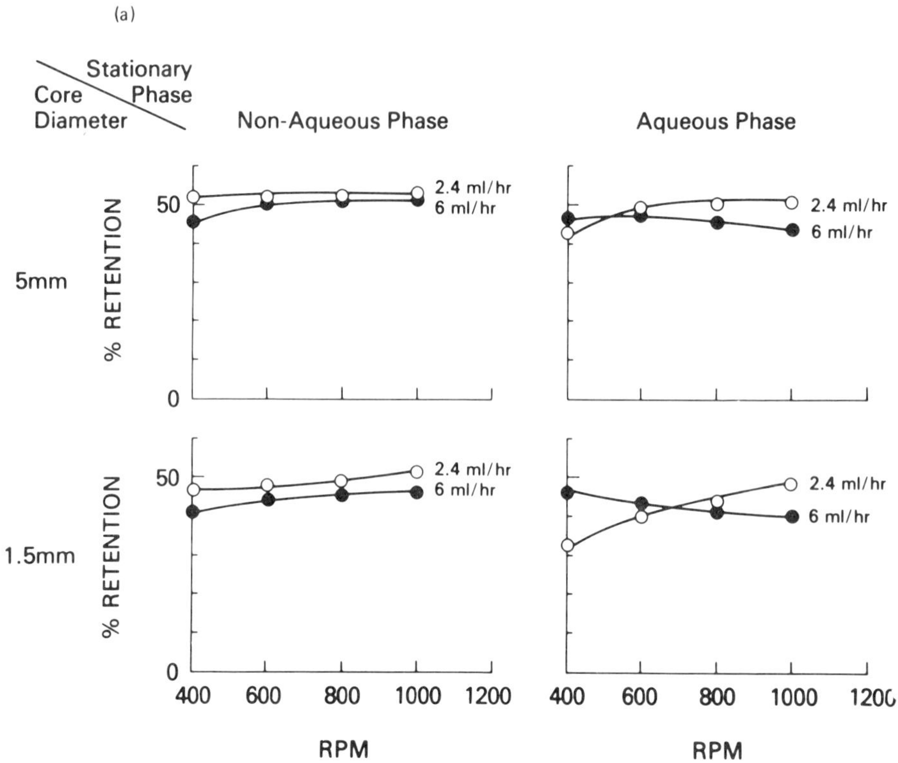

Fig. 89 Phase distribution diagrams obtained from the toroidal coil. (a) Retention of the stationary phase of chloroform/acetic acid/0.1 N HCl (2:2:1) in both 5- and 1.5-mm core columns. (b) Retention of the stationary phase of n-butanol/acetic acid/water (4:1:5) in both 5- and 1.5-mm core columns.

be easily estimated by observing the degree of peak resolution and, if desired, expressed in terms of the number of theoretical plates according to Eq. (30). Overall results indicate that the 1.5-mm core column yields higher peak resolution than the 5-mm core column. Although good peak resolutions are given at 6 ml/hr flow rate in short periods of time, the highest partition efficiency is obtained at 2.4 ml/hr flow rate under revolutional speeds of 600–1000 rpm.

Figure 90b similarly summarizes the results of the oligopeptide separation. Here again the peak resolution obtained by the 1.5-mm core column generally exceeds that obtained by the 5-mm core column. Poor peak resolution observed at 6 ml/hr under 400 rpm was apparently caused by a low retention level of the stationary phase as

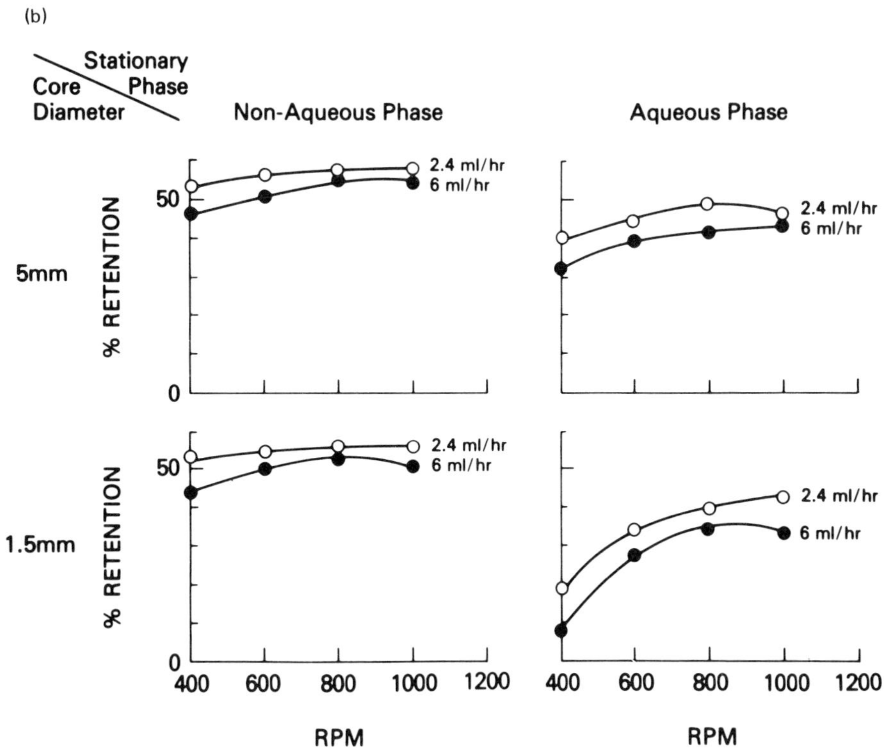

Fig. 89 (Continued)

seen in Fig. 89b. The best results are obtained by the 1.5-mm core column at 2.4 ml/hr under 800–1000 rpm, where impurities present in the sample solution are partly resolved from the major peaks.

Separation of DNP Amino Acids and Oligopeptides with a Long Column

The present method was applied for separation of DNP amino acids and oligopeptides using a long, coiled column. The separation column was prepared from a 50-m × 0.55-mm i.d. PTFE tube by winding it onto a 13-m × 1.5-mm o.d. nylon tube to make approximately 8500 helical turns with a total capacity of about 18 ml. The column was mounted on the column holder with $\beta = 0.75$. Suitable sets of samples were selected each from DNP amino acids and oligopeptides for separation with the two-phase solvent used in the previous experiments. The DNP amino acid mixture was prepared by dissolving a set of samples in the aqueous phase to make a con-

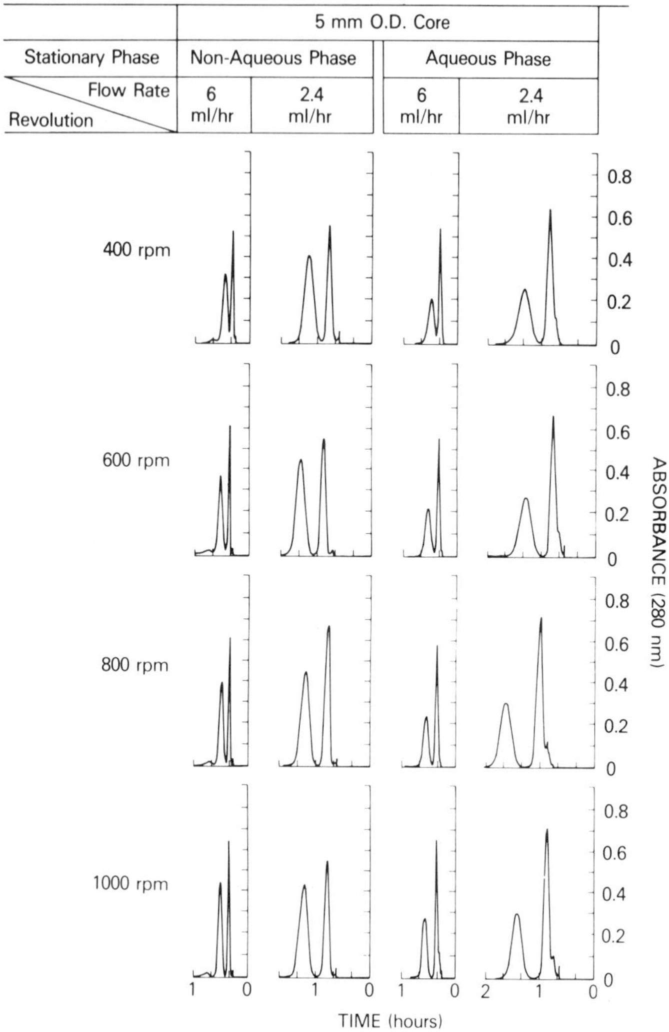

Fig. 90 Effects of revolutional speed and flow rate on partition efficiency of the short toroidal coil (0.55 mm i.d. and 7 m long). (a) Separations of DNP-L-glutamic acid and DNP-L-alanine on chloroform/acetic acid/0.1 N HCl (2:2:1). (b) Separations of L-valyl-L-tyrosine and L-tryptophyl-L-tyrosine on n-butanol/acetic acid/water (4:1:5).

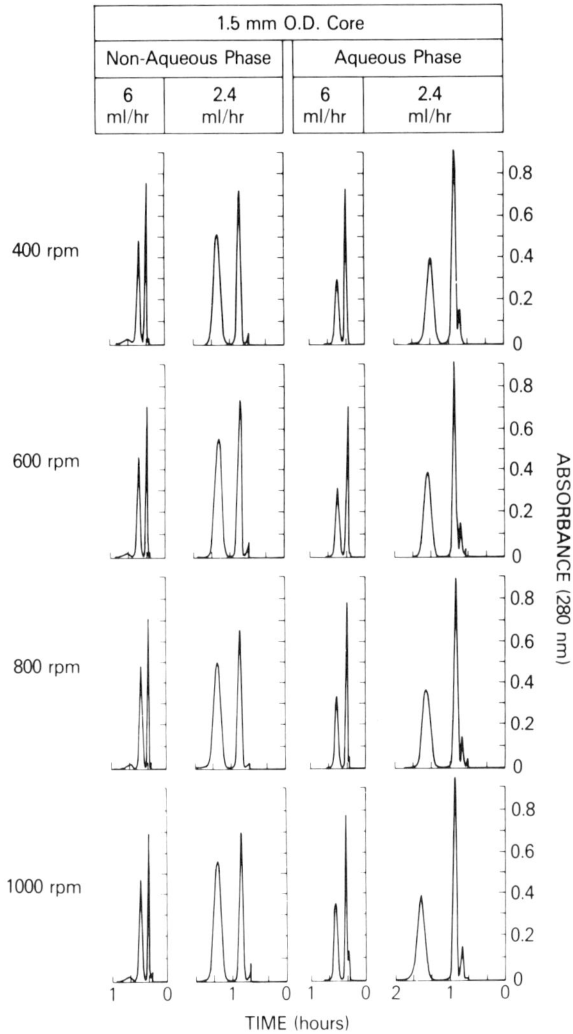

Fig. 90 (Continued)

centration of each component at 0.1–0.4 g%. The oligopeptide sample mixture was similarly prepared to bring the final concentration of each component at 0.05–0.6 g%.

In each separation, the column was first filled with the stationary phase and 50 μl of the above sample solution was injected through the sample port. Then the mobile phase was pumped into the column at a flow rate of 2.4 ml/hr while the apparatus was run at revolution-

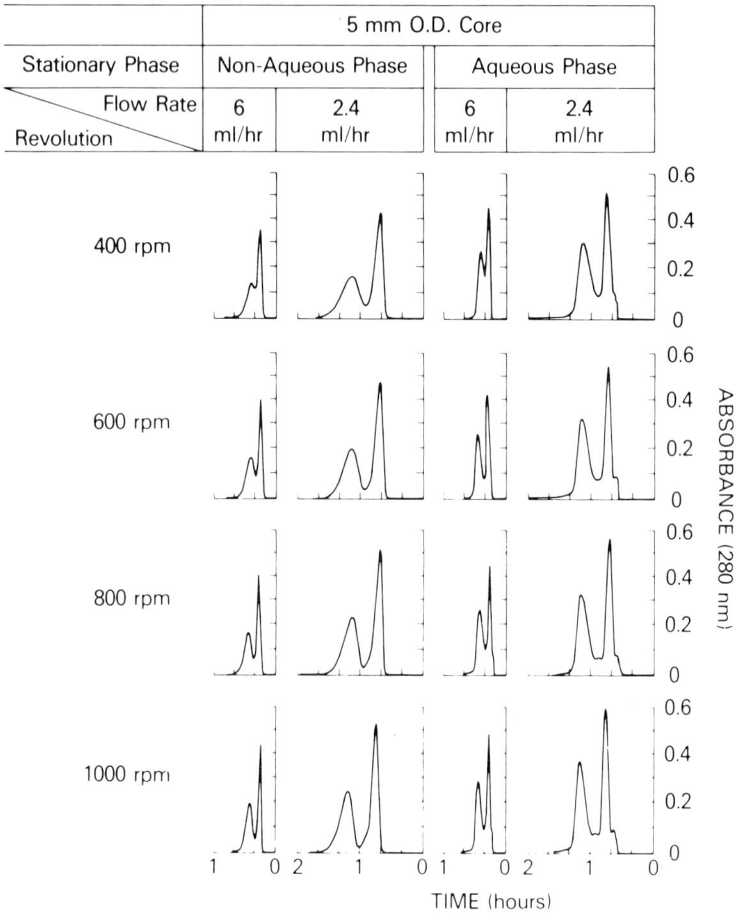

Fig. 90 (Continued)

al speeds ranging between 400 and 600 rpm. The eluate was continuously monitored with an LKB Uvicord III at 280 nm.

Figure 91a shows a chromatogram obtained from a set of DNP amino acids by eluting with the aqueous phase under a revolutional speed of 400 rpm. All components were well resolved as symmetrical peaks and eluted out in 18 hr. Figure 91b shows a similar chromatogram obtained by elution with the nonaqueous phase. Partition efficiencies of these separations estimated from Eq. (30) range between 6000 and 2000 theoretical plates. Figure 92a shows a chromatogram of oligopeptides obtained by eluting with the aqueous phase. Because of the nonlinear distribution isotherm of the samples

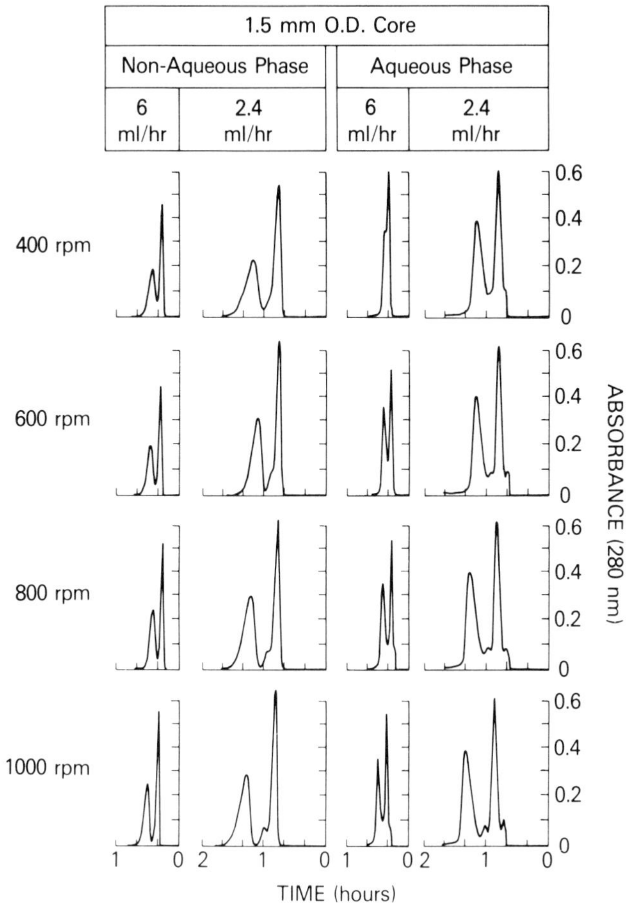

Fig. 90 (Continued)

in the n-butanol phase system, peaks show intensive skewing. The shape of this skewing is reversed when the samples are eluted with the nonaqueous phase as shown in Fig. 92b. The degree of skewing will be decreased as the sample concentration is reduced. As expected from the results of the previous studies, partition efficiency of the oligopeptide separation is lower than that in the DNP amino acid separation mainly due to the higher viscosity of the normal butanol phase system. Partition efficiencies of these oligopeptide separations range between 3000 and 300 theoretical plates.

These chromatograms were obtained under the revolutional speeds considerably below the optimum ranges determined by the previous

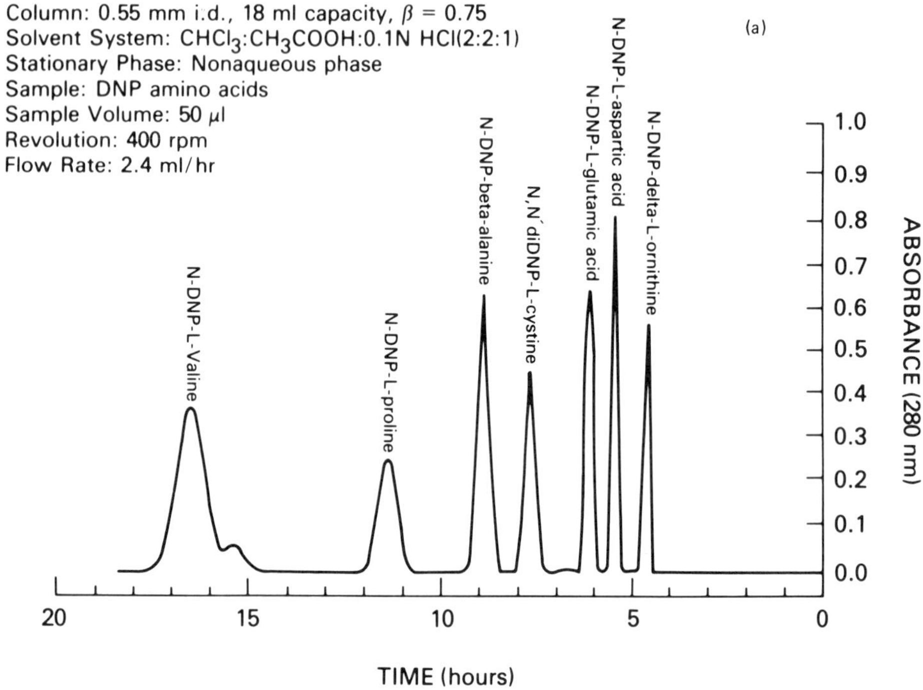

Fig. 91 Separations of a set of DNP amino acids with the long toroidal coil (0.55 mm i.d. and 50 m long). (a) Chromatogram obtained with the upper aqueous phase mobile. (b) Chromatogram obtained with the lower nonaqueous phase mobile.

studies. The use of a long, coiled column causes a pressure buildup in the column which could exceed the maximum capacity of the Chromatronix pump rated at 500 psi. This limited the operation from 400 to 600 rpm to maintain a safety pressure range below 400 psi. However, the PTFE tube used in the column can hold much higher pressures and therefore the use of a metering pump with a higher pressure capacity would yield substantially better resolution by allowing higher revolutional speeds.

Application of the Apparatus

In spite of the excellent analytical capability, the utility of the toroidal CPC has been limited to few applications as described below.

1. *Plant Hormones.* The present method has been successfully applied to separations of various types of plant hormones [72,73].

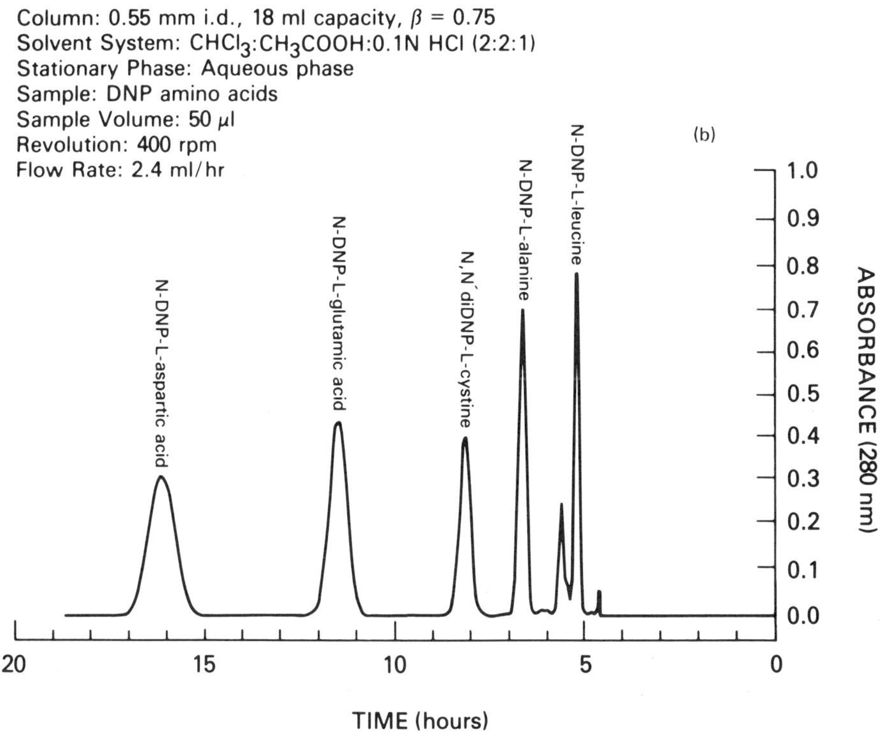

Fig. 91 (Continued)

Indole auxins were separated on either hexane/ethyl acetate/methanol/ water (3:7:5:5) or chloroform/acetic acid/water (2:2:1). The latter solvent system was found to be useful for the separation of abscisic acid from indole-3-acetic acid. Gibberellins (GA_3, GA_4, and GA_7) were separated from each other on ether/methanol/0.5 M phosphate buffer (pH 5.7) (3:1:2) whereas four cytokinins (zeatin, zeatin riboside, 6-isopentenyladenosine, and 6-isopentenyladenine) were separated on a similar solvent system composed of ethyl acetate/ methanol/0.5 M phosphate buffer (pH 7.0) (3:1:3). Details of these separations are discussed by Mandava and Ruth in Chap. 8.

2. *Partition of Cell Organelles with Polymer Phase Systems.* Recently the present method was applied to partition of rat liver cell organelles with aqueous-aqueous polymer phase systems [74]. The size and ease of operation of the toroidal CPC gave significant advantages over conventional CCD systems achieving equivalent resolution in fraction of the times. The results of these separations will be fully described in Chap. 11 by Sutherland.

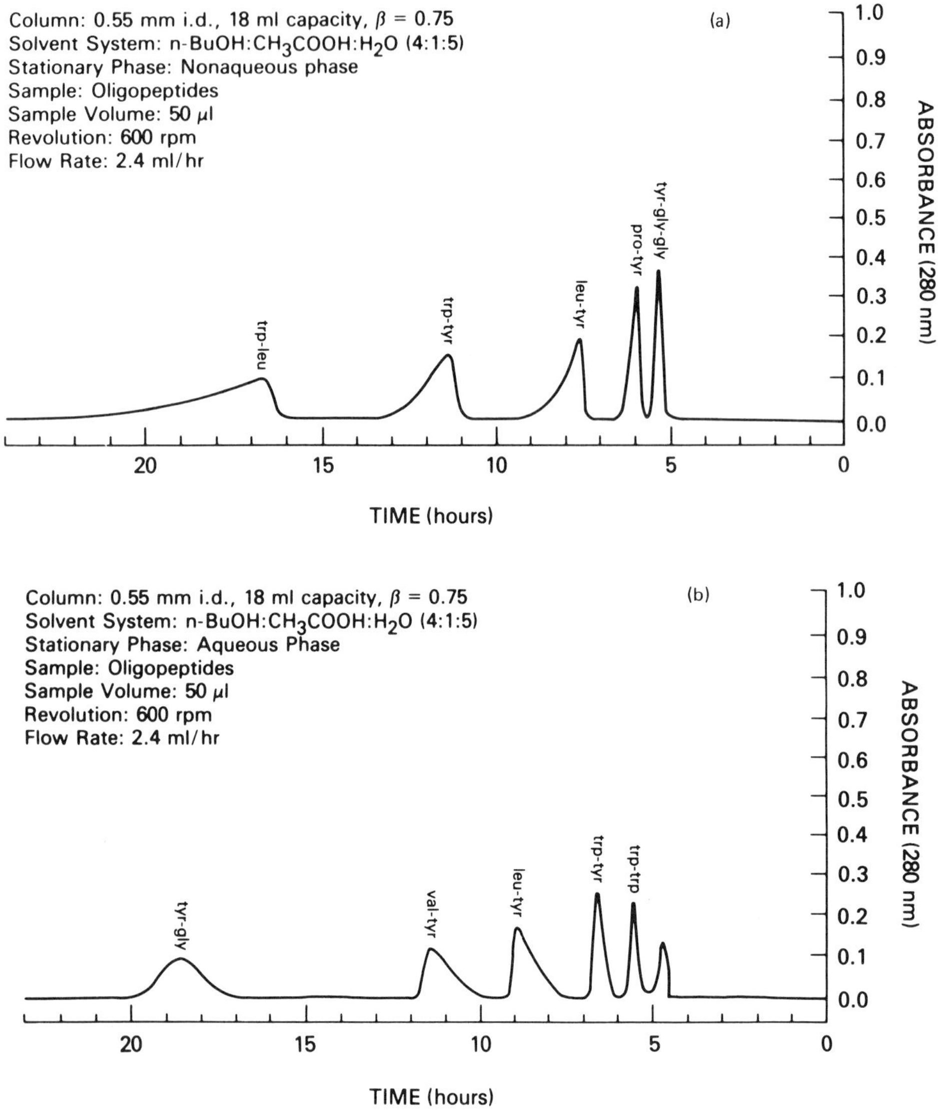

Fig. 92 Separations of a set of oligopeptides with the long toroidal coil (0.55 mm i.d. and 50 m long). (a) Chromatogram obtained with the lower aqueous phase mobile. (b) Chromatogram obtained with the upper nonaqueous phase mobile.

VIII. NONSYNCHRONOUS FLOW-THROUGH COIL PLANET CENTRIFUGES

Among all the CPCs developed for performing CCC, the nonsynchronous flow-through CPCs require the most elaborative mechanical design which provides any desirable combination of rotation (around its own axis) and revolution (around the centrifuge axis) of the coil holder. In our laboratory three types of the nonsynchronous flow-through CPCs were designed, one equipped with a dual-rotary seal [75] and the other two models without rotary seal [37,38] by adapting scheme VI as illustrated in Fig. 17.

The nonsynchronous planetary motion provides a specific advantage in performing partition of various biological samples with aqueous-aqueous polymer phase systems. These polymer phase systems possess extremely low interfacial tension, high viscosity, and relatively small difference in density between the two phases. This unique physical properties of the solvent system produce problems of emulsification under vigorous agitation and high mass transfer resistance under inefficient mixing. In the nonsynchronous flow-through CPCs, the proper degree of phase mixing is provided by adjusting the coil rotation while emulsification is prevented by applying a high revolutional speed to ensure satisfactory retention of the stationary phase. In addition the nonsynchronous flow-through CPCs provide an ideal condition for elutriation of cell particles with a single-phase physiological solution. Constant mixing and dispersing of the particles can be maintained by suitable coil rotation under a strong centrifugal force field generated by a high revolutional rate of the apparatus.

A. Design Principle of the Apparatus

The Nonsynchronous Flow-Through CPC with a Dual Rotary Seal [75]

Figure 93 shows a cross-sectional view through the center of the apparatus. The motor directly drives the central shaft and a pair of rotary wings that support the column holder shaft and the counterweight shaft symmetrically through bearings. The motor also rotates the countershaft by means of toothed pulleys (P_4 and P_3) coupled with a toothed belt. The countershaft is equipped with two identical toothed pulleys (P_2): one (left) drives the column holder through the idler pulleys (P_1) and toothed belts, and the other (right) drives the rotary piece of the dual rotating seal by toothed pulleys (P_1 and P_2) and a toothed belt. Under this arrangement the column holder and the rotary piece of the rotating seal rotates at the same angular velocity to prevent twisting of the flow tubes. The ratio (R) of rotational and revolutional speeds of the column holder is determined by the tooth number of the pulleys P_1 through P_4 as

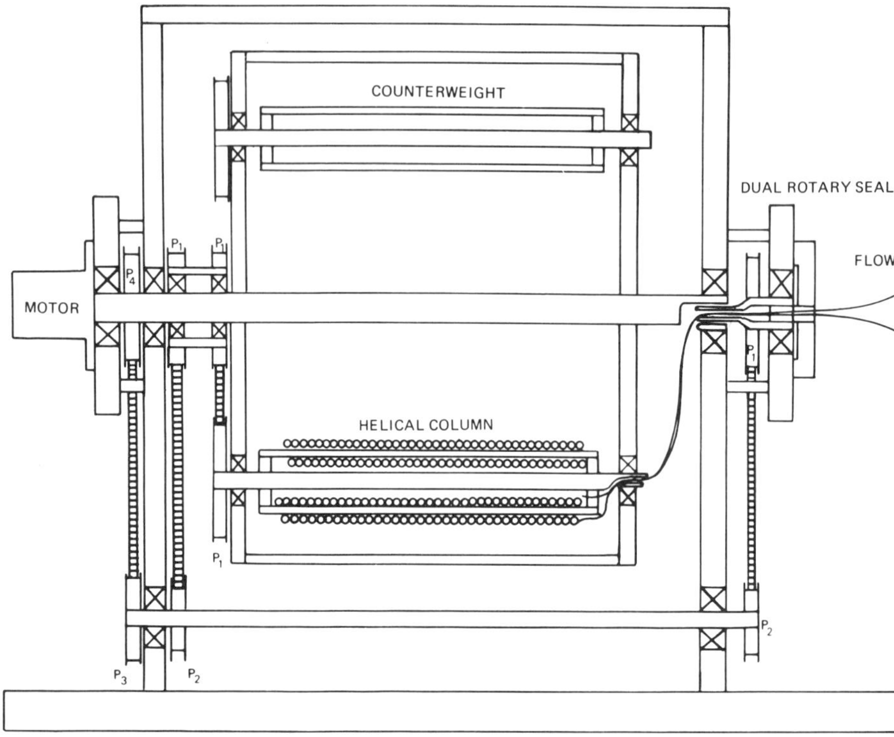

Fig. 93 Cross-sectional view of the nonsynchronous flow-through coil planet centrifuge equipped with a dual rotating seal.

$$R = 1 - \frac{p_2 \times p_4}{p_1 \times p_3} \tag{32}$$

where p_i denotes the tooth number of pulley P_i. In our prototype the rotation/revolution ratio is optimized at 1:143. However, the present system can be modified by using the second motor directly connected to the countershaft to eliminate pulleys P_3 and P_4. In this case the speed of rotation and the speed of revolution can be independently selected by choosing the proper speed for the two motors.

The Nonsynchronous Flow-Through CPC Without Rotary Seals (Scheme VI) [38]

Figure 94a illustrates the principle of the rotating-seal free design of the apparatus. A cylindrical column holder undergoes a planetary

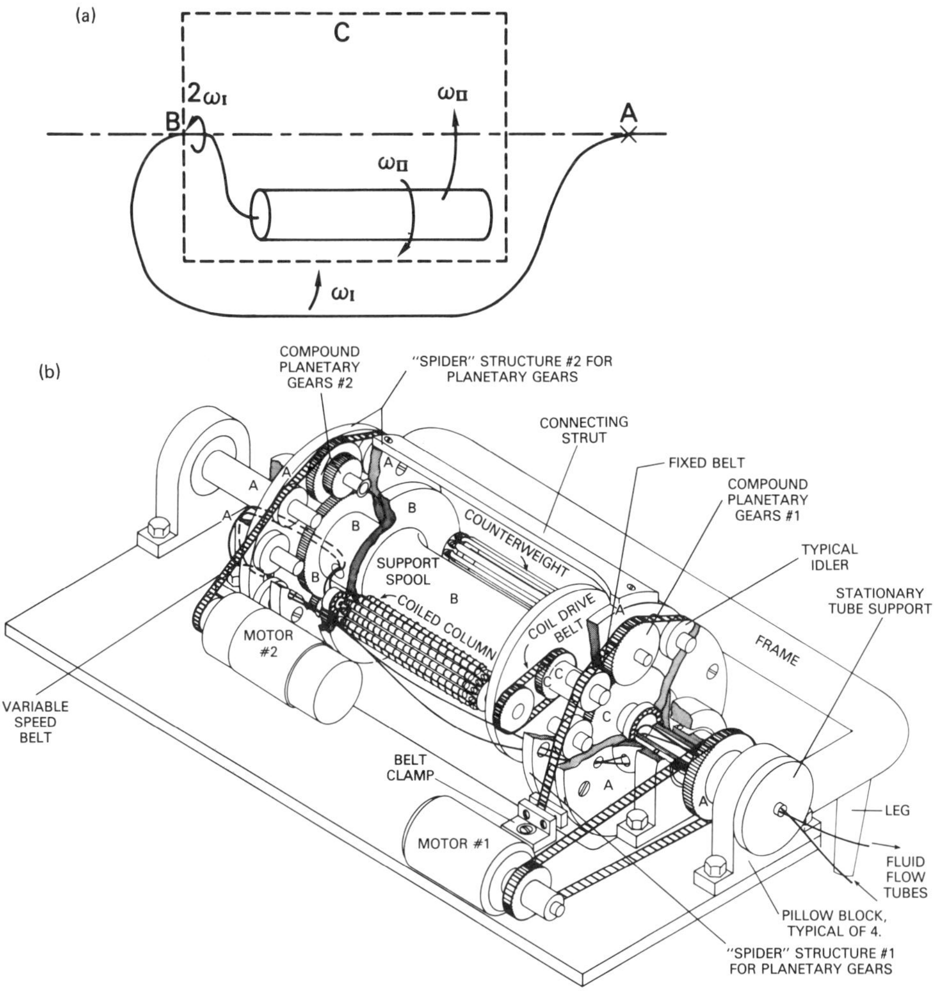

Fig. 94 The nonsynchronous flow-through coil planet centrifuge without rotating seals. (a) Design principle of the apparatus. (b) Sketch of implementation.

motion as indicated by a pair of arrows (ω_{II}). A bundle of flow tubes led from the holder first exits the imaginary compartment C shown by a dotted line at point B located on the central axis of the apparatus and then makes a loop to reach point A located on the central axis of the apparatus on the opposite side. At A the bundle is tightly supported by a stationary member of the apparatus.

For ease in the description of the system, the total pathway of the bundle is divided into two parts, i.e., A to B and within compartment C. Between A and B the loop of the bundle rotates in a circle around the central axis of the apparatus at angular velocity ω_I and at B it rotates about its own axis at $2\omega_I$ in the same direction. The system is identical to that in the flow-through centrifuge described earlier (scheme V in Fig. 17) in that the bundle is free from twisting. Within compartment C, which rotates at $2\omega_I$, the holder undergoes a synchronous planetary motion with respect to the rotating compartment C, i.e., the holder counterrotates about its own axis at ω_{II} and simultaneously revolves around the central axis of the apparatus also at ω_{II}. This system becomes identical to the flow-through coil planet centrifuge system. The direction of this synchronous planetary motion can also be reversed with the same effect. In either case, the bundle is free from twisting within compartment C. The entire length of the bundle therefore becomes free of twisting at any arbitrary, independent values of ω_I and ω_{II}.

Under the above mode of planetary motion, the net revolutional rate of the holder with respect to the outside observer is given by $2\omega_I + \omega_{II}$, the sum of the rotational rate of compartment C ($2\omega_I$) and the revolutional rate of the holder relative to compartment C (ω_{II}). The total acceleration acting on the axis of the holder is given by

$$\alpha = R(2\omega_I + \omega_{II})^2 \tag{33}$$

where α denotes the total acceleration and R is the radius of revolution or the distance from the holder axis to the central axis of the apparatus. The rotational rate of the holder about its own axis is given by ω_{II} and is entirely dependent upon the synchronous planetary motion of the holder within compartment C. Therefore, the ratio of the rotational rate to the revolutional rate of the holder is expressed as $\omega_{II}/(2\omega_I + \omega_{II})$. Because ω_I and ω_{II} are made independently variable, the present system permits any value of the above ratio under a desired acceleration field acting on the axis of the holder. When the coiled separation column is mounted around the holder, the rotation of the holder induces an additional acceleration which acts radially from the axis of the holder to modify the total acceleration expressed in Eq. (33). However, the effect is practically negligible due to the fact that the optimum rotational speed (ω_{II}) for cell separation usually ranges from 3 to 20 rpm (at 3 cm radius) compared with the revolutional rate ($2\omega_I + \omega_{II}$) of 500–1200 rpm (at 15 cm radius). Rotation of the coiled column also produces an undulating centrifugal force field acting on the individual column units mounted eccentrically around the holder as shown in Fig. 94b. Under this arrangement each column unit is subjected to a greater force field at the position remote from the central axis of the apparatus and a smaller force field at the position close to the same axis. While the actual effect of this undulation of the centrifugal force field on cell separation is unknown, it may be desirable to minimize this effect by mounting the column units close to the axis of the holder. This will tend to decrease the eccentric motion caused by the rotation of the column holder.

In the present model the required independent rotations of the coils are provided by two motors and a system of planetary gears to transmit the motions while guiding and not interfering with the motions of the tube bundle.

As shown in Fig. 94b, motor I drives gear supports (spiders) A at both ends of the apparatus at ω_I. The motor drives one spider via a timing belt, and the other spider is connected to the first via two connecting struts. The spiders support the outer loop of tubing and cause the planetary gears to orbit at ω_I.

The internal ring gears of traditional planetary sets have been replaced by timing belts supported by idlers. At one end of the apparatus this belt is fixed, causing the sun gear C to rotate at $2\omega_I$. The effective pitch diameter of the fixed belt is 12 in., that of the sun gear 6 in., and those of the compound gears 2 and 4 in. Since there are two diametrically opposed sets of planet gears at each end, at least one is in mesh at all times, even when the other one is freewheeling as it passes the area where the belt is fixed.

Sun gear C drives the coiled column support via an auxiliary shaft on the centerline and a "coil drive belt" (a timing belt). Thus, with respect to the laboratory frame, the coiled column rotates always at exactly $2\omega_I$. (Study of the kinematics of an orbiting pulley driven by a belt from the central pulley will show that angular rotational velocity with respect to the laboratory frame is independent of orbital velocity.)

At the other end of the apparatus sun gear B also rotates at $2\omega_I$ if motor II is stopped because there is an identical set of planetary gears. However, when motor II imposes an effective angular velocity to the belt acting as the internal gear for the planetary set at that end, then the angular velocity ω_{II} will be superimposed on sun gear B which will revolve at $2\omega_I + \omega_{II}$. Sun gear B is rigidly attached to the support spool and the coiled columns must orbit at $2\omega_I + \omega_{II}$.

The present device is still undergoing mechanical refinement. Standard timing belts (Browning) seem to behave in a trouble-free manner as opposed to a proprietary "sprocket cable" (Winfred M. Berg) which was tried first. There is noticeable vibration due to the polygonal shape of the belts as supported by the finite number of idlers, but it is believed that this could be greatly reduced in another model by using a large number (perhaps double) of smaller idlers. This system does avoid the necessity of custom internal gears which could be substituted.

To avoid the demanding tolerances of alignment of the various bearings for rotating members, the position of each spider assembly is determined by two coaxial bearings, and the support spool is then supported by one bearing in each spider assembly. Thus small amounts of angular misalignment can be tolerated as the system is not overdetermined.

A rectangle formed of 2-in. diameter, 0.125-in. wall aluminum tubing surrounds the rotating elements and supports the bearing assemblies at each end and the motors to the side. Thus rigidity, light weight, and excellent access to all parts are achieved. This stiffness is required to withstand the forces and vibrations produced by 1200 rpm operation (164 g at the 4-in. radius of the axis of the coils).

The Improved Nonsynchronous Flow-Through CPC Without Rotary Seals (Scheme VI) [37]

Figure 95 shows a cross-sectional view through the central axis of the apparatus. The rotor consists of two major rotary structures, i.e., (rotary) frames I and II which are coaxially bridged together with the center piece (dark shade).

Frame I consists of 3 plates rigidly linked together and directly driven by motor I. It holds three rotary elements, the centerpiece (center), countershaft I (bottom), and countershaft II (top), all embedded in ball bearings. A pair of long arms extending symmetrically and perpendicularly from the middle plate forms the tube supporting frame which clears over frame II to reach the central shaft on the right side of frame II. Frame II (light shade) consists of three pairs of arms linked together to rotate around the central shaft. It supports a pair of rotary shafts, one holding a coil holder assembly and the other the counterweight.

There are two motors, motors I and II, to drive the rotor. When motor I drives frame I at ω_I, the stationary pulley 1 introduces counterrotation of pulley 2 through a toothed belt and therefore countershaft 1 rotates at $-\omega_I$ with respect to rotating frame I. This motion is further conveyed to the centerpiece by 1:1 gearing between gears 1 and 2. Thus the centerpiece rotates at $2\omega_I$ or at ω_I with respect to the rotating frame I. The motion of frame II, however, also depends on the motion of motor II.

If motor II is at rest, pulley 5 becomes stationary as is pulley 1 so that countershaft II counterrotates at ω_I as does countershaft I. This motion is similarly conveyed to the rotary arms of frame II by 1:1 gearing between gears 3 and 4, resulting in rotation of frame II at $2\omega_I$ or the same angular velocity as that of the centerpiece. Consequently, coupling of pulleys 6–7 and 8–9 with toothed belts produces no additional motion to the rotary shaft which simply revolves with frame II at $2\omega_I$ around the central axis of the apparatus.

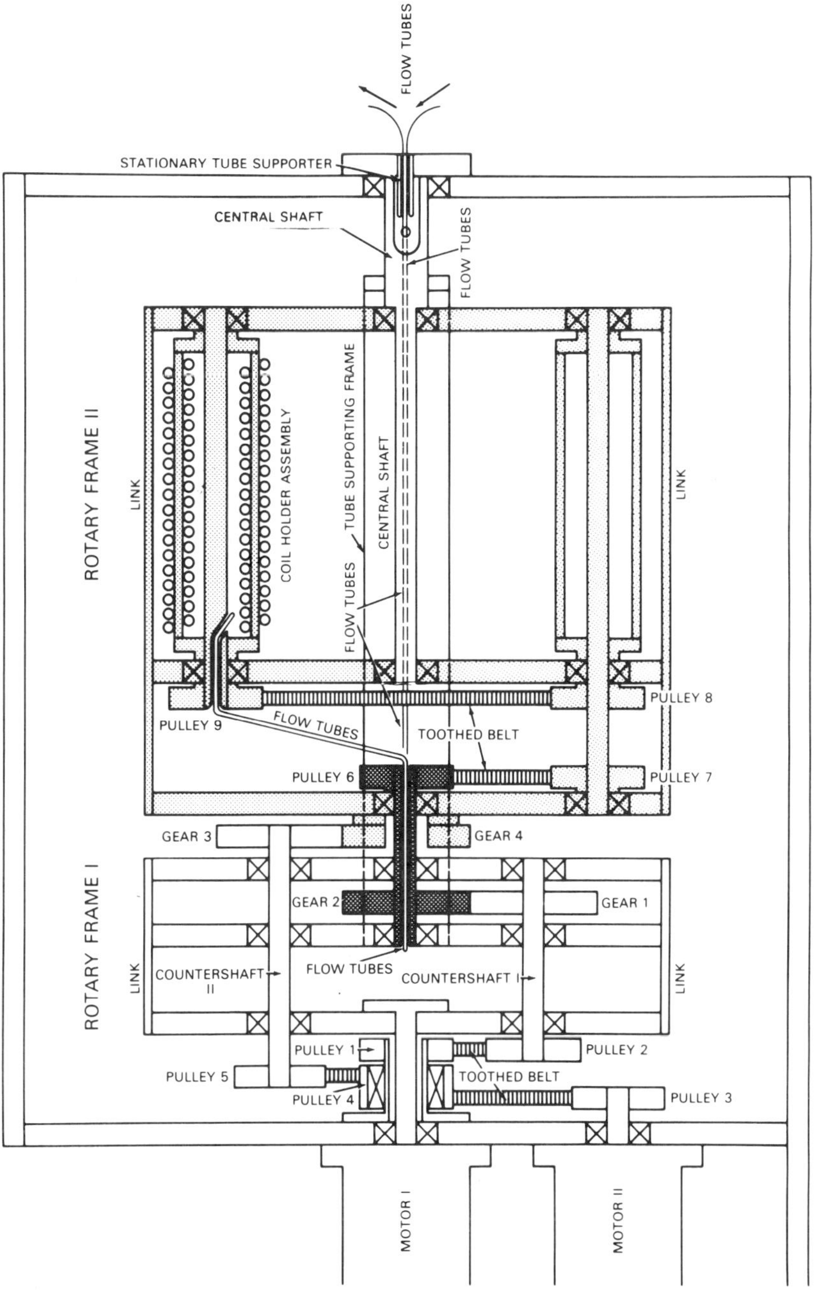

Fig. 95 Improved design of the nonsynchronous flow-through coil planet centrifuge free of rotating seals. Motor I produces a high-speed revolution while motor II gives a slow rotation of the coil assembly around its own axis. This design is based on scheme VI in the nonsynchronous unit shown in Fig. 17.

When motor II rotates at ω_{II}, idler pulley 4 coupled to pulley 3 on the motor shaft rotates at the same rate, which in turn modifies the rotational rate of pulley 5 on countershaft II. Thus countershaft II now counterrotates at $\omega_I - \omega_{II}$ on frame I. This motion further alters the rotational rate of frame II through 1:1 gear coupling between gears 3 and 4. Subsequently, frame II rotates at $2\omega_I - \omega_{II}$ with respect to the earth or at $-\omega_{II}$ relative to the centerpiece which always rotates at $2\omega_I$. The difference in rotational rate between frame II and the centerpiece is conveyed to the rotary shafts through coupling of pulleys 6–7 and 8–9. Consequently, both rotary shafts rotate at ω_{II} about their own axes while revolving around the central axis of the apparatus at $2\omega_I - \omega_{II}$. This gives the rotation/revolution ratio of the rotary shaft:

$$\frac{r}{R} = \frac{-\omega_{II}}{(2\omega_I - \omega_{II})} \tag{34}$$

Therefore, any combination of revolutional and rotational speeds of the coil holder assembly can be achieved by selecting the proper values for ω_I and ω_{II}.

The coiled column was prepared by winding 1-mm i.d. PTFE tubing continually onto six units of 0.68-cm o.d. stainless steel pipe to make about 600 helical turns with a total capacity of 15 ml. The column units were symmetrically mounted around the rotary shaft with screws through the hole at each end of the units. The flow tubes connected to the coiled column were first led through the hole of the rotary shaft and passed through the opening of the centerpiece to exit at the middle portion of frame I. The flow tubes were then led along the tube support frame to clear frame II and then reached the side hole of the central shaft near the right wall of the centrifuge where they were tightly held by the stationary tube supporter. These tubes were lubricated with grease and protected with a piece of flexible plastic tubing at each supported portion to prevent direct contact with the metal parts.

Figure 96 shows the overall picture of the apparatus. The revolutional speed of the coil holder assembly is continuously adjustable up to 1000 rpm combined with any given rotational rate between 0 and 50 rpm in either direction. The improved design of the present apparatus completely eliminated noise and vibration produced by the first model with the spider gear assembly.

B. CCC With Polymer-Polymer Phase Systems

Principle

As described in the hydrodynamic equilibrium system (HDES) (Fig. 13), rotation of a coil in an acceleration field creates an Archimedean

Fig. 96 Photograph of the improved nonsynchronous flow-through coil planet centrifuge free of rotating seals.

screw force which moves all objects in the coil toward the head of the coil. This effect can be used to retain the stationary phase in a rotating coil while the mobile phase continuously elutes through the coil. Using this hydrodynamic effect, CCC has been successfully performed with conventional two-phase solvent systems in a rotating centrifugal force field produced by synchronous planetary motion. However, application of polymer phase systems to the synchronous CPCs produces a problem of emulsification due to the low interfacial tension of the solvent phases combined with vigorous agitation produced by the synchronous rotation of the coil. Successful application of the polymer phase systems to the synchronous CPC systems has been made by modifying the orientation of the column holder to incorporate the feature of the hydrostatic equilibrium system (HSES) by means of the angle rotor CPC or the elution centrifuge described earlier.

Because of its freely adjustable rotation rate with respect to the revolution, the nonsynchronous flow-through CPC enables the use of polymer phase systems by the combined use of slow rotation and high revolution of the coiled column.

Application

Preliminary experiments have been performed to demonstrate the versatility of the present method in separations of various biological samples. Typical examples described below include separation of the plasmid DNA from RNA, partition of mutant bacterial cells, and separation of mammalian erythrocytes.

1. *Plasmid DNA [37]*. Crude preparation of the plasmid pBR 322 was obtained from a lysate of *E. coli* C600 r^-m^- according to the procedure used by Falkow et al. [76]. The polymer phase system composed of 5% (w/w) dextran 500, 4% (w/w) polyethylene glycol 6000, and 10 mM sodium phosphate (pH 6.8) was equilibrated in a separatory funnel and separated before use. In each separation the coiled column was first filled with the stationary lower phase and the sample solution containing 100 μg DNA was injected through the sample port located on the flow line between the inlet of the column and the outlet of the pump. Then the mobile upper phase was pumped through the head of the column at 8.5 ml/hr (FMI Lab Pump, Fluid Metering), while the apparatus was run at 1000 rpm combined with 5 rpm coil rotation. The effluent from the outlet of the column was continuously monitored with an LKB Uvicord S at 256 nm and fractionated with an LKB fraction collector. Aliquots of the fractions were then electrophoreses on 1% agarose gels run at 5 V/cm and, after completion, the gel was stained with ethidium bromide to visualize the DNA and RNA bands.

Figure 97 illustrates the typical results obtained in separation of *E. coli* plasmid DNA with the polymer phase system using the improved

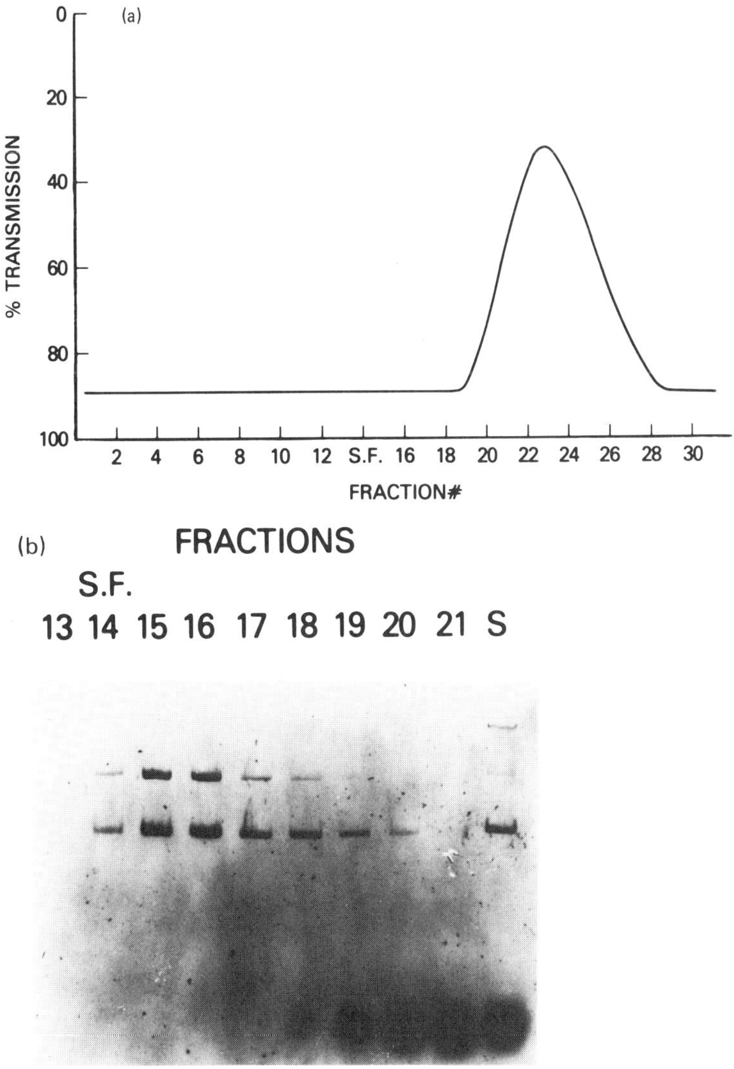

Fig. 97 Separation of *E. coli* plasmid DNA with the nonsynchronous flow-through coil planet centrifuge. (a) Countercurrent chromatogram of the crude sample on the polymer phase system. The chart shows a broad peak of RNAs. The plasmid DNA eluted immediately after the solvent front is not detected. (b) Agarose gel electrophoresis of the fractions. Fractions obtained with countercurrent chromatography above are further analyzed with gel electrophoresis. The plasmid DNA is detected in the early fractions and fairly well separated from the RNAs.

model of the nonsynchronous flow-through CPC. Figure 97a is an elution profile of the sample monitored with the Uvicord S at 256 nm. A broad peak recorded on the chart represents various RNA components which are partitioned rather evenly between the mobile and stationary phases. However, this monitoring system failed to detect a small amount of plasmid DNA in the sample which almost entirely partitioned in the upper mobile phase and therefore eluted immediately after the solvent front (fraction 14) marked on the chart. In order to analyze plasmid DNA, aliquots of early fractions, 14–21, were electrophoresed on 1% agarose gels run at 5 V/cm and after completion gels were stained with ethidium bromide to visualize the DNA and RNA bands (Fig. 97b). The plasmid DNA (sharp bands) is seen to run at the solvent front whereas the RNA (diffuse band near bottom of gel) first appears three to four fractions later. Although there is some degree of overlap, pooling of fractions containing DNA but not RNA, as determined on agarose gels, yielded recoveries of 74%. Boiling of the sample for 2 min followed by quick cooling in a Dry Ice–ethanol bath, prior to loading, resulted in the separation of closed circular plasmid DNA from denatured linear DNA, of which greater than 90% is retained in the stationary phase. The largest sample loading attempted was 400 μg of DNA, with equal resolution obtained.

2. *Salmonella typhimurium strain [37].* Samples of *Salmonella typhimurium* differing in cell surface lipopolysaccharide were prepared by growing *S. typhimurium* strain G30 which lacks UDP glucose-4-epimerase in proteose peptone-beef extract medium, either with galactose (0.01–0.1 M) or in its absence. In the presence of galactose, this strain makes a normal lipopolysaccharide; in its absence, it makes a short chain terminated at the point of addition of the first galactose in the molecule [77]. Bacteria with these different lipopolysaccharides have previously been shown to differ in partition in countercurrent distribution [78].

The polymer phase system used for separation was composed of 6.2% (w/w) dextran 500 (Sigma), 4.4% (w/w) polyethylene glycol 6000 (Union Carbide Corp.), 0.05 M Tris-Cl (pH 7.0), 10 mM potassium phosphate (pH 7.0), and 0.01% sodium azide. In each separation the column was initially filled with the polymer phase system consisting of about equal amounts of the upper and lower phases, and galactose (+) and galactose (−) cell mixture suspended in 1 ml of the upper phase was injected through the sample port. Then the upper mobile phase was pumped through the head of the column at 8.5 ml/hr while the apparatus was run at 1000 rpm with the optimum coil rotation of 5 rpm. The eluates from the tail of the column were continuously monitored with an LKB Uvicord S at 280 nm and fractionated into test tubes. An aliquot of each fraction was diluted with distilled water and the elution profile of cells was determined by measuring the turbidity at 280 nm with a Beckman DU spectrophotometer.

Figure 98 shows three partition chromatograms of *S. typhimurium* obtained with the improved model of the nonsynchronous CPC. The cells grown without galactose are mostly partitioned either at the interface or in the lower stationary phase and therefore retained in the column for a long period of time, these cells being recovered by eluting the column contents after stopping the centrifuge run (top). The cells grown with galactose develop the outer layer with long polysaccharide which alters their partition behavior. As a result, they are mostly distributed in the upper mobile phase and therefore eluted from the column much earlier (middle). The chromatogram obtained from the mixture of these two types of cells reproduces the above two peaks at the respective locations (bottom).

The experiments have been continued with radioisotope-labeling techniques which facilitate the analysis of the cell fractions with a high specificity. The recovery rate of the cells are also accurately measured in this way. It has been found that the column cleaned with 1 N NaOH solution overnight gives near 100% recovery of the activity and the column can be reused without the risk of contamination. With the combined use of radiolabeled galactose (^{14}C) and leucine (^{3}H), the process of lipopolysaccharide formation and translocation to the outer membrane of these cells has been successfully studied by Leive et al. using the improved model of the nonsynchronous flow-through CPC [79].

3. *Mammalian Erythrocytes [80].* Mammalian erythrocytes exhibit characteristic partition behavior on the aqueous-aqueous polymer phase systems composed of polyethylene glycol 6000, dextran 500, and phosphate buffer containing NaCl at various concentrations. Figure 99 shows partition diagram of three species of erythrocytes (dog, human, and sheep). At zero salt concentration, all cells distribute toward the upper, polyethylene glycol-rich phase. As the salt concentration increases the cells become increasingly adsorbed at the interface between the two phases until at 0.15 M salt concentration there are hardly any cells left in the upper phase. The largest differences between the cells are at 0.075 M salt concentration where 61% dog, 40% human, and 25% sheep distribute in the upper phase. This difference in partition coefficeint (K = 1.56, 0.67, and 0.33, respectively) has been used to separate populations of these cells using the nonsynchronous flow-through coil planet centrifuge equipped with a dual rotating seal.

The separation of a 20% mixture of packed sheep and dog RBCs using an 0.075 M NaCl phase system is shown in Fig. 100. The flow rate was 14 ml/hr with a centrifuge speed of 750 rpm and coil rotation 10 rpm. Alternate fractions were diluted and hemolysed to give the results shown in Fig. 100, while the remainder were kept for analysis using a blood cell counter. The cells separated into two populations and the analysis of the peak fractions is shown in Table 5.

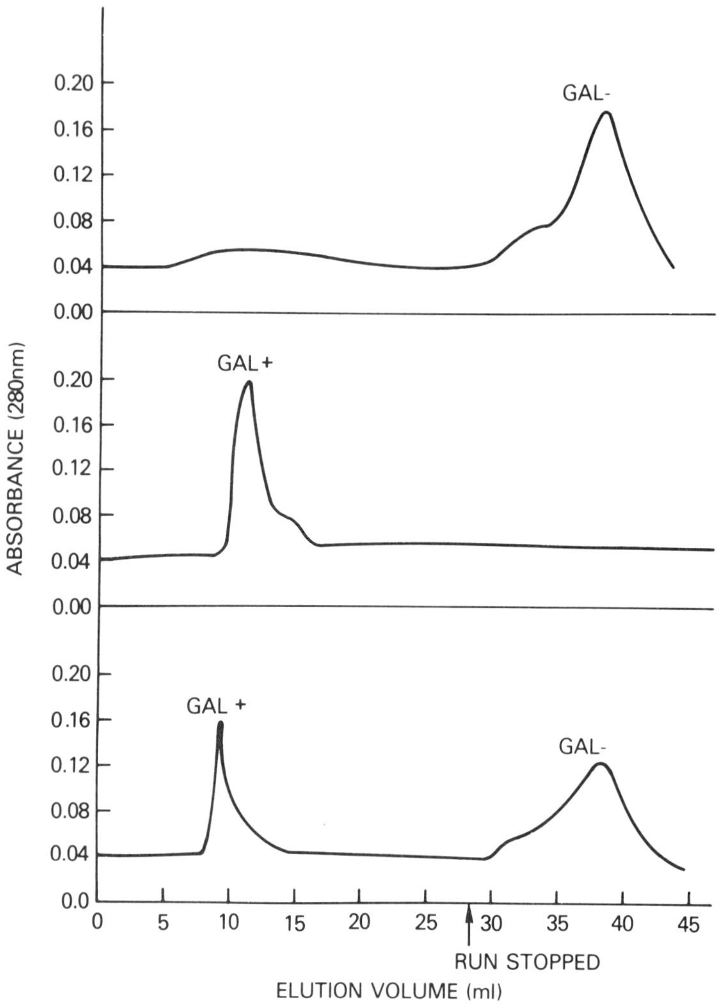

Fig. 98 Separation of *Salmonella typhimurium* strain on the polymer phase system with the nonsynchronous flow-through coil planet centrifuge. *S. typhimurium* G30 strain cells cultured on the galactose-deficient media (gal −) are mostly distributed either at the interface or in the lower stationary phase and therefore retained in the column (top). The same strain of cells grown with galactose (gal +) develops lipopolysaccharide outer layer which alters their partition behavior. These cells are mostly distributed in the upper mobile phase and eluted out much earlier (middle). Countercurrent chromatography applied to the sample mixture containing these two types of cells produces two peaks each at the identical location (bottom).

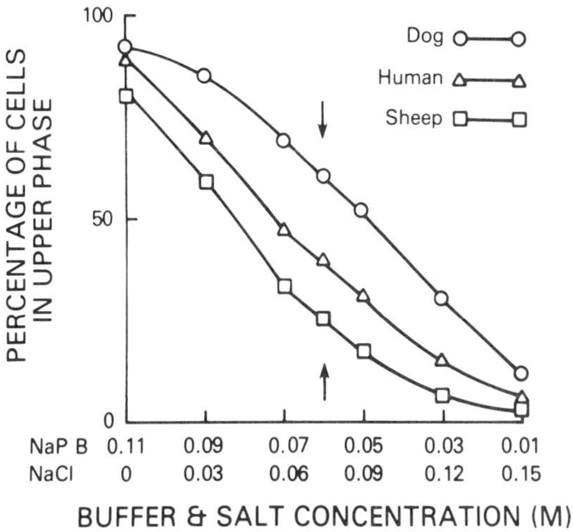

Fig. 99 Partition of dog (○-○), human (△-△), and sheep (□-□) RBCs in a two-phase polymer system at different ionic compositions. Phase system 5% (w/w) dextran 500 and 4% (w/w) PEG 6000. Settling time 2 min. As the salt concentraiton increases the distribution of three species of cells changes from the upper phase to the interface. The difference in partition at the concentrations arrowed is used as a basis for separation.

The partition coefficients of human and dog erythrocytes are much closer and so it was decided to reduce the flow to 7 ml/hr to give better retention, leaving the other parameters the same. The result is shown in Fig. 101 where two populations were separated after injecting a 20% mixture of equal quantities of packed human and dog erythrocytes. The distribution of stationary phase throughout the column contents is improved and this is reflected in the increased retention time of the second peak. As the cell sizes were so similar the blood cell counter could not be used. Instead the in-

Table 5 Proportions of Cells in Peak Fractions

Fraction no.	Proportion of cell species (%)	
	Dog	*Sheep*
3	92	8
15	2	98

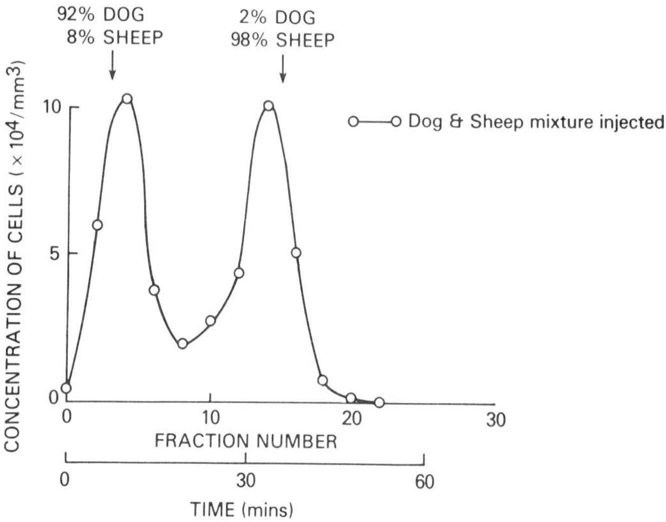

Fig. 100 An equal mixture of dog and sheep RBCs are separated in 60 min using the phase system of Fig. 99 with a salt concentration of 0.075 M NaCl and 0.06 M sodium phosphate buffer. The first peak was found to be 92% dog cells and the second peak 98% sheep cells. A 0.2-ml sample was injected at a concentration of 20% packed cell volume diluted with upper phase. Centrifuge parameters: Rotation (Ω) = 750 rpm; centrifuge radius (R) = 16 cm; coil progression (ω) = 10 rpm; number of coils = 600; coil internal diameter = 1.07 mm; helix diameter (Z) = 7.5 mm; total coil volume = 16.2 ml; flow rates 14 ml/hr; fraction time = 2.5 min.

dividual peak fractions (four highest points in the figure) were bulked and rerun. The results of rerunning the peak material under otherwise identical conditions are plotted for comparison on the same graph. Each peak occurs in the position expected with a reproducibility of 99% for the dog cells and 75% for the human.

C. Cell Elutriation with Physiological Solutions [37]

Principle

Figure 102 illustrates the principle of particle elutriation in a rotating coil [5]. The coil is initially filled with water and then sample particles are introduced at the tail (top), followed by the rotation of the coil. Under a slow rotation (left), all particles always stay together at the bottom of the coil and move toward the head of the coil

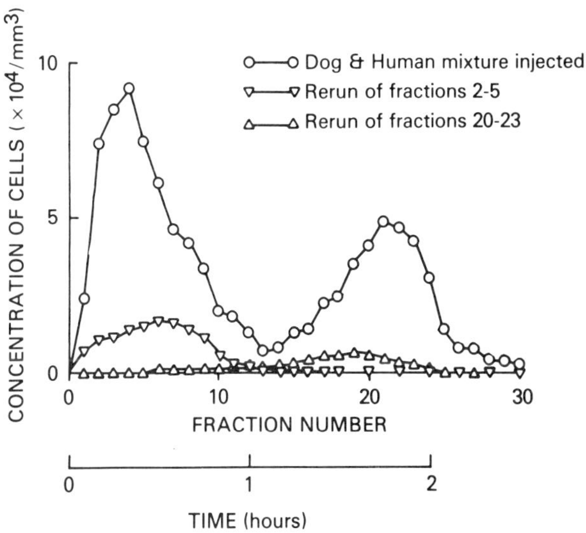

Fig. 101 An equal mixture of dog and human RBCs are separated in 2 hr using the phase system of Fig. 99 with a salt concentration of 0.075 M NaCl and 0.06 M sodium phosphate buffer. The individual peak fractions indicated in the diagram were bulked and rerun, giving 99% reproducibility for the dog cells and 74% for the human. The slower flow has given an improved retention of the stationary phase resulting in better retention of the second peak. A 0.2-ml sample was injected at a concentration of 20% packed cell volume diluted with upper phase. Centrifuge parameters: As for Fig. 100, except flow rate = 7 ml/hr. Fraction time = 5 min.

at a uniform rate of one helical turn per one rotation of the coil. In this case no separation is observed. When the rotational speed is increased (right), the particles fail to remain at the bottom of the coil and they are consequently retarded in their movement toward the head of the coil. Because the degree of this retardation depends on the sedimentation rate of the particles, the particles are chromatographically separated along the length of the coil according to their relative size and density. The fractionation of the particles becomes possible by introducing continuous flow through the column during separation. This method can be applied to the separation of small particles with the nonsynchronous flow-through coil planet centrifuge.

Preliminary Applications

1. *Mammalian Erythrocytes.* Preliminary experiments have been performed to demonstrate the potential capability of the present ap-

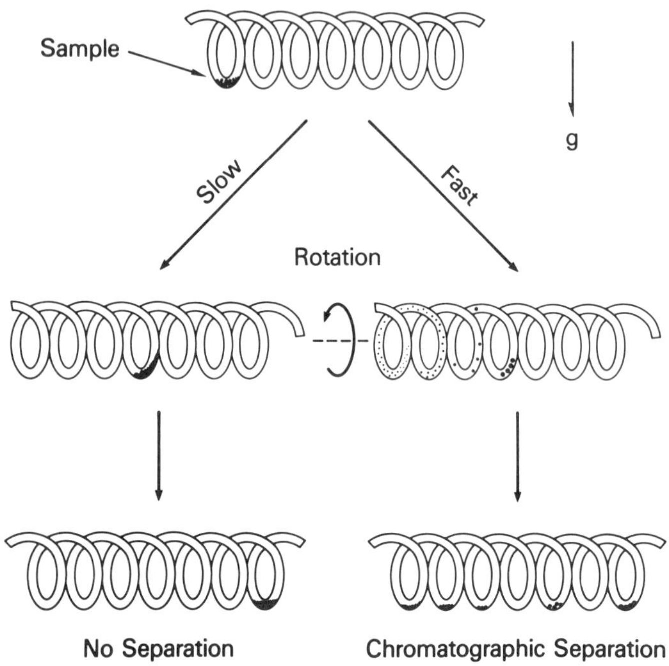

Fig. 102 Principle of particle separation with a rotating coiled tube. The coil is filled with water and the particle mixture is introduced in the tail end, both ends of the coil then being closed (top). Slow rotation of the coil carries all particles together toward the head without separation (left). As the rotational rate is increased, the larger particles move faster than the smaller particles and the particles are chromatographically separated along the length of the coil according to their sedimentation rates.

paratus by the separation of human and sheep erythrocytes according to their size difference. Isotonic buffered saline solution (pH 7.4) was prepared by dissolving NaCl, 90 g; Na_2HPO_4, 13.65 g; and $NaH_2PO_4 \cdot H_2O$, 2.15 g in 1 L of distilled water and diluting 85 ml of this stock solution with distilled water to bring the final volume to 1 L. The human erythrocyte suspension was prepared from EDTA-treated, fresh adult blood by washing with the buffered saline solution three times, repeating centrifugation and decanting the supernatant. One part of the loosely packed cells obtained was finally mixed with four parts of the same solution. The sheep erythrocyte suspension was similarly prepared from ACD-treated blood stored at 4°C. Equal volumes of human and sheep erythrocyte suspension were combined and 0.2 ml of this mixture was used for each separation.

The separation was performed as follows: The column was first filled with the buffered saline solution and the sample mixture was injected through the sample port. Then the column was eluted with the buffered saline solution while the apparatus was run at 1000 rpm combined with a slow column rotation in either direction. The effluent from the outlet of the column was continuously monitored with an LKB Uvicord S at 280 nm and fractionated in the test tubes. The original sample mixture and some fractions were analyzed with a Coulter analyzer to obtain size distribution curves.

Figure 103 shows the preliminary results of elutriation of sheep and human erythrocytes mixture with the improved nonsynchronous flow-through CPC. The experiments were performed in two different ways. In the first method the coiled column was slowly rotated (12.5 rpm) in such a direction that the cells were held back toward the inlet of the column against the flowing stream (22 ml/hr) (head–tail elution). Under this mode of elution the small sheep cells were eluted with a sharp peak which was followed by a broader peak of the large human cells as shown in Fig. 103 (top). This elution profile closely resembles the size distribution curve of the original sample mixture obtained with a Coulter analyzer (Fig. 103, bottom).

In the second method the rotational direction of the coiled column was reversed so that all cells are slowly carried toward the outlet of the column by rotation (4 rpm) while the flowing stream (45 ml/hr) further promotes the traveling rate of the smaller cells (tail–head elution). This mode of elution yielded a different elution curve characterized by a broad peak of the sheep cells and a sharp peak of the human cells as shown in Fig. 103 (middle). This method favors the recovery of the sample and also facilitates the programming of the separation times.

2. *Rat Liver Cells.* The present method was applied to the separation of rat liver cells. Sample cell suspension was prepared by means of intravenous collagenase infusion through the fresh rat liver. Figure 104 shows preliminary results of rat liver cell elutriation with a culture medium obtained by the head–tail elution mode. The elution profile of the total cell population revealed a large early peak followed by a low broad peak while a large population of the cells was still retained in the column. A dye exclusion microscopic viability test revealed that almost all cells in the first peak were damaged cells. The rest of the eluted cells were mostly viable and increased in size with the fraction number. The retained cells were the largest in size and the last few fractions consisted of aggregated cells probably resulting from incomplete digestion of the hepatic cords.

The fact that the nonviable cells were eluted early as a sharp peak strongly suggests that the cells were already damaged before being subjected to centrifugation; this was confirmed by a viability test of the original sample. The PTFE coiled column used for separation is easily sterilized by treating with either acid or alkaline solu-

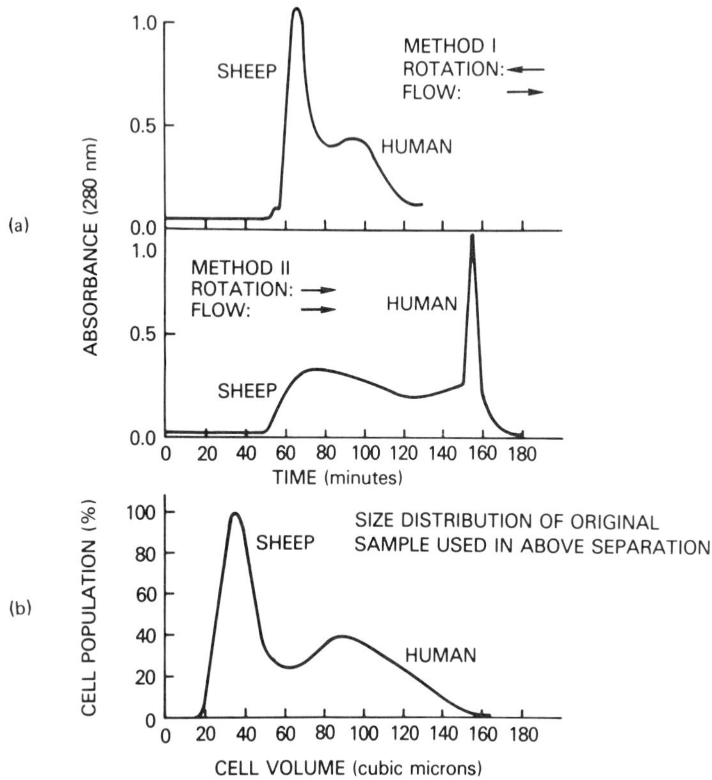

Fig. 103 Separation of human and sheep erythrocytes by elutriation with the nonsynchronous flow-through coil planet centrifuge. In the top figure (a), a mixture of human and sheep cells was eluted with the physiological solution in two different ways. In method I (a, top), cells were held by coil rotation against the flowing stream, which produced an elution profile of cells quite similar to the size distribution curve of the original sample obtained with a Coulter analyzer (bottom figure b). In method II (a, bottom), cells were slowly carried toward the outlet of the column by rotation while the flowing stream promotes the traveling rate of the smaller cells. This method produced a broad peak of the sheep cells and a sharp peak of the human cells.

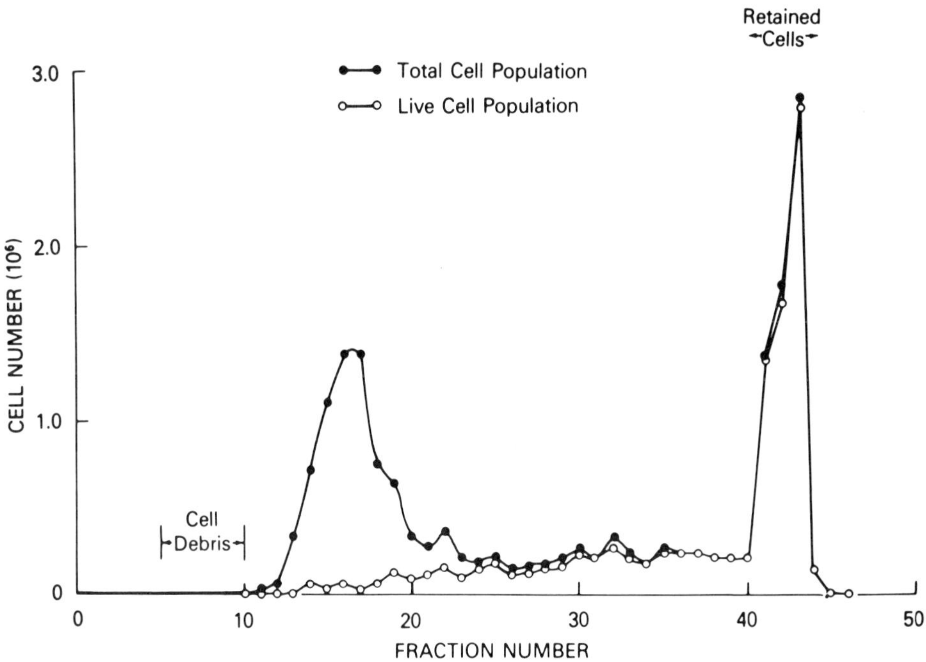

Fig. 104 Elutriation profile of rat liver cells with the nonsynchronous flow-through coil planet centrifuge. Damaged cells eluted first at a sharp peak while most of the viable cells were retained in the coiled column.

tion and the fractions thus obtained may be further used for cell culture or other experiments without risk of contamination.

IX. SLOWLY ROTATING COIL ASSEMBLY

This simple preparative CCC scheme has stemmed directly from the basic hydrodynamic equilibrium system (HDES) described in Sec. II. In the original model, a simple rotary shaft carried a bundle of coiled tubes which was eluted through a rotating seal located at each end of the rotary shaft. The rotating seals were eliminated in the recent model by adapting the scheme V design illustrated in Fig. 17. These devices have yielded efficient preparative-scale separations comparable to those obtained with droplet CCC, but in much shorter elution times.

In the course of development it has been found that the coil, if directly wound around the rotary shaft (coaxial orientation), produces a unique hydrodynamic effect at the critical rotational speed; two immiscible solvent phases confined in the coil are unilaterally distributed in such a way that one phase entirely occupies the head side of the coil and the other phase the tail side of the coil. As described earlier, this unilateral phase distribution can be efficiently utilized for performing CCC. High retention of the stationary phase and excellent peak resolution were achieved under a relatively high flow rate of the mobile phase.

Although the present apparatus has a limited sample-loading capacity of 1-g quantity, the system would be amenable to be scaled up for industrial separations of multigram to kilogram quantities simply by increasing the size and the length of the coiled column. Because the separations are performed under slow coil rotation not exceeding 80 rpm, a large apparatus can be operated without a risk of serious accident caused by the centrifugal force.

A. Design of the Apparatus

Original Apparatus Equipped with Rotating Seals [15,16]

Figure 105 shows a diagram of the first prototype. A helical column is tightly mounted around a hollow rotary shaft (60 cm long and 1.3 cm o.d.) equipped with a rotating seal at each end. The rotary shaft is driven by a motor with a speed reducer, if necessary, through a pair of pulleys coupled with a toothed belt. The aluminum frame holding both the rotary shaft and its driving mechanism can be

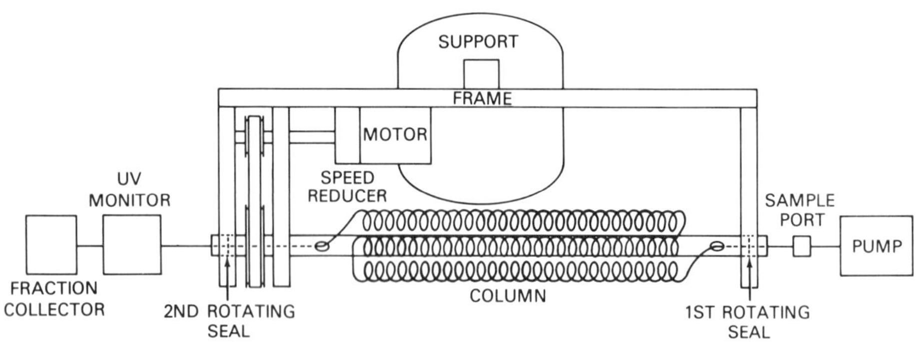

Fig. 105 Top view of the original apparatus. The helical column is mounted around the rotary shaft which is driven by the motor through the pulleys coupled with a belt. Solvent is pumped into the rotating column through the first rotating seal, and the eluate is collected through the second rotating seal. The column angle is adjusted by rotating and then tightening the frame against the support.

positioned at a desirable angle by rotating and tightening it against the standing support. The solvent is pumped through the sample port and first rotating seal into the head end of the column while the eluate from the tail end of the column is collected through the second rotating seal and continuously monitored with an LKB Uvicord III at 280 nm and then fractionated with a fraction collector. Each rotating seal was made from a pair of Kel-F blocks and a PTFE O-ring to minimize corrosion by the solvents.

Each column unit was prepared by winding PTFE tubing, 2.6 mm i.d. and about 5 m in length, onto a plastic or aluminium pipe, 1.25 cm o.d. and 50 cm in length, to make about 110 helical turns. The capacity of the single-column unit was approximately 25 ml. A long column was made by connecting a desirable number of column units in series by tail-to-head connection. The long column can also be made from one piece of tubing, thereby eliminating the need for multiple connections. The column was mounted symmetrically around the rotary shaft to obtain a good balance of the apparatus. Figure 106 shows a photograph of the original apparatus equipped with a long preparative column which consists of a series of nine column units with a total of approximately 1000 helical turns.

Recent Model Free of Rotating Seals [17]

The original model described above is equipped with rotating seals which are the potential sources of complications such as leakage and obstruction. In the second model described below, the use of rotating seals was eliminated by adapting the nonplanetary seal-free design illustrated in Fig. 17 (scheme V).

Figure 107 shows a simplified cross-sectional view through the central axis of the apparatus. The motor drives a rotary frame which consists of three aluminum arms rigidly bridged together with links. The frame holds two rotary elements, the countershaft and the centrally located column holder assembly. The countershaft is equipped with a toothed pulley at one end and a plastic gear at the other end. The pulley of the countershaft is coupled with a toothed belt to an identical stationary pulley mounted on the stationary wall member of the apparatus. This coupling causes a counterrotation of the countershaft on the rotary frame. This motion is further conveyed to the central column holder assembly by 1:1 gear coupling. Consequently, the column holder assembly rotates around its own axis at a rate twice that of the rotary frame in the same direction. This particular design gives a great advantage in that the scheme allows the flow in and out of the rotating column without the use of rotating seals.

Figure 108 shows the photograph of the apparatus holding multiple glass coils, each 0.5 cm i.d., 2.5 cm core diameter, and 90 ml capacity. Each column is supported by a hollow aluminum core

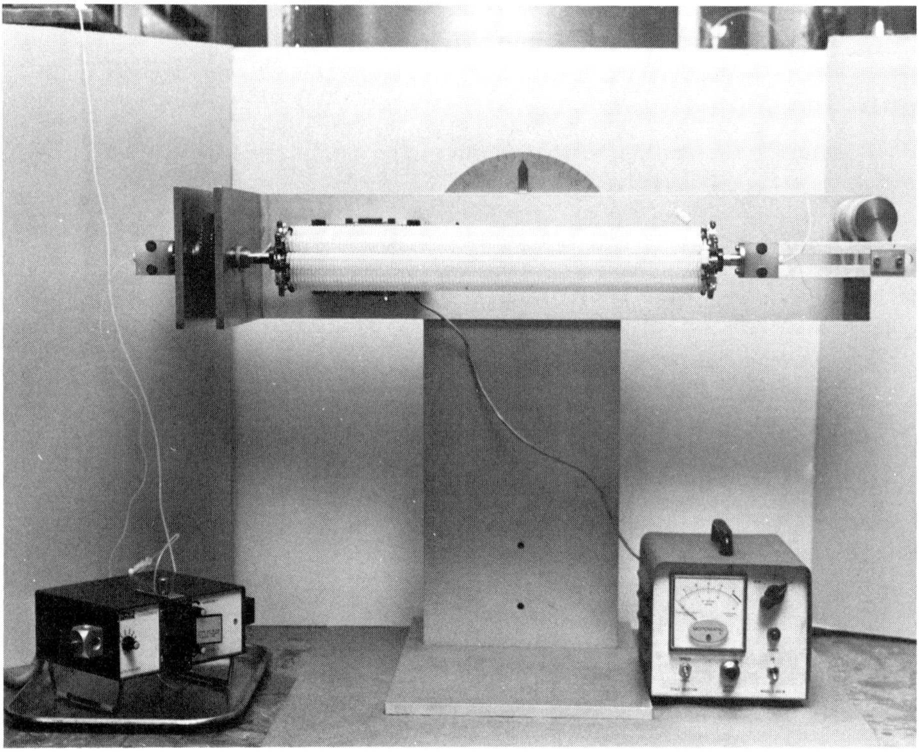

Fig. 106 Preparative countercurrent chromatograph. The apparatus consists of a helical column mounted around the rotary shaft equipped with a rotating seal at each end and driven by a motor through a pair of pulleys coupled with a toothed belt. The solvent is introduced into the column through the first rotating seal (right) and the eluate collected through the second rotating seal (left) is monitored with an LKB Uvicord III at 280 nm. The frame holding the rotary shaft is made adjustable to any desired angle.

which is in turn mounted onto the column holder by a screw at each end. The column holder is equipped with two different levels for mounting columns, the first level being located 6.5 cm from the central axis of the apparatus and the second level 13 cm from the same axis. A maximum of 30 columns can be mounted to the holder, 10 columns at the first level and 20 columns at the second level. The desired number of columns can be connected in series with a short piece of heat-shrinkable PTFE tubing at each junction. Flow tubes from the column are first led through the center hole of the column

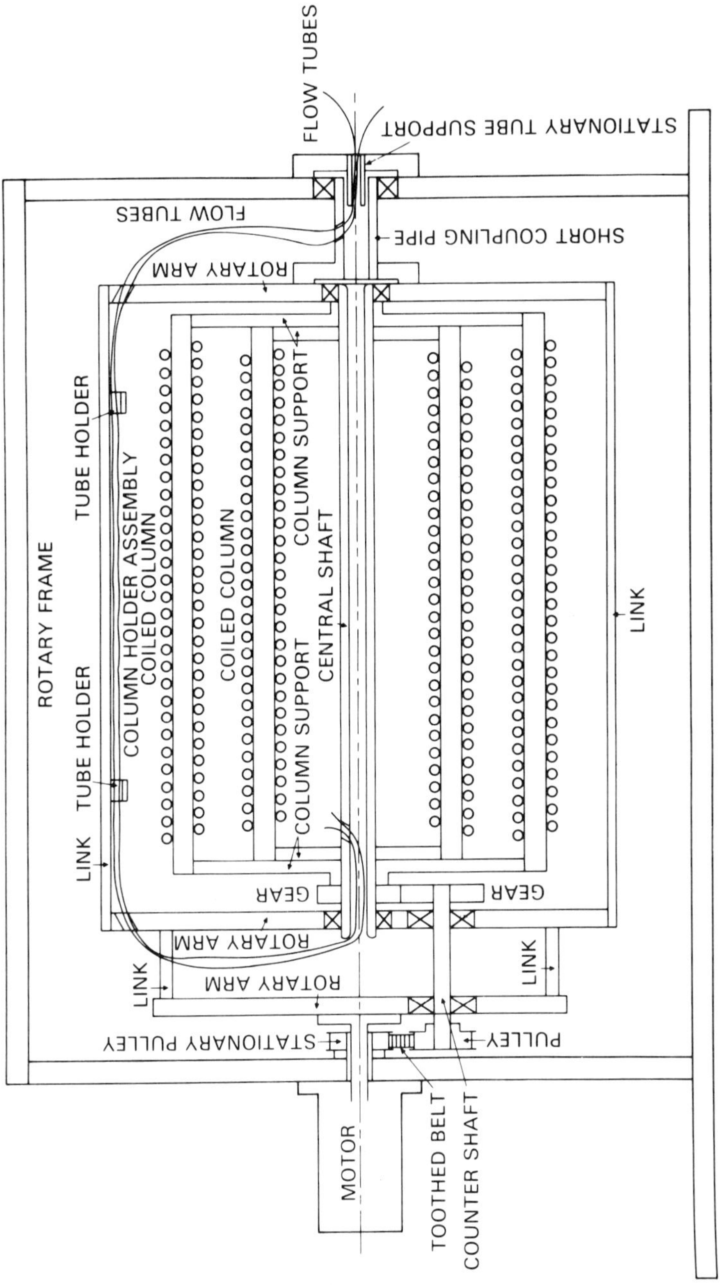

Fig. 107 A simplified cross-sectional view through the central axis of the rotating coil assembly free of rotary seals.

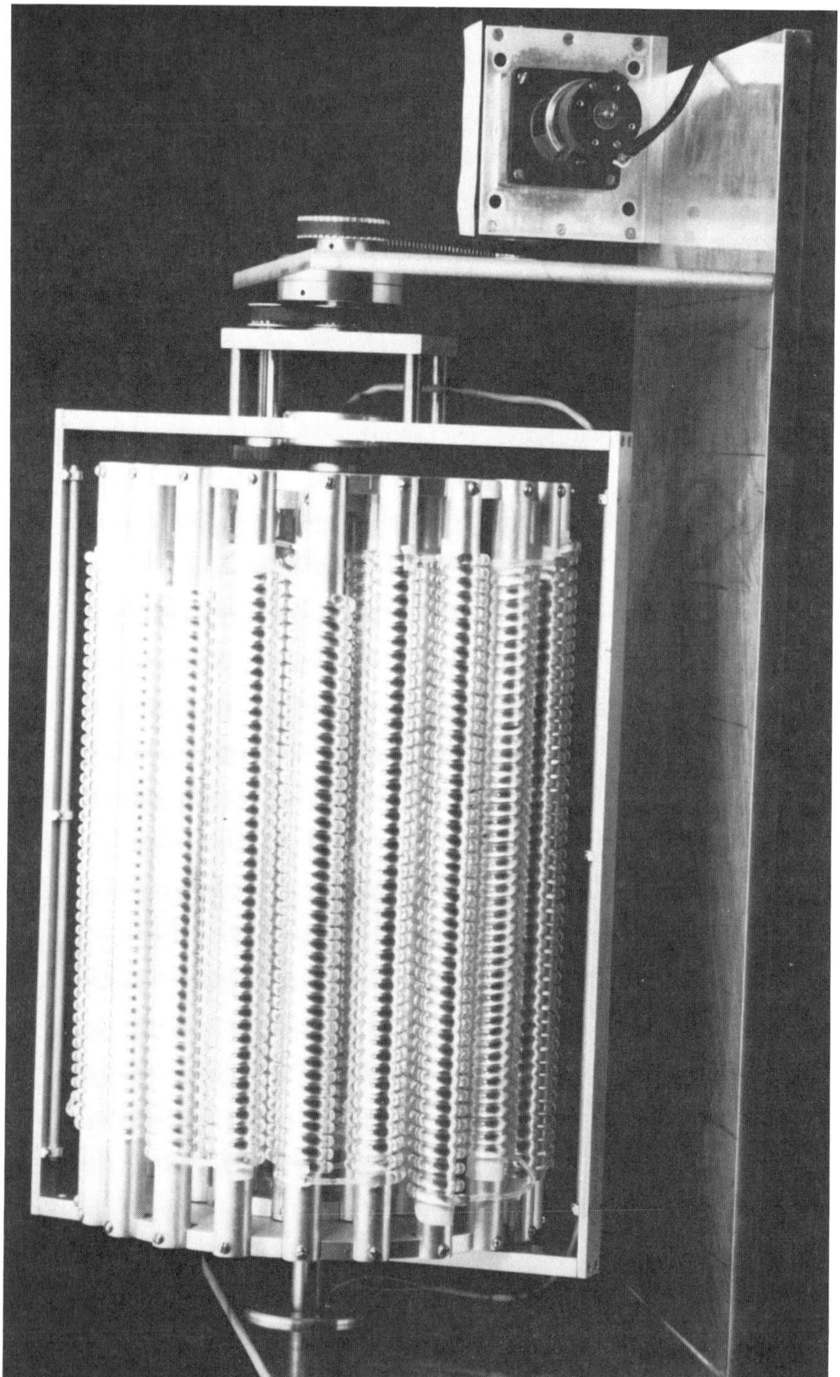

Fig. 108 Photograph of the improved model of the rotating coil assembly.

holder shaft, then passed through a pair of holes at the periphery of the rotary arms, and finally are supported by a stationary tube support located at the central axis of the apparatus. These tubes are thoroughly lubricated with silicone grease and protected with a piece of plastic tubing to prevent contact with metal parts.

The rotational speed of the column assembly can be regulated up to 300 rpm. However, in the present studies, fragility of the glass column limits the maximum rate down to approximately 100 rpm. A Beckman Accu-Flo pump and Chromatronix pump are used to eluate the solvents and an LKB Uvicord III to monitor the eluate at 280 nm.

B. Performance of Small-Bore Coils

Series of preliminary experiments have been performed to evaluate the performance of small-bore coils mounted on the original apparatus (Fig. 106). For these studies two different types of two-phase solvent systems were selected; chloroform/acetic acid/0.1 N HCl (2:2:1) for separations of DNP amino acids and n-butanol/acetic acid/water (4:1:5) for separations of oligopeptides. Performance studies were divided into two parts, i.e., studies on the retention of the stationary phase and partition efficiency, both being made with a short coil mounted parallel to and at a distance of about 2 cm from the rotation axis (eccentric parallel orientation).

Phase Retention Studies

For each solvent system, the retention of the stationary phase was measured under various conditions with respect to column angle, rotational speed, and flow rate. Both upper and lower phases were tested as a stationary phase using a single-column unit. In order to study the phase distribution and flow pattern, it is convenient to color the stationary phase with a dye which partitions almost entirely to the stationary phase. For a chloroform/acetic acid/0.1 N HCl (2:2:1) phase system, Sudan black B was used to color the lower organic phase and basic fuchsin for the upper aqueous phase. For a n-butanol/acetic acid/water (4:1:5) phase system, acid fuchsin was used to color the lower aqueous phase and basic fuchsin the upper nonaqueous phase. The column was first filled with the mobile phase, and about 5 ml of the lightly colored stationary phase was introduced into the column, which was held vertical so that the stationary phase was completely separated from the mobile phase. The length of the column occupied by the colored stationary phase was then measured as A cm. Then the column was set to the desired angle, rotated, and eluted with the mobile phase at the desired rates. After a hydrodynamic equilibrium state of the two phases was established, the length of the column containing the colored stationary phase was measured as B cm. The retention percentage of the stationary phase (R) was then calculated as $R = (A/B) \times 100$.

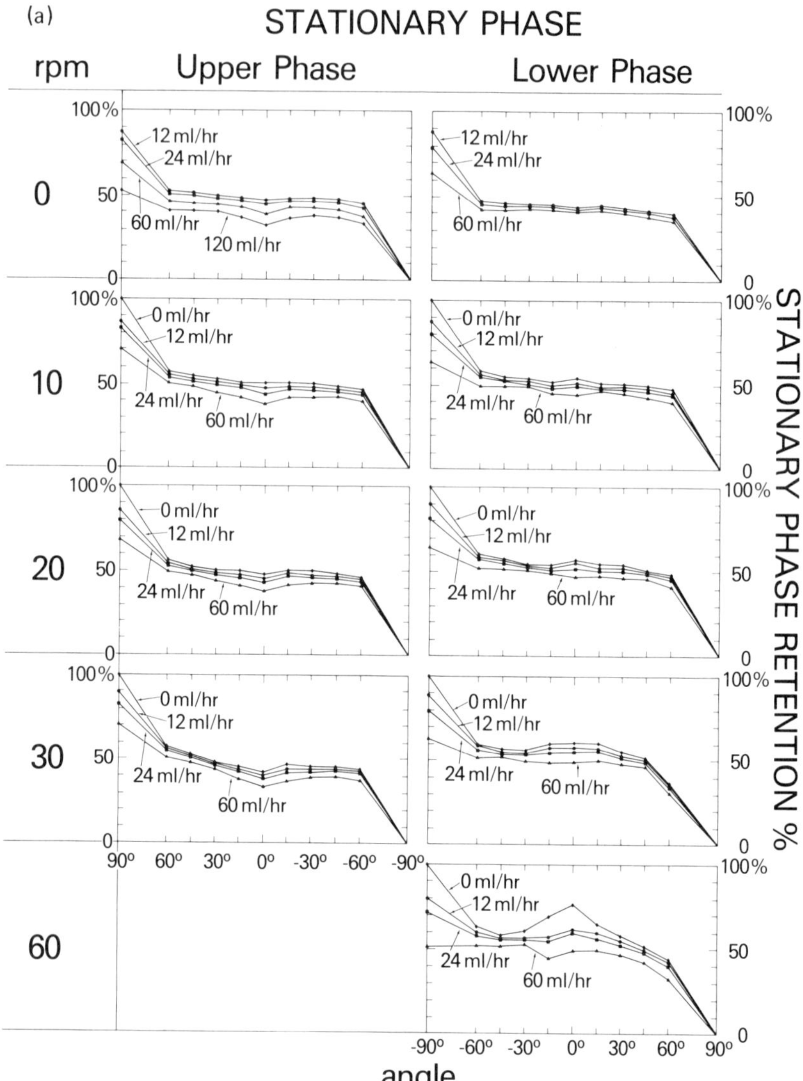

Fig. 109 Retention of the stationary phase with respect to column angle, rotational speed, and flow rate. The column angle is expressed as 0° being horizontal. The abrupt change of the percentage of retention at 0° is caused by the segmental flow of the aqueous phase. (a) Chloroform/acetic acid/0.1 N HCl (2:2:1). (b) n-butanol/acetic acid/water (4:1:5).

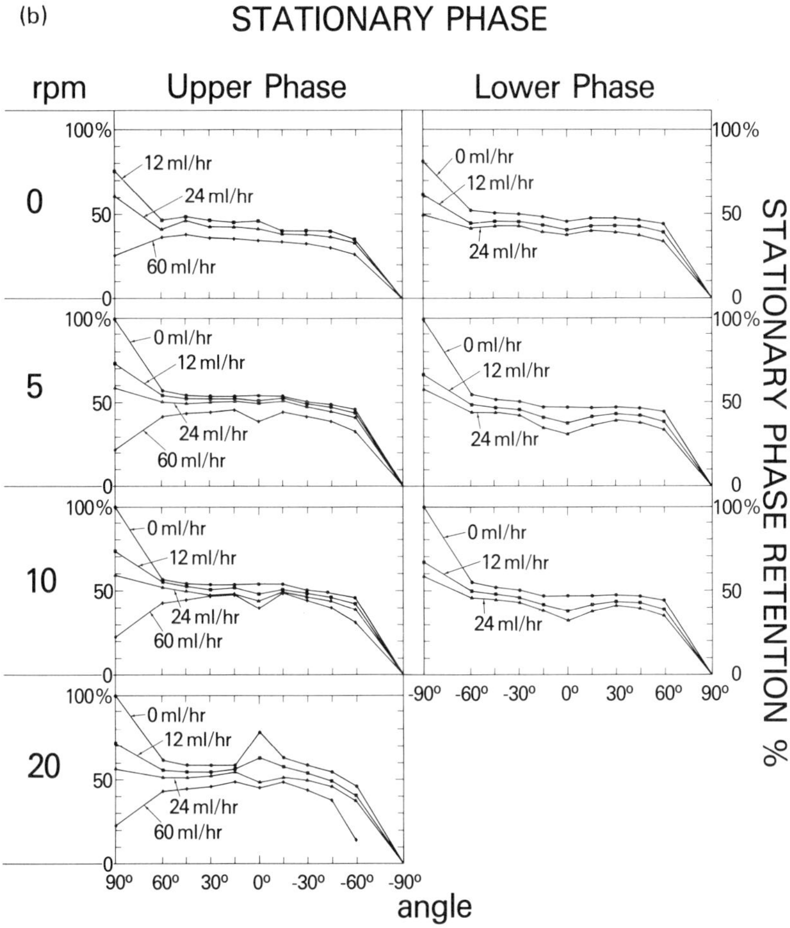

Fig. 109 (Continued)

Retention of the stationary phase in the rotating coiled tube examined under various operative conditions is illustrated in Fig. 109a for the chloroform/acetic acid/0.1 N HCl (2:2:1) and b for the n-butanol/acetic acid/water (4:1:5). In each diagram, the percentage of stationary phase volume occupying the column space is plotted against the applied column angle, where 0° indicates the horizontal column position, +90° and −90° being the vertical column positions with the inlet (head) end upward and downward, respectively. Several curves drawn in each diagram show the effects of flow rate. Ideal phase retention close to 50% is observed in a wide range of the column angle between 0° ± 60° in most of the cases.

Each curve exhibits a characteristic shape which is similar in both phase systems used. Most of the curves show an abrupt decrease in retention at around 0°, suggesting a change of the two-phase flow pattern at this angle. It was observed that at angles over +30° or below −30°, each phase underwent a laminary flow along the side wall of the coiled tube favored by the gravity. When the angle became close to 0°, the pattern changed into a segmental flow where the aqueous phase, having less affinity to the wall surface, was broken down into droplets. In the chloroform phase system, the size of these droplets was much smaller than the tube diameter, while in the n-butanol phase system each droplet could occupy the column space across the diameter of the tube as shown in Fig. 14 (right). Further observation revealed that individual droplets moved locally back and forth in the column synchronously with rotation mostly without coalescence, especially in the latter n-butanol phase system. This suggested that a high partition efficiency may be obtained under this condition. In both phase systems, aqueous phase, which formed droplets in the column would allow higher rotational speeds and/or flow rates if used as the mobile phase. The chloroform phase system, which had a lower viscosity and a greater density difference between the upper and the lower phases, permitted the operation at higher rotational speeds and flow rates than the n-butanol phase system. This also indicated that the smaller bore column could be used for the chloroform phase system to obtain a higher partition efficiency.

Several other two-phase solvent systems have also been tested for phase retention. Satisfactory retention was found with extremely low interfacial tension phase systems such as ethylene glycol ethers/various salt solutions which are suitable for separations of proteins [81]. However, high interfacial tension phase systems such as hexane/water and ethyl acetate/water produced a plug flow in the presently used column, and it is necessary to increase the diameter of the column for successful operation.

Partition Efficiency Studies

Suitable samples which enables detection with a single wavelength of 280 nm were chosen for each two-phase solvent system to study partition efficiency. For the chloroform phase system, a pair of DNP amino acids—DNP-L-glutamic acid (1.9) and DNP-L-alanine (0.56)—were selected on a basis of their desirable partition coefficients (C_u/C_l: solute concentration in the upper phase divided by that in the lower phase) indicated in the parentheses above. Because of the reciprocal relationship between their partition coefficient values, the mixture of these samples gave similar elution curves regardless of the choice of the mobile phase, thus facilitating fair evaluation of the partition efficiency. The sample mixture was dissolved into the upper aqueous phase at a concentration of 0.5 g% for each component. For the n-butanol phase system, two pairs of oligopeptides were selected on the basis of the partition coefficients: L-tyrosylglycylglycine (0.21) and L-leucyl-L-tyrosine (1.0) for studies on elution with the lower phase and L-tryptophyl-L-tryptophan (5.2) and L-leucyl-L-tyrosine (1.0) on elution with the upper phase where their partition coefficients (C_u/C_l) on the n-butanol phase system are indicated in the parentheses. Each sample mixture was dissolved into the lower phase.

In each separation the column was filled with the stationary phase and 0.2 ml of sample solution injected through the sample port followed by elution with the mobile phase under a given set of conditions of column angle, rotational speeds, and flow rates. The eluate was continuously monitored with an LKB Uvicord III at 280 nm to record the elution curve. The partition efficiency in each separation was evaluated from the relative height of the trough between the two peaks and/or by calculating the number of theoretical plates for each component [see Eq. (30)] if the two peaks were well resolved.

Effects of the column angle and rotational speed on separation are shown in Fig. 110a for the DNP amino acid separation and Fig. 110b for the oligopeptide separation. In DNP amino acid separation, the upper aqueous phase was used as the mobile phase at a flow rate of 60 ml/hr, while in the oligopeptide separation the lower aqueous phase was used as the mobile phase at 24 ml/hr. In both a and b, the partition efficiency sharply increases with rotation up to 15 (b) and 30 (a) rpm, which clearly indicates the superiority of this hydrodynamic equilibrium scheme over the basic hydrostatic equilibrium scheme (separation shown at 0 rpm) as explained earlier. In Fig. 110a, the highest peak resolution was obtained at 45 rpm near the horizontal column position where the partition efficiency was rather insensitive to the column angle. In Fig. 110b, the best peak resolution was found at 15 rpm in the horizontal column position where a slight shift of the angle from the horizontal position in either direction substantially lowered the peak resolution.

Effects of flow rates and rotation speeds on peak resolution were further investigated using the horizontal column position. The results of the experiments on the DNP amino acid separation with the chloroform phase system is shown in Fig. 111. In Fig. 111a, where the upper aqueous phase was used as the mobile phase, the best peak resolution is seen at 60 rpm with a slow flow rate of 12 ml/hr. When the partition efficiency is expressed in terms of theoretical plates, the first peak gives 180 TP and the second peak 77 TP, the former exceeding the number of the coil units in the applied column. The partition efficiency may also be expressed as the time required for yielding one theoretical plate or somewhat similar to one transfer time of the Craig's countercurrent distribution apparatus. This value is

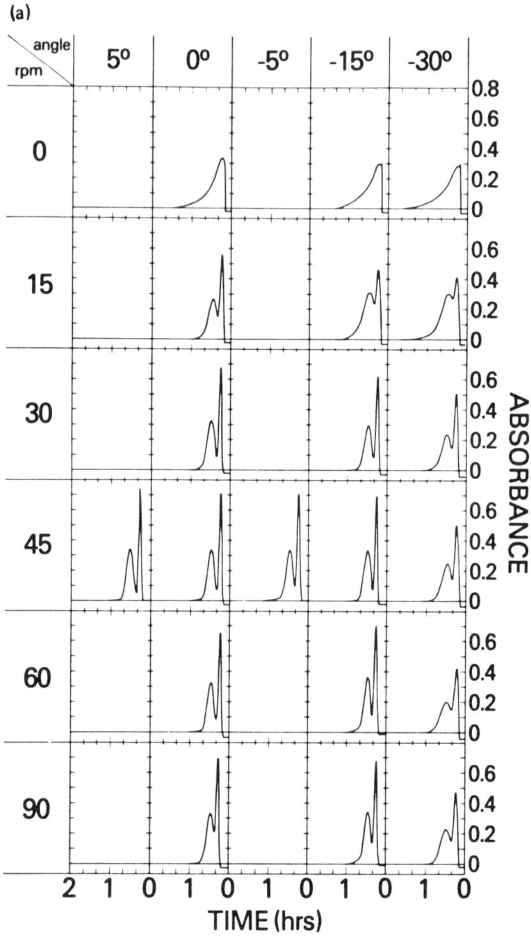

Fig. 110 Effects of column angle and rotational speed on peak resolution. (a) Separation of DNP amino acid. The best peak resolution is obtained at 45 rpm around the horizontal coil position. The separation is sharply improved from 0 to 30 rpm. The first peak: DNP-L-glutamic acid; the second peak: DNP-L-alanine; mobile phase: upper phase; flow rate: 60 ml/hr. (b) Separation of oligopeptides. The highest peak resolution is obtained at 0° at 15 rpm. The separation is sharply improved as the rotational speed is increased from 0 to 10 rpm. The first peak: L-tyrosylglycylglycine; the second peak: L-leucyl-L-tyrosine; mobile phase: lower phase; flow rate: 24 ml/hr.

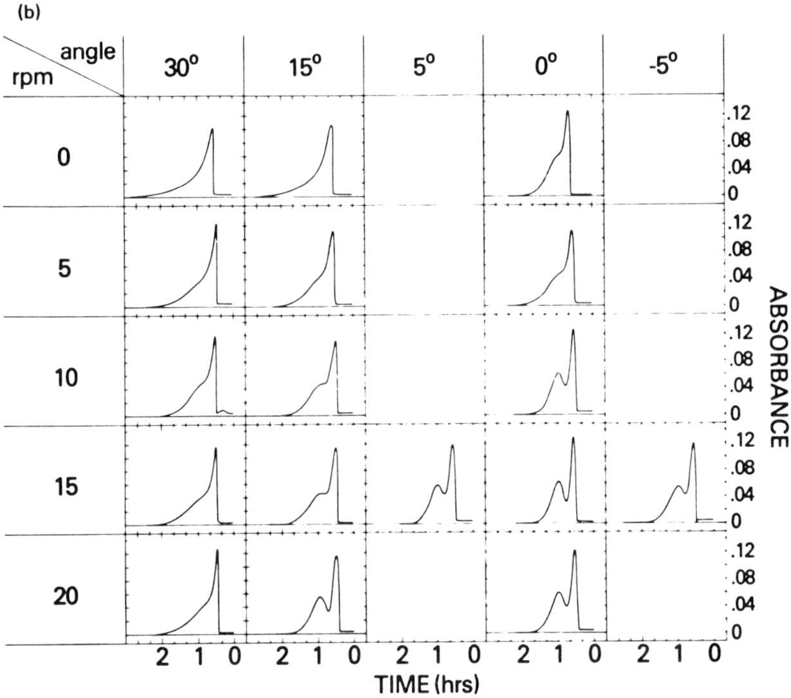

Fig. 110 (Continued)

actually obtained by dividing the retention time of the solvent front by the yielded theoretical plate number. In this case, the separation obtained by the flow rate of 60 ml/hr at 45 rpm gives the best values of 12 sec/TP for the first peak and 20 sec/TP for the second peak. In Fig. 111b, where the lower nonaqueous phase was used as the mobile phase, the best peak resolution is given at 20 rpm with a flow rate of 24 ml/hr yielding the partition efficiency of 86 TP for the first peak and 58 TP for the second peak. The separation time figures are further improved by a higher flow rate of 60 ml/hr at 20 rpm, the values being approximated as 16 sec/TP for the first peak and 24 sec/TP for the second.

Similar experiments were performed on separation of oligopeptides with the n-butanol phase system. The results are summarized in Fig. 112. In Fig. 112a, where the lower aqueous phase was the mobile phase, the highest peak resolution was obtained by a slow flow rate of 6 ml/hr at 15 rpm. The partition efficiency in this separation was estimated as 185 TP for the first peak and 64 TP for the second. The separation time figures are further improved by a faster flow of

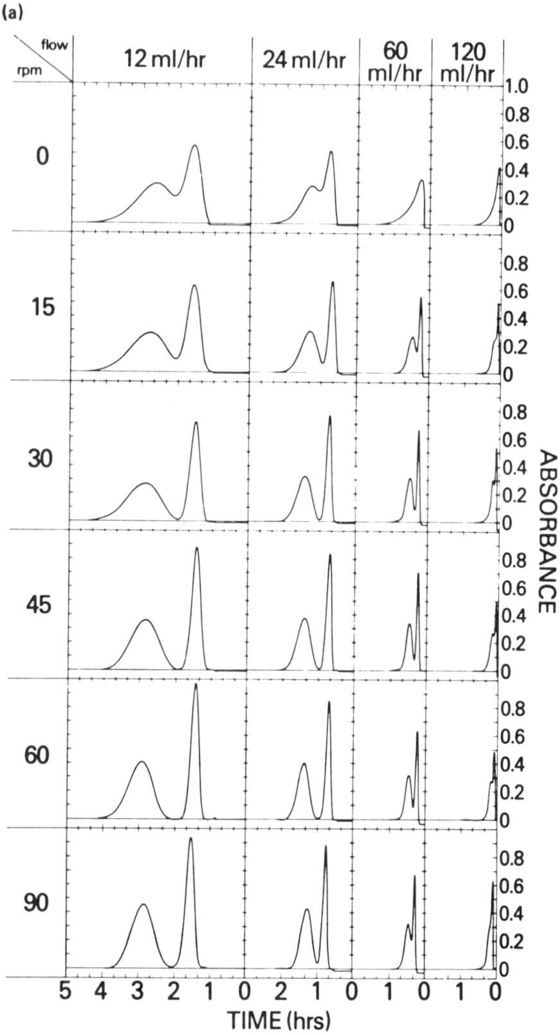

Fig. 111 Effects of flow rate and rotational speed on DNP amino acid separation. (a) Upper aqueous phase mobile. The first peak: DNP-L-glutamic acid; the second peak: DNP-L-alanine; column angle: 0°. (b) Lower nonaqueous phase mobile. The first peak: DNP-L-alanine; the second peak: DNP-L-glutamic acid; column angle: 0°.

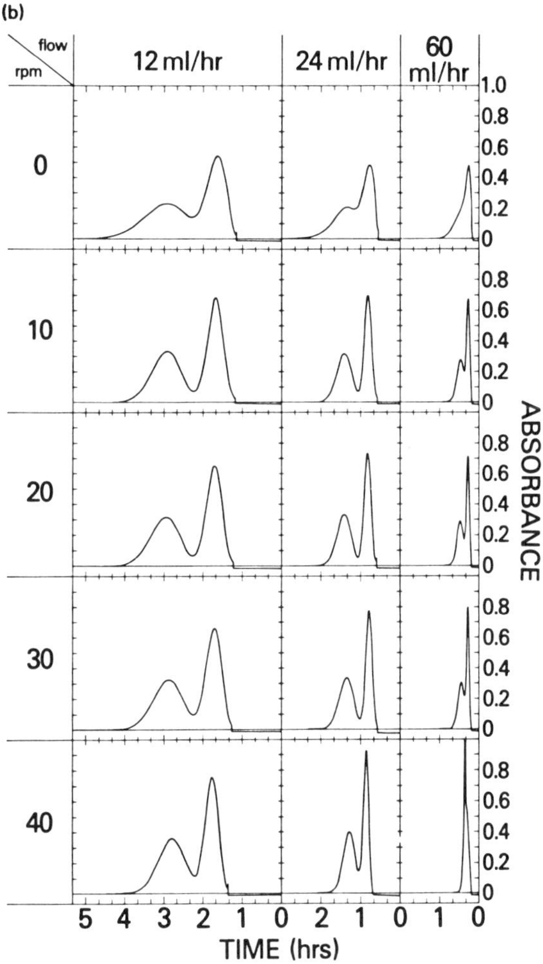

Fig. 111 (Continued)

24 ml/hr at 15 rpm, the figures being 18 sec/TP for the first peak and 54 sec/TP for the second. Similar results were obtained when the upper nonaqueous phase was used as the mobile phase as shown in Fig. 112b. The highest peak resolution was achieved by a 6 ml/hr flow rate at 10 rpm with efficiencies of 240 TP for the first peak and 154 TP for the second. On the other hand, the separation yielded by a 12 ml/hr at 10 rpm gave the improved separation time figures of 30 sec/TP for the first peak and 47 sec/TP for the second.

A series of partition efficiency studies described above clearly indicate that the highest peak resolution can be obtained at a slow

(a)

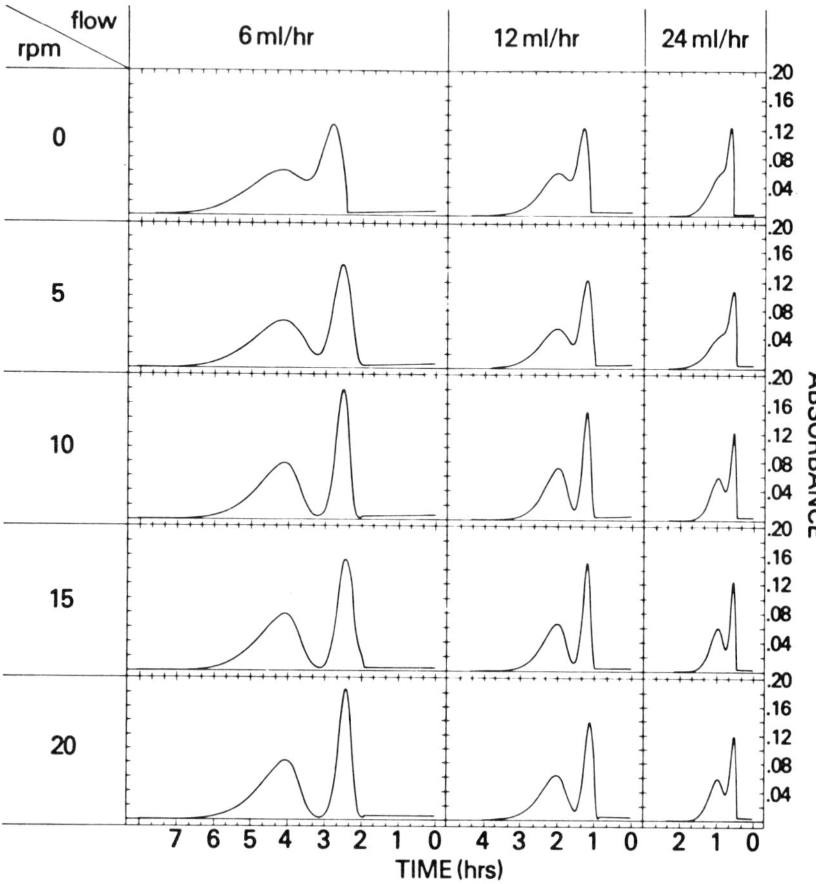

Fig. 112 Effects of flow rate and rotational speed on oligopeptide separation. The highest peak resolution is obtained by slow flow rates whereas the time required for yielding one TP is shortened by a faster flow. (A) Lower aqueous phase mobile. The first peak: L-tyrosylglycylglycine; the second peak: L-leucyl-L-tyrosine. (b) Upper nonaqueous phase mobile. The first peak: L-tryptophyl-L-tryptophan; the second peak: L-leucyl-L-tyrosine.

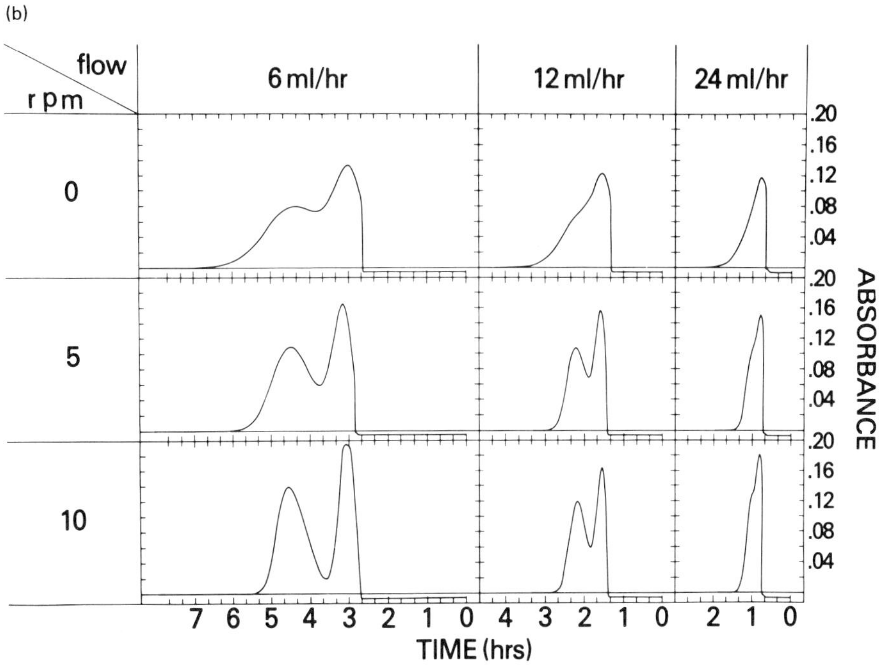

Fig. 112 (Continued)

flow rate with near horizontal column positions. The optimum rotational speed, however, varies with the physical properties of the solvent system and with the choice of the mobile phase in a given phase system. When separation time is the limiting factor in separation, it may be shortened by applying a higher flow rate where sec/TP value is minimized, provided that the retention of the stationary phase is within the optimum range. Comparison of the results obtained from the above two solvent systems may lead to a conclusion that the use of the aqueous phase as the mobile phase, regardless of the relative viscosity, allows a higher rotational speed and flow rate to yield a higher partition efficiency. This fact may be explained on the basis of solvent–wall interaction: the nonaqueous phase has an affinity to the PTFE tube wall and therefore, if used as the stationary phase, permits more stable retention under high rotational speeds. This hypothesis is strongly supported by the experiments with a coiled glass tube using the same solvent systems (see Fig. 116). The physical properties of the solvent system may also play an important role in partition process. The partition efficiencies produced by the chloroform phase system were always

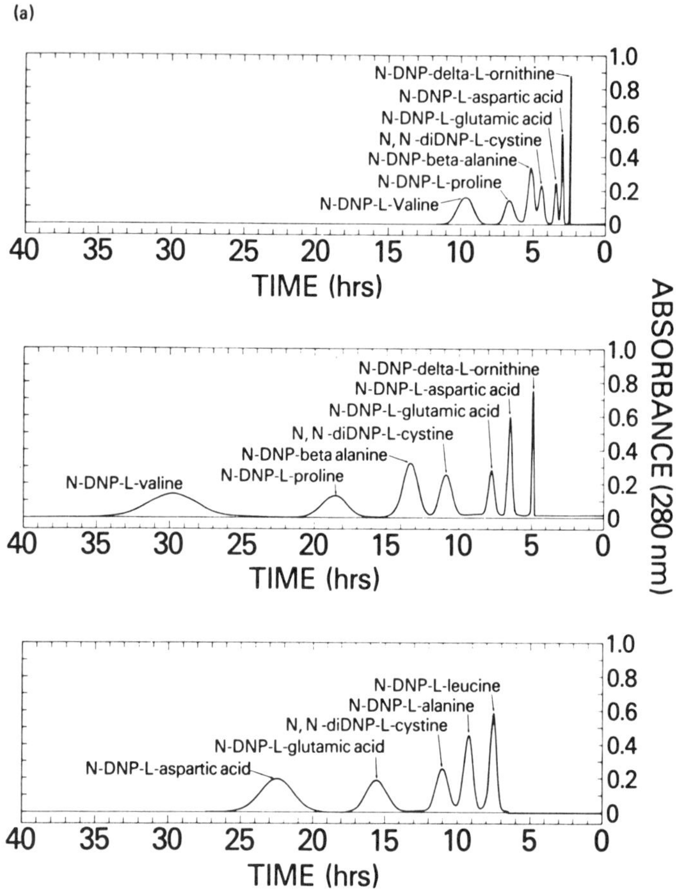

Fig. 113 Preparative chromatograms obtained with a long column. (a) DNP amino acid separations on chloroform/acetic acid/0.1 N HCl (2:2:1). Top: mobile phase, upper phase; flow rate, 60 ml/hr; rotation, 45 rpm; column angle, 0°. Middle: mobile phase, upper phase; flow rate, 24 ml/hr; rotation, 45 rpm; column angle, 0°. Bottom; mobile phase, lower phase; flow rate, 24 ml/hr; rotation, 20 rpm; column angle, 0°. (b) Oligopeptide separations on n-butanol/acetic acid/water (4:1:5). Top: mobile phase, lower phase; flow rate, 12 ml/hr; rotation, 15 rpm; column angle, 0°. Bottom: mobile phase, upper phase; flow rate 12 ml/hr; rotation, 10 rpm; column angle, 0°.

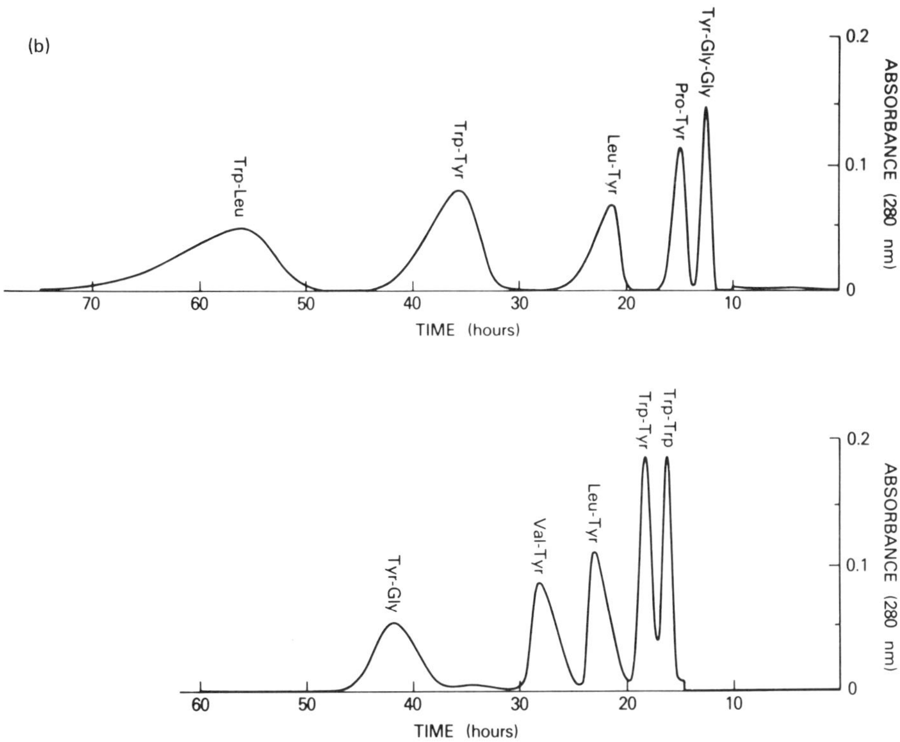

Fig. 113 (Continued)

higher than those produced by the n-butanol phase system. This may be attributed to the physical properties of the chloroform system such as the lower phase viscosity and the greater density difference between the upper and the lower phases.

Preparative Countercurrent Chromatograms

Preparative-scale separations were performed with a long column consisting of nine column units connected in series which has approximately 1000 helical turns with a total capacity of 240 ml. The sample mixture was dissolved in the stationary phase and a sample volume of 10 ml was applied in each separation.

Figure 113a shows three chromatograms of DNP amino acids obtained on the phase system composed of chloroform/acetic acid/0.1 N HCl (2:2:1). The top and middle chromatograms were produced by using the upper aqueous phase as the mobile phase under the optimum operational conditions of 45 rpm at flow rates of 60 and 24 ml/hr, respectively. The bottom chromatogram was produced by using the lower nonaqueous phase as the mobile phase at 20 rpm and a flow rate of 24 ml/hr.

Figure 113b shows preparative chromatograms of oligopeptides on the phase system composed of n-butanol/acetic acid/water (4:1:5) with the same preparative column used for DNP amino acid separations. The top chromatogram was obtained by using the lower aqueous phase as the mobile phase at 15 rpm and 12 ml/hr of flow rate while the bottom was obtained by using the upper nonaqueous phase as the mobile phase at 10 rpm and 12 ml/hr.

These preparative separations were produced on the two commonly used solvent systems which differ greatly in viscosity and phase density difference. These data suggest that other low interfacial tension phase systems used for extraction or partition chromatography may be successfully applied to the present method by using the above operational conditions with minor modifications. The present countercurrent chromatographic scheme allows a 10-ml sample charge to yield high partition efficiencies ranging from one thousand to several hundred theoretical plates where either aqueous or nonaqueous phase can be used as the mobile phase. The method has been successfully applied for fractionation of glycosaminoglycans with n-butanol/aqueous NaCl containing hexadecylpyridinium chloride. Chondroitin 4-sulfate was fractionated according to degree of sulfation and could be completely resolved from heparin. A heparin sample was shown to contain three discrete components differing with respect to sulfaminohexose and sulfate substitution [82].

C. Performance of Large-Bore Coils Eccentrically Mounted on the Holder

In order to scale up the present CCC scheme, series of studies have been performed on large-bore coils (0.5 cm i.d.) by changing various parameters which might significantly affect the performance of the coiled column. The tested parameters include tubing material (glass and FEP), core diameter (1.25 and 2.5 cm), column position on the holder (6.5–13 cm from the axis of rotation), skewed orientation (0–15°), inclination of the apparatus (0–15° from the horizontal), and coaxial orientation to the holder axis. Some of these coil orientations are illustrated in Fig. 114. Among these variations of the coiled columns, the coaxially mounted coils have produced a unique hydrodynamic effect resulting in a radical improvement of peak resolution and therefore are separately described later under the respective heading.

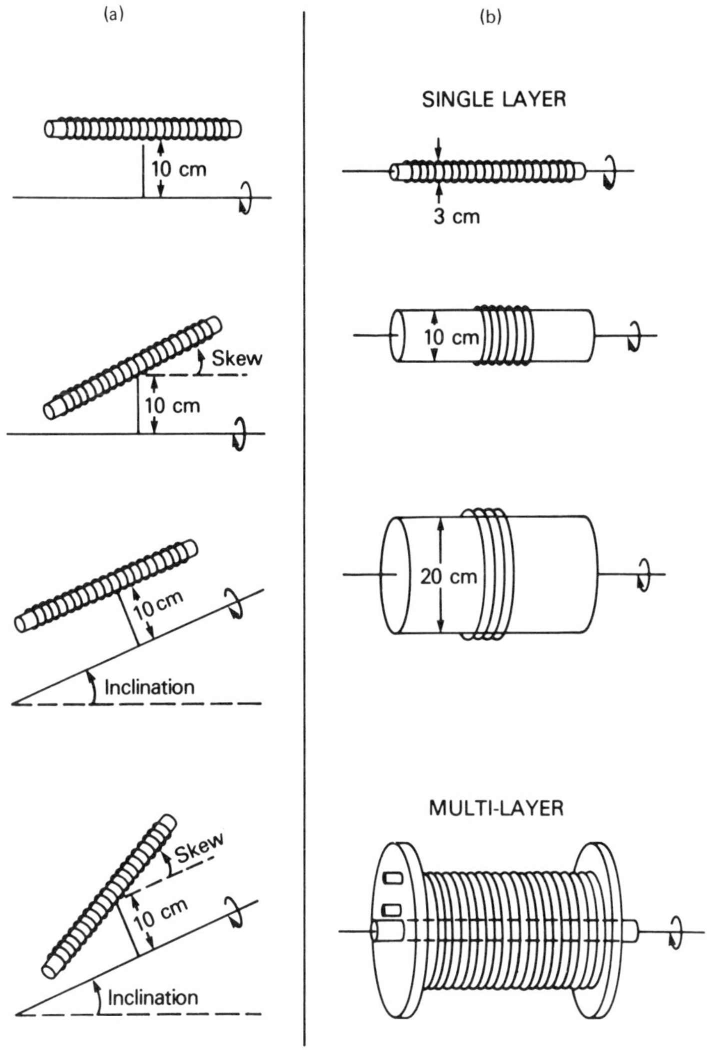

Fig. 114 Various orientations and configurations of the coiled separation columns. (a) Eccentric orientation. (b) Coaxial orientation.

Effects of Core Diameter and Coil Position on the Holder [17]

Performance of glass coils with two different core diameters of 1.25 and 2.5 cm was compared by mounting each coil at the inner and outer positions of the column holder in the improved model of the rotating coil assembly illustrated in Figs. 107 and 108.

1. *Phase Retention Studies.* A set of phase retention diagrams was obtained from a two-phase solvent system composed of chloroform/acetic acid/water (2:2:1) according to the similar procedure previously applied to the small-bore coils. The results obtained with the coils mounted in the outer position of the holder are shown in Fig. 115, where the percentage retention of the stationary phase relative to the total column capacity is plotted against the rotational speed of the column. The several lines drawn in each diagram indicate the effect of different flow rates. Retention near 50% is considered to be ideal although lower levels of retention can be suitable for separations if the inclination of the curve is near horizontal which ensures a stable retention of the stationary phase upon fluctuation of the rotational speed.

Figures 115a and b show retention in the large coil (2.5 cm o.d. core) mounted in the outer position of the column holder for the lower nonaqueous phase (a) and the upper aqueous phase (b). As is clearly observed, the retention levels of the nonaqueous phase are substantially lower than those of the aqueous phase throughout the applied rotational speeds. This higher level of retention produced by the aqueous phase may be largely attributed to its greater affinity to the glass wall of the column. The effects of the flow rate on the retention of the stationary phase are also clearly shown in these diagrams; the slower the flow, the higher the retention levels. Figures 115c and d similarly show the retention levels in the small coil (1.25 cm o.d. core) mounted in the outer position of the column holder for both stationary phases. The data clearly show that the retention levels produced by the small coil is substantially higher than those by the large coil for both nonaqueous and aqueous stationary phases. This may indicate that in the small-core coil the linear velocity of rotation relative to the gravity becomes smaller resulting in less violent mixing of the two phases and therefore a greater amount of the stationary phase is retained in the column at a given rotational speed.

Data obtained with the same columns mounted in the inner position of the column holder gave similar results. Overall results indicate that satisfactory retention levels can be obtained with either type of coils under a wide range of rotational speeds and flow rates.

2. *Partition Efficiency Studies.* The partition efficiency of each coil was evaluated with a two-phase solvent system composed of

chloroform/acetic acid/0.1 N HCl (2:2:1) and a pair of DNP amino acids, DNP-D,L-glutamic acid and DNP-L-alanine.

Figure 116 summarizes the results of DNP amino acid separation with a single coil mounted in the outer position of the column holder under a flow rate of 120 ml/hr. In each diagram, partition efficiency can be easily estimated by the resolution of the two peaks. In all groups the efficiency sharply increases as the rotational speed increases from 0 to 40–60 rpm, where the peak resolution becomes maximum. Further increase of the rotational speed to 80 rpm results in the loss of peak resolution. The optimum rotational speed thus ranges from 40 to 60 rpm for all groups. The highest peak resolution is given by the large coil while the small coil could yield much higher resolution if two coils are connected to make the capacity equal to that of the single large coil. The results obtained with a higher flow rate of 240 ml/hr yielded less efficient separations in both small and large coils compared with those produced at 120 ml/hr. The results obtained from the coils mounted in the inner position of the holder gave separations similar to those produced by respective coils mounted in the outer position of the column holder.

3. *Preparative-Scale Separations with a Long Coil.* The preparative capability of the present CCC scheme was demonstrated by the separation of 1-g samples with a long column consisting of 10 large coils (2.5 cm i.d. core) connected in series in the outside position. The separations were performed at a 120 ml/hr flow rate eluting with both nonaqueous and aqueous mobile phases. Figure 117a shows a chromatogram obtained at the optimum rotational speed of 40 rpm by using the aqueous phase as the mobile phase. The two DNP amino acids were completely resolved as symmetrical peaks and eluted out within 9 hr. The partition efficiency calculated according to the standard formula [Eq. (30)] gives 1250 TP for the first peak and 880 TP for the second peak. Figure 117b shows a similar chromatogram obtained at 60 rpm using the nonaqueous phase as the mobile phase. Because of higher aqueous phase retention, the peak resolution is much greater than that of the separation using the aqueous phase as the mobile phase. The partition efficiency in the latter separation gives lower figures of 1000 TP for the first peak and 830 TP for the second peak.

The above results are comparable with those obtained with the similar glass coils mounted on the original horizontal flow-through CPC [30]. However, the present scheme can hold the greater number of coiled columns compactly around the central axis of the apparatus and therefore is more suitable for scaling up the sample-loading capacity by increasing the size and the number of coiled columns.

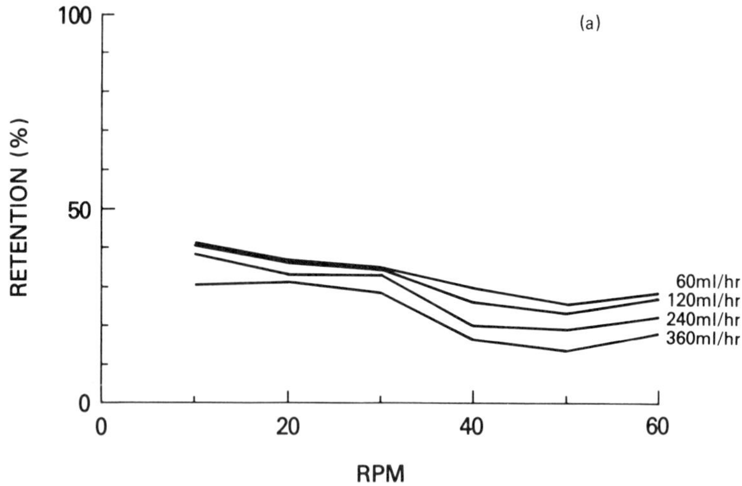

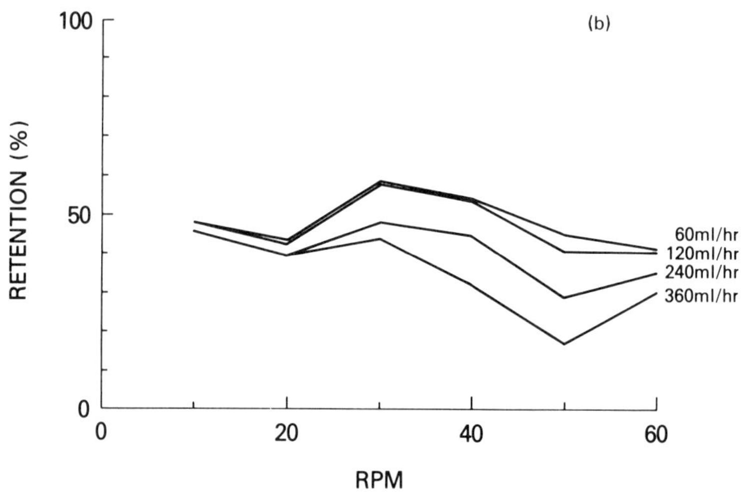

Fig. 115 The effect of rotational speed and flow rate on stationary phase retention. Large coil in outside position: (a) nonaqueous stationary phase; (b) aqueous stationary phase. Small coil in outside position: (c) nonaqueous stationary phase; (d) aqueous stationary phase.

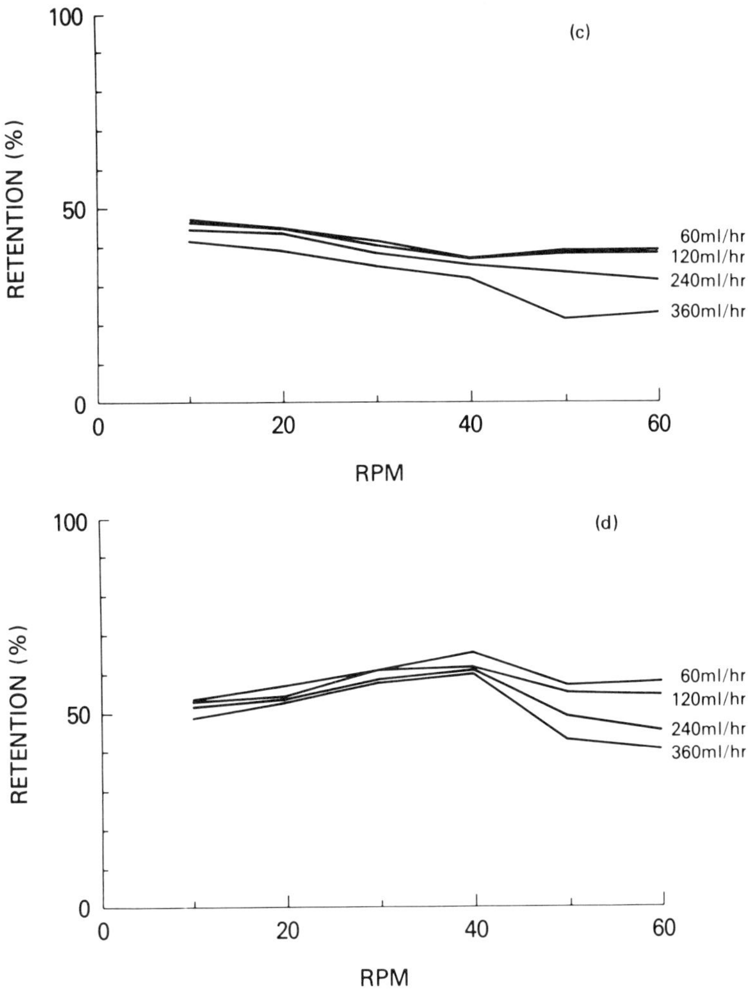

Fig. 115 (Continued)

Effects of Inclination and Skewed Orientation in Glass and Plastic Coils [120]

The above experiments have been expanded to study the effects of inclination and skewed orientation of coils on partition efficiency. In order to test the effect of solvent–wall interaction, glass and FEP (fluorinated ethylene propylene) coils were employed, both having similar dimensions as follows: glass coil: 0.5 cm i.d., 2.5 cm core diameter, 50 helical turns, 90-ml capacity; FEP coil: 0.55 cm i.d.,

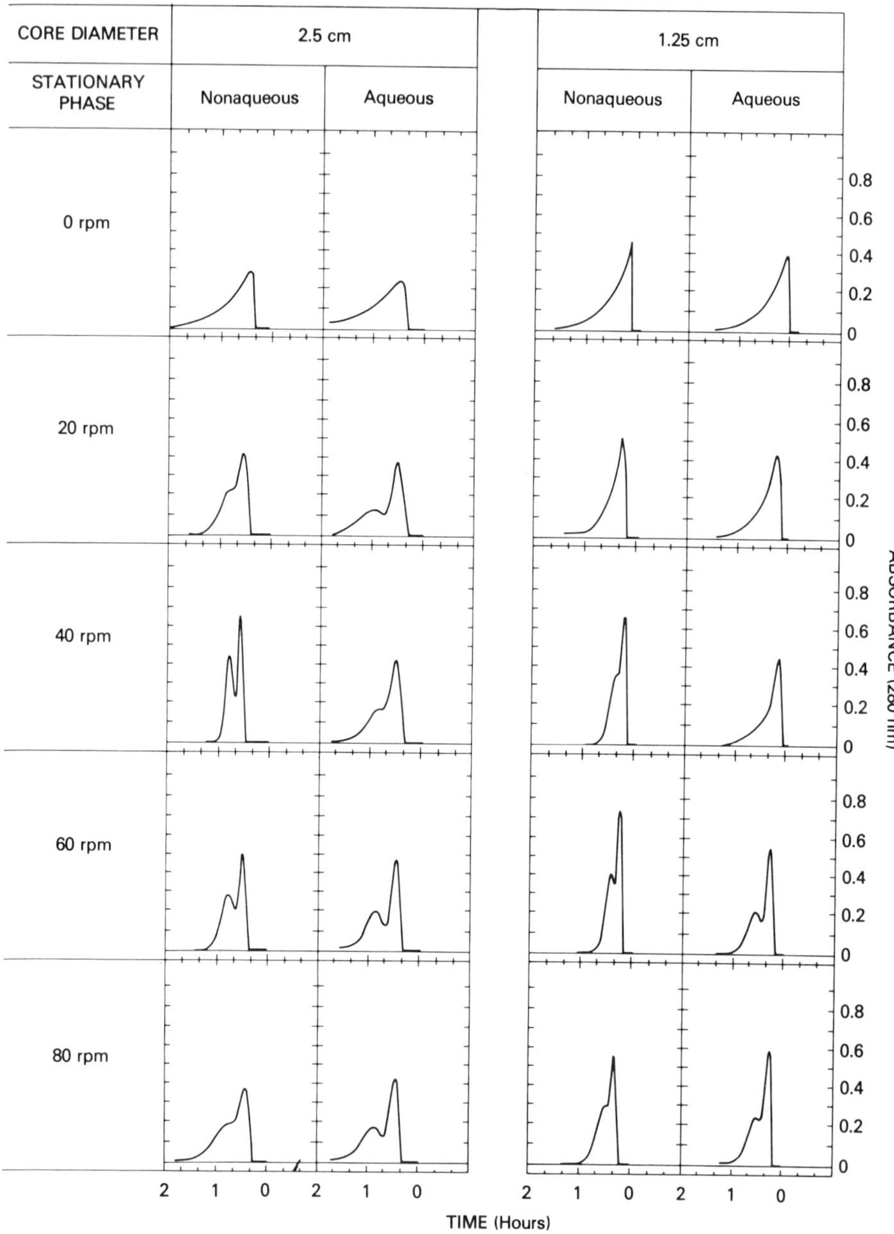

Fig. 116 The effect of rotational speed on the partition efficiency in the separation of DNP amino acids.

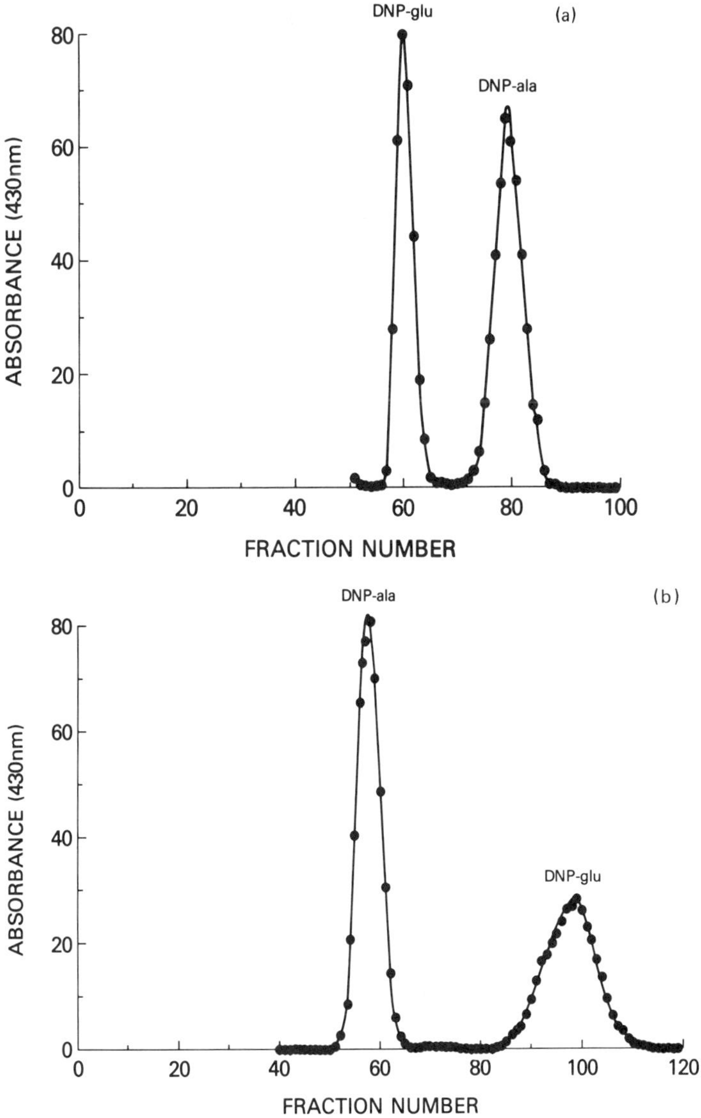

Fig. 117 Preparative-scale separations of DNP amino acids with a long column. (a) Nonaqueous stationary phase. (b) Aqueous stationary phase.

2.5 cm core diameter, 46 helical turns, 110-ml capacity. Each coil was mounted on the rotary shaft of the original apparatus (Fig. 106) in eccentric positions 10 cm away from the axis of rotation by the aid of a clamp rod which could support the coil at the desired skewed position at a given inclination of the rotary shaft (see Fig. 114, left).

Partition efficiency of coils was evaluated in separation of DNP-D,L-glutamic acid and DNP-L-alanine on a two-phase solvent system composed of chloroform/acetic acid/0.1 N HCl (2:2:1), using the similar procedure applied for the previous studies. The DNP amino acid separations obtained with glass and FEP coils are summarized in Fig. 118 where individual charts are arranged according to the applied operational conditions of coil orientation, choice of the stationary phase, and rotational speed.

In the first experimental group with the glass coil (Fig. 118a), the results obtained with the 0° skew, 0° inclination (left column) were almost identical to those obtained in the previous studies (see Fig. 116, left column). These separations were considerably improved by changing either skew or inclination angle. The best peak resolutions were found in 0° skew, 15° inclination at 40 rpm for both nonaqueous stationary phase groups (middle column). In the second experimental group with the FEP coil (Fig. 118b) the best separations were observed at 40 rpm for the stationary nonaqueous phase and 60 rpm for the stationary aqueous phase, both under 0° skew and 0° inclination (left column). Application of either 15° skew or 15° inclination failed to give any appreciable improvement in the peak resolution. Although skew and inclination produced some significant improvement on the performance of the glass coil, the FEP coil positioned at 0° skew and 0° inclination yielded the best peak resolutions among all experimental groups.

D. Performance of Large-Bore Coils Coaxially Mounted on the Holder [7, 120]

While the above studies revealed no significant improvement in performance by modifying the orientation of the eccentrically mounted coils, our efforts have been rewarded in a discovery of a unique performance of the coaxially mounted coils which was incidentally tested on the course of these studies. Further experiments have been performed along the line of this new coil orientation using the improved model of the rotary coil assembly free of rotating seals. In order to mount the coils coaxially with the holder axis at various helical diameters, a spool-shaped coil support was coaxially mounted around the rotary shaft of the apparatus. Coaxial coil orientations examined in the present study are schematically illustrated in Fig. 114 (right).

Phase Distribution Diagrams

Using a two-phase solvent system composed of chloroform/acetic acid/0.1 N HCl (2:2:1), the distribution of the two solvent phases in the rotating coil was studied with a 0.55 cm i.d. FEP coiled tube coaxially mounted on the rotary support in helical diameters of 3, 10, and 20 cm. In order to facilitate observation, the lower nonaqueous phase was colored with Sudan III. The column was first filled with equal volumes of the upper and the lower phases and both inlet and outlet tubes were clamped. Then the apparatus was rotated at a given rate until the two solvent phases established the hydrodynamic equilibrium, in which one of the phases predominantly occupied the head side leaving the excess volume of the other phase at the tail side of the coil. Upon stopping the rotation, the number of helical turns n containing the predominant phase was noted. From the total number of helical turns N, the percentage volume occupied by the predominant phase in the equilibrated coil was calculated from the expression 50N/n. Alternatively, for a coil with large helical diameters, the length of the segment of each phase occupying one helical turn was directly measured to obtain percentage figures for each phase. The above procedure was repeated without renewing the column contents while changing the rotational speed of the coil to obtain a series of measurements for each helical diameter of the coil.

The distribution of the two phases in the coaxially rotated coils measured at their hydrodynamic equilibrium is shown in Fig. 119. In each diagram, the volume percentage of each phase at the head side of the coil is plotted against the applied rotational speed. The three diagrams obtained from different helical diameters show common features characteristic of the coaxial orientation. In the slow rotational speed between 0 and 30 rpm, the two phases distribute fairly evenly in the coil (stage I). When the rotational speed is increased, the lower nonaqueous phase tends to occupy more space on the head side of the coil and, at a critical speed between 60 and 100 rpm, the two phases are almost completely separated along the length of the coil, the lower phase occupying the head side and the upper phase the tail side of the coil (stage II). After this critical range, the amount of lower phase on the head side tends to decrease rather sharply, crossing below the 50% line (stage III). A further increase in the rotational speed again yields an even distribution of the two phases in the coil (stage IV). As the helical diameter increases all these stages tend to shift toward the lower rpm range apparently due to the enhanced centrifugal force field. For performing preparative CCC, stage II is considered to be of greatest interest because the system permits retention of a large amount of the stationary phase for either phase under the proper mode of elution as described earlier. The uneven distribution of the two phases in the coil at the

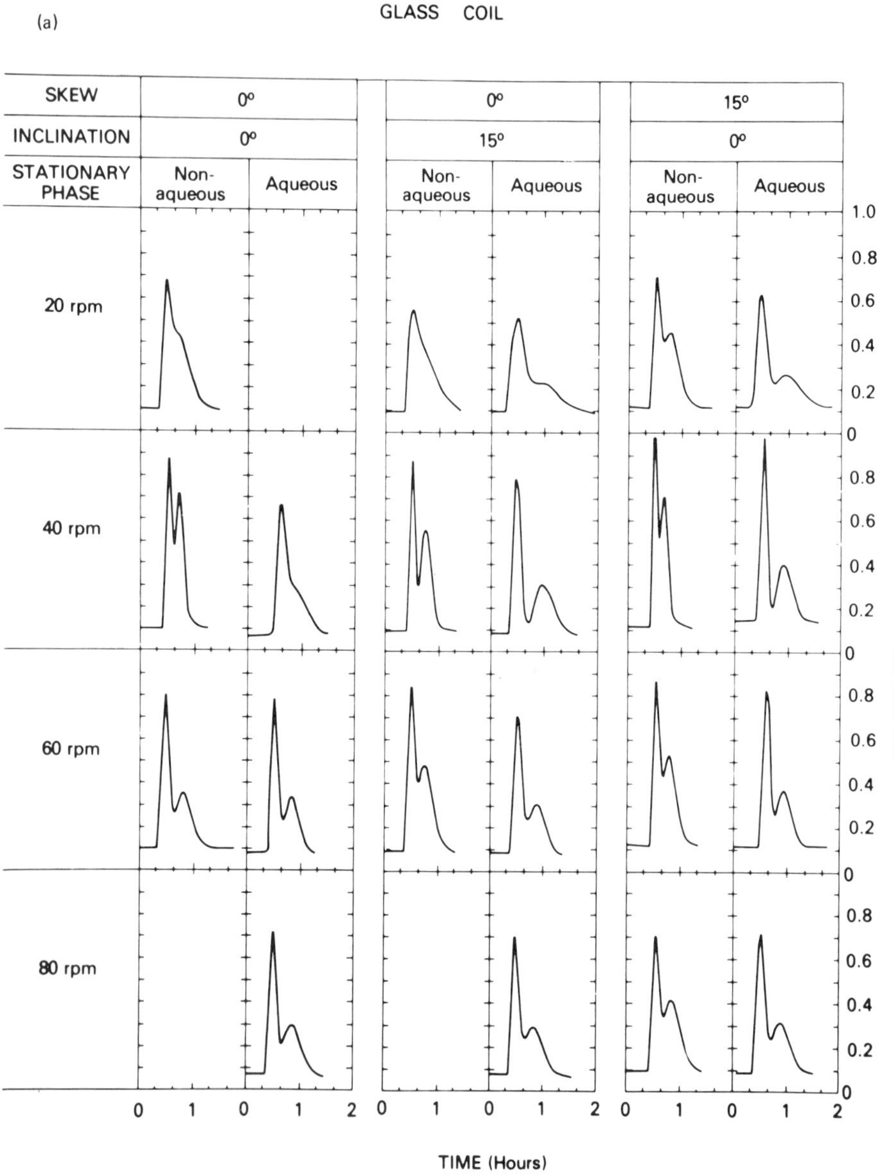

Fig. 118 Effects of skew angle and inclination of the eccentrically rotated coils on the separations of DNP amino acids at various rotational speeds. (a) Glass coil. (b) FEP coil.

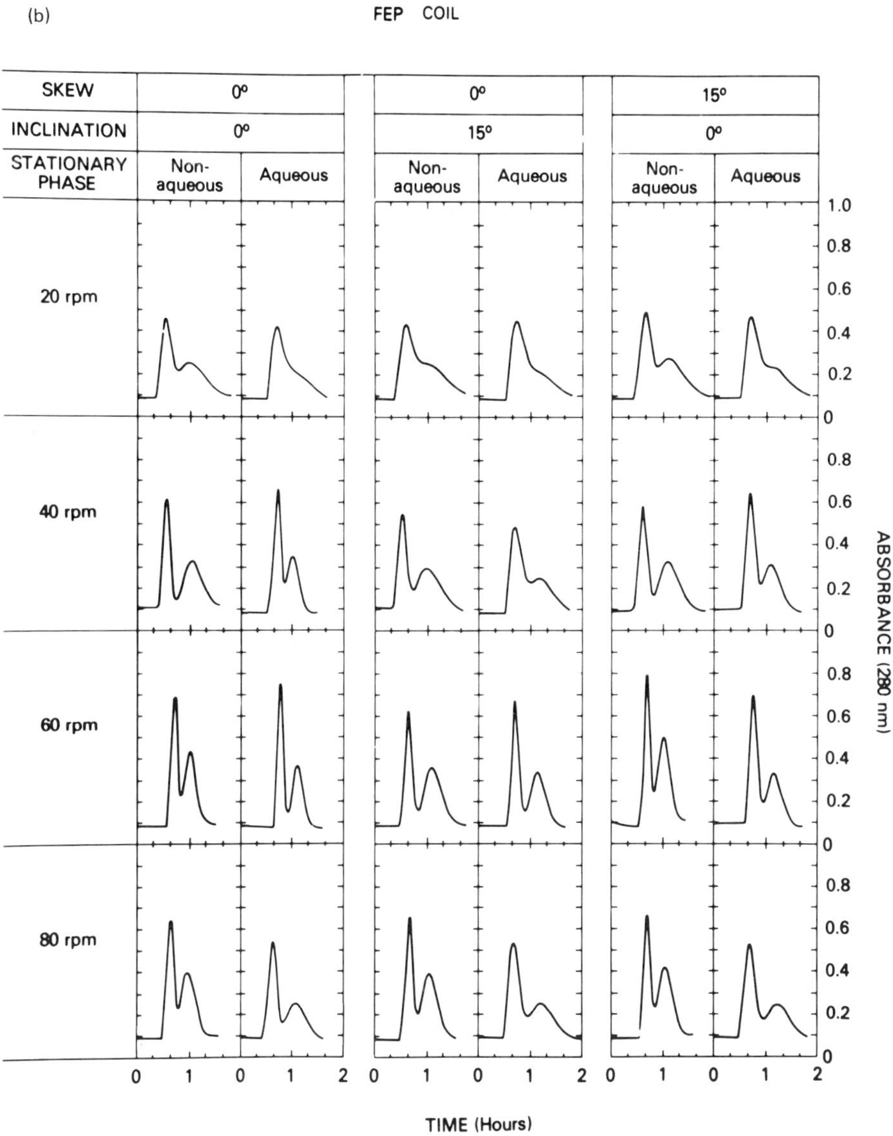

Fig. 118 (Continued)

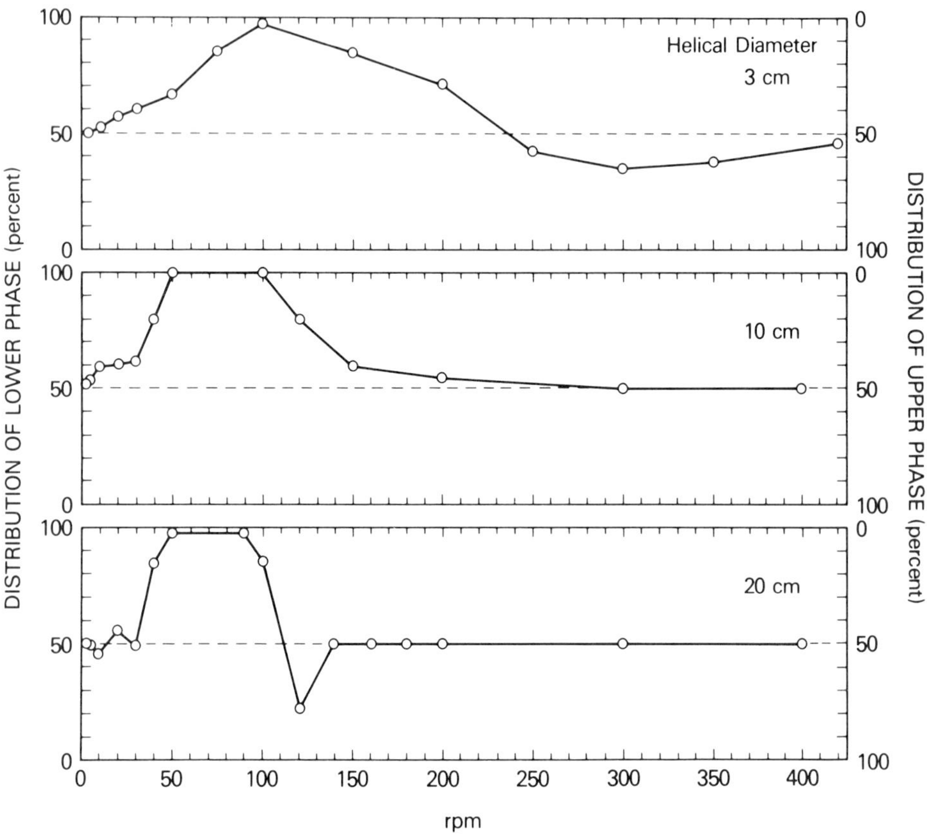

Fig. 119 Phase distribution diagrams for coaxially rotated coils with three different helical diameters.

critical rpm permits an extremely high level of stationary phase retention when either the aqueous phase is eluted in the normal mode (head to tail) or the nonaqueous phase in the reversed mode (tail to head).

Partition Efficiency Studies

The partition efficiency of the coaxially rotated coils was measured by the set of DNP amino acid samples and the two-phase solvent system which were used in the studies on the eccentric orientation. The experimental results with the three different helical diameters are summarized in Fig. 120 in which individual charts are arranged according to various operational conditions. In addition to the normal

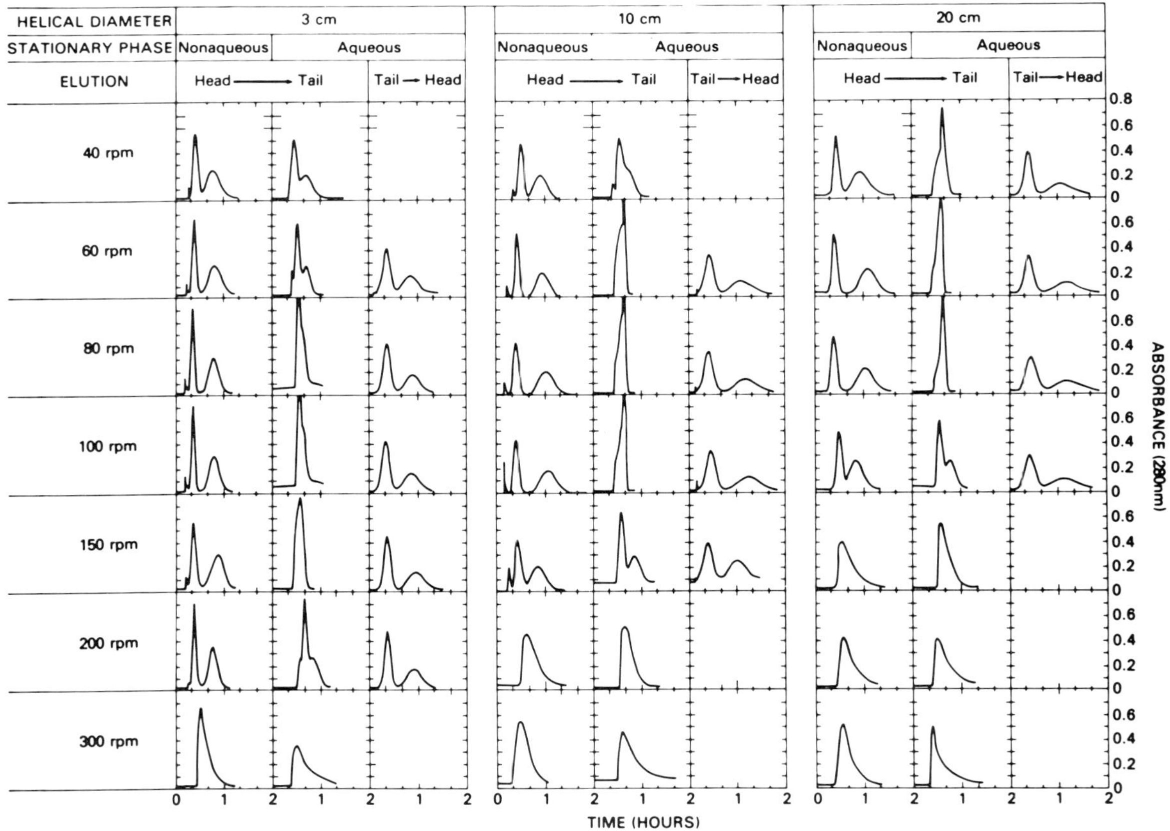

Fig. 120 Effects of helical diameters of the coaxially rotated coil on the separations of the DNP amino acids at various rotational speeds.

elution mode (head-to-tail), the reversed elution mode (tail-to-head) was also applied for the nonaqueous mobile phase at the critical rpm range (stage II) to achieve a high level of stationary aqueous phase retention as is expected from the phase distribution diagrams shown in Fig. 119.

The overall results of these experiments, with the coaxially rotated coils, clearly indicate that the best peak resolutions are obtained at the critical rotational speed ranging between 60 and 100 rpm, when either the aqueous phase was eluted in the normal mode (head-to-tail) or the nonaqueous phase in the reversed mode (tail-to-head). As predicted from the phase distribution diagrams in Fig. 119, the normal mode of elution with the stationary aqueous phase (middle column in each helical diameter in Fig. 120) gave unsatisfactory peak resolution due to a low level of stationary phase retention. Under the proper elution mode the resultant peak resolutions from the coaxially rotated coils are much higher than those from the eccentrically rotated coils shown in Fig. 118.

Comparison of the results obtained from the three different helical diameters illustrated in Fig. 120 reveals the considerable shift of the optimal conditions toward the lower rpm range in the larger helical diameter coil, this being consistent with the finding from the phase distribution diagrams in Fig. 119. However, the satisfactory peak resolution observed in a wide range of rpm in each helical diameter suggests that the choice of a proper rotational speed such as 80 rpm would produce good separations in the coil with any helical diameter between 3 and 20 cm. In light of these experimental data with the coaxially rotated coils, a new column configuration called the multilayer coil (Fig. 114, right bottom) has been made for large-scale preparative CCC.

Preparative CCC with a Multilayer Coil

The preparative capability of the multilayer coil has been demonstrated in the separations of 1-g quantity of the DNP amino acid mixture under the optimum operational conditions determined by the preliminary studies on short coils as described above. Figure 121 shows the chromatograms obtained with the multilayer coil in which both the upper aqueous phase (a) and the lower nonaqueous phase (b) are used as the mobile phase in the suitable elution mode. The two DNP amino acids are well resolved and eluted out as symmetrical peaks. The retention of the stationary phase measured after the separation was 68% in a and 84% in b. The separations obtained with this multilayer coil with a 750-ml capacity are much better than those from the eccentrically mounted glass coils with a similar internal diameter and having a total capacity of 900 ml (Fig. 117).

The multilayer coil assembly described above has a number of advantages over the previous CCC schemes utilizing a slowly rotating

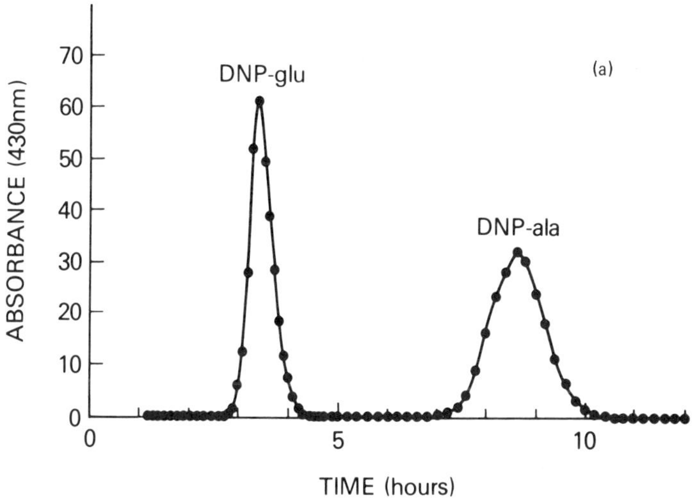

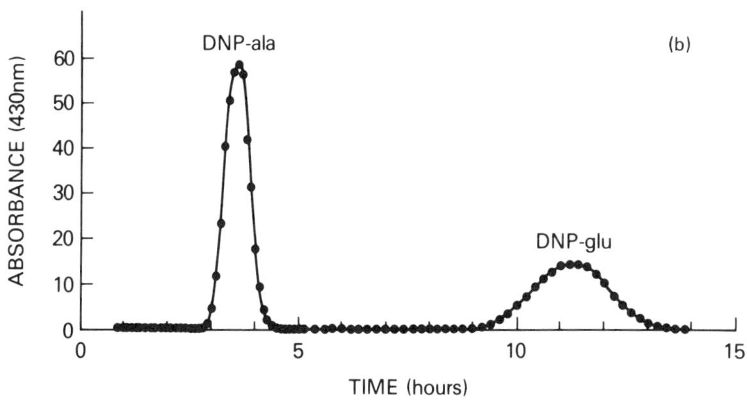

Fig. 121 Chromatograms of DNP amino acids (1g quantity) obtained with the multilayer coil. (a) Elution with the upper aqueous phase in the normal mode (head-to-tail). (b) Elution with the lower non-aqueous phase in the reversed mode (tail-to-head).

coil assembly. Because of the extremely high retention of the stationary phase, both peak resolution and the sample-loading capacity of the column are much increased. The column is leak-free, easily prepared, and relatively inexpensive. The compactness of the column reduces the size of the apparatus saving cost and space in the research laboratory. The present scheme is amenable to be scaled up for large-scale industrial applications.

X. HIGH-PERFORMANCE CCC

The high-performance CCC is considered to be the most advanced form of CCC including a family of efficient CCC schemes such as continuous countercurrent extraction [8,29], high-speed CCC [9,28], and dual CCC [10]. All these schemes utilize a unique hydrodynamic phenomenon called unilateral hydrodynamic equilibrium in which two immiscible solvent phases distribute unilaterally in a rotating coil, one phase entirely occupying the head side and the other phase the tail side of the coil. As described in the foregoing section (coaxially mounted coil), this hydrodynamic effect can be utilized for performing CCC in a unit gravitational field. In this method, the stationary phase is retained in the coil under a subtle balance between the two forces: an Archimedean screw effect generated by the combination of the unit gravity and coil rotation forces the stationary phase to move toward the inlet of the coil while the flowing mobile phase constantly pushes the stationary phase toward the outlet of the coil. In other words, the retention of the stationary phase in the rotating coil entirely relies on the weak unit gravity. Consequently, application of high flow rates of the mobile phase would produce carryover of the stationary phase resulting in loss of peak resolution. For achieving more stable retention of the stationary phase, it is natural to consider the enhancement of the acceleration by centrifugation.

A systematic search for the suitable centrifuge scheme to effect the unilateral hydrodynamic equilibrium has been successfully made among the synchronous planetary motions (Fig. 17) as illustrated in Fig. 122. Scheme I synchronous planetary motion has produced an even distribution of the two solvent phases in the coil as in the basic HDES (see Fig. 16 left) regardless of the orientation of the coil on the holder. Then the search was directed to the scheme IV synchronous planetary motion combined with various configuration and orientation of the coil on the holder. The eccentric orientation of the coil, either parallel or toroidal form, produced alternating segments of the two solvent phases throughout the coil as in the HSES (see Fig. 11). Finally, when the tube was directly wound around the holder (coaxial orientation), the ideal unilateral phase distribution was produced. A long coil can be compactly accommodated in a form of multilayer coil to produce the same effect. Thus, the high-performance CCC schemes have been developed by the combination of the coaxial coil orientation and a particular type of the synchronous planetary motion of the holder.

A. Principle

The principle of the CCC schemes utilizing the unilateral hydrodynamic equilibrium has been described in Sec. II. For elucidating

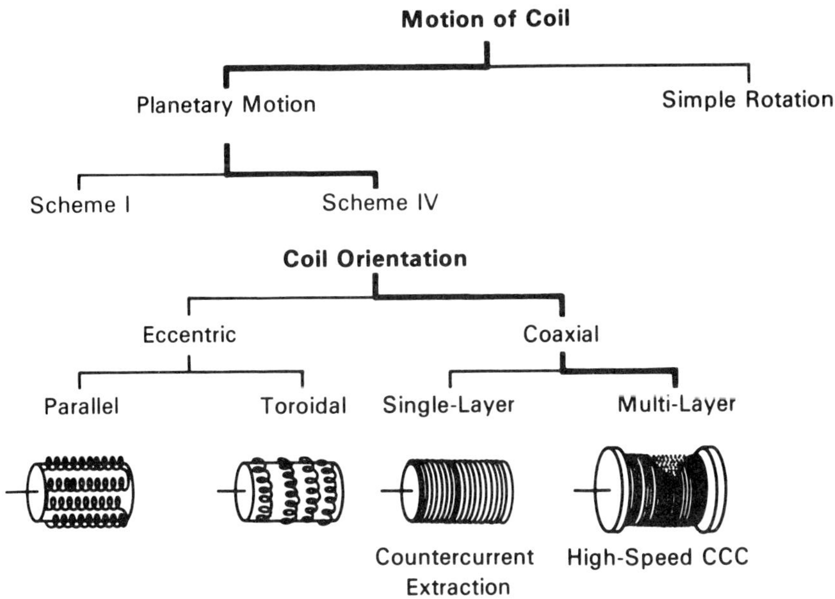

Fig. 122 Search for the high-performance CCC systems.

the principle of the high-performance CCC, the right half of the previous illustration (Fig. 16) is shown in Fig. 123 where the rotating coils are each drawn as a straight tube to indicate the overall distribution of the two solvent phases in the coil.

In Fig. 123a, the two solvent phases confined in the coil have established a unilateral hydrodynamic equilibrium under the scheme IV synchronous planetary motion. The white phase (head phase) is entirely occupying the head side and the black phase (tail phase) the tail side of the coil. This hydrodynamic equilibrium condition clearly indicates that a drop of the tail phase, if introduced into the head of the coil, would move through the head phase toward the tail and similarly a drop of the head phase, if introduced into the tail of the coil, would move through the tail phase toward the head of the coil. This hydrodynamic motion can be effectively utilized for continuous elution with the mobile phase to retain a large volume of the stationary phase in the coil in the following two different ways:

1. Head-to-tail elution mode (Fig. 123b, top). The coil is first entirely filled with the head phase followed by elution with the tail phase from the head toward the tail of the coil.

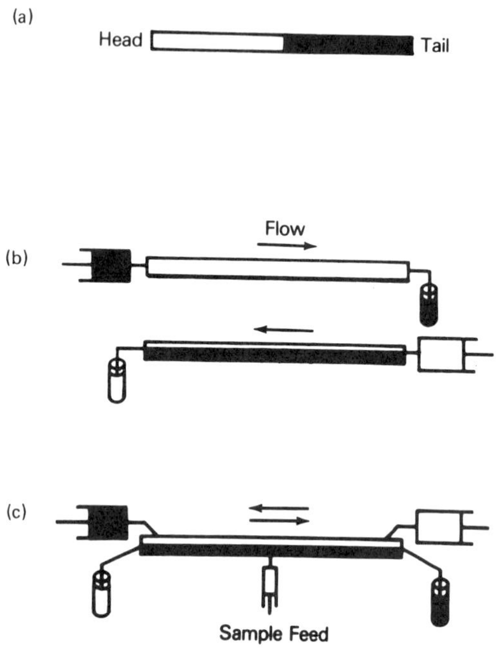

Fig. 123 Principle of the unilateral hydrodynamic equilibrium system. (a) Unilateral hydrodynamic equilibrium in a closed coil. (b) One-way elution modes. (c) Dual-countercurrent system.

2. Tail-to-head elution mode (Fig. 123b, bottom). The coil is first entirely filled with the tail phase followed by elution with the head phase from the tail toward the head of the coil.

In both cases, the system enables a high retention level of the stationary phase in the coil against fast flow of the mobile phase while an undulating centrifugal force field produced by the planetary motion causes efficient local mixing of two phases in each helical turn. As a result the system yields highly efficient chromatographic separations of solutes in short periods of time.

The system also allows simultaneous elution of the two solvent phases through the respective terminal of the coil to effect the true countercurrent flow of the two phases (Fig. 123c). This requires an additional flow line at each terminal of the coil to collect the effluents and a sample feed line at the middle portion of the coil. The dual-countercurrent system has been successfully applied to foam separation as described later in detail [10].

B. Design of the Apparatus

All apparatus for high-performance CCC share a common basic design which produces the scheme IV synchronous planetary motion of the holder (Fig. 17). The design principle of these coil planet centrifuges are illustrated in Fig. 124 where the orientation of the centrifuge axis is placed in the horizontal position. A large cylindrical coil holder coaxially holds a planetary gear which is coupled to an identical stationary sun gear (shaded) mounted on the central axis of the centrifuge. This gear arrangement produces a synchronous planetary motion to the coil holder: The holder revolves around the central axis of the apparatus and simultaneously rotates about its own axis at the same angular velocity in the same direction. In doing so the holder maintains its position parallel to and at distance R from the central axis of the apparatus. This synchronous planetary motion of the holder prevents twisting the flow tubes as previously illustrated by successive positions of the planetary and sun gears during one complete revolution (see Fig. 72b).

The column is prepared by winding a piece of flexible plastic tubing, typically PTFE, around the holder. The column design may be classified into three categories: the single layer or short coil for extraction purposes, the long multilayer coil for high-speed CCC, and the column equipped with five flow channels for dual-countercurrent operation. In all these columns, the ratio between the radii of rotation and revolution ($\beta = r/R$) is an important parameter to determine the pattern of the centrifugal force field as illustrated in Fig. 18d (right).

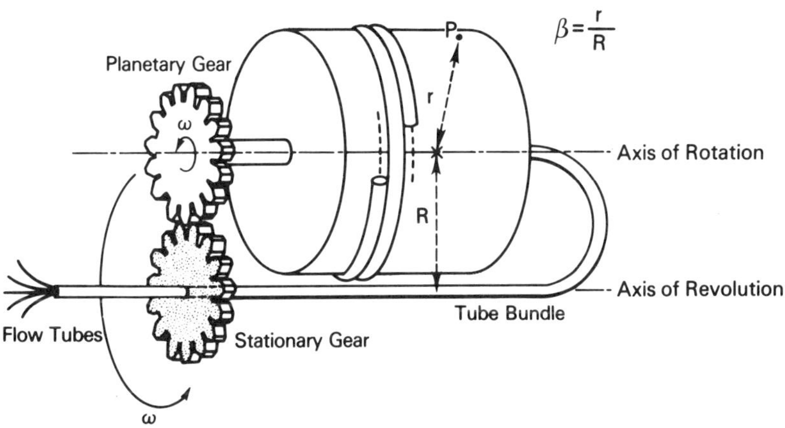

Fig. 124 Design principle of the coil planet centrifuge for high-performance CCC.

Continuous Extraction Apparatus

Figure 125 shows a cross-sectional view of the coil planet centrifuge for continuous extraction [8,29]. The motor drives the rotary frame around the horizontal stationary pipe (shaded) mounted on the axis of the centrifuge. The rotary frame consists of a pair of aluminum disks rigidly bridged together with multiple links (not shown in the diagram) and holds a pair of rotary column holders in the symmetrical positions 10 cm away from the central axis of the centrifuge. The bottom holder has a diameter of 15 cm ($\beta = 0.75$) and the top holder 10 cm ($\beta = 0.5$). The shaft of each holder is equipped with a plastic planetary gear which is coupled to an identical sun gear (shaded) mounted around the central stationary pipe. In order to provide mechanical stability, a short coupling pipe is coaxially mounted to the free end (right side) of the rotary frame while the other end of the coupling pipe is supported by a stationary wall member of the centrifuge through a ball bearing. The coiled column was made by winding the desired length of a PTFE tube around one of the holders while a counterweight was applied on the other holder to balance the centrifuge. A pair of flow tubes from the coiled column is first passed through the center hole of the holder shaft and led through the side hole of the short coupling pipe to reach the opening of the central stationary pipe. These flow tubes are thoroughly lubricated with silicone grease and protected with a piece of plastic tubing at each supported portion to prevent direct contact with metal parts. The revolutional speed can be regulated up to 1000 rpm. The apparatus is a compact table top model whose dimensions are about 16 in. × 16 in. × 17 in.

Multilayer Coil Planet Centrifuges for High-Speed CCC [9,28]

The original multilayer coil planet centrifuge for high-speed CCC (Fig. 126) has been developed from the above model by modifying the column holder. In order to accommodate a long coiled column, a pair of flanges was mounted around the smaller column holder to support multiple layers of coil. This apparatus has produced a rapid chromatographic separation of DNP amino acid samples [28]. The partition efficiency of high-speed CCC was further increased by mounting a longer multilayer coil around a large spool-shaped holder as shown in Fig. 127 [9]. In order to facilitate the column preparation, the column holder was made removable from the rotary frame simply by loosening a pair of screws on each bearing block. Figure 128 shows the multilayer coil planet centrifuges constructed in our laboratory. A compact table top model of the coil planet centrifuge shown in Fig. 128a has a multilayer coil consisting of a 130-m long, 1.6-mm i.d. PTFE tube with a total capacity of about 280 ml. Several hundred milligrams of samples can be loaded in each separation.

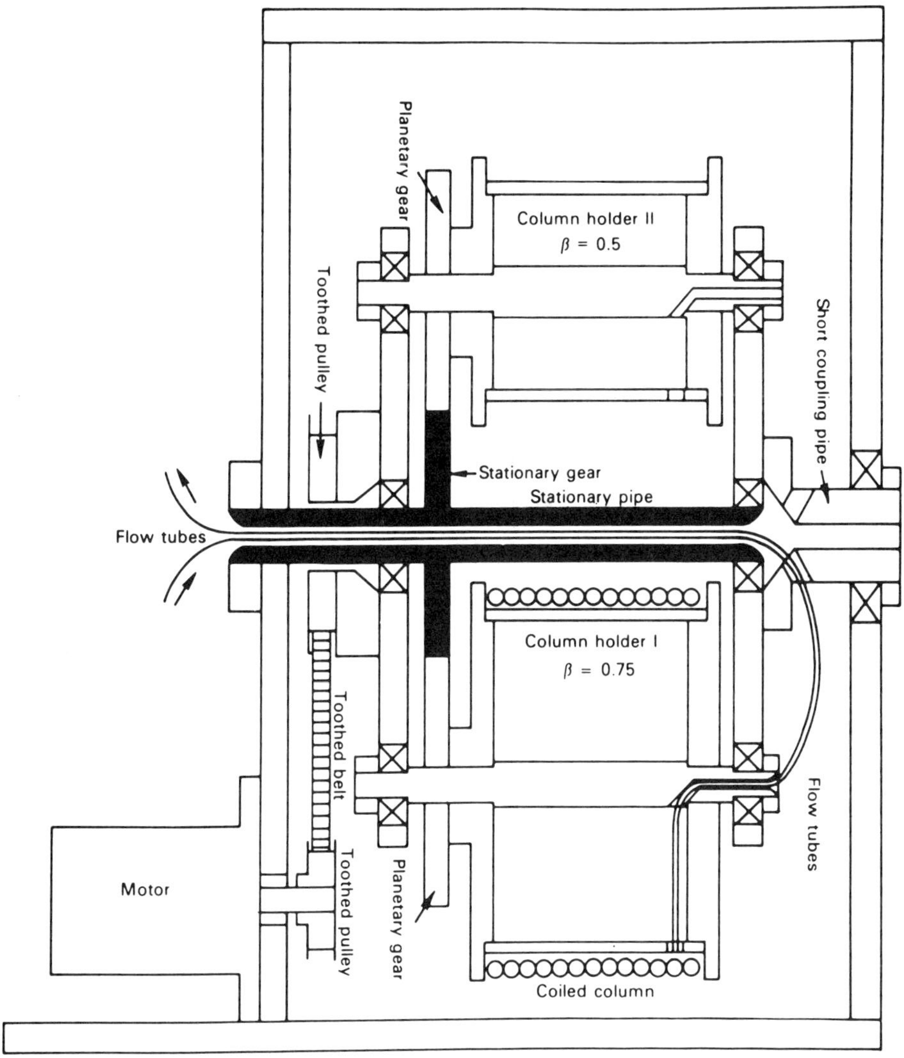

Fig. 125 Cross-sectional view of the coil planet centrifuge for continuous extraction.

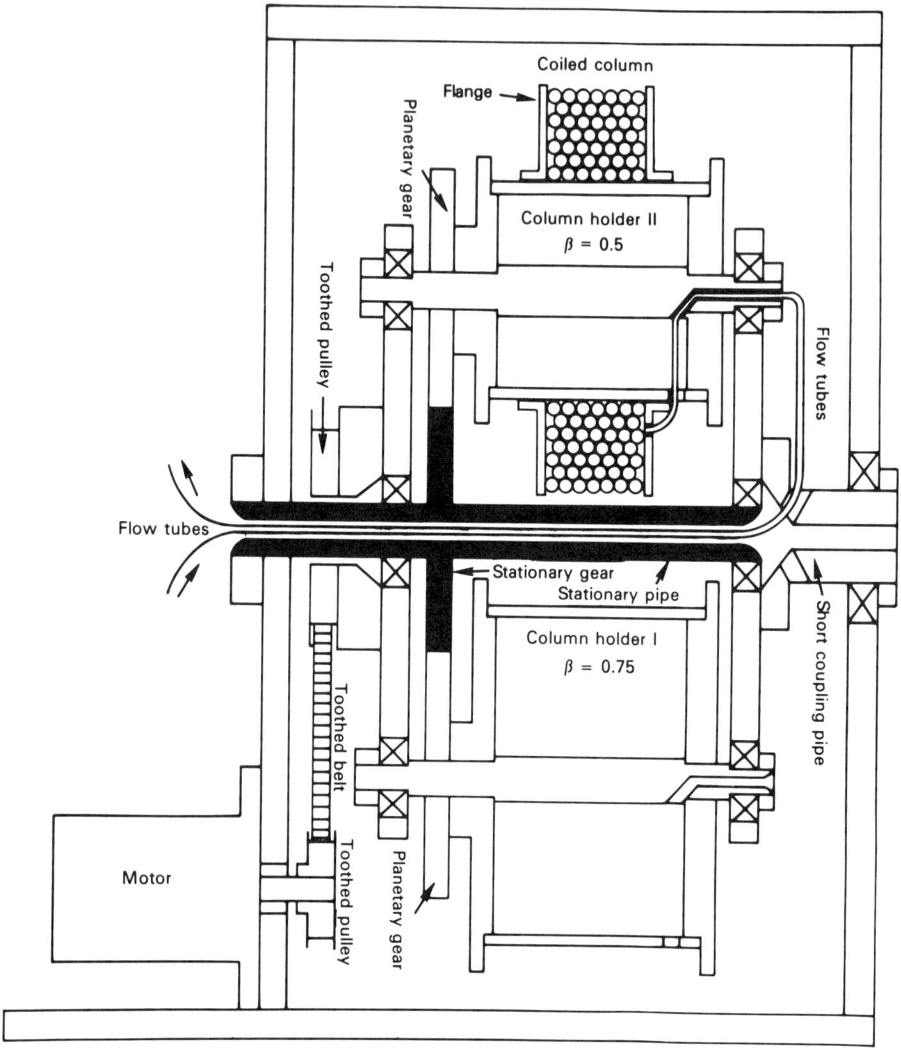

Fig. 126 Cross-sectional view of the original multilayer coil planet centrifuge for high-speed CCC.

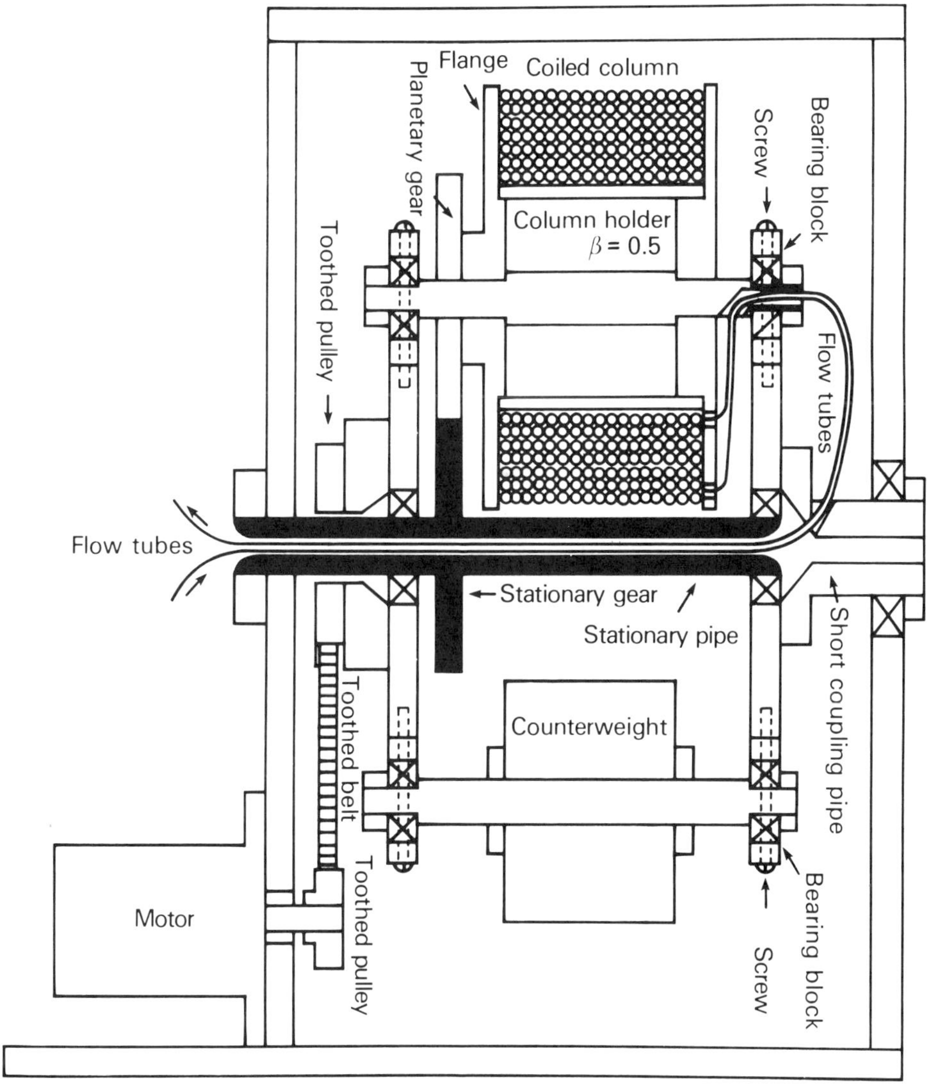

Fig. 127 Cross-sectional view of the coil planet centrifuge with a large multilayer coil.

Fig. 128 Photograph (above) of our prototype of the coil planet centrifuge for high-speed CCC. Photograph (opposite page) of our prototype of the coil planet centrifuge for large preparative-scale separations.

Recently, a similar multilayer coil planet centrifuge became commercially available through P.C. Inc. (Potomac, MD) and also from Pharma-Tech Research Corp. (Baltimore, MD). Figure 128b shows a bench top model of the coil planet centrifuge with a 5.5-mm i.d. multilayer coil for large-scale preparative separations [84,85].

Preparation of the multilayer coil requires caution to prevent dislocation of the column on the holder during the run. Because the holder is constantly subjected to a planetary motion, the resulting acceleration tends to rotate the column around the holder. This problem becomes serious when a large column is subjected to a high revolutional speed. In our laboratory the multilayer coil is prepared in the following manner with a satisfactory result [86]: A sheet of emery cloth is glued around the holder core and the PTFE tubing is tightly wound over it between the two large flanges which provide boundaries for the multilayer configuration. To prevent the movement

Fig. 128 (Continued)

of the column with respect to the holder, each layer of coil is attached to the flanges with a piece of fiberglass-reinforced adhesive tape across the width of the coil. To further enhance greater stability, the same fiberglass tape is wrapped around the circumference of the coil. Around this layer of tape is wound a single piece of wire in which the ends are anchored to each flange. Around this wire another single layer of tape is wrapped, in effect sandwiching the wire between the two layers of tape. This arrangement firmly anchors the multilayer coil to the holder. Recently we tried a different approach by modifying the column holder as follows: Multiple radial grooves (6 to 8 for each flange, measuring 1 mm deep and 2 mm wide, were made on the inside surface of each flange. Similar grooves were also made on the holder core across its width in such a way that each groove met the grooves made on the flanges. A hole fitting to the tip of the plastic syringe was made through the flanges at each point where the grooves made on the core and flanges meet together. The multilayer coil was prepared by simply winding the tube onto the modified flange support and wrapped with the fiberglass adhesive tape over the circumference of the entire column. Then, using a plastic syringe, RTV silicone rubber adhesive sealant was injected into each hole to fill the grooves on the core and on each flange. The sealant was cured for 24 hr at room temperature and the coil was ready for use. In this way preparation of the multilayer coil became less laborious and the process can be automated in the future. Several test runs produced a promising result in our laboratory.

Combined Horizontal Flow-Through Coil Planet Centrifuge [87,88]

In addition to the above scheme IV coil planet centrifuges, a new table top model of the combined horizontal flow-through CPC was used for various hydrodynamic studies. The apparatus produces two different types of synchronous planetary motions, schemes I and IV. The design principle of the centrifuge was previously illustrated in Fig. 76.

Figure 129 shows a cross-sectional view through the central axis of the apparatus. The motor drives the rotary frame around the central stationary pipe (shaded) by a pair of toothed pulleys coupled with a toothed belt. To stabilize the centrifuge system, the other end of the rotary frame is coaxially connected to a short coupling pipe which is in turn supported by a ball bearing mounted in a stationary wall member of the centrifuge. The rotary frame holds column holders and countershaft on each side in symmetrical positions. On the lower side of the rotary frame (pulley drive), a toothed pulley mounted on the central stationary pipe is coupled with a toothed belt to an identical pulley on the countershaft. This coupling

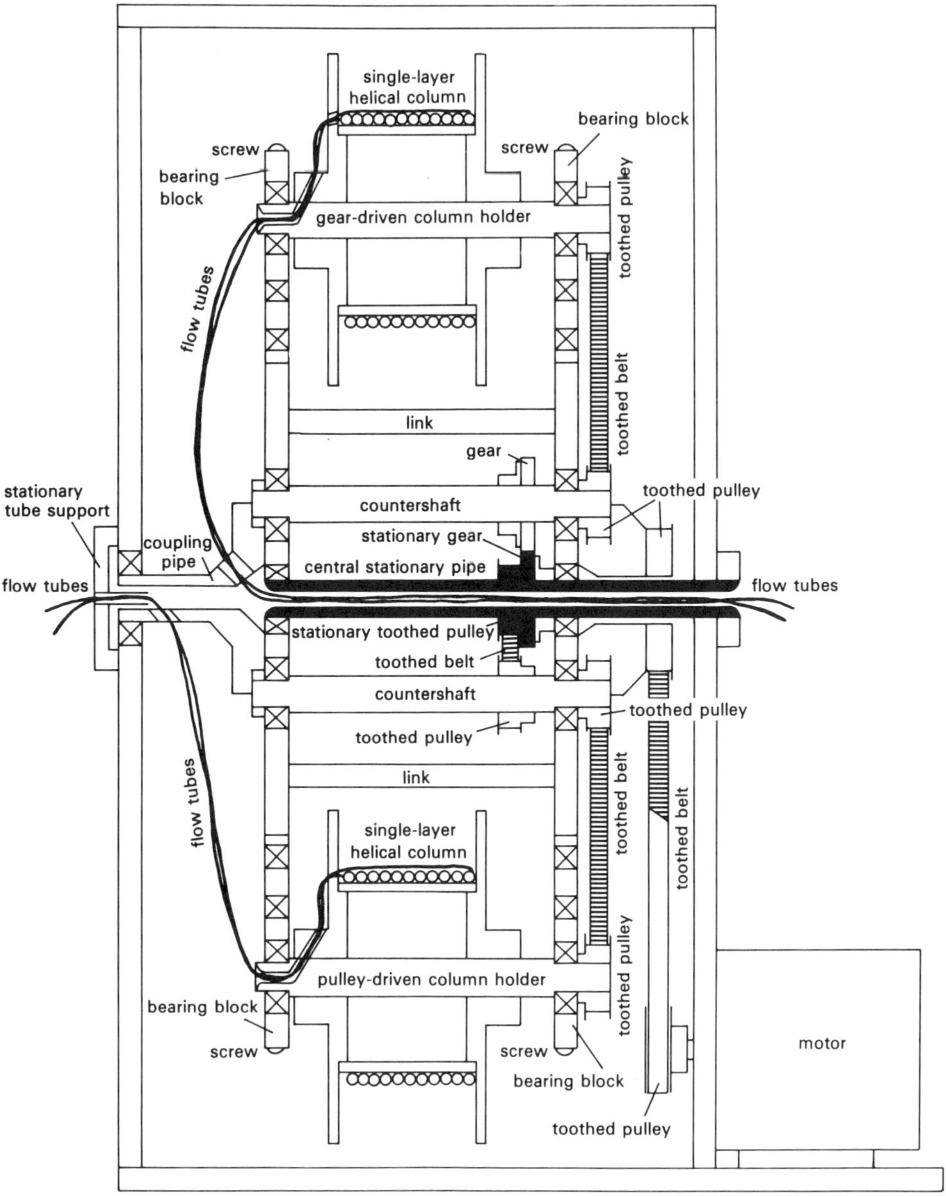

Fig. 129 Cross-sectional view of the combined horizontal flow-through coil planet centrifuge.

produces synchronous counterrotation of the countershaft with respect to the rotary frame. This motion is conveyed to the column holder by a belt coupling a pair of identical toothed pulleys, one mounted on the countershaft and the other on the column holder shaft. Consequently, the column holder undergoes synchronous planetary motion identical to that of scheme I illustrated in Fig. 17. On the upper side of the rotary frame (gear drive), the stationary gear mounted around the central stationary pipe is coupled to an identical gear on the countershaft to produce synchronous rotation of the countershaft with respect to the rotating frame. This motion is similarly conveyed to the holder by a belt and a pair of identical timing pulleys. The gear-driven column holder, then, rotates with respect to the rotary frame once about its own axis per each revolution of the rotary frame around the central axis of the centrifuge as in scheme IV illustrated in Fig. 17.

The column holder on each side is removable from the rotary frame and may be positioned at 15 or 20 cm from the axis of the centrifuge by selecting the respective bearing holes on the pair of bearing blocks. For each position of the holder the β-value can be varied by changing the holder-hub diameter up to a maximum β-value of 0.5, as limited by the presence of the countershaft on each side of the rotary frame. The helical column is prepared by winding a piece of PTFE tubing directly onto the holder making the desired number of helical turns. The flow tubes from the pulley-driven column holder (bottom) are first passed through the center hole of the holder shaft and then led through a side hole of the short coupling pipe to reach the stationary tube support as illustrated. The flow tubes from the gear-driven column holder (top) are similarly passed through the center hole of the holder shaft and then led through another side hole of the short coupling pipe to enter the opening of the central stationary pipe. Each pair of flow tubes is lubricated with silicone grease and the portions rotating with the frame are protected by a flexible plastic sheath to prevent abrasion by direct contact with metal parts. The revolutional speed of the apparatus is continuously adjustable up to 1000 rpm with a speed control unit.

Coil Planet Centrifuge with Large β-Values [88]

In all the coil planet centrifuges described above the diameter of the coil holder mounted on the rotary frame is limited by the presence of the countershaft or the central stationary pipe, the maximum β-values available in these apparatus always being less than 1. In this particular type of the coil planet centrifuge illustrated in Fig. 130 [89], the range of β-values is further extended by eliminating the central stationary pipe to accommodate a large holder extending over the central axis of the apparatus. The coil holder is driven through a short countershaft mounted on the opposite side of the rotary frame. Two bearing holes provide revolutional radii adjustable to 2 and 4 in. The apparatus has been used for phase retention studies with helical columns [88].

Coil Planet Centrifuge for Stroboscopic Observation [90]

This particular type of the apparatus shown in Fig. 131 has been constructed for stroboscopic observation and photography of hydrodynamic motion of the solvent phases in the rotating coiled column. To facilitate observation, one side of the column was widely exposed to illumination by eliminating the bearing support. The separation column consisting of either a single-layer spiral or multilayer coil was mounted over the aluminum plate and covered with a transparent plastic plate. A piece of magnet (not shown in the diagram) mounted onto the lower part of the central rotary shaft activated the magnetic switch to produce synchronous light discharge from the stroboscope.

C. Studies on Hydrodynamic Distribution of Two Solvent Phases in the Rotating Column

As described earlier, the high-performance CCC utilizes a unique hydrodynamic motion of the two solvent phases in the coiled column subjected to a particular type of synchronous planetary motion. The mathematical analysis on the acceleration has been performed to elucidate the characteristic centrifugal force field produced by this synchronous planetary motion [5]. However, the effects of the centrifugal force field on the hydrodynamic distribution of the solvent phases in the coiled column cannot be predicted from the results of these analyses. As a matter of fact, all the knowledge on the unilateral hydrodynamic equilibrium has been exclusively derived from direct or indirect observation and measurement of the two-phase distribution in the coiled column by experimentation.

Stroboscopic Observation on Hydrodynamics [90,91]

A series of experiments has been performed by Conway et al. [90, 91] to study hydrodynamic motion of the two immiscible solvents in the rotating column. Using a specially designed scheme IV coil planet centrifuge with a 10 cm revolutional radius (Fig. 131), phase distribution and mixing of the chloroform/acetic acid/water (2:2:1) system in a single-layer spiral and a multilayer helical coil was observed and photographically recorded under stroboscopic illumination.

The columns (Fig. 132), mounted concentric with the axis of the column holder, were driven in either direction to reverse the head

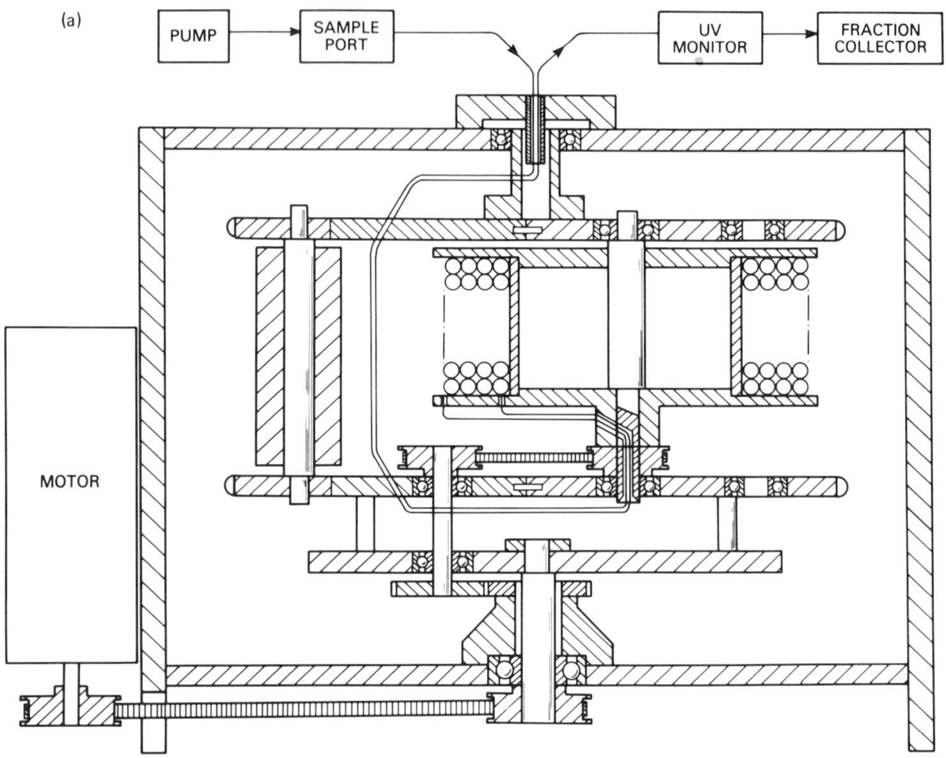

Fig. 130 Coil planet centrifuge providing large β-values. (a) Cross-sectional view through the central axis of the apparatus. (b) Cross-sectional view through the upper plate of the rotary frame.

and tail relationship of the peripheral and central ends of the coil. The centrifugal field (−α), shown for β = 0.5, is unsymmetrical in intensity and direction, where β is the ratio of the coil radius r to the orbital radius R of the column holder. All vectors radiate from a common point, β = 0.25 and α = 0.

The single-layer spiral column consisted of about 3.5 m of 2.5-mm i.d. PTFE tubing, volume 20.9 ml, with 10 coils from 11.5 cm (β = 0.57) to 18 cm (β = 0.9) in diameter. The multilayer helical coil consisted of about 17 m of the same tubing, volume 102 ml, 5 rows wide, 10 helical layers, ranging from 10 cm (β = 0.5) to 17 cm (β = 0.85) in diameter. The column was first filled with stationary phase and mobile phase pumped at 240 ml/hr into either end while rotating the apparatus at 800 rpm. A stroboscope was magnetically synchronized for observation and photography of the phases dyed yellow and red.

(b)

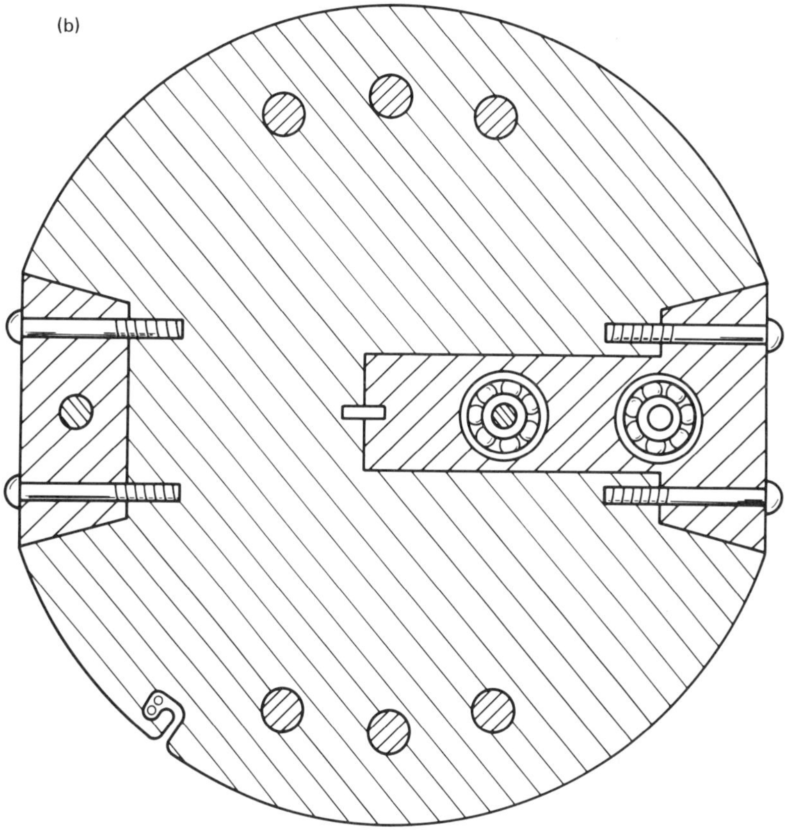

Fig. 130 (Continued)

Phase volume equilibrium was quickly established immediately behind the mobile phase front. Stationary phase retention (S_F = fraction of column volume occupied by stationary phase, tabulated as percentage) for various conditions is summarized in Table 6. Flow direciton is symbolized by the arrow between H (head) and T (tail), located at the coil center (c) or periphery (p). The best retention is obtained with the column head at the center and heavier phase pumped from head to tail (condition A) or lighter phase pumped from tail to head (condition b). However, only slightly lower retention is obtained in the single-layer spiral with reverse rotation (conditions C and d).

Under conditions with good retention and with the heavier phase mobile (A and C), the column was divided into an agitated zone, to the left of the vertical line through $\beta = 0$, in which good mixing was

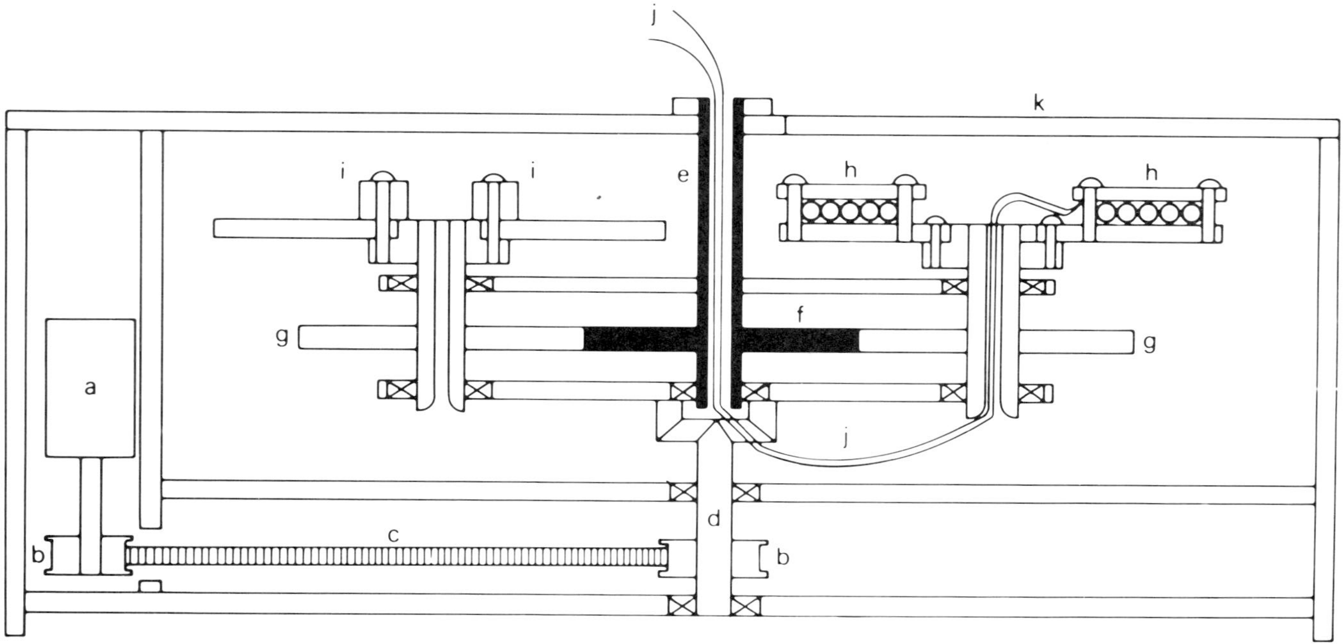

Fig. 131 Cross-sectional view of the coil planet centrifuge for stroboscopic observation of hydrodynamics. (a) Motor; (b) toothed pulley; (c) toothed belt; (d) central rotary shaft; (e) central stationary shaft; (f) stationary sun gear; (g) planetary gear; (h) spiral column with transparent plastic cover; (i) counterweight; (j) flow tubes; (k) transparent plastic plate.

Table 6 Stationary Phase Retention for the $CHCl_3$:HOAc:H_2O, 2:2:1, v/v/v System

		Single-layer spiral, S_F(%) mobile phase		Multilayer coil, S_F(%) mobile phase	
Rotation	Flow	Light	Heavy	Light	Heavy
CW	Hc → Tp	a, 5	A, 82	—	A, 78,78
CW	Hc ← Tp	b, 84,82	B, 3,0.5	b, 70,70	—
CCW	Hp → Tc	c, 12	C, 77	—	—
CCW	Hp ← Tc	d, 78,72	D, 2,0.5	—	—

observed (Fig. 132) and a quiescent zone in the remainder of the column where heavier phase migrated along the periphery of each coil with little apparent mixing. With a mobile lighter phase, mixing was also observed in the left part of the column (conditions b and d) but it was not as uniform, and additional agitated zones were observed in the right side of the coil. Two zones of apparent column flooding were observed with the light mobile phase in the multilayer coil (condition b), one of which was associated with the passage of tubing from the end row of coils to the adjacent inner row.

More recently, Sutherland et al. [91] extended the above study with a similar type of the coil planet centrifuge with a spiral column covering a lower range of β-values from 0.35 to 0.82. The results showed, in addition to the hydrodynamic trends previously reported, that the reversed volume distribution of the two phases at the inner portion of the spiral column with small β-values. The above finding

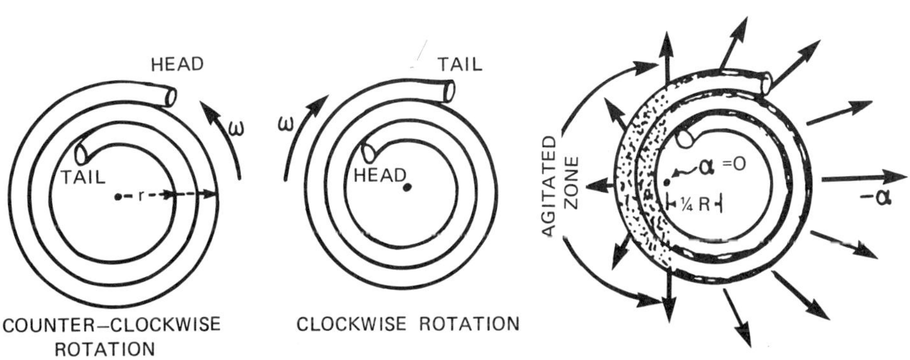

Fig. 132 Head and tail relationships and centrifugal force field (−α) for a spiral column.

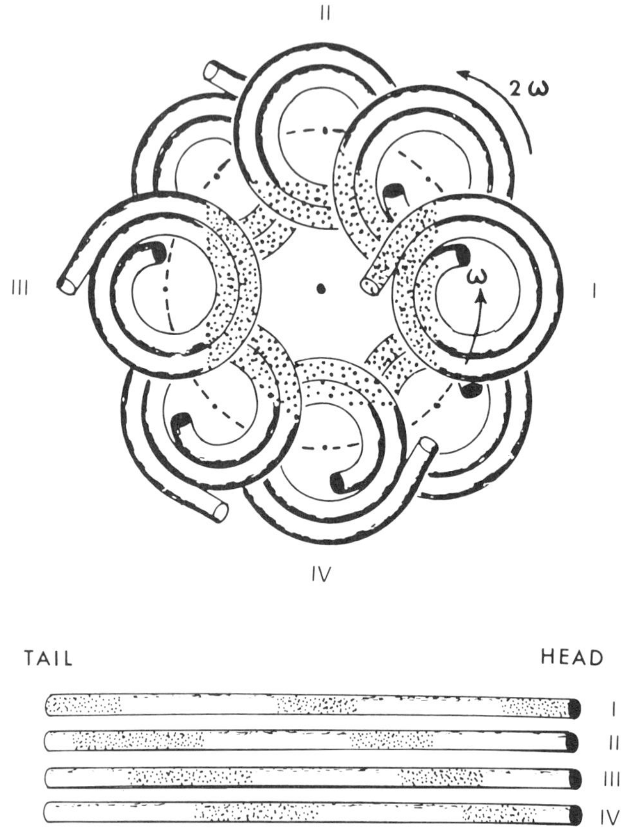

Fig. 133 Motion of mixing zones through a spiral column concentrically mounted on the holder of the scheme IV coil planet centrifuge.

is consistent with the results of hydrodynamic studies described in the following subsection.

These results clearly demonstrate a unique hydrodynamic behavior of the two solvent phases in the present system as schematically illustrated in Fig. 133. The top diagram illustrates successive positions of the spiral column during one complete revolution around the central axis of the centrifuge. The mixing zone is always located near the center of the centrifuge while the spiral column rotates about its own axis twice during one revolutional cycle. At the bottom uncoiled columns numbered I–IV correspond to the column position in the top diagram and show movement of the mixing zone through the spiral column. Each mixing zone is traveling toward the head of the spiral column at a rate equal to the revolutional speed of the column. This indicates an extremely interesting fact that the two solvent phases at any portion of the column are subjected to a typical partition process of repetitive mixing and settling at an enormously high rate of over 13 times per second while the mobile phase is steadily passing through the stationary phase. Consequently, this finding promises a high partition efficiency of the present scheme under a high flow rate of the mobile phase.

While the above described stroboscopic observation can shed light on the hydrodynamic interaction of the two solvent phases taking place in the column, it leaves a number of questions to be answered in the future. Some examples are (a) how the mixing and settling zones are formed, (b) why two solvent phases are unilaterally distributed, (c) which phase becomes the head phase and how it is determined, etc. Although the unique distribution of the centrifugal force vectors described earlier (Fig. 18d, right) seems to provide some rationale to the formation of the mixing and settling zones, further studies are required to establish the theory.

It is most likely that the hydrodynamic interaction of the two solvent phases in the rotating column is caused and affected by a number of physical factors involved in the present system. If this is the case, it becomes necessary to investigate the effects of these physical factors on the hydrodynamic distribution of the two solvent phases in the present system.

Phase Distribution Diagrams

Series of experiments have been systematically conducted to study the hydrodynamic distribution of two immiscible solvent phases in the rotating column utilizing various types of the coil planet centrifuges described earlier. These studies were performed with two types of columns, a single-layer coil and a single-layer spiral. The configuration, handedness, and head–tail orientation of these columns are illustrated in Fig. 134. Phase distribution diagrams were produced for each column by measuring the retention of the stationary phase under various operational conditions with wide ranges of revolutional speeds and flow rates.

For each measurement, the column was first entirely filled with the stationary phase. Then the apparatus was run at a desired revolutional speed while the mobile phase was pumped through the column at a given flow rate. The effluent from the outlet of the column was collected in a graduated cylinder to measure the volume of the stationary phase eluted from the column as well as the total volume of the mobile phase eluted. From the data obtained from the above experiments, phase distribution diagrams were produced. The volume of the stationary phase retained in the column was first calculated from the volume of the stationary phase eluted from the column and expressed as a percentage of the total column capacity ac-

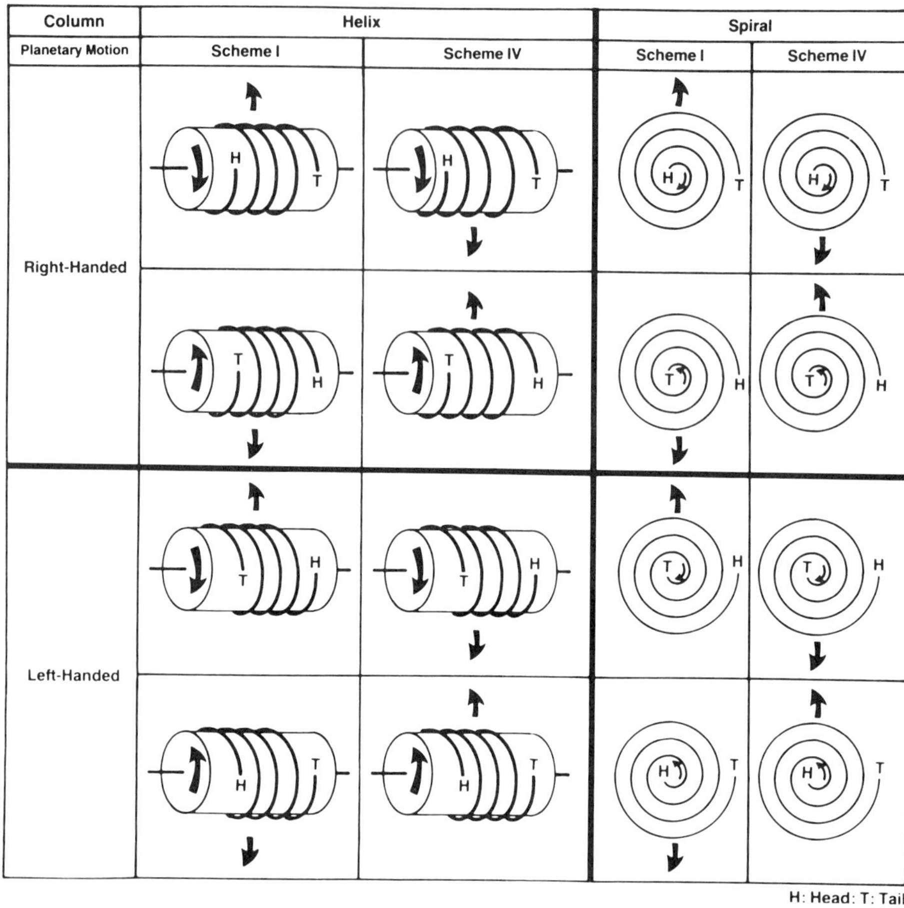

Fig. 134 Head–tail orientation of coiled columns in synchronous planetary motion. H = Head, T = Tail. Under the centrifugal force fields produced by the two typical synchronous planetary motions, the head–tail orientation of both columns is determined by the handedness of the coil and column rotation but is irrelevant to column revolution. Note that the handedness of a flat spiral is altered if viewed from the other side.

Table 7 Symbolic Design of the Retention Curves in the Phase Distribution Diagram for Helical and Spiral Columns

	Elution mode		Retention curve
(A) Helical column	H → T		Solid line
	H ← T		Broken line
	Elution mode		Retention curve
	Upper phase mobile	Lower phase mobile	
(B) Sprial column	EH → IT	IH → ET	Thick solid line
	IH → ET	EH → IT	Thin solid line
	IH ← ET	EH ← IT	Thick broken line
	EH ← IT	IH ← ET	Thin broken line

Abbreviations: H = head; T = tail; I = internal; E = external.

cording to the expression $100(V_c + V_f - V_s)/V_c$, where V_c denotes total column capacity; V_f, the volume of the flow tubes; and V_s, the volume of the stationary phase eluted from the column. The hydrodynamic distribution of the two solvent phases in the column was summarized by a phase distribution diagram constructed by plotting the percentage retention of stationary phase as a function of revolutional speed for a particular mobile phase flow rate. A group of retention curves produced by different elution modes but otherwise idential experimental conditions can be illustrated in the same phase distribution diagram with a set of symbolic designs (Table 7).

1. *Phase Distribution Diagrams Obtained by Two Typical Synchronous Planetary Motions [87].* In the first series of experiments a set of phase distribution diagrams was obtained from two different modes of synchronous planetary motions using the combined horizontal flow-through coil planet centrifuge with a revolutional radius set at 20 cm. As shown in Fig. 120, this apparatus holds two column holders: the pulley-driven holder produces the scheme I planetary motion and the gear-driven holder the scheme IV planetary motion.

Experiments were performed with helical columns, 5 m × 1.6 mm i.d. PTFE tubing wound around a 10-cm diameter column holder mounted on each side. For the present study nine different types of volatile two-phase solvent systems, covering a broad spectrum in hydrophobicity, were selected. They include hexane/water, hexane/

methanol, ethyl acetate/water, ethyl acetate/acetic acid/water (4:1:4), chloroform/water, chloroform/acetic acid/water (2:2:1), n-butanol/water, n-butanol/acetic acid/water (4:1:5) and sec-butanol/water. Each solvent system was prepared by equilibrating the solvent mixture in a separatory funnel at room temperature and separated before use.

The experimental data are summarized in Fig. 135. Nine different solvent systems were examined by using each phase as the mobile phase at a flow rate of 120 ml/hr except for the butanol phase systems in which, because of its high viscosity, the flow rate was reduced at 60 ml/hr. Other important experimental conditions, such as column i.d., R, r, and β values, and the revolutional speeds, etc., are indicated in the diagram. Two retention curves are drawn in each diagram, the solid line for head-to-tail elution, and the broken line for tail-to-head elution. These phase distribution diagrams are arranged, from left to right, in the order of hydrophobicity of the major organic solvent, i.e., hexane, ethyl acetate, chloroform, and butanol. Within each organic solvent group, the phase composition is modified by addition of mutually miscible solvent to moderate the interfacial tension of the solvent system.

The phase distribution diagram of the hexane/water system obtained by the scheme IV planetary motion, upper left diagram in Fig. 135, will serve to illustrate the interpretation of these diagrams. In the upper diagram, plotting lower phase retention on the ordinate implies that the upper phase is mobile. The retention of the lower phase sharply increases with revolutional speed, reaching over 90% at 800 rpm with the tail-to-head elution mode (broken line), while no retention occurs in the head-to-tail elution mode (solid line). This retention behavior can be described as unilateral, to indicate that the upper phase has a strong hydrodynamic tendency to move to the head and the lower phase to the tail of the column. Thus with tail-to-head elution of the upper phase, the lower phase is well retained and the flowing upper phase quickly moves through it. On the other hand, with head-to-tail elution of the upper phase, the natural tendency of the lower phase to move to the tail is enhanced and the lower phase is simply washed out of the column. In the diagram second from the top, where the lower phase is mobile, high stationary phase retention results from the head-to-tail elution mode (solid line) because the natural tendency of the upper phase to flow toward the head is opposed by the mobile phase flow. Again poor retention of the upper phase is obtained if the lower phase is pumped from tail to head. These findings agree with the previous observation [8] that the hexane/water system, if confined in a similar column, establishes a unilateral hydrodynamic distribution where the two phases are completely separated in the direction of the column axis, the upper phase entirely occupying the head side and the lower phase the tail side of the column. This unilateral hydrodynamic dis-

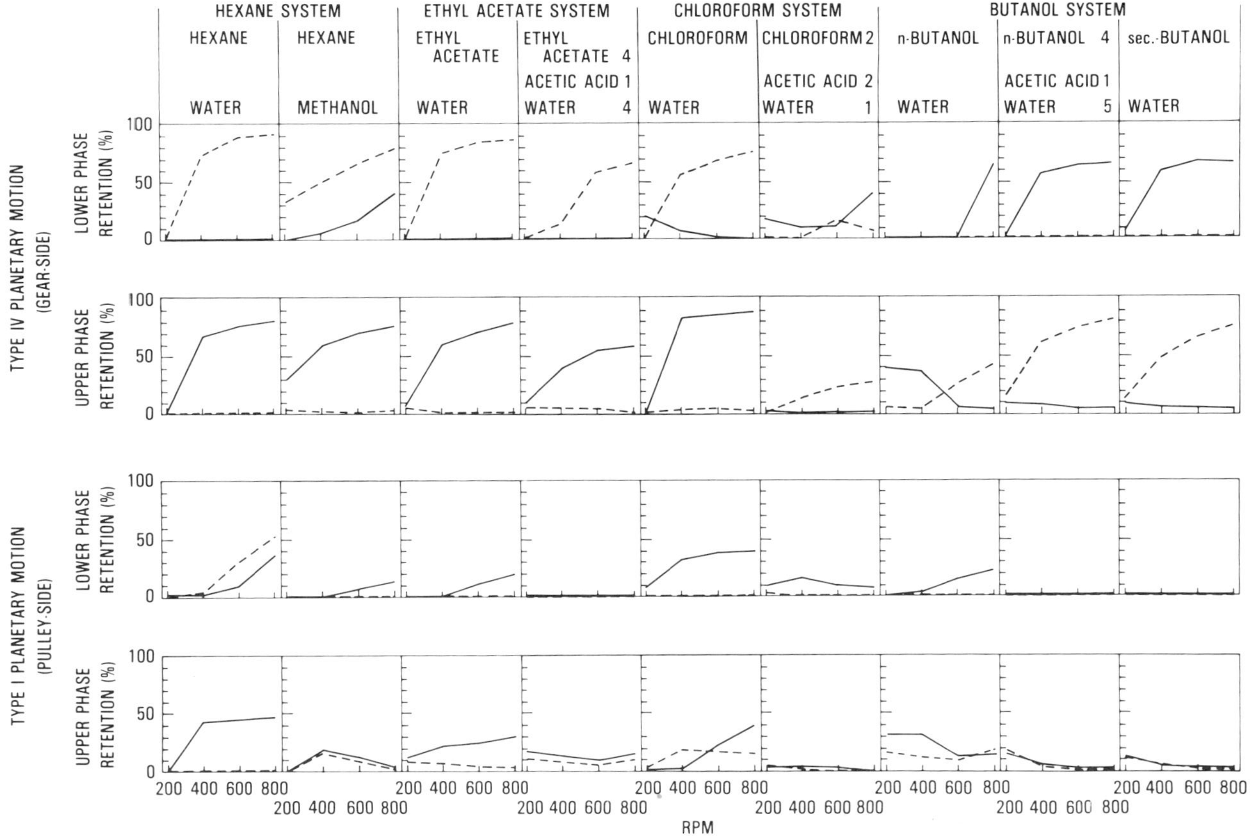

Fig. 135 Phase distribution diagrams in single-layer coil of various two-phase solvent systems obtained by the schemes I and IV planetary motion (1.6 mm i.d., R = 20 cm, r = 5 cm, β = 0.25). (—) Head-to-tail elution. (---) Tail-to-head elution.

tribution of the two phases is ideal for performing CCC because higher retention of the stationary phases is attainable with either phase used as the mobile phase by selecting the proper mode of elution. Two other solvent systems, ethyl acetate/water and chloroform/water, also exhibit the ideal retention curves similar to those of hexane/water, when subjected to the scheme IV planetary motion. These three solvent systems are characterized by high interfacial tension and a strongly hydrophobic organic phase.

Ideal retention curves, but with completely reversed hydrodynamic trends, are observed in the two butanol solvent systems, n-butanol/acetic acid/water (4:1:5) and sec-butanol/water. Both organic phases are very hydrophilic and possess high viscosity and low interfacial tension to the aqueous phase. As seen from the retention curves, these solvent systems display a reversed unilateral hydrodynamic motion in the column; the upper phase moves toward the tail and the lower phase toward the head. If the two phases are confined in the column without introduction of external flow, they establish a unilateral hydrodynamic distribution or complete separation along the axis of the column with the upper phase on the tail side and the lower phase on the head side.

The remaining solvent systems in the scheme IV group show phase distributions intermediate between those of the above two extreme groups. These intermediate solvent systems are usually prepared by adding methanol or acetic acid to a hydrophobic solvent system to moderate the interfacial tension and hydrophobicity of the organic phase. The relatively hydrophilic n-butanol/water also falls in this intermediate group. All solvent systems treated with the scheme IV planetary motion display a trend of the unilateral hydrodynamic distribution in which one phase moves toward the head and the other phase toward the tail of the column, although some are similar to hexane/water while others resemble the reversed hydrodynamic equilibrium of sec-butanol/water.

Phase distribution diagrams obtained from the scheme I planetary motion under the otherwise identical experimental conditions are illustrated in the lower half of Fig. 135. As these phase distribution diagrams clearly indicate, the scheme I planetary motion fails to yield a satisfactory retention level of the stationary phase in all solvent systems. The fact that the solid curve is generally higher than the broken curve indicates that higher retention levels are attained by the head to tail elution mode regardless of which phase is used as the mobile phase. These findings strongly suggest that the two solvent phases establish a basic hydrodynamic equilibrium in the column in which each phase occupies nearly equal volume on the head side as previously demonstrated in the coiled column with much smaller helical diameters [19]. Under this basic hydrodynamic equilibrium, the head-to-tail elution mode always favors the retention of the stationary phase regardless of the choice of the mobile phase, and the maximum attainable retention is usually less than 50% of the total column capacity. The reversed mode of elution from the tail toward the head of the column in these circumstances is usually not feasible because of poor stationary phase retention and steady carryover of the stationary phase resulting in loss of peak resolution.

Comparison of the phase distribution diagrams for the two different types of the synchronous planetary motion clearly reveals that the scheme IV planetary motion produces much higher stationary phase retention for all solvent systems under otherwise identical experimental conditions. This surprising contrast between the retention capability of these two schemes must be attributed to the difference in the centrifugal force field produced by their planetary motions illustrated in Fig. 18d. As stated earlier, the scheme I planetary motion produces a homogeneous centrifugal force field which synchronously rotates with the high revolutional rate of the column. This steadily rotating force field would introduce constant and vigorous agitation of the two solvent phases in a coiled column, especially with large internal diameters.

In the scheme IV synchronous planetary motion, the centrifugal force vectors always directed outwardly from the holder provide a force component which steadily acts perpendicularly to the coiled tube. Under these circumstances the two solvent phases are separated to form two layers along the helical flow path with the heavier phase in the outer portion and the lighter phase in the inner portion except for one local mixing zone for each helical turn as observed under the stroboscopic illumination (Figs. 132 and 133) [90]. Thus, the scheme IV planetary motion provides a suitable condition for the two solvent phases to flow smoothly against each other with no risk of emulsification.

2. *Phase Distribution Diagrams for Helical Column Under Scheme IV Planetary Motion [88].* The second series of experiments was conducted to investigate the effects of various physical factors on the phase distribution diagrams produced by the helical column under the scheme IV planetary motion. Two scheme IV coil planet centrifuges were employed: the apparatus with a 20-cm revolutional radius used in the previous studies and the apparatus with a 10-cm revolutional radius (Fig. 125). In addition to the nine volatile solvent systems previously tested, six different types of butanol solvent systems were newly examined. The data obtained from these experiments are summarized in Figs. 136 and 137.

Overall results clearly indicate that phase distribution curves display diverse patterns according to the hydrophobicity of the solvent systems and that within each solvent system the distribution pattern significantly changes with β-values. As described earlier, these solvent systems may be classified into three groups according to their characteristic retention profile in the phase distribution diagram.

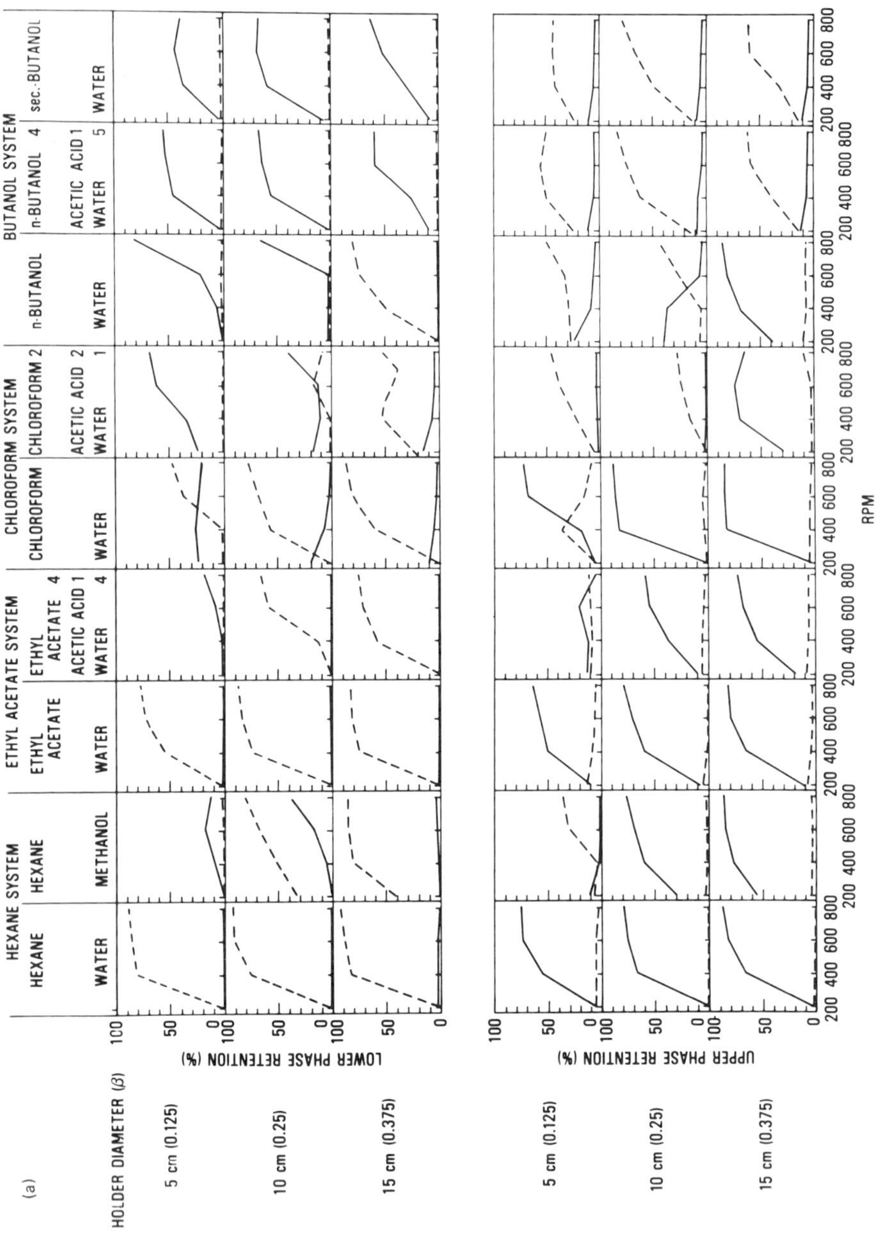
(a)
HOLDER DIAMETER (β)
5 cm (0.125)
10 cm (0.25)
15 cm (0.375)
LOWER PHASE RETENTION (%)
UPPER PHASE RETENTION (%)
HEXANE SYSTEM
HEXANE
WATER
HEXANE
METHANOL
ETHYL ACETATE SYSTEM
ETHYL ACETATE
WATER
ETHYL ACETATE 4
ACETIC ACID 1
WATER 4
CHLOROFORM SYSTEM
CHLOROFORM
WATER
CHLOROFORM 2
ACETIC ACID 2
WATER 1
BUTANOL SYSTEM
n-BUTANOL
WATER
n-BUTANOL 4
ACETIC ACID 1
WATER 5
sec.-BUTANOL
WATER
100
50
0
200 400 600 800
RPM

Fig. 136

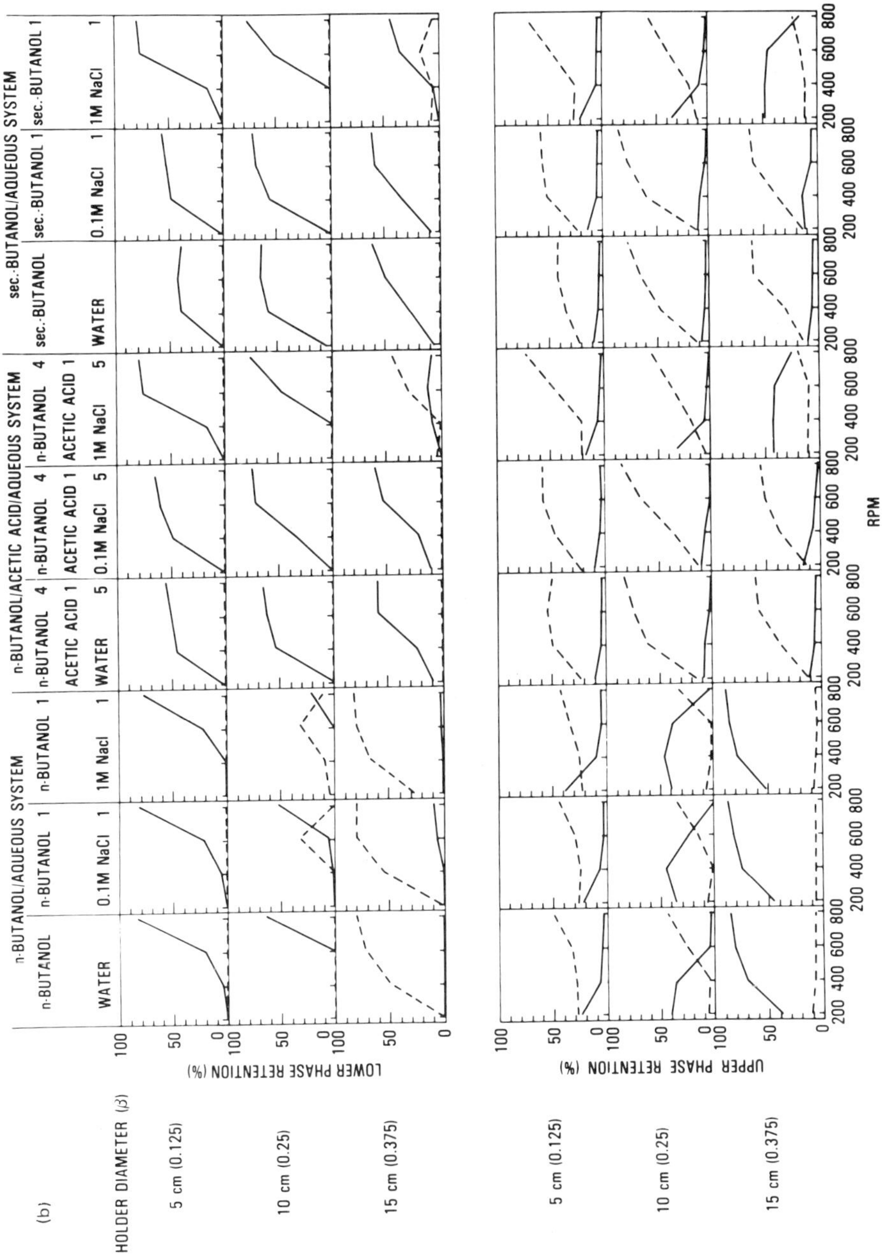

Fig. 136 (Continued)

The hydrophobic solvent group including hexane/water, ethyl acetate/water, and chloroform/water displays high retention of the stationary phase if the upper phase is eluted in the tail-to-head mode (broken line) or the lower phase is eluted in the head-to-tail mode (solid line). The reversed elution modes result in almost no retention except for chloroform/water which produces 20–30% retention at small β-values. This retention profile of the solvent group indicates that the upper phase has a strong hydrodynamic tendency to move toward the head and the lower phase toward the tail of the helical column regardless of the β-values within the range investigated.

The hydrophilic solvent group represented by n-butanol/acetic acid/water (4:1:5) and sec-butanol/water displays a retention profile contrasting with that of the hydrophobic solvent group. These solvent systems yield high retention of the stationary phase when eluting with the upper phase in the head-to-tail mode (solid line) or the lower phase in the tail-to-head mode (broken line), while no retention occurs with the reversed modes of elution. This retention profile of the hydrophilic solvent systems implies a reversed hydrodynamic tendency with the lower phase moving toward the head and the upper phase toward the tail within the range of β-values studied.

The rest of the solvent systems belong to the intermediate solvent group which is characterized by moderate hydrophobicity of the organic phase. These solvent systems exhibit a very complex retention profile which is sensitively affected by a change of the β-value. At a small β-value of 0.125, better retention of the lower phase is obtained under conditions represented by the solid curve while the broken curve indicates better retention of the upper phases. This pattern is analogous to that observed in the hydrophilic solvent group and indicates that the upper phase moves toward the tail and the lower phase toward the head under these conditions. At a larger β-value of 0.375, this retention profile is completely reversed, i.e., the broken curve shows greater lower phase retention and the solid curve shows greater upper phase retention analogous to the hydro-

---

Fig. 136 Phase distribution in helical columns obtained under the scheme IV planetary motion with a 20-cm revolutional radius. (a) Phase distribution diagrams of nine volatile solvent systems. (b) Phase distribution diagrams of butanol solvent systems. Experimental conditions are as follows: Planetary motion: scheme IV. Revolutional speed: 200–800 rpm. Revolutional radius: 20 cm. Holder diameter (β): 5 cm (0.125), 10 cm (0.25), 15 cm (0.375). Column: 1.6-mm i.d. single-layer helix. Flow rate: 120 ml/hr for hexane, ethyl acetate, and chloroform systems and 60 ml/hr for butanol systems. Mobile phase: upper phase (top three rows) and lower phase (bottom three rows). Elution mode: head to tail, solid line; tail to head, broken line (see Table 7A).

phobic solvent group indicating that the upper phase moves toward the head and the lower phase toward the tail. In brief, the hydrodynamic motion and resulting retention profiles of the intermediate solvent group approach those of the hydrophilic solvent group at small β-values and become quite similar to those of the hydrophobic solvent group at large β-values. Addition of NaCl to the hydrophilic butanol solvent systems at 1 M concentration produced a substantial change of the phase distribution profile at large β-values in which the solvent systems displayed retention curves similar to those in the intermediate solvent group.

Phase distribution diagrams obtained from the helical columns under various experimental conditions (Figs. 136 and 137) show complex patterns of the retention curves which are at present unpredictable on a theoretical basis. The retention profile of the two-phase solvent systems is certainly based on the hydrodynamic interaction of the two solvent phases in the rotating coiled column. As described earlier, for scheme IV planetary motion the two solvent phases exhibit unilateral distribution along the axis of the coiled column in such a way that one phase entirely occupies the head side and the other phase the tail side. Which phase is directed toward the head of the column is determined by the combined effects of various factors characteristic of the particular experimental condition. Overall results of the present studies suggest that conditions favoring the upper phase movement toward the head and the lower phase toward the tail include (a) increase in hydrophobicity of the nonaqueous phase; (b) increase in interfacial tension between the two solvent phases; (c) decrease in viscosity of the solvent phases; (d) increase in the density difference between the solvent phases; (e) increase in β-value; (f) decrease in revolutional radius; and (g) increase in the helical diameter of the column.

The effects produced by the physical properties of the solvent system (a–d) are intermingled and difficult to isolate from each other and, in addition, the effects of solvent–wall interaction should

---

Fig. 137 Phase distribution in helical columns obtained under the scheme IV planetary motion with a 10-cm revolutional radius. (a) Phase distribution diagrams of nine volatile solvent systems. (b) Phase distribution diagrams of butanol solvent systems. Experimental conditions are as follows: Planetary motion: scheme IV. Revolutional speed: 400–1000 rpm. Revolutional radius: 10 cm. Holder diameter (β): 5 cm (0.25), 10 cm (0.5), 15 cm (0.75). Column: 1.6-mm i.d. single-layer helix. Flow rate: 120 ml/hr for hexane, ethyl acetate, and chloroform systems and 60 ml/hr for butanol systems. Mobile phase: upper phase (top three rows) and lower phase (bottom three rows). Elution mode: head to tail, solid line; tail to head, broken line (see Table 7A).

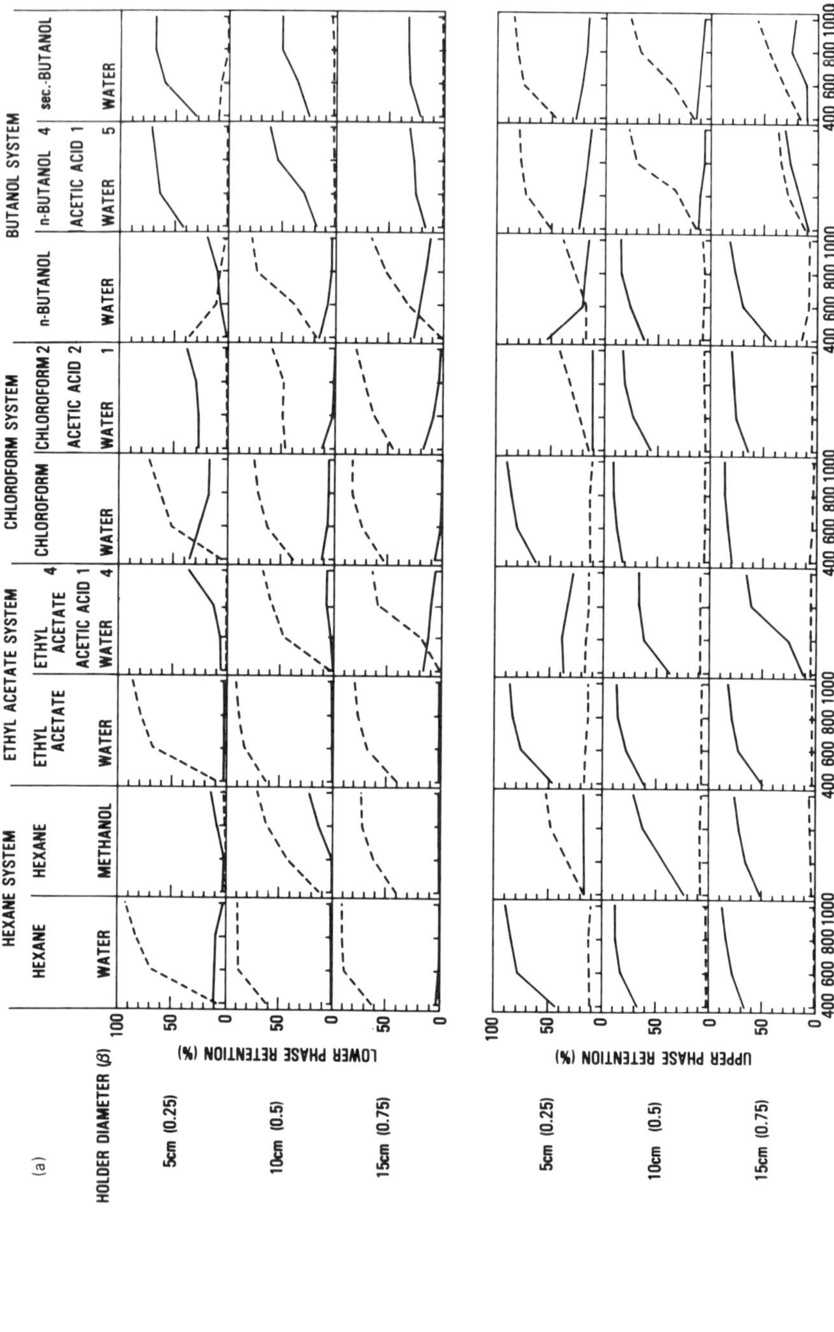
(a)
HOLDER DIAMETER (β)
5cm (0.25)
10cm (0.5)
15cm (0.75)
LOWER PHASE RETENTION (%)
UPPER PHASE RETENTION (%)
HEXANE SYSTEM
HEXANE WATER
HEXANE METHANOL
ETHYL ACETATE SYSTEM
ETHYL ACETATE WATER
ETHYL ACETATE 4 ACETIC ACID 1 WATER 4
CHLOROFORM SYSTEM
CHLOROFORM WATER
CHLOROFORM 2 ACETIC ACID 2 WATER 1
BUTANOL SYSTEM
n-BUTANOL WATER
n-BUTANOL 4 ACETIC ACID 1 WATER 5
sec.-BUTANOL WATER
RPM
100
50
0
400 600 800 1000

Fig. 137

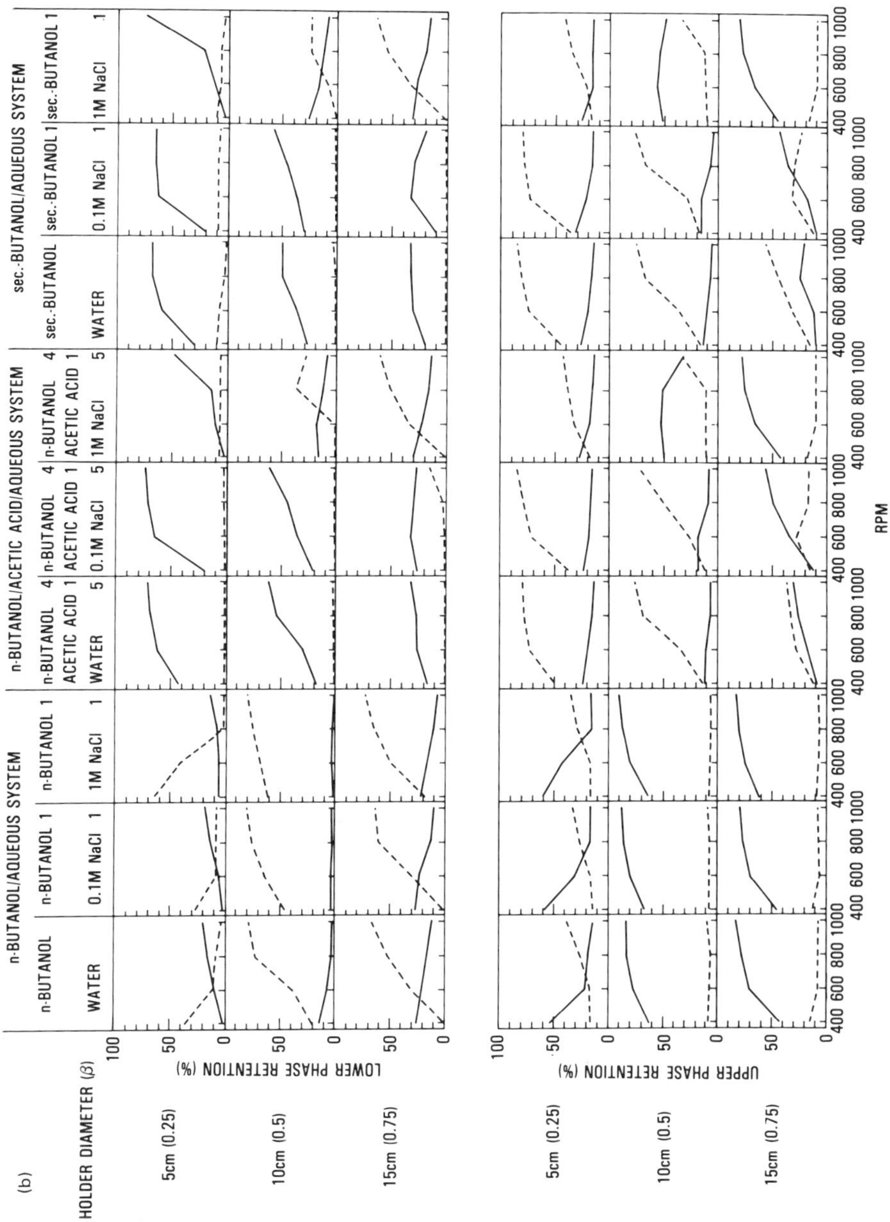

Fig. 137 (Continued)

be further investigated with a column which provides a hydrophilic surface. The effects of the column factors (e–g) are also complex, but the β-value may be the most sensitive parameter of the three. The revolutional speed produces highly complex effects on the interaction of the two solvent phases, often causing a reversal of the unilateral hydrodynamic motion of the two phases which occurs at the crossing point of the two retention curves in the phase distribution diagram. In most cases increasing the revolutional speed tends to direct the upper phase toward the head of the coiled column, i.e., for lower phase retention the broken curve crosses upward over the solid curve and for upper phase retention the solid curve crosses over the broken curve. However, reversal of the retention curves is also occasionally observed with butanol solvent systems. Although the flow rate of the mobile phase is an important factor affecting the retention of the stationary phase, it is not likely to cause reversal of the unilateral hydrodynamic motion.

Among the various factors mentioned above, the β-value may play a universal role in the unilateral hydrodynamic distribution of the two solvent phases. The phase distribution diagrams obtained at various β-values strongly suggest that further increase of the β-value by the use of a large-diameter holder hub would eventually reverse the direction of the unilateral hydrodynamic separation of the hydrophilic solvent systems. The decrease of the β-value by the use of a smaller diameter holder hub may revese the unilateral hydrodynamic motion of the hydrophobic solvent systems, as illustrated by the retention curves of chloroform/water at $\beta = 0.125$ in Fig. 136a.

3. *Phase Distribution Diagrams for Spiral Columns Under Scheme IV Planetary Motion [88].* The third series of experiments has been performed to study the phase distribution in the spiral columns subjected to the scheme IV synchronous planetary motion. In the spiral column retention of the stationary phase is affected by two forces; one is the Archimedean screw force governed by the head–tail relationship as in the helical column and the other, the centrifugal force field gradient created by the spiral pitch of the column. The separate effects of these two forces are graphically presented in the phase distribution diagrams by the choice of symbols for drawing the retention curves as indicated in Table 7B. The Archimedean screw effects are shown by the difference between the solid and broken curves as in the previous diagrams for the helical column while the centrifugal force gradient effects are indicated by the difference between the thick and thin curves. For simplification the choice of stationary and mobile phases giving the greater stationary phase retention as a result of the force gradient is shown by the thick line and the reversed elution mode by the thin line, so that the thick curve always appears higher in the diagram regardless of the choice of the lower or upper layer as the mobile phase.

The phase distribution diagrams for the nine volatile solvent systems for single-layer spiral columns mounted at a 10-cm revolutional radius are illustrated in Fig. 138a where the top two rows show the data for the 1.6-mm i.d. column and the bottom two rows the data for the 2.6-mm i.d. column. In order to provide similar linear flow velocity of the mobile phase in each of the two columns, the flow rate for the 2.6-mm i.d. column was made double that of the smaller column, i.e., 240 ml/hr for the hexane, ethyl acetate, and chloroform systems and 120 ml/hr for the butanol systems.

The effects of the centrifugal force gradient in the spiral column on the retention curves are quite different in each solvent group. In the hydrophobic solvent group including hexane/water, ethyl acetate/water, and chloroform/water, the force gradient contributes little effect in either the 1.6- or 2.6-mm i.d. column as indicated by the small difference in retention level between the thick and thin curves for each pair of the solid and broken curves. In the intermediate solvent group including hexane/methanol, ethyl acetate/acetic acid/water (4:1:4), chloroform/acetic acid/water (2:2:1), and n-butanol/water, the effects of the force gradient becomes significant, especially in the 2.6-mm i.d. column where much wider gaps were formed between the thick and thin curves. In the hydrophilic solvent group of n-butanol/acetic acid/water (4:1:5) and sec-butanol/water, the force gradient effects become further enhanced and, except for the top diagrams (lower phase retention in the 1.6-mm i.d. column), both thick curves are higher than the pair of thin curves indicating that the force gradient plays a more important role in retention than the head–tail relationship. In the 2.6-mm i.d. column, both hydrophilic solvent systems show a radical change in retention curves at the revolutional speeds between 700 and 900 rpm where the two thick curves cross each other indicating that at higher revolutional speeds the upper phase is directed toward the head and the lower phase

---

Fig. 138 Phase distribution in spiral columns obtained under the scheme IV planetary motion with 10-cm revolutional radius. (a) Phase distribution diagrams of nine volatile solvent systems. (b) Phase distribution diagrams of butanol solvent systems. Experimental conditions are as follows: Planetary motion: scheme IV. Revolutional speed: 400–1000 rpm. Holder diameter (β): 10 cm [0.5 (for the internal terminal) to 0.8 (for the external terminal)]. Column: 1.6-mm i.d. (2.4-mm o.d.) and 2.6-mm i.d. (3.6-mm o.d.) single-layer spiral. Flow rate: 120 ml/hr for hexane, ethyl acetate, and chloroform systems and 60 ml/hr for butanol systems in 1.6-mm i.d. column (top two rows); the above rates are doubled for each solvent group in 2.6-mm i.d. column (bottom two rows). Mobile phase: upper phase (first row) and lower phase (second row) in each column. Elution mode: see Table 7B.

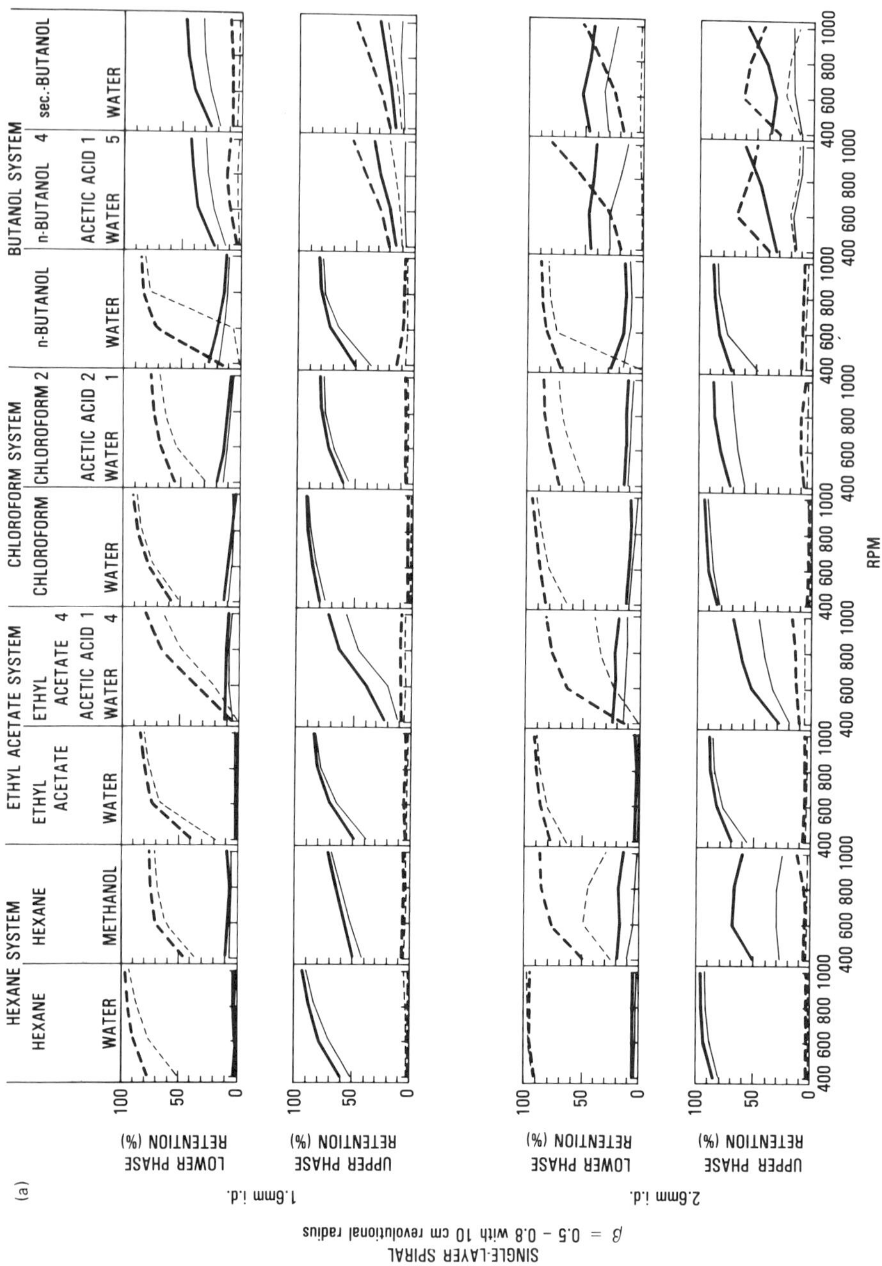
(a)
SINGLE-LAYER SPIRAL
β = 0.5 – 0.8 with 10 cm revolutional radius
1.6mm i.d.
2.6mm i.d.
LOWER PHASE RETENTION (%)
UPPER PHASE RETENTION (%)
100
50
0
HEXANE SYSTEM
HEXANE WATER
HEXANE METHANOL
ETHYL ACETATE SYSTEM
ETHYL ACETATE WATER
ETHYL ACETATE 4 ACETIC ACID 1 WATER 4
CHLOROFORM SYSTEM
CHLOROFORM WATER
CHLOROFORM 2 ACETIC ACID 2 WATER 1
BUTANOL SYSTEM
n-BUTANOL WATER
n-BUTANOL 4 ACETIC ACID 1 WATER 5
sec.-BUTANOL WATER
400 600 800 1000
RPM

Fig. 138

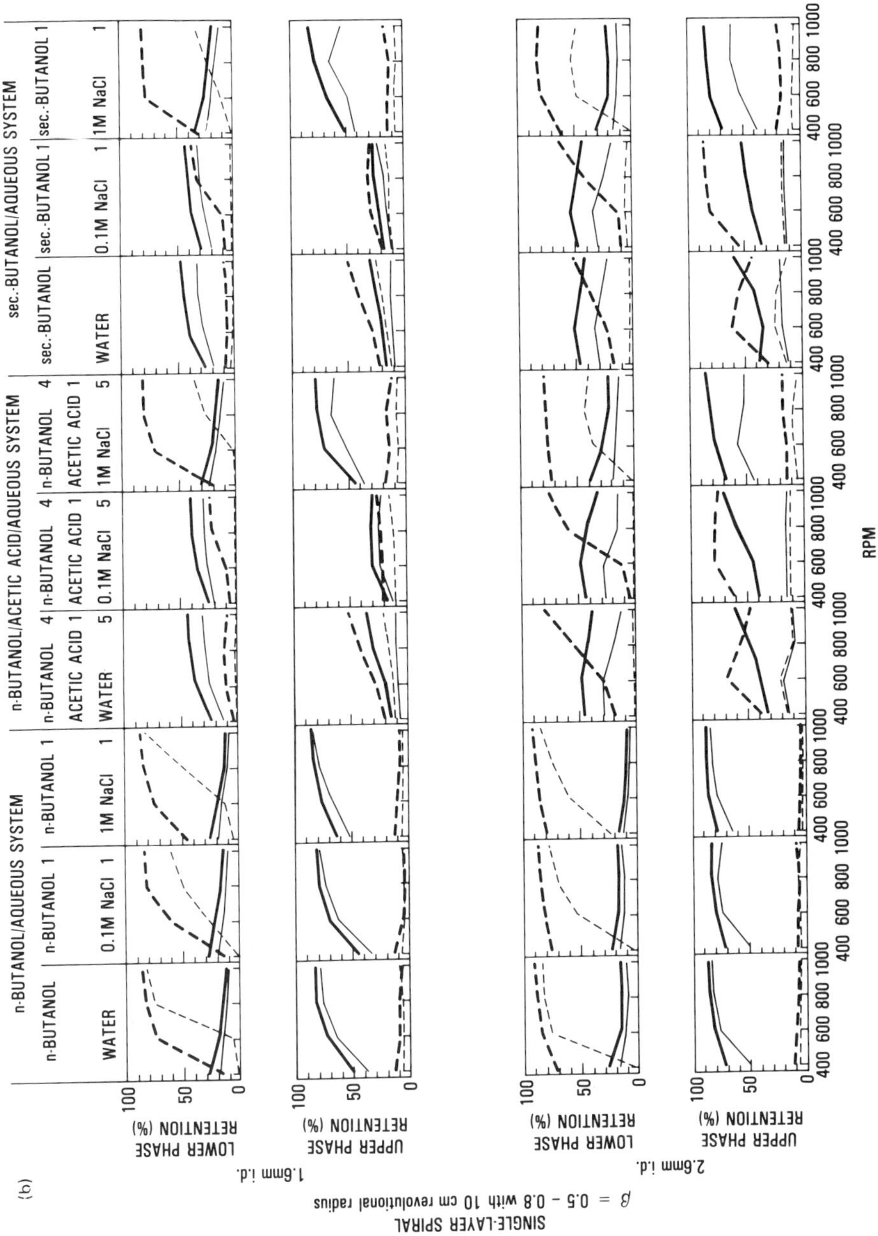

Fig. 138 (Continued)

toward the tail in the spiral column as is the case with other solvent groups.

Figure 138b shows sets of phase distribution diagrams of butanol solvent systems obtained from the 1.6-mm i.d. and 2.6-mm i.d. spiral columns. Addition of NaCl to n-butanol/water produces no significant change in the stationary phase retention profile regardless of the internal diameter of the column. However, the retention curves for the n-butanol/acetic acid/water (4:1:5) and sec-butanol/water are significantly altered at the 1 M NaCl concentration where the phase distribution profile becomes similar to those for the n-butanol/water and other intermediate solvent systems.

Performance of the spiral columns has also been examined with the third type of the coil planet centrifuge which provides greater β-values by further reducing the revolutional radius and/or increasing the holder diameter. The retention data were obtained from a 1.6-mm i.d. spiral column wound around a 15-cm diameter holder which was placed at 10 or 5 cm from the central axis of the centrifuge, providing the β-values of 0.75–0.95 or 1.5–1.9, respectively. Figure 139 shows a set of phase distribution diagrams obtained from four different solvent pairs with a 5-cm revolutional radius (top two rows) and a 10-cm revolutional radius (bottom two rows).

Comparison between the two top and the two bottom rows of the phase distribution diagrams in Fig. 139 shows the effects of reducing the revolutional radius from 10 to 5 cm in the same spiral column on the 15-cm diameter holder. This reduction of the revolutional radius not only causes an increase in β-values but also halves the net centrifugal force field at a given revolutional speed. The results clearly show that the retention levels are further decreased especially in the intermediate solvent systems, although widened gaps between the thick and thin curves in both hydrophobic and intermediate solvent systems indicate the enhanced effects of the force gradient on the retention of the stationary phase in the spiral column. The hydrophilic solvent system of n-butanol/acetic acid/water (4:1:5) shows no significant change when the revolutional radius is reduced. Overall results indicate that the large β-values in the spiral column ob-

---

Fig. 139 Phase distribution in a spiral column obtained under the scheme IV planetary motion with two different revolutional radii. Experimental conditions are as follows: Planetary motion: scheme IV. Revolutional speed: 400–1000 rpm. Revolutional radius: 5 cm (top two rows) and 10 cm (bottom two rows). Holder diameter (β): 15 cm [1.5 (internal terminal)-1.9 (external terminal) and 0.75 (internal terminal)-0.95 (external terminal)]. Column: 1.6-mm i.d. (2.4-mm o.d.) single-layer spiral. Flow rate: 120 ml/hr for chloroform systems and 60 ml/hr for butanol systems. Mobile phase: upper phase (first row) and lower phase (second row) in each group. Elution mode: see Table 7B.

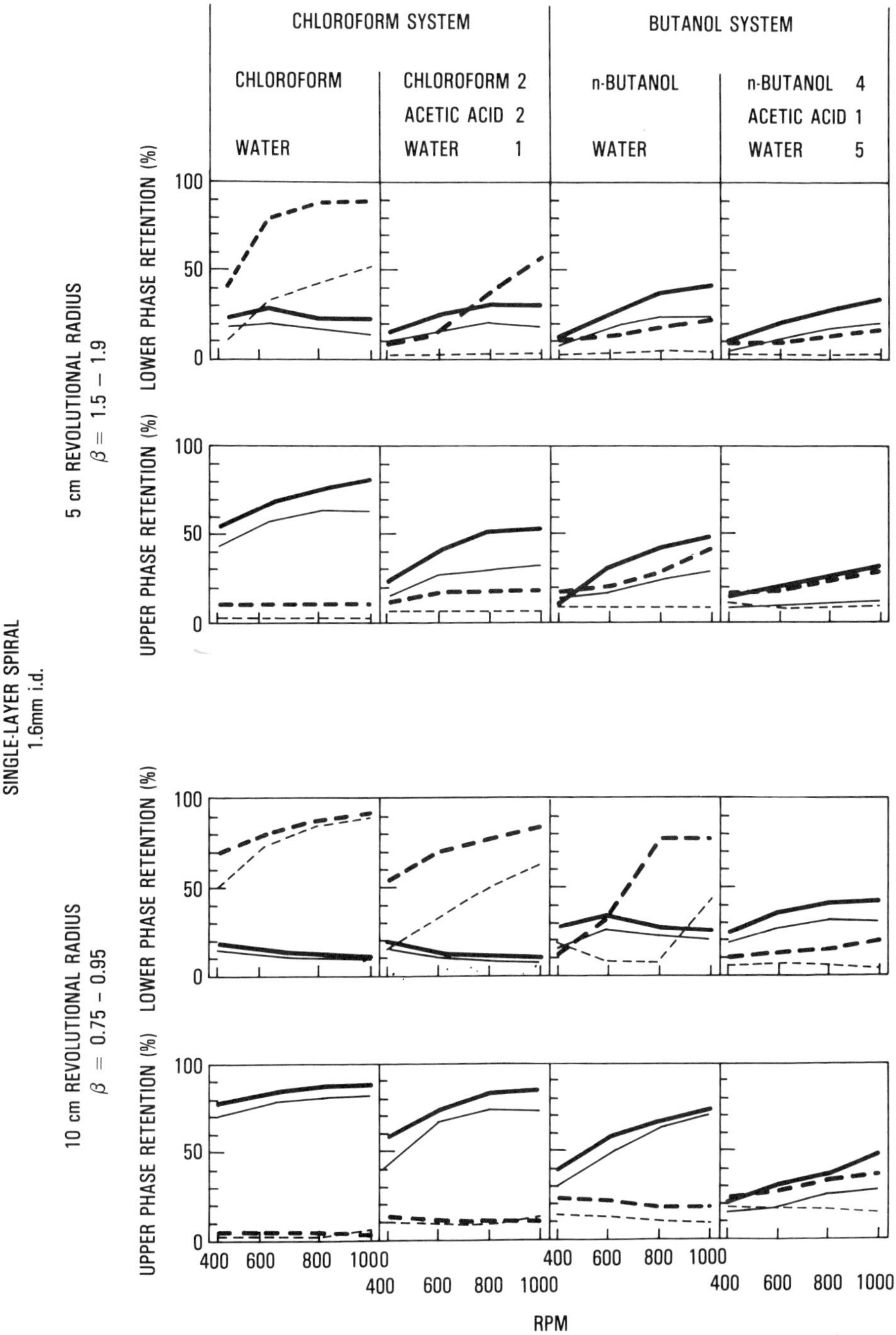
CHLOROFORM SYSTEM
BUTANOL SYSTEM
CHLOROFORM
WATER
CHLOROFORM 2
ACETIC ACID 2
WATER 1
n-BUTANOL
WATER
n-BUTANOL 4
ACETIC ACID 1
WATER 5
SINGLE-LAYER SPIRAL
1.6mm i.d.
5 cm REVOLUTIONAL RADIUS
β = 1.5 – 1.9
10 cm REVOLUTIONAL RADIUS
β = 0.75 – 0.95
LOWER PHASE RETENTION (%)
UPPER PHASE RETENTION (%)
100
50
0
400 600 800 1000
RPM

Fig. 139

tained by increasing the holder diameter and decreasing the revolutional radius produces substantial enhancement of the force gradient effects but fails to improve the retention of the stationary phase apparently due to the loss of the Archimedean screw effects produced by the head–tail hydrodynamic interaction of the two solvent phases.

Various Physical Factors Governing the Phase Distribution [92]

Systematic studies on phase distribution diagrams described above have demonstrated that the two-phase solvent systems display characteristic behavior which is closely related to the hydrophobicity of the nonaqueous phases. The hydrophobic solvent systems exhibit a strong hydrodynamic trend in the coiled column where the two solvent phases are unilaterally distributed in such a way that the upper phase is always directed toward the head and the lower phase toward the tail. The hydrophilic solvent systems show the opposite hydrodynamic tendency, with the upper phase being directed toward the tail and the lower phase toward the head. In the solvent systems with moderate hydrophobicity intermediate between the above two extremes, the hydrodynamic tendency is sensitive to the centrifugal conditions. In a system having a column of small helical diameter with a large radius of revolution, the hydrodynamics of the intermediate solvent group approaches that of the hydrophilic solvent group while in the reversed conditions (large helical diameter and small radius of revolution) the hydrodynamics become quite similar to those of the hydrophobic solvent group.

In the present studies, efforts were made to correlate the hydrodynamics of the two-phase solvent systems with three major physical properties: interfacial tension, viscosity, and the difference in density between the two phases. Measurement of the settling time of the two-phase solvent systems under unit gravity gave a parameter which provided a numerical index closely correlating with the hydrodynamics observed for the solvent systems in a centrifugal force field.

1. *Physical Properties of Solvent Systems and Settling Times.* Major physical properties of the solvent systems including density, viscosity, and interfacial tension were measured using conventional methods as described below. The densities of the upper and the lower phases in each solvent system were measured using a set of hydrometers (Fisher Scientific Company, Pittsburgh) and the density difference between the two phases was calculated. The viscosity of each solvent phase was measured using a falling-ball-type viscometer (VWR Scientific Inc., Baltimore). A modified Wilhelmy balance was used to determine the surface tension of each solvent phase against

air, using water (73 dynes/cm) and hexane (18.4 dynes/cm) for standardization of the instrument for aqueous and organic phases, respectively. The interfacial tension was calculated as the difference in surface tension between the upper and the lower phases of each solvent pair. All measurements above were performed at room temperature of 22°C ± 1°C. The results are summarized in Table 8. Among three physical properties measured for the solvent system, interfacial tension ($\Delta\gamma$) shows extremely high correlation with both the hydrophobicity and the hydrodynamic behavior of the solvent system in that the value is highest for the hydrophobic solvent group and lowest for the hydrophilic solvent group. The viscosity of each phase (η_U and η_L) or the mean viscosity of both phases ($\tilde{\eta}$) is also very well correlated with the solvent group, being lowest in the hydrophobic solvent systems and highest in the hydrophilic solvent systems. The density difference between the upper and the lower phases ($\Delta\rho$) appears to be less well correlated with the solvent groups. Although the value is generally high in the hydrophobic solvent group and low in the hydrophilic solvent group, a discrepancy is presented by ethyl acetate/water in the hydrophobic solvent group which shows the second lowest value among all solvent systems. Thus all three physical properties examined appear to bear some significant relationship to the hydrodynamic behavior of the solvent systems in high-performance CCC.

In order to correlate the hydrodynamic phase distribution in the rotating coiled column to the physical properties of the solvent system, simple tests were performed to measure the settling time of the two solvent phases, i.e., the time required for the solvent mixture to be completely separated into two layers in a unit gravitational field. Each two-phase solvent system was first equilibrated in a separatory funnel at room temperature and separated into two phases. A 2-ml aliquot for each phase, total volume 4 ml, was delivered to a 5-ml graduated cylinder which was then sealed with a glass stopper. The solvents in the cylinder were gently mixed by inverting the cylinder five times and the cylinder was immediately placed on a flat table in an upright position and the time required for the two phases to settle was measured. The measurement was repeated several times to obtain a mean value. A second set of tests was performed in a similar manner except that the cylinder was vigorously shaken five times to see the tendency for emulsification. The test was again repeated several times to obtain a mean value for each solvent system.

The settling times of the two-phase solvent systems are also shown in Table 8 where T indicates the settling time after gentle mixing by inverting the cylinder five times and T' after vigorous shaking of the cylinder five times. The results clearly show a close correlation between the settling times in a unit gravitational field and the hydrodynamic behavior of the solvent phases observed in a centrifugal field. The settling times of the hydrophobic solvent sys-

Table 8 Physical Properties of the Two-Phase Solvent System

Solvent group	Two-phase solvent system (volume ratio)	Interfacial tension (dyne/cm)	Viscosity (cp)		Density (g/cm^3)		Settling time (sec)	
		$\Delta\gamma$	η_U/η_L	$\tilde{\eta}$	ρ_U/ρ_L	$\Delta\rho$	T	T'
Hydrophobic	Hexane/water	52	0.41/0.95	0.68	0.66/1.00	0.34	<1	8
	Ethyl acetate/water	31	0.47/0.89	0.68	0.92/0.99	0.07	15.5	21
	Chloroform/water	42	0.95/0.57	0.76	1.00/1.50	0.50	3.5	5.5
Intermediate	Hexane/methanol	4	0.50/0.68	0.59	0.67/0.74	0.07	5.5	6
	Ethyl acetate/acetic acid/water (4:1:4)	16	0.76/0.81	0.79	0.94/1.01	0.07	15	16
	Chloroform/acetic acid/water (2:2:1)	12	1.16/0.77	0.97	1.12/1.35	0.24	29	27.5
	n-Butanol/water	3	1.72/1.06	1.40	0.85/0.99	0.14	18	14
	n-Butanol/0.1 M NaCl (1:1)	4	1.66/1.04	1.35	0.85/0.99	0.15	16	14.5
	n-Butanol/1 M NaCl (1:1)	5	1.75/1.04	1.40	0.84/1.04	0.20	23.5	21.5
Hydrophilic	n-Butanol/acetic acid/water (4:1:5)	<1	1.63/1.40	1.52	0.90/0.95	0.05	38.5	37.5
	n-Butanol/acetic acid/0.1 M NaCl (4:1:5)	<1	1.68/1.25	1.47	0.89/1.01	0.11	32	30.5
	n-Butanol/acetic acid/1 M NaCl (4:1:5)	1	1.69/1.26	1.48	0.88/1.05	0.16	26.5	24.5
	sec-Butanol/water	<1	2.7/1.67	2.19	0.87/0.97	0.10	57	58
	sec-Butanol/0.1 M NaCl (1:1)	<1	1.96/1.26	1.61	0.86/0.98	0.12	46.5	49.5
	sec-Butanol/1 M NaCl (1:1)	3	1.91/1.29	1.60	0.84/1.03	0.19	34	33.5

Abbreviations: η_U/η_L = upper phase viscosity/lower phase viscosity; $\tilde{\eta}$ = mean viscosity of the two phases; ρ_U/ρ_L = upper phase density/lower phase density; $\Delta\rho$ = density difference between the two phases; T = settling time after gentle mixing, T' = settling time after vigorous shaking.

tems are shortest ranging from 5 to 16 sec while those of the hydrophilic solvent systems are longest ranging from 38 to 59 sec. The intermediate solvent systems gave moderate values ranging from 15 to 29 sec with the exception of the nonaqueous system, hexane/methanol, which had a short settling time of 5.5 sec. The difference between T and T' for each solvent system was found to be quite small in both the intermediate and hydrophilic solvent systems. However, in the hydrophobic solvent group T' becomes much greater than T, especially in hexane/water.

Because of the parallel trend observed with the hydrophobicity of the solvent group, the settling time may serve as a useful parameter for correlating the physical properties of the solvent system with the hydrodynamic behavior of the solvent systems in a centrifugal force field. The values for $\Delta\gamma$, $\Delta\rho$, and $\tilde{\eta}$ (mean viscosity of the upper and the lower phases) in Table 8 are respectively plotted against the settling time (T) of the solvent system in Figs. 140a–c where the open circle represents the hydrophobic solvent system, the solid circle the hydrophilic solvent system, and the semisolid circle the intermediate solvent system. The correlation coefficient (r) for each plot of physical property with settling time (T) was computed for the set of 15 solvent systems. The density difference ($\Delta\rho$) gave a weak negative correlation ($r = -0.45$) (Fig. 140a) and the interfacial tension ($\Delta\gamma$) a moderate negative correlation ($r = -0.65$) (Fig. 140b). A strong positive correlation ($r = +0.88$) was observed between the settling time and the mean viscosity ($\tilde{\eta}$) as illustrated in Fig. 140c. When the correlation coefficient between the viscosity and the settling time is computed individually for each phase, the aqueous phase including the lower phase of hexane/methanol (solid circle) gave the highest value of 0.89 and the nonaqueous phase (open circle) a lower value of 0.83 (Fig. 140d).

The above findings suggest that among various physical properties of the solvent system viscosity may play a major role in retention of the stationary phase. The less viscous the solvent system, the higher level of retention to be expected. Although the interfacial tension and density difference between the two phases appear to be less important factors than the viscosity, they would exert a strong influence on the settling times near the plait point of the solvent as described later in detail.

2. *Effects of the Temperature on the Settling Time.* As described earlier, hydrophilic solvent systems such as n-butanol/acetic acid/water (4:1:5) and sec-butanol/water exhibit unique hydrodynamic behavior: the upper phase is distributed toward the tail and the lower phase toward the head. Therefore the reversed elution mode is required to obtain retention of the stationary phase in the helical column. These solvent systems are characterized by high viscosity and low interfacial tension associated with the longest settling times among all solvent systems examined. The highly significant correla-

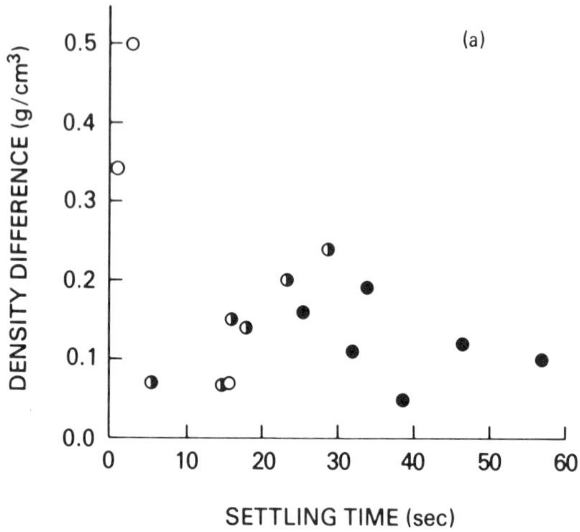

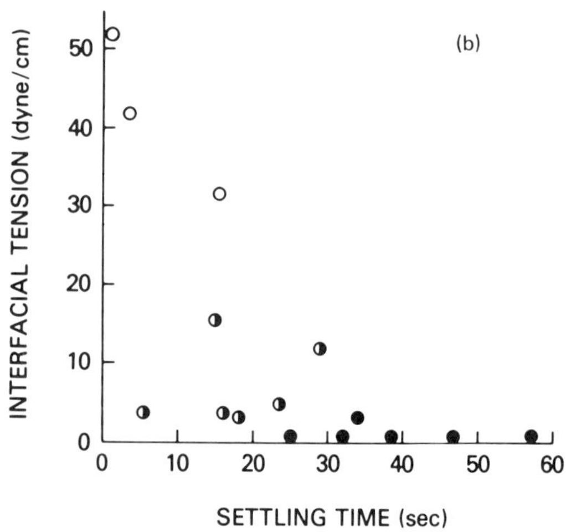

Fig. 140 Correlation between the settling time and three major physical properties of the two-phase solvent system. a–c: open circle indicates the hydrophobic solvent system; semisolid circle, the intermediate solvent system; and solid circle, the hydrophilic solvent system. (a) Density difference of the two phases vs. settling time (r = −0.45). (b) Interfacial tension vs. settling time (r = −0.65). (c) Mean viscosity vs. settling time (r = +0.88). (d) Correlation between the settling time and viscosity of each solvent phase. Open circle: viscosity of nonaqueous phase vs. settling time (r = +0.83). Solid circle: viscosity of aqueous phase including the lower phase of hexane/methanol vs. settling time (r = +0.89).

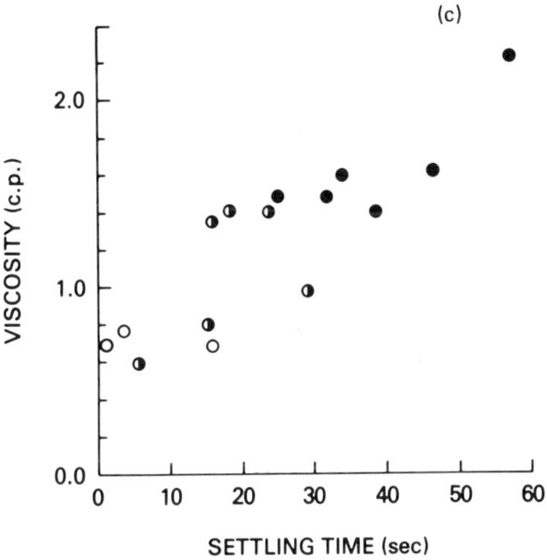

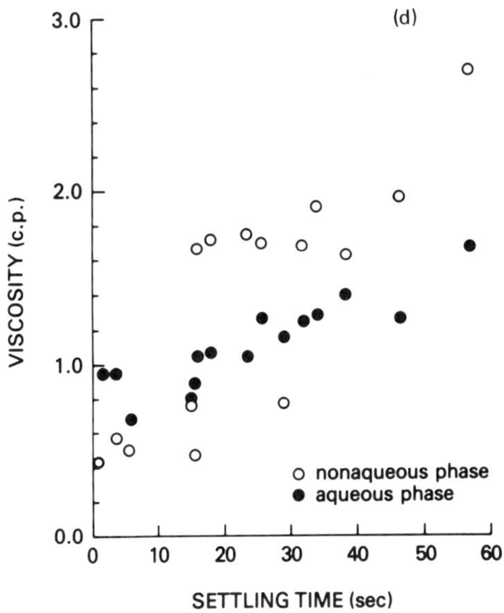

Fig. 140 (Continued)

tion found between viscosity and settling times of various solvent systems strongly suggests that the above unique hydrodynamic behavior of the hydrophilic solvent systems may be attributed to their high viscosity and that reduction of the viscosity by raising the temperature of these solvent systems would reduce their settling times to the range observed for other solvent systems, thus converting their reversed hydrodynamic mode to the normal mode. This reversal of the hydrodynamic behavior of the butanol solvent systems will provide an obvious advantage in performing high-speed CCC in that all conventional solvent systems could then be eluted with the normal mode with high flow rates of the mobile phase.

In order to demonstrate the feasibility of this reversal of hydrodynamics, settling times of the butanol solvent systems were measured at various temperatures ranging from room temperature to around 80°C. Figure 141a shows the effects of temperature on the settling times of n-butanol/water and sec-butanol/water. The settling time of the sec-butanol/water system sharply declines to less than 20 sec on increasing the temperature to 50°C. The settling time of the less viscous n-butanol/water system exhibits a similar trend, dropping to less than 10 sec at 50°C. As judged from the settling times, it is expected that at a temperature of 50°C these solvent systems would produce phase distribution diagrams similar to those of the intermediate and hydrophobic solvent systems which allow high retention levels of the stationary phase with normal elution mode. It is interesting to note that the curves of the settling time vs. temperature for these binary butanol systems (Fig. 141a) are quite similar to that of the viscosity vs. temperature plot for n-butanol (Fig. 141b), which is consistent with the high positive correlation between the settling time and viscosity described earlier.

Figure 142a shows the effects of temperature on the settling times of the ternary butanol systems. In n-butanol/acetic acid/water (4:1:5) the settling times decline with an increase in temperature to 50°C reaching a minimum of 23 sec. Increase in the temperature further results in increase of the settling times up to 65°C, where the system forms a single phase. The increase of the settling times near the plait point is apparently caused by the extremely low interfacial tension and the small difference in density between the two phases. This anomaly can be corrected by reducing the amount of acetic acid in the solvent system as shown in the settling time curve obtained for n-butanol/acetic acid/water (9:1:10), which decreases to about 10 sec at 50°C.

Alternatively, the above problem may be solved by the addition of a salt in the aqueous solution. Figure 142b illustrates effect of salt concentration on the settling time vs. temperature curve in n-butanol/acetic acid/aqueous solution (4:1:5). At 0.1 M NaCl concentration the solvent system still retains the anomalous zone although it has shifted about 15° toward the right on the temperature

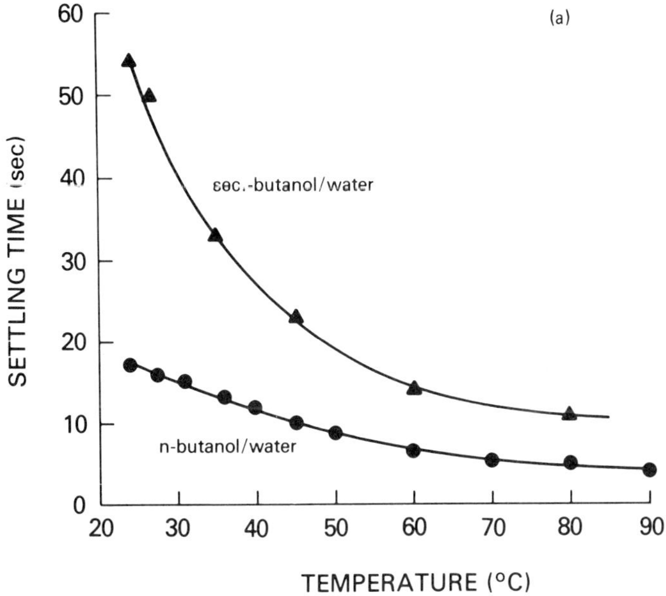

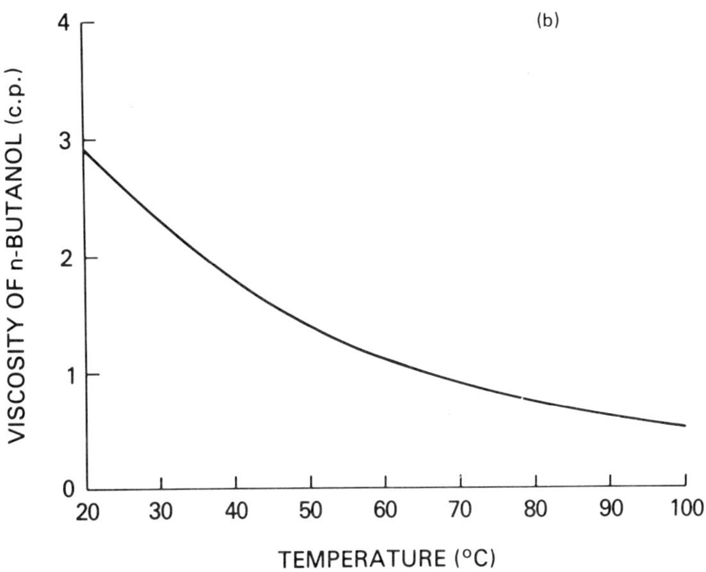

Fig. 141 (a) Effects of temperature on the settling time of binary butanol systems. Circles for n-butanol/water and triangles for sec-butanol/water. (b) Effects of temperature on the viscosity of n-butanol.

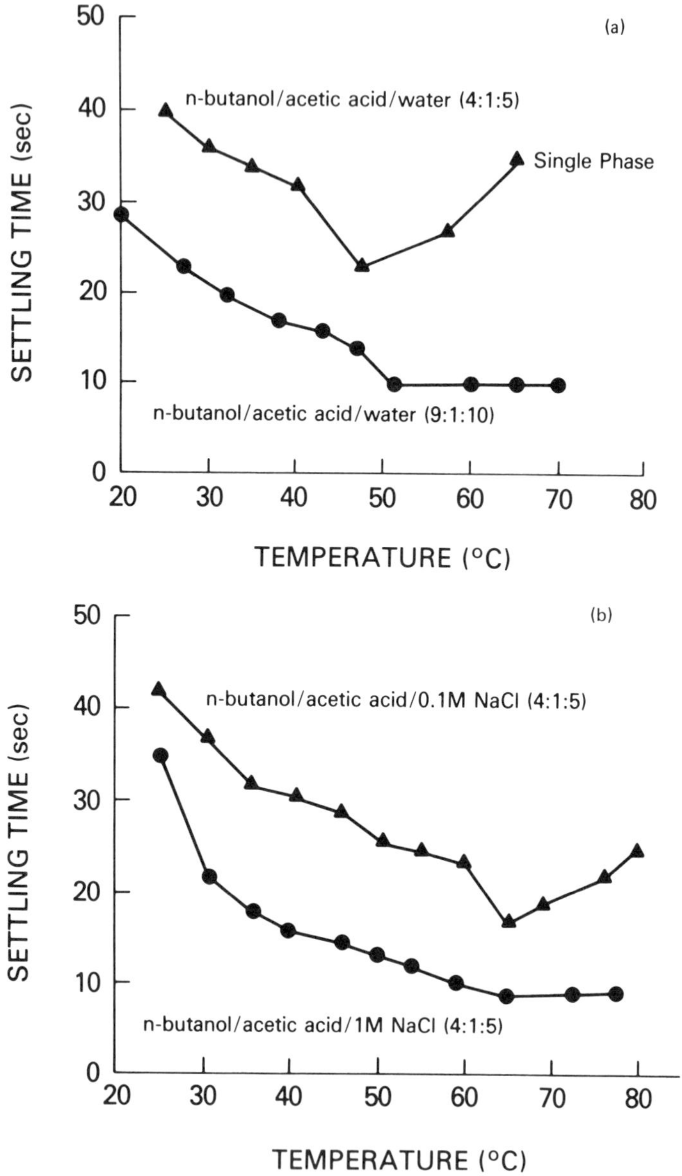

Fig. 142 (a) Effects of temperature on the settling time of ternary n-butanol systems. Circles for n-butanol/acetic acid/water (9:1:10) and triangles for n-butanol/acetic acid/water (4:1:5). (b) Effects of temperature on the settling time of quadruple n-butanol systems containing NaCl. Circles for n-butanol/acetic acid/1 M NaCl (4:1:5) and triangles for n-butanol/acetic acid/0.1 M NaCl (4:1:5).

axis. Increase of the NaCl concentration to 1 M normalizes the settling time curve which declines to 15 sec at 50°C.

The above findings clearly indicate that at higher temperature the hydrophilic solvent systems may display the normal hydrodynamic behavior observed in the hydrophobic and intermediate solvent systems, provided that the phase composition is kept away from the vicinity of the plait point. Raising the temperature slightly above room temperature may also reduce the settling time of the intermediate solvent systems and therefore would improve the retention of stationary phase and increase partition efficiency during the separation.

Performing CCC at elevated temperature will provide a number of beneficial effects including (a) conversion of the hydrodynamic behavior of the hydrophilic solvent systems to the normal mode, (b) improved retention of the stationary phase resulting in higher peak resolution, (c) higher partition efficiency due to reduction of mass transfer resistance, (d) increased sample loading capacity due to higher solubility, etc. These expectations have been confirmed recently using a high-speed CCC apparatus equipped with a temperature control system which has been constructed in our laboratory.

D. Continuous Extraction

The unique capability of the present scheme to retain a large volume of the stationary phase in the column under efficient mixing with the mobile phase provides an ideal condition for performing continuous extraction. Because no solid support is used in the column, a minute amount of the sample can be recovered with minimum adsorption loss onto the liquid–solid interface.

Preliminary Studies on Extraction of DNP Amino Acids [8]

1. *Studies on Retention of the Stationary Phase.* In the first series of experiments, the capability of the present scheme in retaining a large amount of stationary phase was demonstrated with a coiled column prepared from a 2.5-m long, 2.6-mm i.d. PTFE tube which was coiled around the larger holder of the coil planet centrifuge illustrated in Fig. 125. The column consisted of five helical turns at $\beta = 0.75$ and had a total capacity of about 15 ml. Typical two-phase solvent systems composed of ethyl acetate/water and ethyl acetate/0.5 M sodium phosphate (pH 4.4) at a volume ratio of 1:1 were selected.

In each operation the coiled column and the free space in the flow path were entirely filled with the stationary phase and the mobile phase was pumped into the column in the proper direction (head–tail elution for the aqueous phase and tail–head elution for the nonaqueous phase) while the apparatus was run at a given revolutional speed. The eluate through the outlet of the column was pooled in a graduated cylinder to measure the volume of the stationary phase retained in the column. The experiments were performed under various revolutional speeds and flow rates using both nonaqueous and aqueous phases as the stationary phase.

The results of the retention studies are summarized in Fig. 143 where the percentage retention of the stationary phase was plotted against the applied revolutional speeds. The three lines drawn in each diagram indicate the effects of the different flow rates applied. The flow rate of 516 ml/hr is the maximum rate available with the Beckman Accu-Flo pump employed. The ideal retention level for extraction may be considered to be over 70% at or near the plateau of the curve, although much lower retention levels can be applied for extraction unless carryover of the stationary phase occurs. Figures 143a and b show the retention curves of the solvent system composed of ethyl acetate and water, where both nonaqueous (a) and aqueous (b) phases were used as the stationary phase. In both cases the ideal retention levels are provided at the revolutional speed of over 600 rpm at all flow rates applied. Figures 143c and d show similar retention curves for the phase system composed of ethyl acetate/0.5 M NaH_2PO_4 (1:1). In obth c and d, retention levels show much improvement over the previous phase system. Addition of salt to the phase system results in a greater density difference which promotes countercurrent movement of the two phases in the coil. The overall results indicate that the system provides excellent retention under a broad range of operational conditions for both aqueous and nonaqueous stationary phases. The results also suggest that much higher flow rates are applicable with high revolutional speeds.

2. *Studies on Efficiency of Extraction.* The second series of experiments has been performed to demonstrate the capability of the present scheme to extract a solute present in a large volume of the mobile phase into a small volume of the stationary phase retained in the coiled column. This requires a set of conditions such that the solute must favor partition to the stationary phase. With commonly used extraction media such as an ethyl acetate/aqueous system, partition coefficients of various biological materials can be conveniently adjusted by modifying the pH and/or ionic strength of the aqueous phase to meet the above requirement. For the present studies, a pair of DNP amino acids, N-DNP-L-leucine (DNP-leu) and delta-N-DNP-L-ornithine (DNP-orn), were selected as samples because they are readily observed through the column wall during the extraction process under stroboscopic illumination and also provide suitable partition coefficients for this present solvent system. The experiments were performed with the coiled column used in the previous retention studies. The overall experimental conditions in the following studies are summarized in Table 9.

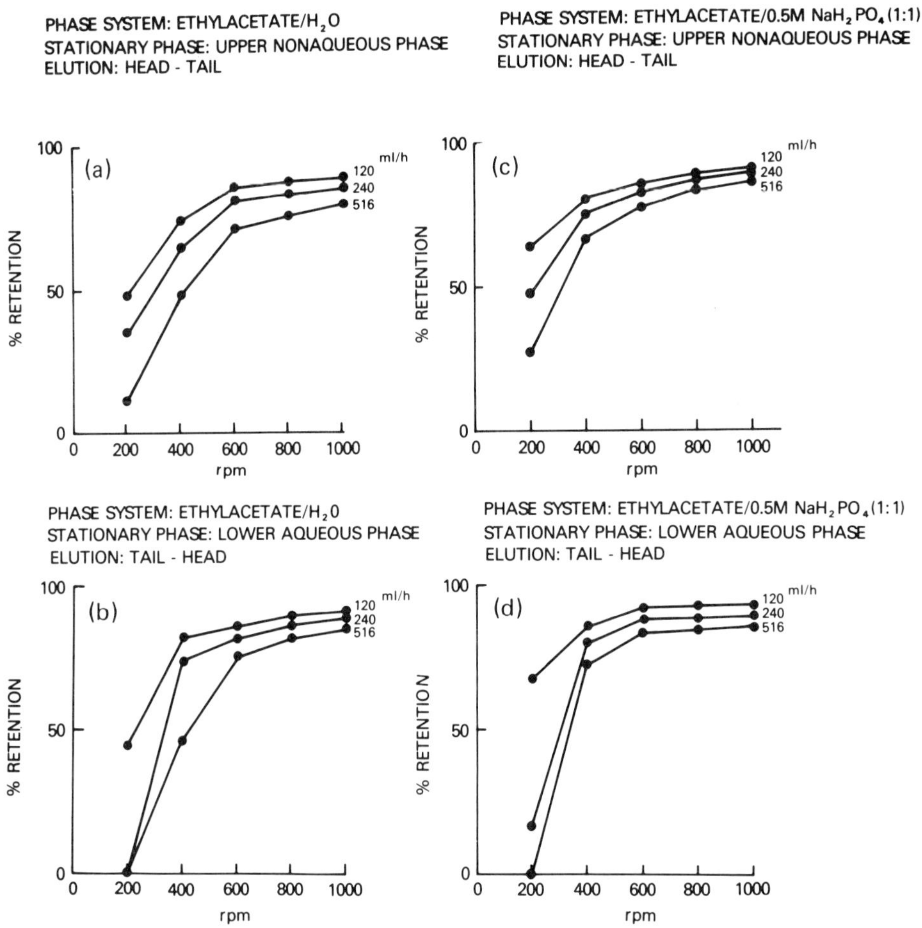

Fig. 143 Phase distribution diagrams of ethyl acetate/aqueous systems in a helical column for continuous extraction.

A typical extraction procedure may be divided into three steps, i.e., extraction, cleaning, and collection. In each operation, the column was filled with the stationary phase and the mobile phase containing the sample was eluted through the column in the proper direction while the apparatus was run at 600 rpm. The extraction process was continued until 400 ml of the mobile phase was eluted. Then the mobile phase was replaced by the same phase but free of solute to wash the column contents. This cleaning process was continued until the additional 100 ml of the mobile phase was eluted.

Table 9 Experimental Conditions and Results of Extraction

Exp. no.	Solvent sample		Mobile phase	Stationary phase	Sample (P.C.)[a]
1	Ethyl acetate 0.5 M NaH_2PO_4	1 2	Aqueous	Nonaqueous	DNP-leu (<0.01)
2	Ethyl acetate 0.5 M NaH_2PO_4	1 2	Aqueous	Nonaqueous	DNP-leu (<0.01)
3	Ehtyl acetate 0.5 M NaH_2PO_4	1 2	Aqueous	Nonaqueous	DNP-leu (<0.01)
4	Ethyl acetate 0.5 M NaH_2PO_4	2 1	Nonaqueous	Aqueous	DNP-orn (<0.01)
5	Ethyl acetate 0.5 M NaH_2PO_4	2 1	Nonaqueous	Aqueous	DNP-orn (<0.01)
6	Nonequilibrium system		5% ethyl acetate in 0.5 M NaH_2PO_4	Ethyl acetate	DNP-leu (<0.01)

[a]Partition coefficient is defined as solute concentration in the mobile phase divided by that in the stationary phase.

This would elute out all impurities having partition coefficients of 0.1 or greater. The sample extracted into the stationary phase in the coiled column was collected by eluting with the mobile phase in the opposite direction. This was done by switching the feed and return flow lines either by simply disconnecting the flow lines or the use of a four-way slide valve. The sample still remaining in the column was then washed out by eluting the column with the other phase originally used as the stationary phase.

The results were summarized in Table 9. In experiments 1–3 in Table 9, DNP-leu was dissolved in 400 ml of the aqueous phase at various concentrations and extracted into the nonaqueous phase retained in the column. The extracted sample was then cleaned by eluting the column with 100 ml of the clean aqueous phase and then collected from the column. The harvested stationary phase volume measured approximately 10 ml, containing over 90% of the original sample. A small amount of the sample still remaining in the column, usually a few percent of the total, was conveniently recovered by

Sample conc. in mobile phase (mg%)	Extracted mobile phase vol. (ml)	Flow rate (direction)	rpm	Collected stationary phase vol. (ml)	Sample recovery (%)
4	400	516 ml/hr (head–tail)	600	10.5	94
0.4	400	516 ml/hr (head–tail)	600	10.0	97
0.04	400	516 ml/hr (head–tail)	600	10.4	100
0.4	400	516 ml/hr (tail–head)	600	11.8	97
0.04	400	516 ml/hr (tail–head)	600	11.8	100
0.4	400	516 ml/hr (head–tail)	600	6.1	99

eluting the column with several milliliters of the nonaqueous phase. The total sample recovery is always well over 90%, as shown in the table. The reduction of the sample concentration from 4 mg% to 0.04 mg% somewhat improved the recovery rate, indicating that no significant sample loss occurs due to the adsorption effects and that further reduction of the sample concentration is feasible with high levels of recovery. The mode of elution that uses the nonaqueous phase as the stationary phase renders a great advantage in practical extraction in that the collected solvent is highly volatile and free of salts, facilitating further concentration. It also permits the stepwise or gradient elution of the sample by eluting the column with a modified aqueous phase to achieve further purification.

In experiments 4 and 5 in Table 9, the DNP-orn sample was dissolved in 400 ml of the mobile nonaqueous phase and extracted into the stationary aqueous phase by eluting the column from the tail toward the head. The retained aqueous phase in the column was then similarly cleaned with 100 ml of nonaqueous phase free of sample.

The collected stationary aqueous phase measured approximately 12 ml in volume. This exceeds the volumes in experiments 1–3, as expected from the results of the retention studies. The sample still remaining in the column was eluted out with several milliliters of the aqueous phase. The sample recovery ranged over 95% with an improved figure at the reduced sample concentration as observed in the previous experiments.

In practice, application of the method to aqueous crude extracts or physiological fluids requires a preliminary adjustment of the solvent composition to provide a suitable partition coefficient of the desired material for the applied pair of solvents. In this case preequilibration of the two phases is not essential. Experiment 6 in Table 9 shows an example of operation with such nonequilibrated solvents. The sample DNP-leu was first dissolved in 400 ml of 0.5 M NaH_2PO_4 aqueous solution containing ethyl acetate at 5%, which is slightly below the saturation level of approximately 7%. The column was filled with ethyl acetate followed by elution with the above sample solution. Both extraction and cleaning processes were performed as in other experiments. The sample solution collected from the column measured slightly over 6 ml. This depletion of the stationary phase apparently resulted from the use of the nonequilibrated solvent pair but without any effect on the sample recovery.

The overall results indicate a potential usefulness of the present method in processing a large amount of crude extracts or biological fluids in research laboratories. A small amount of the sample present in several hundred milliliters of the original solution can be enriched in 10 ml of the nonaqueous phase almost free of salt in 1 hr at a high recovery rate.

Extraction of Urinary Drug Metabolites [29]

Application of the present method to extraction of urinary metabolites of two anthracycline antitumor drugs, adriamycin (ADR) and daunorubicin (DNR), has been conducted by Nakazawa et al. at the National Cancer Institute, Baltimore.

1. *Selection of Extraction Medium.* Search for the suitable extraction medium was curried out by measuring the partition coefficients of these drugs with three different two-phase solvent systems. Chloroform/isopropanol (3:1) mixture, ethyl acetate, and n-butanol were individually tested against 1.0 M solutions of various salts. In each measurement equal volumes of organic and aqueous phases were placed in a test tube, 100 nmol of drug was added, and the tube was vigorously mixed for 30 sec. Quantitites of drug in each phase were measured by fluorometry (SPF 125, American Instrument Co., excitation, 470 nm; emission, 585 nm).

The results of the partition experiments are summarized in Tables 10A and 10B for ADR and DNR, respectively. In these tables partition coefficients were expressed in terms of concentration of drug in the nonaqueous phase divided by that in the aqueous phase or C_{nonaq}/C_{aqu}. In nearly all cases, n-butanol systems yielded highest partition coefficient values among all three solvent systems examined. DNR showed substantially higher partition coefficient values than ADR in nearly every solvent system, apparently due to the higher polarity of ADR over DNR. Although the highest coefficient values were obtained in the n-butanol/Na_2HPO_4 system, precipitation of salt crystals was observed. Further studies showed that crystal formation was avoided by reducing the salt concentration below 0.4 M while maintaining sufficiently high partition coefficient values. Therefore, the following studies were performed exclusively with the n-butanol/ 0.3 M Na_2HPO_4 system. At this Na_2HPO_4 concentration, partition coefficients for ADR and DNR were 31 and 78, respectively.

2. *Extraction of Urinary Drug Metabolites.* Sample urine was collected from patients or healthy volunteers for 24 hr during control periods and patients for the first 24–48 hr following anthracyline chemotherapy. A sufficient amount of Na_2HPO_4 to yield a concentration of 0.3 M was added to several liters of urine. Salted urine sample was filtered to remove precipitated material and then saturated with n-butanol to prevent leaching the stationary phase retained in the column. In some cases, it was necessary to repeat filtration of the urine after saturation with n-butanol. In order to test the efficiency of the countercurrent extraction, known quantities of ADR and DNR were added to control urines. These spiked samples were processed in an identical fashion to posttherapy urines.

Extraction experiments were performed with a table top model of the coil planet centrifuge (Fig. 125) equipped with a 60-ml capacity column consisting of 18 helical turns of a 10-m long, 2.6-mm i.d. PTFE tube mounted around the 15-cm diameter holder (β = 0.75). The column was first entirely filled with n-butanol (stationary phase) and the apparatus was rotated at 650 rpm. Then, 0.3 M Na_2HPO_4 solution saturated with n-butanol was pumped through the column in the head-to-tail elution mode to establish phase equilibrium conditions where the stationary n-butanol phase volume retained in the column was 32 ml or slightly over 50% of the total column capacity. The above prepared urine samples, 1–2 L in volume, were then pumped through the column at flow rates of 500–700 ml/hr using a metering pump. After the extraction process was completed, the column contents were cleaned by eluting the column with 100 ml of 0.3 M Na_2HPO_4 solution saturated with n-butanol. Then the stationary phase containing drug metabolites was drained from the column; several milliliters of n-butanol was passed through the column to recover a small amount of sample still remaining in the column. The n-butanol extracts were combined and evaporated to

Table 10A Partition Coefficients for Adriamycin[a]

Salt	n-Butanol/water (1:1)	Chloroform/ isopropanol/water (3:1:2)	Ethyl acetate/water (1:1)
None	2.1	2.3	0.1
$(NH_4)_2SO_4$	2.2	0.4	0.1
NaH_2PO_4	4.3	0.3	0.4
NH_4Cl	7.6	1.1	0.1
KCl	12.3	2.4	0.8
LiCl	10.0	1.8	0.7
CH_3COONH_4	10.1	4.1	0.4
NaCl	11.7	1.8	0.9
Na_2SO_4	10.8	7.7	0.1
Na_2HPO_4	>40.0	—	>1.2

[a]Adriamycin (100 nmol) was added to each 4.0-ml phase system. The concentration of aqueous salt solution was 1.0 M in all cases. The systems were vigorously mixed and allowed to equilibrate. Concentration of drug in each phase was measured by fluorometry, and partition coefficients were calculated as the ratio of the concentration in the nonaqueous phase to that in the aqeuous phase. Insoluble salt crystals were noted in mixing 1.0 M Na_2HPO_4 with all nonaqueous phases, and no coefficient could be determined for the chloroform/isopropanol/Na_2NPO_4 system.

dryness by flash evaporation. The deep-red residue remaining after drying was redissolved in methanol. Aliquots of prechromatography urine, column eluate (mobile phase), and the final extract from the column in methanol were analyzed for anthracyline content by fluorescence. A small volume of the aliquot was added to 2 ml of 1.8 M hydrochloric acid ethanol (1:3), and the concentration determined against known ADR standards in an Aminco-Bowman spectrophotofluorometer.

Figure 144 shows the experimental results of DNR extraction. The top chromatogram (a) was obtained from a reversed phase HPLC analysis of the urine from a cancer patient who has received DNR. Only two anthracyline species, D_1 and D_2, were detected. The bottom chromatogram (b) was similarly obtained from HPLC analysis of the same urine sample after extraction with the coil planet centri-

Table 10B Partition Coefficeints for Daunorubicin[a]

Salt	n-Butanol/water (1:1)	Chloroform/ isopropanol/water (3:1:2)	Ethyl acetate/water (1:1)
None	2.7	3.2	0.1
$(NH_4)_2SO_4$	5.9	1.6	0.1
NaH_2PO_4	12.4	0.6	0.2
NH_4Cl	16.8	2.3	0.3
KCl	27.3	4.7	0.4
LiCl	20.9	4.4	0.3
CH_3COONH_4	21.6	12.1	1.3
NaCl	24.4	4.6	0.3
Na_2SO_4	36.3	12.5	0.6
Na_2HPO_4	>85.5	—	>6.9

[a]Experimental conditions were as described in Table 10A. Daunorubicin (100 nmol) was added to each system. Insoluble crystals of Na_2HPO_4 were encountered with all nonaqueous phases, and no coefficient could be determined for the chloroform/isopropanol/ Na_2HPO_4 system.

fuge method. It shows enriched D_1 and D_2 peaks and three previously undetectable metabolite peaks, dD_3, demethyl-dD_3, and 4-O-glucuronide. Figure 145 shows a similar enrichment obtained for ADR and its metabolites where relatively greater amounts of interfering materials were recovered due to the higher polarity of ADR.

Comparative studies revealed that the standard method with an XAD-2 resin yielded more efficient extraction of ADR than did the coil planet centrifuge from spiked samples, but the methods were equivalent in their recoveries of the less polar DNR. The reverse was seen in the case of treated patients' urines, with about 50% improvement in recovery of ADR and metabolites using the coil planet centrifuge. This difference may have been due to altered partitioning of drug metabolites between the two phases in comparison with the parent drug. In particular, aglycone metabolites would be expected to distribute well into the n-butanol phase. Both TLC and reversed phase HPLC confirmed that the coil planet centrifuge allowed recovery of relatively greater quantities of aglycones from urine than the XAD-2 column for ADR (60% aglycones with coil planet

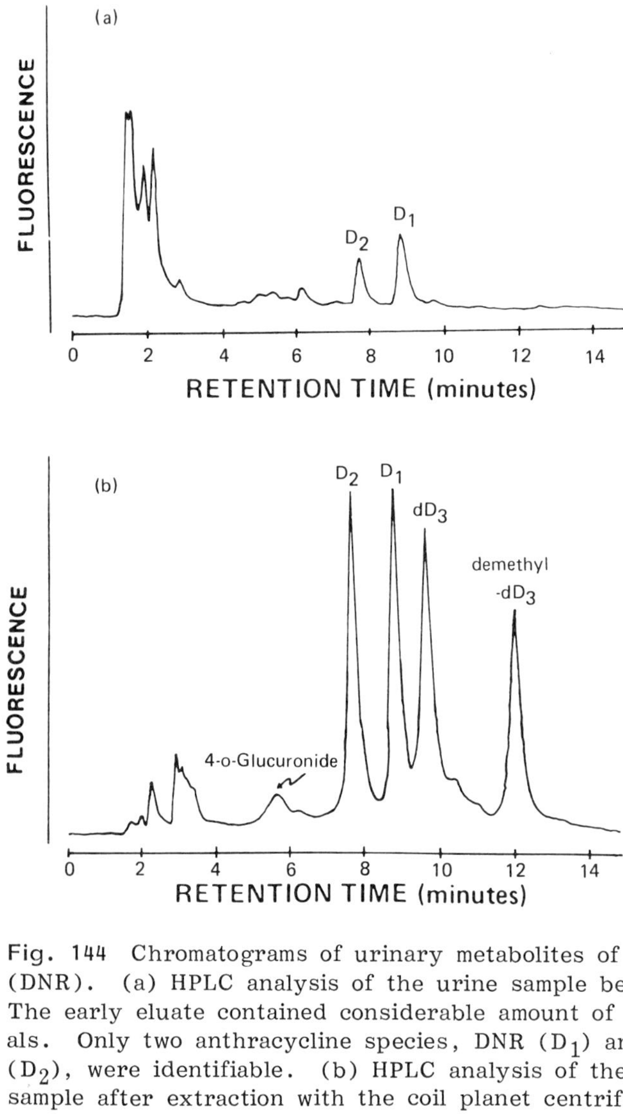

Fig. 144 Chromatograms of urinary metabolites of daunorubicin (DNR). (a) HPLC analysis of the urine sample before extraction. The early eluate contained considerable amount of interfering materials. Only two anthracycline species, DNR (D_1) and 13-OH-DNR (D_2), were identifiable. (b) HPLC analysis of the same urine sample after extraction with the coil planet centrifuge. The early interfering materials were largely excluded from the extract. Sharp, enriched peaks for D_1 and D_2 were evident, and three previously undetectable metabolites of DNR-1 conjugated species and two aglycones, dD_3 and demethyl-dD_3, were made apparent.

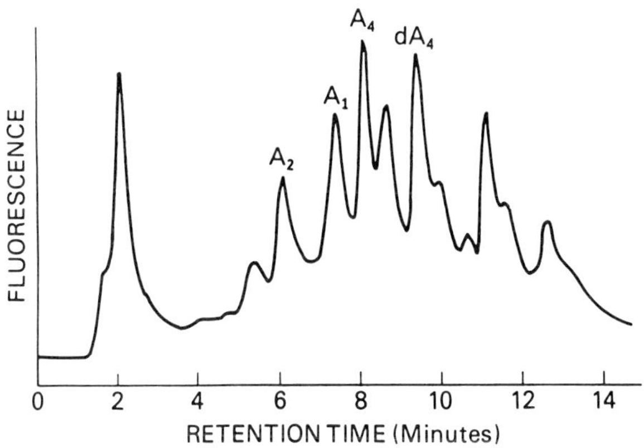

Fig. 145 Chromatogram of urinary metabolites of adriamycin. HPLC analysis of urine extract with the coil planet centrifuge showed enriched adriamycin (A_1) and three metabolites (A_2, A_4, and dA_4).

centrifuge vs. 12% with XAD-2 column) and DNR (48% aglycones vs. 5%).

E. High-Speed CCC

High-speed CCC, which is characterized by high peak resolution and speedy separation, enables highly efficient preparative separations comparable to those obtained with preparative HPLC. Because of its distinctive feature of utilizing the unilateral hydrodynamic equilibrium, the method requires special attention to selection of two-phase solvent systems and various operational conditions as compared with other CCC schemes.

Two-Phase Solvent System

The selection of the solvent systems in high-speed CCC should be focused on the following points:

1. No decomposition or denaturation of the sample
2. Sufficient sample solubility
3. Suitable partition coefficient values
4. Satisfactory retention of the stationary phase

Among those, items 1 and 2 are common to all CCC schemes while items 3 and 4 are particularly important for high-speed CCC and merit further discussion.

In order to clarify the following description, the partition coefficient (P.C.) is defined here as solute concentration in the stationary phase (C_s) divided by that in the mobile phase (C_m) or $K = C_s/C_m$. The P.C. value is accurately determined by a simple test tube experiment with a pure compound or the standard [65]. When the standard is not available, the P.C. of each component in a crude sample mixture can be accurately determined by HPLC analysis on each layer of the two-phase solvent system equilibrated with the sample mixture [93]. TLC is similarly used for evaluation of P.C. values.

As reported elsewhere [94], the optimum P.C. values in CCC depend on the retention of the stationary phase in the column. When the retention of the stationary phase substantially exceeds 50% of the total column capacity (as in the case of high-speed CCC), rapid and effecient separations can be attained by adjusting the P.C. values at somewhere between 1 and 0.5. When the stationary phase retention becomes lower than 50% (as in the many other CCC schemes), the peak resolution is effectively improved by increasing the P.C. values in a range between 1 and 2 or even greater. Since the present method usually permits either upper or lower phase to be used as the mobile phase, the P.C. values between 2 and 0.5 will produce satisfactory separations. More detailed description on this subject is given in Chap. 4.

Once the P.C. (K) of the solute is determined, the retention volume (R) or the peak location on the chromatogram is predicted from the following equation:

$$R = (V_C - R_{SF})K + R_{SF} \tag{35}$$

where V_C is the total column capacity and R_{SF} the retention volume of the solvent front. From the above equation, it is apparent that the solute with $K = 1$ is always eluted at $R = V_C$ or the elution volume equal to the total column capacity, regardless of the position of the solvent front. The solutes with $K < 1$ and $K > 1$ are eluted earlier and later than V_C, respectively. When no stationary phase is retained in the column ($R_{SF} = V_C$), all components would be eluted together at V_C. The P.C. value (K) of each component on the chromatogram can also be determined from a modified equation [62]:

$$K = \frac{R - R_{SF}}{V_C - R_{SF}} \tag{36}$$

In the course of a solvent search, care should be taken to avoid the use of solvent composition at the vicinity of the plait point where the solvent mixture forms a single phase. Such solvent systems possess low interfacial tension and a small difference in density between the two phases which would adversely affect the retention of the stationary phase as described elsewhere. This caution becomes particularly important in large-scale preparative separations where introduction of a large amount of sample may significantly alter the two-phase composition and their relative volumes in the sample compartment at the beginning of the column to adversely affect the retention of the stationary phase and the peak resolution.

Various two-phase solvent systems successfully used in high-speed CCC and other CCC schemes are listed in Part III, Appendix, Tables 1–4.

Sample Solution

Preparation of sample solution requires some considerations on several factors which would affect the partition efficiency. In order to maintain the normal phase composition, samples should be dissolved in the solvents, upper and/or lower phases, to be used for the separation. The maximum concentration of solutes in the sample solution, effectively applied to the column, may be limited by several factors such as solubility, nonlinear isotherm, and alteration of the two-phase composition. The nonlinear isotherm produces skewed peaks causing a loss of peak resolution, while the altered phase composition results in decrease of the stationary phase volume retained in the column. Sample volume also affects the partition efficiency as in other chromatographic methods.

A series of experiments has been carried out to study the effects of sample dose, sample volume, and choice of sample diluents on the stationary phase retention and peak resolution in high-speed CCC [86]. All data were obtained from a 400-ml capacity multilayer coil consisting of a 70-m long, 2.6-mm i.d. PTFE tube mounted on the coil planet centrifuge (Fig. 127) at $\beta = 0.5-0.8$. The performance of the apparatus was evaluated on separation of a set of five DNP amino acid samples with a two-phase solvent system composed of chloroform, acetic acid, and 0.1 N hydrochloric acid (2:2:1). Both lower nonaqueous and upper aqueous phases were used as the mobile phase at a flow rate of 500 ml/hr under a revolutional speed of 800 rpm.

1. *Effects of Sample Dose.* Figure 146 shows effects of the sample dose on the retention of the stationary phase where the retention volume expressed in percentages relative to the total column capacity is plotted against the applied sample volume. The results clearly indicate a general trend that the retention of the stationary phase decreases with the increased sample size for both mobile phase groups. The results also show that the retention level of the non-

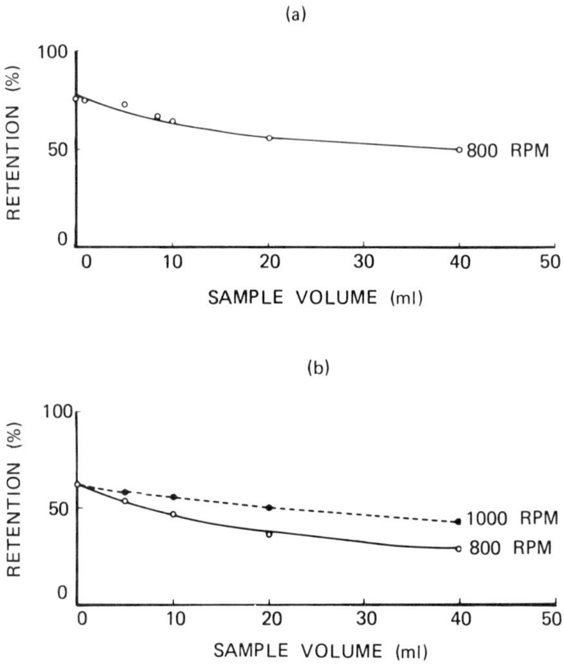

Fig. 146 Effects of sample volume on retention level of the stationary phase. (a) Lower phase mobile. (b) Upper phase mobile.

aqueous phase (b) is substantially lower than that of the aqueous phase (a). However, this low retention level of the nonaqueous phase is improved by applying a higher revolutional speed at 1000 rpm as indicated by the broken line (b).

Chromatograms obtained from these experiments are illustrated in Fig. 147 where individual charts are arranged according to the sample dose and the mobile phase. Although the peak resolution gradually decreases with the increased sample dose, the integrity of the individual peaks is well preserved in all charts except for the 2-g run with the mobile upper phase. This lowest peak resolution coincides with the lowest retention level of the stationary phase at 28%. The results strongly suggest that the retention level of the stationary phase plays a critical role in partition efficiency in the present method.

2. *Effects of Sample Volume and Choice of Dilutents.* Effects of the sample volume and the choice of the sample diluents on the peak resolution were studied on separations on 1-g quantity of the DNP amino acid mixture. The results are shown in Fig. 148 where both the lower nonaqueous phase (A) and the upper aqueous phase (B) are used as the mobile phase.

These chromatograms clearly show that the choice of the sample diluent makes little difference in resolution for both early and late appearing peaks until the sample volume is increase to 40 ml. With a large sample volume a significant decrease in peak resolution is observed in early appearing peaks, especially when the sample is dissolved entirely in the mobile phase (Fig. 148, i and p). The loss of peak resolution becomes minimized if the sample is dissolved in the stationary phase (Fig. 148, g and r). Further observation reveals that chromatograms i and p (Fig. 148) with the lowest resolution in the early peaks show the best resolved late peaks among all 40-ml sample groups. This peculiar elution profile of the early and late peaks may be clearly understood on the basis of the two-phase interaction within the sample compartment at the beginning of the partition process.

When the sample mixture is introduced with the stationary phase, the eluting mobile phase will pick up each individual component at a different rate according to the partition coefficient. The components which favor the partition in the mobile phase, hence producing the early appearing peaks, are quickly depleted from the sample compartment of the stationary phase and concentrated in a small volume of the mobile phase, resulting in a sharp sample band. Therefore, a large volume of sample can be injected into the column without causing significant peak broadening of early appearing peaks. Although the components favoring the partition in the stationary phase are not concentrated by the mobile phase, they are subjected to the partition process in the column for longer periods of time to produce broader but better resolved peaks. Therefore, the initial band width for these components becomes less significant for separation. The above effects become reversed when the sample mixture is introduced with a large volume of the mobile phase. In this case the components favoring the partition in the stationary phase, thus producing the late appearing peaks, are concentrated in the stationary phase at the beginning of the column, while other components tend to produce broader peaks affected by the sample volume. Because these early appearing peaks are eluted close together, the loss of resolution among those peaks becomes serious compared with the late eluted peaks. By dissolving the sample in equal volumes of the upper and lower phases, resolution can be relatively maintained for both early and late appearing peaks. There is a decrease in resolution when the sample is dissolved in a large volume of the upper and lower phase mixture but not to the degree seen when the sample is dissolved in a large quantity of the mobile phase.

As discussed above, in large-scale separations the best results are usually attained by dissolving the sample mixture entirely in the stationary phase. However, in some instances the sample mixture contains a component or components having low solubility in the stationary phase and therefore an enormously large volume of the sta-

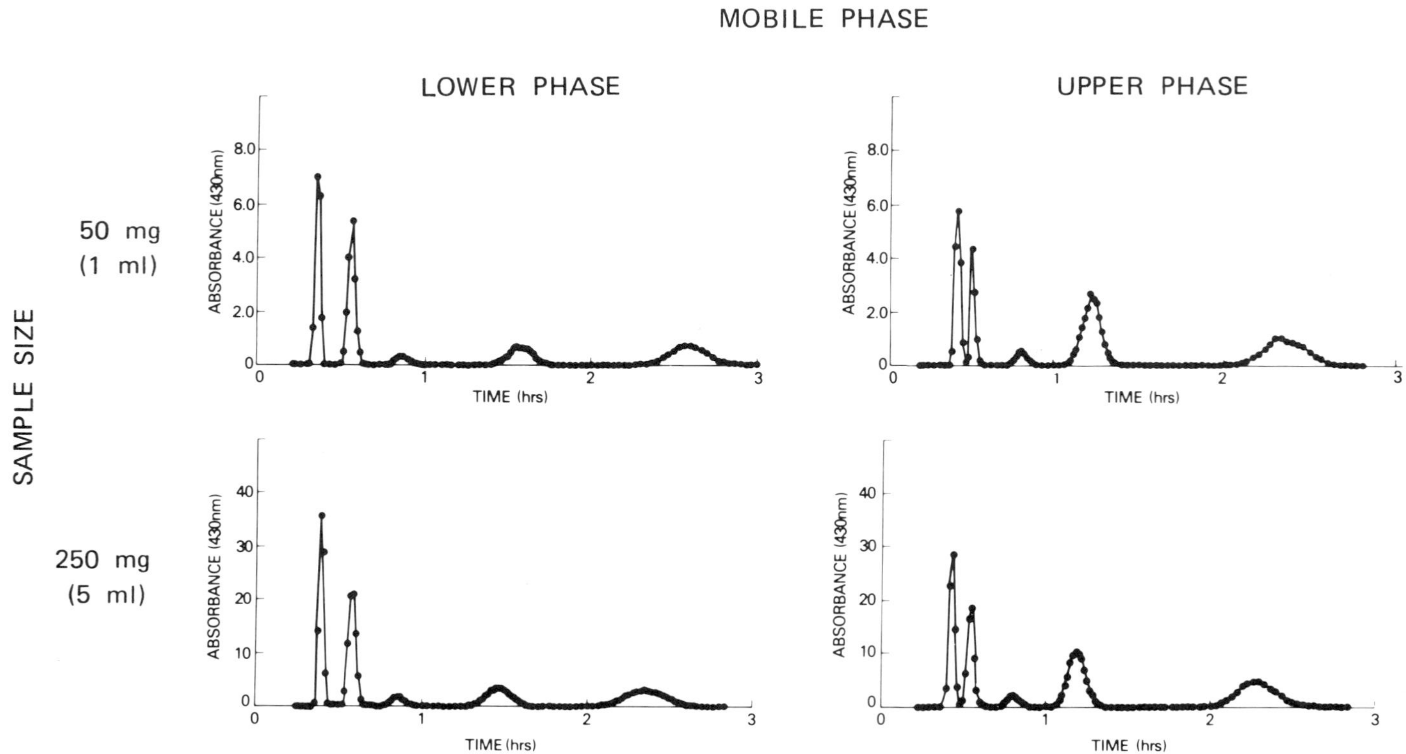

Fig. 147 Effects of sample size on separation of a set of DNP amino acids. Order of elution: lower phase mobile (left column): DNP-valine, DNP-alanine, diDNP-cystine, DNP-glutamic acid, DNP-aspartic acid. Upper phase mobile (right column): DNP-aspartic acid, DNP-glutamic acid, diDNP-cystine, DNP-alanine, DNP-valine.

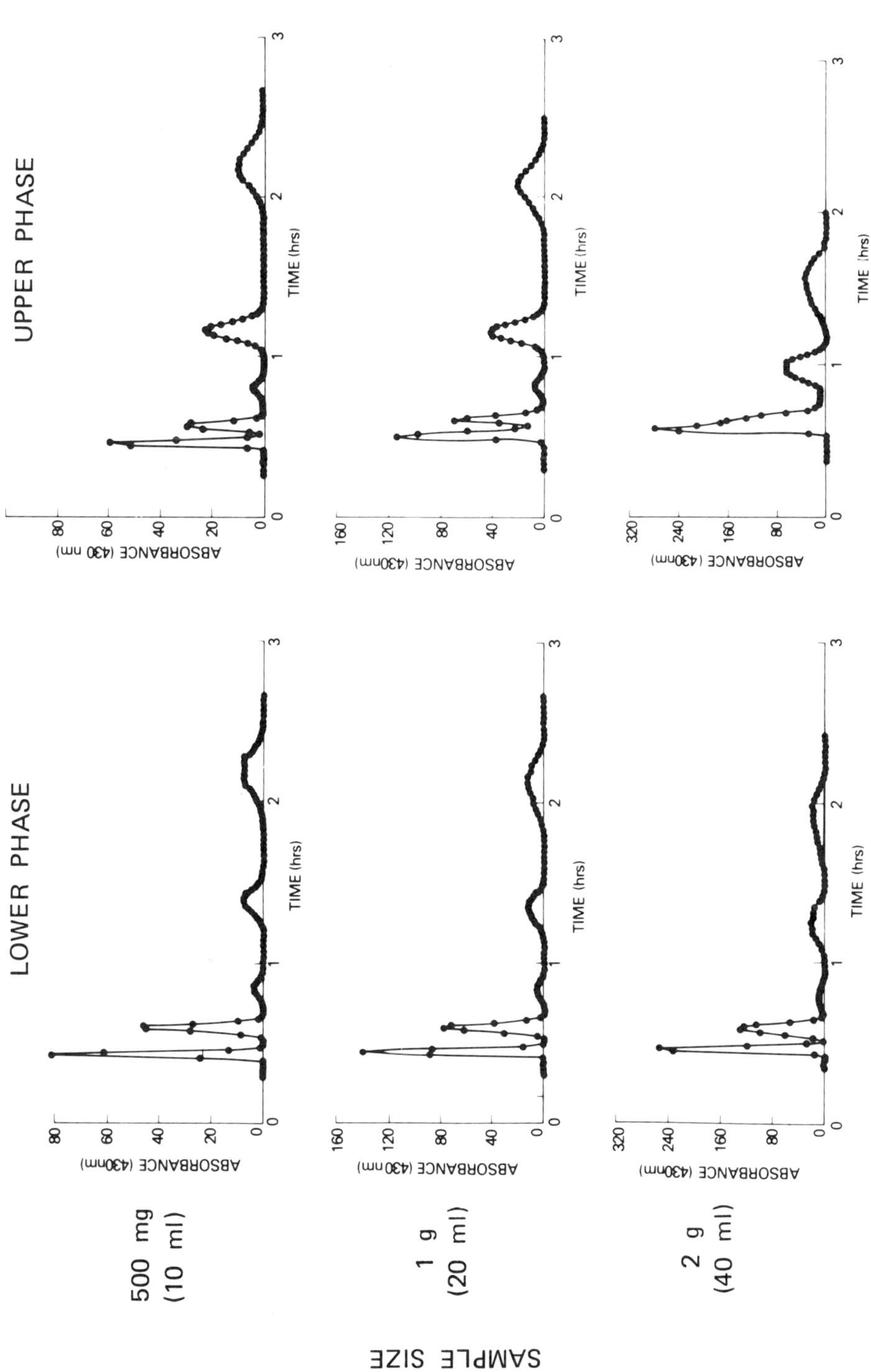

MOBILE PHASE
LOWER PHASE
UPPER PHASE
SAMPLE SIZE
500 mg
(10 ml)
1 g
(20 ml)
2 g
(40 ml)
ABSORBANCE (430nm)
ABSORBANCE (430 nm)
TIME (hrs)
0
1
2
3
20
40
60
80
120
160
240
320

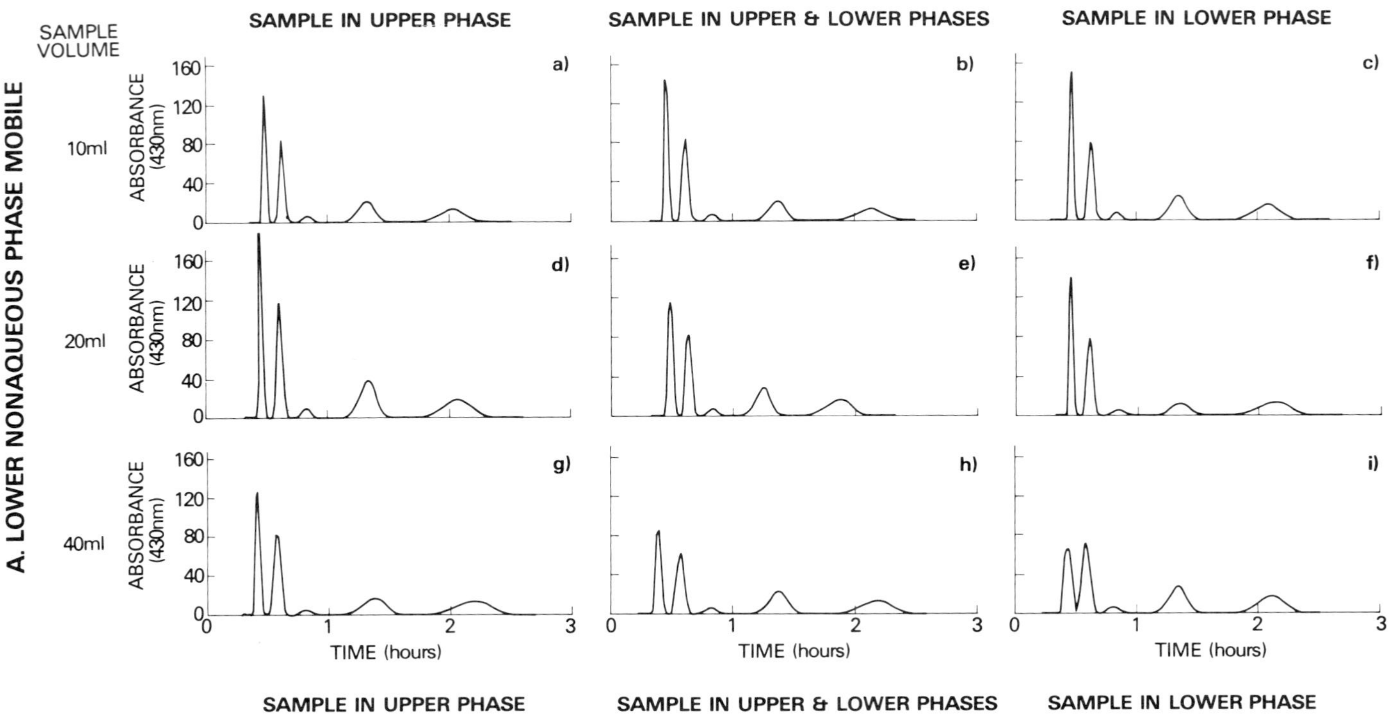

Fig. 148 Effects of sample volume and sample diluent on separation of a set of DNP amino acids. For order of elution see Fig. 147

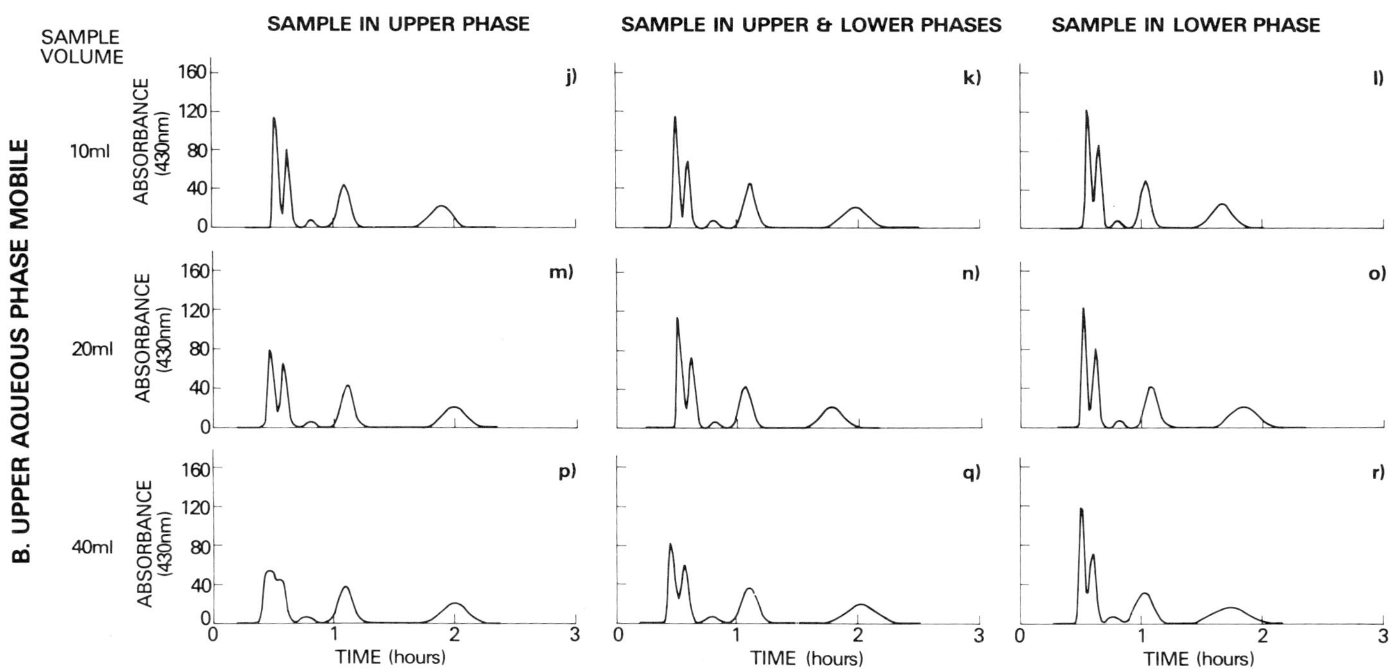

Fig. 148 (Continued)

tionary phase is required to dissolve the sample. In this situation the addition of the mobile phase to the sample solution will give beneficial effects in reducing the sample volume. The use of the two phases for the sample solution provides an additional advantage in that one can ensure formation of two phases in the sample solution. When the sample concentration exceeds the critical range, solvent system may form a single phase in the sample compartment at the beginning of the column. Even though the sample solution forms the two layers, the excessive alteration of the phase composition may adversely affect the retention of the stationary phase. As described elsewhere, this can be predicted by measuring the settling time of the sample solution. When the settling time exceeds a critical range of 30 sec, it is most likely that the phase distribution mode in the sample compartment is reversed and in this situation the application of the normal elution mode may cause a large loss of the stationary phase. Because of these reasons, the use of two solvent phases for preparation of sample solution is strongly recommended.

Separation Procedure

As in many other CCC methods, high-speed CCC is performed in the following steps: The column is first entirely filled with the stationary phase. This is followed by sample injection through the sample port. Then, the apparatus is rotated at the desired rpm while the mobile phase is pumped into the column at the optimum flow rate. The effluent from the outlet of the column is continously monitored with a UV monitor and fractionated into test tubes with a fraction collector. After the separation is completed, the apparatus is stopped and the column contents are collected by either eluting the solvents to fractionate into test tubes or flushing with pressured N_2 into a container. The column is then washed by introducing a small volume (20–30 ml) of intermediate solvent miscible with either phase such as methanol and flushing it with N_2. This process may be repeated until the washings become clean. Then the column is dried by passing N_2, if desired. Beside the above general procedure common with other CCC schemes, high-speed CCC requires special attention to the following points to achieve the best results.

The present system produces vigorous mixing of the solvents in the column and permits application of a high flow rate of the mobile phase without a loss in peak resolution provided that a satisfactor volume of the stationary phase (over 50% of the total column capacity) is retained in the column. Although the hydrodynamic behavior of the two solvent phases in the rotating coil is extremely complex, a simple settling test described earlier may provide useful guidance for optimizing the operational conditions to attain a satisfactory retention of the stationary phase [92].

In a standard multilayer coiled column (β = 0.5–0.8, 10-cm revolutional radius) [9,86], a solvent system with a settling time less than 30 sec would display normal hydrodynamic behavior which distributes the upper phase toward the head and the lower phase toward the tail. Accordingly, the elution should be performed by pumping the upper phase from tail to head (broken curve in the phase distribution diagram, Fig. 137a) or the lower phase from head to tail (solid curve in the phase distribution diagram, Fig. 137a). The flow rate of the mobile phase may also be judged from the settling times. For a 1.6-mm i.d. column, flow rates over 240 ml/hr may be applicable if the settling time is less than 10 sec, and a reduced flow rate should be used for solvent systems with longer settling times. When the settling time of the solvent system considerably exceeds 30 sec, it is most likely that the hydrodynamic behavior of the two phases is reversed, the upper phase now being distributed toward the tail and the lower phase toward the head. Accordingly, the upper phase should be pumped from the head toward the tail (solid curve) and the lower phase from the tail toward the head (broken curve) at the reduced flow rates (typically 60 ml/hr or less for the 1.6-mm i.d. column).

Alternatively, the separation can be performed at elevated temperature which permits the normal mode of elution for the viscous polar solvent systems [92]. Preliminary experiments have been successfully conducted at 40–50°C using a coil planet centrifuge equipped with a temperature control system [70,95,96]. In order to maintain the original two-phase composition in the column, the solvents should be equilibrated and kept at the elevated temperature in a water bath. In many occasions, the partition coefficient of the compound is sensitively altered with temperature and therefore the solvent system should provide a suitable partition coefficient of the aimed compound at the elevated temperature.

In addition to the isocratic elution described above, CCC enables stepwise and gradient elutions [9,19,20,54]. However, the successful application requires a particular choice of the solvent system as follows: the key element producing the gradient (such as acid, neutral salt, etc.) should be partitioned almost entirely into the mobile phase and at the same time should not significantly alter the volume ratio of the two solvent phases in the column. Solvent systems such as n-butanol/aqueous solution for the gradient of dichloroacetic acid or trifluoroacetic acid concentration and n-butanol/phosphate buffer for a pH gradient will suffice the above requirements if the lower aqueous phase is used as the mobile phase.

While the method permits a choice in the two elution modes, the head-to-tail elution usually gives superior results. In some solvent systems, the tail-to-head elution mode produces a steady carryover of the stationary phase, resulting in gradual loss of the stationary phase and reduced peak resolution. Even though quite often this carryover is minor and does not significantly affect the separation, the presence of stationary phase droplets in the flow cell disturbs the absorbance monitoring of solute peaks by producing a large noise in recording. The tail-to-head elution mode also requires an additional consideration to maintain a uniform flow rate. Countercurrent flow of the two solvent phases through the rotating coil creates a pressure gradient along the length of the coil in such a way that the pressure on the head side becomes higher than that on the tail side. Consequently, introduction of the mobile phase through the tail may produce a negative pressure at the inlet of the column, resulting in an increased flow rate of the mobile phase by sucking an extra amount of solvent from the reservoir through the metering pump equipped with one-way check valves. This adverse phenomenon is enhanced by several factors such as a large density difference between the two solvent phases, application of high revolutional speed, and slow flow rate, etc. However, this problem can be easily solved by restricting the flow through the column with a piece of narrow-bore tubing, typically 0.5 mm i.d. × 50 cm, inserted at the outlet of the column [9]. For monitoring the pressure, the use of a metering pump equipped with a pressure gauge is recommended.

Examples of Application

During the past few years, application data of high-speed CCC have been sharply increasing as the apparatus became commercially available in the United States. Although the number of publications of high-speed CCC is still limited, the method has been applied to a broad spectrum of samples as listed in Part III, Appendix, Table 1.

Performance of high-speed CCC has been evaluated on separations of the standard set of DNP amino acid samples on a two-phase solvent system composed of chloroform/acetic acid/0.1 N HCl (2:2:1). The results obtained with a 1.6-mm i.d. and 280-ml capacity multilayer coil are illustrated in Fig. 149. The first chromatogram (Fig. 149a) was obtained by using the upper aqueous phase as the mobile phase at an optimum flow rate of 240 ml/hr in a tail-to-head elution mode. All seven components were well resolved into symmetrical peaks and eluted out in about 4 hr. Comparative studies show that the present method can produce similar peak resolution 18 times faster than the droplet CCC previously described (see Fig. 24). The second chromatogram (Fig. 149b) was obtained by using the lower nonaqueous phase as the mobile phase under a head-to-tail elution. As mentioned earlier, the head-to-tail elution mode permitted application of higher flow rate of 420 ml/hr to shorten the separation time required. Sample size in these separations can be increased to several hundred milligrams without causing detrimental loss of peak resolution [9].

Figure 150 shows chromatograms of purines and pyrimidines obtained with a two-phase solvent system composed of n-butanol/1 M potassium phosphate (pH 6.5) (1:1). Separations were similarly performed by using both the lower phase (a) and the upper phase (b) as the mobile phase at the same flow rate of 240 ml/hr in the respective elution modes. Although the upper butanol phase has a higher viscosity than the lower aqueous phase, these separations demonstrate that either phase can be chosen as the mobile phase to obtain similar peak resolutions. Because of a relatively large difference in density between the two phases, the tail-to-head elution mode (b) caused a negative pressure which resulted in increased flow by sucking the solvent from the reservoir through the metering pump. This complication was eliminated by inserting a narrow-bore tubing at the outlet of the column to create a positive pressure of 30 psi.

Figure 151 shows a chromatogram of indole plant hormones on a two-phase solvent system composed of n-hexane/ethyl acetate/methanol/water (3:7:5:5). The separation was performed at a flow rate of 240 ml/hr using the lower aqueous phase as the mobile phase in a head-to-tail elution mode. This solvent system is ideal for separations of moderately hydrophobic compounds since the partition coefficient values can be optimized simply by adjusting the volume ratio between hexane and ethyl acetate. Separations of various plant hormones [73] and other agrochemicals [97–99] are described in detail in Chap. 8.

Figure 152 shows an example of antibiotics separations using a similar solvent system reported by Brill et al. [100]. Two components, tirandamycin A and B, with a minor structural difference are completely separated, tirandamycin B being recovered from the column long after tirandamycin A is eluted out from the column. Recently, varieties of antibiotics have been successfully isolated by high-speed CCC as summarized in Part III, Appendix, Table 1. Various advantages of the present method in isolation of antibiotics are instructively discussed in Chap. 9.

Tannins, which have produced poor peak resolution in HPLC, have been successfully separated by high-speed CCC by Putman and Butler [101]. Partition coefficient of tannins in n-butanol/aqueous systems was found to be insensitive to pH but conveniently adjusted by manipulating the ionic strength of neutral salts. Figure 152 shows a chromatogram of crude tannin extracted from IS8768 sorghum obtained with n-butanol/0.1 M NaCl (1:1) using the lower aqueous phase as the mobile phase. For larger-scale separations, addition of 1 mM phytic acid (pH 4.0) was required to protect tannin from oxidation and preserve the fractions for further analysis.

Figure 153 shows a chromatogram of a natural gramicidin mixture with a two-phase solvent system composed of chloroform/benzene/methanol/water (15:15:23:7) using the upper aqueous phase as the mobile phase [9]. All major components were well separated and each component was further partially resolved into valine and isoleucine analogs. With the same solvent system, peaks of gramicidin C

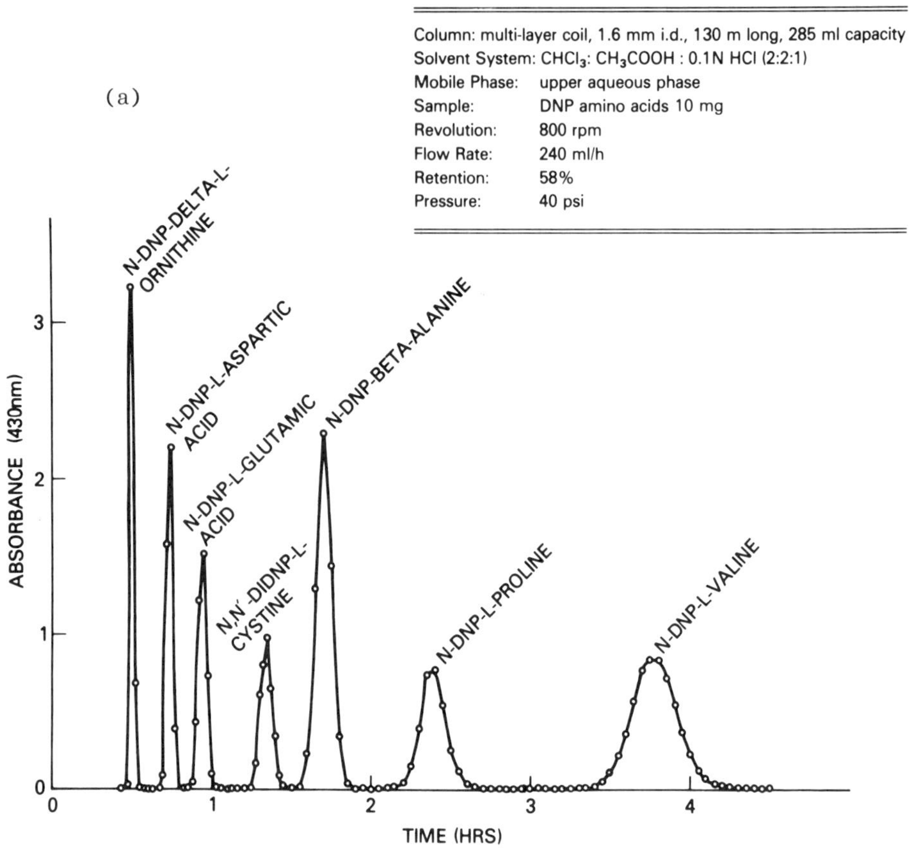

Fig. 149 Chromatograms of DNP amino acids. (a) Upper aqueous phase mobile. (b) Lower nonaqueous phase mobile.

analogs may be improved by using the lower nonaqueous phase as the mobile phase. Many hydrophilic peptides can be conveniently separated with n-butanol/aqueous systems where partition coefficient values are adjustable by addition of a small amount of trifluoroacetic acid or dichloroacetic acid. It has been demonstrated by Knight et al. that both the partition efficiency and retention of the stationary phase of the viscous butanol systems can be greatly improved by performing the separation at elevated temperature of 45–50°C and one-step purification of crude synthetic peptides becomes feasible with the present method [70,95]. Separations of various peptides with CCC will be fully discussed in Chap. 10.

As described earlier, high-speed CCC permits the use of gradient elution as frequently used in HPLC. Figure 155 shows chromatograms

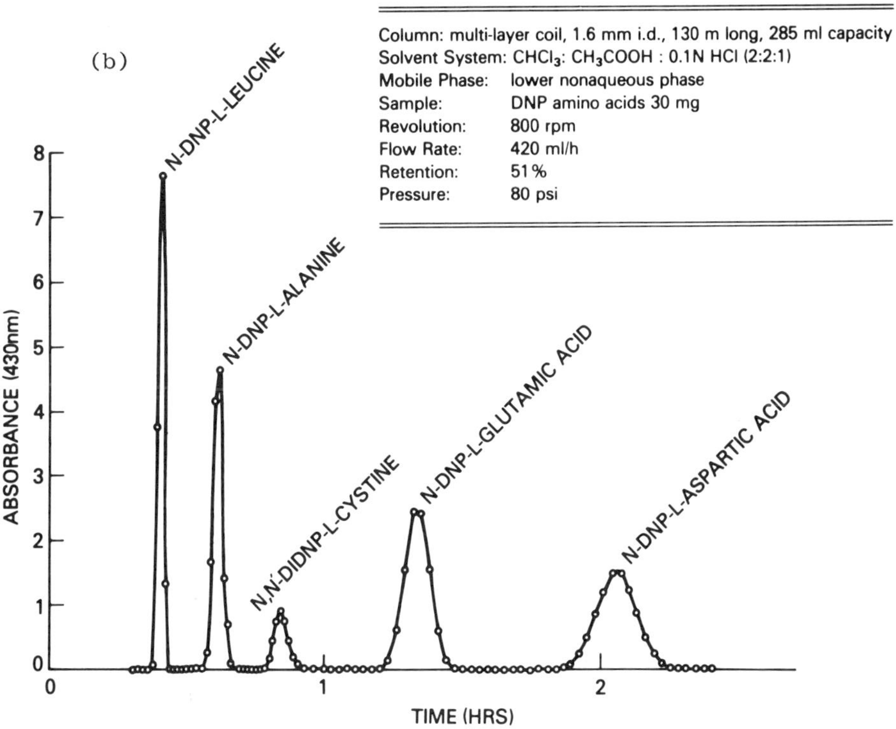

Fig. 149 (Continued)

of seven dipeptides obtained by gradient elution [9]. In both chromatograms a linear gradient of dichloroacetic acid concentration (1%–0%) was applied on the base of n-butanol/0.1 M ammonium formate (1:1) with 2-hr (a) and 4-hr (b) durations. Two pairs of isomers, tyr-val vs. val-tyr and tyr-leu vs. leu-tyr, are well resolved in both chromatograms.

In the past, development of high-speed CCC had been mainly directed toward preparative-scale separations where on-line UV monitors were almost exclusively utilized for sample detection. Recently, development of analytical high-speed CCC has been initiated by two research groups, each adapting a new detection system to high-speed CCC. Romanach et al. investigated a possiblity of interfacing high-speed CCC to a Fourier transform infrared (FT-IR) spectrometer [102]. Their preliminary studies revealed that a high solute concentration characteristic of CCC fractions could produce very high signal-to-noise ratios while the use of chloroform as the mobile phase gave no interference bands from the stationary phase

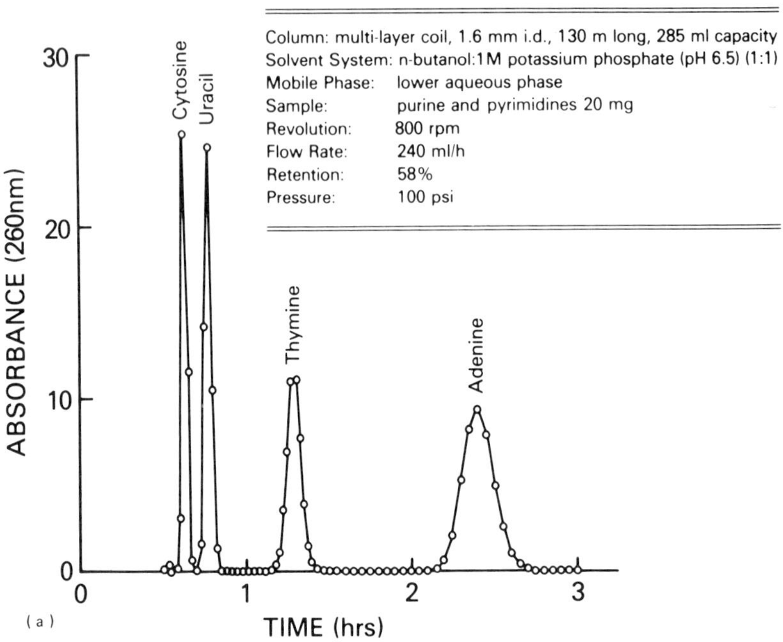

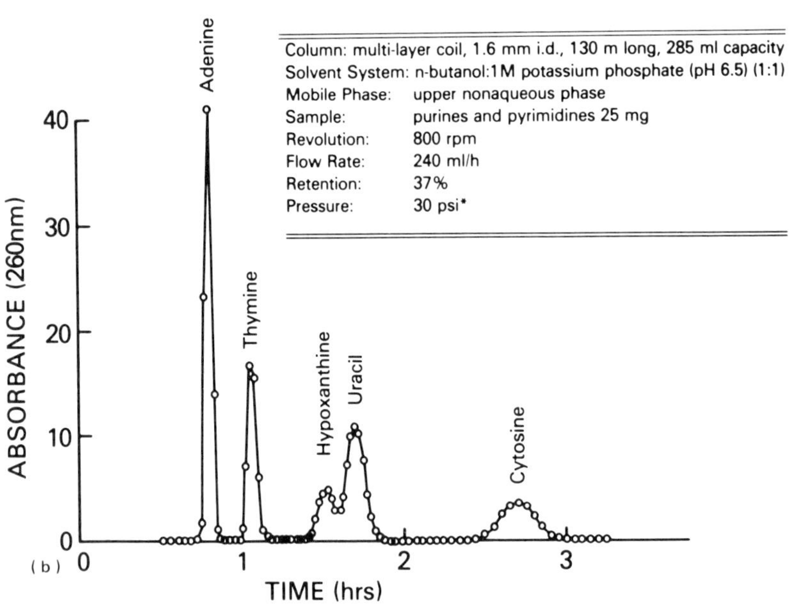

Fig. 150 Chromatograms of purines and pyrimidines. (a) Lower aqueous phase mobile. (b) Upper nonaqueous phase mobile.

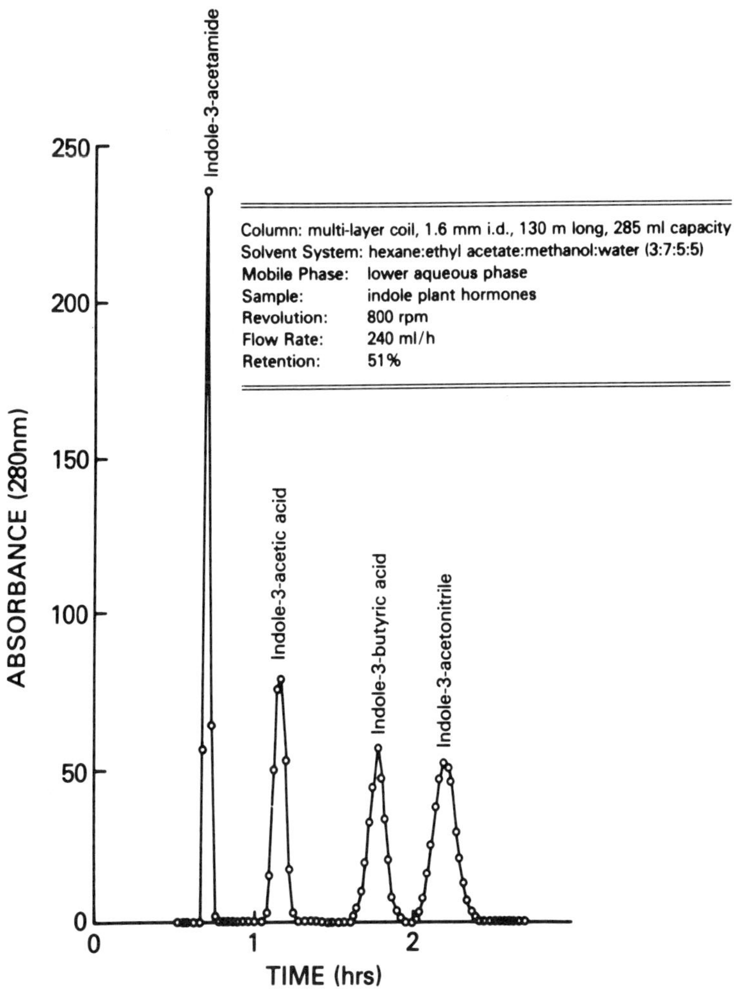

Fig. 151 Chromatogram of indole plant hormones.

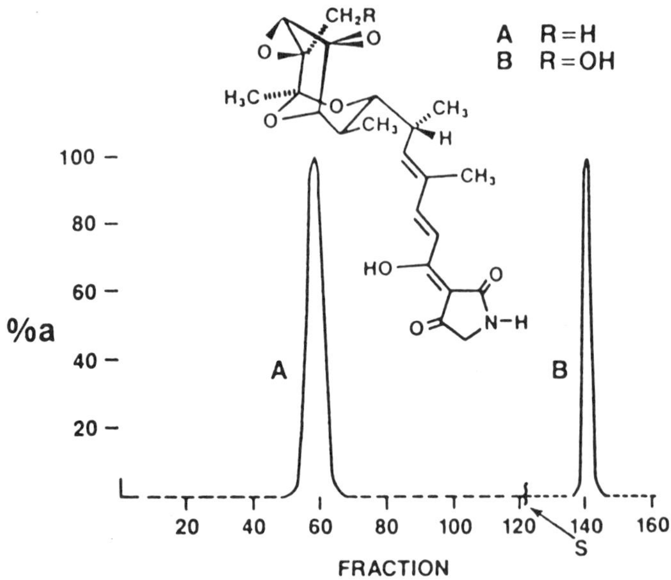

Fig. 152 CCC separation of tirandamycin A and B with n-hexane/ethyl acetate/methanol/water (70:30:15:6).

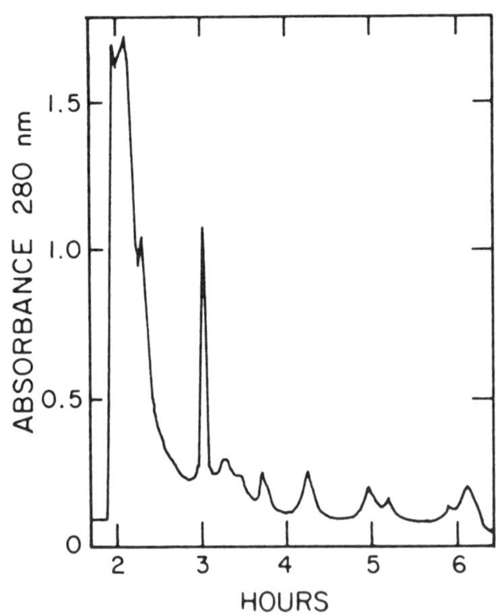

Fig. 153 CCC separation of crude tannin extracted from 20 g of IS8768 sorghum. Sample volume: 8 ml (4 ml of each phase); solvent system: n-butanol/0.1 M sodium chloride (1:1); stationary phase: upper nonaqueous phase; flow rate: 38 ml/hr; retention of stationary phase: 68%.

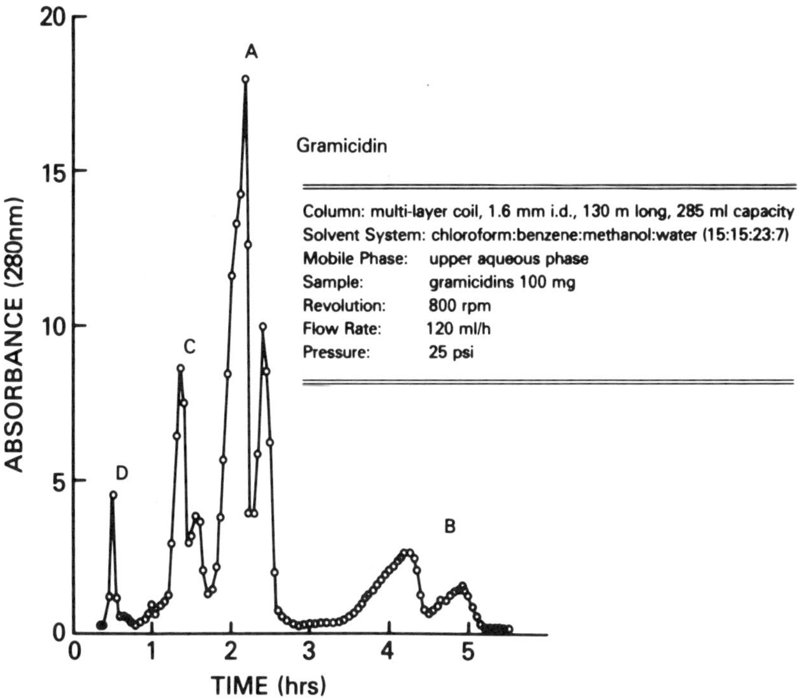

Fig. 154 Chromatogram of gramicidins.

mainly composed of methanol and water. These results clearly demonstrate the feasibility of a HSCCC/FT-IR spectrometry interface. Recent studies on digitonin fractionation with HPLC and CCC by Yang et al. showed that, despite an apparently lower peak resolution, CCC fractions produce more intense single ion for each peak in californium-252 plasma desorption mass spectrometry [103]. More recently, Chou et al. constructed a new angle rotor coil planet centrifuge equipped with a multilayer coil for the purpose of direct interfacing with a mass spectrometer equipped with a thermospray device [104]. These new developments would greatly expand the analytical capability of high-speed CCC in the near future.

F. Foam CCC [10]

Unilateral hydrodynamic equilibrium, illustrated in Fig. 123, permits true countercurrent flow of two immiscible phases through a rotating coil. This dual CCC system has been successfully applied to foam separation by using foam to carry foam-active materials. The method

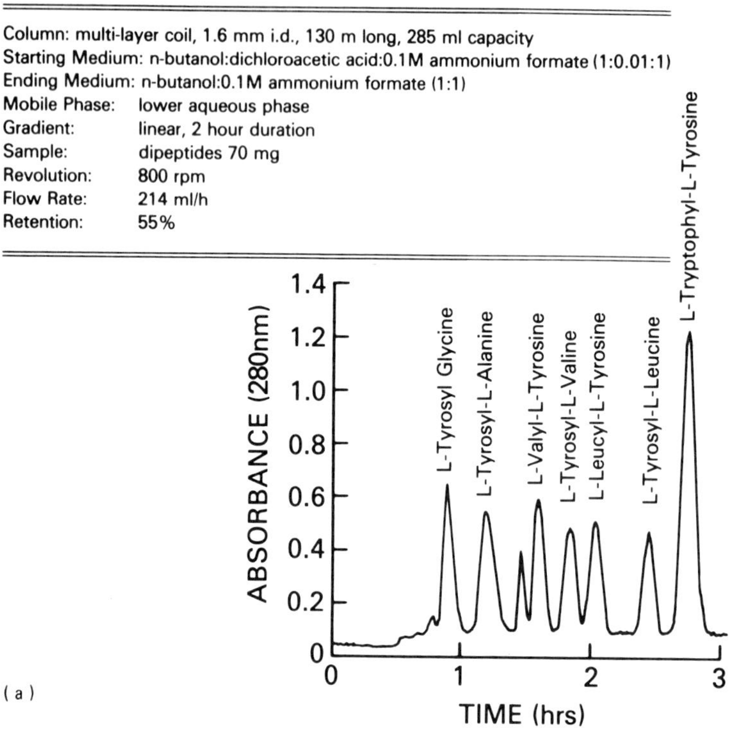

(a)

Fig. 155 Chromatograms of dipeptides obtained with gradient elution. Duration of gradient: 2 hr (a) and 4 hr (b).

has a number of advantages over the conventional foam separation methods [105] such as high efficiency, speedy separation, versatility, etc., and therefore will become a useful tool in research laboratories. Preliminary studies have demonstrated the potential capability of the method to separate macromolecules and biological particles.

Mechanism of Foam CCC

Figure 156 illustrates a schematic column design for foam CCC. The coiled column is equipped with five flow channels. The liquid phase is introduced through the liquid feed line located near the tail and drained through the liquid collection line at the head of the coil. The gas phase is similarly introduced through the gas feed line located near the head and the generated foams are harvested through the foam collection line at the tail of the coil. The sample solution may be introduced through the sample feed line opened at the middle portion of the coil.

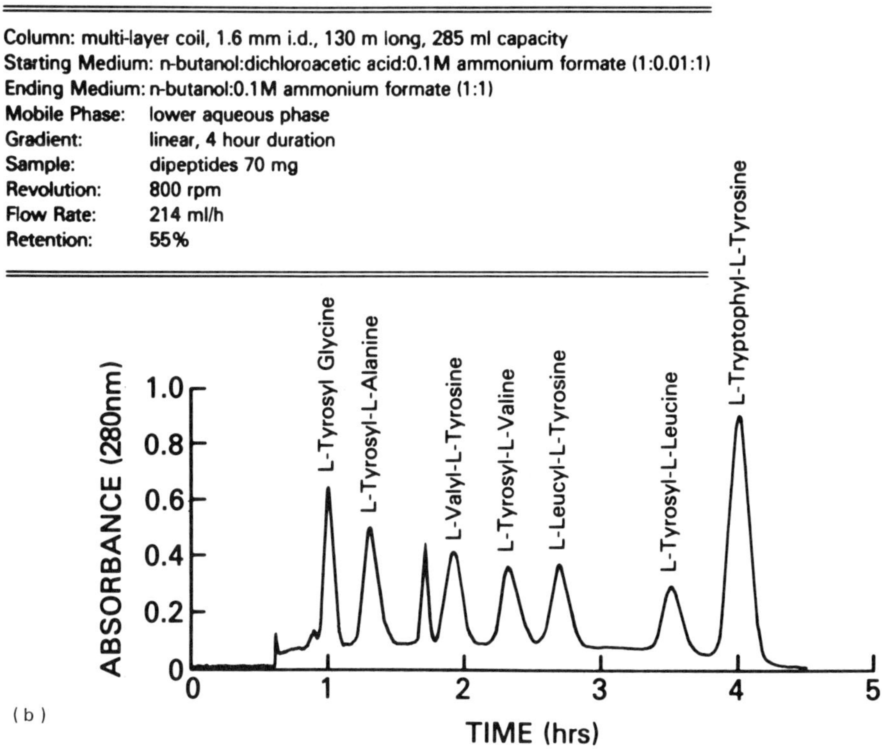

Fig. 155 (Continued)

In a typical operation mode the liquid phase containing surfactant and N_2 gas are simultaneously introduced through the respective lines into the rotating column while the sample solution is continuously fed through the sample feed line. Consequently, the samples are separated according to their foam affinity. Solutes or particulates having an affinity to the foam are quickly carried with the foaming stream toward the tail and harvested through the foam collection line while other materials in the sample solution are carried with the liquid stream in the opposite direction toward the head and eluted out through the liquid collection line.

This foam CCC method can be applied to a broad spectrum of samples having a foam affinity which may be classified into the following two categories: (a) Direct affinity to the gas–liquid interface. Detergents and many other foam-producing materials can be harvested through the foam collection line without any special treatment. (b) Affinity to the foam-producing carrier or collector molecule. Samples which lack a direct affinity to the gas–liquid interface can be indi-

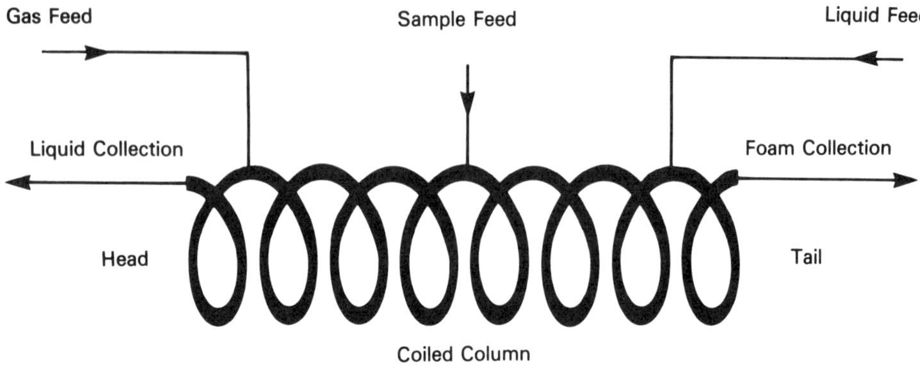

Fig. 156 Schematic column design for foam CCC.

rectly adsorbed to the foam if they have an affinity to the foam-producing agents lining the gas—liquid interface in the foams. This type of affinity effectively used for foam separation may vary in a wide range from a nonspecific form such as surface electric charges to a highly specific form such as enzyme-substrate or enzyme-inhibitor binding provided that one of the pair has either direct or indirect foam affinity.

Preliminary Studies

Series of preliminary experiments have been performed by using a combined horizontal flow-through coil planet centrifuge with a 20-cm revolutional radius (Fig. 129). The coiled column, consisting of a 10-m long, 2.6-mm i.d. PTFE tube with a total capacity of 50 ml, was mounted on the gear-driven holder of 12.5-cm diameter. The column was equipped with five flow channels as illustrated in Fig. 157. The liquid feed line was connected to a Milton Roy mini-pump (flow rate: 214 ml/hr) and the gas feed line to an N_2 cylinder pressured at 80 psi. Flow through the liquid collection line was regulated with a needle valve while the foam collection line was directly left open to the air. The sample solution was fed through the sample feed line either with a syringe (batch separation) or a metering pump (continuous separation). Effluents through each collection line were either manually fractionated into test tubes or pooled into a graduated cylinder. Preliminary studies were performed in three different elution modes using rhodamine B as a test sample and sodium dodecyl sulfate (SDS) as a collector. Experimental conditions used in the present studies are summarized in Table 11.

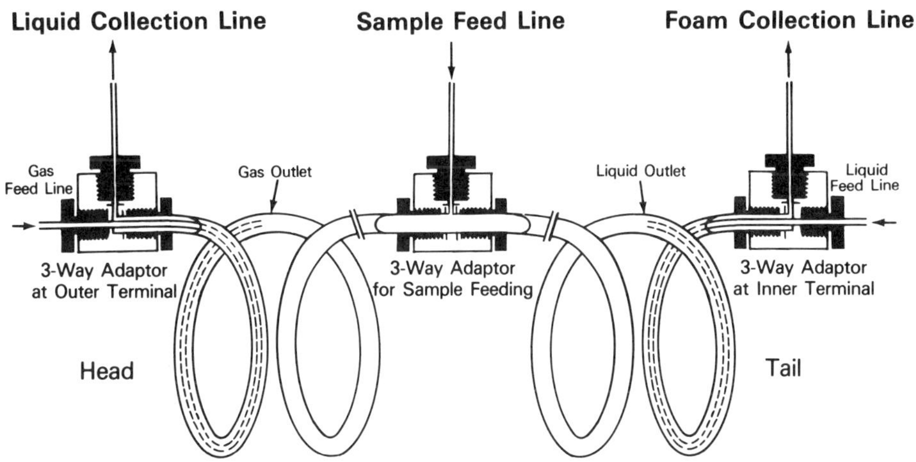

Fig. 157 Layout of five flow channels on the foam CCC column. Note: In the actual configuration of the coiled column consisting of the double layers, the first and the second layers have different handedness of the coil.

1. *Continuous Enrichment and Stripping.* This experiment was performed to demonstrate the capability of the method to concentrate and/or eliminate a minute amount of material present in a large volume of sample solution. The sample solution containing rhodamine B at a 10^{-6} M concentration and SDS at 10^{-3} M as a collector was introduced through the liquid feed line at 214 ml/hr against N_2 flow through the gas feed line at 80 psi, while the apparatus was run at 500 rpm. The sample feed line was closed and not used in this experiment. The liquid collection rate was adjusted at a level slightly below the liquid feed rate so that the foam collection rate became as small as several hundred microliters per hour which yielded the foam highly enriched with rhodamine B. After 1 L of the sample solution was eluted, the liquid collection line was closed to elute rhodamine B remaining in the column through the foam collection line. The stripped liquid collected through the liquid collection line was fluorometrically analyzed to determine the concentration of rhodamine B. The results showed that the dye concentration in the stripped solution was 1.3×10^{-9} M while over 99% of rhodamine B was recovered through the foam collection line with a 2-ml volume resulting in over 500-fold enrichment.

2. *Batch Separation with Sample Injection Through Sample Feed Line.* This experiment was initiated by establishing a liquid-gas countercurrent flow equilibrium through the coiled column. At the

Table 11 Experimental Conditions of Foam CCC Applied to the Present Study

Factors	Variable items	Applied conditions
Centrifuge	Planetary motion	Synchronous planetary motion on gear-driven holder
	Revolutional radius (R)	20 cm
	Rotational radius (r)	6.4–6.7 cm
	β-value (r/R)	0.32–0.34
	Revolutional speed	500 rpm
Column	Diameter	2.6 mm i.d.
	Length	10 m
	Configuration	Coaxial to holder axis, two layers of coil
Flow	Liquid feed rate	214 ml/hr
	Gas feed rate	600 ml/min (80 psi)
	Sample feed rate	Varied
	Foam collection rate	Varied
	Liquid collection rate	Varied

revolutional speed of 500 rpm a surfactant solution containing SDS at 10^{-3} M was pumped through the liquid feed line while the N_2 gas flow was introduced through the gas feed line at 80 psi. After the hydrodynamic equilibrium was reached, 0.5 ml of sample solution containing rhodamine B and Evans blue, each at 5×10^{-4} M, was injected through the sample feed line. The needle valve on the liquid collection line was adjusted to make a 1:3 volume ratio between the foam and liquid fractions. Effluents from both collection lines were separately fractionated into a series of test tubes at 30-sec intervals. The concentration of each dye in the fractions was spectrophotometrically determined using 556 nm for rhodamine B and 620 nm for Evans blue. Figure 158 shows the typical experimental result obtained with the present method. The upper chromatogram obtained through the foam collection line shows a sharp single peak consisting

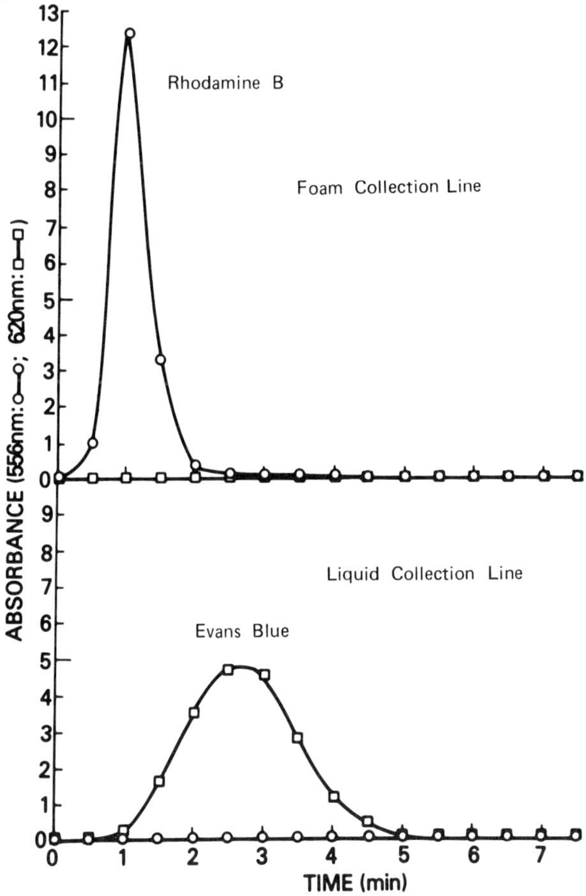

Fig. 158 Batch separation of rhodamine B and Evans blue with foam CCC. Rhodamine B having foam affinity was quickly eluted through the foam collection line (upper chromatogram) while Evans blue was carried with the liquid stream in the opposite direction and eluted through the liquid collection line (lower chromatogram). Note: In analysis, Evans blue absorbed the light at both wavelengths at a ratio of 0.6 (566 nm/620 nm). Therefore, the 556-nm curve in the lower chromatogram, which represents the elution profile of rhodamine B, was given by the total absorbance at 556 nm less the absorbance at 620 nm multiplied by 0.6.

entirely of rhodamine B with the peak maximum 1 min after sample injection. The lower chromatogram obtained through the liquid collection line shows a broad symmetrical peak of Evans blue with the peak maximum at 2.75 min after sample injection. These results are quite reproducible and injection of the single component produced the similar peak through the respective collection line. The volume of the liquid phase present in the column under a steady-state hydrodynamic equilibrium in these experiments ranged between 4 and 5 ml, which amounted to approximately 10% of the total column capacity.

3. *Continuous Separation by Continuous Sample Feeding Through Sample Feed Line.* Rapid and clean separation of the two dyes in the batch separation method described above indicated the feasibility of continuous separation by steadily feeding the sample mixture at a proper rate. Under otherwise identical experimental conditions used in the batch separation, the sample solution was continuously introduced through the sample feed line at various flow rates. The satisfactory separations were obtained at sample feed rates of 0.36 ml/min or less, which separated each sample at the maximum rate of 1.8×10^{-7} mol/min. The application of higher flow rates resulted in an initial accumulation of rhodamine B in the column which was later followed by elution of rhodamine B through the liquid collection line.

The experiments were continued to study the effects of SDS concentration on the hydrodynamic behavior of rhodamine B in the foam separation column under a high sample feed rate of 3.8×10^{-7} mol/min. The results are summarized in Fig. 159 where the percentage distribution of the rhodamine sample is plotted for the applied SDS concentration expressed in a log scale in abscissa. Among three curves drawn in the diagram, the thick solid curve indicates the amount of dye eluted through the foam collection line; the thin solid curve, the amount of dye retained with the column; and the broken curve, the amount of dye eluted through the liquid collection line. At a high SDS concentration of 10^{-2} M, the dye exhibited little affinity to the foam and mostly eluted through the liquid collection line. As the SDS concentration was decreased, the dye rapidly developed the foam affinity and, at 10^{-3} to 2×10^{-3} M SDS concentrations, over 90% of the dye was collected through the foam collection line while the liquid collection line eluted clear liquid almost free of rhodamine B. Further decrease of the SDS concentration resulted in a decreased foam recovery rate of rhodamine B causing the retention within the separation column. If the sample feeding is stopped at this stage, elution of the dye through the foam collection line continues at the same rate until the retained dye is completely recovered. If the sample feeding is continued, the dye continuously accumulates in the column and finally appears through the liquid collection line. The results clearly indicate that the foam recovery rate of rhodamine B is largely governed by the SDS concentration which yields the highest recovery rate at 10^{-3} M to 2×10^{-3} M under the present experimental condition.

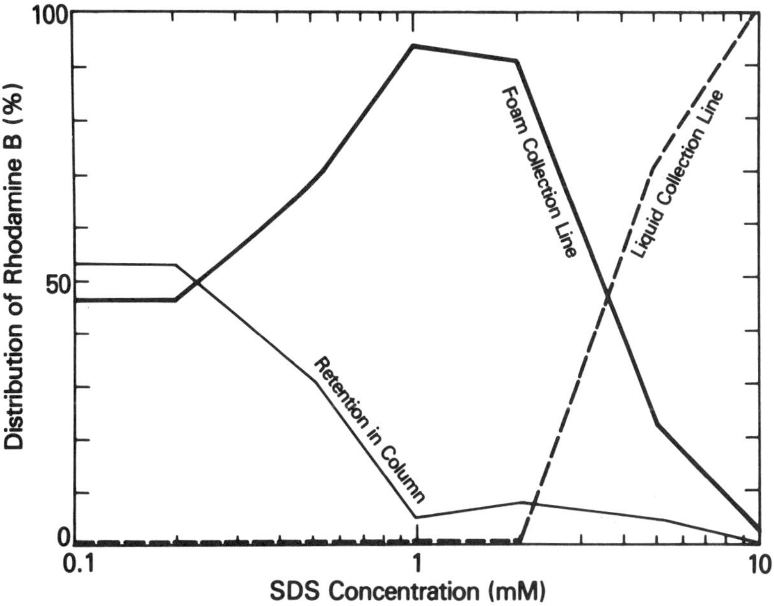

Fig. 159 Distribution of rhodamine B at various SDS concentrations in the liquid phases.

Feasibility Studies on Separation of Macromolecules

The present method has been applied to the separation of proteins without the use of a surfactant collector. As is well known, exposure of proteins such as BSA to a gas–liquid interface may cause denaturation which alters the physiological function of the molecule. The preliminary studies were conducted to test vulnerability and foam-producing capacity of proteins with the present system by injecting the sample solution into the running column through the sample feed line. Several kinds of proteins including BSA, human and sheep hemoglobins, and ovalbumin were examined. Among these only BSA showed an active foam-producing ability and was collected through the foam collection lines whereas other proteins were mostly eluted through the liquid collection line without any visible evidence of denaturation. BSA fractions eluted through the foam collection line showed various degrees of turbidity apparently due to the denaturation of the molecule. Further experiments revealed that the intensity of turbidity depended highly on the composition of the applied liquid phase. The use of salt-free distilled water or dilute acid solution caused the most intensive turbidity. Addition of a surfactant to the liquid phase decreased the degree of turbidity but at the same time lowered the foam recovery rate of BSA. Sodium phosphate solution of slightly alkaline pH (7.2–8.9) at a relatively high ionic strength (0.2–0.5 M) produced minimum turbidity with a high BSA recovery of over 90% through the foam collection line.

Figure 160 illustrates a preliminary result of the batch separation of BSA and sheep hemoglobin mixture obtained with a liquid phase composed of 0.2 M dibasic sodium phosphate solution (pH 8.9) under the standard experimental condition previously applied to the dye separation. BSA with a foam-producing capacity was quickly eluted through the foam collection line with 6 min while sheep hemoglobin was entirely recovered through the liquid collection line in about 10 min.

The preliminary studies on foam separation of proteins described above furnish a useful guidance for further development of the present method. Although simple adjustment of pH and ionic concentration of the liquid phase worked out well for separation of BSA, many other proteins lack an active foam-producing capability and therefore require the selective collectors to acquire the foam affinity. Hopefully, the use of such collectors would also prevent the protein molecules from direct contact with the gas–liquid interface, thus reducing the possibility of denaturation. One useful future application of the present method may be the separation of macromolecules or particulates with foam affinity CCC using highly specific collector molecules.

XI. SUMMARY

In this chapter a variety of existing CCC schemes have been reviewed in the approximately chronological order of their development. Development of the CCC technology during the past 20 years is summarized in Fig. 161, where a simple separation method using a rotating coil branches out to form an enormous variety of efficient CCC schemes. In the middle of the 1960s the coil planet centrifuge had first demonstrated the potential capability of CCC by separation of dyes, proteins, and cell particles in the end-closed coils. Several years later, introduction of flow-through mechanisms to the coiled column had established two basic CCC systems, both being capable of continuous elution, monitoring, and fractionation as routinely performed in liquid chromatography. The relatively simple hydrostatic equilibrium system (HSES) was quickly improved to produce several efficient CCC schemes for both analytical (helix CCC) and preparative (droplet CCC and locular CCC) separations. Development of hydrodynamic equilibrium system (HDES) has been conducted in two different directions, i.e., large preparative scale separations under a unit gravitational field and analytical to semipreparative

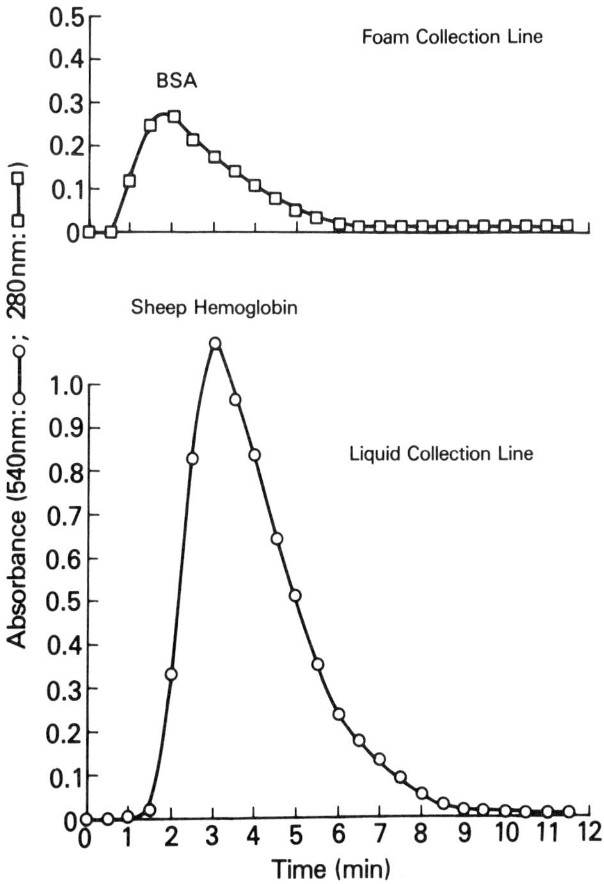

Fig. 160 Preliminary separation of proteins with the foam CCC method. BSA having a foam-producing capacity was collected through the foam collection line while sheep hemoglobin was entirely recovered through the liquid collection line.

separations utilizing a centrifugal force field. Remarkable technical progress in CCC has been brought forth by introduction of the flow-through systems to the coil planet centrifuge without the use of rotary seals. Among various types of synchronous coil planet centrifuges, the scheme IV centrifuge has proven most versatile producing a variety of efficient CCC schemes, whereas the nonsynchronous coil planet centrifuges were found to be particularly useful for partition of macromolecules and biological particulates with aqueous-aqueous polymer phase systems. Most recently, an epoch-making ad-

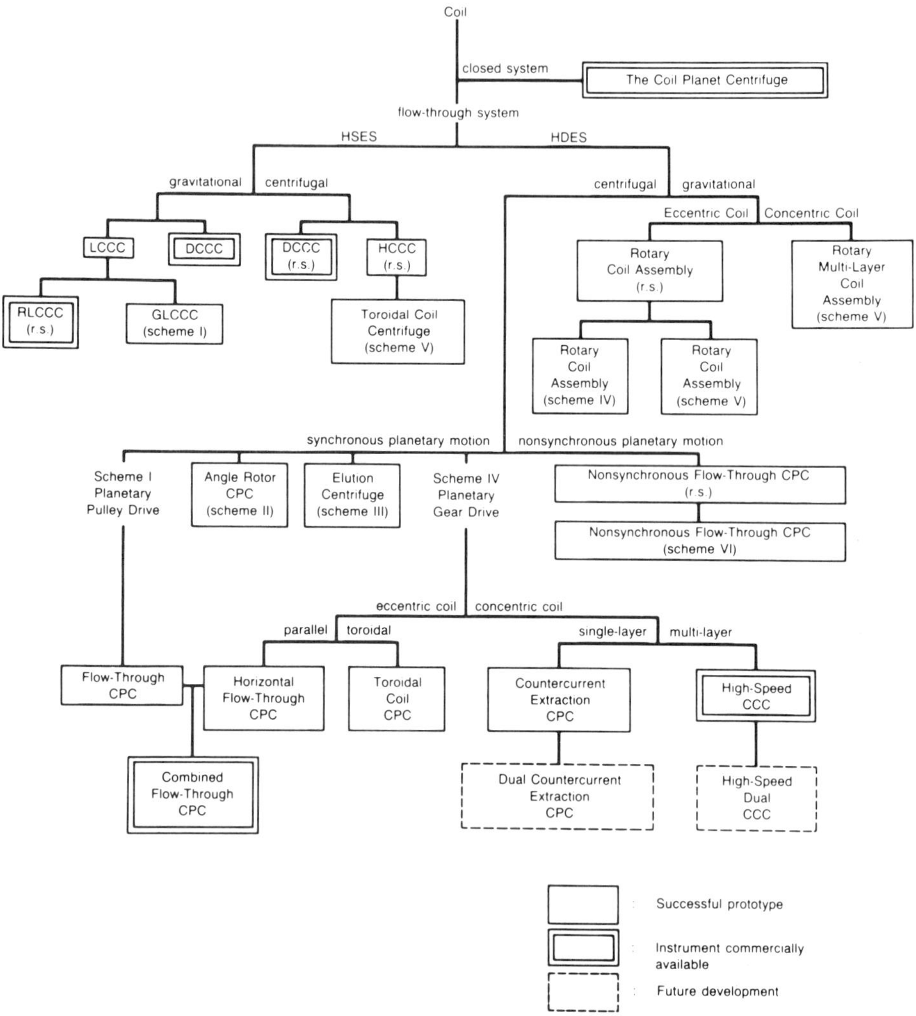

Fig. 161 Summary of the development of CCC technology over the past 20 years. HSES, hydrostatic equilibrium system; HDES, hydrodynamic equilibrium system; CCC, countercurrent chromatography; DCCC, droplet CCC; HCCC, helix CCC; LCCC, locular CCC; RLCCC, rotation LCCC; GLCCC, gyration LCCC; CPC, coil planet centrifuge; r.s., rotating seal.

vance in the CCC technology has been achieved by combination of the scheme IV planetary motion and concentric coil orientation on the holder which permitted extremely high retention of the stationary phase under a fast flow rate of the mobile phase. This high-performance CCC has been successfully applied to continuous extraction, high-speed CCC, and dual CCC with highly promising results. Innovations and progress in the CCC technology described above have been reviewed from time to time by a number of publications including review articles in technical journals [4–6,13,106–113], comprehensive book chapters [7,61,114], and a monograph [115].

Although later developed schemes generally yielded superior performance, all CCC schemes have their own specific features as summarized in Table 12. Here special emphasis should be made for some CCC schemes which possess enormous potential for future development and application. The rotating coil assembly, the most primitive form of the hydrodynamic equilibrium system (HDES), is simple and yet produces relatively rapid and efficient separations. The simplicity of the design, slow rotational motion required, compact column configuration, and the well-balanced system all favor scaling up the column capacity for large industrial-scale separations without mechanical problems. The capability of the system to allow dual countercurrent flow also suggests that the system may be automated for continuous operation to be efficiently used for purification or enrichment of isotopes, recycling of nuclear wastes, etc. The high-performance CCC schemes produce high retention of the stationary phase and efficient mixing which altogether yield high peak resolution in a short period of time. The fact that the system utilizes a rotary seal-free elution system and requires relatively low column pressure suggests that the partition efficiency of the system may be easily increased several fold by applying a longer multilayer coil mounted on a wider column holder. The system also permits dual countercurrent separation for continuous sample feeding as demonstrated in foam CCC.

Compared with other chromatographic methods, all CCC schemes share the common advantages of being free from various complications caused by the use of solid supports. Adverse adsorption effects which often result in loss and alteration of sensitive samples would not be entirely eliminated from HPLC as long as the method uses the solid support matrix providing a huge liquid–solid interface area. Because of high sample recovery and mild separation environment, CCC will provide a powerful tool for isolating a minute amount of biologically active materials present in a large volume of the sample solution. In preparative-scale separations where liquid chromatography suffers from serious loss of partition efficiency and expenditure of solvents, CCC can maintain excellent peak resolution with minimum need for the solvent. A large quantity of crude extracts or synthetic mixture can be directly loaded into the column to yield highly pure fractions which are sufficient for mass spectrometric analysis. In

Table 12 Various CCC Schemes and Their Applications

Apparatus	Preparative CCC	Analytical CCC	Partition with polymer phase systems	Other applications	Special features
Original CPC[a]		+	+	Osmotic fragility test	Closed system
Droplet CCC[b]	+				Column stationary simple
Locular CCC	+				Reliable retention of stationary phase
Slowly rotating coil	+			Continuous extraction, dual CCC	Large preparative separations
Toroidal coil centrifuge (scheme V)		+	+	Cell elutriation	
Centrifugal droplet CCC	+	+	+		High partition efficiency, rapid separation

Flow-through CPC (scheme I)	+	+		Particle separation, foam CCC	High partition efficiency
Angle rotor CPC (scheme II)		+	+		
Elution centrifuge (scheme III)		+	+		
Horizontal flow-through CPC (scheme IV)	+	+			High partition efficiency, reliable phase retention
Toroidal CPC (scheme IV)		+	+		High partition efficiency
Nonsynchronous CPC (scheme VI)		+	+	Cell elutriation	Freely adjustable coil rotation
Multilayer CPC (scheme IV)	+	+		Continuous extraction, foam CCC, dual CCC	High phase retention, high partition efficiency, rapid separation

[a] CPC, coil planet centrifuge.
[b] CCC, countercurrent chromatograph.

short, CCC has a number of desirable features which offer supplementary or alternative choice to HPLC for improving future separation methodology.

APPENDIX

A. New Nomenclature for the Synchronous Planetary Motion

As described in Sec. III, various types of synchronous planetary motions have been utilized for performing countercurrent chromatography (CCC). Previously, these rotary seal-free flow-through planetary centrifuge schemes were simply assigned numerals in the order of simplicity from scheme I to scheme IV as illustrated in Fig. 17 (Sec. III). Recently, two new CCC schemes have been introduced: one is called the improved angle rotor coil planet centrifuge [116,117] and the other the cross-axis synchronous flow-through coil planet centrifuge [118,119]. These two schemes are each based on a new member of the synchronous planetary motion which also belongs to the above family. Consequently, all these planetary motions can be sequentially arranged as illustrated in Fig. 162a, where each diagram is assigned a simple descriptive term according to the relative orientation of the holder to the main axis of the centrifuge.

The I-L-J series (left column) is formed by tilting the holder of type I (top) toward the central axis of the centrifuge while the new series I-X-J (right column) is made by rotating the holder sideways so that the distance between the two axes always remains the same. The hybrid type J-L had been applied to the improved angle rotor countercurrent chromatograph by setting the column angle at 25° against the centrifuge axis [116,117]. The type X planetary motion was utilized in the cross-axis synchronous flow-through coil planet centrifuge for large-scale preparative separations [118,119].

Similar nomenclature may also be applied to a series of the nonsynchronous planetary motions (Fig. 162b).

B. General Formulas for Computing Acceleration for I-L-J Series of Synchronous Planetary Motions

In order to facilitate comparative studies, efforts have been made to derive a set of generalized formulas which can be used to compute acceleration generated by all types of synchronous planetary motions in the I-L-J series. On the basis of the data thus obtained, the centrifugal force field of all these planetary motions can be expressed in the same format for comparison.

We consider a discoid body with radius r and its axis being held at angle Ψ from the central axis of the stationary frame (Fig. 163).

The discoid body undergoes a synchronous planetary motion in such a way that it revolves around the central axis at an angular velocity ω with a revolutional radius R and simultaneously rotates about its own axis at the same angular velocity. Then, we wish to study the motion of an arbitrary point present on the periphery of the discoid body.

Figure 164a shows an x-y-z coordinate system where the center of the discoid body initially locates at point $Q_0(R,0,0)$ and the arbitrary point at $P_0(R-r\cos\Psi, 0, r\sin\Psi)$. The center of the discoid body circles around point O in the x-y plane and at time t moves angle $\theta = \omega t$ to reach point $Q(R\cos\theta, R\sin\theta, 0)$. By the aid of a diagram illustrated in Fig. 164b, the location of the arbitrary point on the discoid body is expressed by P(x,y,z) according to the following equations:

$$x = R\cos\theta + r[1 - (1 + \cos\Psi)\cos^2\theta] \tag{37}$$

$$y = R\sin\theta - r(1 + \cos\Psi)\sin\theta\cos\theta \tag{38}$$

$$z = r\sin\Psi\cos\theta \tag{39}$$

where R is the distance $\overline{OQ}$; r, the distance $\overline{QP}$; and Ψ, the angle formed between the z-axis and the body axis $\overline{QA}$.

The acceleration α produced by the planetary motion is then computed from the second derivatives of the above equations as

$$\alpha_x = \frac{d^2x}{dt^2} = -R\omega^2[\cos\theta - 2\beta(1 + \cos\Psi)\cos 2\theta] \tag{40}$$

$$\alpha_y = \frac{d^2y}{dt^2} = -R\omega^2[\sin\theta - 2\beta(1 + \cos\Psi)\sin 2\theta] \tag{41}$$

$$\alpha_z = \frac{d^2z}{dt^2} = -r\omega^2\sin\Psi\cos\theta \tag{42}$$

where $\beta = r/R$ and $R \neq 0$. Using Eqs. (40)–(42), the x, y, z components of the acceleration vector can be computed and displayed each separately or in a combined form in the x-y-z coordinate system of the stationary reference frame.

However, the effects of the acceleration on the subjects moving with the discoid body can be more effectively visualized if the acceleration is expressed with respect to the revolving body frame. Figure 165 shows the relationship between the original x-y-z coordinate system and the x_b-y_b-z_b body coordinate system in which the z_b-axis coincides with the axis of the discoid body. The

(a)

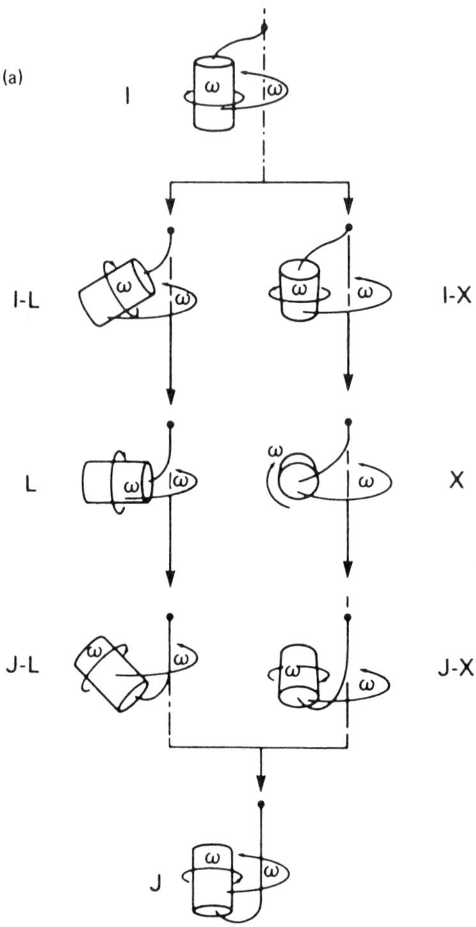

Fig. 162 New nomenclature for the flow-through centrifuge schemes free of rotary seals. (a) Synchronous schemes. (b) Synchronous, nonplanetary, and nonsynchronous schemes.

x_b-y_b-z_b coordinate system revolves around the z-axis of the original coordinate system in such a way that the radially directed x_b-axis forms angle Ψ against the x-y plane while the y_b-axis is always confined in the same plane. Consequently, the discoid body coaxially rotates around the z_b-axis at angular velocity ω in the x_b-y_b plane.

Transformation of the acceleration from the original x-y-z coordinate system to the x_b-y_b-z_b body coordinate system can be performed by the following equations:

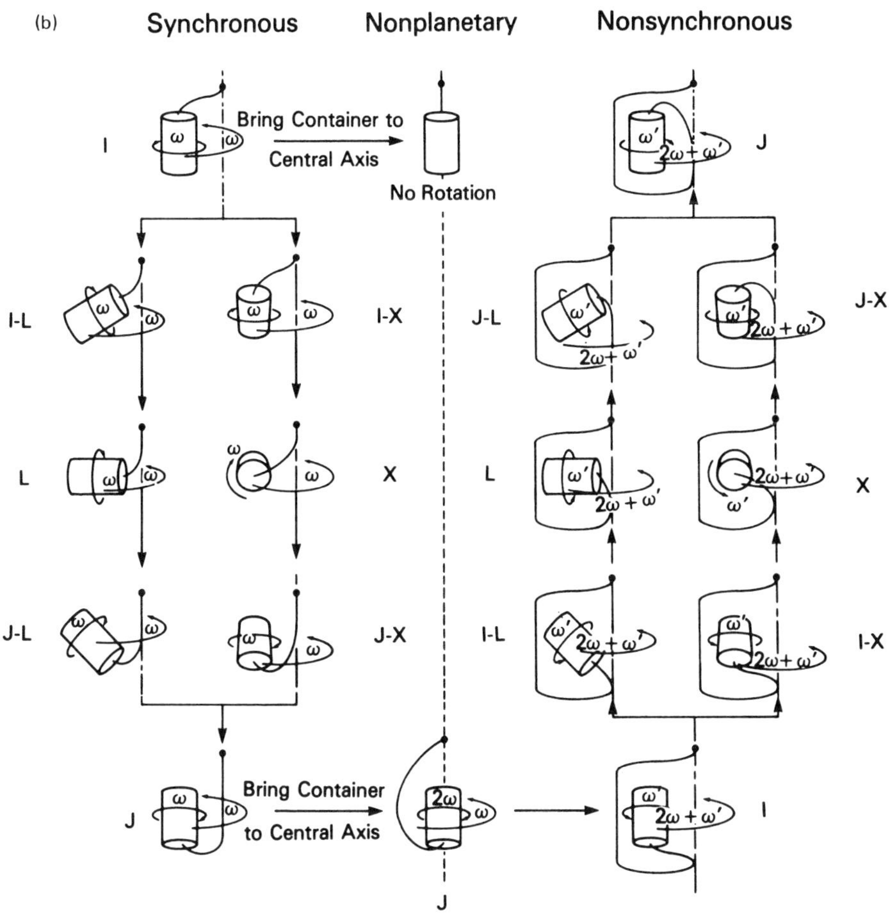

Fig. 162 (Continued)

$$\alpha_{x_b} = (\alpha_x \cos\theta + \alpha_y \sin\theta)\cos\Psi - \alpha_z \sin\Psi \tag{43}$$

$$\alpha_{y_b} = -\alpha_x \sin\theta + \alpha_y \cos\theta \tag{44}$$

$$\alpha_{z_b} = (\alpha_x \cos\theta + \alpha_y \sin\theta)\sin\Psi + \alpha_z \cos\Psi \tag{45}$$

which becomes

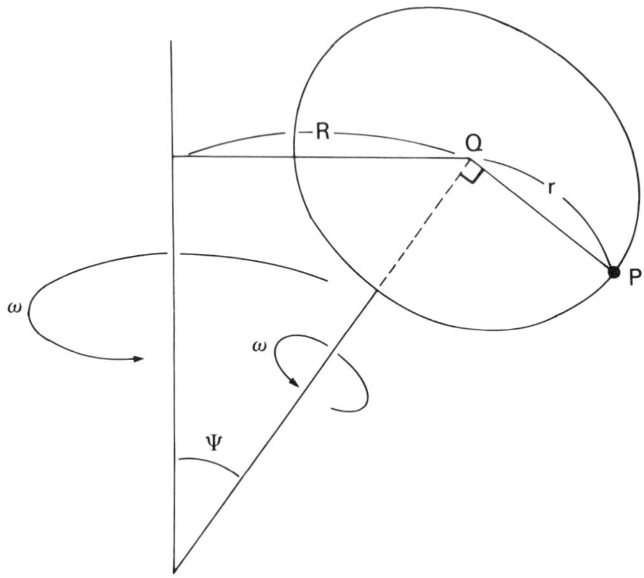

Fig. 163 Synchronous planetary motion of the discoid body.

$$\alpha_{x_b} = - R\omega^2[\cos \Psi - \beta(1 + \cos \Psi)^2 \cos \theta] \quad (46)$$

$$\alpha_{y_b} = R\omega^2(1 + \cos \Psi)2\beta \sin \theta \quad (47)$$

$$\alpha_{z_b} = - R\omega^2 \sin \Psi[1 - \beta(2 + \cos \Psi)\cos \theta] \quad (48)$$

For convenience of the diagrammatical expression of acceleration vectors, the first two components α_{x_b} and α_{y_b} are combined into a single vector and expressed as an arrow in the x_b-y_b plane, while the third vector α_{z_b} is separately expressed as a column. Figure 166 illustrates distribution of centrifugal force vectors (the same magnitude as the acceleration but acting in the opposite direction) produced by the planetary motion at $\Psi = 25°$.

In the diagram the center of the discoid body coincides with the center of the x_b-y_b body coordinate system where five concentric circles indicate location of arbitrary points at the labeled β-values ($\beta = r/R$) of 0.1, 0.25, 0.5, 0.75, and 1.0 from the center ($\beta = 0$) toward the periphery. On each circle, the centrifugal force vectors are drawn at every 30° interval as two components. As mentioned earlier, the force component acting in the x_b-y_b plane is indicated by an arrow while the second component ($-\alpha_{z_b}$) acting perpendicularly to the x_b-y_b plane is expressed as a column in which the force pointing above the plane is shown by an ascending column and the force pointing below the plane by a descending column. Although the present diagram (Fig. 166) illustrates the centrifugal force field acting at a given moment, successive change of the force vector at a given point during revolution can be easily visualized by either moving the point counterclockwise against the force field or rotating the force field clockwise with respect to the point.

Equations (46)–(48) derived from the above analysis can be universally applied for computing the acceleration produced from all types of I-L-J synchronous planetary motions by substituting the corresponding angles for Ψ, e.g., 180° for type I (scheme I), 90° for type L (scheme III), and 0° for type J (scheme IV). Then distribution of the centrifugal force vectors can be computed for each scheme and expressed in the same format as illustrated in Fig. 166 for $\Psi = 25°$.

Figure 167 illustrates a set of the centrifugal force distribution diagrams for five types of the synchronous planetary motion. In the first column (left), the new descriptive terms are introduced for each type of the synchronous planetary motion to replace the old terms indicated in parentheses. The second column gives angle Ψ characteristic of each type of the synchronous planetary motion. In the two hybrid types, type I-L and type J-L, Ψ is chosen at 135° and 45°, respectively, to facilitate comparative studies.

The third column in Fig. 167 illustrates orientation and planetary motion of the discoid body. It also indicates orientation of the x_b-y_b-z_b coordinate system which serves as a reference frame for the force distribution diagram shown in the next column. The orientation of the x_b-y_b-z_b body coordinate system is selected in such a way that the z_b force component acting upward from the discoid body is always indicated as an ascending column in the force distribution diagram except for the transitional angle of $\Psi = 90°$ where the z_b force component acting radially from the center of revolution toward the periphery (right) is expressed as the ascending column. Accordingly, the orientation of the reference frame is corrected by rotating the x_b-y_b-z_b coordinate frame around the y_b-axis by 180° for $\Psi > 90°$. This transposition also causes conversion of the +/– signs for the x_b force component to normalize the direction of the x_b/y_b force vector expressed as an arrow in the force distribution diagram.

In the fourth and the final column in Fig. 167 (right), the distribution of the centrifugal force vectors is displayed for various Ψ-values in the order from 180° (top) to 0° (bottom) using the reference body coordinate frame indicated in the third column. As described earlier, all these force distribution diagrams are computed

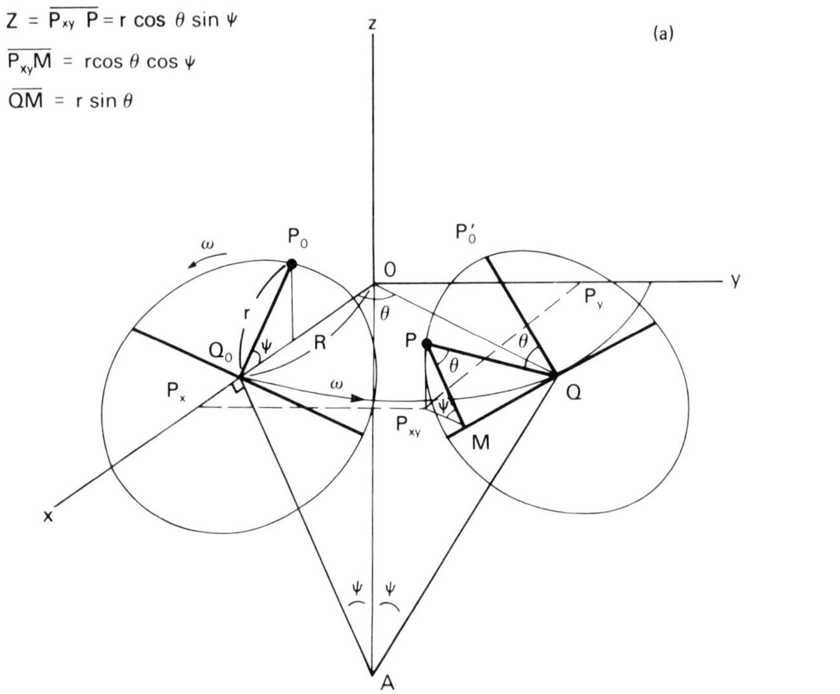

Fig. 164 Coordinate systems for analysis of acceleration. (a) x-y-z coordinate system. (b) x-y coordinate system for assisting the analysis.

from a set of Eqs. (46)–(48) by inserting the respective Ψ-values indicated in the second column. In order to facilitate comparison, the length of the arrows and the columns which represent the magnitude of the force vector is normalized on the basis of the unit ($R\omega^2$) used in the first diagram (Ψ = 180°), so that the lengths of vectors drawn in these diagrams accurately represent the relative magnitude of the force generated by the respective planetary motion at the same revolutional radius (R) and angular velocity (ω).

Observation of the force distribution diagrams from the top through the bottom reveals successive changes in distribution patterns of the first force component (arrows) while distribution of the second force component acting along the z_b-axis (columns) shows disruption of the continuity between Ψ = 135° and 90° due to the transposition of the x_b-y_b-z_b reference frame as described earlier. At Ψ = 180° the diagram shows homogeneous two-dimensional distribution of the centrifugal force vectors each measuring a unit length and acting in

$$x = \overline{OP_x} = \overline{OQ_x} + \overline{Q_xM_x} - \overline{P_xM_x}$$
$$\overline{OQ_x} = R\cos\theta$$
$$\overline{Q_xM_x} = \overline{QM}\sin\theta = r\sin^2\theta$$
$$\overline{P_xM_x} = \overline{P_{xy}M}\cos\theta = r\cos^2\theta\cos\Psi$$

$$y = \overline{OP_y} = \overline{OQ_y} - \overline{M_yQ_y} - \overline{P_yM_y}$$
$$\overline{OQ_y} = R\sin\theta$$
$$\overline{M_yQ_y} = \overline{QM}\cos\theta = r\sin\theta\cos\theta$$
$$\overline{P_yM_y} = \overline{P_{xy}M}\sin\theta = r\cos\theta\sin\theta\cos\Psi$$

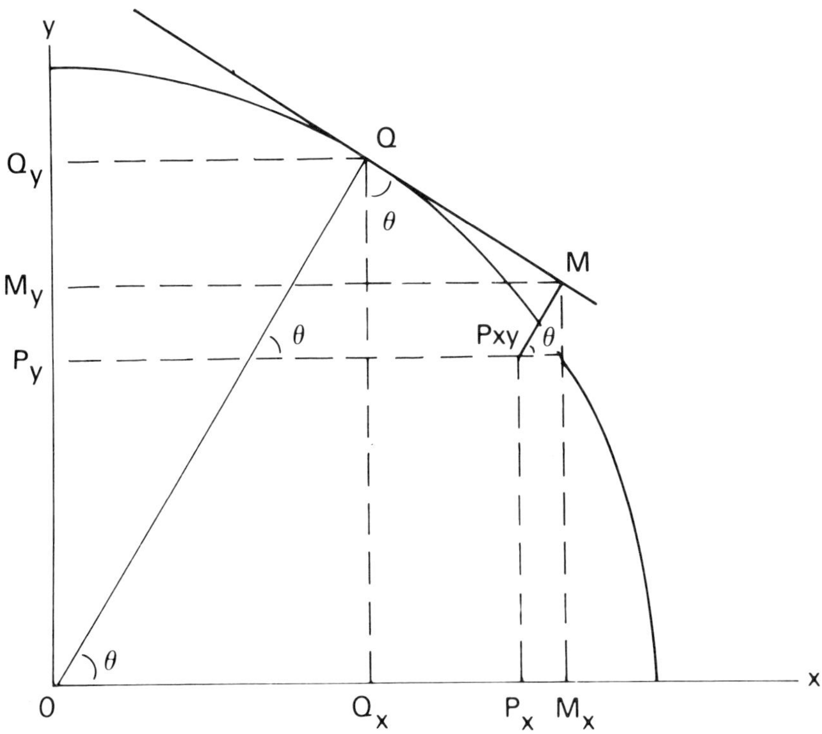

(b)

Fig. 164 (Continued)

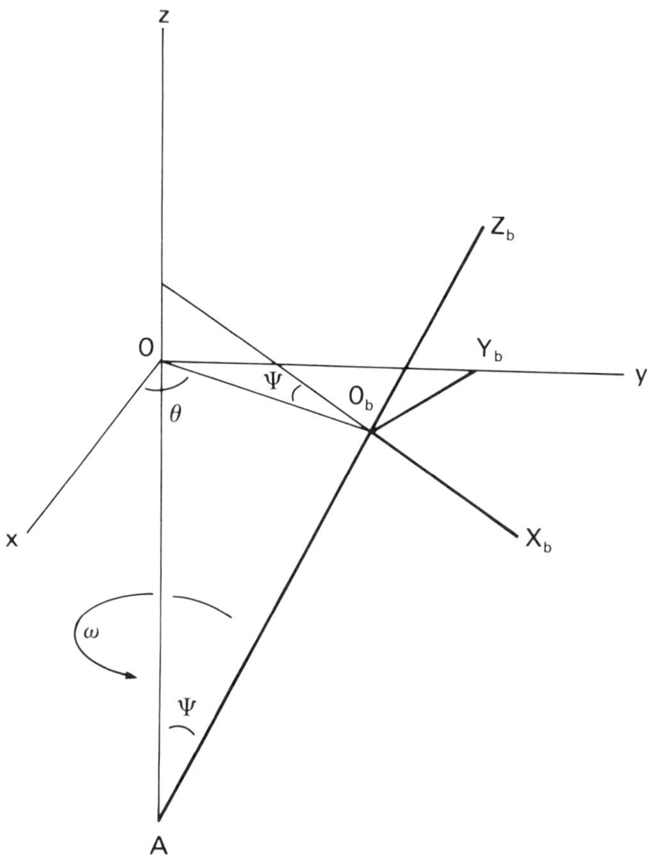

Fig. 165 The x_b-y_b-z_b coordinate system and its spacial relation to the original x-y-z coordinate system.

the same direction parallel to the radius of revolution. At $\Psi = 135°$ the force distribution becomes three-dimensional: the first component (arrow) shows somewhat diverged distribution with slight decrease in magnitude while the second component (column) acts strongly downward on the left and slightly upward on the right. When Ψ reaches the transitional angle of 90°, the first force component (arrow) forms symmetrical distribution always acting outwardly from the circle while the second force component (column) acts very strongly upward (or toward the periphery along the radius of revolution) on the right. Here it is interesting to note that the second force component also acts downward on the left indicating that the centrifugal force is momentarily directed toward the

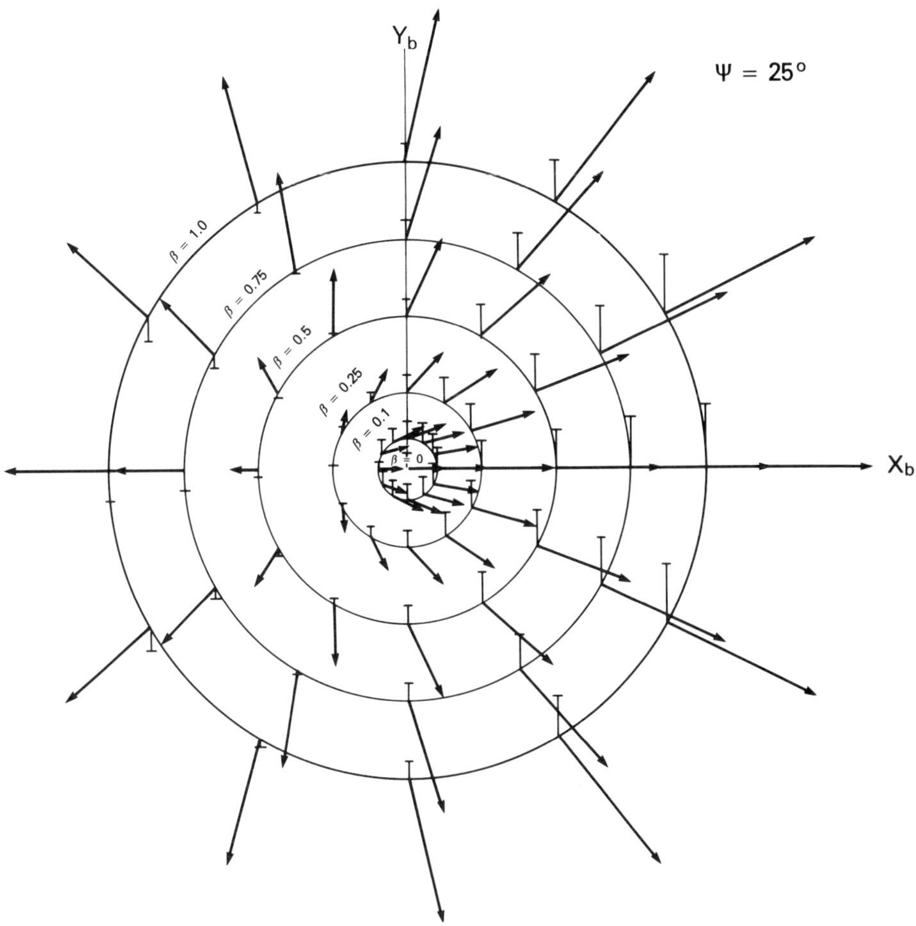

Fig. 166 Distribution of the centrifugal force vectors produced by the new synchronous planetary motion at $\Psi = 25°$.

center of revolution at the very top of the discoid body. This strange phenomenon, which has been reported elsewhere [21], is explained on the basis of the Coriolis force produced by the planetary motion. As Ψ is further decreased to 45°, the first force component (arrow) remarkably gains the strength especially on the right, while the second component (column) maintains the similar profile. Finally, at $\Psi = 0°$ the distribution of the force vectors returns to the two-dimensional pattern as in the first diagram ($\Psi = 180°$) but it consists of a complex arrangement of force vectors

New Nomenclature of Synchronous Planetary Motion (Old Terms)	Angle ψ	Planetary Motion and Reference Coordinate Frame	Force Distribution Diagram
Type I (Scheme I)	180°	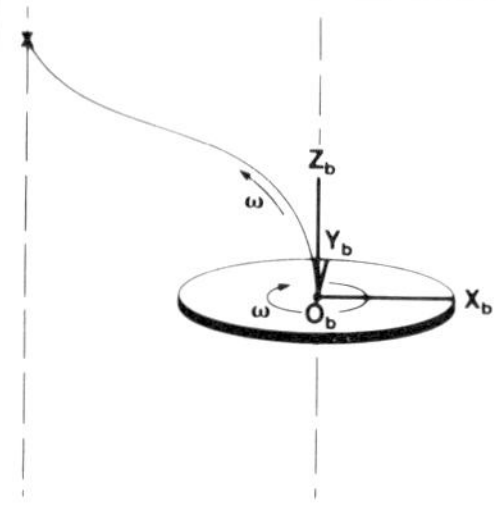	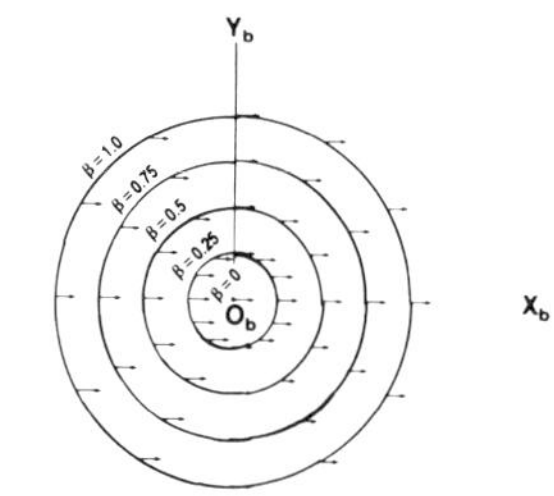
Type I-L (Scheme II)	135°	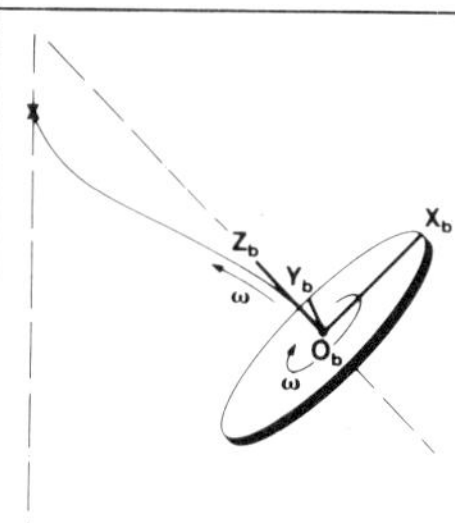	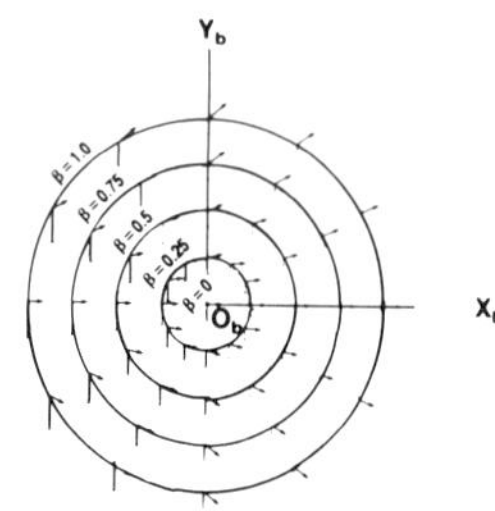
Type L (Scheme III)	90°	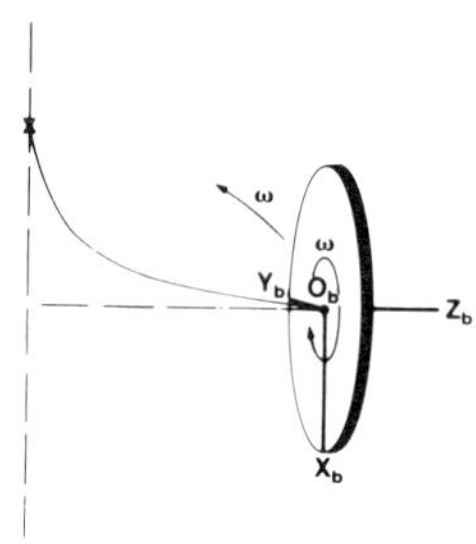	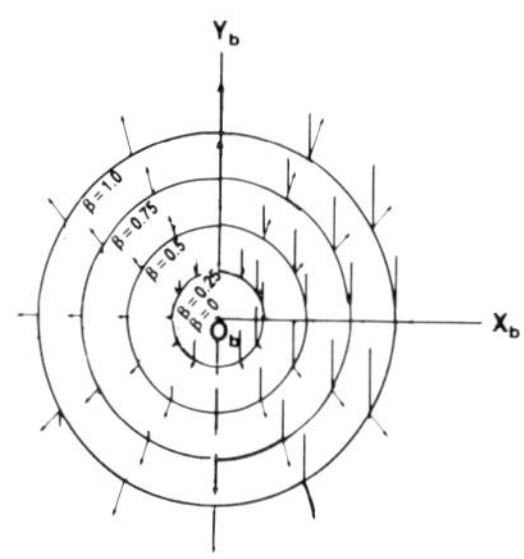

Fig. 167 A set of the centrifugal force distribution diagrams for five types of synchronous planetary motion.

diverging from each circle with a great enhancement of the magnitude especially at the remote location from the center of the discoid body.

C. Analysis of Acceleration for the Type X Synchronous Planetary Motion

The mathematical analysis of acceleration produced by the Type X synchronous planetary motion (Fig. 162a) is performed in the follow-

Type J-L	45°	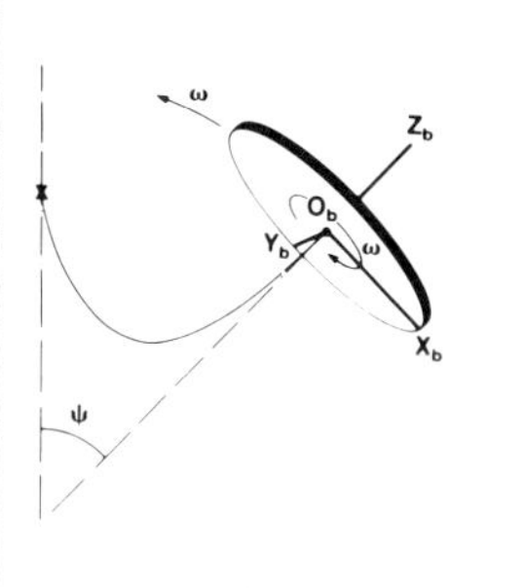	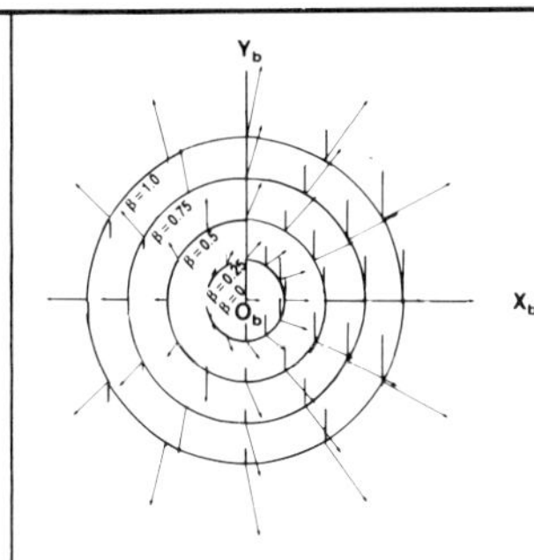
Type J (Scheme IV)	0°	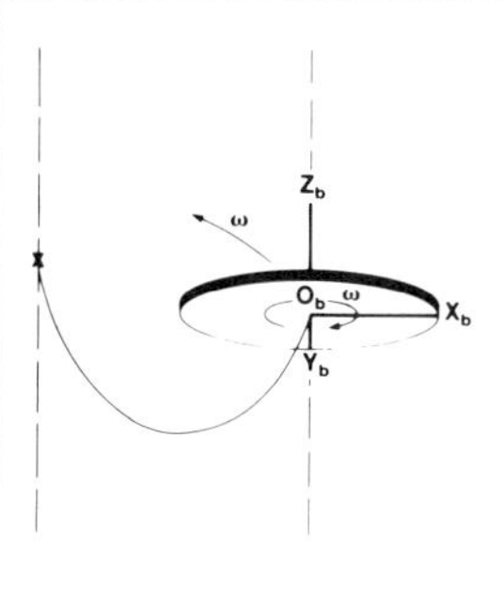	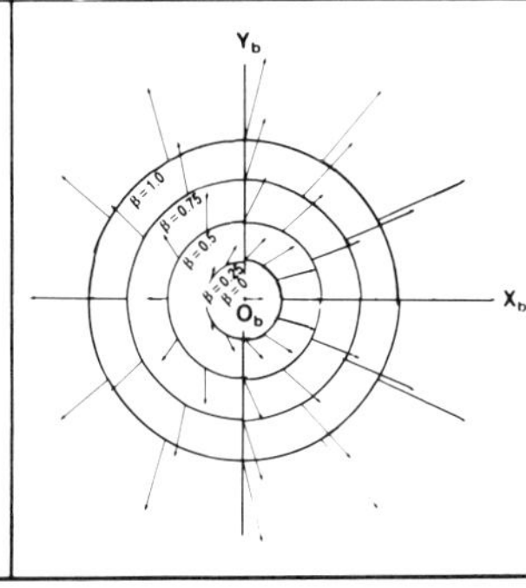

Fig. 167 (Continued)

ing two steps. The first analysis is performed on an arbitrary point located in the central plane of the holder across the holder axis. In this particular case, the arbitrary point, the radius of revolution, and the central axis of the centrifuge are all confined in the same plane. In the second analysis, the arbitrary point can be chosen at any location on the holder, thus yielding generalization formulas for computing acceleration.

Acceleration Acting on the Central Plane of the Holder

We consider a circular body with radius r and undergoing a planetary motion in such a way that it revolves around the central axis of the centrifuge ($\overline{AB}$) and simultaneously rotates about its own axis (Q) at the same angular velocity (ω) as illustrated in Fig. 168. In so doing, the circular body always maintains its axis perpendicular to and at the same distance ($\overline{OQ}$ = R) away from the central axis. Then, we wish to study the motion of an arbitrary point (P) located at the periphery of the circular body ($\overline{QP}$ = r).

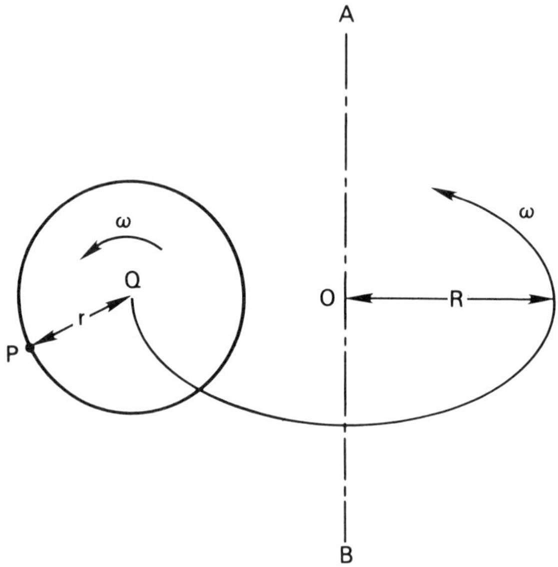

Fig. 168 Orientation and motion of the circular body for analysis of acceleration.

Figure 169 shows an x-y-z coordinate system where the z-axis coincides with the central axis and the center of the circular body (Q) circles around point O in the x-y plane. The arbitrary point on the circular body and the center of the circular body are initially located on the x-axis at P_0 and Q_0, respectively. The circular body undergoes a synchronous planetary motion as described above, i.e., it revolves around the z-axis and rotates about its own axis at the same angular velocity ω. Consequently, after a lapse of time t, the arbitrary point moves to point P and the center of the circular body to point Q. Then the location of point P (x, y, z) is expressed in the following equations:

$$x = R \cos \theta - r \cos^2 \theta \tag{49}$$

$$y = R \sin \theta - r \sin \theta \cos \theta \tag{50}$$

$$z = r \sin \theta \tag{51}$$

where $R = \overline{OQ}$, $r = \overline{QP}$, and $\theta = \omega t$.

The acceleration acting on the arbitrary point is then given by the second derivatives of these equations and

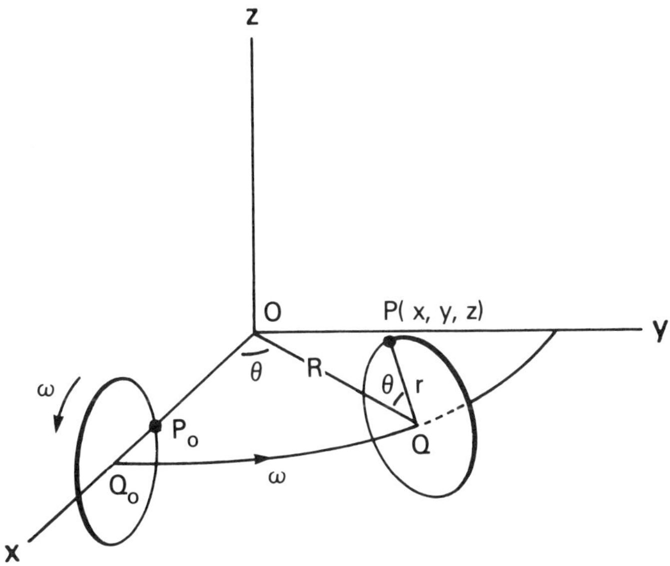

Fig. 169 The x-y-z coordinate system for analysis of acceleration acting on the circular body.

$$\frac{d^2x}{dt^2} = -R\omega^2 \cos\theta + 2r\omega^2 \cos 2\theta = -R\omega^2(\cos\theta - 2\beta \cos 2\theta) \tag{52}$$

$$\frac{d^2y}{dt^2} = -R\omega^2 \sin\theta + 2r\omega^2 \sin 2\theta = -R\omega^2(\sin\theta - 2\beta \sin 2\theta) \tag{53}$$

$$\frac{d^2z}{dt^2} = -r\omega^2 \sin\theta = -R\omega^2 \beta \sin\theta \tag{54}$$

where $\beta = r/R$.

In order to visualize the effects of acceleration on the objects rotating with the circular body, it is more convenient to express the acceleration vectors with respect to the body frame or the X_b-Y_b-Z_b coordinate system illustrated in Fig. 170. Transformation of the vectors from the original x-y-z coordinate system to the X_b-Y_b-Z_b coordinate system may be performed according to the following equations:

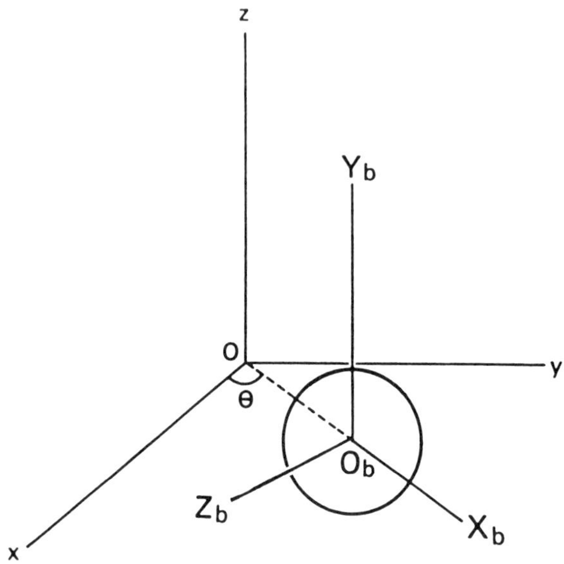

Fig. 170 Relationship between the reference x-y-z coordinate system and the X_b-Y_b-Z_b body coordinate system.

$$\alpha_{X_b} = \left(\frac{d^2x}{dt^2}\right)\cos\theta + \left(\frac{d^2y}{dt^2}\right)\sin\theta = -R\omega^2(1 - 2\beta\cos\theta) \qquad (55)$$

$$\alpha_{Y_b} = \frac{d^2z}{dt^2} = -R\omega^2\beta\sin\theta \qquad (56)$$

$$\alpha_{Z_b} = \left(\frac{d^2x}{dt^2}\right)\sin\theta - \left(\frac{d^2y}{dt^2}\right)\cos\theta = -R\omega^2 2\beta\sin\theta \qquad (57)$$

where αX_b, αY_b, and αZ_b indicate the acceleration vectors acting along the corresponding coordinate axes.

From Eqs. (55)–(57), the centrifugal force vectors (same magnitude with the acceleration but acting in the opposite direction) at various points on the circular body are computed and diagrammatically illustrated in Fig. 171. In order to illustrate three-dimensional pattern of the centrifugal force vectors in a two-dimensional diagram, two force vectors $-\alpha X_b$ and $-\alpha Y_b$ are combined into a single arrow forming various angles from the X_b-axis, whereas the third force vector $-\alpha Z_b$, which acts perpendicular to the X_b-Y_b plane, is drawn

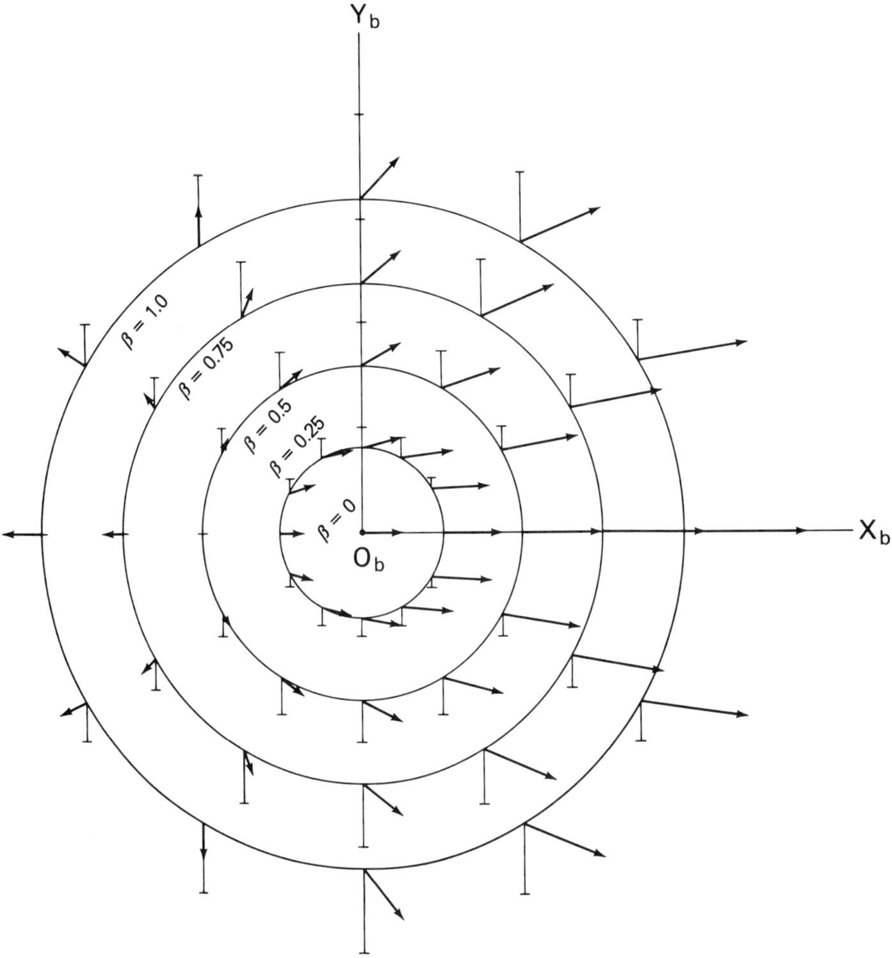

Fig. 171 Force distribution diagram for the cross-axis synchronous flow-through coil planet centrifuge (type X).

as vertical columns along the Y_b-axis. The ascending column indicates the force acting upward ($Z_b > 0$) and the descending column, the force acting downward ($Z_b < 0$). Several concentric circles around point O_b (the center of the circular body) indicate the location on the circular body corresponding to various β-values labeled in the diagram.

While the force distribution diagram thus produced (Fig. 171) indicates the distribution of centrifugal force vectors acting on the circular body at any given moment, an arbitrary point on the circular body is rotating clockwise around the center of the body coordinate system (O_b) at a uniform angular velocity equal to revolution. Consequently, during one revolutional cycle, the arbitrary point is subjected to a series of centrifugal force vectors indicated by the force distribution diagram.

As illustrated in Fig. 171, the force distribution diagram for the cross-axis synchronous coil planet centrifuge reveals complex three-dimensional fluctuation of the centrifugal force field during each revolutional cycle where both magnitude and direction of force vectors change with β or the location of the point on the circular body. The X_b-Y_b component (expressed as arrows) shows a mirror image between the upper and the lower halves of the circular body but it displays a great asymmetry along the X_b-axis. At small β or the locations close to the center of the circular body, all force vectors are directed toward the right, indicating that the force rotates around each point as the circular body revolves. When β becomes greater than 0.5, however, the force vectors are always directed outwardly from the circle indicating that the rotation of the force around each point changes into the oscillatory motion. On the other hand, the Z_b-component (expressed as columns) displays perfect symmetry around the center of the circular body while it gradually increases both the magnitude and amplitude of oscillation as β becomes greater.

Acceleration Acting on the Arbitrary Point on the Holder

Figure 172 shows a x-y-z coordinate system where a cylindrical body (column holder) revolves around the z-axis and synchronously rotates about its own axis positioned in the x-y plane. The cylindrical body contains the circular body drawn as a thin circle at the middle portion together with points P_o and P, which were subjected to the preceding analysis. An arbitrary point, for the present analysis, is located at distance ℓ away from point P along the axis of the cylindrical body. As shown in Fig. 172, the arbitrary point initially locates at P'_o (R-r, ℓ, 0) and after a lapse of t moves to point P' (x', y', z') which is expressed in the following equations:

$$x' = R \cos \theta - r \cos^2 \theta - \ell \sin \theta \tag{58}$$

$$y' = R \sin \theta - r \sin \theta \cos \theta + \ell \cos \theta \tag{59}$$

$$z' = r \sin \theta \tag{60}$$

The acceleration acting on point P' is similarly obtained from the second derivatives of the above equations and

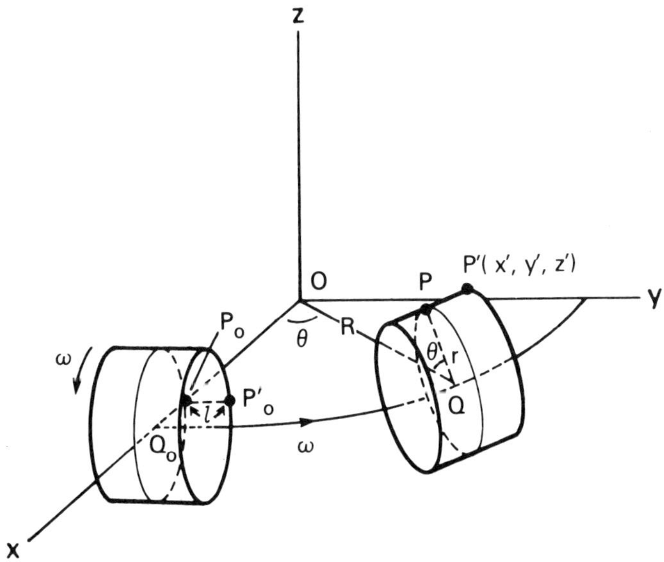

Fig. 172 The x-y-z coordinate system for analysis of acceleration acting on the cylindrical body.

$$\frac{d^2x'}{dt^2} = - R\omega^2(\cos\theta - 2\beta\cos 2\theta) + \ell\omega^2\sin\theta \qquad (61)$$

$$\frac{d^2y'}{dt^2} = - R\omega^2(\sin\theta - 2\beta\sin 2\theta) - \ell\omega^2\cos\theta \qquad (62)$$

$$\frac{d^2z'}{dt^2} = - R\omega^2\beta\sin\theta \qquad (63)$$

where $\beta = r/R$.

According to the procedure used in the preceding analysis, these acceleration vectors expressed in reference to the original x-y-z coordinate system are further transformed into the X_b-Y_b-Z_b body coordinate system (see Fig. 170) by the following equations:

$$\alpha'_{X_b} = \left(\frac{d^2x'}{dt^2}\right)\cos\theta + \left(\frac{d^2y'}{dt^2}\right)\sin\theta = - R\omega^2(1 - 2\beta\cos\theta) \qquad (64)$$

$$\alpha'_{Y_b} = \frac{d^2z'}{dt^2} = -R\omega^2\beta \sin\theta \tag{65}$$

$$\alpha'_{Z_b} = \left(\frac{d^2x'}{dt^2}\right)\sin\theta - \left(\frac{d^2y'}{dt^2}\right)\cos\theta = -R\omega^2 2\beta \sin\theta + \ell\omega^2 \tag{66}$$

Comparison between the two sets of Eqs. (55)–(57) and (64)–(66) above clearly shows that deviation of the point from the central plane of the holder has no effect on the acceleration components acting in the X_b-Y_b plane. The above effect is only found on the acceleration vectors acting along the Z_b-axis where the magnitude of all vectors is uniformly affected by $-\ell\omega^2$ regardless of the values for R and r. Consequently, deviation of the point along the holder axis causes asymmetric distribution of the laterally acting force field in such a way that the degree of asymmetry increases as r becomes smaller.

Equation (66) may be written as

$$\alpha'_{Z_b} = -\omega^2(2r \sin\theta - \ell) \tag{67}$$

indicating that when $\ell \geqslant 2r$, $\alpha'_{Z_b} \geqslant 0$, that is, the direction of the acceleration (or the centrifugal force) vector acting along the Z_b-axis becomes unilateral.

REFERENCES

1. Y. Ito, M. A. Weinstein, I. Aoki, R. Harada, E. Kimura, and K. Nunogaki, *Nature, 212,* 985 (1966).
2. Y. Ito, I. Aoki, and E. Kimura, *Anal. Chem., 41,* 1579 (1969).
3. P. Å. Albertsson, *Partition of Cell Particles and Macromolecules,* Almqvist and Wiksell, Stockholm, and John Wiley and Sons, New York, 1960.
4. Y. Ito, R. E. Hurst, R. L. Bowman, and E. K. Achter, *Sep. Purif. Methods, 3,* No. 1, 133 (1974).
5. Y. Ito, *J. Biochem. Biophys. Methods, 5,* 105 (1981).
6. Y. Ito and W. D. Conway, *Anal. Chem., 56,* 534A (1984).
7. Y. Ito, in *Advances in Chromatography,* Vol. 24, Marcel Dekker, New York, 1984, p. 181.
8. Y. Ito, *J. Chromatogr., 207,* 161 (1981).
9. Y. Ito, J. Sandlin and W. G. Bowers, *J. Chromatogr., 244,* 247 (1982).
10. Y. Ito, *J. Liq. Chromatogr., 8,* 2131 (1985).
11. Y. Ito and R. L. Bowman, *J. Chromatogr. Sci., 8,* 315 (1970).

12. T. Tanimura, J. J. Pisano, Y. Ito, and R. L. Bowman, *Science*, *169*, 54 (1970).
13. Y. Ito and R. L. Bowman, *Anal. Chem.*, *43*, 69A (1971).
14. Y. Ito and R. L. Bowman, *Science*, *167*, 281 (1970).
15. Y. Ito and R. L. Bowman, *Anal. Biochem.*, *78*, 506 (1977).
16. Y. Ito and R. L. Bowman, *J. Chromatogr.*, *136*, 189 (1977).
17. Y. Ito and R. Bhatnagar, *J. Chromatogr.*, *207*, 171 (1981).
18. Y. Ito and R. L. Bowman, *Science*, *173*, 420 (1971).
19. Y. Ito and R. L. Bowman, *J. Chromatogr. Sci.*, *11*, 284 (1973).
20. Y. Ito and R. L. Bowman, *Anal. Biochem.*, *65*, 310 (1975).
21. Y. Ito, R. L. Bowman, and F. W. Noble, *Anal. Biochem.*, *49*, 1 (1972).
22. Y. Ito, *Anal. Biochem.*, *100*, 271 (1979).
23. Y. Ito, *J. Chromatogr.*, *188*, 43 (1980).
24. Y. Ito, *Anal. Biochem.*, *102*, 150 (1980).
25. Y. Ito, *J. Chromatogr.*, *192*, 75 (1980).
26. Y. Ito and R. L. Bowman, *Anal. Biochem.*, *82*, 63 (1977).
27. Y. Ito and R. L. Bowman, *J. Chromatogr.*, *147*, 221 (1978).
28. Y. Ito, *J. Chromatogr.*, *214*, 122 (1981).
29. H. Nakazawa, C. E. Riggs, Jr., M. J. Egorin, S. M. Redwood, N. R. Bachur, R. Bhatnagar, and Y. Ito, *J. Chromatogr.*, *307*, 323 (1984).
30. Y. Ito, *J. Chromatogr.*, *196*, 295 (1980).
31. Y. Ito, J. Suaudeau, and R. L. Bowman, *Science*, *189*, 999 (1975).
32. J. Suaudeau, T. Kolobow, R. Vaillancourt, A. Carvalho, Y. Ito, and A. J. Erdman III, *Transfusion*, *18*, No. 3, 312 (1978).
33. Y. Ito and R. L. Bowman, *Anal. Biochem.*, *85*, 614 (1978).
34. I. A. Sutherland and Y. Ito, *J. High Res. Chromatogr. Chromatogr. Com.*, *10019*, 171 (1978).
35. I. A. Sutherland, D. Heywood-Waddington, and T. J. Peters, *J. Liq. Chromatogr.*, *7*, No. 2, 363 (1984).
36. S. D. Flanagan, G. Johansson, B. Yost, Y. Ito, and I. A. Sutherland, *J. Liq. Chromatogr.*, *7*, No. 2, 385 (1984).
37. Y. Ito, G. T. Bramblett, R. Bhatnagar, M. Huberman, L. Leive, L. M. Cullinane, and W. Groves, *Sep. Sci. Tech.*, *18*, No. 1, 33 (1983).
38. Y. Ito, P. Carmeci, R. Bhatnagar, S. Leighton, and R. Seldon, *Sep. Sci. Tech.*, *15*, 1589 (1980).
39. Y. Ito, *J. Chromatogr.*, *188*, 33 (1980).
40. Y. Ito and G. J. Putterman, *J. Chromatogr.*, *193*, 37 (1980).
41. A. I. M. Keulemans, in *Gas Chromatography* (C. G. Verver, ed.), Reinhold, New York, 1957, p. 113.
42. W. Murayama, T. Kobayashi, Y. Kosuge, H. Yano, Y. Nunogaki, and K. Nunogaki, *J. Chromatogr.*, *239*, 643 (1982).

43. T. Kolobow, Y. Ito, I. Mychkovsky, P. Peters, and J. Morabito, *J. Liq. Chromatogr.*, *8*, 2173 (1985).
44. E. H. Wrench, U.S. Patent 3,358,072, 1967.
45. D. A. Adams, U.S. Patent 3,586,413, 1971.
46. Y. Ito, U.S. Patent 4,425,112, 1984.
47. Y. Ito, unpublished experiments.
48. T. Kolobow, J. L. Sandlin, and Y. Ito, unpublished data.
49. R. E. Hurst and Y. Ito, *Clin. Chem.*, *18*, 814 (1972).
50. J. W. Lightbown, P. Newland, I. A. Sutherland, and J. W. A. Dymond, *Proc. Analyst. Div. Chem. Soc.*, *14*, 34 (1977).
51. I. A. Sutherland, J. S. Lee, and D. J. Gauvreau, *Anal. Biochem.*, *89*, 213 (1978).
52. Y. Ito and R. L. Bowman, *Sep. Sci.*, *11*, 201 (1976).
53. Y. Ito and R. L. Bowman, *Anal. Biochem.*, *61*, 288 (1974).
54. Y. Ito and R. L. Bowman, *Science*, *182*, 391 (1973).
55. W. Hausmann, J. R. Weisiger, and L. C. Craig, *J. Am. Chem. Soc.*, *77*, 723 (1955).
56. J. F. Tinker and G. B. Bown, *J. Biol. Chem.*, *173*, 585 (1948).
57. E. J. Harfenist and L. C. Craig, *J. Am. Chem. Soc.*, *74*, 3083 (1952).
58. G. Jonansson, A. Hartman and P. Å. Albertsson, *Eur. J. Biochem.*, *33*, 379 (1973).
59. D. Metcalf and M. A. S. Moor, in *Haemopoietic Cells* (A. Neuberger and E. L. Tatum, eds.), Elsevier/North-Holland, Amsterdam, 1971, p. 383.
60. T. Y. Zhang, *J. Chromatogr.*, *315*, 287 (1984).
61. Y. Ito, in *Methods in Enzymology, Vol. 91, Enzyme Structure*, Part I, (C. H. W. Hirs and Serge N. Timasheff, eds.), Academic, New York, 1983, p. 335.
62. E. A. B. Brown and Y. Ito, *J. Biochem. Biophys. Met.*, *3*, 77 (1980).
63. Y. Ito, *Protein, Nucleic Acid, Enzyme*, *26*, 1021 (1981) (Japanese).
64. R. D. Friedman, H. Nakazawa, F. E. Chou, N. R. Bachur, and Y. Ito, *J. Chromatogr.*, *252*, 283 (1982).
65. W. D. Conway and Y. Ito, *J. Liq. Chromatogr.*, *7*, 275 (1984).
66. G. J. Putterman, F. Perini, E. L. White, and Y. Ito, *Proc. Sixth Am. Pep. Symp.*, Pierce Chemical Co., 1979, p. 113.
67. G. J. Putterman, M. B. Spear, and F. Perini, *J. Liq. Chromatogr.*, *7*, 341 (1984).
68. M. Knight, Y. Ito, and T. N. Chase, *J. Chromatogr.*, *212*, 356 (1981).
69. M. Knight, C. A. Tamminga, Y. Ito, J. D. Gardner, and T. N. Chase, *J. Liq. Chromatogr.*, *301*, 277 (1984).
70. M. Knight, Y. Ito, A. M. Kask, C. A. Tamminga, and T. N. Chase, *J. Liq. Chromatogr.*, *7*, 2525 (1984).

71. M. Knight, A. M. Kask, and C. A. Tamminga, *J. Liq. Chromatogr.*, *7*, 351 (1984).
72. N. B. Mandava and Y. Ito, *J. Chromatogr.*, *247*, 315 (1982).
73. N. B. Mandava and Y. Ito, *J. Liq. Chromatogr.*, *7*, 303 (1984).
74. I. A. Sutherland, D. Heywood-Waddington, and T. J. Peters, *J. Liq. Chromatogr.*, *8*, 2315 (1985).
75. Y. Ito, P. Carmeci, and I. A. Sutherland, *Anal. Biochem.*, *94*, 249 (1979).
76. P. Guerry, D. J. LeBlanc, and S. Falkow, *J. Bacteriol.*, *116*, 1064 (1978).
77. M. J. Osborn, S. M. Rosen, L. Rothfield, and B. L. Horecker, *Proc. Natl. Acad. Sci. USA*, *48*, 1831 (1962).
78. O. Stendal, K. E. Magnusson, C. Tagesson, R. Cunningham, and L. Edebo, *Infect. Immun.*, *7*, 573 (1973).
79. L. Leive, L. M. Cullinane, Y. Ito, and G. T. Bramblett, *J. Liq. Chromatogr.*, *7*, 403 (1984).
80. I. A. Sutherland and Y. Ito, *Anal. Biochem.*, *108*, 367 (1980).
81. R. R. Porter, *Biochem. J.*, *53*, 320 (1953).
82. R. E. Hurst, J. Y.-p. Sheng, and Y. Ito, *Anal. Biochem.*, *85*, 230 (1978).
83. H. Nakazawa, P. A. Andrews, N. R. Bachur, and Y. Ito, *J. Chromatogr.*, *205*, 482 (1981).
84. J. L. Sandlin and Y. Ito, *J. Liq. Chromatogr.*, *8*, 2153 (1985).
85. J. L. Sandlin and Y. Ito, *J. Chromatogr.*, *348*, 131 (1985).
86. J. L. Sandlin and Y. Ito, *J. Liq. Chromatogr.*, *7*, 323 (1984).
87. Y. Ito, *J. Chromatogr.*, *301*, 377 (1984).
88. Y. Ito, *J. Chromatogr.*, *301*, 387 (1984).
89. Y. Ito, U.S. Patent 4,487,693, 1984.
90. W. D. Conway and Y. Ito, Abstract of the 1984 Pittsburgh Conference and Exposition on Analytical Chemistry and Applied Spectroscopy, 471 (1984).
91. I. A. Sutherland and D. Heywood-Waddington, Abstract of the 1985 Pittsburgh Conference and Exposition on Analytical Chemistry and Applied Spectroscopy, 302 (1985).
92. Y. Ito and W. D. Conway, *J. Chromatogr.*, *301*, 405 (1984).
93. W. D. Conway and Y. Ito, *J. Liq. Chromatogr.*, *7*, 291 (1984).
94. W. D. Conway and Y. Ito, *J. Liq. Chromatogr.*, *8*, 2195 (1985).
95. M. Knight, Y. Ito, P. Peters, and C. diBello, *J. Liq. Chromatogr.*, *8*, 2281 (1985).
96. M. Knight, Y. Ito, J. L. Sandlin, and A. M. Kask, *J. Liq. Chromatogr.*, *9*, 791 (1986).
97. N. B. Mandava, Y. Ito, and J. M. Ruth, *J. Liq. Chromatogr.*, *8*, 2221 (1985).
98. J. M. Ruth, Y. Ito, and N. B. Mandava, *J. Liq. Chromatogr.*, *8*, 2239 (1985).
99. Y. W. Lee and C. E. Cook, *J. Liq. Chromatogr.*, *8*, 2253 (1985).

100. G. M. Brill, J. B. McAlpine, and J. E. Hochlowski, *J. Liq. Chromatogr.*, *8*, 2259 (1985).
101. L. J. Putman and L. G. Butler, *J. Chromatogr.*, *318*, 85 (1985).
102. R. J. Romanach, J. A. de Haseth, and Y. Ito, *J. Liq. Chromatogr.*, *8*, 2209 (1985).
103. Y. M. Yang, H. A. Lloid, L. Pannell, H. M. Fales, R. A. Macfarlane, and Y. Ito, *Biomed. Environ. Mass Spectrom.*, *13*, 439 (1986).
104. F. E. Chou, A. Yergey, and Y. Ito, Abstract of the 1986 Pittsburgh Conference and Exposition on Analytical Chemistry and Applied Spectroscopy, 1046 (1986).
105. P. Somasundaran, *Sep. Purif. Methods*, *1*, 117 (1972).
106. Y. Ito, *Trends Biomed. Sci.*, *7*, 47 (1982).
107. Y. Ito, *Industr. Res. Devel.*, *24*, 108 (1982).
108. N. B. Mandava, Y. Ito, and W. D. Conway, *Am. Lab.*, *14*, No. 10, 62 (1982).
109. N. B. Mandava, Y. Ito, and W. D. Conway, *Am. Lab.*, *14*, No. 11, 48 (1982).
110. Y. Ito, *Farumashia* (Pharmaceutical Society of Japan), *FARUAW* *19*, 1054 (1983) (Japanese).
111. W. D. Conway and Y. Ito, *LC Magazine*, *2*, 368 (1984).
112. Y. Ito, *Chromatogr. Int.*, *7*, 4 (1985).
113. Y. Ito, *CRC Crit. Rev. Anal. Chem.*, *17*, 65 (1986).
114. W. D. Conway, Countercurrent Chromatography, in *Handbook of Naturally Occurring Pesticides, Vol. 1, Methods, Theory and Practice*, (N. B. Mandava, ed.), CRC Press, Boca Raton, 1983.
115. W. D. Conway, *Countercurrent Chromatography*, in preparation.
116. Y. Ito, *J. Chromatogr.*, *358*, 313 (1986).
117. Y. Ito, *J. Chromatogr.*, *358*, 325 (1986).
118. Y. Ito, *Sep. Sci. Tech.*, Special Issue for Preparative Chromatography, in press.
119. Y. Ito, *Sep. Sci. Tech.*, Special Issue for Preparative Chromatography, in press.
120. Y. Ito and R. Bhatnagar, *J. Liq. Chromatogr.*, *7*, 257 (1984). 1 (1972).

chapter four

Theoretical Aspects of Countercurrent Chromatography

Walter D. Conway
State University of New York at Buffalo
Amherst, New York

Generally speaking, the mathematical relationships employed in other forms of chromatography also apply to CCC. It is convenient to summarize these relationships with reference to the typical elution profile illustrated in Fig. 1. Definitions of the symbols are summarized in the Appendix at the end of the book.

I. FUNDAMENTAL PARAMETERS

A. The Capacity Factor

Perhaps the most fundamental parameter in chromatography is the capacity factor k', which is defined as

$$k' = \frac{Q_s}{Q_m} = \frac{\text{quantity of solute in stationary phase}}{\text{quantity of solute in mobile phase}} \tag{1}$$

This is related to the partition coefficient K by the phase volume ratio:

$$k' = \frac{Q_s}{Q_m} = \frac{C_s V_s}{C_m V_m} = K \frac{V_s}{V_m} \tag{2}$$

where V_s and V_m are the volumes of stationary and mobile phase retained in the column and K is defined as

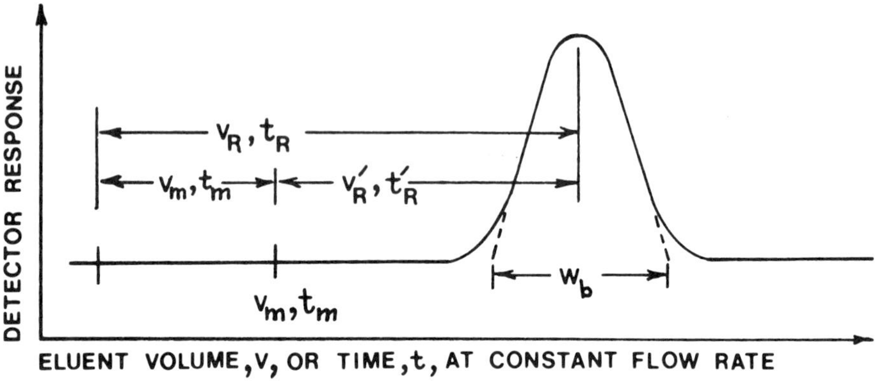

Fig. 1 Relationship of various chromatographic parameters to the elution profile.

$$K = \frac{C_s}{C_m} = \frac{\text{concentration of solute in stationary phase}}{\text{concentration of solute in mobile phase}} \quad (3)$$

Note that the ratio used to define K in chromatography is the reciprocal of that conventionally used to describe CCD.

If the flow rate is constant, the horizontal axis in Fig. 1 expressed in units of volume will be proportional to the corresponding units of time. The initial point on the profile corresponds to the beginning of the chromatogram and assumes that the dead volume in the feed lines is negligible. That is usually true in preparative columns of 100 ml or more in volume. If the dead volume is not negligible, a correction must be made as illustrated in Fig. 2.

B. The Stationary Phase Volume

The second point on the profile in Fig. 1, V_m or t_m, represents the elution of the "solvent front." The front may be an imaginary point on the chromatogram, the position of which can be calculated, or it may be visible if suitable markers are present in the sample. The most common CCC methodology involves first filling the column completely with stationary phase; the sample is then injected, following which the mobile phase is pumped into the column at a constant flow rate. Assuming the sample volume is negligibly small, the leading edge of mobile phase may be regarded as the solvent front. A certain volume of stationary phase is collected in a receiver or fraction collector as the mobile phase proceeds through the column. If it is assumed that hydrodynamic equilibrium is established immediately

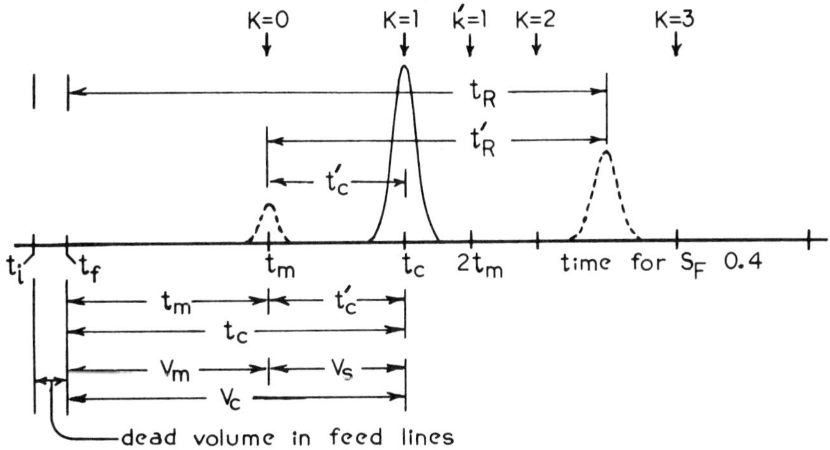

Fig. 2 Relationship of retention time and partition coefficient in a CCC system. All features of the diagram are general except the time axis, which corresponds to S_F of 0.4. t_i is the time of injection. $t_f - t_i$ is a correction for feed line dead volume. Therefore, all parameters are measured from t_f as the start of the chromatogram.

behind the solvent front, then the stationary phase which has been displaced is a meaure of the volume of mobile phase V_m in the column. This method is not always suited to detection with an in-stream monitor because the cell may become fouled with droplets when the interface passes through. However, it often provides a reasonable estimate of V_m where the volume is measured directly. The estimate may be somewhat low if the initial breakthrough of mobile phase is used as an indicator, but it is generally satisfactory if collection is continued for a suitable time after the initial breakthrough.

An alternative method of estimating V_m is to empty the column contents by gently blowing out the contents with compressed gas at the end of the run and to measure the volume of each phase directly. This approach may evaporate some of the more volatile phase. It may also underestimate the effective volume of stationary phase in relation to peaks eluted early, if there was appreciable additional bleed of stationary phase subsequent to elution of the peaks. Other methods of estimating V_m include introduction of a nonretained marker substance with the sample, or calculation of V_m from the retention time of a standard substance for which the partition coefficient is known in the solvent system being used.

The volume of stationary phase V_s can be calculated by subtracting V_m from the column volume V_c. The latter is usually measured by determining the increase in weight when the column is filled with water. It is convenient to express the volume of stationary phase retained under a particular set of conditions as the fraction S_F of column volume.

$$V_s = V_c - V_m \tag{4}$$

$$S_F = \frac{V_s}{V_c} \tag{5}$$

It is often convenient to express the phase volume ratio in terms of S_F as

$$\frac{V_s}{V_m} = \frac{S_F}{1 - S_F} \tag{6}$$

The previous relationship between k' and K in Eq. (2) then becomes

$$k' = K\left(\frac{S_F}{1 - S_F}\right) \tag{7}$$

C. Relationship of k' and Retention Time

The time t_m actually represents the time spent by the solute in the mobile phase during its transit time through the column. This time is the same for all solutes. The total time spent by the solute in the column is t_R, the retention time. Therefore, the solute must spend the difference between these two times in the stationary phase. This difference is the corrected retention time, t'_R on the diagram in Fig. 1. Detailed discussion of this reasoning, which leads to Eq. (8), may be found in standard texts on chromatography [1–3].

$$t'_R = T_R - t_m \tag{8}$$

An alternative definition of the capacity factor as the ratio of the time spent by the solute in the stationary phase t_s so that in the mobile phase t_m permits the retention parameters to be expressed as

$$k' = \frac{t_s}{t_m} = \frac{t'_R}{t_m} = \frac{t_R - t_m}{t_m} \tag{9}$$

which allows k' to be evaluated by direct measurement of distances on the chromatogram. The corresponding relationships in terms of volume are

$$k' = \frac{V'_R}{V_m} = \frac{V_R - V_m}{V_m} \tag{10}$$

Equations (9) and (10) are the fundamental retention equations in chromatography. While the parameters on the chromatogram may be expressed in units of time or volume, it is usually more convenient to simply measure the appropriate distances on the chromatogram in millimeters. Since k' is a dimensionless number, the units cancel in Eqs. (9) and (10).

D. The Partition Coefficient

Conventional chromatography is usually described in terms of the capacity factor k' rather than by the partition coefficient K. The stationary phase in conventional systems is retained by adsorption or chemical bonding to a solid supporting matrix. Solute adsorption to the matrix surface often contributes to the solute retention. Therefore, k' in conventional chromatography is determined by a mixed mechanism and is not always directly predictable from the partition coefficient. In addition, the value of the phase volume ratio is usually not directly available in conventional chromatography.

A major advantage of CCC lies in the absence of a solid supporting matrix. Separation in CCC is therefore entirely dependent on liquid-liquid partition and the elution profile is predictable from the more fundamental partition coefficient K, which can be determined by means independent from the chromatogram. In addition, the phase volume ratio is directly measurable in CCC and can be used along with the partition coefficeint to calculate k' using Eq. (2).

In terms of K, Eqs. (9) and (10) become

$$K = \frac{t'_R}{t_m}\left(\frac{V_m}{V_s}\right) = \frac{t'_R}{t_m}\left(\frac{1 - S_F}{S_F}\right) = \frac{t_R - t_m}{t_m}\left(\frac{1 - S_F}{S_F}\right) \tag{11}$$

and

$$K = \frac{V'_R}{V_m}\left(\frac{V_m}{V_s}\right) = \frac{V'_R}{V_m}\left(\frac{1 - S_F}{S_F}\right) = \frac{V_R - V_m}{V_m}\left(\frac{1 - S_F}{S_F}\right) \tag{12}$$

which can be used to calculate K directly from a chromatogram.

In examining CCC chromatograms, it is often convenient to calculate the position at which a solute with K = 1 will elute. This is shown as the peak drawn with a solid line in Fig. 2. The relationships in the diagram are general, but distances on the time axis correspond specificially to a system with S_F of 0.4. The retention time for K = 1 can be calculated from Eq. (13), which is simply a rearrangement of Eq. (11).

$$t_R = t_m + \frac{K t_m S_F}{1 - S_F} \tag{13}$$

The second term on the right of Eq. (13) is, of course, the time spent by the solute in the stationary phase. For K = 1, this time corresponds to the time interval t'_c required to elute a volume of mobile phase equivalent to the volume of stationary phase V_s in the column. The retention time t_R for a solute with K = 1, therefore, corresponds to the time t_c required to elute a volume of mobile phase equal to the column volume V_c. The retention time can be calculated from V_c and the mobile phase flow rate f using Eq. (14):

$$t_{R,K=1} = t_c = \frac{V_c}{f} \tag{14}$$

This relationship can be clearly seen from the notation of the various parameters beneath the chromatogram in Fig. 2. The notation in Fig. 2 allows for a correction for feed line dead volume, where it is not negligible. The choice of symbols—t_i, t_f, and t_m—does not conflict with customary diagrams (Fig. 1) which don't illustrate the dead volume correction.

In a case such as that illustrated in Fig. 2, where flow rate is not specified but S_F and the position of t_m are known, the retention time for a solute with K = 1 can be calculated from Eq. (13). For S_F = 0.4, the result is

$$t_{R,K=1,S_F=0.4} = t_c = t_m + 0.67t_m = 1.67t_m \tag{15}$$

and the distance $0.67t_m$ can be laid off on the time axis. Solutes with K less than 1 will elute between t_f and t_c. Subsequent peaks with K = 2, 3, etc., will be eluted at successive time intervals of $0.67t_m$, in this specific case, or $[S_F/(1 - S_F)]t_m$ in general.

The partition coefficeint of the solute shown by the dotted peak in Fig. 2 can be calculated to be 2.5, from measurement of t'_R or t_R and using Eq. (11). It can also be calculated from the general relationship

$$K = \frac{t'_R}{t'_c} \tag{16}$$

using the appropriate distances measured on the chromatogram, where t'_c is the corrected retention time $t_c - t_m$ for the elution of one column volume. The effect of K on elution is probably seen most easily from Eq. (16) which indicates the elution sequence for K = 1, 2, 3, etc., at successive multiples of the time t'_c by K.

A similar relationship exists for multiple values of the capacity factor k'. Elution of a solute with k' = 1 occurs at the time $2t_m$ indicated in Fig. 2 and successive solutes with k' = 2, 3, 4, etc., elute at t_R values which increase by t_m for each unit increase in k'.

E. Retention

The retention volume V_R and the retention time t_R in CCC can be expressed in various ways by rearrangement and recombination of the equations presented thus far. The following are convenient:

$$V_R = V_c[1 + S_F(K - 1)] \tag{17}$$

$$t_R = t_c[1 + S_F(K - 1)] \tag{18}$$

Equations (17) and (18) show that V_R and t_R are increased by increases in V_c, S_F, and K. They also demonstrate that at K = 1, V_R and t_R equal V_c and t_c, respectively.

F. Column Efficiency

The shape of the chromatographic elution profile follows a Poisson distribution. For columns having a large number of theoretical plates, this may be approximated by a normal distribution. The efficiency, or number of theoretical plates N, is commonly estimated by the equation:

$$N = 16\left(\frac{V_R}{W_b}\right)^2 = 16\left(\frac{t_R}{W_b}\right)^2 \tag{19}$$

where W_b is the 4σ base width. W_b is estimated by drawing tangents to the sides of the peak as illustrated in Fig. 1. Other means of estimating N are discussed in standard works on chromatography [3,4].

G. Separation Factor

The separation factor α is defined as the ratio of the capacity factors k', the partition coefficients K, or the corrected retention times t'_R of two solutes, the ratio always being expressed as a value greater than unity [Eq. (20)].

$$\alpha = \frac{K_2}{K_1} = \frac{k'_2}{k'_1} = \frac{t'_2}{t'_1} \qquad K_2 > K_1,\ k'_2 > k'_1,\ t'_2 > t'_1 \tag{20}$$

The separation factor is a measure of the selectivity or ability of a particular solvent system to separate a particular pair of solutes. It may be determined nonchromatographically from known partition coefficients or from measurements of K, k', or t' taken directly from chromatograms. The value of α is independent of the means of expressing the set of partition coefficients (upper/lower or lower/upper), as long as the ratio of the two values of K or k' (each individually expressed with the same convention) is calculated to be greater than 1. Likewise it does not matter whether the chromatographic system used employs a mobile lighter phase or a mobile heavier phase.

II. RESOLUTION

The separation of the peaks 1 and 2 in Fig. 3 is conventionally expressed by the resolution, defined as

$$R_s = \frac{2(V_{R2} - V_{R1})}{W_{b2} + W_{b1}} = \frac{2\ \Delta V}{W_{b2} + W_{b1}} = \frac{2\ \Delta t}{W_{b2} + W_{b1}} \tag{21}$$

At a constant flow rate, the corresponding retention times may also be used. A resolution of 1.5 corresponds to a 6σ separation or baseline resolution. This results from the fact that W_b corresponds to a width of 4σ on the peak, whereas the total peak is essentially contained in a width of $\pm 3\sigma$ or 6σ. A resolution of 1.0 corresponds to a 4σ separation, each peak then containing about 2% of the other substance [5]. If a chromatogram is generated by monitoring light absorption at a fixed wavelength, the response of different substances will not be the same unless the absorptivity of both substances is the same. Consequently, an accurate estimation of resolution cannot usually be obtained from a chromatogram based on absorbance.

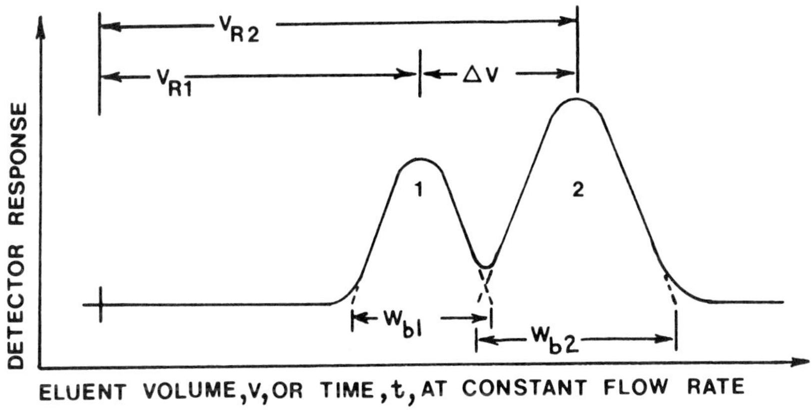

Fig. 3 Resolution of chromatographic peaks 1 and 2.

A. The Knox Equation

The major factors contributing to resolution have been summarized by Knox [4] in the form of the equation:

$$R_s = \underbrace{\frac{1}{4}(\alpha - 1)}_{\text{separation factor term}} \underbrace{\sqrt{N}}_{\text{efficiency term}} \underbrace{\left(\frac{k'_1}{k'_1 + 1}\right)}_{\text{capacity factor term}} \qquad (22)$$

The equation is an approximation which applies only when α is close to unity and hence $\alpha - 1$; and consequently, R_s is almost zero. In this case, k'_1 and k'_2 are almost identical and can be replaced by the value of k'_1. The equation is widely used as an aid to optimization of column chromatography because it can be partitioned into the three terms indicated in Eq. (22), each of which can be independently evaluated from the chromatogram.

The factor α, which is largely determined by the solvent system, should be as large as possible to maximize R_s. Likewise, the efficiency N, which is largely determined by the nature of the column, should be as large as possible. But the capacity factor term approaches an asymptote of unity as k'_1 is increased. The increase in the capacity factor term drops from 34%, when k'_1 is increased from 1 to 2, to about 4% for an increase in k'_1 from 4 to 5 and drops to about 1% and less for increases above 8. Values of k'_1 above about

4, therefore, contribute very little to resolution whereas they increase retention time significantly.

B. An Approximate Resolution Equation for CCC

Substitution of the following equivalence:

$$k' = K_1 \left(\frac{S_F}{1 - S_F} \right) \tag{23}$$

for k' in Eq. (22) provides the following equivalent of the Knox equation which is more applicable to CCC:

$$R_s = \frac{1}{4} (\alpha - 1)\sqrt{N} \left[\underbrace{\frac{K_1}{K_1 + \left(\frac{1 - S_F}{S_F} \right)}}_{\text{approximate partition coefficient term}} \right] \tag{24}$$

The first two terms are the same as in Eq. (22), but the S_F factor in the denominator of the partition coefficient term shows that resolution will be increased by increasing S_F. The partition coefficient term approaches an asymptote of unity, but the contribution of increasing K_1 continually decreases as S_F increases. Thus, when S_F is 0.4, the value of the approximate partition coefficient term reaches 0.8 at a K_1 of 6, whereas when S_F is 0.8, the term reaches 0.8 at a K_1 of 1. The approximate Eq. (24) actually overestimates the contribution of K_1 to resolution, since it fails to include a further interaction of α with K_1, but it does provide some insight into the nature of the resolution process in CCC.

C. An Exact Resolution Equation for CCC

Starting from the definition of resolution [Eq. (21)], Conway and Ito [6] derived an exact resolution equation, which may be presented [7] in the format of Eq. (24) as

$$R_s = \frac{1}{4} (\alpha - 1)\sqrt{N} \left[\underbrace{\frac{K_1}{K_1 \left(\frac{\alpha + 1}{2} \right) + \left(\frac{1 - S_F}{S_F} \right)}}_{\text{exact partition coefficient term}} \right] \tag{25}$$

The partition coefficient term in Eq. (25) contains no approximations. When $\alpha = 1$, indicating zero resolution, Eq. (25) reduces to Eq. (24). Clearly there is an interaction of α with K_1 in the denominator of the partition coefficient term, which serves to reduce the contribution of K_1 to resolution as α increases.

The contribution of the partition coefficient term' to resolution for an S_F of 0.4 is summarized in Fig. 4. An S_F of 0.4 is typical for the horizontal flow-through coil planet centrifuge. As mentioned earlier, the approximate term overestimates the contribution of the partition coefficient term to resolution. Whereas the approximate partition coefficeint term [Eq. (24)] approaches an asymptote of 1, the asymptotic value for the exact term [Eq. (25)] decreases with increasing α. The K_1 value corresponding to 80% of the term's contribution to resolution also decreases significantly with increasing α. Thus for α of 1.5, an 80% K_1 contribution is reached at K_1 of 4.8, whereas when α is 10, the value is reached at a K_1 of 1.0.

Similar plots for an S_F of 0.8, which is frequently attained with the multilayer coil planet centrifuge, are given in Fig. 5. Here the approximate term indicates an 80% contribution of K_1 at a K_1 value of 1.0. The 80% values calculated from the exact equation range

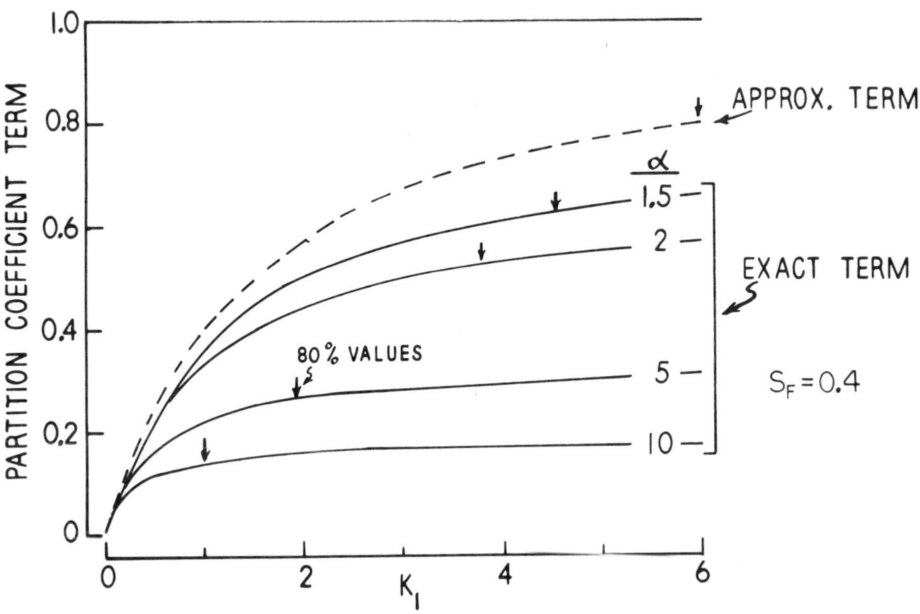

Fig. 4 Plots of the approximate and exact partition coefficient terms from Eqs. (24) and (25), respectively, as a function of K_1 and various α values, for S_F of 0.4. Points where the terms reach 80% of their asymptotic values are indicated.

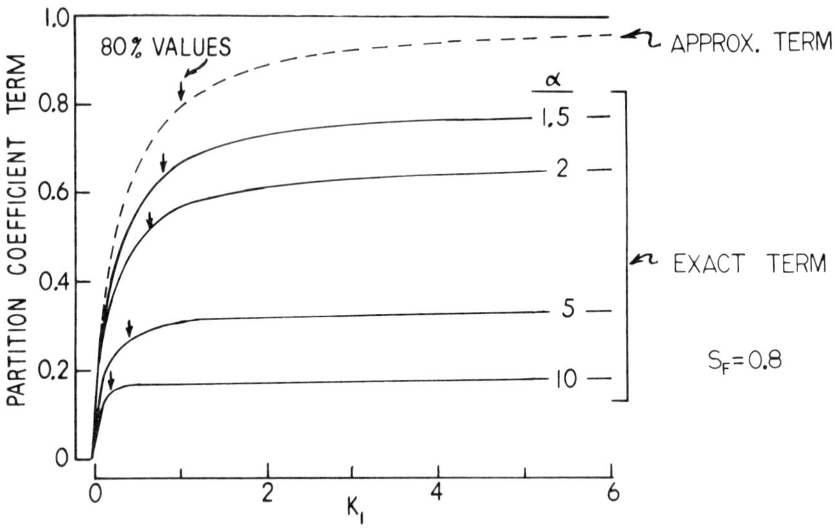

Fig. 5 Plots as described in Fig. 4 except $S_F = 0.8$.

from a K_1 of 0.8 for α of 0.5 to a K_1 of about 0.2 for an α of 10. These K_1 values for S_F of 0.8 are about one-sixth of the corresponding values for S_F of 0.4. Since K_1 is directly related to retention time [Eq. (18)], in effect, Figs. 4 and 5 show that the same resolution can be achieved with an S_F of 0.8 in about one-sixth the time required with an S_F of 0.4. This assumes that the same mobile phase flow rate is used in both cases. A difference in speed is observed experimentally when the performance of the multilayer coil planet centrifuge is compared with the horizontal flow-through coil planet centrifuge, although it appears to be less than sixfold. A corresponding saving in solvent is also observed.

The interaction of α and K_1 to decrease the effectiveness of the partition coefficient term, as seen in Figs. 4 and 5, should not be perceived as a disadvantage to increasing α. The effect is more than compensated by the contribution of the $\alpha - 1$ term in Eq. (25). For instance, in Fig. 5 (for S_F of 0.8), when α increases from 1.5 to 2.0, the partition coefficient term at $K_1 = 1$ decreases from 0.67 to 0.57, while the $\alpha - 1$ term doubles from 0.5 to 1.0, making the overall contribution of the combined terms to resolution rise from 0.34 to 0.57. Increases in α always overcome the decrease in the partition coefficient term and produce an enhancement of resolution.

The complexity of the interaction of K_1, α, and S_F is illustrated in Fig. 6, which is drawn for an α of 2 [6]. The diagram indicates that substances with high partition coefficients K_1 are best separated

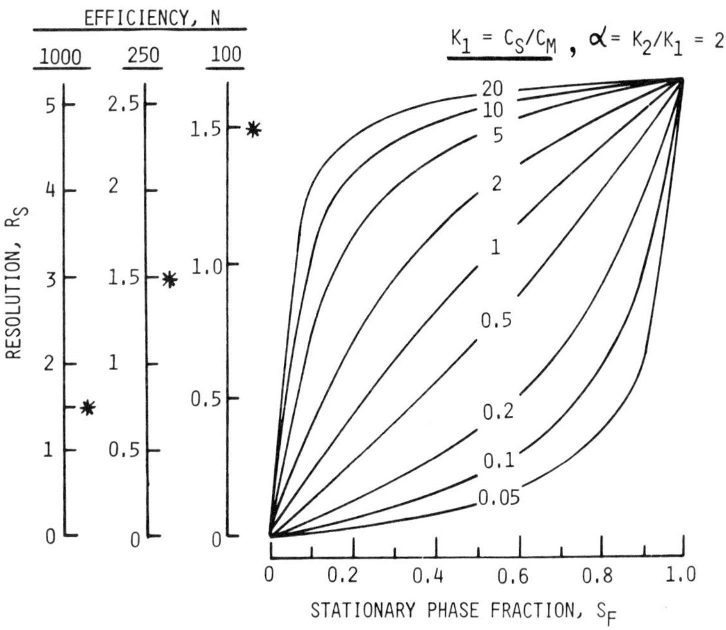

Fig. 6 Predicted resolution as a function of S_F and K_1 with an α value of 2, for columns providing 100, 250, and 1000 theoretical plates. Baseline resolution, R_s = 1.5, is indicated by an asterisk. (Reprinted from Ref. 6 with permission.)

in systems with low S_F. Increasing S_F improves the resolution of solute pairs with low partition coefficients. For instance, an S_F of of about 0.8 would be required to resolve a solute pair with K_1 of 0.1 and α of 2 on a column of 1000 theoretical plates. This illustrates that the overall effect of increasing S_F is to adjust the k' of the system to its optimum range of effectiveness.

The optimum range for k' in HPLC is usually taken to be from about 1 to 5, the ideal value being 2, in terms of minimizing the time required for separation [6]. These values are based on the approximate resolution equation which assumes an α of 1, and therefore apply particularly to low α values. Calculating the equivalent K_1 values with Eq. (7) shows that this optimum range of k' corresponds, for S_F of 0.4, to K values of 1.5–7.5 with an optimum of 3, and for S_F of 0.8, to a range of 0.25–1.25 with an optimum of 0.5. These may be compared with the $K_1^{80\%}$ values, for α of 1.5 (Figs. 4 and 5), of 4.8 for S_F of 0.4, and 0.8 for S_F of 0.8. The $K_1^{80\%}$ values correspond in both cases to a k' of 3, compared to the ideal of 2 recommended for HPLC. This suggests that while

Table 1 N_{REQ}- and $K_1^{80\%}$ Values for Baseline Resolution of a Solute Pair for Various α Values in Systems with S_F of 0.8 and 0.4

	$K_1 80\%$		
α	S_F 0.4	S_F 0.8	N_{REQ} for R_S 1.5
1.1	5.7	0.95	6200
1.2	5.4	0.91	1700
1.25	5.3	0.89	1140
1.5	4.8	0.80	350
2	4.0	0.67	130
5	2.0	0.33	32
10	1.0	0.17	22

little additional resolution will be obtained in CCC by employing K_1 values greater than the $K_1^{80\%}$ values, good separation might be achieved in a shorter time using somewhat lower K_1 values.

D. N Required for Separation

The exact resolution equation [Eq. (25)] can be rearranged as shown here to permit calculation of the number of theoretical plates required N_{REQ} for any desired resolution for various values of K_1, α, and S_F [6].

$$N_{REQ} = \left(\frac{4R_s}{\alpha - 1}\right)^2 \left[\frac{K_1\left(\frac{\alpha + 1}{2}\right) + \left(\frac{1 - S_F}{S_F}\right)}{K_1}\right]^2 \qquad (26)$$

A summary of values of N_{REQ} AND $K_1^{80\%}$ values for baseline resolution (R_s of 1.5) of a solute pair for various values of α in systems with S_F of 0.8 and 0.4 is given in Table 1.

With an α value of 1.1, the theoretical plate requirement is too high for separation with currently available CCC devices, which provide about 1000 theoretical plates. Analytically oriented CCC devices which use small-bore columns and operate at very slow flow rate have, however, provided efficiencies of this magnitude.

A minimum α value of 1.25 is required for baseline resolution in currently available CCC devices providing about 1000 theoretical

Table 2 Separation Factor α Required for Resolution

	K_1		α_{REQ}, R_S 1.5			
k'_1	S_F 0.4	S_F 0.8	N 1000	N 500	N 250	N 100
0.10	0.15	0.025	3.31	4.41	6.15	10.4
0.20	0.30	0.05	2.26	2.86	3.81	6.14
0.40	0.60	0.10	1.73	2.08	2.64	4.00
0.67	1.00	0.17	1.52	1.77	2.17	3.14
1.00	1.50	0.25	1.42	1.62	1.94	2.71
1.33	2.00	0.33	1.37	1.54	1.82	2.50
2.00	3.00	0.50	1.31	1.46	1.70	2.28
2.68	4.00	0.67	1.29	1.43	1.64	2.18
4.00	6.00	1.00	1.26	1.39	1.58	2.07
6.67	10.00	1.67	1.24	1.36	1.54	1.98
8.00	12.00	2.00	1.23	1.35	1.53	1.96

plates. For this, a K_1 value of about 0.9 is optimum using a system providing an S_F of 0.8, or a K_1 value of about 5 can be employed in a system where S_F is 0.4. The latter separation will take about six times as long and will consume a correspondingly larger volume of solvent. It is interesitng to note that in each of these situations (S_F of 0.4 and 0.8), the optimum partition coefficient corresponds to exactly the same capacity factor. The effect of increasing S_F is to increase the capacity factor [Eq. (23)], thereby shifting the optimum partition coefficeint to a lower value. Table 1 should be very helpful in guiding the search for optimum solvent systems for CCC.

E. α Required for Separation

If the numerator and denominator of Eq. (25) are multiplied by $S_F/(1 - S_F)$, the resulting terms replaced by k' according to Eq. (7), and the result solved for α, the following equation is obtained [6]:

$$\alpha_{REQ} = \left(\frac{4R_S}{\sqrt{N} - 2R_S}\right)\frac{1}{k'_1} + \frac{\sqrt{N} + 2R_S}{\sqrt{N} - 2R_2} \qquad (27)$$

The α term is designated α_{REQ} and it is convenient to use Eq. (27) to calculate the separation factor required for resolution, as a function of k'_1, for arbitrarily chosen values of R_s and N. A plot of α_{REQ} vs. $1/k'_1$ has been published [6], but it is not convenient for use with CCC. The summary of α_{REQ} values presented in Table 2 is more readily applied to CCC. It is useful to answer the question: given a solute pair with K_1 of 1.0, or other value, what value of α is required to achieve baseline separation on a column providing 1000 theoretical plates and having an S_F of 0.4? Entering the table under S_F 0.4 for K_1 of 1 and moving over to the N 1000 column, one finds α_{REQ} is 1.52. For shorter columns, providing fewer theoretical plates, α_{REQ} is larger. If S_F is increased, α_{REQ} drops.

As indicated by Eq. (27), α_{REQ} is determined by k'_1. The values of K_1 listed in Table 2 are equivalent to the listed k'_1 values, but in applying the table to CCC it is more convenient to enter the K_1 column, since in searching for solvent systems it is more expedient to measure K. Additional columns of K_1 for other values of S_F can be calculated with Eq. (7). In applying Table 2, one should keep in mind that it is not advantageous to employ partition coefficients greater than the $K_1^{80\%}$ values for Table 1. If a solvent system providing a high α value is available for a separation and it is desired to recover the second solute as well, a shorter column might be considered in order to reduce the time required for elution of the second solute. The corrected retention time of the second solute t'_2 is directly related to α [Eq. (28)].

$$\alpha = \frac{t'_2}{t'_1} \tag{28}$$

An alternative to consider, if only the second solute is of interest, is to invert the solvent system, thereby reversing the elution order. Still another possibility is to recover the second solute by pumping out the column contents after elution of the first solute. These optional approaches are major conveniences of CCC.

F. Validation of the Resolution Equation

The validity of Eq. (25) was tested using a series of 30 chromatograms of the separation of DNP-glutamic acid and DNP-alanine in the solvent system $CHCl_3$/HOAc/0.1 N HCl (2:2:1) [6]. These were obtained on a slowly rotating, nonplanetary coaxial column of 5.5-mm bore FEP tubing, 4.2 m long, wound in helical diameters of 3, 10, and 20 cm. The various modes of operation are summarized in Fig. 7. Measured resolution was calculated from the chromatograms using Eq. (21). Predicted resolution was obtained using Eq. (25) with the mean N determined from the chromatogram, S_F determined from the emptied column contents, and the value of α determined from partition

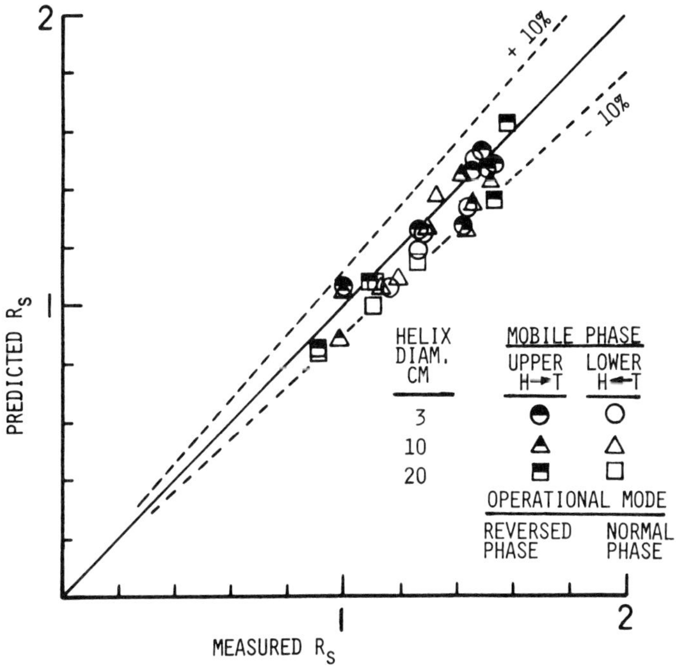

Fig. 7 Predicted vs. measured resolution for CCC separation of DNP-glutamic acid and DNP-alanine with the system $CHCl_3$/HOAc/0.1 N HCl (2:2:1). (Reprinted from Ref. 6 with permission.)

coefficients (1.9 and 0.56 for DNP-glu and DNP-ala, respectively, as C_{upper}/C_{lower}) measured nonchromatographically by partitioning and spectrophotometry.

The plot shows high correlation (r = 0.929), with predicted values being usually slightly lower (slope 0.919, y intercept 0.07) than measured values. The largest deviations varied only 10% from the theoretical unit slope. The most likely source of error was thought to be incomplete recovery of the column contents, which would result in a slightly low estimate of S_F. For practical purposes, there is excellent agreement between the predicted and measured values of resolution. The use of Eqs. (25)–(27) should facilitate the choice of solvent systems or column size when partition coefficients are known for the solutes to be separated.

G. Explanation of Symbols

An attempt has been made to assign readily identifiable symbols to parameters associated with CCC but at the same time to avoid conflict with symbolism generally employed in conventional chromatography. The symbols employed are summarized in the Appendix at the end of the book. Considerable work in CCC at the present time employs short columns, where a correction for dead volume in the feed lines must be employed. The symbolism adopted in Fig. 2 appears to facilitate discussion of the correction while avoiding symbols, particularly t_0, which might conflict with some usage in column chromatography.

The time of injection is specified by t_i and the time delay associated with elution of the feed line contents by t_f. The CCC chromatogram proper then starts from t_f. Most texts omit symbolic designation of these points and they do not appear in most formulas for chromatography. The symbol t_0 was avoided since some texts use t_0 to designate the time of emergence of an unretained peak, for which k' is zero. The symbol t_m is used for that purpose here. It presents no conflict and has the advantage of emphasizing that it is the time of emergence of one mobile phase volume, and is also the time spent by all solutes in the mobile phase.

In a similar manner, the symbol t_c is used to designate the time of emergence of a solute for which K = 1. It emphasizes that this occurs at the emergence of one column volume of mobile phase, the corresponding symbol V_c being a natural choice for column volume. The corresponding corrected retention time $(t_c - t_m)$ is then designated t_c'. The choice of t_s is avoided for designation of $t_c - t_m$, because t_s is often used as a general term to designate the time spend by a solute in the stationary phase as it passes through the column, and in this sense $t_s = t_R'$.

S_F is a convenient choice for the stationary phase fraction, or fraction of column volume occupied by stationary phase. The symbol t_R is used here for retention time, but it is convenient to drop the subscript when specific peaks are referred to, such as t_1, t_2, etc. Likewise, it is often convenient to drop the subscript b from W_b.

Other symbols in the Appendix do not require specific justification and are simply listed for convenience.

REFERENCES

1. A. I. M. Keulemans, *Gas Chromatography*, Reinhold, New York, 1975.
2. J. C. Giddings, *Dynamics of Chromatography*, Marcel Dekker, New York, 1965.
3. L. R. Snyder and J. J. Kirkland, *Introduction to Modern Liquid Chromatography*, 2nd ed., John Wiley and Sons, New York, 1979.
4. A. S. Said, *Theory and Mathematics of Chromatography*, Dr. Alfred Huthig Verlag, New York, 1981.

5. B. L. Karger, L. R. Snyder, and C. Horváth, *An Introduction to Separation Science,* John Wiley and Sons, New York, 1973.
6. W. D. Conway and Y. Ito, *J. Liq. Chromatog.*, *8*, 2195 (1985).
7. W. D. Conway, Theoretical aspects and recent applications of countercurrent chromatography, paper 162, Division of Analytical Chemistry, 189th National Meeting, American Chemical Society, Miami Beach, April 1985.

Part 2

APPLICATIONS

chapter five

Natural Products Isolation of Droplet Countercurrent Chromatography

Kurt Hostettmann and Andrew Marston
University of Lausanne
Lausanne, Switzerland

I. INTRODUCTION

The support-free separation technique, droplet countercurrent chromatography (DCCC), as its name implies, involves the passage of droplets of a mobile phase through a liquid stationary phase. By this means, a solute in the system is continuously partitioned between the two phases. A suitable choice of operating conditions enables the separation of mixtures of compounds present in the solute. The inherent simplicity of operation of a droplet countercurrent chromatograph and the ability to separate anything from milligram to gram quantities has led to the widespread acceptance of the technique, especially in the field of natural products [1–4]. The compatibility of the apparatus even with acid- and base-containing solvent systems is a great advantage and increases the choice of solvent combinations.

II. APPARATUS

Droplet countercurrent chromatographs are commercially available from two manufacturers: Büchi A. G. (Flawil, Switzerland) and Tokyo Rikakikai Co. Ltd. (Tokyo, Japan). These instruments consist of 200–600 vertical columns (20–60 cm in length) of narrow-bore silanized glass or Teflon tubing, interconnected in series by Teflon capillary tubes. Droplets of mobile phase are passed through the

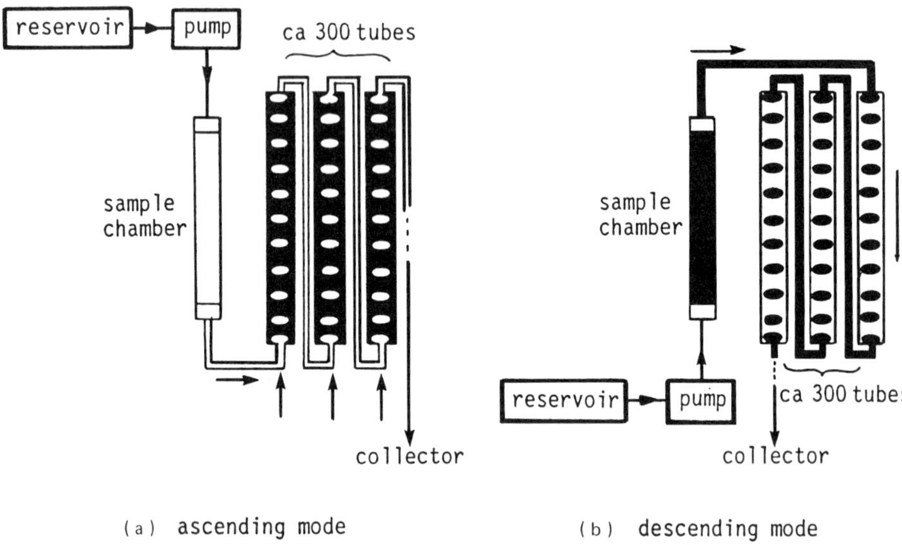

Fig. 1 The principle of DCCC: (a) ascending mode and (b) descending mode. (Adapted from Ref. 1.)

columns full of stationary phase and collection of mobile phase is effected at the end of the battery of columns (Fig. 1).

A suitable two-phase solvent system for the separation of the mixture is chosen and the whole apparatus is then filled with the selected stationary phase. The sample, dissolved in the lighter phase or the heavier phase or in a mixture of both phases, is injected into the sample chamber. Mobile phase is subsequently pumped via the sample chamber into the first of the columns, displacing the sample and forming a stream of droplets in the immiscible stationary phase. Depending on the choice of solvents for the mobile and stationary phases, these droplets are made to either ascend or descend through the columns. The capillary tubing connecting the columns does not allow entry of the stationary phase, except at the very beginning of the separation.

As the mobile phase moves through the column in the form of droplets, turbulence within the droplet promotes efficient partitioning of the solute between the two phases. Separation occurs according to the difference in the partition coefficients of the components of the sample. The size and mobility of the droplets are functions of the column bore, the flow rate of the mobile phase, the diameter of the inlet tip, the difference in specific gravities of the two liquid phases, the viscosity of the solvent, and the surface tension.

Columns with an internal diameter of 2 mm are most often used for separations but internal diameters of 2.7, 3.0, and 3.4 mm are now commercially available [4]. The columns with a wider bore have the advantages that higher rates of flow are possible and there is a higher sample loading capacity. For example, sample loadings of up to 6.4 g of crude extract on columns of 3.4 mm i.d. have been carried out [4] and even sample sizes of 16 g have been possible with 10-mm i.d. columns [5]. Care must be taken, however, to avoid a very high flow rate, as this can lead to elution of the stationary phase. In general, columns of less than 1 mm i.d. produce flow plugs in which the entire contents of the tube are displaced.

Resolution can be improved by increasing the number of columns but separation times are lengthened accordingly.

Since the glass tubes and interconnecting Teflon capillaries are chemically inert, the DCCC apparatus is compatible with solvent systems containing acids, alkalis, and all organic solvents.

III. CHOICE OF SOLVENTS

The formation of droplets, necessary for DCCC, depends on such factors as the difference in density between the two liquid phases, the viscosity of the solvents, and the interfacial tension. Crucial, therefore, for successful DCCC separations is the choice of two-phase solvent system. Binary solvent systems are impractical for the formation of suitable droplets because of the large difference in polarity between the two components. Ternary (or quaternary) systems are therefore recommended for the preparation of the two phases, such that the addition of a third (or fourth) component, miscible with the other components, diminishes the difference in polarity between the two phases (Fig. 2). The selectivity of the system is thus increased, allowing the separation of closely related substances. Furthermore, the role of the third component is to diminish the interfacial tension and the viscosity of the system.

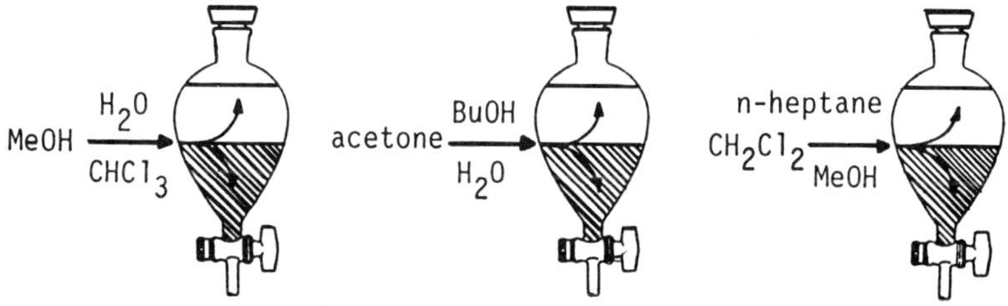

Fig. 2 Diminution of the polarity difference between two immiscible liquids by the addition of a third component, soluble in both solvents.

Table 1 Aqueous Solvent Systems for Droplet Countercurrent Chromatography

Sample	Basic binary system	Auxiliary solvent
Very polar	n-Butanol/water	Methanol, ethanol, acetone, acetic acid, (pyridine)
Polar	Chloroform/water	Methanol, n-propanol, i-propanol

A. Aqueous Systems

The most frequently employed components of aqueous DCCC two-phase systems are indicated in Table 1.

Methanol is often used as auxiliary solvent because a small change in the amount of methanol leads to a large change in the polarity of the system.

The well-known Craig countercurrent distribution system methanol/hexane/water is unsuitable because of a tendency to form plugs in the narrow-bore columns.

A rapid TLC method exists for the choice of two-phase aqueous solvent system that is most applicable to the separation of a given mixture or extract [8,9]. A thin-layer chromatogram on silica gel is run, with the water-saturated organic layer (i.e., the organic phase of the two-phase solvent system) as eluent. If the R_f values of the compounds to be separated are higher than about 0.50 (less polar solutes), given that there is a suitable separation of the components of the mixture, the less polar layer is suitable for use as the mobile phase. For more polar solutes ($R_f < 0.50$), the more polar layer should be used as the mobile phase. When the R_f values lie in the range 0.40–0.60, the separation can be achieved by using either the more polar layer or the less polar layer as the mobile phase.

If the R_f values of the compounds to be separated are either too low (<0.20–0.30) or too high (>0.70–0.80), no elution will take place in a reasonable time, although the compounds can be recovered quantitatively from the remaining stationary phase.

An alternative method of selecting suitable solvents is to monitor the distribution of 5–10 mg of sample between 5 and 10 ml of each of the two phases constituting the solvent system [10]. A system in which 15–25% of the sample is distributed in one of the phases is chosen. The stationary phase is chosen as the phase that contains the most sample.

Numerous solvent systems have been used for the separation of natural products by DCCC [1,4,7] and some representative aqueous systems are listed in Table 2.

Table 2 Aqueous DCCC Solvent Systems for the Separation of Natural Products

Solvent system	Substances separated
$CHCl_3$/MeOH/H_2O (7:13:8 and other proportions)	Glycosides, tannins, alkaloids, antibiotics, etc.
$CHCl_3$/MeOH/acetate buffer, pH 3.6 (9:12:8)	Alkaloids
$CHCl_3$/MeOH/1% aq. NH_3 (7:12:8)	Basic steroid saponins
$CHCl_3$/MeOH/5% HCl (5:5:3)	Alkaloids
$CHCl_3$/MeOH/0.5% aq. $CaCl_2$ n-PrOH (50:60:40:6)	Glycolipids
$CHCl_3$/MeOH/H_2O/n-PrOH (9:12:8:1)	Saponins, iridoid glycosides, xanthone glycosides, flavonoid glycosides, anthraquinone glycosides, diterpene glycosides
$CHCl_3$/MeOH/H_2O/i-PrOH (5:6:4:1)	Saponins
$CHCl_3$/MeOH/H_2O/n-BuOH (10:10:6:1)	Flavonoid glycosides
$CHCl_3$/MeOH/H_2O/n-PrOH/EtOH (9:6:8:1:8)	Saponins
$CHCl_3$/MeOH/H_2O/C_6H_6 (3:5:2:1)	Quassinoids
$CHCl_3$/MeOH/H_2O/C_6H_5Me (5:7:2:5)	Alkaloids
$CHCl_3$/HOAc/H_2O (2:2:1)	Gibberellins
CH_2Cl_2/MeOH/H_2O (8:13:7)	Saponins
n-BuOH/HOAc/H_2O (4:1:5)	Anthocyanins, flavonoid glycosides, peptides
n-BuOH/MeOH/H_2O (5:1:5)	Naphthalide glycosides
n-BuOH/n-PrOH/H_2O (2:1:3)	Tannins
n-BuOH/EtOH/H_2O (4:1:5)	Iridoid and secoiridoid glycosides
n-BuOH/Me_2CO/H_2O (33:10:50)	Sennosides
i-BuOH/TFA/H_2O (120:1:160)	Peptides
Petrol ether/EtOH/H_2O/EtOAc (5:4:1:2)	Gibberellins, anthraquinones
C_6H_{14}/Et_2O/n-PrOH/95% EtOH/H_2O (4:8:3:5:4)	Diterpenoids, coumarins

B. Nonaqueous Systems

The use of aqueous solvent systems introduces limitations concerning the range of compounds which may be separated by DCCC. Weakly polar or nonpolar substances cannot generally be chromatographed under these conditions because they tend to elute with the solvent front, even when the organic phase is the mobile phase (there are exceptions, notably in the diterpene field; these are described under the relevant section). In addition, precipitation of weakly polar compounds may occur in the DCCC apparatus. However, nonaqueous systems capable of droplet formation do exist [6]. Basically the same criteria hold as for aqueous systems, i.e., a third or fourth component of the solvent is required, to decrease the difference in polarity between the two immiscible solvents and decrease the interfacial tension, leading to better droplets. Mixtures containing n-heptane and methanol (or acetonitrile) are often employed, with a chlorinated hydrocarbon or acetone as the third component.

Nonaqueous systems are important for the separation of weakly polar compounds which are unstable in the presence of water or which decompose during chromatography on silica gel.

For the choice of nonaqueous solvent systems, TLC on silica gel is not suitable as the separation is based entirely on adsorption, so chemically bonded phases, e.g., RP-8, may be substituted for silica gel plates [11]. When nonaqueous DCCC solvent systems containing n-heptane are being investigated, chromatography on RP-8 HPTLC plates with the lower layer is employed for the analysis of samples [11].

The number of known nonaqueous DCCC solvent systems is not large but a representative selection is shown in Table 3.

Table 3 Nonaqueous DCCC Solvent Systems for the Separation of Natural Products

Solvent system	Substances separated
C_6H_{14}/EtOAc/$MeNO_2$/MeOH (8:2:2:3)	Essential oils
C_7H_{16}/$C_2H_4Cl_2$/MeOH (47:6:72)	Triterpenes, steroids
C_7H_{16}/Me_2CO/MeOH (5:1:4)	Triterpenes, steroids, depsides
C_7H_{16}/CH_2Cl_2/MeCN (10:3:7)	Triterpenes, steroids
C_7H_{16}/C_6H_{12}/CH_2Cl_2/C_6H_5Me/MeCN (8:2:3:1:5:7)	Alkaloids

IV. APPLICATIONS

The first documented use of DCCC was in the baseline separation of an artificial mixture of seven DNP amino acids, with $CHCl_3$/HOAc/0.1 M HCl (2:2:1) as the solvent system, in the ascending mode (Fig. 3). Resolution was high (900 theoretical plates for dinitrophenyl alanine) [12]. Although subsequent applications have, in the main, involved the particular suitability of the method for the preparative separation of polar compounds, DCCC has also found application as an *analytical* tool. The estimation of saponins in *Panax* species (Araliaceae), using just 25 mg of each total saponin mixture, has been possible by DCCC [13], as has the detection of 20-hydroxyecdysone (12 mg) in a 2.05-g portion of the crude methanol extract of fresh fruits of *Vitex thyrsiflora* (Verbenaceae) [14].

The separation of polar natural products by DCCC is now a routine laboratory technique and has met with considerable success especially in the fields of polyphenols and saponins. The main reasons are as follows:

A solid support is not required.
The direct introduction of crude extracts can always be performed.
No loss of material occurs.
A wide range of solvent systems can be used.

Although DCCC is not confined to the separation of polar compounds, certain classes of substances, such as glycosides, often present particular isolation problems using conventional techniques

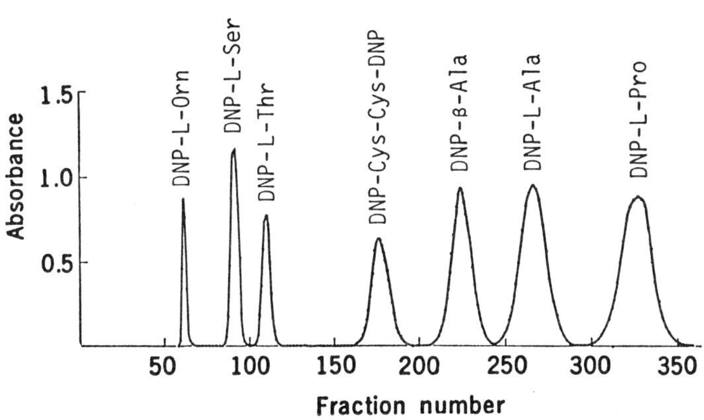

Fig. 3 DCCC separation of DNP amino acids. (Reprinted with permission from Ref. 12.)

and it is here that DCCC alone, or in combination with other methods, has proved so worthwhile.

Chloroform/methanol/water systems of varying compositions are the most widely used (see Table 2) in view of their good droplet-forming characteristics and convenient viscosities. Separations are also relatively fast. In addition, as previously mentioned, small differences in the methanol contents lead to large polarity differences of the systems. Chlorinated solvents, which have high densities and low viscosities, are particularly suited for use as descending mobile phases.

The separation of the components of a mixture having a wide range of polarities is possible by DCCC if first the organic phase is used as mobile phase to elute less polar compounds. The polar components can then be recovered from the apparatus and the aqueous phase used as the mobile phase for their separation. For example, this method has been used to effect the separation of ethyl acetate-soluble acids from seeds of *Phaseolus coccineus* (Leguminosae) [15]. Alternatively, combinations of separation methods may be employed for complex problems; DCCC may then constitute one of the earlier prefractionation steps or, indeed, one of the final steps for obtaining pure compounds.

The simplicity of operation of a droplet countercurrent chromatograph and the very performant separations obtained with minimal quantities of solvent has led to the preponderance of applications of the method, greatly outnumbering those of other modern countercurrent techniques such as RLCCC.

A. Aqueous Systems

Polyphenols

The acidic hydroxyl groups in polyphenols often cause irreversible adsorption on solid stationary phases during column chromatographic procedures and it would appear that DCCC is ideally suited to the separation of this class of compounds. Furthermore, polyphenols often occur as polar glycosides and these are readily soluble in the solvent systems employed in DCCC. Indeed, many polyphenol glycosides have been successfully purified by DCCC, the flavonoid glycosides providing a large number of applications of the method [16]. In Table 4, several solvent systems for polyphenols are listed; it is immediately obvious that $CHCl_3/MeOH/H_2O$ mixtures of different compositions are the most widely used.

Each solvent system is not exclusive to a certain class of polyphenols and, indeed, there is frequent duplication of the solvent for different separation problems.

Flavonoids. In the case of flavonoid glycosides, it has been possible to isolate pure compounds directly from DCCC of the crude extracts.

Table 4 Selected DCCC Solvent Systems for Polyphenols

Class	Solvent
Flavonol aglycones	$CHCl_3/MeOH/n\text{-}BuOH/H_2O$ (10:10:1:6)
Flavonol glycosides	$CHCl_3/MeOH/H_2O$ (7:13:8)
Flavone glycosides	$CHCl_3/MeOH/n\text{-}PrOH/H_2O$ (5:6:1:4)
C-glycosylflavonoids	$CHCl_3/MeOH/n\text{-}BuOH/H_2O$ (10:10:1:6)
Isoflavone glycosides	$CHCl_3/MeOH/H_2O$ (7:13:8)
Anthocyanins	$n\text{-}BuOH/AcOH/H_2O$ (4:1:5)
Tannins	$n\text{-}BuOH/n\text{-}PrOH/H_2O$ (2:1:3)
Anthraquinones	$n\text{-}C_6H_{14}/EtOH/H_2O/EtOAc$ (5:4:1:2)
Anthraquinone glycosides	$CHCl_3/MeOH/H_2O$ (5:5:3)
Phenol glycosides	$CHCl_3/MeOH/H_2O$ (7:13:8)
Biphenyl glycosides	$CHCl_3/MeOH/H_2O$ (5:5:3)

For example, using the solvent system $CHCl_3/MeOH/n\text{-}PrOH/H_2O$ (5:6:1:4), a variant of the basic $CHCl_3/MeOH/H_2O$ system, in the descending mode, with 3.4-mm i.d. columns, a methanol extract of *Lomatogonium carinthiacum* (Gentianaceae) yielded the glycosides 6-hydroxyluteolin 7-O-glucoside (1) and 6-hydroxyluteolin 7-O-gentiobioside (2) [17,18].

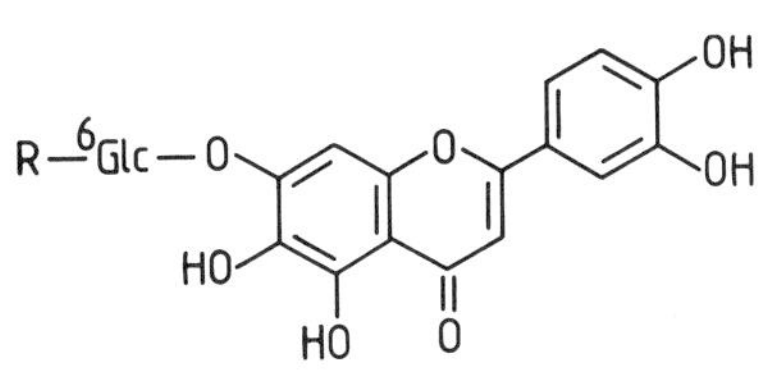

1 R = H

2 R = Glc

A further useful solvent system for the separation of flavonoid glycosides is $CHCl_3/MeOH/n\text{-}BuOH/H_2O$ (10:10:1:6). Employed in the ascending mode, this system was used in the isolation of 12 flavonol glycosides (3–14) from the leaves of *Securidaca diversifolia* (Polygalaceae) [19]. The chromatogram obtained is shown in Fig. 4. Final purifications of the glycosides were achieved by low-pressure reversed phase liquid chromatography and/or Sephadex LH-20 gel filtration.

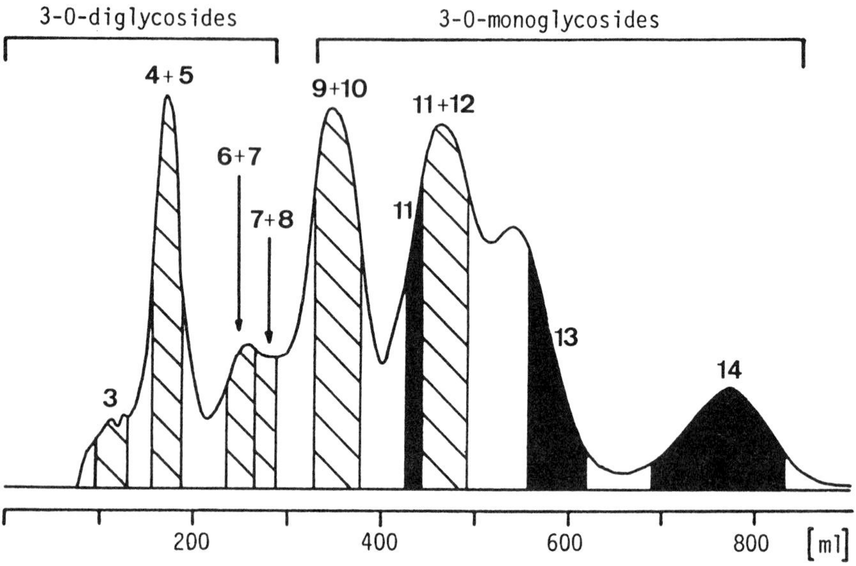

Fig. 4 DCCC separation of flavonoid glycosides from *Securidaca diversifolia* ($CHCl_3$/MeOH/n-BuOH/H_2O 10:10:1:6; ascending mode; 288 columns; 2.7 mm i.d.; sample: 2 g of MeOH extract).

For other practical examples, the field is well covered by a number of reviews; see Refs. [4,16,20].

Xanthones. Xanthones, similar to flavones, have been isolated by DCCC. Thus a crude chloroform extract (200 mg) of *Gentiana strictiflora* (Gentianaceae) gave bellidifolin (29 mg) and desmethylbellidifolin (10 mg), eluting with just 160 ml of descending mobile phase of the solvent system $CHCl_3$/MeOH/H_2O (13:7:4) [9].

Tannins. Tannins often exist as complicated mixtures in plants and are very difficult to purify. The operation of DCCC with solvents of high polarity has meant that this techniques lends itself particularly well to the separation of tannins. DCCC has been exploited by Okuda and coworkers for the isolation of a series of ellagitannins from *Geranium, Casuarina, Stachyurus,* and *Geum* species [21–23]. Included in these is the novel C-glucosidic ellagitannin casuarinin. Columns of 3.2 mm i.d. were used, with n-BuOH/n-PrOH/H_2O (2:1:3) (ascending) and (4:1:5) (ascending) as the solvents. DCCC was generally followed by gel filtration on Sephadex LH-20 as the final purification step.

The antiallergic ellagitannin agrimoniin has been isolated by a very similar method (n-BuOH/n-PrOH/H_2O 2:1:3, ascending) from rhizomes of *Potentilla erecta* (Rosaceae) [24].

Anthocyanins. Anthocyanins form another class of polyphenolics which gives separation problems. This is a result of their instability under any but acidic conditions and their irreversible adsorption to conventional column-packing materials. Recently, however, DCCC has been applied to the separation of anthocyanins from black currants (*Ribes nigrum*, Saxifragaceae) and raspberries (*Rubus idaeus*), using an acidic solvent system (n-BuOH/AcOH/H_2O, 4:1:5), the upper layer as mobile phase and columns of 2 mm i.d., illustrating well the versatility of the DCCC technique under conditions which deviate from neutrality. By this means, cyanidin 3-O-glucoside (15), delphinidin 3-O-glucoside (16), cyanidin 3-O-rutinoside (17), and delphinidin 3-O-rutinoside (18), extracted from blackcurrants, were separated very efficiently [25].

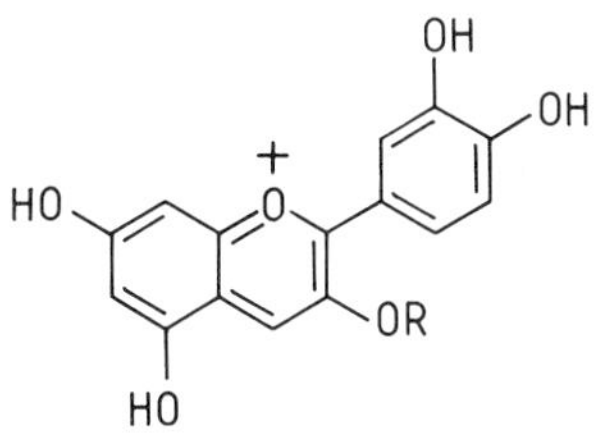

15 R = glucosyl

16 R = rutinosyl

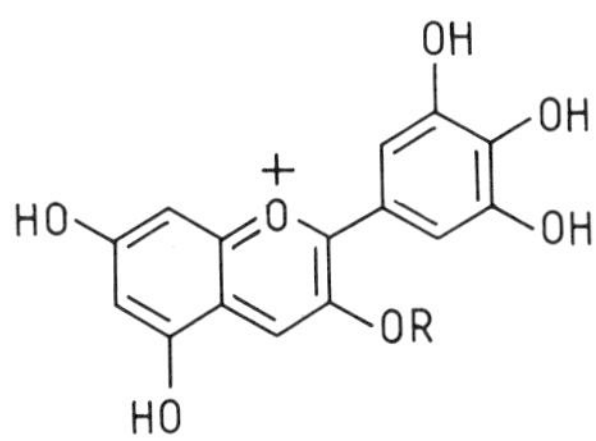

17 R = glucosyl

18 R = rutinosyl

Anthraquinones. Anthraquinones are often conveniently separated by DCCC. As well as the separation of sennosides from rhubarb [7], emodin 8-O-primeveroside, emodin 8-O-gentiobioside, and emodin 8-O-glucoside have been isolated from the stem bark of *Rhamnus catharticus* (Rhamnaceae) by repeated DCCC ($CHCl_3$/MeOH/H_2O/HCOOH, 70:130:80:1) in both ascending and descending modes [26].

Another interesting application of DCCC is in the separation of anthraquinone glycoside diastereoisomers. The separation of the diastereoisomers aloin A and aloin B of barbaloin (19), formerly accomplished by HPLC or by fractional crystallization, has now proved possible by DCCC [27]. The choice of solvent ($CHCl_3$/MeOH/H_2O, 7:13:8, on columns of 2 mm i.d.) was governed by consideration of the behavior of the diastereoisomers on TLC and their partition coefficeints, using HPLC to determine the partition of aloins A and B between the two solvent phases. Despite the conversion of aloin A into aloin B in aqueous solutions, the DCCC method was sufficiently rapid to avoid this problem.

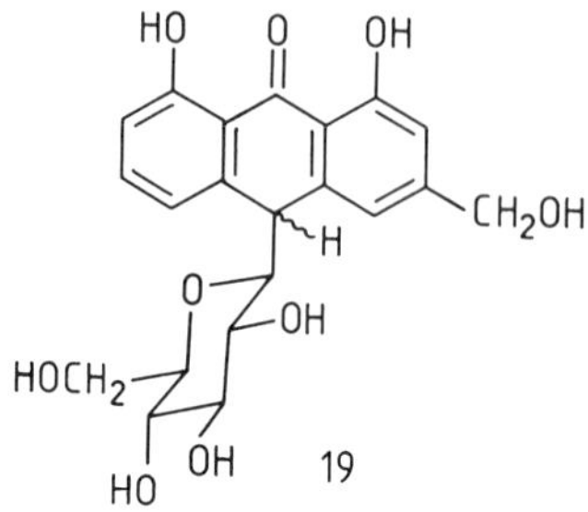

19

DCCC has been used for the examination of anthraquinones from cell suspension cultures. In the case of cultures of *Morinda citrifolia* (Rubiaceae), both the aglycones and glycosides of anthraquinones have been separated by DCCC [28], thus providing an example whereby weakly polar substances are also separated with an aqueous solvent system. Extraction of the cells was carried out with benzene, THF, and MeOH. The benzene extract was subjected to DCCC (n-C_6H_{14}/EtOH/H_2O/EtOAc, 5:4:1:2, ascending) and the weakly polar anthraquinones were eluted with the mobile phase. DCCC of the THF extract, under the same conditions, gave the anthraquinone aglycones in the mobile phase. The remaining stationary phase, containing the anthraquinone glycosides, was rechromatographed by DCCC using $CHCl_3$/MeOH/H_2O (5:5:3, descending) in order to separate the glycosides. Further anthraquinone glycosides were obtained from the methanol extract by DCCC with EtOAc/n-PrOH/H_2O (7:3:9) in the ascending mode.

Saponins

The saponins represent a very large group of complex natural products, often with important biological activities. Interest in these glycosidic compounds is growing rapidly, especially as the introduction of modern separation techniques, such as DCCC, has greatly simplified the problems associated with their isolation. DCCC is not a universal solution to the separation of saponins but, in conjunction with other separation techniques, provides a powerful tool for obtaining pure compounds. Saponins differ not only in the nature of the aglycone but also in the nature and length of the sugar chains. Sometimes the structural differences are very small and DCCC is of value in separating these closely related compounds, especially as the recoveries of chromatographed products can be as high as 98% [7]. Droplet countercurrent chromatography has found numerous applications in the separation of triterpene glycosides, steroid glycosides, and basic steroid saponins, but of these three groups, triterpene glycosides provide the most examples (Table 5).

From the table, it is immediately obvious that solvent systems formed of $CHCl_3/MeOH/H_2O$ mixtures are also well suited to the separation of various types of saponins [1,4] and can be used either in the ascending mode (very polar saponins) or in the descending mode (saponins possessing one or two sugars and few free hydroxyl groups).

A number of plants containing triterpene saponins have been found to exhibit strong molluscicidal activity and are therefore of great importance as potential agents for the control of schistosomiasis [29]. DCCC has been used for the isolation of four triterpenoid saponins (20–23) with strong molluscicidal activity from berries of the common ivy *Hedera helix* (Araliaceae) [30]. After extraction of the berries with 70% ethanol and partitioning the extract between n-butanol and water, the n-butanol layer, exhibiting the biological activity, was submitted to DCCC, using $CHCl_3/MeOH/H_2O$ (7:13:8, descending) as solvent (Fig. 5). The first peak to elute contained a mixture of fatty acids, chlorophylls etc. The pure compounds

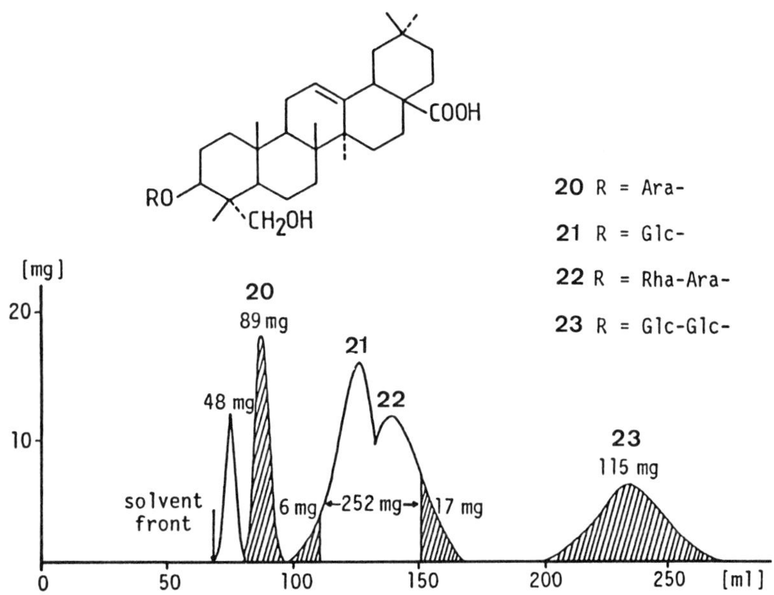

Fig. 5 DCCC of a crude extract of *Hedera helix* berries (1.2 g) with $CHCl_3/MeOH/H_2O$ 7:13:8; mobile phase: lower layer. (Reprinted with permission from Ref. 8.)

Table 5 Some Separations of Saponins by DCCC

Class	Source	Solvent system	Ref.
Triterpene glycosides	*Hedra helix* (Araliaceae)	$CHCl_3/MeOH/H_2O$ 7:13:8 (descending)	[30]
	Celmisia petriei (Compositae)	$CHCl_3/MeOH/H_2O$ 7:13:8 (ascending)	[31]
	Tetrapanax papyriferum (Rhamnaceae)	$CHCl_3/MeOH/H_2O$ 7:13:8 (ascending)	[32]
	Xanthoceras sorbifolia (Sapindaceae)	$CHCl_3/MeOH/H_2O$ 7:13:8 (ascending)	[33–35]
	Lonicera nigra (Caprifoliaceae)	$CHCl_3/MeOH/H_2O$ 5:9:6 (descending)	[36]
	Gleditsia japonica (Leguminosae)	$CHCl_3$/MeOH/n-PrOH/H_2O 9:12:1:8 (ascending)	[37]
	Gundelia tournefortii (Compositae)	$CHCl_3/MeOH/H_2O$/n-PrOH/EtOH 9:6:8:1:8 (ascending)	[38]
		$CHCl_3/MeOH/H_2O$/n-PrOH/EtOH 9:10:8:1:2 (descending)	
	Platycodon grandiflorum (Campanulaceae)	$CHCl_3$/MeOH/n-PrOH/H_2O 5:6:1·2:4	[39]
Dammarane saponins	*Hovenia dulcis* (Rhamnaceae)	$CHCl_3/MeOH/H_2O$ 7:13:8 (ascending)	[40,41]
		$CHCl_3/MeOH/H_2O$ 5:6:4 (ascending)	
	Zizphyus jujuba (Rhamnaceae)	$CHCl_3/MeOH/H_2O$ 5:6:4 (ascending)	[42]
	Colubrina asiatica (Rhamnaceae)	$CHCl_3$/MeOH/n-PrOH/H_2O 9:12:1:8	[43]
	Bacopa monniera (Scrophulariaceae)	$CHCl_3/MeOH/H_2O$ 5:6:4	[44]

Table 5 (Continued)

Class	Source	Solvent system	Ref.
Spirostanol saponins	*Balanites aegyptiaca* (Balanitaceae)	CH_2Cl_2/MeOH/H_2O 8:13:7 (ascending)	[45,46]
Furostanol glycosides	*Trigonella foenum-gracecum* (Leguminosae)	$CHCl_3$/MeOH/H_2O 7:13:8 (descending)	[47]
Steroid glycosides from marine organisms	*Acanthaster planci* (Acanthasteridae)	EtOAc/n-BuOH/H_2O 1:2:3 (ascending) $CHCl_3$/MeOH/n-PrOH/H_2O 7:13:1:8 (descending) $CHCl_3$/MeOH/H_2O 7:13:8 (descending)	[48]
	Luidia maculata (Asteroidea)	n-BuOH/H_2O	[5]
	Hacelia attenuata (Ophidiasteridae)	$CHCl_3$/MeOH/H_2O 10:14:7 (descending)	[49]
	Ophidiaster ophidianus (Opidiasteridae)	n-BuOH/Me_2CO/H_2O 3:1:5 (descending)	[50]
	Linckia laevigata	n-BuOH/Me_2CO/H_2O 3:1:5 (ascending)	[51]
	Protoreaster nodosus (Oreasteridae)	n-BuOH/Me_2CO/H_2O 3:1:5 (ascending)	[52]
Basic steroid saponins	*Fritillaria thunbergii* (*Liliaceae*)	CHCl/MeOH/1% aq. NH_3 7:12:8 (ascending)	[53]

(20–23) (shaded areas in Fig. 5) were obtained in order of increasing polarity, as expected by use of the less polar layer as mobile phase, and were shown to be glycosides of hederagenin. The most polar components (free sugars, polysaccharides, etc.) were recovered from the stationary phase of the apparatus. In one separation step, therefore, major plant constituents were obtained directly in pure form from a crude extract with low solvent consumption.

Extensive use of DCCC, with $CHCl_3/MeOH/n\text{-}PrOH/H_2O$ (5:6:1.2:4) as solvent system, in the separation of 15 bidesmosidic oleanane-type saponins from the roots of *Platycodon grandiflorum* (Campanulaceae) has been reported [39]. In some cases, isomeric saponins differing only in the positions of substitution of the acetate groups on the sugar residues were successfully separated by this particular DCCC solvent system.

Marine organisms are currently being widely investigated and have furnished a variety of exotic new compounds. Among these are a number of steroid glycoside sulfates which have been successfully separated by DCCC [48]. Since these glycosides have a tendency to decompose on preservation or in solution, procedures for their rapid separation are required; DCCC is ideally suited to this particular problem. Furthermore, the use of DCCC has led to the isolation of pure products from very complex mixtures. For example, the starfish *Acanthaster planci* (Acanthasteridae) yielded a pregn-9(11)-ene derivative and two related glycoside sulfates after an extensive purification procedure which included several DCCC steps. These were (a) $EtOAc/n\text{-}BuOH/H_2O$ solvent system, using 18 columns, each 10 mm i.d.; (b) $CHCl_3/MeOH/n\text{-}PrOH/H_2O$ (7:13:1:8, descending); and (c) $CHCl_3/MeOH/H_2O$ (7:13:8, descending) [48].

A large-scale DCCC procedure for preliminary purification, using 19 columns (10 mm i.d.; 30 cm long) with n-BuOH-saturated H_2O as the stationary phase and H_2O-saturated n-BuOH as the mobile phase, has been applied to the isolation of a cholest-9(11)-en-23-one glycoside sulfate from the starfish *Luidia maculata* (Asteroidea) [5].

Cardenolides

These are structurally related to saponins and the same solvent systems can often be used to effect separations.

In the case of *Digitalis lanata* (Scrophulariaceae), more than 70 cardenolides have been found and there are often only subtle differences (mainly in the nature of the sugar moiety) in the structures of these pharmacologically important heart glycosides. Their separation is usually problematical. Despite the large number of known cardenolides, three new glycosides have been isolated from *Digitalis lanata*, largely by means of DCCC [54]. The DCCC solvent systems $CHCl_3/MeOH/H_2O$ (5:6:4) and $CH_2Cl_2/MeOH/H_2O$ (5:6:4) (both descending) were used for the separation of the glycosides from a leaf extract.

Further applications of DCCC to the separation of cardenolides have been reported in the isolation of affinosides from *Anodendron affine* (Apocynaceae) [55,56], in the isolation of cardenolide glycosides and an aglycone from *Adonis aleppica* (Ranunculaceae; $CHCl_3$/MeOH/H_2O 5:6:4 and CH_2Cl_2/MeOH/H_2O 5:6:4) [57], in the isolation of thevetins from *Thevetia peruviana* (Apocynaceae; $CHCl_3$/MeOH/H_2O 7:13:9) [58] and of closely related strophanthidin glycosides from *Lophopetalum toxicum* (Celastraceae) [59].

Diterpenes

A number of very efficient separations of weakly polar diterpenes have been performed with aqueous DCCC solvent systems. Droplet countercurrent chromatography is particularly suitable for the separation of, e.g., phorbol, ingenol, and ingol derivatives for several reasons: (a) the technicalities of the separation method are simpler than the previously employed Craig countercurrent distribution (as mentioned previously); (b) separation times are not too long; (c) autoxidation is kept to a minimum for these sensitive and sometimes unstable substances; (d) conventional chromatographic techniques are often not sufficient to separate the components of the complex mixtures that are obtained. By means of DCCC, phorbol, the parent alcohol of the potent tumor promoter 12-O-tetradecanoyl phorbol 13-acetate (TPA), was separated from 4α-phorbol, using n-hexane/diethyl ether/n-propanol/95% ethanol/water 4:8:3:5:4 (descending) [60,61].

Eleven short-chain esters of phorbol and 4-deoxy-4α-phorbol were isolated from croton oil (*Croton tiglium*, Euphorbiaceae) by combinations of DCCC (n-hexane/diethyl ether/n-propanol/ethanol/water (17:40:15:25:40, ascending), low-pressure column chromatography and preparative TLC [62].

An antileukemic quassinoid, simalikalactone D, was obtained from an ethanol extract of the wood of *Simaba multiflora* (Simaroubaceae) with $CHCl_3$/C_6H_6/MeOH/H_2O (3:1:5:2, descending) as the DCCC solvent system [63].

Iridoids

The technique of DCCC has been reasonably well exploited for the iridoid glycosides (see Table 6 for representative examples).

In general, $CHCl_3$/MeOH/H_2O mixtures are very suitable for the separation of iridoid glycosides but for iridoid glycoside esters it is desirable to use the less polar lower layer as mobile phase. However, final purification of three iridoid glycosides from *Randia canthioides* (Rubiaceae) using another solvent system, n-BuOH/EtOH/H_2O (4:1:5, ascending) and preparative scale TLC has been reported [73]. Columns of 3.4 mm i.d. were employed.

Table 6 Separations of Iridoid Glycosides by DCCC

Source	Solvent system	Ref.
Ajuga pyramidalis (Labiatae)	$CHCl_3/MeOH/H_2O$ 43:37:20 (ascending)	[9]
Lonicera coerulea (Caprifoliaceae)	$CHCl_3/MeOH/H_2O$ 43:37:20 (descending)	[64]
Nepeta cataria (Labiatae)	$CHCl_3/MeOH/H_2O$ 5:5:3 (descending)	[65]
Actinidia polygama (Actinidiaceae)	$CHCl_3/MeOH/H_2O$ 5:5:3 (descending)	[69]
Allamanda cathartica (Apocynaceae)	$CHCl_3/MeOH/H_2O$ 5:6:4 (ascending)	[66]
Allamanda neriifolia (Apocynaceae)	$CHCl_3/MeOH/H_2O$ 5:6:4 (ascending)	[67]
Sambucus ebulus (Caprifoliaceae)	$CHCl_3/MeOH/H_2O$ 5:6:4 (descending)	[100]
Olea europaea (Oleaceae)	$CHCl_3/MeOH/H_2O$ 13:7:4 (ascending)	[68]
Penstemon serrulatus (Scrophulariaceae)	$CHCl_3/MeOH/H_2O$ 7:13:8 (ascending)	[70,71]
Penstemon whippleanus (Scrophulariaceae)	$CHCl_3/MeOH/n\text{-}PrOH/H_2O$ 7:12:1:8 (ascending)	[72]
Randia canthoides (Rubiaceae)	$n\text{-}BuOH/EtOH/H_2O$ 4:1:5 (ascending)	[73]
Ligustrum japonicum (Oleaceae)	$n\text{-}BuOH\text{-}EtOH/H_2O$ 4:1:5 (ascending)	[74]

DCCC ($CHCl_3$/MeOH/H_2O 5:5:3, descending) has been used in the separation of iridoid enol glucosides from *Actinida polygama* (Actinidiaceae) [69]. In one instance, the DCCC apparatus was charged with a 14.4-g sample and pure iridodialo-β-D-gentiobioside was obtained from the separation.

Coumarins

Two DCCC separations of coumarins have appeared in the literature. Yang and Kinghorn [75] successfully isolated scopoletin, 6,7,8-trimethoxycoumarin and 5,6,7,8-tetramethoxycoumarin from the Chinese tallow tree *Sapium sebiferum* (Euphorbiaceae) with the aid of the solvent system employed for the separation of diterpenes, C_6H_6/Et_2O/n-PrOH/EtOH/H_2O (7:16:6:10:8, ascending). Scopoletin was also isolated from *Gundelia tournefortii* (Compositae) by DCCC with $CHCl_3$/C_6H_6/MeOH/H_2O (15:15:23:7, descending) [76].

Alkaloids

Opium alkaloids have been successfully separated from *Papaver somniferum* (Papaveraceae) by DCCC [77]. Using the more polar upper layer of a chloroform/methanol/water system as the mobile phase, crude morphine, pure thebaine, and a mixture of narceine and codeine were obtained. After elution of thebaine, the stationary phase was recovered and rechromatographed with the less polar layer as mobile phase. By this means, narcotine and papaverine were separated. A one-step procedure can be achieved if the pH of the mobile phase is modified during the elution. In this way, Tani et al. [78] were able to resolve highly complex alkaloid mixtures from Papaveraceae, including the isolation of two new protoberberine alkaloids from *Corydalis ophiocarpa* (Papaveraceae).

The versatility of DCCC techniques is ably demonstrated by the use of a buffer-containing solvent system (0.5 M phosphate buffer/MeOH/$CHCl_3$/n-PrOH adjusted to pH 5) for the purification of commercially available veratrine [79]. A measure of pH control during the separation is thus achieved. Since the organic layer was used as the mobile phase, problems of removing the buffer from the purified sample were minimal. Veratridine, which is a stimulator of cellular Na^+ influx, was completely separated from cevadine and other more rapidly eluting contaminants.

A series of papers on the application of DCCC to the isolation of alkaloids from *Strychnos usambarensis* (Loganiaceae) has been published [80–82]. Previously, difficulties had been encountered in obtaining pure alkaloids from this plant, mainly due to irreversible adsorptions on solid supports during chromatography [80]. Fluorocurarine was obtained by two successive DCCC runs (solvent systems $CHCl_3$/MeOH/n-PrOH/H_2O 9:12:1:8, descending and $CHCl_3$/

$MeOH/H_2O$ 5:10:6, descending) after preliminary purification by column chromatography [80]. Significantly, the 10-hydroxy and 11-hydroxy isomers of unsambarine were successfully separated as their highly polar chloride salts by DCCC ($CHCl_3$/MeOH/pH 3·6 acetate buffer 9:12:8, descending mode) [82].

Several closely related pyrrolizidine alkaloids have been isolated from methanol extracts of *Senecio cacaliaster* (Compositae) and *Alkanna tinctoria* (Boraginaceae), employing DCCC (toluene/chloroform/methanol/water 5:5:7:2, ascending) as one of the purification steps [83,84].

Other separations of alkaloids by DCCC, together with the relevant operating conditions, include:

Quaternary alkaloids from *Corydalis tashiroi* (Papaveraceae) [85] $CHCl_3/MeOH/H_2O$ (5:5.7:3, descending)
Ipecoside [86] $CHCl_3$/MeOH/n-PrOH/H_2O (45:60:2:40, descending)
Benzylisoquinoline alkaloid from *Zizyphus jujuba* (Rhamnaceae) [87] $C_6H_6/CHCl_3/MeOH/H_2O$ (5:5:7:2, ascending)

Carbohydrates

Ogihara et al. [7] were able to obtain separations of galactose, xylose, rhamnose and glucose, arabinose, fucose monosaccharide mixtures by simply introducing their solutions into a DCCC apparatus running with $CHCl_3/MeOH/H_2O$ (7:13:8, ascending; 600 tubes; 40 cm × 1.65 mm i.d.).

As an extension of this method, monosaccharides resulting from the hydrolysis of saponins have been purified by DCCC, also using as solvent $CHCl_3/MeOH/H_2O$ (7:13:8) [88]. Furthermore, oligosaccharides obtained as partial hydrolysis products of saponins have been separated by DCCC. Using $CHCl_3/MeOH/H_2O$ 7:12;:8 in the ascending mode, two disaccharides and a trisaccharide were separated from the hydrolysis products of a hexaglycosidic saponin, after preliminary purification on a Sephadex LH-20 column [89].

Amino Acids and Peptides

As previously mentioned, the first reported separation with droplet countercurrent chromatography involved dinitrophenyl amino acids [12]. Since then, however, relatively few applications of DCCC in the field of amino acids and peptides have been described, despite the potential for the use of acidic, highly polar solvent systems (Table 7).

In a rather novel area of DCCC, the complete resolution of DL-isoleucine into its enantiomers has been achieved in a two-phase buffered n-butanol/water system containing Cu^{2+} and N-n-dodecyl-

Table 7 Solvents for the Separation of Amino Acids and Peptides by DCCC

$CHCl_3$/MeOH/0.1 M HCl (19:19:12)
$CHCl_3$/MeOH/H_2O/C_6H_6 (15:23:7:15)
$CHCl_3$/MeOH/0.1 M HCl/C_6H_6 (5:10:4:11)
$CHCl_3$/HOAc/0.1 M HCl (2:2:1)
i-BuOH/TFA/H_2O (120:1:160)
i-BuOH/HOAc/H_2O (4:1:5)

L-proline [90]. Teflon tubes of 4 mm i.d. and a flow rate (descending mode) of 1 ml/min were employed. The copper ions were extracted into the organic phase as a complex with the long chain proline derivative, thus facilitating the distribution of the enantiomers into the organic phase.

A combination of DCCC and chromatography on Amberlite XAD-2, silica gel, and Sephadex LH-20 has been used in the isolation of antitumor cyclic hexapeptides from *Rubia cordifolia* (Rubiaceae) [91]. These hexapeptides have been shown to possess activity against a number of experimental animal tumors, including solid tumors [91].

Miscellaneous

Lipids. When dealing with complex lipids, Craig counterucrrent distributions tend to form stable emulsions which are hard to separate into two phases. However, DCCC is ideally suited to the problems of separating lipids, examples of separations having been reported by Otsuka and Yamakawa [92]. They isolated glycolipids from a variety of sources, using several different DCCC solvent systems, most notably $CHCl_3$/MeOH/0.5% aq. $CaCl_2$/n-PrOH (50:60:40:6, ascending) and $CHCl_3$/MeOH/H_2O/n-PrOH (50:60:40:2, ascending), with 500 standard columns. Alternative column chromatographic procedures proved much more tedious and more time and solvent consuming.

A fractionation of human serum lipids has been described by Konobu et al. [93]. With heptane/n-butanol/chloroform/methanol/60% acetic acid (3:2:2:3:5), a baseline separation of tricolein, palmitic acid, capric acid, and lecithin was achieved by the ascending method.

Steroids. Phytoecdysones have successfully been isolated by DCCC and, in the case of phytoecdysones of *Ajuga remota* (Labiatae), separation by DCCC was in fact preferred to separation by preparative scale HPLC [94]. Direct fractionation of an ethyl acetate extract of the roots with $CHCl_3/MeOH/H_2O$ (13:7:4, ascending) gave rapid and nondestructive separations of β-ecdysone, ajugasterone C, and cyasterone, in the decigram amounts required for biological testing.

Polyhydroxylated sterols have been obtained from the starfish *Protoreaster nodosus* (Oreasteridae) [95] and sulfated steroids from *Ophioderma longicaudum* (Ophiuroidae) [96] by a combination of DCCC ($CHCl_3/MeOH/H_2O$ 7:13:8, ascending mode) and preparative HPLC.

Cyanogenic Glycosides

As well as a flavone glucuronide, scutellarin, rosmarinic acid, and caffeic acid, two cyanogenic glycosides [prunasin and (R)-2-(2-O-β-D-glucopyranosyl-β-D-glucopyranosyloxy)phenylacetonitrile] have been isolated from leaves of *Perilla frutescens* (Labiatae). The procedure involved a preliminary separation on Amberlite XAD-2, DCCC with the solvent $CHCl_3/MeOH/H_2O$/n-BuOH (9:12:8:2, ascending), Diaion HP 20 AG column chromatography, and silica gel chromatography [97].

B. Nonaqueous Systems

The use of nonaqueous solvent systems in DCCC has enormous potential for the separation of weakly polar or apolar compounds, especially those which are unstable in the presence of water or decompose during chromatography on silica gel.

One of the first applications of such a solvent system was in the direct isolation of vulpinic acid (24) from an extract of the lichen *Letharia vulpina* on columns of 2.7 mm i.d. using methanol/acetone/n-heptane 4:1:5 in the descending mode [11].

24

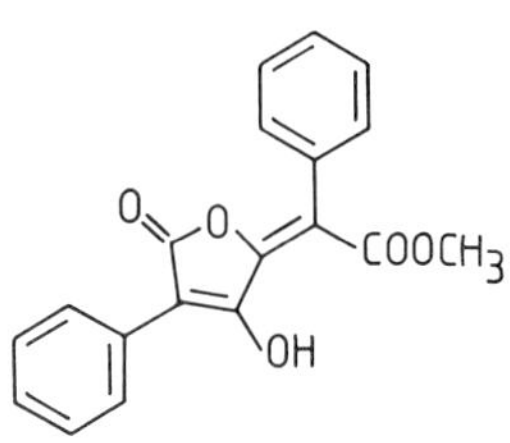

The triterpenes oleanolic acid and hederagenin could be separated from an acid-hydrolyzed extract of *Hedera helix* (Araliaceae) (methanol/1,2-dichloroethane/n-heptane 57.6:4.8:37.6, descending), while a mixture of betulinic acid, betulin, β-amyrin, and cholesterol could be separated into its components by DCCC with acetonitrile/dichloromethane/n-heptane 7:3:10, in the descending mode [11]. However, with methanol/acetone/n-heptane 4:1:5, betulin and betulinic acid were not separated.

The iridoid isovaltrate was obtained in a pure state from *Valeriana edulis* (Valerianaceae) after DCCC with the solvent n-hexane/ethyl acetate/nitromethane/methanol 8:2:2:3, first in the ascending and then in the descending mode [98]. However, caution should be exercised with nitromethane, which is incompatible with UV detection and may cause oxidation.

Geranyl isobutyrate was isolated from the essential oil of *Echinacea angustifolia* (Compositae) by a similar procedure to that above, using n-heptane/ethyl acetate/nitromethane/methanol 8:2:2:3 in the descending mode [98].

By using another nitromethane system, the essential oil of chamomile (*Matricaria chamomilla,* Compositae) could be split into its terpenoid constituents farnesene, chamazulene, spathulenole, bisabolole, and bisabololoxides A and B by DCCC, first with the solvent system n-hexane/ethyl acetate/nitromethane/methanol 9:2:2:3 in the ascending mode and then in the proportions 30:5:15:5, also in the ascending mode [98].

Recently another nonaqueous solvent has been described—and applied to the isolation of furoacridone alkaloids from tissue cultures of *Ruta graveolens* (Rutaceae) [99]. A dichloromethane extract of callus tissue was chromatographed by DCCC, with the system n-heptane/cyclohexane/dichloromethane/toluene/acetonitrile (8:2:3:1:5:7, descending) and the acridone alkaloids, including the new compounds 1-hydroxyrutacridone epoxide (25) and rutagravin (26), were finally purified by silica gel column chromatography and preparative TLC.

25

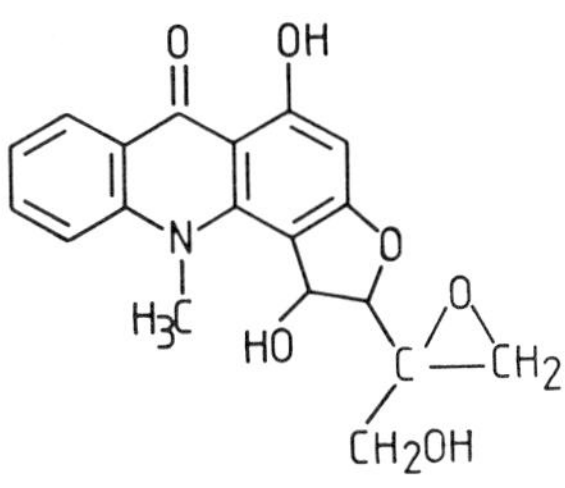

26

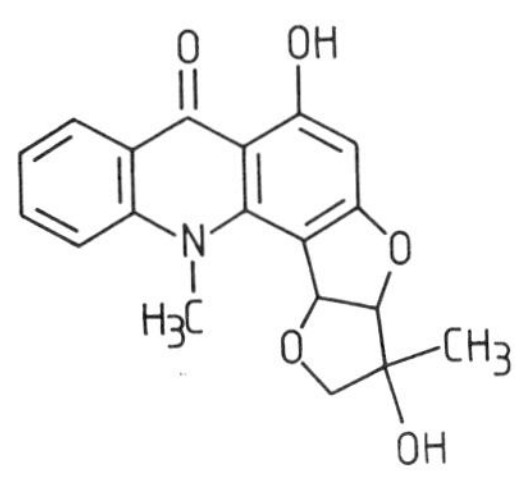

V. OUTLOOK

As illustrated in this chapter, DCCC has found the widest range of applications of all countercurrent chromatographic methods. A large number of natural products, many of them highly polar, have been isolated by procedures which include at least one DCCC step. The reasons for this are numerous but one of the most obvious is the ease of manipulation of droplet countercurrent chromatographs. There are no rotating parts, thus immediately eliminating one complication. Although droplet formation is essential to the success of the method, establishment of the initial conditions is straightforward and no preliminary purification of crude extracts is necessary. Multigram quantities can be loaded at one time and capacities are considerably higher than those achieved with other countercurrent techniques. Efforts are being made to discover new solvent systems (notably nonaqueous mixtures), shorten separation times, and further increase capacity, but nevertheless DCCC is a very useful tool in the hands of separation scientists.

While the need to have solvent systems that produce reasonable droplets is a disadvantage of DCCC, the biggest drawback is that separation times are so long—often a matter of many hours or even days. This is a familiar problem with gravity-driven chromatographic methods. However, other countercurrent techniques, such as centrifugal countercurrent chromatography, have greatly increased speeds of separation; in some cases resolution is also improved. Consequently, these new developments in countercurrent chromatography, once they become cheaper and more accessible, will probably represent the future of liquid-liquid partition methods for the separation of natural products.

REFERENCES

1. K. Hostettmann, *Planta Med.*, *39*, 1 (1980).
2. K. Hostettmann, in *Natural Products as Medicinal Agents* (J. L. Beal and E. Reinhard, eds.), Hippokrates-Verlag, Stuttgart, 1981, p. 79.
3. K. Hostettmann, *Adv. Chromatogr.*, *21*, 165 (1983).
4. K. Hostettmann, M. Hostettmann, and A. Marston, *Nat. Prod. Rep.*, *1*, 471 (1984).
5. T. Komori, H. C. Krebs, Y. Itakura, R. Higuchi, K. Sakamoto, S. Taguchi, and T. Kawasaki, *Liebigs Ann. Chem.*, 2092 (1983).
6. K. Hostettmann, C. Appolonia, B. Domon, and M. Hostettmann, *J. Liq. Chromatogr.*, *7*, 231 (1984).
7. Y. Ogihara, O. Inoue, H. Otsuka, K. Kawai, T. Tanimura, and S. Shibata, *J. Chromatogr.*, *128*, 218 (1976).
8. K. Hostettmann, M. Hostettmann-Kaldas, and K. Nakanishi, *J. Chromatogr.*, *170*, 355 (1979).
9. K. Hostettmann, M. Hostettmann-Kaldas, and O. Sticher, *Helv. Chim. Acta*, *62*, 2079 (1979).
10. J. K. Snyder, K. Nakanishi, K. Hostettmann, and M. Hostettmann, *J. Liq. Chromatogr.*, *7*, 243 (1984).
11. B. Domon, M. Hostettmann, and K. Hostettmann, *J. Chromatogr.*, *246*, 133 (1982).

12. T. Tanimura, J. J. Pisano, Y. Ito, and R. L. Bowman, *Science*, *169*, 54 (1970).
13. H. Otsuka, Y. Morita, Y. Ogihara, and S. Shibata, *Planta Med.*, *32*, 9 (1977).
14. I. Kubo, A. Matsumoto, F. J. Hanke, and J. F. Ayafor, *J. Chromatogr.*, *321*, 246 (1985).
15. J. R. Bearder and J. MacMillan, in *"Gibberellins: Chemistry, Physiology and Use"* Monograph 5, (J. R. Lenton, ed.) British Plant Growth Regulator Group, Wantage, 1980, p. 25.
16. K. Hostettmann and M. Hostettmann, in *The Flavonoids: Advances in Research* (J. B. Harborne and T. J. Mabry, eds.) Chapman and Hall, London, 1982, p. 1.
17. D. Schaufelberger, M. Hamburger, and K. Hostettmann, *Bull. Liais. Groupe Polyphénols* (Narbonne), 11 (1983).
18. D. Schaufelberger and K. Hostettmann, *Phytochemistry*, *23*, 787 (1984).
19. M. Hamburger, M. Gupta, and K. Hostettmann, *Phytochemistry*, *24*, 2689 (1985).
20. K. Hostettmann, in *Annual Proceedings of the Phytochemical Society of Europe*, Vol. 25, (C. F. van Sumere and P. J. Lea, eds.), Clarendon, Oxford, 1985, p. 1.
21. T. Okuda, T. Yoshida, M. Ashida, and K. Yazaki, *J. Chem. Soc. Perkin I*, 1765 (1983).
22. T. Okuda, T. Yoshida, M. Kuwahara, M. U. Memon, and T. Shingu, *Chem. Pharm. Bull.*, *32*, 2165 (1984).
23. T. Yoshida, Y. Maruyama, M. U. Memon, T. Shingu, and T. Okuda, *Phytochemistry*, *24*, 1041 (1985).
24. K. Lund and H. Rimpler, *Deutsche Apotheker Zeitung*, *125*, 105 (1985).
25. G. W. Francis and Ø. M. Andersen, *J. Chromatogr.*, *283*, 445 (1984).
26. H.-W. Rauwald, *Z. Naturforsch. C.*, *38*, 170 (1983).
27. H.-W. Rauwald, *Arch. Pharm.* (Weinheim), *315*, 769 (1982).
28. K. Inoue, H. Nayeshiro, H. Inouye, and M. Zenk, *Phytochemistry*, *20*, 1693 (1981).
29. A. Marston and K. Hostettmann, *Phytochemistry*, *24*, 639 (1985).
30. K. Hostettmann, *Helv. Chim. Acta*, *63*, 606 (1980).
31. D. D. Rowan and R. H. Newman, *Phytochemistry*, *23*, 639 (1984).
32. M. Takai, S. Amagaya, and Y. Ogihara, *J. Chem. Soc. Perkin I*, 1801 (1977).
33. Y. Chen, T. Takeda, and Y. Ogihara, *Chem. Pharm. Bull.*, *33*, 127 (1985).
34. Y. Chen, T. Takeda, and Y. Ogihara, *Chem. Pharm. Bull.*, *33*, 1043 (1985).

35. Y. Chen, T. Takeda, and Y. Ogihara, *Chem. Pharm. Bull.*, *33*, 1387 (1985).
36. B. Domon and K. Hostettmann, *Helv. Chim. Acta.*, *66*, 422 (1983).
37. Y. Okada, K. Koyama, K. Takahashi, T. Okuyama, and S. Shibata, *Planta Med.*, *40*, 185 (1980).
38. H. Wagner, H. Nickl, and Y. Aynehchi, *Phytochemistry*, *23*, 2505 (1984).
39. H. Ishii, K. Tori, T. Tozyo, and Y. Yoshimura, *J. Chem. Soc. Perkin I*, 661 (1984).
40. O. Inoue, T. Takeda, and Y. Ogihara, *J. Chem. Soc. Perkin I*, 1289 (1978).
41. Y. Kimura, Y. Kobayashi, T. Takeda, and Y. Ogihara, *J. Chem. Soc. Perkin I*, 1923 (1981).
42. N. Okamura, T. Nohara, A. Yagi, and I. Nishioka, *Chem. Pharm. Bull.*, *29*, 676 (1981).
43. H. Wagner, S. Ott, K. Juric, J. Morton, and A. Neszmelyi, *Planta Med.*, *48*, 136 (1983).
44. K.-I. Kawai and S. Shibata, *Phytochemistry*, *17*, 287 (1978).
45. H.-W. Liu and K. Nakanishi, *J. Am. Chem. Soc.*, *103*, 7005 (1981).
46. H.-W. Liu and K. Nakanishi, *Tetrahedron*, *38*, 513 (1982).
47. R. K. Gupta, D. C. Jain, and R. S. Thakur, *Phytochemistry*, *24*, 2399 (1985).
48. T. Komori, J. Matsuo, Y. Itakura, K. Shakamoto, Y. Ito, S. Taguchi, and T. Kawasaki, *Liebigs Ann. Chem.*, 24 (1983).
49. L. Minale, C. Pizza, F. Zollo, and R. Riccio, *J. Nat. Prod.*, *46*, 736 (1983).
50. R. Riccio, C. Pizza, O. Squillace-Greco, and L. Minale, *J. Chem. Soc. Perkin I*, 655 (1985).
51. R. Riccio, O. S. Greco, and L. Minale, *J. Nat. Prod.*, *48*, 97 (1985).
52. R. Riccio, F. Zollo, E. Finamore, L. Minale, D. Laurent, G. Bargibant, and J. Pusset, *J. Nat. Prod.*, *48*, 266 (1985).
53. J. Kitajima, T. Komori, T. Kamasaki, and H.-R. Schulten, *Phytochemistry*, *21*, 187 (1982).
54. D. Krüger, P. Junior, and M. Wichtl, *Planta Med.*, *49*, 74 (1983).
55. F. Abe and T. Yamauchi, *Chem. Pharm. Bull.*, *30*, 1183 (1982).
56. F. Abe and T. Yamauchi, *Chem. Pharm. Bull.*, *31*, 1199 (1983).
57. P. Junior, D. Krüger, and C. Winkler, *Deutsche Apotheker Zeitung*, *125*, 1945 (1985).
58. I. M. Said, *Planta Med.*, *51*, 536 (1985).
59. H. Wagner, H. Habermeier, and H.-R. Schulten, *Helv. Chim. Acta*, *67*, 54 (1984).
60. G. T. Marshall and A. D. Kinghorn, *J. Chromatogr.*, *206*, 421 (1981).

61. A. D. Kinghorn and G. T. Marshall, U.S. Patent 4,468,328 (1984).
62. G. T. Marshall and A. D. Kinghorn, *J. Am. Oil Chem. Soc.*, *61*, 1220 (1984).
63. M. Arisawa, A. Fujita, N. Morita, A. D. Kinghorn, G. A. Cordell, and N. R. Farnsworth, *Planta Med.*, *51*, 348 (1985).
64. I. Calis and O. Sticher, *J. Nat. Prod.*, *48*, 108 (1985).
65. M. Tagawa and F. Murai, *Planta Med.*, *39*, 144 (1980).
66. T. Yamauchi, F. Abe, and M. Taki, *Chem. Pharm. Bull.*, *29*, 3051 (1981).
67. F. Abe, T. Mori, and T. Yamauchi, *Chem. Pharm. Bull.*, *32*, 2947 (1984).
68. I. Kubo and A. Matsumoto, *J. Agric. Food Chem.*, *32*, 687 (1984).
69. F. Murai and M. Tagawa, *Planta Med.*, *37*, 234 (1979).
70. P. Junior, *Planta Med.*, *50*, 438 (1984).
71. P. Junior, *Planta Med.*, *50*, 417 (1984).
72. P. Junior, *Planta Med.*, *50*, 444 (1984).
73. S. Uesato, E. Ali, H. Nishimura, I. Kawamura, and H. Inouye, *Phytochemistry*, *21*, 353 (1982).
74. K. Inoue, T. Nishioka, T. Tanahashi, and H. Inouye, *Phytochemistry*, *21*, 2305 (1982).
75. P. Yang and A. D. Kinghorn, *J. Nat. Prod.*, *48*, 486 (1985).
76. A. Kery, A. M. Hussein and A. Al-Jeboory, *Fitoterapia*, *56*, 42 (1985).
77. K. Hostettmann, M. Hostettmann-Kaldas, and O. Sticher, *J. Chromatogr.*, *186*, 529 (1979).
78. C. Tani, N. Nagakura, and C. Kuriyama, *Yakugaku Zasshi*, *98*, 1658 (1978).
79. G. Holan, W. M. P. Johnson, and K. Rihs, *J. Chromatogr.*, *288*, 479 (1984).
80. M. Caprasse, C. Coune, and L. Angenot, *J. Pharm. Belg.*, *36*, 243 (1981).
81. M. Caprasse, C. Coune, and L. Angenot, *J. Pharm. Belg.*, *38*, 135 (1983).
82. M. Caprasse, D. Tavenier, and L. Angenot, *J. Pharm. Belg.*, *38*, 211 (1983).
83. E. Roeder, H. Wiedenfeld, and R. Britz-Kirstgen, *Phytochemistry*, *23*, 1761 (1984).
84. E. Roeder, H. Wiedenfeld, and R. Schraut, *Phytochemistry*, *23*, 2125 (1984).
85. C. Tani, N. Nagakura, S. Saeki, and M. T. Kao, *Planta Med.*, *41*, 403 (1981).
86. G. Höfle, N. Nagakura, and M. H. Zenk, *Chem. Ber.*, *113*, 566 (1980).

87. H. Otsuka, Y. Ogihara, and S. Shibata, *Phytochemistry, 13*, 2016 (1974).
88. Y. Kobayashi, T. Takeda, and Y. Ogihara, *Chem. Pharm. Bull., 29*, 2222 (1981).
89. H. Nickl, PhD. thesis, Munich, 1982.
90. T. Takeuchi, R. Horikawa, and T. Tanimura, *J. Chromatogr., 284*, 285 (1984).
91. H. Itokawa, K. Takeya, N. Mori, T. Hamanaka, T. Sonobe, and K. Mihara, *Chem. Pharm. Bull., 32*, 284 (1984).
92. H. Otsuka and T. Yamakawa, *J. Biochem., 90*, 247 (1981).
93. K. J. Konubu, H. Takao, and M. Kuroda, *Yakugaku Zasshi, 96*, 1379 (1976).
94. I. Kubo, J. A. Klocke, I. Ganjian, N. Ichikawa, and T. Matsumoto, *J. Chromatogr., 257*, 157 (1983).
95. L. Minale, C. Pizza, R. Riccio, C. Sorrentino. and F. Zollo, *J. Nat. Prod., 47*, 790 (1984).
96. R. Riccio, M. V. D'Auria, and L. Minale, *Tetrahedron, 41*, 6041 (1985).
97. M. Aritomi, T. Kumori, and T. Kawasaki, *Phytochemistry, 24*, 2438 (1985).
98. H. Becker, J. Reichling, and W. C. Hsieh, *J. Chromatogr., 237*, 307 (1982).
99. A. Nahrstedt, V. Wray, B. Engel, and E. Reinhard, *Planta Med., 51*, 517 (1985).
100. G.-A. Gross, O. Sticher, and C. Anklin, *Helv. Chim. Acta, 69*, 156 (1986).

chapter six

Rotation Locular Countercurrent Chromatography for Natural Products Isolation

Isao Kubo, Gary T. Marshall, and Frederick J. Hanke
University of California at Berkeley
Berkeley, California

I. INTRODUCTION

Rotation locular countercurrent chromatography (RLCCC) is a liquid-liquid countercurrent chromatographic technique currently receiving interest for the separation of a variety of natural products. As an all-liquid technique, it offers the advantages of another recently devised liquid partition system, droplet countercurrent chromatography (DCCC, reviewed elsewhere in this volume). These are an inherently large sample capacity, small solvent consumption, a simple apparatus design which operates at low pressure, an inert environment, and the assurance of complete recovery of an applied sample as there are not solid supports which might allow for irreversible binding or chemical degradation. In contrast to DCCC, the design of the RLCCC apparatus doesn't require droplet formation. Thus RLCCC can accommodate a wider variety of solvent systems than DCCC. In addition, a gradient elution of mobile phase solvents is possible in RLCCC, formation of emulsions is not problematical, and chromatographic runs can be somewhat shorter for RLCCC as compared to DCCC. However, the number of theoretical plates in the commercially available RLCCC model is less (approximately 300) than for the currently available DCCC instruments (approximately 1000).

The RLCCC became commercially available in 1981 [1]. The currently available device consists of 16 glass columns (11 mm interior diameter and 500 mm in length) interconnected in series by Teflon capillary tubing. Within each glass column are 37 chambers or locules formed by Teflon disk and spacers. Solvent can pass between locules via openings in the center of each disk. In addition, the 16 columns are mounted around a central axis which spins at 60–120 rpm during operation. This rotation facilitates the mixing of the biphasic solvents in each locule and is not used to induce any centrifugal force. Furthermore, the angle of this axis may be varied from a vertical position for loading stationary phase to a horizontal position for loading mobile phase. An angle of 20–40° is suggested as optimum for partition efficiency during separations [2] (see Fig. 1).

Following the selection of the biphasic solvent system, the stationary phase is loaded into the apparatus with the column assembly in the vertical position to dispel any previous contents. The column assembly is then placed in a horizontal position and the mobile phase is pumped into the system displacing 50% of the stationary phase in each locule. The columns are then placed at 20–40° to the horizontal. If, for example, the mobile phase is the lighter phase, the system will be placed in the ascending mode. A volume of fresh mobile phase entering the first locule will wash through the heavier stationary phase and displace an equal volume of the old mobile phase through the center hole leading to the next locule. The sample, dissolved in one or both of the two phases, is introduced via a sample loop placed between the pump and the column assembly. The flow of mobile phase displaces the sample from the sample loop into the first locule where it can be separated by partition between the stationary and mobile phases. Mobile phase containing solute is then recovered from the end locule of the sixteenth column. Sample material not eluting with the mobile phase is reclaimed by washing the column with additional stationary phase and/or methanol.

Although the RLCCC design was first published in 1970 [2], applications of RLCCC have been limited. The first RLCCC separation reported was, in 1982, of the partial resolution of racemic norephedrine by partition of enantiomeric salts [3]. Subsequent RLCCC separations of flavonoids, chromomycin antibiotics, and a plant growth regulator have recently been reviewed [4].

We have been using RLCCC and DCCC instruments for the isolation of biologically active natural products since 1982. Some of this work has not been previously published. Hence, our experience in the application of RLCCC to the analysis of natural products will be reviewed herein. In addition, comparisons of separations by RLCCC and DCCC shall be described as well as a modification of the RLCCC apparatus which allows a rapid separation of crude extracts within several hours.

II. APPLICATIONS

A. Dihydrochalcone

A methanol extract of the leaves of the apple, *Malus pumila* var. *dulcissima* Mill. (Rosaceae), was recently found to have antifeedant

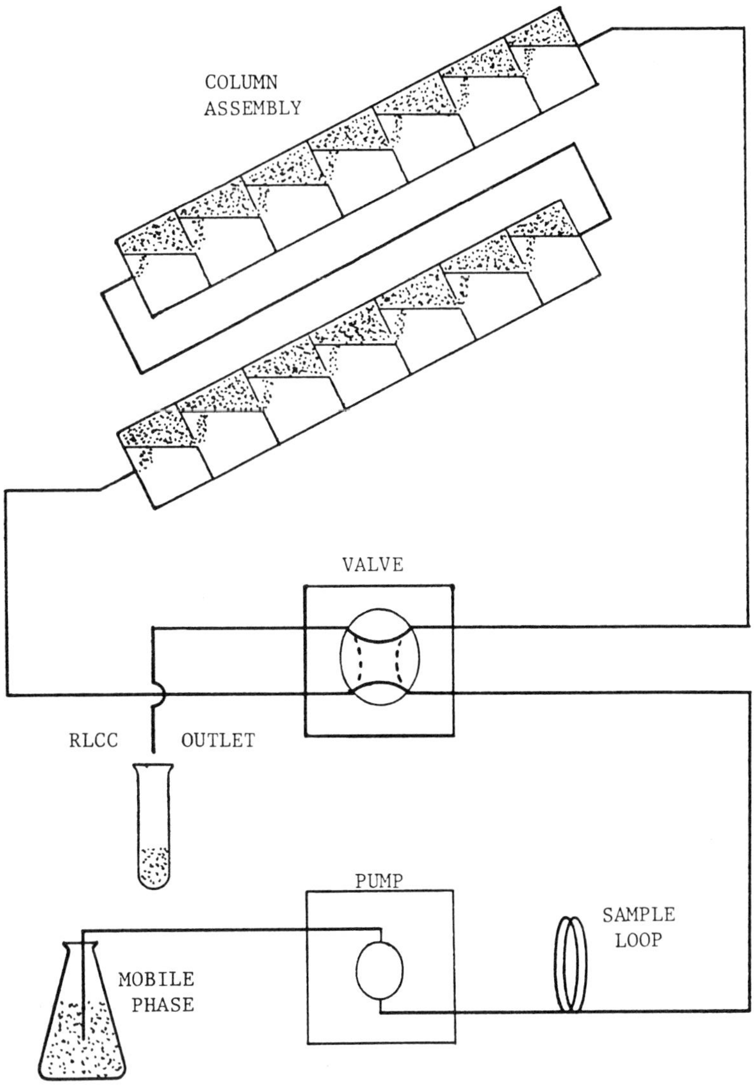

Fig. 1 Schematic of the rotation locular countercurrent chromatograph shown in the ascending mode.

activity against a peast aphid. *Schizaphis graminum* Rondani, in an artificial diet feeding bioassy [5]. The crude extract was separated into hexane, ether, ethyl acetate, and water-soluble portions. The biological activity was found in the ethyl acetate fraction. RLCCC was chosen for further separation of this polar material using $CHCl_3/MeOH/H_2O$ (13:7:4, v/v) as the solvent system in the ascending mode. The major component of the extract, phloretin-4'-O-β-D-glucopyranoside, proved to be the active constituent [5]. Elution of the mobile phase was monitored by UV detection and the chromatogram obtained is represented in Fig. 2.

B. Coumarins

The rutaceous plant *Clausena anisata* (Willd.) Benth. is used in parts of Africa as an insecticide as well as medicinally. The nonpolar ether-soluble fraction retained the activity of the original extract following a solvent partition. An application of 1 g of this fraction to RLCCC using a new nonaqueous solvent system we developed, n-heptane/$CHCl_3/CH_3CN$ (50:15:35, v/v) in the ascending mode, resulted in the isolation of three known alkylcoumarins [6]; xanthoxyletin (530 mg), xanthyletin (25 mg), and 3-(1,1-dimethylallyl)-xanthyletin (41 mg) (Fig. 3). The major component, xanthoxyletin, was the most active as an insect antifeedant while xanthyletin was almost as strong. The third coumarin was only a weak antifeedant. The antifeedant activity of xanthoxyletin has been previously reported [7]. We found this new solvent system not applicable to DCCC (2 mm by 400-mm columns) due to poor droplet formation.

C. Flavonoid Glucoside and Monoterpene

A byproduct in the production of the spice saffron, which consists of the dried flower stigmas and style tips of *Crocus sativus* L. (Iridaceae), are the flower petals. The application of RLCCC to a methanol extract of these flower petals resulted in the isolation of a flavonoid glucoside and a monoterpene [6]. The methanol extract, 500 mg, was submitted to RLCCC using the highly polar solvent system EtOAc/n-BuOH/CH_3CN/H_2O (5:1:7:10, v/v) in the ascending mode. The results of this fractionation (Fig. 4) yielded 130 mg of the flavonoid kaempferol-3-O-sophoroside, and a smaller amount of the monoterpene safranal. Similar results were obtained by DCCC using the same solvent system in the ascending mode (Fig. 5).

D. Seco-iridoid Glycosides

The common black olive, *Olea europaea* L. (Oleaceae), produces fruit which contain bitter substances having molluscicidal activity against a schistosomiasis intermediate host snail *Biomphalaria glabrata*

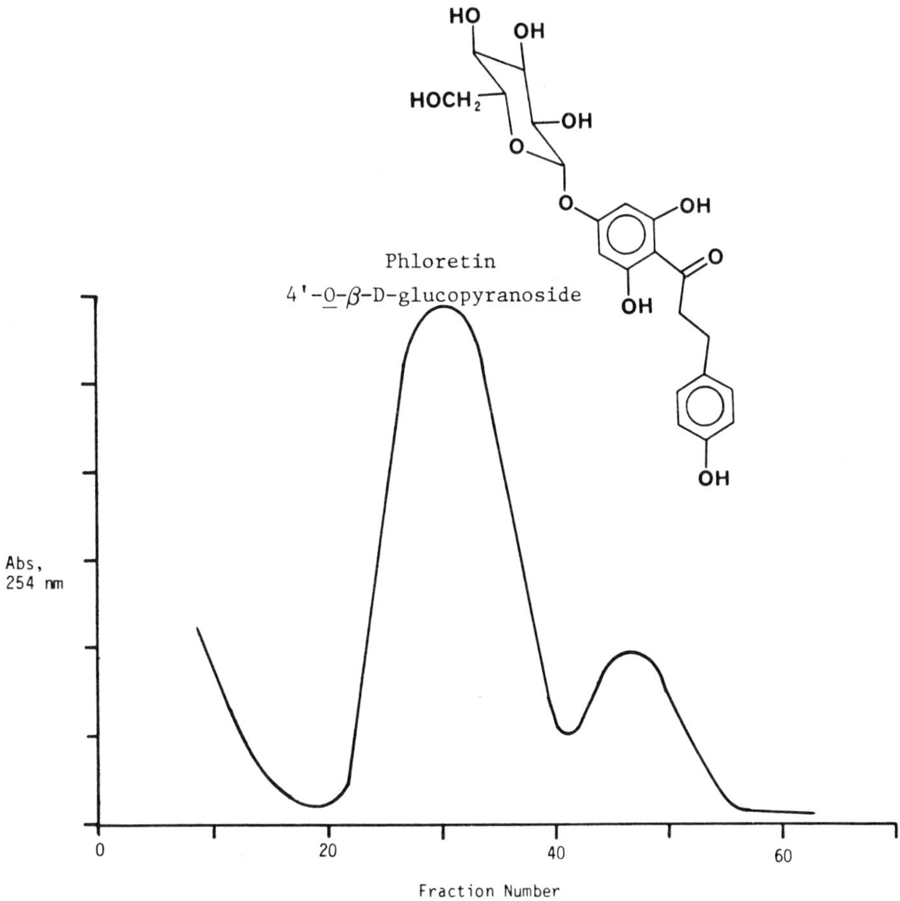

Fig. 2 RLCCC of the methanol extract (1.5 g) of *Malus pumila* var. *dulcissima* with $CHCl_3/MeOH/H_2O$ (13:7:4, v/v) by the ascending method; detection at 254 nm, 1.5 days.

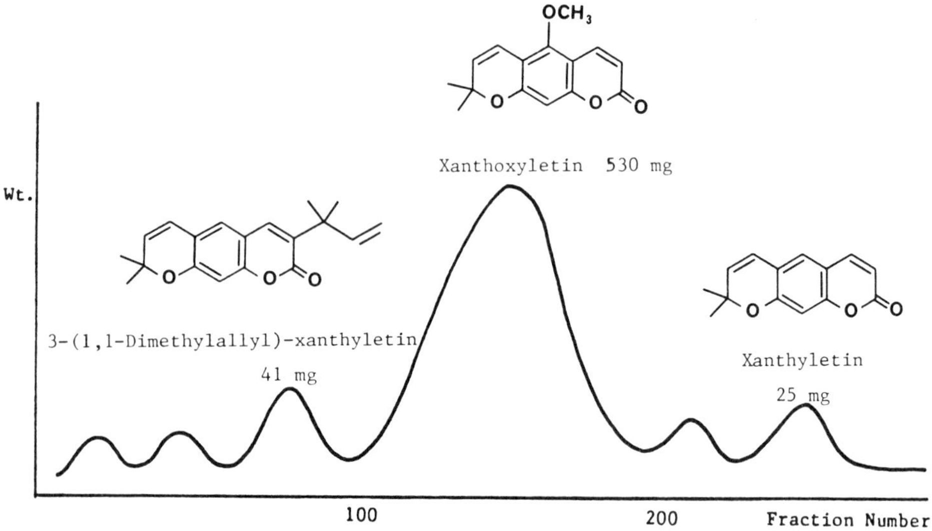

Fig. 3 RLCCC of the ether extract (0.95 g) of *Clausena anisata* with n-heptane/$CHCl_3$/CH_3CN (50:15:35, v/v) by the ascending method; 5 ml per fraction.

[8]. Separation of the crude extract into ether, ethyl acetate, and water-soluble portions resulted in the activity concentrated in the ethyl acetate fraction. Due to the polar nature of the ethyl acetate solubles, further chromatographic separation was performed using both RLCCC and DCCC. The same solvent system was used in each case, $CHCl_3$/MeOH/H_2O (13:7:4, v/v), in the ascending mode. Efficient separation of two molluscicidal seco-iridoids glycosides, oleuropein and ligstroside, was achieved using both chromatographic strategies and the results are represented in Figs. 6 and 7 [8]. This separation using RLCCC was completed in 3 days. However, 5 days was required for similar results to be achieved using DCCC.

E. Saponins

Aqueous extracts of the fruits of the East African plant *Phytolacca dodecandra* l'Herit (Phytolaccaceae) have shown potent molluscicidal activity against the water snail *Biomphalaria glabrata* [9]. Two portions of a methanolic extract, 4 g each, was submitted to RLCCC using EtOAc/EtOH/H_2O (40:20:40, v/v) as the solvent system in the ascending mode. Fractions were recovered which contained three active saponins, diglucosides of bayogenin, oleanolic acid, and

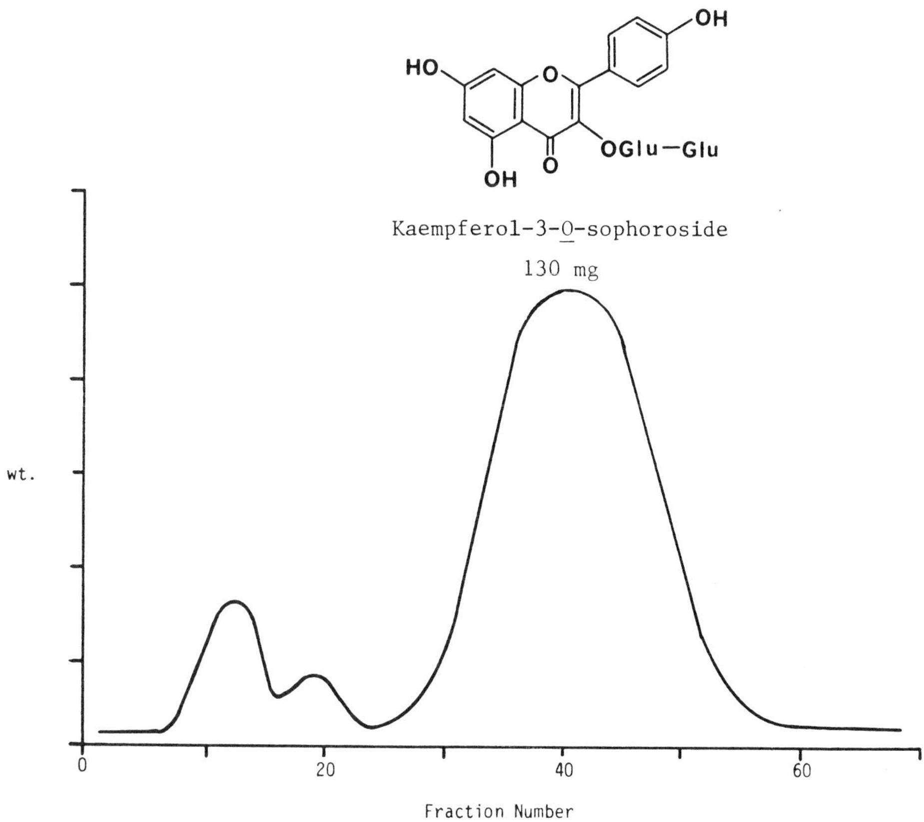

Fig. 4 RLCCC of the methanol extract (0.5 g) of *Crocus sativus* petals with EtOAc/n-BuOH/CH_3CN/H_2O (5:1:7:10, v/v) by the ascending method; 12 ml per fraction, 2.5 days.

hederagenin, but required some final purification by reversed phase octylsilyl-bonded silica gel and Sephadex LH-20 chromatography.

The methanol extract of *Pachyelasma tessmannii* (Sabaceae), a tree native to Camaroon, was shown to be molluscicidal against the water snail *Biomphalaria glabrata* [6]. The constituents involved appeared to be fairly polar in nature, thus the solvent system $CHCl_2$/MeOH/H_2O (7:13:8, v/v) was chosen using the organic phase as the mobile phase in the descending mode. An injection of 200 mg, monitored by UV at 284 nm, resulted in the isolation of three triterpenoid saponins differing in their sugar moieties (Fig. 8). Although RLCCC could not achieve complete resolution of these saponins, final purification was performed on NH_2-bonded HPLC. The

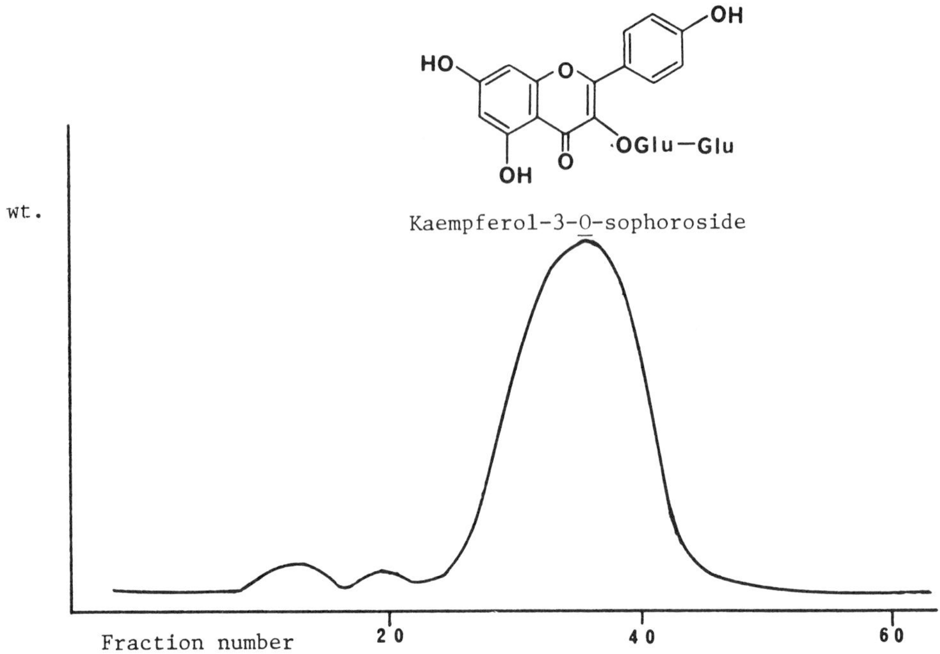

Fig. 5 DCCC of the methanol extract (0.7 g) of *Crocus sativus* petals with EtOAc/n-BuOH/CH_3/CN/H_2O (5:1:7:10, v/v) by the ascending method; 12 ml per fraction, 2.5 days.

use of RLCCC for preliminary fractionation as effective as reversed phase octyldecylsilyl-bonded HPLC but with a much greater sample capacity of up to 3 g per injection.

F. Phytoecdysteroids

Large amounts of the phytoecdysteroids 20-hydroxyecdysone and ajugasterone C were isolated from the fresh root bark of the West African medicinal plant *Vitex madiensis* (Verbenaceae). These compounds are potent disrupters of the insect molting cycle [10]. Prior to their isolation and identification, the ethyl acetate soluble portion showed one major constituent by silica TLC. However, HPLC suggested two components. To separate these compounds, RLCCC was tried using $CHCl_3$/MeOH/H_2O (13:7:4, v/v) as the solvent system in the ascending mode. Complete resolution was not achieved (Fig. 9). The application of the same sample and solvent system to DCCC was successful in resolving these two phytoecdysteroids (Fig. 10).

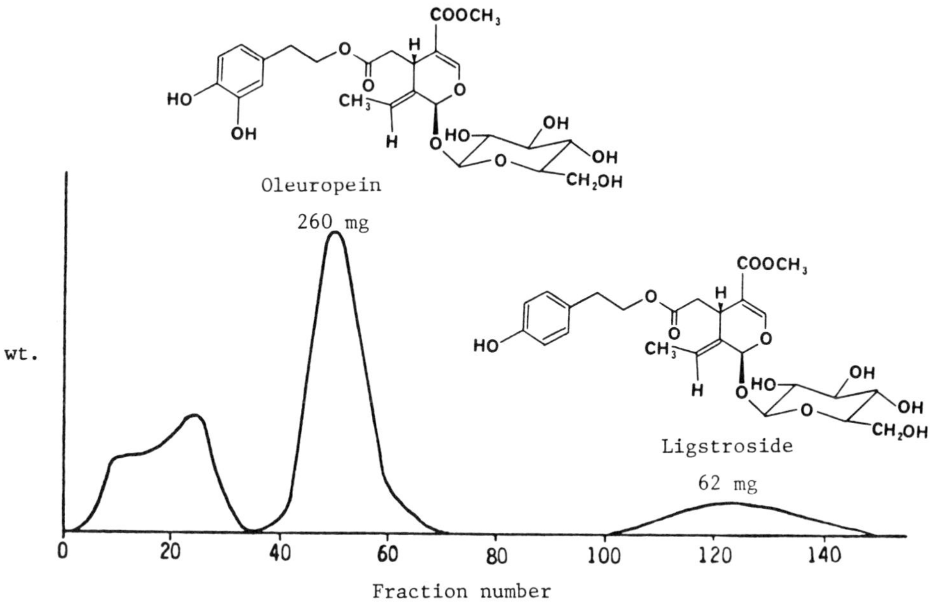

Fig. 6 RLCCC of the methanol extract (1.0 g) of *Olea europaea* with $CHCl_3/MeOH/H_2O$ (13:7:4, v/v) by the ascending method; 1.4 ml per fraction, 3 days.

G. "Flash" RLCCC

Frequently in the analysis of natural products having biological activity, initial partition of an extract into petroleum ether or hexane, diethyl ether, ethyl acetate, and water-soluble fractions is performed to narrow the spectrum of chemical constituents thus concentrating the biological activity. We believed that RLCCC, with its ability to utilize gradient elution and having a sample capacity of up to several grams, would be more effective than simple partitioning for the initial fractionation of plant extracts for bioassay. The normal operation of RLCCC dictates a flow rate of about 0.2 ml/min. Since we were interested in obtaining a rapid, initial separation of a crude extract, some sacrifice of chromatographic efficiency for fast elution rates was acceptable. The commercially available RLCCC unit comes equipped with a pump head which allows for a maximum flow rate of approximately 3 ml/min. However, replacement of this pumping unit with one designed for use on a DCCC effectively quadruples the maximum flow rate to approximately 12.5 ml/min. This modification increases the system pressure substantially and requires resetting

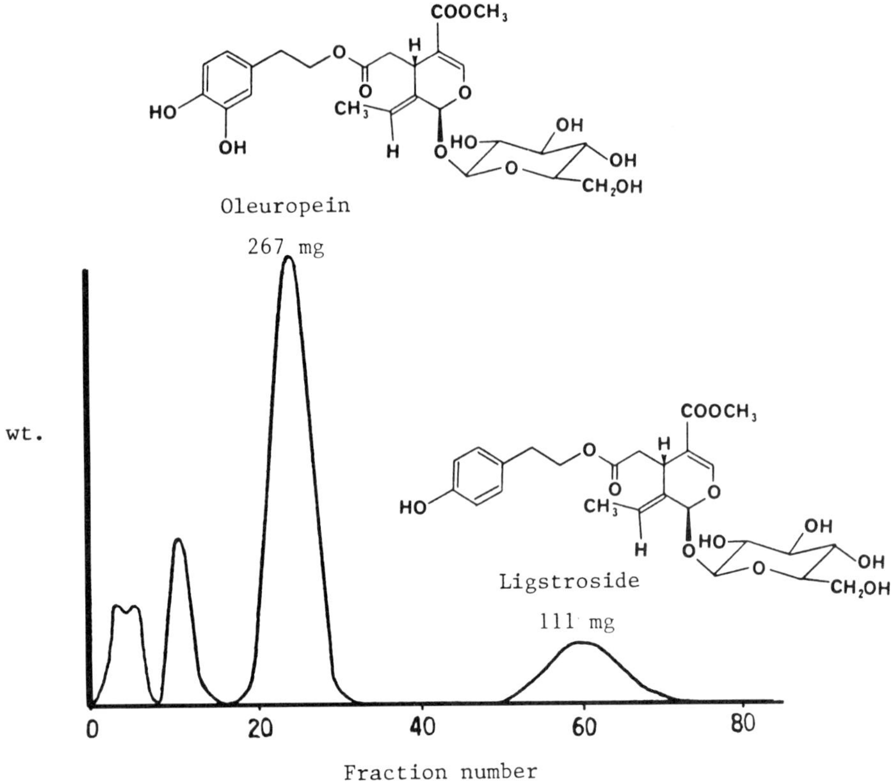

Fig. 7 DCCC of the methanol extract (1.0 g) of *Olea europaea* with $CHCl_3/MeOH/H_2O$ (13:7:4, v/v) by the ascending method; 1.4 ml per fraction, 5 days.

the automatic pressure regulator to shut the pump off at the higher pressure of 100 psi. In our testing, the system still functions normally and no leaking of solvent occurs. As the minimum flow rate for the new pump head is similar to that used for most normal separations, around 0.1 ml/min for the commercial RLCCC pump head and around 0.15 ml/min for the newly attached DCCC pump head, normal separations are still possible after this modification. At the new maximum flow rate, only 15 min is required to completely load a mobile phase onto a previously filled aqueous stationary phase. An extract of several grams can be loaded onto the RLCCC and eluted sequentially with hexanes, ether, ethyl acetate (approximately 500 ml each) and finally a methanol/water wash. This potentially yields about 10 fractions of widely differing polarities within a few hours.

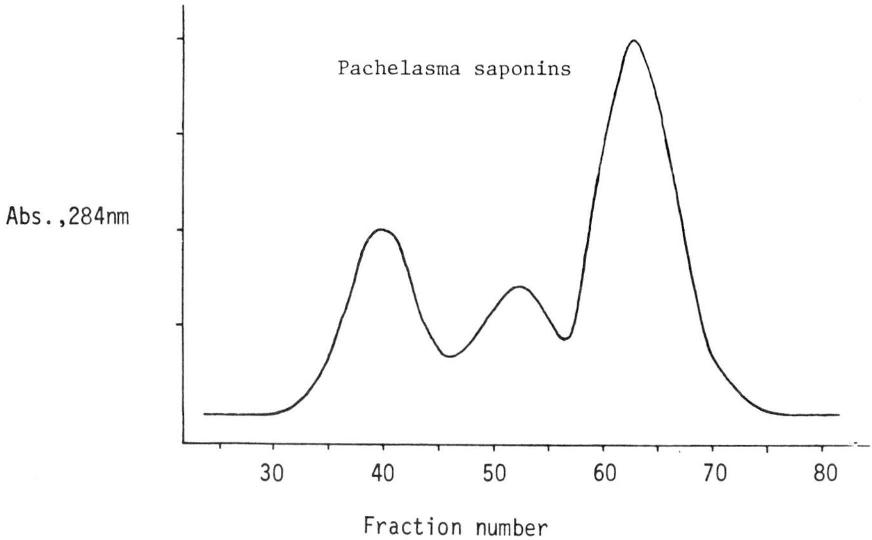

Fig. 8 RLCCC of the methanol extract (0.2 g) of *Pachyelasma tessmannii* with $CHCl_3/MeOH/H_2O$ (7:13:8, v/v) by the descending method; 5 ml per fraction, 3 days.

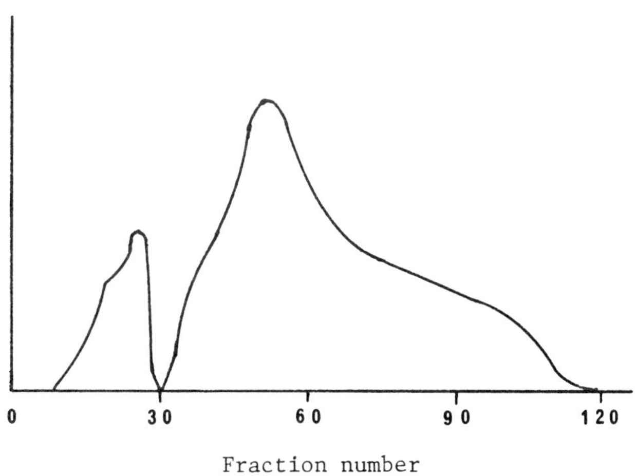

Fig. 9 RLCCC of the ethyl acetate extract (2.2 g) of *Vitex madiensis* with $CHCl_3/MeOH/H_2O$ (13:7:4, v/v) by the ascending method; 1.4 ml per fraction.

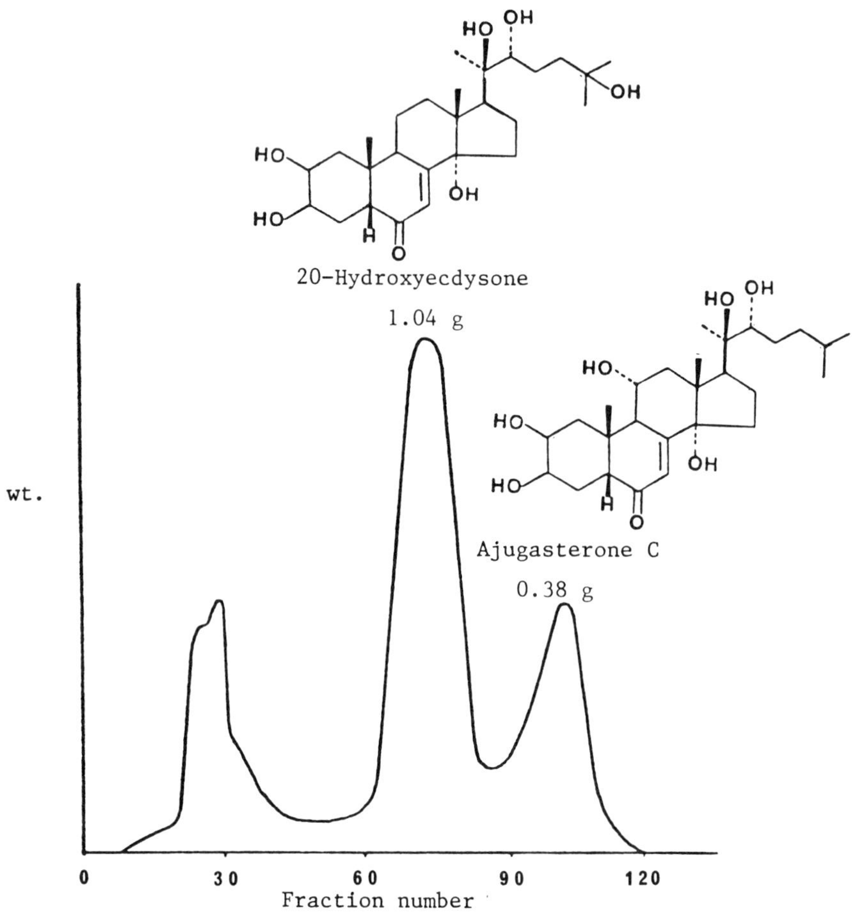

Fig. 10 DCCC of the ethyl acetate extract (2.7 g) of *Vitex madiensis* with $CHCl_3/MeOH/H_2O$ (13:7:4, v/v) by the ascending method; 1.5 ml per fraction.

Furthermore, the greater resolution of this type of initial separation can be achieved without forming troublesome emulsions that are frequently encountered in separatory funnel-type solvent partitions.

This general fractionation scheme has recently been applied as the initial partition step in the separation of some biologically active plant extracts currently receiving interest in this laboratory. In one example, 1.5 g of a methanolic plant extract was applied to "flash" RLCCC using water as the stationary phase and eluted sequentially with 500-ml portions of hexanes, ether, and methyl ethyl

Table 1 Solvent Systems Used for Rotation Locular Countercurrent Chromatography

Compound class	Solvent system (volume ratios)	Ref.
Dihydrochalcone	$CHCl_3/MeOH/H_2O$ (13:7:4)	[5]
Coumarins	n-Heptane/$CHCl_3/CH_3CN$ (50:15:35)	[6]
Flavonoid glycosides	EtOAc/n-BuOH/CH_3CN/H_2O (5:1:7:10)	[6]
	EtOAc/n-PrOH/H_2O (40:20:40)	[4]
aglycones	$CHCl_3/MeOH/H_2O$ (33:40:27)	[4]
Iridoid glycosides	$CHCl_3/MeOH/H_2O$ (13:7:4)	[8]
Saponins	$CHCl_3/MeOH/H_2O$ (13:7:4)	[6]
	EtOAc/EtOH/H_2O (40:20:40)	[9]
Phytoecdysteroids	$CHCl_3/MeOH/H_2O$ (13:7:4)	[10]
Phenylpropanoid[a]	A. $CH_2Cl_2/MeOH/H_2O$ (35:65:45)	[4]
	B. $CHCl_3/MeOH/H_2O$ (35:65:45)	[4]
Chromomycin antibiotics	$CHCl_3$/EtOAc/MeOH/H_2O (2:4:2:1)	[4]
(±) norephedrine[b]	0.5 M aq. $NaPF_6$	
	0.3 M (R,R)-di-5-nonyltartrate in CH_2Cl-CH_2Cl	[3]

[a]Solvent system A was used initially with the organic phase as the mobile phase. After 20 hr, the organic phase from solvent system B was used to elute the sample from the RLCCC.
[b]The aqueous phase was used as the stationary phase while the organic layer was the mobile phase.

ketone. This was followed by an aqueous wash. Nine fractions of 200–250 ml each were collected. The second fraction representing only 2.2% of the weight of the applied sample was biologically active while the first and third through sixth fractions were inactive. Activity again appeared in fractions 7–9 with this activity due to different compounds from those found in fraction 2. This initial separation was achieved very quickly, in less than 4 hr, and no emulsions were formed although the RLCCC column assembly was rotating at full mixing speed (around 120 rpm).

III. CONCLUSIONS

Rotation locular countercurrent chromatography (RLCCC) is subject to a growing range of applications in natural product separations. Although primarily involved with the isolation of polar compounds, its utility in the resolution of nonpolar constituents using nonaqueous solvent systems is also apparent. Thus, to date, 10 solvent systems have been reported for the separation of natural products by RLCCC (Table 1).

Few direct comparisons of the two liquid partition systems RLCCC and DCCC have been made. Applications using identical solvent systems for the RLCCC and DCCC separation of two plant extracts, *Crocus sativus* L. and *Olea europaea* L., achieved similar separations of a flavonoid glucoside (and monoterpene) and of secoiridoid glycosides, respectively, yet RLCCC required less time. However, RLCCC was not successful at achieving the complete resolution of the phytoecdysteroids, 20-hydroxyecdysone and ajugasterone C, while DCCC using the same solvent system could effect their separation. The ability of either instrument to resolve a mixture depends on the complexity and partition coefficients of the constituents to be separated and the suitability of the solvent system. Many factors must be weighed, including the separation time required, in choosing a preference of either RLCCC or DCCC for a particular separation.

Perhaps the application of RLCCC is most appropriate in tandem with other chromatographic systems, such as DCCC and reversed phase HPLC, for the fractionation of polar compounds. The ability of "flash" RLCCC to a rapidly and efficiently resolve a crude extract into a narrower range of compound polarities makes it an ideal method for initial fractionation of an extract for subsequent bioassay.

REFERENCES

1. D. Schaufelberger, M. Hamburger, and K. Hostettmann, Bulletin de Liason no. 11 du Groupe Polyphenols, Narbonne, France, *Compte-Rendu des Journals Internetionales d'Etude et de l'Assemblee General,* Toulouse, September 1982, pp. 29–30.
2. Y. Ito and R. L. Bowman, *J. Chromatogr. Sci., 8,* 315 (1970).
3. B. Domon and K. Hostettmann, *J. Chromatogr., 250,* 149 (1982).
4. J. K. Snyder, K. Nakanishi, K. Hostettmann, and M. Hostettmann, *J. Liq. Chromatogr., 7,* 243 (1984).
5. I. Kubo and A. Matsumoto, *Chem. Pharm. Bull., 33,* 3817 (1985).
6. I. Kubo et al., unpublished.
7. T. Gebreyesus and A. Chapya, in *Natural Products for Innovative Pest Management,* Vol. 2 (D. L. Whitehead and W. S. Bowers, eds.), Pergamon, New York, 1983, pp. 237–242.

8. I. Kubo and A. Matsumoto, *J. Agric. Food Chem.*, *32*, 687 (1984).
9. B. Domon and K. Hostettmann, *Helv. Chim. Acta*, *67*, 1310 (1984).
10. I. Kubo, A. Matsumoto, and J. F. Ayafor, *Agric. Biol. Chem.*, *48*, 1683 (1984).

chapter seven

Centrifugal Droplet Countercurrent Chromatography in Natural Products Chemistry

Reimar C. Bruening*, Fadila Derguini, and Koji Nakanishi
Columbia University
New York, New York

The development of increasingly sophisticated instrumentation and methodology for the isolation and structure elucidation of compounds of biological origin enables contemporary natural products chemistry to answer biological problems on a molecular level which often puzzled scientists for decades. While physicochemical methods for the elucidation of molecular structure rely more and more on the help of the computer and therefore become increasingly straightforward in their application, the successful isolation of pure substances from biological material remains largely empirical. Since chromatographic methods play the major role in obtaining a pure compound from complex mixtures, any new addition to the arsenal of chromatographic tools is highly welcome. The rediscovery of countercurrent chromatography in the form of droplet countercurrent chromatography (DCCC) about 15 years ago provided such an addition [1]. The development of compact size instrumentation based on dynamic liquid-liquid partition of a solute between two immiscible phases created a true alternative to other highly developed chromatographic techniques, such as high-performance liquid chromatography (HPLC), gel filtration, or electrophoresis. Centrifugal droplet countercurrent chromatography (CDCCC), or better centrifugal partition chromatography (CPC) with only one phase mobile, marks a rather recent development in liquid-liquid partition methodology. However, it shares with its other relatives, such as DCCC [2], rotation locular countercurrent chromatography (RLCCC) [3] or the various coil planet centrifuge techniques [4], the basic characteristics of liquid-liquid partition which makes them so distinct from all the other chromatographic

**Current affiliation:* University of Hawaii at Manoa, Honolulu, Hawaii methods. While the latter are based on the interaction of the solute between a solid support and the liquid mobile phase, the modern countercurrent methodologies, in absence of any solid support, score the following advantages:

In terms of column chromatography, the volume ratio of stationary phase to total column volume is higher in partition chromatography than in conventional chromatography. This means that larger quantities of sample can be retained in the stationary phase, which makes these techniques suitable for preparative scale separations.

Since there is no solid stationary phase, compounds with otherwise poor retention characteristics, such as glycosides, oligo- or polysaccharides, peptides and proteins, or compounds which would chemically interact with the solid phase and thereby be irreversibly retained or even decomposed can easily be handled by a countercurrent separation. This means that without the risk of decomposition or denaturation all the components in the sample can be recovered either with the mobile phase or after the chromatographic run by removing them from the remaining stationary phase.

Compared to the sometimes very specialized and costly solid supports used in bonded phase liquid chromatography, countercurrent techniques use only common, easily available solvents which makes this kind of separation, especially on a large scale, economically very attractive.

Restrictions in the choice of solvents apply, however, for DCCC since the mobile phase has to form distinct droplets that can pass through the stationary phase without replacing it [5]. It is here that centrifugal partition chromatography has its first advantage because virtually all kinds of solvents can be used as long as their mixture forms a stable two-phase system, since the two phases are continuously demixed by centrifugal action. This means that even the so-called aqueous polymer two-phase systems can be applied, e.g., for the separation of biopolymers, cell organelles, or even whole cells [6]. In centrifugal partition chromatography the stationary phase is securely kept in the column by centrifugal force so that bleeding from the column during the run is greatly reduced. Even more important is the fact that separations, which often take a long time with the other paritition chromatography methods, can be performed without loss of resolution in a few hours. This makes centrifugal partition chromatography the method of choice for the isolation and separation of unstable or highly sensitive compounds where the speed of the chromatographic run is of crucial importance. It was this speed of separation together with the unique possibility of switching phases or elution modes during the separation that allowed the isolation of the tunichromes, the rather unstable polyphenolic blood pigments from vanadium-collecting tunicates (see below) [7].

Since centrifugal partition chromatography's introduction in the international literature several years ago went largely unnoticed [8], a brief survey over its basic principles and technical aspects shall be given herewith. In droplet countercurrent chromatography the droplet of the mobile phase either rises through the heavier stationary phase which is kept in a narrow-bore glass column thus giving the ascending mode, or if it is the heavier phase falls through the lighter stationary phase in the glass column resulting in descending mode. In both cases, earth's gravity plays the crucial role retaining the heavier stationary phase in the column in the case of ascending mode or letting the heavier mobile phase move through the stationary phase descending mode. Since natural gravity is a comparatively weak force, special care has to be taken not to replace the stationary phase from the columns by moving the mobile phase too fast. This drastically curtails the speed of droplet countercurrent separations making runs of several days not an exception. By applying artificial gravity centrifugal partition chromatography overcomes this limitation. Turning the column by 90° and placing it parallel to the radius of a rotor, a strong variable centrifugal force can be applied to a liquid inside the column by turning the rotor at various speeds. This results in a much stronger retention of the stationary phase in the column and thus allows the mobile phase to be pumped against the centrifugal force at much higher speeds. In descending mode, on the other hand, the heavier mobile phase is allowed to jet parallel to the centrifugal force vector through the stationary phase and to successfully replace it at the column outlet (Fig. 1). Naturally, this setup leads to much higher back pressures which calls for drastic technical modifications in designing a functional centrifugal instrument. The presently available unit has a back pressure limit of 60 atmospheres and it could be demonstrated that the best separations are indeed obtained at this pressure [8]. Figure 2a shows that the corresponding columns are actually holes 4 cm long and 3 mm in diameter which are drilled into a block of Tefzel (= poly-C_2F_3Cl) and connected with each other by small holes of 1 mm diameter which are drilled parallel to the columns. Fifty such columns are placed in one block which is called a cartridge.* Figure 2b shows how 12 of these cartridges are assembled in a rotor and connected with each other by small-bore pressure-resistant Teflon tubing. Since this rotor is in motion during the chromatographic run, a special connection, through a so-called rotary seal joint, had to be constructed which allows the leak-free

*This describes the standard-type cartridge 30E. Two new designs, the high-resolution type 200E with 2400 columns and 1000E with 480 wide-bore columns for cell-particle separation, recently became available.

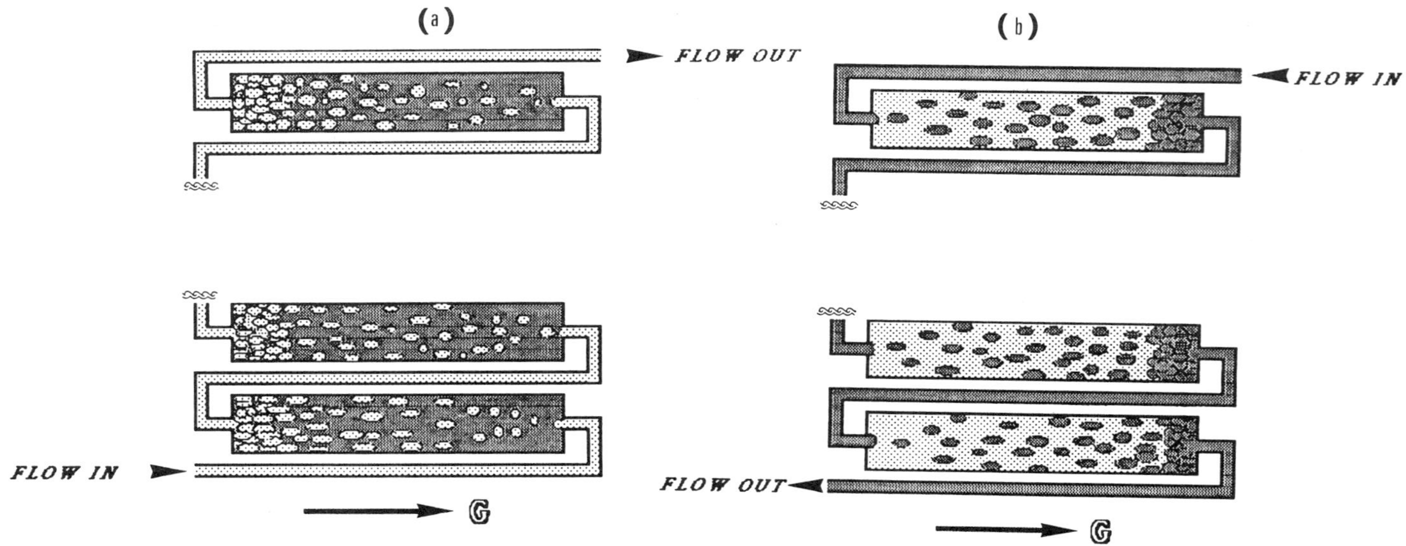

Fig. 1 Schematic representation of mobile phase flow in a centrifugal partition chromatography cartridge. (a) "Ascending mode": the lighter phase (light color) is mobile and is pumped against the centrifugal force G through the stationary, heavier phase (dark color). (b) "Descending mode": the heavier phase is mobile and pumped parallel to the G vector; the lighter, stationary phase is continuously replaced at the column outlet.

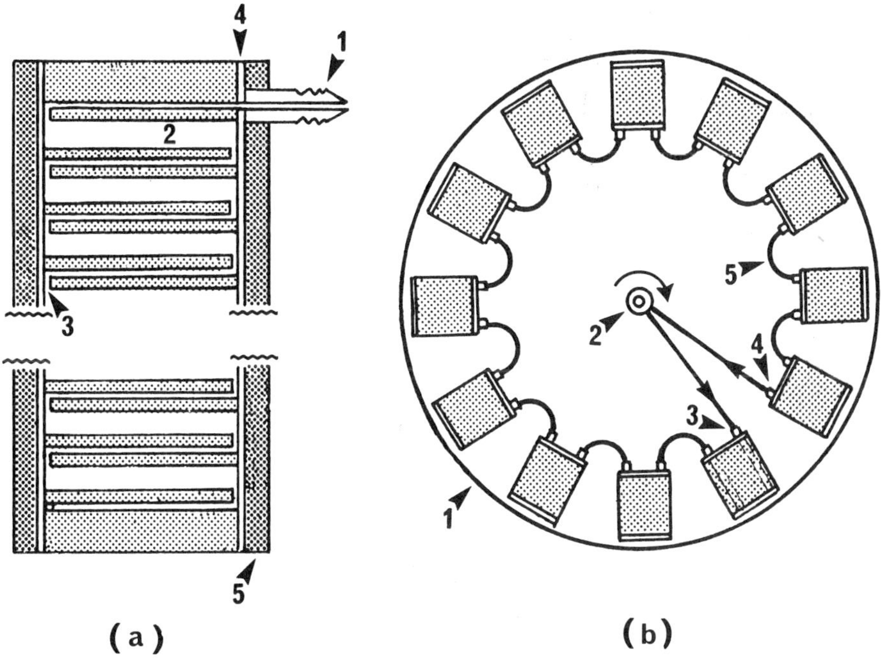

Fig. 2 (a) Cross section through a centrifugal partition chromatography cartridge type 30E; 1 = conical connectors for solvent inlet and outlet; 2 = "columns" (3 × 40 mm); 3 = connecting channels (1 × 40 mm); 4 = Teflon gasket; 5 = aluminum walls, pressed together by steel bolts. (b) Assembly of 12 cartridges in the rotor (= 1) for "ascending mode"; 2 = rotary seal joint; 3 = solvent inlet; 4 = solvent outlet; 5 = connecting tubing. In "descending mode" 3 and 4 are reversed. (After Ref. 8, modified.)

transit of the mobile phase from the static part of the instrument into the moving rotor, through the cartridge, and out of the moving rotor back into the static part. The rotary seal joint is constructed after the rotor-stator principle, where a moving graphite disk is pressed by a spring against a highly polished stationary ceramic surface; this allows a leak-free operation up to the pressure limit. The centrifugal partition chromatography system flow sheet in Fig. 3 shows how three different solvents can be selected, how the sample is introduced through a sample loop, and especially noteworthy, how the phases can be switched in order to change from ascending to descending mode or vice versa without interrupting the motion of the rotor. In any case, however, the eluate is usually contaminated with small amounts of coeluting stationary phase. It is

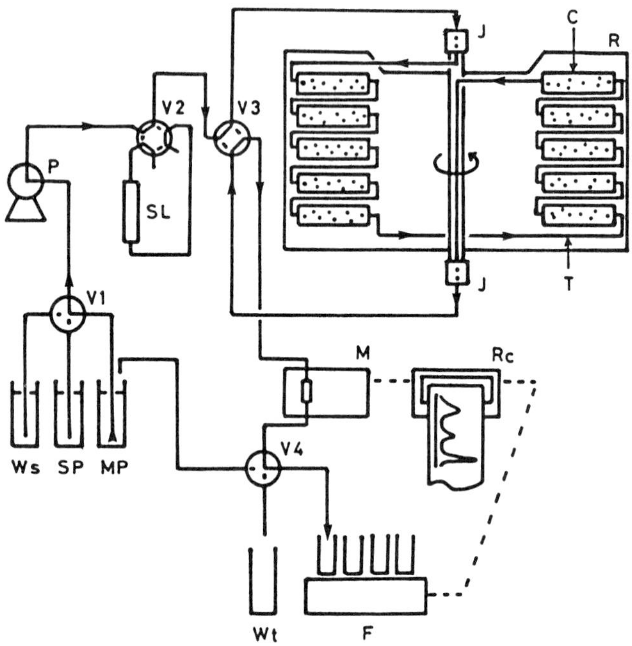

Fig. 3 Flow sheet of a laboratory scale centrifugal partition chromatography unit; WS = washing solvent, SP = stationary phase, MP = mobile phase, Wt = waste; V1 = solvent selection valve, V2 = sample injection valve, V3 = elution mode selection valve (depicted for "ascending mode"), V4 = collecting/recycling valve; P = pump; SL = sample loop; C = cartridges; R = rotor; J = rotary seal joints; T = connecting tubing; M = flow cell and UV monitor; Rc = recorder; F = fraction collector. (After Ref. 6.)

therefore important to provide any subsequent flow-through detector with a preparative flow cell of very short pathlengh or to automatically dilute the eluate with a solvent which allows the two phases to remix prior to detection [9]. Depending of the viscosity of the mobile phase, any feasible flow rate can be applied. As shown in Fig. 4, the original authors could indeed demonstrate that the resolution of the chromatogram is largely independent from the *applied* flow rate as long as the pumping rate, which is the *apparent* flow rate, does not exceed the *true* flow rate; this true rate is a function of the hydrodynamic equilibrium inside the column and therefore depends on the rotational speed and the resulting back pressure (Fig. 5a). The relationship between resulting back pressure and rotational speed on the other side depends on the difference in den-

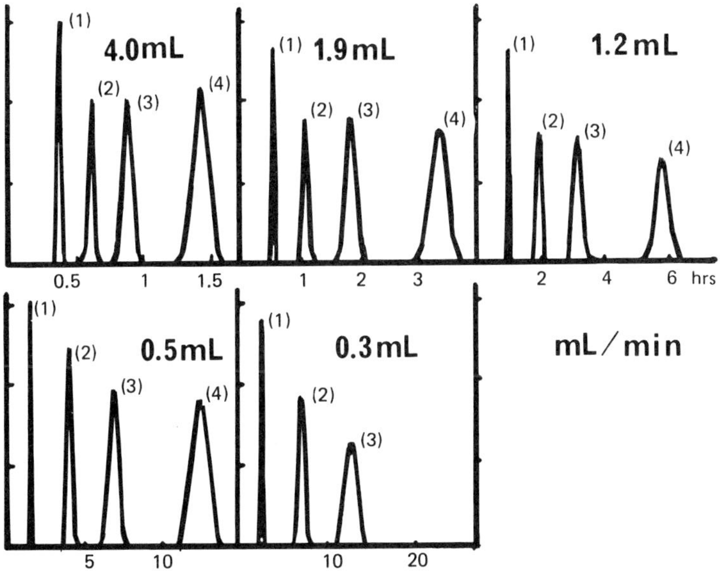

Fig. 4 Separation of DNP amino acids at different flow rates. Solvent system: chloroform/0.1 N hydrochloric acid/acetic acid (2:1:2). Stationary phase = lower phase ("ascending mode"). Peaks: 1 = NS-DNP-L-ornithine hydrochloride, 4 mg; 2 = DNP-L-threonine, 4 mg; 3 = N,N-di-DNP-L-cystine, 8 mg; 4 = DNP-L-proline, 12 mg. Rotational speed: 700 rpm. (From Ref. 8.)

sities between the two phases (Fig. 5b). The final separation parameters are therefore a compromise between pumping rate and rotational speed in order to achieve the highest possible back pressure with a given solvent system; the possibility of stepless regulation of the rotational speed hereby becomes mandatory (500–2000 rpm in modern centrifugal partition chromatography instruments).

Since its introduction, quite a few applications of centrifugal partition chromatography to problems of isolation of natural products appeared in the Hapanese literature ranging from the separation of several sugars [8], saponins [8], tannins [6,10], fatty acids [8], iridoid glucosides [6,11], and other natural products [6,11] to the separation of lysozyme from egg white or the preparation of L-leucine dehydrogenase from *Bacillus sphericus* [11]. The latter two separations are examples of the usage of so-called polymeric aqueous phases where normally mixtures of polyethylene glycols and methylcellulose in a buffered system provide two phases which can be used in centrifugal partition chromatography in spite of their high viscosity;

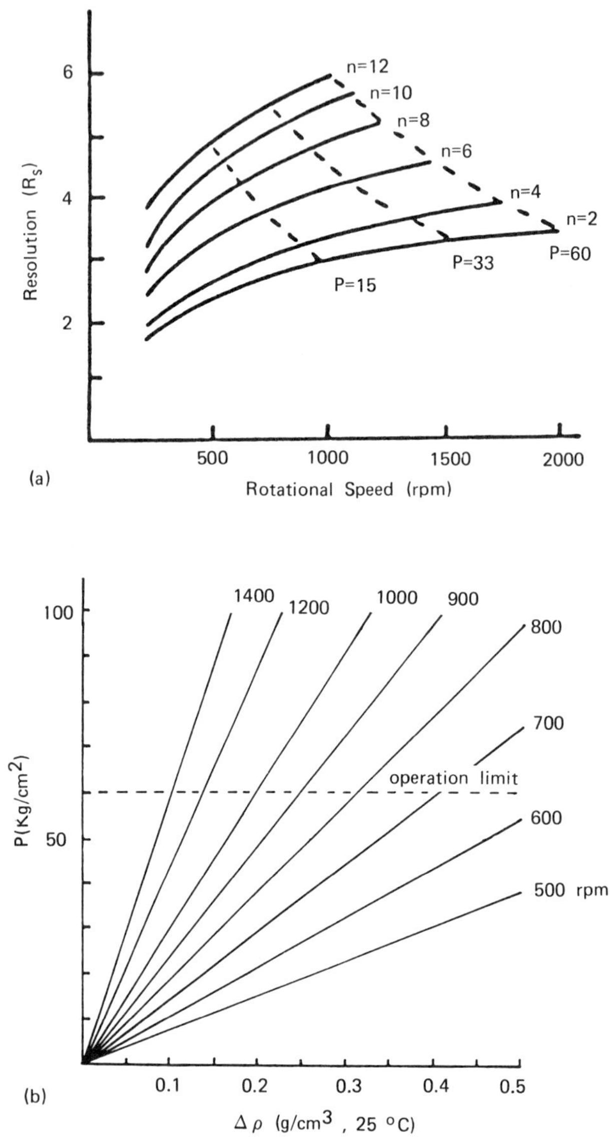

Fig. 5 (a) Rotational speed vs. resolution (R_S). The numbers at the solid lines indicate the numbers of cartridges used; P values indicate the corresponding pumping pressures (dotted lines). Sample: DNP amino acids; other conditions see Fig. 4; flow rate: 1.0 mL/min. (b) Density difference between upper and lower phase ($\Delta\rho$) vs. back pressure (P) at various rotation speeds (12 cartridges type 30E). (From Ref. 8.)

this might illustrate the usefulness of the technique for biochemical separations. The following three separations were obtained from the first centrifugal partition chromatography instrument prototype in the United States and illustrate each of the above mentioned advantages of centrifugal partition chromatography over the other countercurrent techniques. In the first two cases, the suitability of a two-phase solvent system was determined on the base of a separation obtained on normal phase TLC with the organic or upper phase as developing solvent [5]: separation of the major spots in a Rf region of 0.25–0.5 was considered ideal. In the case of the tunichrome separation, the distribution of the colored pigments was measured colorimetrically, and the solvent composition was tuned to give a ratio of approx. 30% tunichrome in the upper phase.

The chromomycins are a class of antitumor antibiotics with anthracyclin-glycoside structures which are produced by *Streptomyces griseus* #7 [12]. The main components chromomycin A2, A3, and A4 are structurally distinguished by subtle changes in the glycoside portion of the molecule [13]. The original separation was conducted on silica gel and suffered large losses due to irreversible retention on the solid support. A breakthrough in separation proved to be the application of rotation locular countercurrent chromatography (RLCCC), which allowed the separation of all three major components on a 700-mg scale with 97% recovery [14]. However, with the flow rate of 0.5 mL/min the separation took 2½ days. Figure 6 shows the separation of chromomycins by centrifugal partition chromatography. With a different solvent system and a flow rate of 1.6 mL/min 700 mg of crude chromomycin mixture could be separated in approximately 4 hr. Even if the chromatographic trace shows anything but baseline separation, the corresponding TLC analsyis verifies that the isolates could be obtained in a purity higher than 97%. While this demonstrates the usefulness of centrifugal partition chromatography for the separation of structurally very similar compounds from a complex mixture in a relatively short time, the following example illustrates how centrifugal partition chromatography renders possible the separation of cis-trans isomers of the heat- and light-sensitive retinal series [15]. The challenge was the preparation of the pure 11-cis isomer from a mixture of other photoisomers obtained from irradiated trans-retinal [16]. The solvent system is entirely anhydrous and based on the polarity difference between cyclohexane-pentane and acetonitrile-methanol (Fig. 7). Compared to other adsorption chromatographic separaitons of retinal mixtures, it is remarkable that this solvent system allows the elution of 11-cis retinal, the naturally occurring visual pigment, prior to all the other isomers. It was thus possible to prepare from 160 mg of photoisomerization mixture 50 mg of pure 11-cis isomer and to recover 40 mg of the starting material trans-retinal. The most remarkable feature of this separation, however, is its short time; the whole

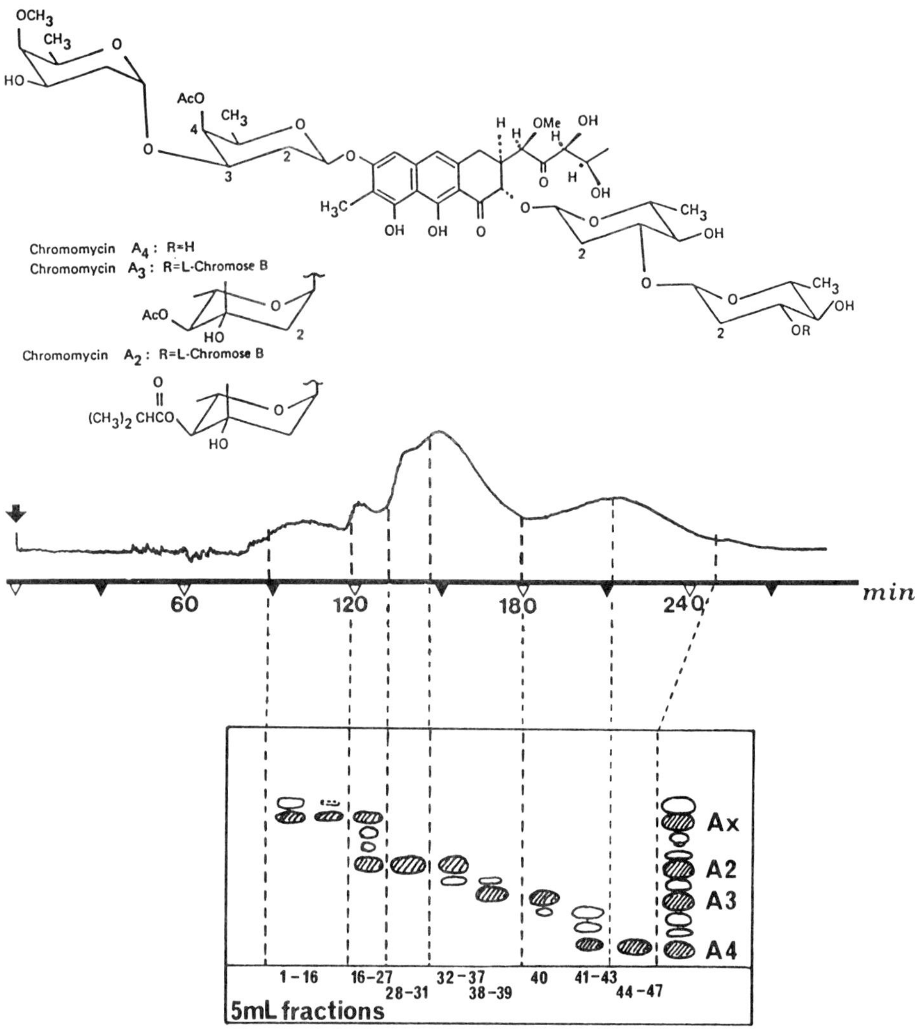

Fig. 6 Centrifugal partition chromatography separation of chromomucins (700 mg). Solvent system: toluene/2-butanone/methanol/water (2:2:2:1). Solvent system: toluene 2-butanone methanol water (2:2:2:1). Stationary phase = lower phase ("ascending mode"). Flow rate: 1.6 mL/min. Rotation speed: 700 rpm. Pressure: 35 atm (500 psi). Detection: UV 450 nm, 1.0 AUFS. TLC analysis: HPTLC plates LPH-KF (silica, Whatman), 100 × 200 mm; mobile phase = upper phase of CPC solvent; detection: UV 366 nm.

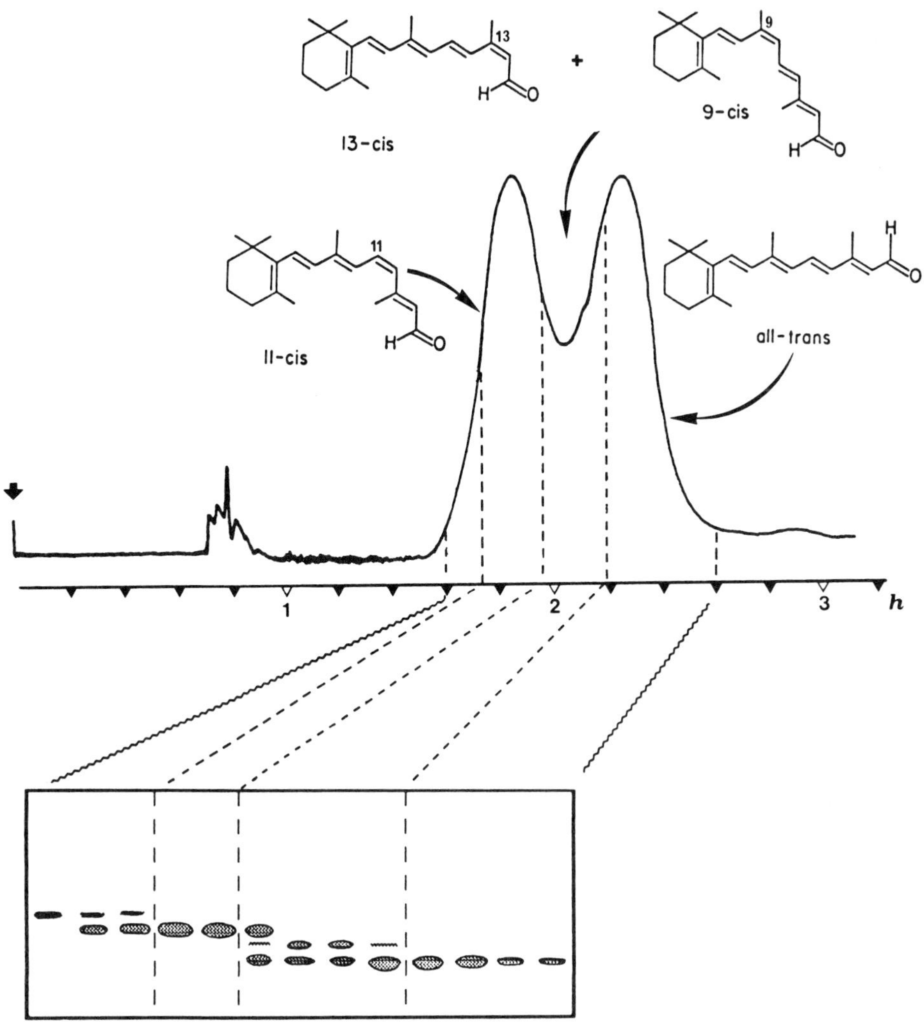

Fig. 7 Centrifugal partition chromatographic separation of cis-trans photoisomers of all trans-retinal (160 mg). Solvent system: cyclohexane/n-pentane/acetonitrile (5:2:5) + 0.1% methanol. Stationary phase: lower ("ascending mode"). Flow rate: 2.5 mL/min. Rotational speed: 1500 rpm. Pressure: 49 atm (700 psi). Detection: UV 420 nm, 2.0 AUFS. TLC analysis: HPTLC plates LHP-KF (silica, Whatman); mobile phase = upper phase of CPC solvent; visualized with vanillin-sulfuric acid.

run was finished in less than 2½ hrs, a time short enough to avoid any thermal decomposition. The third experiment illustrates the advantage of the possibility to switch the elution modes in centrifugal partition chromatography while an actual separation is going on. If in any two-phase solvent system the lighter phase is also the less polar phase, then an "ascending" elution mode in centrifugal partition chromatography becomes comparable to a normal phase separation in liquid chromatography. Analogously, if the lower, more polar phase becomes the mobile phase as it is in "descending" mode, a separation obtained in this way becomes comparable with a reversed phase separation. In other words, to switch elution modes in centrifugal partition chromatography during the chromatographic run would be comparable to the changing of columns and eluent during a liquid chromatography experiment. While the latter is practically impossible, the usefulness of this concept in centrifugal partition chromatography is illustrated in Fig. 8 [11]. Here a hypothetical five-component mixture is introduced with components E, D, and C being the less polar compounds while A and B are more polar. If elution starts in ascending or normal phase mode E, C, and D will eventually be separated and elute as distinct peaks; A and B, however, are merely separated and remain still mostly in the stationary phase; it would take a considerable amount of time and solvent to transport them totally through the stationary phase into the mobile phase prior to elution. Any pair of unstable compounds would hardly survive such a long exposure to solvents or electrolytes. The switching of elution mode after a short "recycling" period instead allows transporting of these just separated components form the column by having the lower phase changed from stationary to mobile. A and B will now appear from the inlet of the rotor as two separate peaks where they can be collected. This special feature of centrifugal partition chromatography is best illustrated by the separation of the tunichromes, the blood pigments from the sea squirt *Ascidia nigra* L. (Fig. 9) [7]. Tunichromes as conjugated phenols are sensitive toward oxidation and persists, even in degassed aqueous systems, only for a short time. The tunichromes of the B series differ only in their phenolic hydroxyl groups. Since their polarities are therefore very similar, a solvent system had to be designed which showed a comparatively subtle difference in polarity between the two phases. The unusually complex two-phase system shown in Fig. 10 fulfills this demand. Since tunichromes slowly decompose even in degassed, buffered, and deoxigenized water, it became mandatory to cut the retention time down to a minimum. Therefore, after elution of the two less polar component pairs tunichrome B5,6 and tunichrome, B3,4, the elution mode was switched form ascending to descending mode with the stationary phase now becoming the mobile phase. After initial elution of polymeric material, a sign of decomposition during the separation, tunichrome B1,2, the major component in the tunichrome

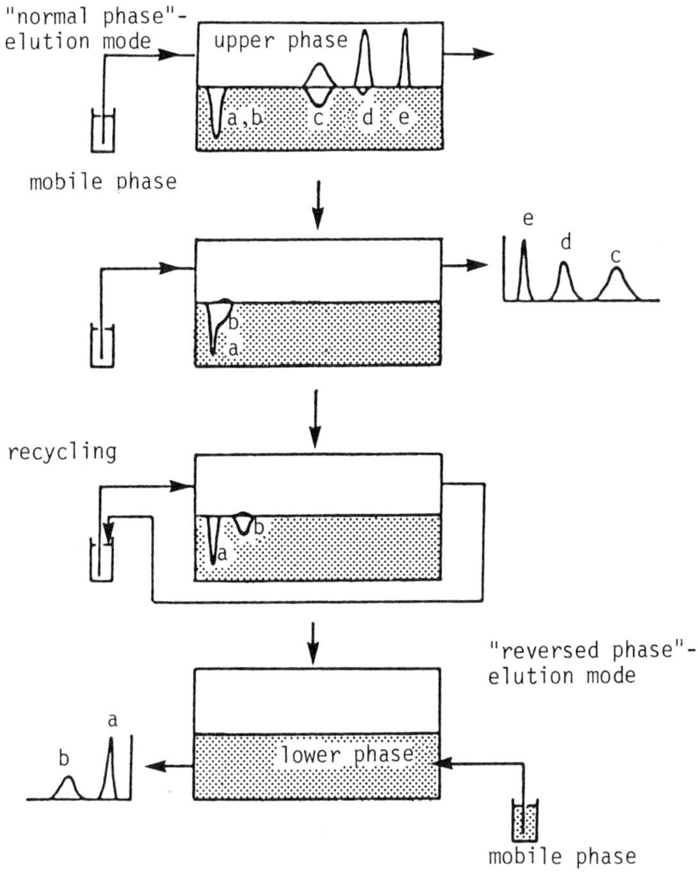

Fig. 8 Effect of mobile phase switching in centrifugal partition chromatography on the separation of a hypothetical five-component mixture (see text).

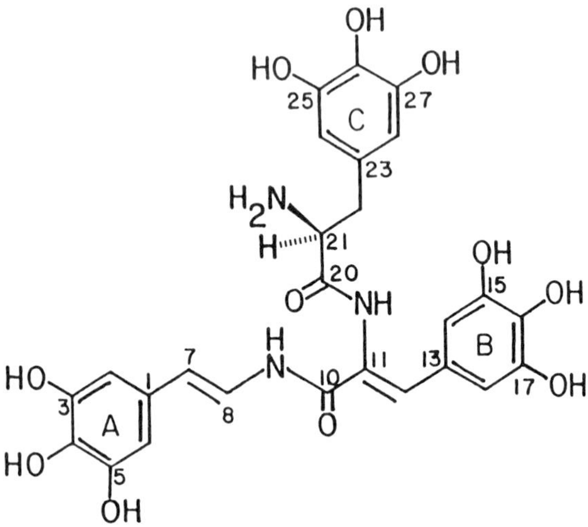

Fig. 9 Chemical structure of tunichrome B-1.

mixture, eluted as a narrow peak after 8 hr. The elution of these more polar compounds as very sharp bands after switching the elution mode is a further benefit of this technique.* The total recovery of tunichromes after centrifugal partition chromatography was better than 80% compared to a recovery of less than 5% from a ODS2 reversed phase column in the subsequent HPLC separation [7].

The previous examples show the distinct advantages of centrifugal partition chromatography over other chromatographic methods, especially in terms of smooth and fast operation. It becomes obvious that centrifugal partition chromatography as a separation technique is just at the beginning of its career. Its great potential, especially towards the separation of labile components from biological sources, for the preparation of biopolymers, or for the separation of stereo or optical isomers, remains largely to be explored. Even industrial scale application has recently been realized by construction of a pilot plant-size machine for the separation of carotenoids [11]. It has to be mentioned, however, that compared to other chromatographic techniques, especially to other liquid-liquid partition methodologies, centrifugal partition chromatography has one essential drawback, its high price tag [18].

---

*The theoretical plate number for the TB-3,4 peak equals approximately 700, while the calculation for TB-1,2, after phase switch, gives a value of N ~ 1800! (See next page, Fig. 10.)

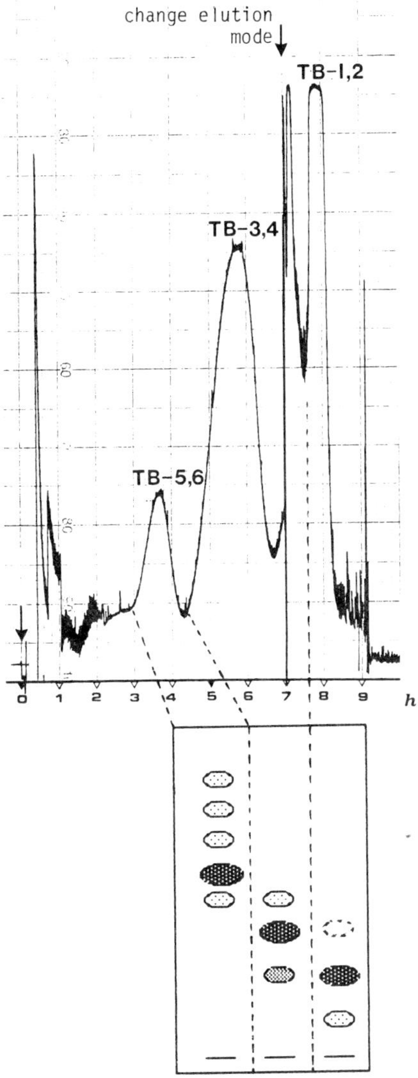

Fig. 10 Centrifugal partition chromatography separation of tunichromes B (200 mg). Solvent system: 3-methylbutanol/n-butanol/n-propanol/water (8:12:10:30) with 0.25% formic acid and 1% tert-butylsulfide. Stationary phase: lower ("ascending mode") for 7 hr; then mobile phase switched to "descending mode." Flow rate: 2.4 mL/min. Rotation speed: 1000 rpm. Pressure: 42 atm (600 psi). Detection: UV 380 nm, 2.56 AUFS. TLC analysis: HPTLC plates LHP-KF (silica, Whatman): mobile phase: ethylacetate/2-butanone/toluene/formic acid/water (50:24:6:10:10); detection:triphenylboric acid-ethauolamine complex ("natural products reagent") [17], UV 366 nm.

REFERENCES

1. T. Tanimura, J. J. Pisano, Y. Ito, and R. L. Bowman, *Science, 169,* 57 (1970).
2. (a) K. Hostettmann, M. Hostettmann-Kaldas, and K. Nakanishi, *J. Chromatog., 170*, 335 (1979); (b) K. Hostettmann, M. Hostettmann, and A. Marston, *Nat. Prod. Reports*, 471 (1984) and rerefences therein.
3. J. K. Snyder, K. Nakanishi, K. Hostettmann, and M. Hostettmann, *J. Liq. Chromatogr., 7*, 243 (1984).
4. (a) Y. Ito and W. D. Conway, *J. Anal. Chem., 56*, 534A (1984); (b) Y. Ito, *J. Biochem. Biophys. Methods, 5*, 105 (1981) and references therein; (c) Y. Ito, I. Aoki, E. Kimura, K. Nunogaki, and Y. Nunogaki, *Anal. Chem., 41*, 1579 (1969); (d) Y. Ito, M. A. Weinstein, I. Aoki, R. Harada, E. Kimura, and K. Nunogaki, *Nature, 212*, 985 (1966).
5. (a) K. Hostettmann, *Planta Med., 39*, 1 (1980); (b) K. Hostettmann, *Adv. Chromatogr., 21*, 165 (1983); (c) F. J. Hanke and I. Kubo, *J. Chromatogr., 329*, 395 (1985).
6. "Centrifugal Continuous Flow Extraction and Fractionation System CPC-L and CPC-I," Sanki Engineering Ltd., Kyoto, 1984.
7. R. C. Bruening, E. M. Oltz, J. Furukawa, K. Nakanishi, and K. Kustin, *J. Am. Chem. Soc., 107*, 5298 (1985); *J. Nat. Prod., 49*, 193 (1986).
8. W. Murayama, T. Kobayashi, Y. Kosuge, H. Yano, Y. Nunogaki, and K. Nunogaki, *J. Chromatogr., 239*, 643 (1982).
9. T. Araki and Y. Kubo, 1st CPC Colloquium, Abstract #2, Sanki Engineering Ltd., Kyoto, 1986.
10. T. Okuda, T. Hatano, and K. Yazaki, *Chem. Pharm. Bull., 31*, 333 (1983).
11. Sanki Engineering Technical Manual, "Continuous Flow Extraction and Fractionation System. Centrifugal Countercurrent Chromatograph. CPC-N90 CPC-A90," Kyoto, 1983.
12. M. Miyamoto, Y. Kawamatsu, K. Kawashima, M. Shinohara, and K. Nakanishi, *Tetrahedron Lett.*, 2355 (1964); M. Miyamoto, K. Morita, Y. Kawamatsu, K. Kawashima, and K. Nakanishi, *Tetrahedron, 23,* 421 (1967).
13. J. Thiem and B. J. Meyer, *J. Chem. Soc., Perkin Trans. II,* 1331 (1979).
14. J. K. Snyder, K. Nakanishi, K. Hostettmann, and M. Hostettmann, *J. Liq. Chromatogr., 7*, 243 (1984).
15. R. C. Bruening, F. Derguini, and K. Nakanishi, *J. Chromatogr., 357*, 340 (1986).
16. M. Denny, M. Chun, and R. S. H. Liu, *Photochem. Photobiol., 33,* 267 (1981).
17. R. Neu, *Naturwissenschaften, 44*, 181 (1957).
18. The instrument is avaiable from Sanki Engineering, Ltd., 2-6-10 Imazato, Nagaokakyo, Kyoto, Japan.

chapter eight

Applications of Countercurrent Chromatography in Agricultural Chemistry

John M. Ruth
Agricultural Environmental Quality Institute
Beltsville Agricultural Research Center, ARS, USDA
Beltsville, Maryland

N. Bhushan Mandava*
Office of Pesticides and Toxic Substances
U.S. Environmental Protection Agency
Washington, D.C.

I. INTRODUCTION

This chapter reviews applications of countercurrent chromatography to the separation and analysis of agricultural chemicals (what are sometimes called "agrochemicals") and to some areas of plant and animal biochemical research of especial interest in agriculture and nutrition. The latter aspect represents a sampling of such topics and does not discuss those that happen to be covered in other chapters of this volume.

Agricultural chemicals in the more restricted sense ("agrochemicals") are for some purposes classified into two categories: (a) crop production chemicals and (b) crop protection chemicals. Crop production chemicals (all of those chemical materials that contribute to productivity), in the language of pesticide regulation, include herbicides, plant growth regulators, and growth modifiers such as desiccants applied during the crop cycle. Crop protection chemicals are those used, before, during, or after harvest, to protect crops or stored agricultural products from pests of all kinds, such as bacteria, fungi, nematodes, insects, and rodents.

According to the regulations of the United States Environmental Protection Agency (EPA), any chemical, whether it is natural or synthetic, when released into the environment to control pests or

Current affiliation: Todhunter, Mandava, & Associates, Washington, D.C.

plants is called a pesticide and is subjected to registration under the Federal Insecticide, Fungicide, and Rodenticide Act before it is marketed for that use.

Pesticides contribute not only to increased agricultural productivity but also to public health protection. However, the same pesticides may be hazards to human health and to the environment. In order to minimize the risks, the EPA under Section 408 of the Federal Food, Drug, and Cosmetic Act sets tolerance limits for pesticides that are used on human food and animal feed crops. Analytical methods are the necessary means for determining pesticide residues in food and feed products.

Modern analytical methods used for detection, isolation, identification, and analysis of agrochemical residues present in plants, animals, raw and processed foods, and animal feeds are mainly chromatographic and spectrometric. They are capable of detecting and analyzing residues at parts-per-million and parts-per-billion levels quickly, reliably, and accurately. Commonly used chromatographic methods are gas chromatography (GC) and high-performance liquid chromatography (HPLC). Spectroscopic methods include ultraviolet (UV), infrared (IR), nuclear magnetic resonance (NMR), and mass spectrometry (MS). Mass spectrometry is the method most often used for final identification.

Progress in Fourier transformation techniques has improved the speed and sensitivity of IR and NMR analyses. In recent years, combinations of a chromatographic method with a spectroscopic method have in several cases been successfully interfaced, for the rapid separation, characterization, and measurement of compounds present in complex mixtures. Some of those combinations, such as GC/MS and GC/IR, have been well developed, and others, such as LC/MS, continue advancing. Computer acquisition and processing of data make possible rapid retrieval and display, and assist in many interpretations that would otherwise not be possible.

Along with refinements of the established methods of analysis and progress in their automation, new techniques are always being sought. One such is countercurrent chromatography (CCC), whose development has been reviewed recently by Ito [1]. That paper lists the different kinds of CCC instruments and summarizes their operating characteristics. The earlier forms of CCC were based on the principles of hydrostatic equilibrium. One of them, called droplet countercurrent chromatography, was described by Ito and Bowman [2,3] and Tanimura et al. [4]. Hostettmann [5] applied it to the preparative scale separation of natural products. Bearder and MacMillan [6] separated gibberellins by droplet countercurrent chromatography.

The further development of CCC produced a number of instruments, including

1. The horizontal flow-through coil planet centrifuge
2. The toroidal coil planet centrifuge
3. The preparative countercurrent chromatograph, and
4. The high-speed countercurrent chromatograph

whose mechanism of operation is the technologically much more complex hydrodynamic equilibrium system [1]. These instruments incorporate various forms of rotating coils and depend for their operation on the effects of centrifugal force on the flowing liquid system. Mandava et al. [7,8] reviewed the development of earlier instruments and their applications. As will be seen below, high-speed preparative countercurrent chromatography, described by Ito et al [9], is one of the most recent and most spectacular achievements. The notation for elution and partition coefficients used in this chapter is taken from Chap. 12, Sec. V.D.

This chapter reviews applications of CCC which are at the research stage, and is not concerned with any official methods of analysis. Because CCC is a relatively new and potentially powerful tool, those applications will be presented in some detail. The pioneering work on the interfacing of CCC to Fourier transform infrared spectrometry (FT-IR) by Romanach et al. [10] and by Romanach and de Haseth [11] holds great promise for the location of components in complex mixtures. Chapter 12 will review chromatographic/mass spectrometric techniques as the background for the development of CCC/MS.

II. PLANT GROWTH REGULATORS

The naturally occurring plant growth regulators, or plant hormones, are divided into five groups. The classification (Bearder [12]; MacMillan [13]; Takahashi [14]) is based on both biological activity and structural similarity. The groups (Table 1) are (a) ethylene, (b) the indole auxins, (c) the gibberellins, (d) the cytokinins, and (e) abscisic acid and related compounds. Ethylene, a plant hormone involved in such processes as fruit ripening and parthenocarpy, has been reviewed by Abeles]15]. Since it is a gas, it falls outside the scope of the present chapter. In some types of work, it can be handled by gas chromatography/mass spectrometry, and in others, by the trapping of samples for mass spectrometric analysis (e.g., as by Abeles et al. [16]).

The name "auxin" (from the Greek *auxein*, to increase) refers to a number of compounds isolated from plant material, of which the

Table 1 Classification of Plant Hormones

I.	Ethylene: $CH_2 = CH_2$
II.	Indole auxins: Indole-3-R
	R = CH_2 COOH: Indole-3-acetic acid (IAA)
	R = CH_2CH_2 COOH: Indole-3-propionic acid (IPA)
	R = $CH_2(CH_2)$ COOH: Indole-3-butyric acid (IBA)
	R = COOH: Indole-3-carboxylic acid (ICA)
	R = CH = CH COOH: Indole-3-acrylica acid (IAca)
	R = CH_2 $CONH_2$: Indole-3-acetamide (ICA)
	R = CH_2CN: Indole-3-acetonitrile (ICN)
III.	Gibberell9ns: The IUPAC Nomenclature is based on the *ent*-gibberellane structure. A system of numbers, GA_n, where n is an integer indicating the order of assignment of the numbers, is also used. Important examples are GA_3, GA_4, and GA_7.
IV.	Cytokinins: The basic skeleton is a 6-substituted aminopurine ring system. Important examples are zeatin (Z), zeatin riboside (ZR), 6-isopentenyladenine (IP), and 6-isopentenyladenosine (IPA).
V.	Abscisic acid and related compounds
VI.	Steroidal plant hormones: This group includes brassinolide and more recently discovered compounds.

most important one is indole-3-acetic acid. In general, the auxins have included compounds containing an indole skeleton variously substituted at the C-3 position, and in one subclass having a chlorine atom at the C-4 position. Because of some association of the name "auxin" with other substances, the compound expression "indole auxins" is used to limit discussion to the indole compounds.

Gibberellins belong to one of two groups, C_{20}, possessing all of the diterpenoid carbon atoms, and C_{19}, in which the angular C-20 is missing. In the older literature, the nomenclature of the gibberellins was based on the gibbane skeleton. More recently, it has been replaced by a new system using the *ent*-gibberellane skeleton, as proposed by Rowe et al. [17] to the 1968 meeting of the IUPAC Commission on the Nomenclature of Organic Chemistry [12,18,19]. The everyday system GA_n, where n is an integer indicating the order in which the names have been assigned, was adopted on the recommendation of MacMillan and Takahashi [20]. Bearder [12] comments further on the complications of gibberellin nomenclature. Some 60 or more gibberellins have now been identified [12-14]. Some common ones are gibberellin A_1 (GA_1); gibberellin A_3 (GA_3), or gibberellic acid; gibberellin A_4 (GA_4); and gibberellin A_7 (GA_7). Auxins and gibberellins are considered responsible for elongation of plants, in addition to other roles that they have in plant growth and development. Cytokinins are involved in cell division. The basic skeleton of all cytokinins is a 6-substituted amino purine ring system. Among several cytokinins (Table 1) identified from higher plants, zeatin and 6-isopentenyl adenine and their ribosides occur widely. These three groups—namely, indole auxins, gibberellins, and cytokinins—are generally referred to as plant growth promoters.

Abscisic acid (ABA) and metabolites (group V, Table 1) such as phaseic and dihydrophaseic acids are isolated from plants. These compounds are normally described as plant growth inhibitors. Although ABA's function is to retard the growth of the plant, it is also involved in other aspects of the life of the plant, such as bud development, plant stress, and water retention by the tissue.

A sixth class, steroidal plant hormones, appears to be in the process of establishing itself. The first member of the group was brassinolide. Other naturally occurring steroidal growth promoters have recently been isolated, and a number of synthetic analogs have been shown to possess similar biological activity [21].

Because growth regulators are present only in minute quantities in plant materials, their identification and quantitative estimation have required sensitive analytical techniques, which today involve mainly chromatography and spectrometry. Except for ethylene, which is a gas, all other plant hormones have been shown to be separable by CCC, which was also employed for the quantitative estimation of ABA in plant tissue. Apart from the naturally occurring phytohormones, there are a number of syntheic materials that show similar biological activity. These agrochemicals are registered for use as growth modifiers during the crop cycle or at the end of it. CCC methods are also applicable to the study of the metabolism of these synthetic growth regulators, but that application remains to be carried out.

A. Indole Auxins

Mandava and Ito [22-25] separated mixtures of several indole auxins and metabolites (Table 2) by toroidal coil planet centrifuge. Two solvent systems were employed:

- Solvent system 1: Hexane/ethyl acetate/methanol/water (0.6:1.4:1:1)
- Solvent system 2: Chloroform/acetic acid/water (2:1:1)

The elution coefficients (K_E* = concentration in mobile phase/concentration in stationary phase) in these solvent systems are shown in Table 2. Because of the differences in their K_E values here, it would be expected that these indoles, when present as a mixture, could be separated as individual components by CCC. If the lighter phase is kept stationary on the column and the heavier phase is used to elute the column, the K_E(H/L) values for these compounds (Table 2, columns 5 and 7) serve as a useful guide for predicting the order of elution. When the phases are reversed, the K_E(L/H) values in columns 4 and 6 of Table 2 must then be used. Because the elution pattern follows the decreasing order of the elution coefficients, the separation of the mixture begins with the compound having the largest elution coefficient, followed by compounds with lower ones.

A mixture containing five indole derivatives (IA, K_E(H/L) = 3.83; ICA, 1.61; IAA, 1.01; IBA, 0.57; and IAN, 0.33) whose elution coefficients vary from 3.83 to 0.33 was chromatographed in solvent system 1, keeping the lighter (organic) phase stationary. For convenience, the list in parentheses gives the numerical value of the elution coefficient K_E(H/L) immediately following the abbreviation used for the compound name. With the heavier (aqueous) phase as mobile phase (Fig. 1), the compounds eluted in the predictable order: IA (elution coefficient, 3.83) came first, followed by ICA (1.61), IAA (1.01), IBA (0.57), and IAN (0.33).

Using solvent system 2, a mixture of abscisic acid (ABA) with four indolecarboxylic acids (ICA, 1.07; IAcA, 0.78; ABA, 0.56; IBA, 0.36; IPA, 0.26) was analyzed by CCC. Again keeping the

---

*For preferred usage of terminology in CCC, see Chap. 2, Sec. III.

Table 2 Elution Coefficients of Indoles in Two Different Solvent Systems

Compound	Abbreviation	Concentration (mg/ml)	Partition coefficient[a] Solvent system I $K_E(L/H)$	Solvent system I $K_E(H/L)$	Solvent system II $K_E(L/H)$	Solvent system II $K_E(H/L)$
Indole-3-acetic acid	IAA	20	0.99	1.01	0.94	1.06
Indole-3-acetamide	IA	20	0.26	3.83	1.05	0.95
Indole-3-acetonitrile	IAN	20	3.00	0.33	0.39	2.55
Indole-3-carboxylic acid	ICA	10	0.63	1.61	1.07	0.93
Indole-3-acrylic acid	IAcA	10	0.88	1.44	0.78	1.28
Indole-3-butyric acid	IBA	20	1.74	0.57	0.36	2.81
Indole-3-propionic acid	IPA	20	2.02	0.50	0.26	3.80

[a]Solvent system I, hexane/methanol/water (0.6:1.4:1:1); solvent system II, chloroform/acetic acid/water (2:2:1). H, heavier phase; L, lighter phase. *Source*: Ref. 24.

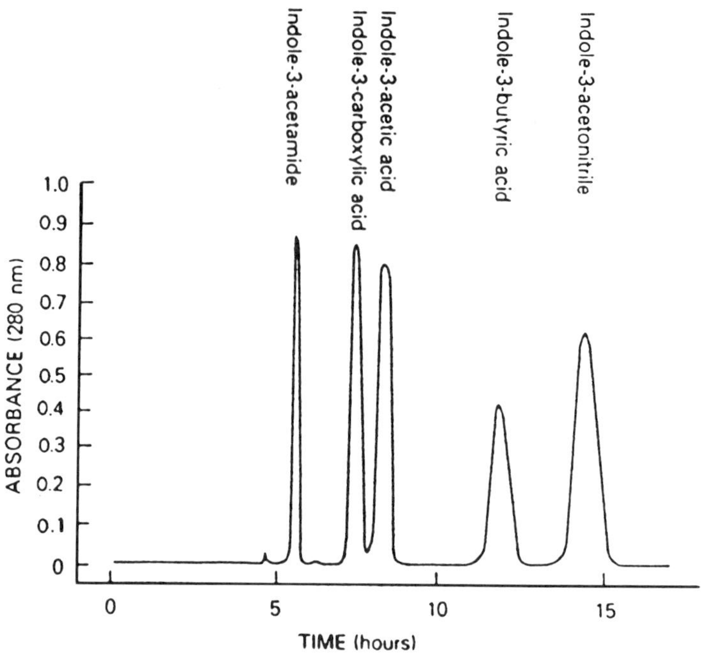

Fig. 1 Separation of indole compounds by CCC in hexane/ethyl acetate/methanol/water (0.6:1.4:1:1); mobile phase aqueous. (From N. B. Mandava and Y. Ito (1982), *J. Chromatogr. 247,* 315.)

organic phase stationary (which is the heavier phase this time), and making the aqueous phase (the lighter phase, this time) mobile, all five compounds were separated (Fig. 2). The elution coefficients $K_E(L/H)$ of the indoles may be found in column 6 of Table 2. If IAA had been present in this mixture, with solvent system 2, it would have fallen between the ICA and IAcA peaks, obscuring their resolution, in which case a different solvent system would have been chosen. The separation of ABA from IAA in the same system is shown in Fig. 3.

B. Gibberellins

The use of droplet countercurrent chromatography (DCCC) by Bearder and MacMillan [6] to separate gibberellins was the first published application of countercurrent chromatography to plant hormone chemistry. In that technique [2-4] (see Sec. I), the stationary phase is confined in a set of tubular columns mounted vertically. The mobile phase, which may be either the upper or the lower phase, passes through it in the form of small droplets. If the

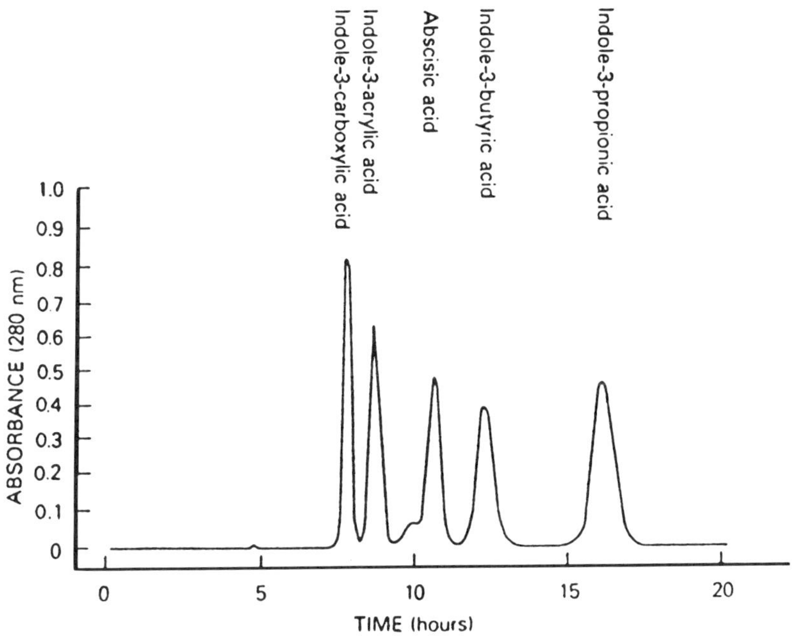

Fig. 2 CCC separation of a mixture of four indolic carboxylic acids and abscisic acid in chloroform/acetic acid/water (2:2:1); organic phase stationary. (From N. B. Mandava and Y. Ito (1982), *J. Chromatogr.* *247*, 315.)

lighter phase is mobile, it is introduced at the bottom of the first tube; if the heavier is mobile, at the top. The connection from the bottom or top of one tube to the top or bottom of the next is made by small diameter PTFE tubing. After the mobile phase has passed through all of the tubes, it is recovered from the last one by a fraction collector.

Bearder and MacMillan [6] employed a Tokyo Rikakikai Co. DCCA containing 350 tubes, each 40 cm long X 2 mm i.d. The volume collected per fraction was about 7 ml, and analyses were done by thin-layer chromatography, gas chromatography, weight, or bioassay. The paper gives six examples as described below.

(1) A mixture containing 30% GA_{13} and 70% GA_3, obtained from a culture of *Gibberella fujikuroi*, was used to isolate GA_{13} required for biosynthetic studies. The solvent system was chloroform/acetic acid/water (2:2:1). The organic (heavier) phase was used as the mobile, descending phase. From 0.7 g of the mixture, pure GA_{13} was collected in fractions 21–24, and pure GA_3, in fractions 55–64.

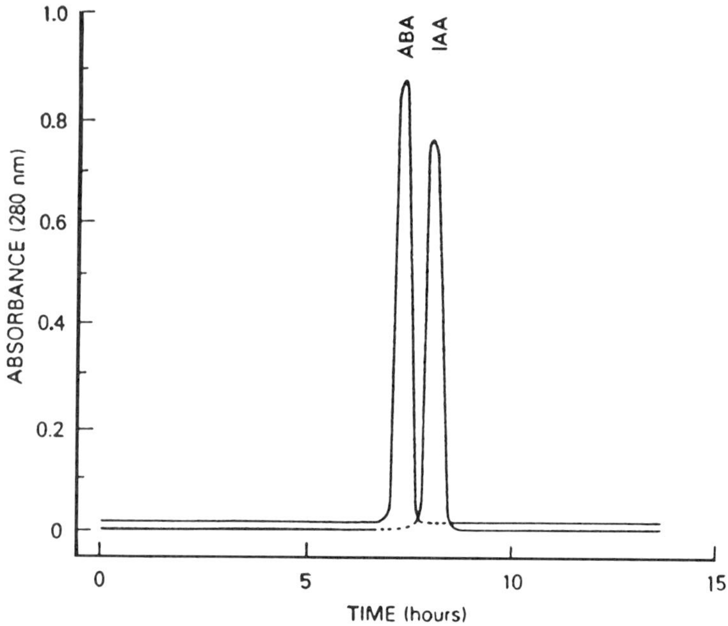

Fig. 3 CCC separation of indole-3-acetic acid and abscisic acid in chloroform/acetic acid/water (2:2:1); organic phase stationary. (From N. B. Mandava and Y. Ito (1982), *J. Chromatogr.* *247*, 315.)

(2) After adsorption chromatography of an extract of *G. fujikuroi*, a resulting mixture of isoGA_7, GA_4, and GA_{14} (1:1:1) was separated to obtain pure GA_{14} and a partial resolution of the remaining two, isoGA_7 and GA_4. The solvent system was methylene chloride/methanol/water (5:6:4), and the organic (heavier) phase was used as the mobile, descending phase. Starting with 1.7 g of the mixture, pure GA_{14} was separated in fractions 52–64. GA_4 and isoGA_7 were partially separated, in fractions 21–26 and fractions 24–32, respectively. Beginning with 5.0 g of the three-component mixture, GA_{14} was still separated in a single run.

(3) The third of the six examples (Bearder and MacMillan [6]) involved samples from synthetic studies to confirm the structures of five new gibberellines that had been detected in extracts of wheat germ by Gaskin et al. [26]. The addition of the elements of water to the Δ-1 double bond of 3-didehydroGA_3 methyl ester yielded a mixture of starting material with the 1β- and 1α-alcohols. The solvent system for DCCC was chloroform/methanol/water (7:15:8), and the organic (heavier) layer was used as the mobile, descending phase. From 600 mg of mixture, the yield was 278 mg starting

material, 190 mg 1β-alcohol, and 13 mg 1α-alcohol, in elution volumes 90–120, 160–210, and 230–265 ml, respectively. At a later stage, the 1α-alcohol was reduced to a mixture of GA_{57} methyl ester and its 3α-epimer. The latter pair were most conveniently separated by DCCC under the same conditions.

(4) The separation of a partially purified extract (35 mg) derived from endosperm tissue of *Cucurbita maxima* was attempted by DCCC, using methylene chloride/methanol/water (5:6:4), with the organic (heavier) layer as the mobile, descending phase. When the required component was eluted in the first 80 fractions, the material in the tubes (26 mg) was recovered, and was rerun with the aqueous layer as the mobile, ascending phase. GA_{58} was then eluted in fractions 20–26 (140–180 ml), and a compound believed to be *ent*-6α,7α,12β-trihydroxykaur-16-en-19-oic acid was obtained from fractions 37–43 (260–300 ml).

(5) The acids (680 mg) soluble in ethyl acetate were extracted from 225 g of 47-day-old seed of *Phaseolus coccineus*. The resulting mixture was partitioned by DCCC in a solvent system consisting of methylene chloride/methanol/water (5:6:4), with the organic phase mobile (in the descending mode). Fractions 1–54 were evaporated to dryness and bioassayed. The material remaining in the tubes was recovered and rerun with the aqueous layer of the same solvent system in the mobile, ascending mode. Bioassay of fractions 1–60 detected biological activity in fractions 16–23 and 45–47, with inhibition in fractions 25–31. The n-butanol-soluble acids from the same source were processed in the same solvent system, with the aqueous phase mobile (ascending). Fractions 1–72 showed no biological activity (no conjugates of bioactive GA's present).

(6) As an illustration of the use of DCCC to isolate nonpolar compounds, the ethyl acetate-soluble portion (1 g) of a methanol extract of *Helianthus decapetalus* flower heads was processed by DCCC with the organic (lighter) phase of the solvent system light petroleum (b.p. 60–80°)/ethanol/water/ethyl acetate (5:4:1:2) mobile (ascending mode). The components *ent*-13(S)-angeloyloxyatisir-16-en-19-oic acid (87 mg) and *ent*-12β-acetokaur-16-en-19-oic acid (87 mg) were collected in fractions 40–48 (310–380 ml) and fractions 64–74 (500–590 ml), respectively. Both were found essentially pure when tested by thin-layer chromatography and by NMR.

Although the six examples were done by droplet countercurrent chromatography, they illustrate equally well the potential value of the newer centrifugal methods. The authors [6] point out in the cases of examples 1 and 2 that DCCC is simpler and more economical than the methods usually employed. The same is of course true for CCC, and the resolution will be better. Those authors mention the usual advantages that have also been noted for CCC: wide range of sample sizes, small total solvent volumes, and high recovery (no adsorbing solid phase present). They mention also the disadvantages of low resolution and restricted number of solvent systems suitable for droplet countercurrent chromatography. Those are of course the problems that have been solved by the development of the centrifugal instruments, where the relative specific gravities and the viscosities of the phases are less restrictive because of the greater mixing and separatory power of the centrifuge.

CCC separation was investigated by Mandava and Ito [23,24] for the most commonly occurring gibberellins, GA_3, GA_4, and GA_7, which possess the highest biological activity. As in HPLC, the separation of gibberellins presented problems, mainly because of the lack of a proper detection system. They showed no UV absorption between 220 and 360 nm, and therefore the standard UV monitors with fixed wavelength detectors at 254 nm were not effective. However, all the gibberellins have strong absorption around 210 nm, which was used to monitor them with a variable UV detector. The choice of solvents was also limited because gibberellins do not partition very well in the solvent systems ordinarily employed. Durley and Pharis [27] reported previously that a solvent mixture containing ethyl acetate/phosphate buffer (pH 7) was the ideal system in the countercurrent distribution of gibberellins. Since ethyl acetate (UV cutoff at 230 nm) interferes with UV monitoring at 210 nm, diethyl ether was tried in its place. However, the use of a diethyl ether/0.5 M phosphate buffer system has the disadvantage that it has a very high interfacial tension, which creates a high column pressure. This difficulty was overcome by introducing methanol into the system. Solvent system 3 was found to be suitable for gibberellin separation.

Solvent system 3: Diethyl ether/methanol/0.5 M phosphate buffer (3:1:2).

The separation of gibberellins A_3, A_4, and A_7 in solvent system 3 shows (Fig. 4) that GA_3 elutes first, followed by GA_7 and GA_4, when the lighter (organic) phase is the mobile phase. (The caption of Fig. 5 of [24] contains an error in the specification of the solvent system system, which is correctly described in the text.) To confirm that only the chromatographic peaks contained gibberellins, all the fractions were evaluated by a gibberellin bioassay. By using bioassays of fractions to monitor the chromatography, it would be possible to use CCC for the separation, detection, identification, and analysis of gibberellins from natural sources.

C. Cytokinins

Countercurrent chromatography was found very effective [23-24] for the separation of a mixture of four cytokinins: zeatin (Z),

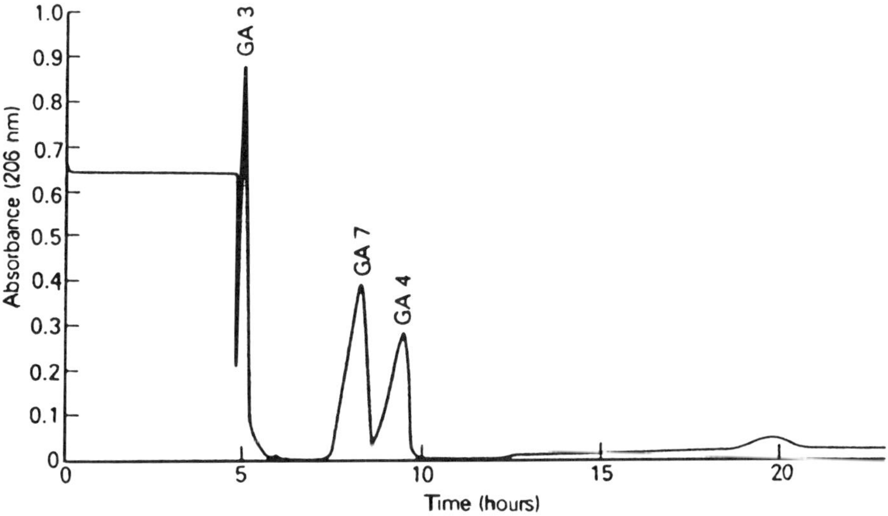

Fig. 4 Separation of gibberellins A_3, A_4, and A_7 in diethyl ether/methanol/0.5 M phosphate buffer, pH 5.9 (3:1:2); organic phase mobile. (From N. B. Mandava and Y. Ito (1982), *J. Chromatogr. 247,* 315.)

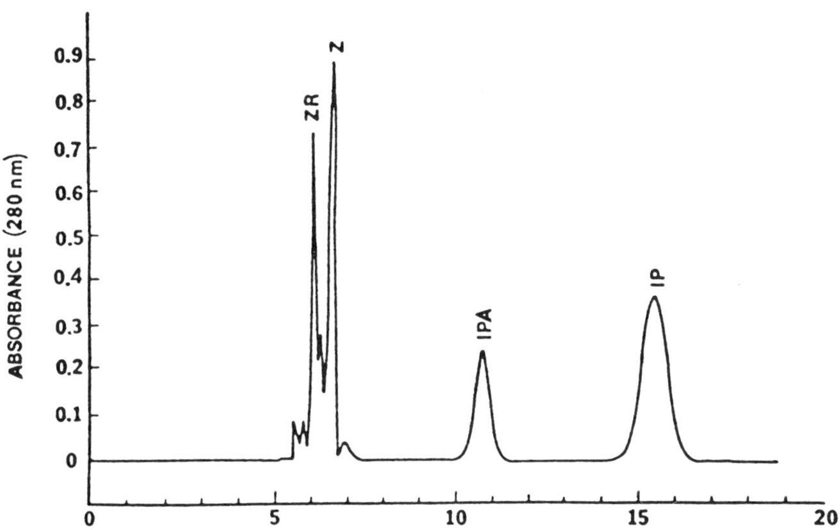

Fig. 5 CCC separation of cytokinins (zeatin, 6-isopentenyl adenine, and their ribosides) in ethyl acetate/methanol/0.5 M phosphate buffer, pH 7 (3:1:3); organic phase stationary. (From N. B. Mandava and Y. Ito (1982), *J. Chromatogr. 247,* 315.)

Table 3 Partition Coefficients of Cytokinins in Ethyl Acetate/Methanol/0.5 M Phosphate Buffer (pH 7) (3:1:3)

		Partition coefficient	
Cytokinin	Abbreviation	C_{aq}/C_{org}	C_{org}/C_{aq}
Zeatin riboside	ZR	4.25	0.24
Zeatin	Z	2.26	0.44
6-Isopentenyladenosine	IPA	0.73	1.37
6-Isopentenyladenine	IP	0.20	5.00

Source: Ref. 24.

zeatin riboside (ZR), 6-isopentenyladenine (IP), and 6-ispentenyladenosine (IPA). They were well resolved in a solvent system containing ethyl acetate, methanol, and phosphate buffer. A change of pH from 5 to 9 affected their elution coefficients considerably when they were partitioned between ethyl acetate and 0.5 M phosphate buffer. For example, ZR showed a $K_E(H/L)$ range of 8–10, while IPA has a range of 0.08–0.1 in this solvent system. Solvent system 4 was the combination which gave the desired range of $K_E(H/L)$ (0.2–5.0) for these compounds (Table 3).

Solvent system 4: Ethyl acetate/methanol/0.5 M phosphate buffer, pH 7 (3:1:3)

The cytokinin mixture, when subjected to CCC separation with the organic phase stationary, was completely resolved, according to the decreasing order of elution coefficients $K_E(H/L)$; i.e., ZR, which has the highest elution coefficient, appeared first, followed by Z, IPA, and IP (Fig. 5).

However, the Z and ZR peaks appeared too close, whereas IP and its riboside (IPA) were separated farther apart. The commercial samples of Z and ZR contained impurities which were found in the chromatogram before, after, and between the Z and ZR peaks. The Z and ZR peaks may be resolved well either by simply reversing the phase in the CCC run or by choosing an entirely different solvent system. When the same solvent system is used, but the phases are reversed, the order of elution of the compounds is reversed (Fig. 6). IP, which had the longest retention time (Fig. 5), is now eluted with the shortest retention time (Fig. 6). ZR, which had the shortest retention time (Fig. 5), now has the longest (Fig. 6). Z and ZR are now well separated, so that the impurities in the mixture do not interfere with the Z and ZR peaks.

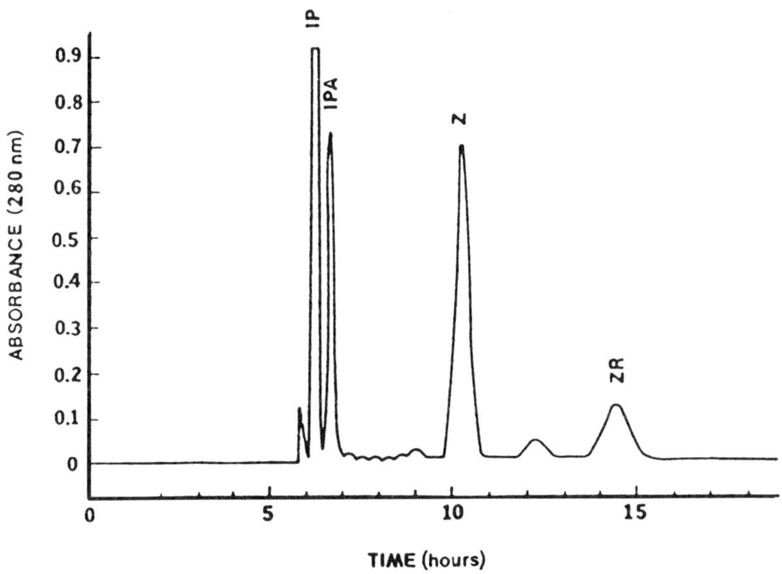

Fig. 6 CCC separation of cytokinins (zeatin, 6-isopentenyl adenine, and their ribosides) in ethyl acetate/methanol/0.5 M phosphate buffer, pH 7 (3:1:3); aqueous phase stationary. (From N. B. Mandava and Y. Ito (1982), *J. Chromatogr.* *247*, 315 (1982).)

In summary, the CCC applications described above illustrate the following basic facts: (a) The elution coefficients $K_E(H/L)$ and $K_E(L/H)$ are reciprocals of each other. (b) Retention times measured from the start of elution to the collection of a given fraction are related directly to the elution coefficient K_E or $K(C_m/C_s)$, where the subscript m identifies the mobile phase and s labels the stationary phase. (c) Thus, in addition to the assistance in the choice of solvent system that the partition coefficients give, they account for the fact that the reversal of the choices for mobile and stationary phases will give new retention times, which are in the inverse order of the retention times measured before the phase reversal.

D. Abscisic Acid

All of the data discussed previously were obtained with a toroidal coil planet centrifuge, which gives good separations but is limited in sample size to 1 mg or less. Included in that collection of data was the separation of ABA from IAA that was presented in Fig. 3 above. The instrument is well suited only for analytical scale applications. When it is desired to analyze for trace organics, such as plant

Table 4 General Operating Conditions for CCC Instruments in Plant Hormone Analysis

CCC instrument	Flow rate (ml/hr)	Revolutional speed (rpm)
1. Toroidal coil planet centrifuge	2.5–4.0	450–500
2. Preparative CCC with rotating coil assembly	180–190	40–50
3. High speed preparative CCC: small column	75–80	800–850
medium column	240	800

Source: Ref. 25.

hormones, present in the parts-per-million range, preliminary cleanup procedures must be employed. Those procedures are often time consuming. To circumvent that problem, other CCC instruments are available. One such is the preparative countercurrent chromatograph (Sec. II.D.1 and II.D.2), which is capable of handling gram quantities.

Although the earlier CCC seemed slower than HPLC, countercurrent chromatography has the advantages of (a) large sample-handling capacity; (b) avoidance of the replacement of expensive HPLC packed columns, which are sometimes damaged during cleanup with strong solvents; (c) replacement of much of the tedious preliminary cleanup procedure for natural products; and (d) a continuing development that is now making it competitive with HPLC in elution speed (Sec. II.D.3 below). Mandava and Ito [25,28] applied both the older preparative CCC and the newer high-speed instrument in further work on abscisic acid, which was also present in one of the mixtures whose analysis was described in Sec. II.A of this chapter. The general operating conditions of the instruments are given in Table 4.

Tests of Preparative CCC with Plant Hormone Mixtures

Before the application of CCC, the general approach to abscisic acid (ABA) analysis was slow, and involved (a) extraction of plant samples by conventional isolation methods (such as partitioning of the extract with acid and base to separate basic and acidic components); (b) repeated purification of the extract containing the ABA by chromatographic methods; (c) following that initial cleanup, the final purification by HPLC; and (d) characterization of ABA, in the

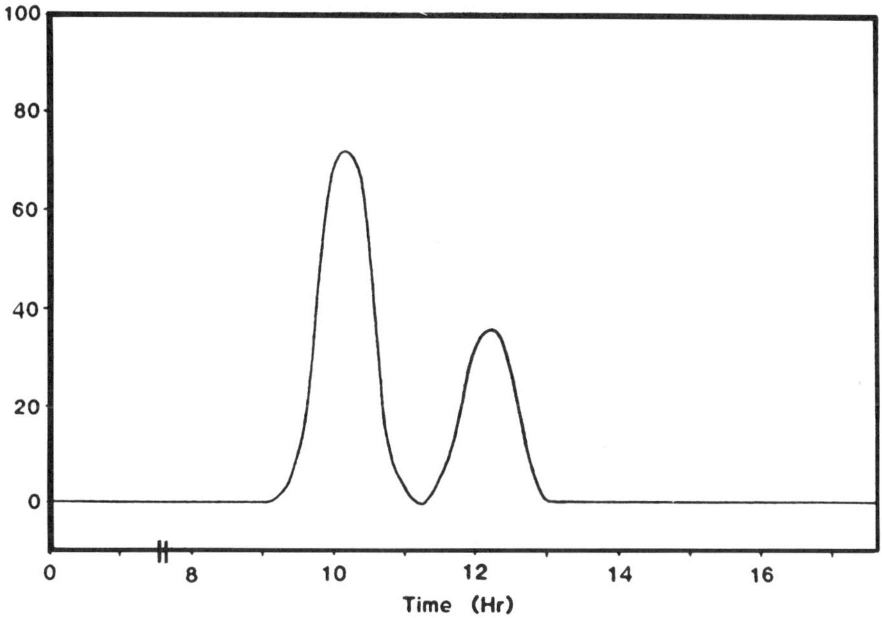

Fig. 7 Separation of ABA and IAA by preparative CCC. (From N. B. Mandava and Y. Ito (1984), *J. Liq. Chromatogr.* 7, 303.)

form of its methyl ester, by gas chromatography/mass spectrometry (GC/MS).

In preparation for the processing of actual samples of plant material, a mixture of standard abscisic acid and indoleacetic acid was put through the original type of preparative countercurrent chromatograph. A typical run with a mixture of ABA (100 mg) and IAA (100 mg) was done with a solvent system obtained by equilibriating 1600 ml chloroform, 1600 ml acetic acid, and 800 ml water. The Chromatograph was operated at 45 rpm. The column was filled with aqueous (lighter) phase and eluted with organic (heavier) phase, maintaining the flow rate at 3.1 ml/min. The UV detection wavelength was 260 nm. The chromatogram (Fig. 7) for a mixture of ABA and IAA is similar to the one obtained in the analytical scale run (Fig. 3.). In the preparative CCC run, the peaks are broader and the retention times are longer, but the separation appears upon visual inspection to be just as good. The solvent system and the choice of mobile phase are the same, which makes the order of elution the same. Other differences between the analytical and preparative runs are in column size, elution rate, centrifugal speed, and sample size [25].

Table 5 Abscisic Acid Analysis by Countercurrent Chromatography[a]

Plant sample	ABA content (ng/g plant tissue)
1. Zoysia grass seed	22
2. Soybean seed	26
3. Avocado fruit	15

[a]The values reported here are only approximate (estimated error = 4.86) and based on the average of three experiments.
Source: Ref. 25.

Preparative CCC of Crude Plant Extracts

After the tests described, crude extracts of plant samples were processed in a preparative countercurrent chromatograph. Gram quantities, obtained from pound quantities of plant tissue by conventional extraction methods used in ABA analysis, were loaded. The CCC instrument was operated under the same conditions used for the standard mixture of ABA and IAA, and the solvent system was the same. The effluent corresponding to the rentention time (9–11 hr) of ABA was collected. After evaporation of the solvent, the residue was subjected to analytical CCC purification by toroidal coil planet centrifuge. The operating conditions and the solvent system were the same described before (Fig. 3). This two-step procedure was said to be very efficient, eliminating the need for two of the usual preliminary purification steps in the analysis of plant samples for ABA. Identity of the ABA obtained by CCC from zoysia grass seed extract was further confirmed by comparing its behavior in HPLC with that of standard ABA [25].

For final verification that the fraction having the retention time of the ABA standard was indeed ABA, the material was converted to the corresponding methyl ester and analyzed by GC/MS. The mass spectrum of the methyl ester (Chap. 12) from the plant product was identical to that of the standard ABA methyl ester. For a further test of the two-stage CCC procedure for ABA analysis, a few more plant samples were analyzed for ABA content (Table 5).

The experiments suggest that CCC is potentially very useful for the analysis of trace compounds, such as plant hormones, that are sometimes quite difficult to separate by other methods. CCC not only offers an alternative analytical method, but also replaces some preliminary procedures, thereby making the analytical task simpler and easier. If some of the earlier CCC instruments require longer

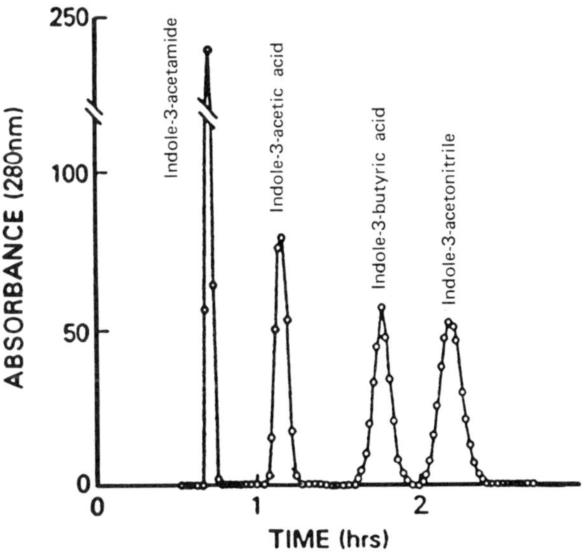

Fig. 8 Separation of indole hormones by high-speed semipreparative CCC. (From N. B. Mandava and Y. Ito (1984), *J. Liq. Chromatogr.* 7, 303.)

operation times than HPLC, it must be remembered that in doing so, they are replacing much of the older cleanup procedures, and at the same time are working under automatic control, releasing the analyst to do other phases of his work simultaneously.

High-Speed Preparative Countercurrent Chromatography

In continuing advances, the problem of long elution time in the separation of complex mixtures was solved by the design of the high-speed preparative countercurrent chromatograph [1,9]. This newer instrument provides much shorter elution times, as compared with older CCC instruments, without sacrificing resolution. It can be used in either analytical or preparative mode by changing the coiled column assembly to fit the sample size, from the microgram to the gram scale.

Mandava and Ito [25] reported running milligram amounts of a mixture of indole auxins with a semipreparative column in the high-speed preparative CCC, as a test of speed and resolution (Fig. 8). The chromatogram was compared with that in Fig. 1, which was the result of running a 25-μg sample [24] of a somewhat similar mixture as an analytical sample in the toroidal coil planet centrifuge. Resolution of the larger sample was adequate. The time for emergence of the last peak, containing the indole-3-acetonitrile, was improved

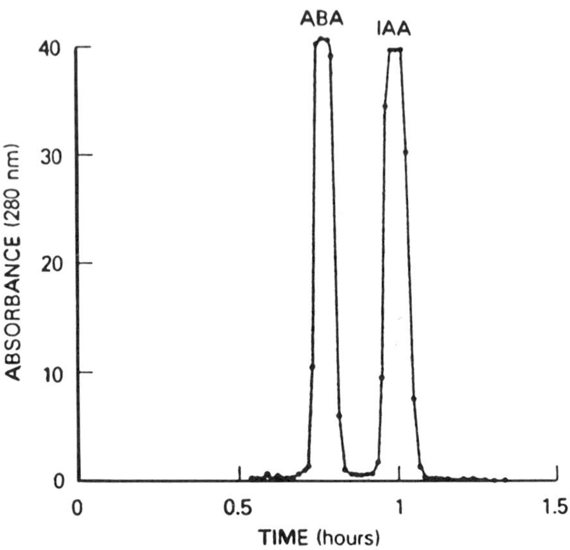

Fig. 9 Separation of ABA and IAA by high-speed semipreparative CCC. (From N. B. Mandava and Y. Ito (1984), *J. Liq. Chromatogr.* 7, 303.)

almost by an order of magnitude, a little over 2 hr in the new machine as opposed to about 14 hr in the older one.

As shown in Fig. 3, the separation of a mixture of indole-3-acetic acid and abscisic acid in solvent system 2, chloroform/acetic acid/water (2:2:1), in the toroidal coil planet centrifuge required about 8 hr [24]. The older preparative CCC (Fig. 7) required about 13 hr [25]. With the new high-speed semipreparative CCC, the same mixture was separated (Fig. 9) in about an hour, and the solvent required was about 200 ml. This is comparable in both run time and solvent use to a typical HPLC run [25]. The commercial availability of the high-speed semipreparative countercurrent chromatograph (P.C. Inc., Potomac, MD and Pharma-Tech Research Corporation, Baltimore, MD) should lead to a rapid increase in applications of CCC.

III. HERBICIDES

Herbicides have acquired great economic importance both to the practitioners of large-scale, specialized agriculture and to the firms which manufacture them. In circumstances in which they are found in significant concentrations outside the area of application, or as persistent residues that influence subsequent activities within the area of application, they become of interest in other ways, both to the general public and to the user. In all phases of manufacture, use, disposition, and in the associated research and developmental activities, they present another set of challenges to the chemist.

Hassall [29] classifies herbicides according to their mode of application:

I. Foliar application

  A. Systemic (translocated)
  B. Contact

II. Soil application

While doing so, he cautions the reader that such a classification is not to be followed too rigorously, quoting the examples of (a) the 2,4-D type, applied to leaves, while the related 2,4-dichlorophenoxyethylsulfate is applied to soils; (b) many carbamates applied to soils, but some used as foliage sprays; and (c) the question of translocation or contact action may depend on the sensitivity of the analytical method used to measure concentrations of organics in aqueous solution.

Another classification assigns herbicides applied in the presence of crop plants to one of two classes of use, "preemergence" or "postemergence," depending on the time of application relative to the time of emergence of the crop plants through the soil surface. Where analytical samples are concerned, the concentration of the herbicide and the nature of the matrix in which it is found will depend on whether it is a residue from application and, if it is, on the mode of application.

A. Phenoxyalkanoic Acid Herbicides

The history of World War II research on the phenoxyalkanoic acid herbicides and of their introduction as commercial weedkillers after the war has been reviewed by Loos [30]. These compounds have been used in large quantities to control broad-leaved weeds, and are especially valued for their selective action, against such weeds but not against grasses. Some of the important ones have been 2,4-dichlorophenoxyacetic acid (2,4-D), 2,4,5-trichlorophenoxyacetic acid (2,4,5-T), and their esters and salts. It was Rachel Carson [31] who summoned public attention to the dangers of environmental pollution and focused it on them. Since then, much research and development have been done on methods of pesticide disposal for use under a variety of circumstances. Ultraviolet oxidation processes are potentially valuable for handling moderate-sized batches of leftover solutions and washwater under the isolated conditions of farm use [32,33].

In photochemistry, many experiments are done to measure the initial rate of formation of a product or the initial rate of degradation of the starting material, where both the concentration of starting material and the extent of reaction are held to low values. The smaller the extent of degradation is, the greater the uncertainties associated with the presence of impurities are. The questions raised include not only whether minor components are degraded, but also whether sensitization or quenching occurs through the interaction of energy states of the desired reagent and an unknown or undesired species.

The commercial "isooctyl esters" of this class of herbicides are mixtures of the esters of many alcohols, mainly branched octanols. Their photodegradation products become correspondingly more complex mixtures. In order to simplify the photoproduct analysis in the case of 2,4,5-trichlorophenoxyacetic acid (III), called 2,4,5-T, the n-octyl ester was prepared [34]. Ruth et al. [35,36] analyzed and purified the product by countercurrent chromatography with the new horizontal flow-through coil planet centrifuge [37]. The column was made up of 10 helical units wound from No. 10 PTFE tubing (2.6 mm i.d.) and connected in series, as described in detail by Ito and Bowman [38]. It had a total of about 1000 helical turns, held a volume of about 260 ml, and was mounted on the gear side holder at β = 0.23 [38]. The speed of rotation of the centrifuge was 400 rpm. The solvent flow rate from the Chromatronix metering pump was 24 ml/hr. The instrument was equipped with an LKB Uvicord S monitor, an LKB recorder, and an LKB fraction collector.

For processing the 2,4,5-T n-octyl ester, a hexane/methanol/water system was investigated. In hexane/methanol, without water, the elution coefficient $K_E(L/H)$ was 0.60. For 100 ml hexane, 90 ml methanol, and 10 ml water, the coefficient was calculated to be 2.43. For 100 ml hexane, 80 ml methanol, and 20 ml water, the elution coefficient was 8.40. The hexane/methanol system, without water, was judged to be satisfactory and was used in the separation. The upper phase was mobile. For the chromatogram in Fig. 10, the sample consisted of 250 μl of the impure ester mixed with 5 ml of the upper phase and 5 ml of the lower phase, which dissolved the esters in a form suitable for injection. The chromatogram shown is similar to the one recorded with preliminary sample one-fifth this size.

The compound separated (Fig. 10) in CCC fractions 105–123, and that in fractions 160–190 were identified by their mass spectra as the n-octyl ester and the methyl ester of 2,4,5-T, respectively. As the chromatogram shows, the methyl ester and a couple of minor impurities were easily removed.

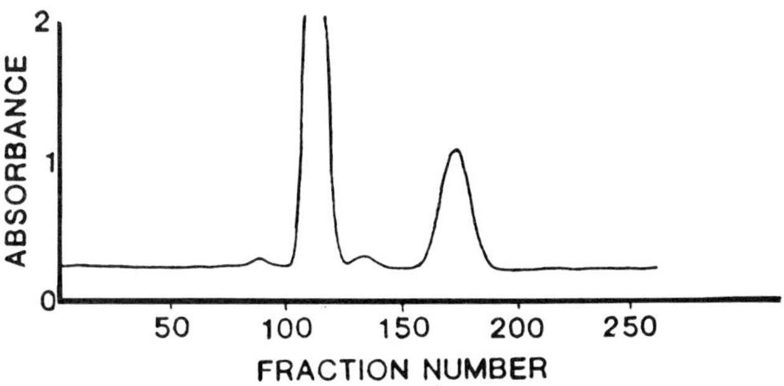

Fig. 10 CCC purification of the n-octyl ester of 2,4,5-trichlorophenoxyacetic acid (2,4,5-T). (From J. M. Ruth, Y. Ito, N. B. Mandava, and V. P. Flanagan (1985), *J. Liq. Chromatogr.* *8*, 2239.)

B. Triazine Herbicides

The symmetrical triazine herbicides are a rather large group, most often applied to the soil in preemergence or immediate postemergence treatment for the control of grass and weed seedlings on cropland, and sometimes in other situations [29]. Some of them are manufactured and used in large quantities. Mandava et al. [39] separated a mixture of four very similar chlorinated s-triazines by countercurrent chromatography.

Where a suitable two-phase solvent system is not already known, it is necessary to make up experimental solvent mixtures and to measure partition coefficients and solubilities of the compounds in the trial systems. The experiments are continued until a system possessing satisfactory partition and solubility data has been found. For the triazines, several kinds of solvent systems were investigated. Among them, systems composed of chloroform/methanol/water and chloroform/acetic acid/water in various volume ratios showed inadequate partition coefficient values, and most of the compounds were partitioned into the lower (nonaqueous) phases. Two types of solvent systems with suitable properties, n-hexane/methanol/water and n-nexane/ethyl acetate/methanol/water, were found.

Two prototype instruments, the high-speed countercurrent chromatograph and the horizontal flow-through coil planet centrifuge, were used in the triazine separations. Both belong to a family of coil planet centrifuges which produce a synchronous planetary motion identical to that in the toroidal coil planet centrifuge [25]. The general operating conditions of the two instruments are

Table 6 General Operating Conditions for CCC Instruments in s-Triazine Herbicides Analysis

CCC instrument	Flow rate (ml/hr)	Revolutional speed (rpm)
1. High speed countercurrent chromatograph	150	800
2. Horizonatl flow-through coil planet centrifuge	100	600

Source: Ref. 39.

summarized in Table 6. The common names and the *Chemical Abstracts* names of the triazines are listed in Table 7.

Elution Coefficients

Each elution coefficient was determined by adding a known amount of the triazine to measured volumes of the solvent system. After thorough mixing, the two phases were separated, and the absorbance was measured at 260 nm. The ratio of the absorbance values of the two phases gave the elution coefficient.

A typical example is provided by the determination of the value for atrazine in the solvent system n-hexane/ethyl acetate/methanol/water (8:2:5:5). To a mixture of 1.6 ml of n-hexane, 0.4 ml of ethyl acetate, and 1 ml of water was added 1 ml of atrazine solution

Table 7 Common Names and Chemical Abstracts Names for Chlorinated s-Triazines

Common name	Chemical Abstracts name
Atrazine	6-Chloro-N-ethyl-N'-(1-methylethyl)-1,3,5-triazine-2,4-diamine
Propazine	6-Chloro-N,N'-bis(1-methylethyl)-1,3,5-triazine-2,4-diamine
Simazine	2-Chloro-N,N'-diethyl-1,3,5-triazine-4,6-diamine
Trietazine	6-Chloro-N,N,N'-triethyl-1,3,5-triazine-2,4-diamine

Table 8 Partition Coefficients of Chlorinated s-Triazines in Two Solvent Systems at Different Volume Ratios

		Partition coefficient $(C_u/C_l)^a$			
Solvent composition	Volume ratio	Trietazine	Propazine	Atrazine	Simazine
1. Hexane/methanol/water	2 : 2 : 0	0.51	0.38	0.30	0.28
	2 : 2 : 1	2.30	0.51	0.22	0.20
	2 : 1 : 1	4.57	1.16	0.43	0.16
	2 : 1 : 2	18.60	3.80	1.08	0.63
	2 : 0 : 2	108.00	14.70	3.09	2.00
2. Hexane/ethyl acetate/ methanol/water	5 : 5 : 5 : 5	9.02	6.10	3.71	2.02
	6 : 4 : 5 : 5	7.64	5.18	2.89	1.19
	7 : 3 : 5 : 5	6.95	4.41	2.16	0.92
	8 : 2 : 5 : 5	6.90	3.95	1.41	0.63
	9 : 1 : 5 : 5	5.41	2.37	1.10	0.35
	10 : 0 : 5 : 5	4.57	1.16	0.43	0.16

[a] C_u: solute concentration in upper phase; C_l: solute concentration in lower phase.
Source: Ref. 39.

Table 9 Partition Coefficients of Chlorinated s-Triazines in Hexane/Ethyl Acetate/Methanol/Water (8:2:5:5)

Compound	Partition coefficient[a]	
	C_u/C_l	C_l/C_u
Trietazine	6.90	0.14
Propazine	3.95	0.25
Atrazine	1.41	0.71
Simazine	0.63	1.60

[a]C_u = solute concentration in upper phase; C_l = solute concentration in lower phase.
Source: Ref. 39.

in methanol (concentration: 1 mg/ml). The atrazine and the solvents were mixed vigorously, and the two phases were allowed to separate. A 0.5-ml volume of each phase was diluted with 3 ml of methanol. The absorbance was measured at 260 nm with a Zeiss UV-visible spectrophotometer, using a 1-cm light path quartz cell. The readings were 0.294 and 0.208 for the lighter (upper) phase and the heavier (lower) phase, respectively. Then the coefficient $K_E(L/H)$ is equal to 0.294/0.208 = 1.41 in that solvent system.

The two solvent systems, n-hexane/methanol/water and n-hexane/ethyl acetate/methanol/water, both yielded suitable ranges of elution coefficients, as presented in Table 8. The system n-hexane/ethyl acetate/methanol/water (8:2:5:5) showed the best range of elution coefficient values and satisfactory solubilities for all of the triazines to be used in the synthetic mixture. Therefore, it was chosen for the chromatography. The elution coefficients $K_E(L/H)$ and $K_E(H/L)$ in this solvent system are listed in Table 9.

The authors point out that with this range of elution coefficients, the choice of mobile phase should be made so as to obtain the best retention of the stationary phase in the column. That choice depends on the characteristics of the instrument to be used. High-speed CCC, which produces high stationary phase retention, will yield rapid and efficient separations by using the lighter (upper) phase as the mobile phase. On the other hand, the horizontal flow-through coil planet centrifuge produces much lower retention of the heavier phase as a stationary phase and will yield better results (higher chromatographic peak resolution) when the heavier phase is used as the mobile phase.

When the correct set of elution coefficients, K_E or $K(C_m/C_s)$, has been selected, where the subscript m identifies the mobile phase and s, the stationary phase, the components of the mixture emerge in the order of decreasing elution coefficients. The one with the largest coefficient comes out first and the one with the smallest emerges last. In Table 9, the coefficients $K_E(L/H)$ in column 2 are the appropriate ones to use when the lighter phase is mobile. If the phases are interchanged, so that the heavier phase becomes the mobile phase, the appropriate coefficients are then the $K_E(H/L)$ found in column 3 of Table 9. Then the order of elution is reversed from what it was with the other phase mobile.

High-Speed Countercurrent Chromatograph

The high-speed countercurrent chromatograph performs fast and efficient separations on both the analytical and the semipreparative scales, depending on the column used. The design is similar to that of the toroidal coil planet centrifuge except that the separation column consists of multiple layers of coiled PTFE tubing, wound coaxially around a spool-shaped holder of 10-cm diameter [1,9]. The columns are easily changed by loosening two screws.

As Ito et al. [9] stated, "Among the existing CCC schemes, the present scheme has the greatest capacity for the retention of the stationary phase against a high flow rate of the mobile phase. This fact enables one to perform efficient separations in short periods of time ranging from 2 to 5 h, instead of overnight or several days as required for other preparative CCC schemes. The low column pressure, ranging from 20 to 100 psi, in the present scheme, suggests that the efficiency can be further improved by the use of a longer column and/or a higher flow rate at an increased revolutional speed. The method also permits almost universal application of the conventional two-phase solvent systems where either phase can be used as the mobile phase." Ito [1] explains, "The multilayer configuration of the coiled column creates a gradual change of the centrifugal force field along the layers of the coil from the internal terminal toward the external terminal. This centrifugal force field gradient can be effectively used to improve the retention of the stationary phase in the column by applying the proper mode of elution.... The force gradient tends to move the heavier phase from the internal terminal toward the external terminal of the multilayer coil and vice versa for the lighter phase. Since the heavier phase usually moves toward the tail and the lighter phase toward the head in each layer of the coil, the maximum retention of the stationary phase and hence the highest peak resolution can be attained by pumping either the heavier phase through the internal head terminal or the lighter phase through the external tail terminal of the multilayer coil."

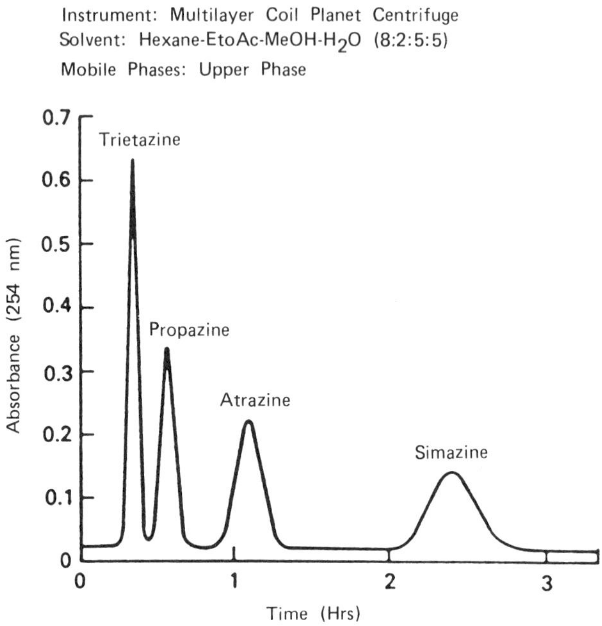

Fig. 11 Separation of chlorinated s-triazines by high-speed countercurrent chromatography in hexane/ethel acetate/methanol/water (8:2:5:5); lighter phase mobile. (From N. B. Mandava and Y. Ito, and J. M. Ruth (1985), *J. Liq. Chromatogr.* *8*, 2221.)

The column used in the triazine separations was 130 m long and 1.6 mm i.d., with a capacity of 285 ml. The coil was first filled with the stationary phase, and the sample was injected through the internal, head end of the column. Then, with the apparatus running at 800 rpm, the mobile phase was introduced, also through the internal head end of the column, at a rate of 150 ml/hr. The eluate was continuously monitored at 254 nm with an LKB Uvicord S and collected with an LKB fraction collector.

In a typical experiment, an 8-mg sample of a mixture, containing 2 mg each of atrazine, propazine, simazine, and trietazine, was chromatographed in the high-speed CCC apparatus in hexane/ethyl acetate/methanol/water (8:2:5:5). The heavier (aqueous) phase was stationary and the lighter phase was mobile. The partition coefficients K_E (Table 9) ranged from 0.63 to 6.90. The compounds eluted in the expected order, trietazine (K_E(L/H) = 6.90) first, followed by propazine (3.95), then atrazine (1.41), and finally simazine (0.63), all within 3 hrs as is seen in Fig. 11. The eluate was collected in 90 fractions (5.2 ml/fraction, at a flow rate of 2.6 ml/min). The first peak was contained in fractions 10–11; the second, in fractions 15–19; the third, in fractions 30–35; and the fourth, in fractions 66–79. Because the four compounds were well separated, only the middle fraction from each peak was analyzed by mass spectrometry. As expected, the mass spectra confirmed trietazine under the first peak, propazine under the second, atrazine under the third, and simazine under the fourth.

Horizontal Flow-Through Coil Planet Centrifuge

The second instrument used in the triazine study was a table top model of the horizontal flow-through coil planet centrifuge. This improved version of the original bench top model has a column holder reduced to a length of 20 cm, giving greater mechanical strength and permitting a speed of 600 rpm. Consequently, the system shows higher partition efficiency and better stationary phase retention. The higher flow rate of the mobile phase provides faster separations. Increasing the column length by using double layers on each unit of the column assembly partially compensated for the reduced length of the column holder. As in the high-speed CCC apparatus, the column holder is easily removed from the rotary frame by loosening two screws on each bearing block.

Each column unit was prepared from a 6-m long, 1.6-mm i.d. piece of PTFE tubing, by winding it tightly onto a 14-cm long 1.25-cm o.d. aluminum pipe core, making two coiled layers with about 100 helical turns. Eight such units were connected head-to-tail in series and mounted symmetrically around the column holder. The total column capacity was about 100 ml.

A mixture of 1 mg each of the same four triazines was chromatographed on the horizontal flow-through coil planet centrifuge. The same solvent system was used, but the phases were interchanged, so that the lighter, nonaqueous phase was stationary, and the heavier, aqueous phase became the mobile phase. The column was first filled with the stationary (lighter) phase and the sample solution was injected through the sample port. Then, with the centrifuge rotating at 600 rpm, the lower phase was pumped into the head of the column at the rate of 100 ml/hr. The effluent from the column outlet was monitored continuously with a Uvicord S at 254 nm and caught with an LKB fraction collector. After the first three peaks were eluted, the phases were switched so that the lighter phase became the mobile phase and quickly eluted the fourth peak from the column (Fig. 12).

While the heavier, aqueous phase was mobile, the pattern of the first three peaks followed the order of the elution coefficients E_H/L in column 3 of Table 9, with simazine emerging first, atrazine second, and propazine third, the opposite of the order seen in Fig. 11.

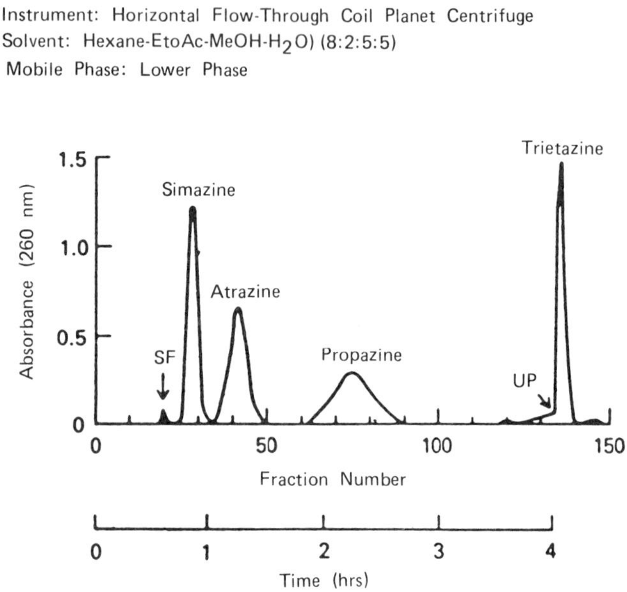

Fig. 12 Separation of chlorinated s-triazines by horizontal flow-through coil planet centrifuge, in hexane/ethyl acetate/methanol/water (8:2:5:5); heavier phase mobile. (From N. B. Mandava, Y. Ito, and J. M. Ruth (1985), *J. Liq. Chromatogr. 8,* 2221.)

When Fig. 11 is compared with Fig. 12, it is immediately noticed that in the former, simazine and atrazine are widely separated, whereas in the latter, propazine and trietazine are similarly isolated. Although the separations are all adequate in the present case, it is obvious that such an interchange of the phases may be of great assistance in the separation of the components of a more complex mixture.

IV. PROSTAGLANDINS

The prostaglandins [40–42] are a group of active biochemical compounds widely distributed in low concentrations in various body fluids and tissues. They have a complex set of effects, including the inhibition of stomach acid production, the stimulation of smooth muscle contraction, the opening of bronchial tubes, the constriction of the pupil of the eye, and the blocking of the breakdown of fats. They appear to be active in the body's defensive mechanisms, such as vomiting, fever, and inflammation, in the regulation of blood

Table 10 Partition Coefficients Calculated Using Eq. (1) from the Experimental Results Obtained with Small-Bore Column at Two Different Flow Rates and with Upper (Acetic Acid) Phase Mobile

	Flow rate (ml/hr)	
Prostaglandin	2.4	6.0
6-Keto-$PGF_{1\alpha}$	1.0	1.0
Thromboxane B_2	0.71	0.71
$PGF_{2\alpha}$	0.64	0.67
PGD_2	0.53	—
PGE_2	0.47	0.49
13,14-Dihydro-15-keto-$PGS_{2\alpha}$	0.25	0.34
13,14-Dihydro-15-keto-PGE_2	0.22	0.24
PGB_2	0.20	0.15
PGA_2	—	0.15

Source: Ref. 43.

pressure, in the control of pituitary hormone production, in the operation of the male and female reproductive systems, and in kidney functions. Evidence accumulates that the primary activity of prostaglandins is concerned with the cell membrane. Aside from their participation in the physiological well-being of domesticated animal species, prostaglandins are of obvious concern in agricultural research through their relationship to the nutritional intake of essential fatty acids.

Brown and Ito [43–44] separated mixtures of several prostaglandins and metabolites by countercurrent chromatography in the new horizontal flow-through coil planet centrifuge. The version of that instrument which was new at that time was one that had two different columns, usable simultaneously. The two columns exhibited different modes of planetary motion, as well as different lengths, diameters, coil geometries, and volumes. One, on the pulley side, was designed for efficient mixing and efficient separations, for analytical use. Each unit of that column consisted of 340 turns of 0.55-mm i.d. PTFE tubing coiled around a 0.68-cm o.d. core, with a total volume of 2.4 ml. Usually 10 such segments were connected in series, to give a column volume of 24 ml. The other column, on the gear side, featured stable retention of the stationary phase in

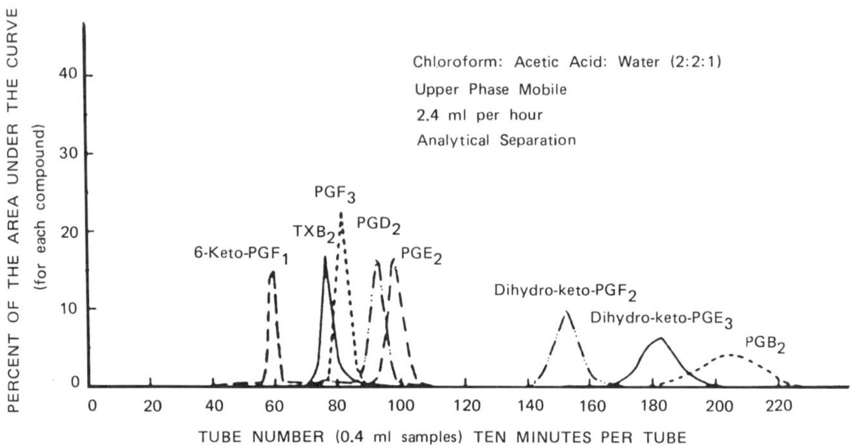

Fig. 13 The separation of prostaglandins using a flow rate of 2.4 ml/hr with the upper (acetic acid) phase mobile. (From E. A. B. Brown and Y. Ito (1980), *J. Biochem. Biophys. Methods 3*, 77.)

large-bore columns for preparative scale separations. In it, each of the 10 units consisted of 100 turns of 2.6-mm i.d. PTFE tubing wound on a 1.25-cm o.d. core, with a capacity of about 26 ml, so that the column had a volume of 260 ml. The speed of revolution was continuously adjustable from 0 to 600 rpm. The solvent system was chloroform/acetic acid/water (2:2:1, v/v). The elution coefficients $K_E(L/H)$ (lighter phase mobile) for the various solutes, shown in Table 10 for two different flow rates in the small-bore column, were calculated from the chromatogram by means of the equation:

$$K_E(L/H) = \frac{V_c - V_S}{V_R - V_S} \qquad (1)$$

in which V_c is the total column capacity, V_S is the retention volume of the solvent front, and V_R is the retention volume of the peak maximum, all in milliliters.

The use of mixtures of C-14- and tritium-labeled compounds make it possible to plot overlapping peaks that were not completely separated. Figure 13 is an example of the results, reported to be equivalent to what had been obtained by others using HPLC and a reversed phase column. The effect of interchanging mobile and stationary phases, as described in earlier sections of this chapter, was also employed in the separation of prostaglandin mixtures.

V. INSECTICIDES

Insecticides have developed an economic significance similar to that of herbicides and along with it a corresponding set of regulations to control the hazards associated with use and residual contamination. The development of analytical techniques is important for that purpose, as well as for application in research on all aspects of the subject. Insecticide types of major importance include organophosphorous compounds, carbamates, organochlorine compounds, and natural and synthetic pyrethroids [29]. Two organophosphorus pesticides have been analyzed by countercurrent chromatography.

A. Fenthion and Impurities

The organophosphorus pesticides are important partly because of their environmental degradability, but that same reactivity contributes to the difficulty of obtaining high-purity materials and of preserving them in storage. Fenthion (I), developed by Schrader and Schlegk [45,46], is a contact and stomach insecticide and has been used against a variety of househould, garden, and crop pests and against animal parasites [46,47]. Its IUPAC name is O,O-dimethyl O-4-methylthio-m-tolyphosphorothioate. In the *Chemical Abstracts* system, it is O,O-dimethyl-O-[3-methyl-4-(methylthio)phenyl]phosphorothioate. The literature contains many code numbers and trade marks [46]. References to various analytical procedures are listed by Ibrahim and Cavagnol [47], who developed a method based on the UV absorption by fenthion at 252 nm, after chromatography on a Florisil column with heptane.

Among the compounds sometimes found in association with fenthion as original impurities or as alteration products are the S-methyl isomer; the oxon, in which the sulfur bonded to the phosphorous is replaced by an oxygen atom; 4-(methylthio)-m-cresol; 3-methyl-4-(methylthio)anisole (II); O,O-dimethyl chloridophosphorothioate; and O,O,O,O-tetramethyl dithiopyrophosphate [47]. The sulfoxides and sulfones of fenthion, of its S-methyl isomer, and of its oxon would be expected to be formed by oxidation of the thioethers [48,49].

Some work on the purification of fenthion and the identification of impurities was reported by Ruth et al. [36,50]. Preliminary examinations by thin-layer chromatography had been done according to the procedure described by Ibrahim and Cavagnol [47]. Fenthion, technical 96% and analytical standard 97.2%, was used without dilution (1 μl/spot). The other compounds were each dissolved in redistilled acetone for spotting (10 μl of 1 g/L solution/spot). The plates were developed with a mixture of heptane and redistilled

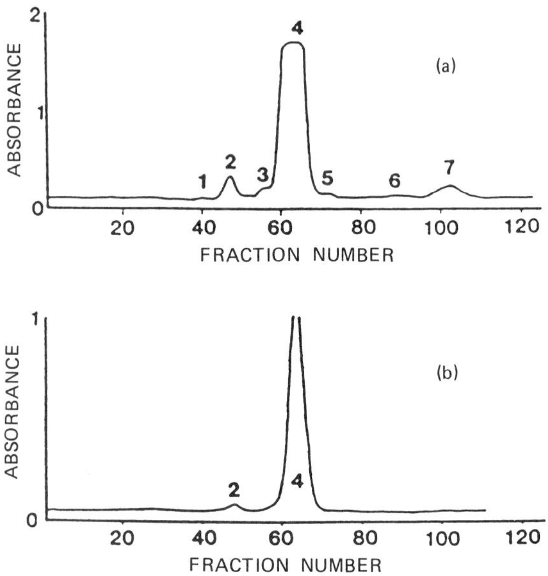

Fig. 14 CCC separation of fenthion and impurities. (a) Technical fenthion (96%). (b) Fenthion analytical standard (97.2%). (From J. M. Ruth, Y. Ito, N. B. Mandava, and V. P. Flanagan (1985), *J. Liq. Chromatogr.* *8*, 2239.)

acetone (7:1 by volume), sprayed with DCQ reagent (2,6-dibromo-N-chloro-p-quinoneimine, 1% in acetone), and heated in the oven at 110°C for 10 min.

The CCC separations were done with the new horizontal flow-through coil planet centrifuge, the same instrument described in Sect. III.A of this chapter. The column and operating conditions were recorded there for the phenoxyacetic acid herbicide separations. The solvent system for fenthion, the same one used for the phenoxyacetate esters, was prepared by equilbrating hexane and methanol (2:1 by volume). The upper phase was mobile. The elution coefficient for fenthion, $K_E(L/H)$ was 0.56. The sample contained 20 μl fenthion mixed with 5 ml of the upper phase for charging. The fraction collector was operated at 6 ml/tube (24 ml/hr).

The countercurrent chromatogram of the technical fenthion is shown in Fig. 14A and that of the analytical standard fenthion in Fig. 14B. The technical grade insecticide yielded traces of six impurities, in addition to the fenthion, which is represented by CCC peak 4. The fenthion analytical standard showed only the fenthion

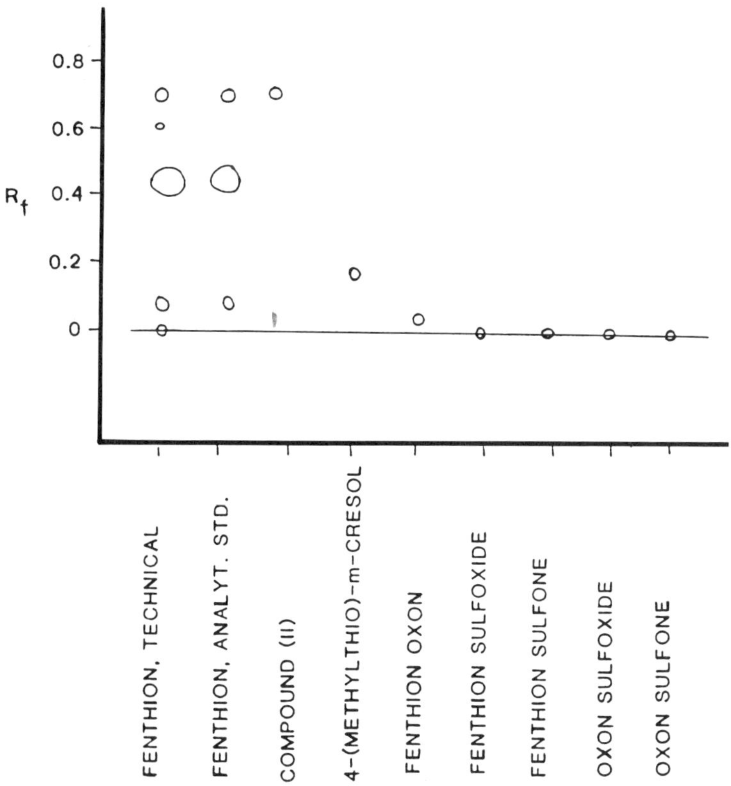

Fig. 15 Results of thin-layer chromatographic tests of fenthion and related compounds. R_f = (distance traveled by spot)/(distance traveled by solvent front). (From J. M. Ruth, Y. Ito, N. B. Mandava, and V. P. Flanagan (1985), *J. Liq. Chromatogr. 8*, 2239.)

and one impurity, CCC peak 2. Thin-layer chromatography (Fig. 15) did not find all of the impurities of the technical fenthion that are clearly seen on the CCC chart. On the other hand, it did show a weak spot corresponding to the slow-moving component of R_f 0.08 found also in the technical grade material. This may represent a trace of the component recorded by CCC peak 7 in Fig. 14A for the technical grade sample, widely separated from the fenthion. The impurity, 3-methyl-4-(methylthio)anisole, present in the analytical standard fenthion (CCC peak 2) was identified by the thin-layer chromatography, and both it and the fenthion (CCC peak 4) were confirmed by mass spectrometry of the CCC fractions.

B. Demeton

The active constituent of the insecticide demeton has been reported to be an isomeric mixture of demeton-O (O,O-diethyl-O-ethylmercapto-ethyl thiophosphate) and demeton-S (O,O-diethyl-S-ethylmercapto-ethyl thioophosphate). Because of isomerization, compounds of the S=P-O- and O=P-S- types are usually obtained as mixtures, and quantities of the pure isomers are difficult to prepare. Gardner and Heath [51] obtained small amounts of the two isomers of demeton by silicia gel partition chromatography, using an isooctane/methanol solvent system. Toxicological study with the purified isomers indicated that demeton S is about 10 times as toxic to insects and mammals as demeton O, which appeared to explain discrepancies in the results of earlier tests in which the composition of the mixture of isomers may have varied. A preparative method was needed for the separation of demeton isomers in the quantities required for further toxicological evaluation. Lee and Cook [52] developed a method which uses the high-speed countercurrent chromatograph.

The operation of the chromatograph, which uses a multilayer coil, has been discussed in Sec. III.B.2 above. The column used for the demeton separation was made from 1.6-mm i.d. PTFE tubing and had a total capacity of 360 ml. The solvent system n-pentane/ethanol/water (6:5:1) was equilibrated in a separatory funnel and separated. The coiled column was filled with the lighter phase at a flow rate of 15 ml/min. The mixture of demeton isomers (600 mg) was dissolved in 3.0 ml of the two-phase solvent system (1.5 ml of each phase) and injected through the head (inlet) of the column. The mobile phase (heavier phase) was then pumped through the inlet at 2.5 ml/min, with the system rotating at 800 rpm, and the column effluent was collected every 2 min. The fractions were concentrated under reduced pressure at 25°C and monitored by thin-layer chromatography on silica gel, with 25% acetone in CH_2Cl_2 as the developing solvent. The isomers were identified at this stage by comparing their thin-layer behavior with that of authentic standards.

Figure 16 is the plot of the chromatographic data obtained by monitoring the CCC effluent by thin-layer chromatography and assuming a Gaussian distribution of weight in the fractions containing the isomers. The more polar demeton S isomer (O=P-S- type) was eluted first, in fractions 15, 16, and 17. The less polar demeton O isomer (S=P-O- type) emerged much later, in fractions 53–59. The R_f-values on the silica gel plates were 0.13 and 0.45 for the demeton S and demeton O, respectively. The structure assignments were confirmed by proton NMR (60 MHz), based on the relative areas of the multiplets centered at 4.1 ppm for oxygen-bonded methylenes and at 2.8 ppm for sulfur-bonded methylenes. Lee and Cook note that the whole operation took less than 3 hr and that the

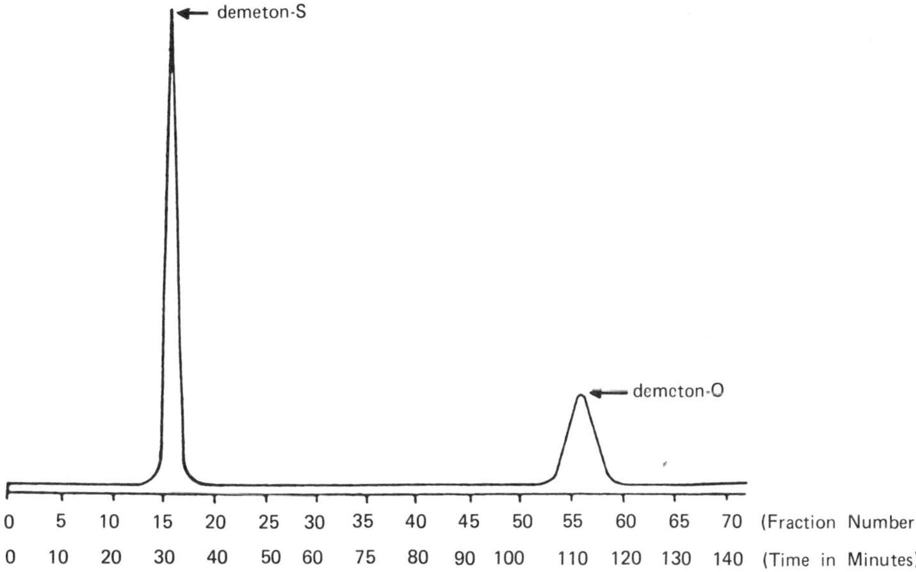

Fig. 16 Plot of countercurrent chromatogram of Demeton-O and Demeton-S, derived from TLC monitoring by assuming Gaussian distribution of weight in the fractions containing the isomers. Fractions 15–17: Demeton-S; 53–59: Demeton-O. Revolution speed: 800 rpm; flow rate: 2.5 ml/min; fraction volume: 5 ml/tube. (From Y. W. Lee and C. E. Cook (1985), *J. Liq. Chromatogr.* *8*, 2253.)

wide separation of the peaks suggests that baseline separation could still be obtained at much greater sample loads.

VI. CONCLUSION

The advantages of countercurrent chromatography over other separative techniques have been pointed out in the literature. The absence of solid adsorbents and the operation of the process at room temperature constitute an especially gentle treatment of labile chemicals and of complex mixtures of natural materials containing low concentrations of physiologically active biochemicals. The capacity for handling such materials on a large scale, competing in time and cost with other processes, is another extremely attractive property. It was therefore natural that CCC should quickly find applications in agricultural research, and it has done so. With improved apparatus becoming commercially available, those applications are likely to multiply rapidly.

REFERENCES

1. Y. Ito, Development of High-Speed Countercurrent Chromatography, *Adv. Chromatog.*, *24*, 181 (1984).
2. Y. Ito and R. L. Bowman, *J. Chromatogr. Sci.*, *8*, 315 (1970).
3. Y. Ito and R. L. Bowman, *Anal. Chem.*, *43*, No. 13, 69A (November 1971).
4. T. Tanimura, J. J. Pisano, Y. Ito, and R. L. Bowman, *Science*, *169*, 54 (1970).
5. K. Hostettmann, *Planta Med.*, *39*, 1 (1980).
6. J. R. Bearder and J. MacMillan, in *Gibberellins: Chemistry, Physiology and Use* (J. R. Lenton, ed.), British Plant Growth Regulators Group, Monograph 5, 1980, pp. 25–30.
7. N. B. Mandava, Y. Ito, and W. D. Conway, *Am. Lab.*, *14*, No. 10, 62 (October 1982).
8. N. B. Mandava, Y. Ito, and W. D. Conway, *Am. Lab.* *14*, No. 11, 48 (November 1982).
9. Y. Ito, J. Sandlin, and W. G. Bowers, *J. Chromatogr.*, *244*, 247 (1982).
10. R. J. Romanach, J. A. de Haseth, and Y. Ito, *J. Liq. Chromatogr.*, *8*, 2209 (1985).
11. R. J. Romanach and J. A. de Haseth, High Speed Countercurrent Chromatography/Fourier Transform Infrared (CCC/FT-IR) Spectrometry, *Abstracts, Pittsburgh Conference on Analytical Chemistry and Applied Spectroscopy*, Atlantic City, NJ, Abstract 1052, 1986.
12. J. R. Bearder, in *Hormonal Regulation of Development. I. Molecular Aspects of Plant Hormones* (J. MacMillan, ed.) (Encyclopedia of Plant Physiology, Vol. 9), Springer-Verlag, Berlin, 1980, pp. 9–112.
13. J. MacMillan (ed.), *Hormonal Regulation of Development. I. Molecular Aspects of Plant Hormones* (Encyclopedia of Plant Physiology, Vol. 9), Springer-Verlag, Berlin, 1980.
14. N. Takahashi (ed.), *Chemistry of Plant Hormones*, CRC, Boca Raton, FL, 1986.
15. F. B. Abeles, *Ethylene in Plant Biology*, Academic, New York, 1973.
16. F. B. Abeles, J. M. Ruth, L. E. Forrence, and G. R. Leather, *Plant Physiol.*, *49*, 669 (1972).
17. J. W. Rowe, *The Common and Systematic Nomenclature of Cyclic Diterpenes*, 3rd revision [Proposal to the IUPAC Commission on the Nomenclature of Organic Chemistry], Forest Products Laboratory, Forest Service, U.S. Department of Agriculture, Madison, WI (October 1968).

18. N. Takahashi, I. Yamaguchi, and H. Yamane, Gibberellins, in *Chemistry of Plant Hormones*, (N. Takahashi, ed.), CRC, Boca Raton, FL, 1986, pp. 57–151.
19. J. MacMillan and R. J. Pryce, in *Phytochemistry*, Vol. 3 (L. P. Miller, ed.), Van Nostrand Reinhold, New York, 1973, pp. 283–326.
20. J. MacMillan and N. Takahashi, *Nature, 217,* 170 (1968).
21. N. B. Mandava, *Ann. Rev. Plant Physiol. Plant Mol. Biol., 39,* 23 (1988).
22. N. B. Mandava and Y. Ito, Application of Countercurrent Chromatography to Separation of Plant Hormones, *14th Middle Atlantic Regional Meeting of the American Chemical Society,* Philadelphia, Abstract No. ANAL 24, 1980, p. 41.
23. N. B. Mandava and Y. Ito, Plant Hormone Analysis by Counter-Current Chromatography, *Abstracts, Pittsburgh Conference on Analytical Chemistry and Applied Spectroscopy,* Atlantic City, NJ, Abstract N. 64, 1981.
24. N. B. Mandava and Y. Ito, *J. Chromatogr., 247,* 315 (1982).
25. N. B. Mandava and Y. Ito, *J. Liq. Chromatogr. 7,* 303 (1984).
26. P. Gaskin, P. S. Kirkwood, J. R. Lenton, J. MacMillan, and M. E. Radley, *Agric. Biol. Chem., 44,* 1589 (1980).
27. R. C. Durley and R. P. Pharis, *Phytochemistry, 11,* 317 (1972).
28. N. B. Mandava, Analysis of Abscisic Acid by Countercurrent Chromatography, *Abstracts, Pittsburgh Conference on Analytical Chemistry and Applied Spectroscopy,* Atlantic City, NJ, Abstract No. 613, 1982.
29. K. A. Hassall, *The Chemistry of Pesticides,* Verlag-Chemie, Weinheim, Deerfield Beach, FL, 1982.
30. M. A. Loos, in *Herbicides: Chemistry, Degradation, and Mode of Action,* 2nd ed. (P. C. Kearney and D. D. Kaufman, eds.), Marcel Dekker, New York, 1975, pp. 1–128.
31. R. Carson, *Silent Spring,* Houghton Mifflin, Boston, 1962.
32. J. M. Ruth, P. C. Kearney, and Q. Zeng, *Chemosphere, 14,* 1181 (1985).
33. P. C. Kearney, Q. Zeng, and J. M. Ruth, *Treatment and Disposal of Pesticide Wastes,* American Chemical Society, Symposium Series No. 258, Washington, D.C., 1984, pp. 185–209.
34. J. M. Ruth, in *Handbook of Natural Pesticides,* Vol. 2, (N. B. Mandava, ed.), CRC, Boca Raton, FL, 1985, pp. 323–438.
35. J. M. Ruth and Y. Ito, Pesticide Analysis by Countercurrent Chromatography, *95th Annual Meeting of the Association of Official Analytical Chemists,* Washington, D.C., Kontes Workshop, Program, 1981, p. 24.
36. J. M. Ruth, Y. Ito, N. B. Mandava, and V. P. Flanagan, *J. Liq. Chromatogr., 8,* 2239 (1985).
37. Y. Ito, *J. Biochem. Biophys. Methods, 5,* 105 (1981).

38. Y. Ito and R. L. Bowman, *J. Chromatogr.*, *147*, 221 (1978).
39. N. B. Mandava, Y. Ito, and J. M. Ruth, *J. Liq. Chromatogr.*, *8*, 2221 (1985).
40. P. B. Curtis-Prior, *Prostaglandins: An Introduction to their Biochemistry, Physiology and Pharmacology*, North-Holland, Amsterdam, 1976.
41. P. Crabbe (ed.), *Prostaglandin Research*, Academic, New York, 1977.
42. F. Berti and G. P. Velo (eds.), *The Prostaglandin System: Endoperoxides, Prostacyclin, and Thromboxanes* (Proceedings of a NATO Advanced Study Institute held September 2–13, 1979, in Erice, Sicily), Plenum, New York, 1981.
43. E. A. B. Brown and Y. Ito, *J. Biochem. Biophys. Methods*, *3*, 77 (1980).
44. E. A. B. Brown and Y. Ito, Separation of Prostaglandins Using the New Horizontal Flow-Through Coil Planet Centrifuge, *Abstracts, Pittsburgh Conference on Analytical Chemistry and Applied Spectroscopy*, Atlantic City, NJ, Abstract No. 63, 1981.
45. G. Schrader, *Hofchen-Briefe (Bayer)*, *13*, 1 (1960).
46. C. S. Worthing and S. B. Walker (eds), *The Pesticide Manual*, 7th ed., British Crop Protection Council, 1983.
47. F. B. Ibrahim and J. C. Cavagnol, *J. Agric. Food Chem.*, *14*, 369 (1966).
48. H. Niessen, H. Tietz, and H. Frehse, *Pflanzenschutz Nachrichten "Bayer,"* *15*, 125 (1962).
49. Eto, M., *Organophosphorus Pesticides: Organic and Biological Chemistry*, CRC, Boca Raton, FL 1974.
50. J. M. Ruth, E. A. B. Brown, and N. B. Mandava, *Abstracts, Pittsburgh Conference on Analytical Chemistry and Applied Spectroscopy*, Atlantic City, NJ, Abstract No. 65, 1981.
51. K. Gardner and D. F. Heath, *Anal. Chem.*, *25*, 1849 (1953).
52. Y. W. Lee and C. E. Cook, *J. Liq. Chromatogr.*, *8*, 2253 (1985).

chapter nine

Countercurrent Chromatography for Drug Discovery and Development

David G. Martin
The Upjohn Company
Kalamazoo, Michigan

Nature has been the source of a magnificent variety of molecular structures. The diverse biological activities of these varied structural types has provided many of our useful drugs, utilizing the natural product itself or a structural modification. The fact that the biologically active principles in nature are frequently present at only a few parts per million in complex mixtures of nonvolatile components has provided those searching for new drugs with an incredible variety of challenging separation problems. Separation techniques fit into broad categories such as adsorption chromatography on a wide range of solid supports, ion exchange procedures, size exclusion techniques, and solvent partition methods. For each problem the physical properties of the desired biologically active components determine which techniques are appropriate. The isolation of the pure biologically interesting components from natural sources often requires the application of all available techniques consistent with their physical properties. This chapter addresses the virtues of countercurrent chromatography (CCC) for the biological activity-monitored isolation of potential new drugs from natural sources.

CCC has already proven advantageous for the isolation and purification of a diverse array of biologically active structural types. CCC greatly facilitated the isolation of a variety of structural types of antimicrobial antibiotics from crude fermentation extracts by procedures which were convenient and less degradative than other chromatographic techniques [1]. It provided a convenient resolution of a spectrum of prostaglandins and their metabolities affording higher recoveries of components than adsorption chromatography [2].

The critical role of CCC in the isolation of the extremely labile tunichromes [3] is noteworthy. CCC provided a convenient resolution of minor components of cortical steroids susceptible to degradation during the course of adsorption chromatography [4]. The reactive antitumor agent diaziquone was separated from its reaction products [5] and the high purity of fractions from reduction products of daunomycin facilitated further instrumental analyses without interference caused by contamination from chromatographic solid supports [6]. Structurally similar alkaloids were readily resolved and isolated from crude plant extracts [7] as were similar triazine herbicides [8] and oxygen and sulfur isomers of the insecticide Dementon [9]. The method proved useful for separating pure organophosphorus insecticide fenthion and phenoxy herbicides from contaminants [10]. Dipeptides (L-tryptophyl-L-leucine and L-valyl-L-tyrosine were nicely resolved on a gram scale in a larger scale coil planet centrifuge [11]. A synthetic heptapeptide fragment and octapeptide analog of cholecystokinin were efficiently and conveniently purified by this method [12]. The physiologically active neutropeptide bombesin, a tetradecapeptide, was synthesized and purified to over 98% purity by CCC [13]. Systems based on two immiscible aqueous polymer phases fractionated enzyme activities from rat liver homogenate with good recovery of enzyme activities [14]. Foam affinity CCC holds considerable potential for the separation of proteins and biopolymers [15]. Other impressive applications of such methods as droplet countercurrent chromatography, centrifugal droplet countercurrent chromatography, rotation locular countercurrent chromatography, and centrifugal countercurrent chromatography have recently been reviewed [16–19] and are found in other chapters of this book.

The high resolving power of multistage countercurrent distribution techniques of Craig and others has long been recognized. However, such methods are sufficiently time consuming and inconvenient to severely limit their application in the search for drugs where many natural sources must be investigated to find the precious few containing novel and useful active agents. This led to the development of CCC and the availability of a variety of greatly improved methods for carrying out fractionations based on phase distributions whithout solid supports. Two excellent reviews on the development of CCC are available [20, 21] and other chapters in this book cover historical development, instrumentation, and theoretical aspects. The outstanding versatility of the method is illustrated by its usefulness in the broad scope of separations mentioned in the applications. Other often quoted advantages include resolution, elimination of binding or deactivation by solid adsorbents, convenience and speed, and inexpensive operation.

This chapter focuses on the more subtle but extremely important advantage of total recovery that CCC offers the search for drugs from natural sources. This advantage is particularly critical for the isolation of biologically active agents that lack convenient monitors. An awareness of some of the problems encountered in the isolation of such less visible agents leads to a greater appreciation of the advantages of CCC. Biological activities and monitors play a crucial role in the isolation of biologically active components from natural sources. To briefly define that role and illustrate the concepts of high and low visibility, a short discussion of antimicrobial "agar," cytotoxic cell culture, and in vivo antitumor activities is presented in the initial section on problems and the correlation approach. Sections follow on the advantages of CCC for the initial fractionation, illustrative examples of agents conveniently isolated in the initial fractionation, and finally examples of the improved resolutions possible when measurable properties are available. These final examples demonstrate the critical nature of solvent selection, desalting a polar, water-soluble substance, and the quantitative recovery permitted by CCC.

I. BACKGROUND PROBLEMS AND THE CORRELATION APPROACH

A. Detectability or Selective Visibility

Detectability or selective visibility of biologically interesting trace components in the complex mixtures comprising their environments is crucial to meaningful isolation efforts. Ideally this visibility arises from the nature of the desired biological activity. For example, the search for antimicrobial antibiotics is greatly facilitated by the visibility derived from their in vitro inhibition of microorganisms cultivated on agar trays. This inhibition provides the basis for bioassay and bioautography schemes which have been invaluable for the isolation of novel antibiotics and for the identification of cultures producing known antibiotics. Using this agar visibility, the natural product chemist can efficiently evaluate potential isolation methods consuming only small amounts of crude natural product. He can readily establish if the desired activity can be recovered from a variety of chromatographic adsorbents. Little time or sample is lost establishing which adsorbents irreversibly bind or deactivate the desired activity. Although CCC has made valuable contributions to the isolation of antimicrobial antibiotics [1], additional critical benefits from CCC are available for the isolation of potential new drugs lacking the high visibility of these antibiotics.

Our efforts at Upjohn to isolate novel agents having in vivo activity against P388 leukemia in mice by screening soil culture fermentations illustrate some of the problems involved in the isolation of less visible agents. Soil cultures were initially prescreened for in vitro cytotoxicity against cultured leukemia cells and several antimicrobial systems to select cultures having some biological activity and therefore more potential for producing in vivo activity [22]. Broth from in vitro active cultures was then evaluated for in vivo activity in mice and isolation work was then carried out on in vivo active cultures. In these cultures the in vivo active principle or principles are usually present in minute amounts in complex mixtures of in vitro active and inactive components. There is no straightforward way of knowing which, if any, of the major in vitro activities present are due to in vivo active components. Information on the behavior of the in vivo activities in broth is not readily determined quickly from small samples. Since only a fraction of the structurally identified antibiotics are reported to have some type of antitumor activity and a growing number of antitumor agents lack significant antimicrobial activity, assumptions about correlating activities are tenuous.

B. Correlation Approach and the Initial Fractionation

One possible approach involves enriching the most promising in vitro activity present by solvent, resin, or carbon extraction and then determining the in vivo activity of the enriched fraction. If the in vitro and in vivo activities are enriched to a similar extent a rough but still tenuous correlation is established. Fractionation of the enriched sample, initially monitored by the appropriate in vitro activity, is then attempted. Such fractionation is usually attempted by some form of chromatography. Selection of an appropriate chromatographic vehicle is far from obvious without information on the behavior and properties of the in vivo active components. The in vivo activity could easily be lost through binding or deactivation by a solid chromatographic adsorbent. Determining if an in vivo activity can be fractionated on a given adsorbent by small-scale prelimianry studies is practically impossible in the absence of an established correlation with a more readily monitored property. Fractionation based on partitioning activities between the phases of immiscible solvent mixtures affords a much better possibility for total recovery of all stable activities, thus facilitating the establishment of a meaningful correlation between the in vivo activity and any more monitorable properties present in the in vivo active fraction.

For the initial fractionation of enriched antitumor extracts, we have utilized the Ito multilayer coil separator-extractor available from P.C., Inc. (Potomac, Md., USA). CCC in the Ito coil has proven an excellent method for this initial fractionation and provided a practical solution to many of our separation problems. In that initial fractionation, a broad range of activity polarities can be accommodated by selecting an appropriate solvent system. The enriched activity is initially partitioned between the preequilibrated phases of several biphasic solvent systems of varying polarities and the in vitro activity of each phase determined. The system yielding a roughly equal distribution of the enriched in vitro activity is then utilized for CCC. Pertinent comments on solvent selection are found in Sec. II.C. Normally 200–500 mg of the enriched sample is dissolved in either or both phases of the selected system and introduced into the coil with a loop valve. At flow rates of 1–5 ml/min, most fractionations require less than a day and collected fractions are monitored and pooled based on the selected in vitro activity. The pools, including the stationary phase, are then investigated for correlating spectral and chromatographic properties as well as in vivo activity if consistent with sample size and potency. Even if the amount of the strongly in vitro enriched fraction is insufficient for in vivo assay, spectral data frequently lead to the identification of known antitumor antibiotics. Early identification of known antibiotics as the source of a culture's antitumor properties allows concentration of isolation efforts on potentially novel cultures. Isolation work is continued on cultures providing fractions of unidentified and enriched in vivo activity from the initial fractionation. This fraction does not necessarily correspond with the enriched fraction of the initially selected in vitro activity. Subsequent isolation efforts benefit from any more readily monitorable properties correlating with the in vivo activity. Such properties greatly facilitate evaluating and monitoring further purification steps including CCC. Discussion of CCC on samples of known properties will follow the stated advantages of CCC for the initial fractionation of a biologically active natural product and same illustrative examples.

II. ADVANTAGES OF CCC FOR THE INITIAL FRACTIONATION

A. Elimination of Binding or Deactivation by Solid Adsorbents

Loss of valuable sample is usually a vital concern of the natural product chemist. In the search for highly visible agents he has the potential to predetermine if or to what extent irreversible binding and/or deactivation by a solid adsorbent will occur. In the search for less visible agents, such as in vivo active antitumor agents, the extent of irreversible binding and/or deactivation by solid adsorbents is not readily determined before a correlation with a more readily monitored property has been established; avoidance of this potential loss with CCC is highly advantageous.

B. Total Recovery of Stable Activities

As in the previous response on binding, this advantage is more critical in the search for less visible agents. With highly visible agents, the recovery of the bioactivity of interest from different chromatographic procedures can be estimated by small-scale preliminary experiments. Before a correlation with a more monitorable property has been established by the initial fractionation, the recovery of less visible activities such as in vivo activity is not predictable. For example, if the in vivo active agent present in an enriched extract is bound or more strongly retained on a given adsorbent than an in vitro activity with which it had been erroneously assumed to correlate, this in vivo active agent would probably be lost. With CCC that in vivo active agent could probably be easily recovered from the stationary phase retained in the coil.

C. Versatility from a Wide Range of Solvent Polarities

Selection of the proper solvent system is critical for any chromatographic resolution. With CCC the restrictions imposed by fluid dynamics [23, 24] replace the restrictions imposed by the properties of the adsorbent in adsorption chromatography. The hydrodynamic behavior of biphasic solvent systems in CCC has been correlated with such physical properties as interfacial tension, viscosity, and density differences of the two phases [25]. Approaches for selecting appropriate solvent systems for CCC have been reviewed [16] but these approaches assume knowledge about the properties of the agent under isolation. Such approaches are therefore not applicable for the initial fractionation efforts on less visible activities. Efficient resolution by CCC requires a significant distribution of the activity of interest in both phases; if the phase used as mobile phase holds most of an activity, that activity will appear in the solvent front and if the stationary phase is strongly favored, the activity will remain in the stationary phase. The most direct approach, partitioning aliquots of the enriched extract residue between the preequilibrated phases of several solvent systems, evaluating the separated phases for the activity of interest, and selecting the system affording the most even distribution of the activity, will usually ensure adequate resolution by CCC. If the enriched extract displays sufficient in vivo potency that distributions could be evaluated without consuming too much of the available sample, in vivo evaluation would be the best way to ensure resolution of the in vivo activity. Often that is not the case, and an in vitro activity is selected to evaluate distributions for determination of the solvent system. In such situations, total recovery of stable activities is particularly important. It is also apparent that a fortuitous mixture of polar and nonpolar activities might resemble an even distribution of a single activity. Activities then recovered from solvent front and stationary phase would require redistributions to find suitable solvent systems for better resolutions.

Although almost limitless number of solvent combinations are available for CCC, logical, empirical approaches can quickly lead to efficient solvent systems. For example, a hypothetical unknown might be distributed between the preequilibrated phases of several systems such as hexane/methanol/water (10:9:1), ethyl acetate/hexane/methanol/water (1:1:1:1), chloroform/methanol/water (5:8:5), and butanol/water (1:1); the layers would then be separated and evaluated by the appropriate assay. Any system distributing 45–55% of the activity in the organic phase could be directly utilized for CCC. If none of these systems met that requirement, the system yielding the most favorable distribution would be modified by changing component ratios or including another modifying solvent. Hexane is a useful modifier of organic upper phase systems for driving components into the lower phase and allowing higher methanol/water ratios. However, hexane is a rather poor solvent for many compounds; if system 2 in the example distributed too much activity into the upper phase, better results might be obtained by replacing ethyl acetate with ether. Cyclohexane is a useful modifier for organic lower phase systems which retain too much activity in the lower phase. However, toluene might be a better modifier for solubility considerations, or substituting carbon tetrachloride for chloroform might improve the distribution. Ether might be added to the fourth system to drive activities toward the lower layer or salts added to drive activities toward the upper layer.

D. Convenience and Speed

Through relatively recent developments, a variety of CCC techniques such as droplet CCC [17], centrifugal droplet CCC [18, 26], rotation locular CCC [18], and planet coil centrifugal CCC [20, 21] are available for convenient resolution of mixtures, and applications are described in other chapters of this book. Our experience with the commercial planet coil, the Ito multilayer coil separator-extractor, has been delightful. The relatively simple, sturdy construction and the absence of rotating seals have provided remarkably reliable service despite heavy usage. At flow rates of 1–5 ml/min, fractionations of 200–500 mg of enriched extract, dissolved in a mixture of the phases and introduced through a loop valve, are usually completed in several hours. Little sample preparation and normally less than a liter of mobile phase are required, simplifying operating procedures.

E. Inexpensive Operation

The expense of chromatographic columns and adsorbents is eliminated and a relatively low volume of solvent is required for each resolution.

The crude mixtures encountered in natural product isolation work tend to spoil chromatographic columns at a relatively high rate.

F. High Resolution

Dramatic resolutions of crude natural products by CCC have been reviewed earlier in this chapter and our search for novel antitumor agents from microbial sources has generated additional examples. Some of these will be briefly described to further illustrate the versatility, solvent selection, operating procedures, and resolving power of CCC in the Ito coil.

III. SOME AGENTS CONVENIENTLY ISOLATED BY CCC

Figure 1 lists some known agents encountered in the course of screening for antitumor agents and solvent systems conveniently yielding

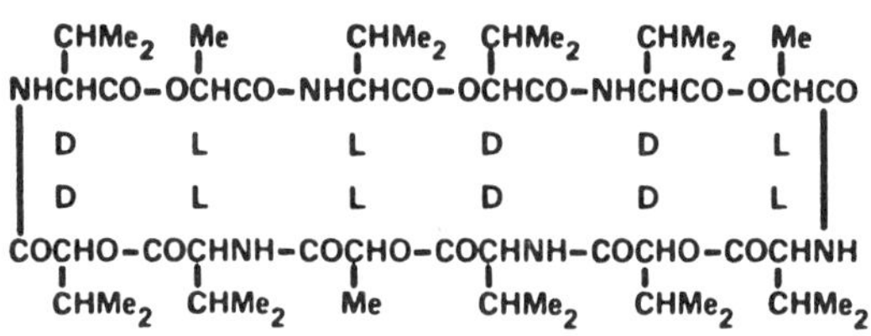

Valinomycin: H–M–W (10:9:1)

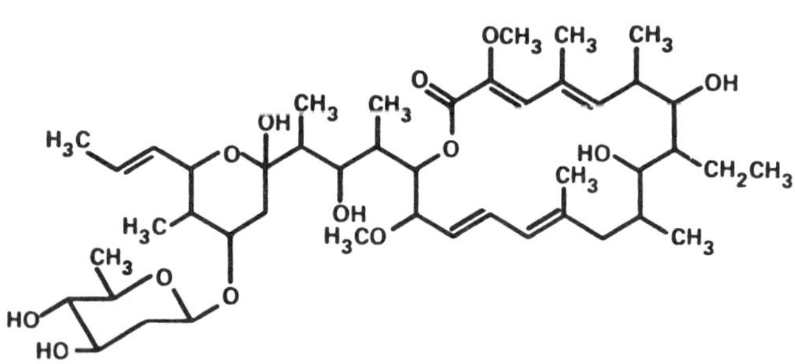

Concanamycin C: H–EA–M–W (1:1:1:1)

Fig. 1 Convenient solvent systems.

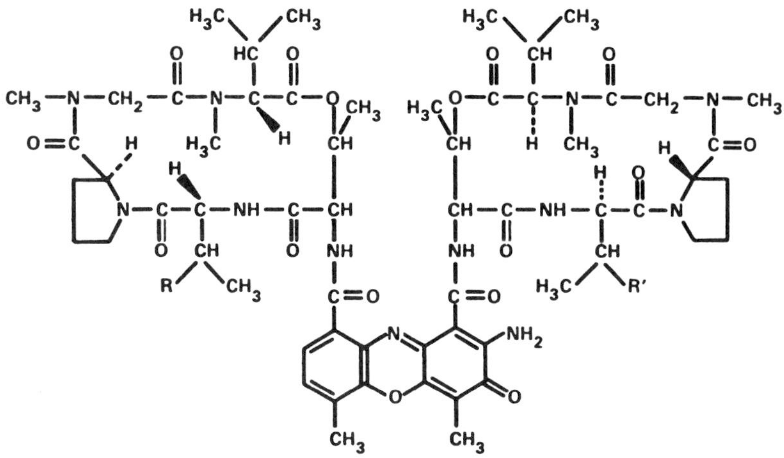

Actinomycin C1 (R, R′=CH_3) C2 (R=CH_3, R′=CH_3CH_2) C3 (R, R′=CH_3CH_2): E–H–M–W (5:1:4:5)

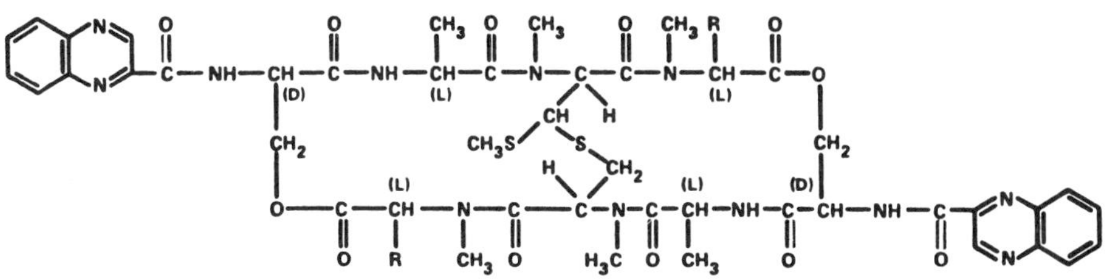

Quinomycin A R=$CH(CH_3)_2$ Quinomycin C R=$CH(CH_3)CH(CH_3)_2$: EA–H–M–W (3:2:2:3)

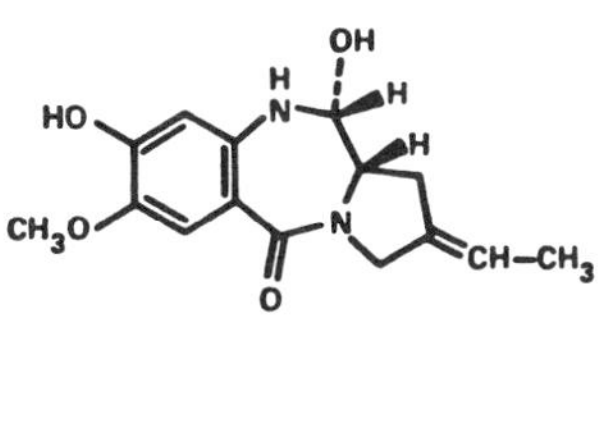

Tomaymycin
EA–H–M–W (3:1:1:3)

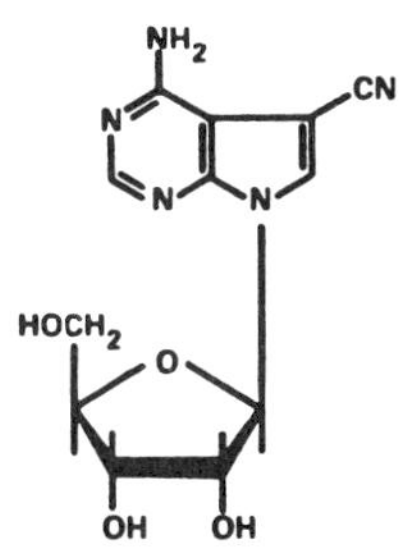

Toyocamycin
B–W (1:1)

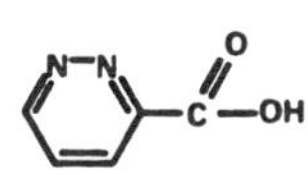

Pyridazine 3 carboxylic acid
B–A (1:1)

H is Hexane, E is Ether, EA is Ethyl Acetate, M is Methanol, B is Butanol, W is Water, and A is 0.2 N Acetic Acid

Fig. 1 (Continued)

identifiable samples of these agents. The lipophilic antitumor antibiotic valinomycin [27] was isolated using the system from hexane and 90% methanol. Macrolide antibiotic concanamycin C[28] was isolated with the system from equal parts of hexane/ethyl acetate/methanol/water. This agent had resisted our isolation attempts from the cytotoxic culture producing it until we isolated it by CCC in the Ito coil. An excellent example of the resolving power of CCC in the Ito coil is the resolution of the actinomycin C complex into its components [29]. Structural considerations illustrate the nontrivial nature of this resolution. Actinomycin C2 contains a single methylene group more than actinomycin C1 (D) and a single methylene group less than actinomycin C3 [30]. The system from equal parts of hexane/ethyl acetate/methanol/water resolved the complex but required 1.7 L of mobile phase to resolve 100 mg of the complex. A TLC densitomteric assay, convenient for rapid estimation of multiple samples, was developed and used to find an improved solvent system for CCC. The system ether/hexane/methanol/water (5:1:4:5) resolved the complex utilizing only 600 ml of mobile phase. Extracts from an initial 2 L of a cytotoxic screening broth afforded well-resolved samples of quinomycin A (levomycin, echinomycin) [31] and quinomycin C[32] on CCC with slightly increased levels of ethyl acetate and water. This isolation by CCC was more efficient and convenient than our previously utilized isolation procedures [33]. Tomaymycin [34] was isolated with still higher levels of ethyl acetate and water. The considerably more polar antitumor agent toyocamycin [35] was isolated with a butanol/water system. The polar, water-soluble nicotinamide antimetabolite pyridazine-3-carboxylic acid [36] was isolated using a system from butanol and dilute acetic acid.

IV. IMPROVED RESOLUTIONS WITH MEASURABLE PROPERTIES

The advantages of CCC for the initial fractionation of a biologically active natural product also apply to subsequent purification of potentially novel agents. As purification progresses, information gained about the physical and chromatographic properties of the actual component of interest allows for better monitoring and quantitation of that component in final purification schemes. As the component becomes truly visible, advantages pertaining to binding and total recovery are less important than those regarding versatility, convenience, inexpensive operation, and especially resolution. The natural product chemist can now readily predict the recovery of the component from various chromatographic procedures by preliminary small-scale experiments. The more accurate estimation of the distribution of the component in prequilibrated biphasic solvent systems allows selection of solvent systems leading to improved resolution in the coil.

A variety of challenging separation problems occur in the development work required to determine if novel antitumor natural products will become useful drugs. The contributions of CCC to three of these separation problems will be described to emphasize solvent selection procedures when measurable physical properties of the substrate are available.

A. Critical Nature of Solvent Selection

Acivicin, 1a (Fig. 2), isolated [37] during the course of screening soil cultures for antimetabolite antitumor activity [38], is a glutamine antagonist in mammalian systems. Its unique enzyme-inhibiting and antitumor effects [39] led to its ongoing clinical trials by the National Cancer Institute. A purified batch of fermentation-derived acivicin was contaminated by a trace of a very polar, ninhydrin-positive substance. The contaminant moved with the dibasic acid cystine on silica gel plates with polar systems, but a more rigorous identification was desired. Direct isolation of the contaminant itself proved difficult. Although the contaminant, like acivicin itself, was not significantly retained by C18, some enrichment was accomplished by percolating an aqueous solution of the contaminated sample through a column of C18. The early acivicin fractions contained approximately 5% levels of the contaminant. Since isolation of the contaminant itself seemed unlikely, the enriched sample was benzoylated affording a mixture of derivatives with better prospects for resolution and increased UV absorption advantageous for monitoring and quantitation. Authentic sampels of benzoyl acivicin [40] and benzoyl cystine were prepared and compared with toluene and ethyl acetate extracts of the acidified benzoylation reaction by TLC densitometry. The toluene extracts contained benzoic acid plus approximately 25% of the total benzoyl acivicin and gave no peaks moving with benzoyl cystine. Most of the benzoyl acivicin and essentially all of the benzoyl contaminant (moving with benzoyl cystine) were in the ethyl acetate extract; the TLC densitometric results of the ethyl acetate extract are summarized in Fig. 2. Baseline resolution of a minor component that close to a major component would be difficult by adsorption chromatography but was readily achieved by CCC with the right solvent system. To select a solvent system with sufficient resolving power, aliquots of the extract residue were partitioned between 1-ml portions of the upper and lower phases of the preequilibrated solvent systems from different ratios of hexane, ethyl acetate, methanol, and dilute acetic acid. The phases were separated and the distributions of both components in the phases of each system were estimated by TLC densitometry. The results (Table 1) dramatically illustrate the importance of selecting the right solvent system. Relatively minor changes in solvent composition caused major shifts in component distributions. CCC using upper phase as the mobile phase with the

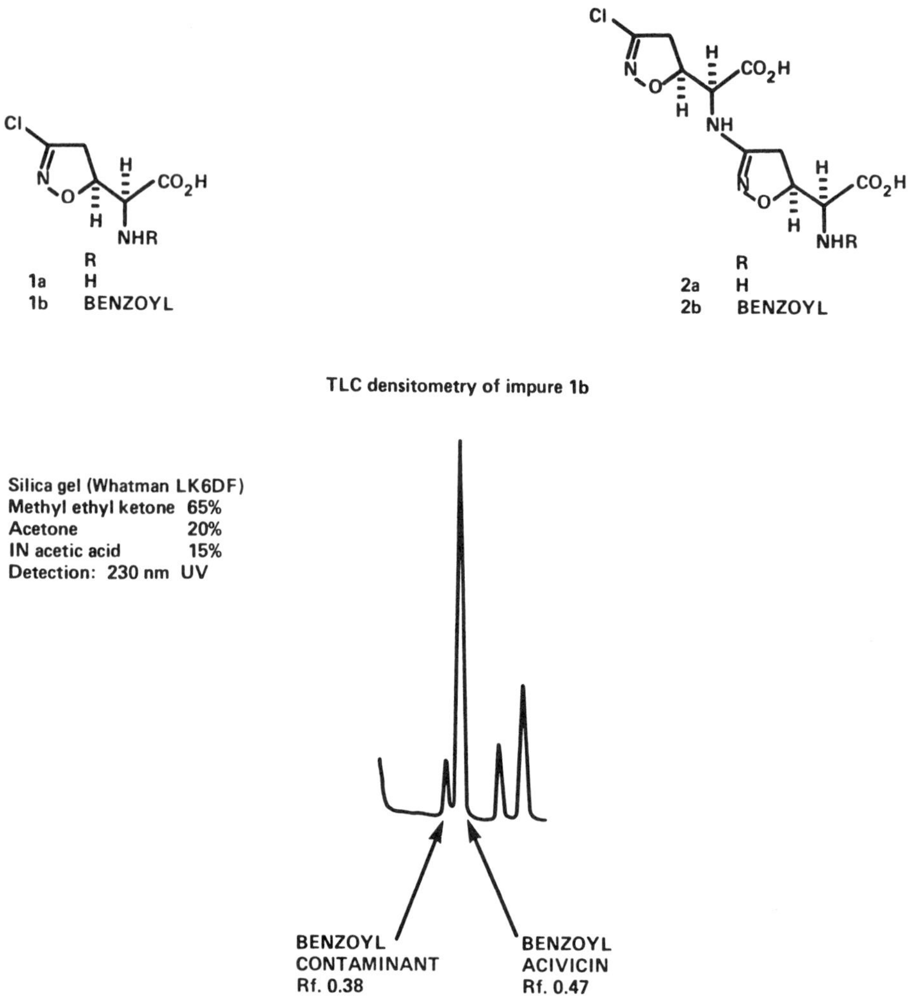

Fig. 2 Acivicin and two contamination problems.

first two systems would readily resolve these benzoyl derivatives but would require more time and solvent than necessary to eluate both components. The third system was selected for CCC because in that system benzoyl acivicin was slightly partial to the upper phase while the benzoyl contaminant was slightly partial to the lower phase. The fourth system illustrates that further changes would eliminate resolution of the desired components. Time spend on optimizing a solvent system for countercurrent chromatography is well spent.

Table 1 Distributions on Impure **1b**, Benzoyl Acivicin

H	E	M	A	H=M	E=A	Benzoyl acivicin	Benzoyl contaminant
1	1	1	1	25%	25%	U-35% L-65%	U- 9% L-91%
2	3	2	3	20%	30%	U-52% L-48%	U-37% L-63%
5	8	5	8	19%	31%	U-55% L-45%	U-47% L-53%
1	2	1	2	17%	33%	U-62% L-38%	U-62% L-38%

Abbreviations: H = hexane; E = ethylacetate; M = methanol; and A = 0.2 N HOAC.

After filling the Ito coil (column 2.6 mm i.d.; volume approximately 330 ml) with the lower phase as the stationary phase, the extract residue, approximately 400 mg dissolved in 8.5 ml of a mixture of the upper and lower phases, was introduced into the coil through a loop valve. Upper phase was then used as the mobile phase at a flow rate of 3 ml/min while the coil was rotated at 800 rpm. After the loss of approximately 26 ml of the stationary phase (retention 304 ml or 92%), collected fractions of the mobile phase were monitored by TLC densitometry. After elution of 190 ml of mobile phase, 367 mg of pure benzoyl acivicin was eluted in the next 190 ml. After an additional 266 ml of compound-free eluate, 19 mg of pure benzoyl contaminant was eluted in the following 380 ml of mobile phase. FAB/MS established the identity of the isolated benzoyl contaminant with benzoyl cystine.

B. Purification of an Alkali-Induced Acivicin Contaminant: Desalting a Water-Soluble Polar Substance

Traces of an alkali-induced contaminant were introduced into a misformulated lot of acivicin under preparation for clinical trials. The recovery of pure acivicin, isolation of the contaminant, and determination of its structure as **2a** (Fig. 2) have been described [40]. Solution of that problem provided another illustrative example of the utility and power of CCC. Efforts to crystallize polar, water-soluble, nonextractable **2a** were unsuccessful and its physical properties were not conducive to further purification. Derivatization by Schotten–Baumann benzoylation led to **2b**, only slightly less polar on silica gel than **2a** itself. Although **2b** was still water-soluble after acidification and was not extractable with ethyl acetate, it was now readily monitored and quantitated by TLC densitometry. Derivative **2b** proved extractable with several protions of butanol but a sample purified to apparent chromatographic homogeneity on silica gel resisted crystallization efforts. By contrast, CCC readily yielded pure, crystalline **2b**. Initially, aliquots of the crude **2b** from derivatizing 50 mg of **2a** were partitioned between 0.5-ml portions of the upper and lower phases of several preequilibrated, acidic biphasic solvent systems and the concentration of **2b** in each phase estimated by TLC densitometry. The system from chloroform/methanol/0.2 N acetic acid (1:1:1) evenly partitioned **2b** between its phases and was used for CCC in the Ito multilayer coil.

After filling the column (2.6 mm i.d.; volume approximately 330 ml) with the upper phase as the stationary phase, viscous oil **2b** was dissolved in 8 ml of a mixture of the phases and introduced into the coil through a loop valve. Lower phase was then used as the mobile phase at a flow rate of 4.6 ml/min while the coil was rotated at 800 rpm. After the loss of 79 ml of stationary phase (retention 251 ml or 76%), collected fractions of the mobile phase were monitored by TLC densitometry. After 444 ml of mobile phase, homogeneous **2b** appeared in the next 185 ml of eluate. Evaporation of center-cut fractions (111 ml containing 89% of the total **2b**) left 40 mg of residue which was readily recrystallized from acetone/cyclohexane. This application of CCC in the Ito coil illustrates its potential for desalting appropriate water-soluble compounds; it is possible that salts, incompletely resolved from water-soluble amino acid **2b** on silica gel, foiled crystallization attempts after adsorption chromatography.

C. Quantitative Recovery of Sample

Screening for cytotoxic antitumor agents uncovered the microorganism *Streptomyces zelensis* [41] producing the novel, potent agent CC-1065, 1 (Fig. 3) [42–44]. Alkaline fragmentation of CC-1065 afforded the cyclopropapyrroloindole (CPI) unit 2 (Fig. 3) [45]. Essentially pure but pigmented CPI was obtained by chromatography on silica gel. Facile resolution of CPI from contaminating pigments by CCC provides the final example selected for this chapter. Aliquots of pigmented CPI were partitioned between the preequilibrated phases of several biphasic solvent systems and its concentration in the phases estimated by TLC densitometry. The system from chloroform/cyclohexane/methanol/water (5:2:10:5) evenly distributed CPI between its phases and was selected for CCC in the Ito coil. After filling the column (2.6 mm i.d.; volume approximately 330 ml) with the upper phase as the stationary phase, 101 mg of pigmented CPI, dissolved in 8 ml of a mixture of the phases, was introduced into the coil through a loop valve. Lower phase was then used as the mobile phase at a flow rate of 2 ml/min while the coil was rotated at 800 rpm. After the loss of 43 ml of stationary phase (retention 287 ml or 87%), collected fractions of the mobile phase were monitored

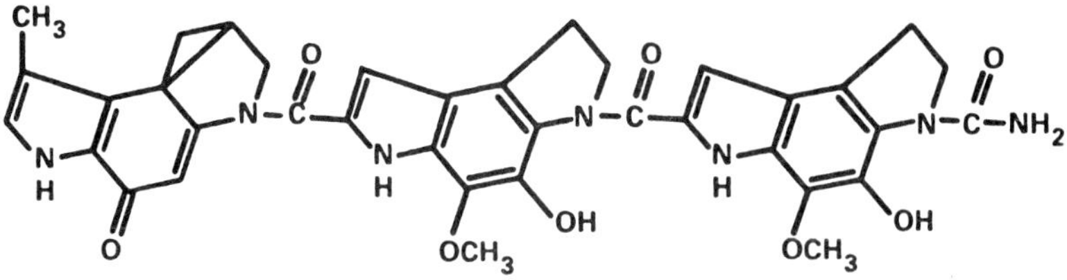

1

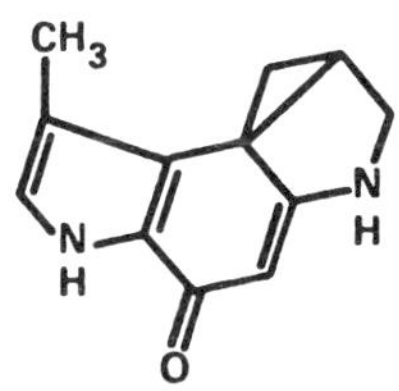

2

Fig. 3 CC-1065 and a fragmentation product.

by TLC densitometry. After 40 ml of mobile phase, the next 40 ml displayed a purplish red color but the initial 280 ml of mobile phase contained only 4 mg of residue. Homogeneous, colorless CPI appeared in the following 120 ml of eluate. Evaporation and trituration afforded 97 mg of pure CPI as cream-colored crystals. This example illustrates the quantitative recoveries permitted by CCC in the coil.

V. CONCLUSION

CCC offers many advantages for the isolation and purification of the biologically active components present in some natural sources so that they can be characterized and evaluated as potential drugs. Abundant applications to a broad spectrum of structural types illustrate the versatility and resolving power of this method. New developments and commercially available equipment provide convenient, rapid, and economical operation. The absence of solid supports allows better isolation of components susceptible to binding or deactivation by solid adsorbents and samples free of contamination from these adsorbents. The absence of solid supports greatly facilitates the isolation of biologically interesting components not rendered highly visible by the nature of their biological activity since it affords total recovery of all stable activities. Optimizing solvent systems for each application is critical for successful resolution by CCC.

ACKNOWLEDGMENTS

The author is grateful to the Upjohn Company for providing the excellent facilities and resources required in the search for novel antitumor drugs. This search is a team effort and the many contributions and support of coworkers are greatly appreciated. He especially thanks his associates, Carolyn Biles and Randy Peltonen, who carried out the initial fractionations described. Finally, he is extremely grateful to Dr. Ito and his colleagues for their extensive pioneering and development efforts on CCC and to Peter Carmeci for manufacturing the planet coil centrifuge.

REFERENCES

1. G. M. Brill, J. M. McAlpine, and J. E. Hochlowski, *J. Liq. Chromatogr.*, *8*:2259 (1985).
2. E. A. B. Brown and Y. Ito, *J. Biochem. Biophys. Meth.*, *3*: 77 (1980).
3. R. C. Bruening, E. M. Oltz, J. Furukawa, and K. Nakanishi, *J. Am. Chem. Soc.*, *107*:5298 (1985).
4. R. G. Williams, 1985 Pittsburgh Conference, New Orleans, La., Feb. 25–Mar. 1, 1985, Abstr. 300.
5. R. D. Friedman, H. Nakazawa, F. E. Chou, N. R. Bachur, and Y. Ito, *J. Chromatogr.*, *252*:283 (1982).
6. H. Nakazawa, P. A. Andrews, and N. R. Bachur, *J. Chromatogr.*, *205*:482 (1981).
7. T. Zhang, *J. Chromatogr.*, *315*:287 (1984).
8. N. B. Mandava, Y. Ito, and J. M. Ruth, *J. Liq. Chromatogr.*, *8*:2221 (1985).
9. Y. W. Lee and C. E. Cook, *J. Liq. Chromatogr.*, *8*:2253 (1985).
10. J. M. Ruth, Y. Ito, N. B. Mandava, and V. P. Flanagan, *J. Liq. Chromatogr.*, *8*:2239 (1985).
11. J. L. Sandlin and Y. Ito, *J. Chromatogr.*, *348*:131 (1985).
12. M. Knight, *LC Magazine*, December 1985, 3 No. 12.
13. M. Knight, Y. Ito, P. Peters, and C. DiBello, *J. Liq. Chromatogr.*, *8*:2281 (1985).
14. I. A. Sutherland, D. Heywood-Waddington, and T. J. Peters, *J. Liq. Chromatogr.*, *8*:2315 (1985).
15. Y. Ito, *J. Liq. Chromatogr.*, *8*:2131 (1985).
16. W. D. Conway and Y. Ito, *LC Magazine*, May 1984, 2 No. 5.

17. K. Hostettmann, G. Appolonia, B. Domon, and M. Hostettmann, *J. Liq. Chromatogr.*, *7*:231 (1984).
18. J. K. Snyder, K. Nakanishi, K. Hostettmann, and M. Hostettmann, *J. Liq. Chromatogr.*, *7*:243 (1984).
19. N. B. Mandava and Y. Ito, *J. Liq. Chromatogr.*, *7*:303 (1984).
20. Y. Ito and W. D. Conway, *Anal. Chem.*, *56*:534A (1984).
21. Y. Ito in *Advances in Chromatography*, Vol. 24 (J. C. Giddings, E. Grushka, J. Cazes, and P. R. Brown, eds.), Marcel Dekker, New York, 1984, p. 181.
22. L. J. Hanka, B. K. Bhuyan, D. G. Martin, G. L. Neil, and J. D. Douros, *Antibiot. Chemother.*, *23*:26 (1978).
23. Y. Ito, *J. Chromatogr.*, *301*:377 (1984).
24. Y. Ito, *J. Chromatogr.*, *301*:387 (1984).
25. Y. Ito and W. D. Conway, *J. Chromatogr.*, *301*:405 (1984).
26. W. Murayama, T. Kobayashi, Y. Kosuge, Y. Nunogaki, and K. Nunogaki, *J. Chromatogr.*, *239*:643 (1982).
27. M. M. Shemyakin, N. A. Aldanova, E. I. Vinogradova, and M. Y. Feigina, *Tetrahedr. Lett.*, *1963*:1921 (1963).
28. H. Kinashi, K. Someno, and K. Sakaguchi, *J. Antibiot.*, *37*: 1333 (1984).
29. D. G. Martin, R. E. Peltonen, and J. W. Nielsen, *J. Antibiot.*, *39*:721 (1986).
30. A. B. Mauger, *Topics Antibiot. Chem.*, *5*:223 (1980).
31. W. Keller-Schierlein, M. Lj. Mihailovic, and V. Prelog, *Helv. Chim. Acta*, *42*:305 (1959).
32. T. Yoshida and K. Katagiri, *J. Antibiot.*, *14A*:330 (1961).
33. D. G. Martin, S. A. Mizsak, C. Biles, J. C. Stewart, L. Baczynskyj, and P. A. Meulman, *J. Antibiot.*, *28*:332 (1975).
34. L. H. Hurley, *J. Antibiot.*, *30*:349 (1977).
35. K. Ohkuma, *J. Antibiot.*, *14A*:343 (1961).
36. S. Gabriel and A. Colman, *Ber.*, *32*:395 (1899).
37. D. G. Martin, D. J. Duchamp, and C. G. Chidester, *Tetrahedr. Lett.*, *1973*:2549 (1973).
38. L J. Hanka, D. G. Martin, and G. L. Neil, *Cancer Chemother. Rep.*, *57*:141 (1973).
39. R. H. Earhart and G. L. Neil, *Adv. Enzyme Regulation*, *24*: 179 (1986).
40. D. G. Martin, S. A. Mizsak, and J. W. Nielsen, *J. Antibiot.*, *39*:603 (1986).
41. L. J. Hanka, A. Dietz, S. A. Gerpheide, S. L. Kuentzel, and D. G. Martin, *J. Antibiot.*, *31*:1211 (1978).
42. D. G. Martin, C. G. Chidester, D. J. Duchamp, and S. A. Mizsak, *J. Antibiot.*, *33*:902 (1980).
43. C. G. Chidester, W. C. Krueger, S. A. Mizsak, D. J. Duchamp, and D. G. Martin, *J. Am. Chem. Soc.*, *103*:7629 (1981).
44. D. G. Martin, C. Biles, S. A. Gerpheide, L. J. Hanka, W. C. Krueger, J. P. McGovern, S. A. Mizsak, G. L. Neil, J. C. Stewart, and J. Visser, *J. Antibiot.*, *34*:1119 (1981).
45. D. G. Martin, S. A. Mizsak, and W. C. Krueger, *J. Antibiot.*, *38*:746 (1985).

chapter ten

Countercurrent Chromatography for Peptides

Martha Knight
Peptide Technologies Corporation
Washington, D.C.

I. INTRODUCTION

Countercurrent chromatography has been a useful method of peptide purification since its inception. An early instrument for this purpose, the countercurrent distribution machine, developed by Lyman Craig in the 1940s was a large glass train of tubes that transferred the upper phase of binary solvent systems [1]. This was the primary means for purifying and isolating peptides. Through those years countercurrent distribution was used for analytical as well as preparative chromatography. Now with the development of high-performance liquid chromatography (HPLC), analytical scale chromatography can be performed on peptides conveniently and, as the technology advances, with increasingly smaller quantities of material. Nevertheless, for preparative scale purification countercurrent chromatography still holds advantages. Chromatographing in a totally liquid open tubing without a solid support, no losses of material can occur by adsorption or irreversible binding. Thus all material is recoverable. Additionally, in suitable solvent systems loads up to 1000 mg of peptide can be truly fractionated in one run. Thus in research laboratories dealing with the need to purify a couple of hundred milligrams of peptides, countercurrent chromatography is turned to using coil planet centrifuges invented by Ito. These pose as modern compact devices able to carry out the task of the former countercurrent distribution machines with more versatility.

As the coil planet instruments were developed, the initial experiments to demonstrate performance were done with amino acid derivatives. The resolving power was well demonstrated with good separations of closely related DNP amino acids [2–4]. The ensuing demonstrations of selectivity were with dipeptides. The effects of different operating conditions were tested with these compounds. The conditions included the effect of rotation rate, flow, and solvent systems, and various mobile phases. Insulin, a peptide hormone long studied and chromatographed by countercurrent chromatography, was chromatographed in an early planet centrifuge apparatus and shown to be better resolved than in countercurrent distribution [5]. Cyclic antibiotics such as gramicidin and echinomycins were chromatographed as well and showed the utility of countercurrent chromatography in the coil planet centrifuge [6,7]. This process being a form of partition chromatography fractionates small compounds and larger compounds such as peptides. However, the application to proteins, especially enzymes, has been limited because of denaturation in organic solvents used such as alcohols. Some progress is being made in this regard with the two-phase dextran, polyethylene glycol systems [8,9]. Nevertheless, a group of biomolecules that have been extensively purified by this technique are the peptides. Most of the peptides that have been studied are considered small peptides below the molecular weight of 2000. Thus more experience with larger peptides is necessary. From the work to date it is hoped that with knowledge of the size and charge of the peptides their behavior in certain solvents will be predictable.

Peptides and related compounds such as antibiotics are presently of high interest in biomedical research and application. For many years oxytocin, vasopressin, and the kinin hormones were studied for their roles in hypertension. Endocrine hormones such as ACTH and insulin have great involvement in disease. Research in these hormones involved achieving chemical synthesis and ensuing structure–activity studies. Physiological studies of metabolism elucidated enzymes for which inhibitors were designed such as those for the angiotensin-converting enzyme. During the late 1960s the hypothalamic releasing factors were discovered and many analogs, especially of LH-RH, have been made for structure–activity studies. The discovery of an endogenous opiate agonist proved to be a pentapeptide, enkephalin. Since then many peptides have been found in the brain, thus structure–activity studies are being done for many peptides with the purpose of designing new specific and potent drugs for many diseases. As a result of rapid gene-sequencing technology the structures of new translation products are being elucidated and the protein sequences are of interest to synthesize for antibody generation. Thus there is a great deal of peptide synthesis activity in research laboratories at present.

As a result of Ito's research in countercurrent chromatography, machines have been designed and some have been built as prototypes. Peptides have been chromatographed on at least four of these. Most of the instruments have not yet gone on to be developed commercially but are useful chromatographs. Described here is the application of coil planet centrifuges in the preparative purification of peptides. It is hoped that ideas for further development of this promising technology will be generated and carried out.

II. HORIZONTAL FLOW-THROUGH COIL PLANET CENTRIFUGE

A. Description of Instrument

The horizontal flow-through coil planet centrifuge was designed and built in 1977 and has been described in various reports (Fig. 1) [2–6]. The instrument was the application of the seal-free continuous coil design with inlet and outlet tubing secured at one end. The coils were mounted parallel and at a distance B from the axis of rotation and on the gear side, according to scheme IV in the design of synchronous flow-through coils (Fig. 2) [6]. On the other side (pulley side) as a counterbalance was another coil of 0.5-mm i.d. tubing whose planetary motion is described in scheme I [6]. This can serve as an analytical coil suitable for separations of small amounts of sample such as radiolabeled compounds. Experiments using this coil separated DNP amino acids and dipeptides [3,4]. Thus this apparatus was the combined flow-through coil planet centrifuge with the I and IV configuration of synchronous motion [6]. The inlet and outlet tubing of both coils passed through the center shaft of rotation. It is the "semipreparative" 2.6- or 1.6-mm i.d. gear side coil that has been used extensively for peptide separations. The total capacity of the coil is 280 ml for 50-m long, 2.6-mm i.d. tubing. Since 1978 this apparatus has been used routinely for the purification of over 100 synthetic peptides generated by the solid phase technique. Commercial prototypes of this machine have been built by Union Carbide and later Kontes. A preparative unit consisting of 5-mm i.d.. glass coils mounted circularly around the central rotational axis was also made by Kontes at that time, but was never used for peptides although it is capable of larger scale preparative chromatography [30].

B. Experimental Methods

All of the peptides were synthesized by the Merrifield solid phase method [12] and the peptides were cleaved from the solid support by reaction with hydrogen fluoride in the presence of anisole. The product was washed with ethyl acetate, dried, and peptide extracted from the resin in water or acetic acid and lyophilized. A determina-

Fig. 1 The horizontal flow-through coil planet centrifuge (Kontes prototype). The column holder (upper side) supports 10 rods wrapped with polytetrafluoroethylene 2.-mm i.d. tubing which is used for preparative chromatography. On the lower or pulley side is another coil of 0.5-mm i.d. tubing suitable for analytical scale chromatography. The tubing is continuous from the pump and a sample slide valve through the central axis (right side) and coiled around the mounted columns beginning at the left and exits at the same side through the central axis out the right side through the flow cell and into the fraction collector.

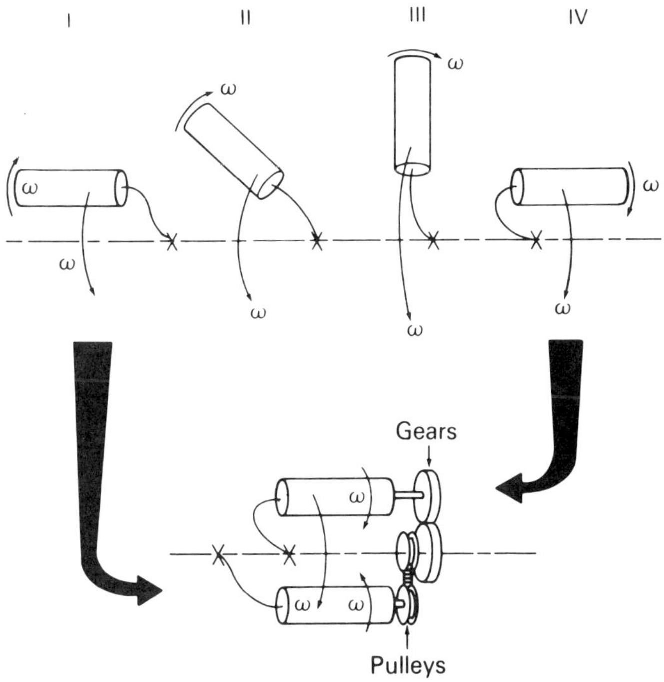

Fig. 2 Basis for the design of the horizontal flow-through coil planet centrifuge. The orientations I—IV are of the cylindrical column coiled tubing holder undergoing synchronous planetary motion. In schemes I and IV, the orientation of the column coil with respect to the center axis of rotation is parallel, and due to the synchronous counterrotation of the holder in I and corotation in IV, the inlet and outlet tubing arranged as shown and clamped down do not twist during the motion. The preparative coil is mounted on the gear side and the tubing enters through the center shaft as diagrammed.

from the resin in water or acetic acid and lyophilized. A determination of the partition coefficient is made by dissolving a small sample in the solvent system, shaking, and, after separation of the phases, measuring the concentration in both phases, thus giving the ratio of upper to lower phase (C_u/C_l) or mobile to stationary phase (C_m/C_s). For the horizontal flow-through coil planet centrifuge, the phase chosen as the mobile phase is that giving a partition coefficient lower than 1 to 0.05. The sample is dissolved in a small volume (not more than one-tenth the total volume of the coil) of both phases and loaded onto the coil previously filled with the stationary phase. The sample can be introduced directly or via a four-way slide valve connected in-line. The rotation is set at 400 rpm and pumping of the mobile phase is begun at a flow rate of 24 ml/hr for the butanol solvent systems or 2–3 times faster for other solvents. The effluent is passed through an LC flow cell detecting at 280 nm. The direction of flow through the cell is upward for a mobile lower phase and downward for the upper phase mobile in order to clear phase droplets. Fractions of 6 ml are usually collected. The chromatography is left to proceed for 2–3 column volumes, then either the other phase is chromatographed or the contents are pumped out and fractions collected. When this is necessary, the sample may have separated somewhat away from slower impurities. This is done by stopping the rotation and pumping faster. In order to move both phases equally it is helpful to rotate the coil slowly in the opposite direction. Between runs, if desired, the coil can be cleaned by pumping acetone and water and drying with a nitrogen stream.

The fractions containing purified peptide if not detected in-line can be determined manually by diluting 0.5–0.1 ml aliquots with 1 ml alcohol and reading the absorbance at a wavelength characteristic of the peptide. Tryptophan- and tyrosine-containing peptides absorb at 280 nm and phenylalanine-containing peptides absorb at 254 nm. This is routinely done in our laboratory because the recorder tracing has a great deal of noise. The LKB chopper recorders, which we do not have, are the least affected by phase droplets. Also colorimetric determination specific for certain amino acids in peptides can be made on the fractions. Examples of assays include Sakaguchi for arginine, Pauly for histidine and tyrosine, and general ones such as biuret, ninhydrin, and Folin Lowry [12]. The fractions containing peptide are concentrated in a rotary evaporator to remove solvent and lyophilized in water, glacial acetic acid, or dilute ammonia according to its solubility. The peptides are subsequently analyzed for purity by analytical HPLC, TLC, and for composition by amino acid analysis. The recent edition of *Solid Phase Peptide Synthesis* [12] is a good source for these analytical methods.

C. Applications of the Horizontal Flow-Through Coil Planet Centrifuge

Preparative Chromatography of Peptides

The first group of peptides chromatographed on the horizontal flow-through coil planet centrifuge were DNP amino acids ranging from polar to neutral amino acids separated in chloroform, acetic acid, and 0.1 N HCl [3,4]. Fractionation occurred with either phase mobile. With the organic phase used as the stationary phase, the polar amino acids were eluted first and the neutral ones later; with

Table 1 Countercurrent Chromatography of Amino Acid Derivatives and Small Peptides

Compound	$K(C_m/C_s)$	Solvent Composition	Volume Ratio	Mobile Phase	Ref.
		Chloroform/ acetic acid/	2:2:1	Upper	[4]
DNP[a] ornithine	>100	0.1 N HCl			
DNP Asp	3.8	"	"	"	"
DNP Glu	1.9	"	"	"	"
di-DNP Cys	.94	"	"	"	"
DNP β Ala	.71	"	"	"	"
DNP Ala	.56	"	"	"	"
DNP Pro	.45	"	"	"	"
DNP Val	.26	"	"	"	"
DNP Leu	5.6	"	"	Lower	"
DNP Ala	1.8	"	"	"	"
di-DNP Cys	1.1	"	"	"	"
DNP Glu	.53	"	"	"	"
DNP Asp	.26	"	"	"	"
		n-Butanol/ acetic acid/	4:1:5	Lower	
Tyr Gly Gly	4.0	Water			"
Pro Tyr	1.9	"	"	"	"
Leu Tyr	1.0	"	"	"	"
Trp Tyr	.45	"	"	"	"
Trp Leu	.33	"	"	"	"
Trp Trp	5.2	"	"	Upper	"
Trp Tyr	2.2	"	"	"	"
Leu Tyr	1.0	"	"	"	"
Val Tyr	.53	"	"	"	"
Tyr Gly Gly	.31	"	"	"	"
Ac Phe D-Glu	.83	"	"	"	[13]
Ac Leu D-Glu	.65	"	"	"	"
Ac Met D-Glu	.45	"	"	"	"
Ac Asp L-Glu	<.29	"	"	"	"

[a]DNP = Dinitrophenyl-L-amino acids.

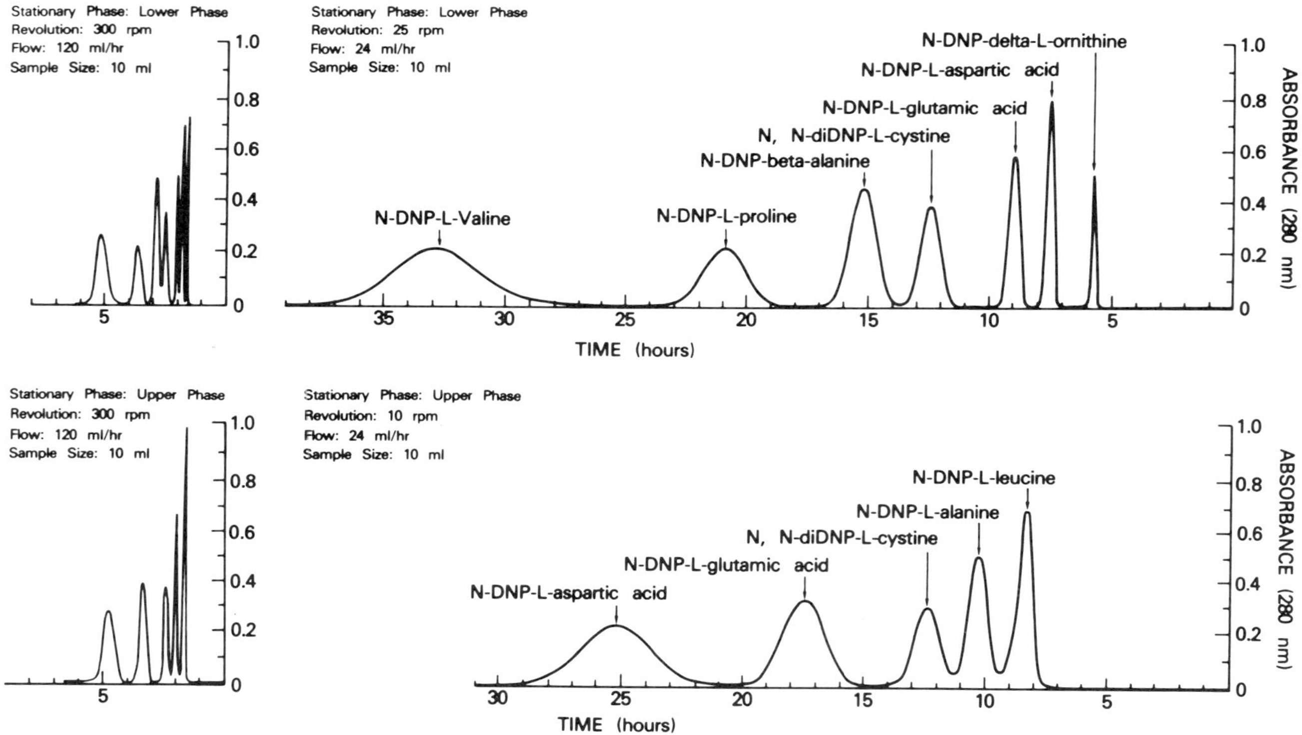

Fig. 3 The effect of flow rate and centrifugal speed on the separation of DNP amino acids and peptides in the horizontal flow-through coil planet centrifuge. Increasing the centrifugal speed decreases elution time without decreasing resolution. Upper two panels: Separation of DNP amino acids in chloroform, acetic acid and 0.1N HCl (2:2:1). Lower two panels: Separation of peptides with varying flow and centrifugal rates and mobile phases. The solvent system used was n-butanol, acetic acid, and water (4:1:5). (Reprinted from Ref. 3.)

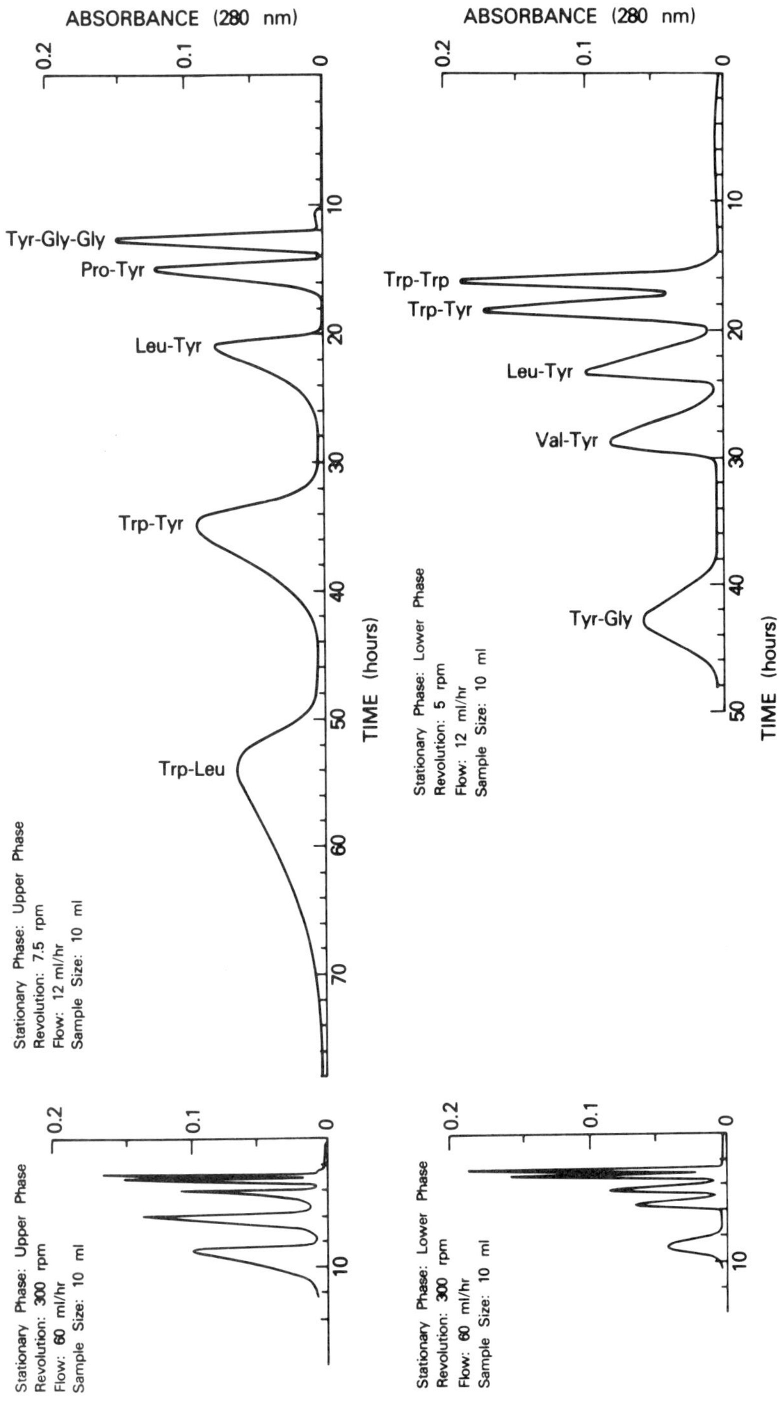
ABSORBANCE (280 nm)
0.2
0.1
0
Tyr-Gly-Gly
Pro-Tyr
Leu-Tyr
Trp-Tyr
Trp-Leu
TIME (hours)
10
20
30
40
50
60
70
Stationary Phase: Upper Phase
Revolution: 7.5 rpm
Flow: 12 ml/hr
Sample Size: 10 ml
Trp-Trp
Trp-Tyr
Leu-Tyr
Val-Tyr
Tyr-Gly
Stationary Phase: Lower Phase
Revolution: 5 rpm
Flow: 12 ml/hr
Sample Size: 10 ml
Stationary Phase: Upper Phase
Revolution: 300 rpm
Flow: 60 ml/hr
Sample Size: 10 ml
Stationary Phase: Lower Phase
Revolution: 300 rpm
Flow: 60 ml/hr
Sample Size: 10 ml

the lower phase stationary, the elution was in the opposite order. The powerful resolution of this method was demonstrated in the chromatography of a series of aromatic dipeptides in n-butanol, acetic acid, and water as shown in Fig. 3. The more hydrophobic peptides were eluted first with the nonaqueous phase mobile. There was good resolution of the neutral peptides with partition coefficients ranging from 5.2 to 1.31 [4]. The dipeptides with differences in one aliphatic side chain were well separated. Other small negatively charged dipeptides, acetyl Asp Glu and a series of D-Glu dipeptides, have been preparatively purified in the same solvent system (Table 1). The order of elution correlated with the polarity.

Much work has been done with peptides that are larger but less than 10 amino acids. Most of these have been cholecystokinin-related peptides. A series of fragments of the active sulfated octapeptide, CCK-8 or CCK 26–33, that were N-acetylated C-terminal amides, both sulfated and unsulfated, were prepared for pharmacological experiments [14]. More recently, analogs of CCK-8 and of the antagonist fragments have been synthesized. During the purification of the fragments, conditions were found suitable for fractionating the synthetic products. The most commonly used organic solvent was n-butanol. The upper phase giving the lower partition coefficient was used as the mobile phase so that faster eluting impurities could be eluted first and avoiding the major product eluting with the solvent front. With experience it was found that from the appearance on silica gel TLC in the n-butanol/acetic acid/water solvent system

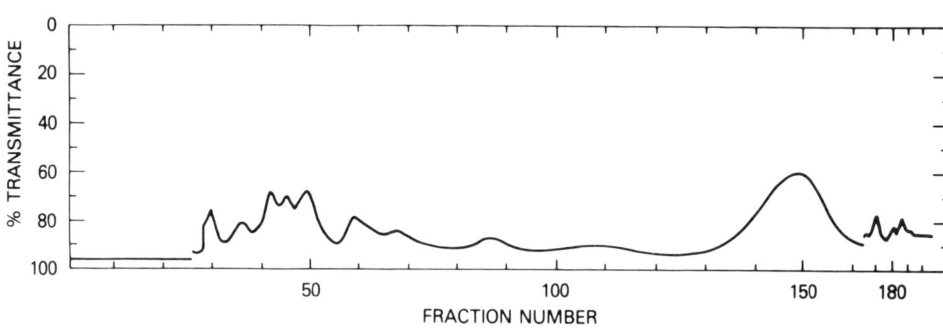

Fig. 4 Purification of Ac Asp Tyr Met Gly Trp NH_2 in the horizontal flow-through coil planet centrifuge in the BAW (4:1:5) system with the lower phase mobile. The solvent front was at fraction 26 and pure peptide, 102 mg (from a 0.5 mmole synthesis), was contained in fractions 130–161. The K is 0.14. The R_f of the peptide in silica gel TLC in BAW (4:1:1) was 0.58. (Reprinted from Ref. 14.)

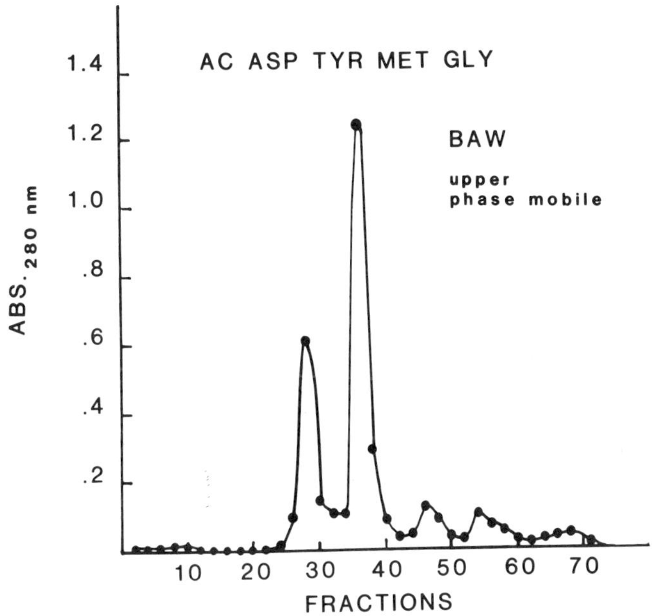

Fig. 5 Purification on the horizontal flow-through coil planet centrifuge of 291 mg synthetic Ac Asp Tyr Met Gly, which has a low R_f on TLC in the BAW system. The upper phase was used as the mobile phase and the solvent front came out of fraction 22 and purified peptide, 176 mg, at fractions 35–40 giving a K of 0.48. (Reprinted from Ref. 15.)

(4:1:1, by volume), the mobility of the peptide would be predicted. Peptides having an R_f of 0.6 or higher would be eluted with the n-butanol solvent front; thus it was better to elute with the lower phase mobile. This was the case with a CCK pentapeptide fragment (Ac Asp Tyr Met Gly Trp NH_2) which has a high R_f and was purified with the lower phase mobile (Fig. 4) [14]. The shorter fragment, Ac Tyr Met Gly, which has a low R_f on TLC, was eluted well behind the solvent front with a mobile upper phase (Fig. 5) [15].

A correlation of behavior on silica gel TLC and in countercurrent chromatography is evident in Fig. 6. The sulfation reaction of Ac Asp Tyr(SO_3H) Met Gly Trp Met NH_2 was chromatographed in 0.2 M NH_4OAc/n-butanol (1:1). The starting material or unsulfated peptide was eluted first followed by a side product and then the peptide. A large amount of salt was retained in the coil and pumped out. Each fraction was concentrated and analyzed by thin-layer chromatography as shown. The order of elution correlated with its mobility in TLC.

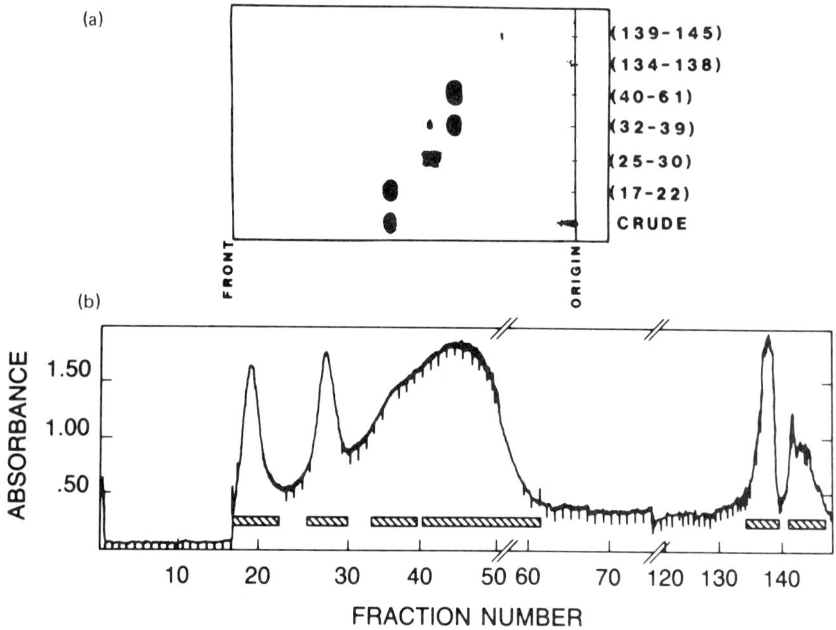

Fig. 6 Analysis by TLC (a) of the countercurrent chromatographic fractionation of the sulfation reaction of Ac Asp Tyr(SO_3H) Met Gly Trp Met NH_2 (b). Unsulfated peptide, 206 mg, was treated with 660 mg pyridine–sulfur trioxide complex; the reaction was neutralized with 15 ml M $NaHCO_3$ and lyophilized. The material was run in the horizontal flow-through coil planet centrifuge in 0.2 M NH_4OAc and n-butanol (1:1) with the upper phase mobile. The solvent front was eluted at fraction 17 and at 118 the contents were pumped out. The first peak contained 102 mg of starting peptide, the second peak 10 mg of side product. The first part of the product peak, fractions 32–39, contained 24 mg and 40–61 had 42 mg of pure sulfated peptide. Fractions 134–138 and 139–145 contained large amounts of salt. The fractions were concentrated and 50 μg was analyzed by TLC in BAW (4:1:1), Ehrlich spray. The mobility of the fractions correlated to their elution volume.

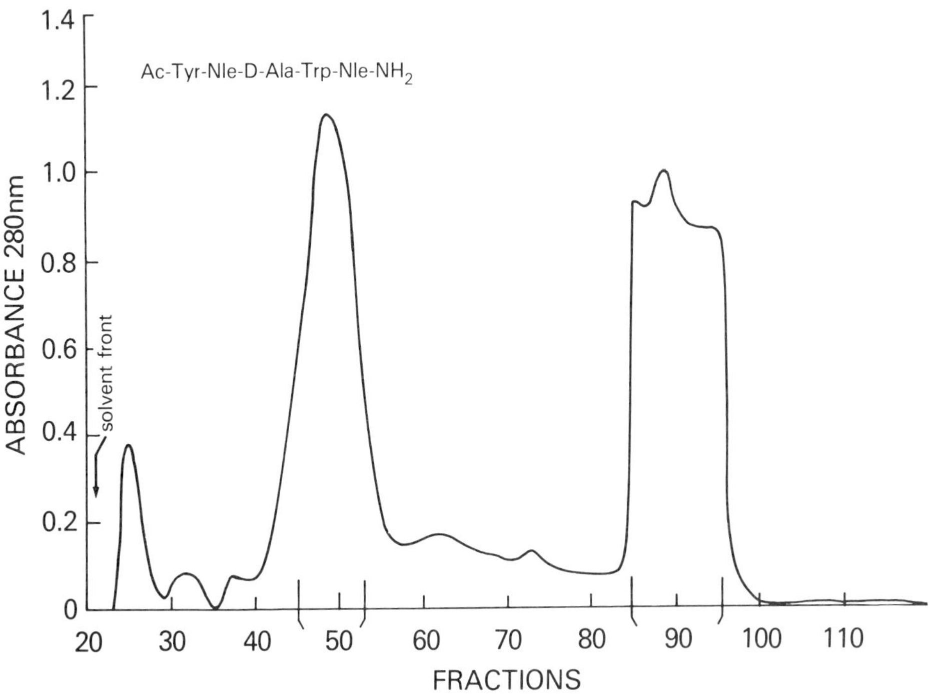

Fig. 7 Purification of Ac Tyr Nle D-Ala Trp Nle NH_2 on the horizontal flow-through coil planet centrifuge in the chloroform, acetic acid, and 0.01 N HCl (2:2:1) solvent system with the upper, aqueous phase mobile. The crude sample, 663 mg, was loaded and at a flow of 60 ml/hr and 400 rpm, 10-ml fractions were collected. The solvent front came out at fraction 23 and the lower phase was chromatographed at fraction 74 and the contents pumped out at fraction 100. There was some carryover of stationary phase throughout the main peak. The purified peptide eluted at fractions 45–53 which contained 234 mg peptide and 89 mg of a side product was collected in fractions 85–95.

Solvent System Selection

Generally for peptides that move in silica gel TLC in the BAW system, the upper phase of n-butanol solvent systems can be used as the mobile phase for peptides having an R_f of 0.4 or less. The lower phase can be used as the mobile phase for peptides having moderate mobility (R_f of 0.4–0.7) in TLC. For even more hydrophobic peptides that display greater mobility in TLC, the solvent system

Table 2 Various Solvent Systems Used for the Countercurrent Chromatography of Synthetic Peptides in the Horizontal Flow-Through Coil Planet Centrifuge

Solvent system	Mobile phase	Peptide	$K(C_m/C_s)$	Ref.
Ethanol n-butanol/ hexane/acetic acid/water (1:3:2:1:5)	Upper	A_c Asp Tyr(SO_3H) Met D-Ala Trp Met Asp Phe NH_2	1.22	[14]
0.2–0.5 M Ammonium acetate pH 7–9/ n-butanol (1:1)	Upper	Ac Asp Tyr Met Gly Trp Met NH_2	8.80	[14]
		Ac Asp Tyr(SO_3H) Met Gly Trp TFA Lys	1.22	[25]
		Ac Asp Tyr(SO_3H) Met Gly Trp Met NH_2	0.90	[14]
		Ac Asp Tyr(SO_3H) Met Gly Trp NH_2	0.16	[14]
		Ac Asp Tyr(SO_3H) Met Gly Trp Met	0.03	[14]
		Ac Asp Tyr(SO_3H) Met Gly Trp Met Asp NH_2	0.03	[14]
		Ac Tyr(SO_3H) Nle D-Ala Trp Nle NH_2	3.30	[16]

0.1% Trifluoroacetic acid/n-butanol (1:1)	Upper	Thr Ser Glu Lys Ser Gln Thr Pro Leu Val Thr	0.19	[26]
	Lower	Tyr Gly Gly Phe Met Thr Ser Glu Lys	1.26	[27]
Dichloro-methane/acetic acid/water (2:1:2)	Upper	Mercaptoisobutyryl Tyr Met Gly Trp NH_2	0.63	[16]
Chloroform/acetic acid/0.01 N HCl (2:2:1)	Upper	Ac Tyr Nle D-Ala Trp Nle NH_2	0.49	[16]
	Upper	Naphthoyl Tyr Nle D-Ala Trp Nle NH_2	0.15	[16]
	Lower	Ac Asp Tyr Met D-Ala Trp Met Asp Phe NH_2	0.57	[16]
Chloroform/benzene/methanol/water (15:15:23:7)	Upper	Gramicidins		[6]

chloroform/acetic acid/0.01 N HCl (2:2:1) with the upper aqueous phase mobile is suitable. Separation of Ac Tyr Nle D-Ala Trp Nle NH_2 and mercaptoisobutyryl Tyr Met Gly Trp NH_2 are shown in Figs. 7 and 8. Other examples are listed in Table 2.

Solvent systems can be modified to change the K of a sample such that it will be retained in the coil and not emerge with the solvent front. To lower the K, (C_u/C_l) ethyl acetate or hexane may be substituted for part of the n-butanol in the BAW system. Means of raising the partitioning into the upper phase are to increase the ionic strength of volatile salts or increase the concentration of volatile acids such as varying the percentage of trifluoroacetic acid from 0.1% to higher. Pyridine acetate in n-butanol affords higher K values for peptides. Two compositions that have been used are 0.1% acetic acid/n-butanol/pyridine (11:5:3) and n-butanol/pyridine/acetic acid/water (150:2:0.2:150). There is interference with UV detection but dual flow cells can be used for background subtraction. This system as well as the ammonium acetate system with a high pH can be used for acidic peptides to maintain their solubility or if acidic conditions are to be avoided. From the limited experience with longer peptides it is characteristic for these peptides to have low partitioning into the upper phase in the BAW system. Thus it is recommended to chromatograph in that system or 1% trifluoroacetic acid with lower phase mobile. Generally, basic peptides which are very water-soluble can be chromatographed in the acidic n-butanol systems with a mobile upper phase. *Solid Phase Peptide Synthesis* [12, p. 97] can be consulted for further suggestions for solvent systems.

Since much work was done with the n-butanol/acetic acid/water solvent system (4:1:5), a compilation of the results with different peptides is presented in Table 3. The peptides are grouped according to the mobile phase used and decreasing order of elution represented by the partition coefficient calculated from the elution volume.

Partition Coefficient Determination

The volume of elution of the compound is related to the partition coefficient (K). The partition coefficient is most conveniently expressed as C_m/C_s or the ratio of the solute concentration in the mobile phase to the solute concentration in the stationary phase, which is equal to the ratio of the stationary phase volume retained in the column to the retention volume of the sample after the solvent front. From the peak fraction and fraction volume, the K can be determined from the total capacity of the coil V_c and excluded volume V_m. In a coil of 2.6-mm i.d. tubing, the total capacity is 280 ml. For runs in solvent systems where the stationary phase is retained approximately 50% (which is the case of most tried at room temperature in the horizontal flow-through coil planet centrifuge) the V_m is

Table 3 Chromatography of Peptides in the Flow-Through Coil Planet Centrifuge in n-Butanol, Acetic Acid, and Water (4:1:5)

Peptide	$K(C_m/C_s)$	Ref.
Upper Phase Mobile		
Ac Asp Tyr Met Gly Trp Met	5.08	[14]
Myristyl Gly Ser Ser Lys Ser Lys Pro Lys Asp Pro Ser Glu Arg Arg Arg Tyr NH_2	4.50	[28]
Ac Asp Tyr Met Gly NH_2	0.95	[14]
Ac Asp Tyr(SO_3H) Met Gly Trp Lys	0.91	[14,25]
Ac Asp Tyr(SO_3H) Met Gly Trp Met Asp	0.88	[14]
Ac Asp Tyr Met Gly	0.48	[14,15]
Tyr Ala Ala Nle Ala Ala Met Arg Asp Val Val Leu Phe Glu Lys	0.46	[4]
Asp Tyr(SO_3H) Met Gly Trp Met Asp Phe NH_2	0.27	[14]
Ac Asp Tyr(SO_3H) Met Gly NH_2	0.23	[16]
Gly Ser Ser Lys Ser Lys Pro Lys Asp Pro Ser Glu Arg Arg Arg Tyr NH_2	0.15	[18]
Ac Asp Tyr(SO_3H) Met Gly	0.12	[15,16]
Lower Phase Mobile		
Ac Asp Tyr Met Gly Trp Met Asp NH_2	0.03	[14,15]
Ac Asp Tyr Met Gly Trp NH_2	0.14	[14,15]
Ac Asp Tyr Met Gly Trp Met	0.17	[14]
Ac Asp Tyr Met Gly Trp Met Asp	0.49	[14]
PCA Pro Pro Gly Gly Ser Lys Val Ile Leu Tyr[a]	1.51	]14]
Ser Ser Ile Ile Arg	4.00	[4]

[a] [Tyr^{11}] Head activator peptide [31].

calculated from the solvent front tube or fraction × the volume of the fractions (usually 6 ml). The retained stationary phase volume is $V_s = V_c - V_m$.

$$K = V_s[(\text{peak fraction} \times \text{fraction volume}) - V_m]$$

$$K = (V_c - V_m)/(V_p - V_m)$$

For example, if the peak tube of a sample is 65 and the solvent front occurs at tube 26 in a 280-ml coil and fractions of 6 ml are collected, the K is calculated from the run as follows:

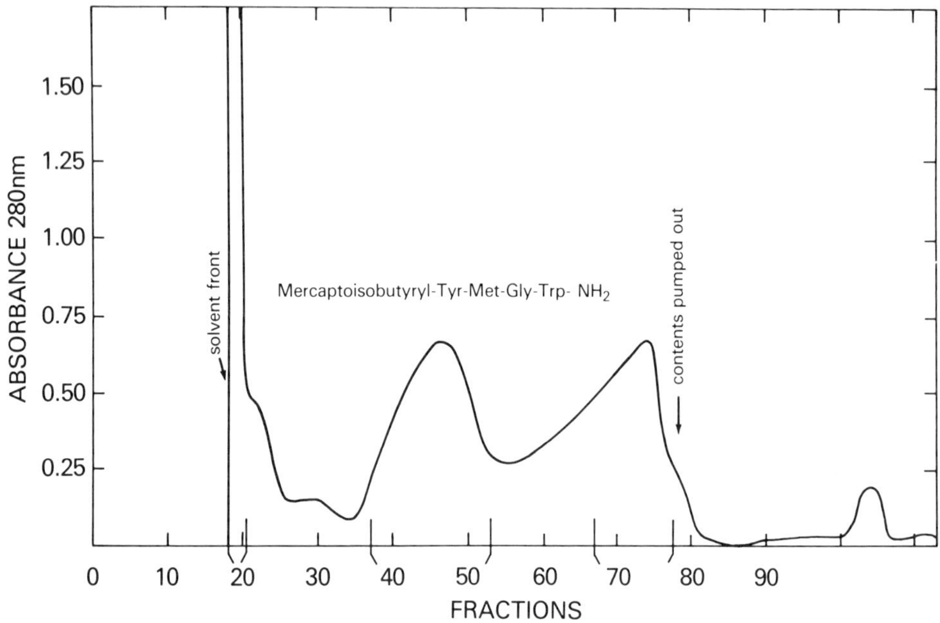

Fig. 8 Chromatography of a hydrophobic peptide, mercaptoisobutyryl Tyr Met Gly Trp NH_2, in the horizontal flow-through coil planet centrifuge in the solvent system dichloromethane/acetic acid/water (2:1:2) with the upper phase mobile. A sample of 460 mg was purified at a flow of 60 ml/hr with 8-ml fractions collected. The solvent front came out at fraction 18 and the purified peptide at fractions 38–53 giving a K of 0.63. Structure was confirmed by mass spectroscopy. Fractions 18–19 contained 45 mg, 38–53 contained 128 mg pure peptide, and fractions 67–76, 22 mg.

$$V_m = 26 \times 6 \text{ ml} = 156 \text{ ml}$$

$$V_s = 280 \text{ ml} - 156 \text{ ml} = 124 \text{ ml}$$

$$V_{65} - V_m = 390 \text{ ml} - 156 \text{ ml} = 234 \text{ ml}$$

$$K = (V_c/V_{65} - V_m) = 124 \text{ ml}/234 \text{ ml} = 0.53$$

Advantages of the Horizontal Flow-Through Coil Planet Centrifuge

This particular instrument, the horizontal flow-through coil planet centrifuge, is unfortunately not commercially available but it continues to be a reliable means of purifying peptides. It is the most versatile of the planet centrifuges at present because all solvent systems can be used at room temperature. Only adjustments in flow rate are necessary for retaining the solvents of higher viscosity. From an analysis of resolution in countercurrent chromatography carried out in coil planet centrifuges [17], it has been concluded that the horizontal flow-through instrument is better for resolving substances of low partitioning, K values less than 1, than the multilayer coil, a later designed instrument to be described. The partition coefficient, defined here as the ratio of solute in the mobile phase to the stationary phase, is simply the inverse to that described in the treatment of the theory by Conway and Ito [17]. This characteristic is useful for fractionating complex mixtures. The multilayer coil planet centrifuge has very high resolving power with compounds of partition coefficients higher than 1. The only disadvantage of the horizontal flow-through instrument is the speed of operation compared to later instruments, but after loading of the sample, the chromatography can be left unattended overnight if necessary. These instruments are still in use in certain laboratories as the first routine purification step in the preparation of peptides.

III. HIGH-SPEED COUNTERCURRENT CHROMATOGRAPHY OF PEPTIDES ON THE MULTILAYER COIL PLANET CENTRIFUGE

A. Principle of Apparatus

Further discoveries about the hydrodynamics of two-phase solvent systems lead to the development of high-speed countercurrent chromatography [18,19]. It was found that in a coaxial multilayer coil at a rapid flow rate and at a critical rotation rate, a unilateral hydrodynamic equilibrium occurs. This means that the two phases separate to the opposite ends, usually the less dense phase to the head and the heavier phase to the tail end of the coil. In order to efficiently utilize the centrifugal force gradient formed in a multilayer coil, the tubing inlet is set on the inner part of the coil and the outlet on the outside. Chromatography can be performed at a high centrifugation rate in the normal elution mode where the column is filled with the stationary head phase and the other phase is pumped from head to tail at a fast rate. The stationary phase is greatly retained (80%). The reverse elution mode is also possible whereby the coil is filled with the tail phase and the head phase is pumped in

Fig. 9 Multilayer coil planet centrifuge with a coaxial coil of PTFE tubing (upper side) mounted 10 cm from central axis of centrifuge. A counterweight is on the lower side. A metal gear (not plastic material which deforms at higher temperature) mounted on the coil holder shaft is coupled to the central axis to produce the planetary motion. Shown are electric heating pads inside the walls of the container and a temperature control unit outside. The motor and controller are also shown. The inlet and outlet tubing is introduced from the left (clamped down on the outside wall) through the central axis and into the center of the coil (upper right). This instrument was built by the Laboratory of Technical Development, National Heart, Lung, and Blood Institute.

the tail-to-head direction. Again the rapid centrifugation rate enhances the acceleration field producing a high stationary phase retention. The multilayer coil possesses high resolution and high sample capacity for compounds of partition coefficients around 1. Low partitioning compounds are retained for an inordinate amount of time. During the early experiments with n-butanol solvent systems, stationary phase retention was low due to the high viscosity of

Table 4 Partition Coefficients of Bombesin in n-Butanol Solvent Systems at 45°C

Solvent system	Volume ratio	$K(C_u/C_l)$	Ref.
n-Butanol/ dichloroacetic acid/ water	1:0:1	0.27	[21]
	1:.01:1	1.00	[21]
	1:.02:1	1.59	[21]
	1:.04:1	2.06	[21]
	1:.06:1	2.78	[21]
n-Butanol/ trifluoroacetic acid/ water	1:.01:1	1.20	[22]
n-Butanol/ acetic acid/ water	4:1:5	0.19	[21]

n-butanol which was decreased by raising the temperature. Thus performing the chromatography at 40–45°C prevented loss of the stationary phase [20].

B. Purification of Bombesin and Cholecystokinin Analogs

Following these principles of operation, a purification of bombesin was performed in a table top model of the multilayer coil planet centrifuge in a heated enclosure (Fig. 9) [21]. The coil was of 1.6-mm i.d. PTFE tubing with a total volume of 280 ml and was mounted 4 in. away from the central axis (B = 0.5–0.8, ratio of the radius of the column coil to radius of the centrifuge). To the inside walls of the cabinet were attached heated pads and the outside was insulated with urethane foam. The partition coefficient of the peptide was measured at 45°C in various n-butanol solvent mixtures. It was found that changing the percentage of dichloroacetic acid in aqueous n-butanol changed the partitioning significantly such that a partition coefficient of 1 could be attained with 1% dichloroacetic acid. The partition coefficients of bombesin with various solvent systems are given in Table 4. For high-speed conditions, the percentage of acetic acid had to be reduced in the n-butanol/acetic acid/water system for the stationary phase to be retained.

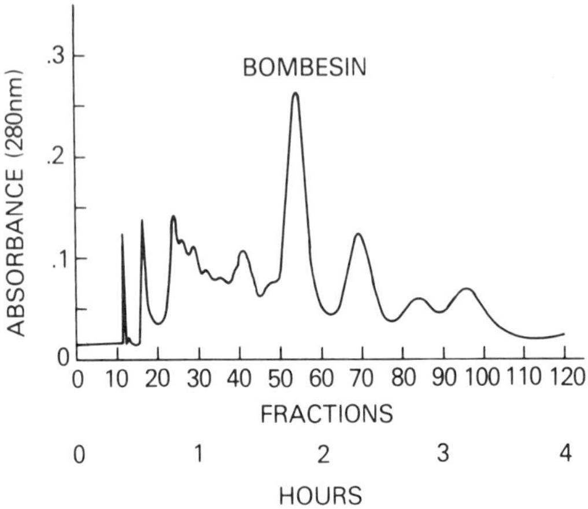

Fig. 10 High-speed chromatography of bombesin in the multilayer coil planet centrifuge. An example of 120 mg was chromatographed in 1% dichloroacetic acid and n-butanol with the lower phase mobile at 150 ml/hr at a rotation rate of 800 rpm. The major peak, fractions 53–58, contained 98 mg of the purified peptide.

Thus samples of 100–120 mg peptide were chromatographed in 1% dichloroacetic acid/n-butanol (1:1) with the lower phase mobile introduced through the inner head of the coil at a flow of 150 ml/hr. The rotation rate was 800 rpm. The effluent was passed through a detector flow cell and fractions of 5 ml were collected every 2 min. The major peak emerged in 80 min (Fig. 10). The aqueous fractions were extracted with ether to remove the dichloroacetic acid and lyophilized. The recovery of purified peptide was approximately 82% by weight and over 98% pure as determined by analytical HPLC (Figs. 11 and 12). Further experiments were done giving similar results with a totally volatile solvent system, 0.5% trifluoroacetic acid/n-butanol (1:1) in which bombesin has a partition coefficient of 1.2. In this system a final extraction step was not necessary [22].

Chromatography of Ac Asp Tyr Met Gly Trp Met Asp NH_2 on the multilayer and the horizontal flow-through coil planet centrifuges shown in Fig. 13 demonstrated a comparison of the resolution of the two instruments [23]. With the use of the lower phase as the mobile phase of ammonium acetate, pH 8.5, n-butanol (1:1) the peptide was retained and better separated from impurities in the multilayer coil planet centrifuge. However, with the upper phase mobile the peptide was well purified in the horizontal flow-through coil planet centrifuge (Fig. 14).

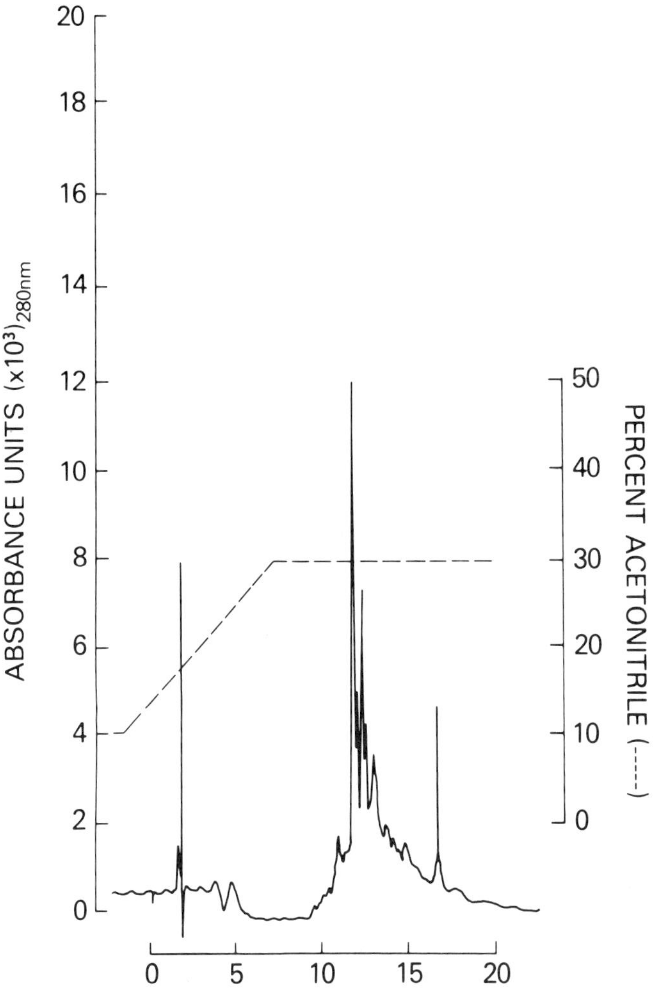

Fig. 11 Analytical HPLC of 7.5 μg of crude synthetic peptide on μ-Bondapak C_{18}(0.4 X 30 cm, Waters, Millipore) in 0.1% phosphoric acid and acetonitrile gradient as shown by dashed line. Flow rate was 2 ml/min and detection was by absorbance at 280 nm at 0.02 unit full scale.

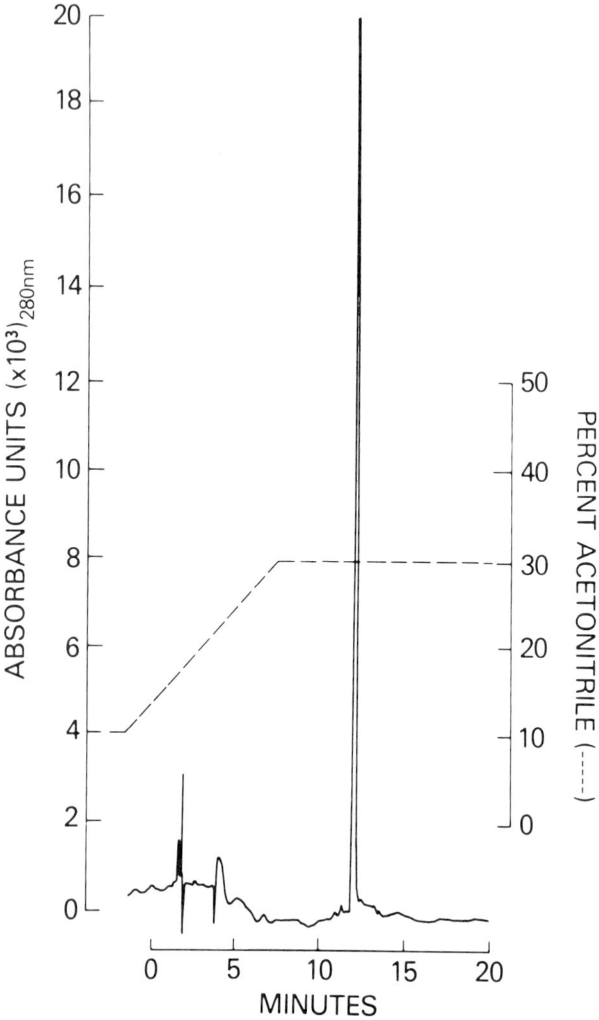

Fig. 12 HPLC analysis of 10 μg of fraction 55 of the countercurrent chromatogram of bombesin. Analytical conditions are the same as in Fig. 11.

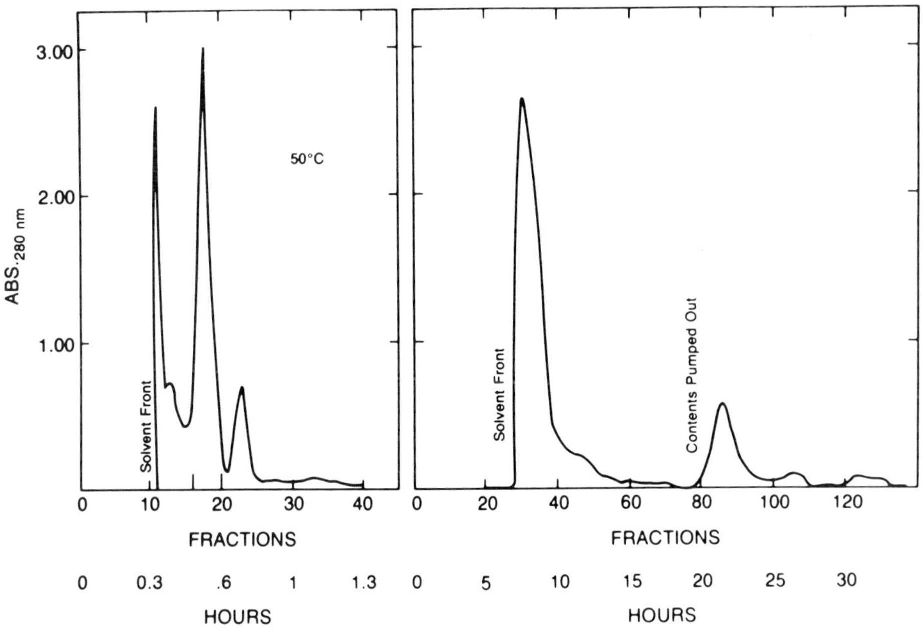

Fig. 13 Comparison of the chromatography of Ac Asp Tyr Met Gly Trp Met Asp NH_2 on the multilayer (left) and horizontal flow-through (right) coil planet centrifuges. Both runs were performed in 0.2 M ammonium acetate, pH 9.5, n-butanol to maintain solubility of the acidic peptide and with the lower phase mobile. The amount of sample loaded in the multilayer instrument was 15 mg and 10 mg pure peptide was recovered in the major peak of fractions 16–20. The flow was 150 ml/hr and rotation rate 800 rpm at 50°C. Fractions of 2 min or 6 ml were collected. The peptide emerged in less than 1 hr. On the horizontal flow-through planet centrifuge, 300 mg crude peptide was run in the usual conditions of 24 ml/hr flow rate and 400 rpm. The major peak was not as well resolved as in the other chromatography as well as when the upper phase is used as the mobile phase (Fig. 14). However, 170 mg pure peptide was recovered in fractions 33–39 and the first part of the peak (28–32 contained impure peptide. (Reprinted from Ref. 23.)

CCK-8 analogs prior to sulfation were passed on the multilayer coil planet centrifuge in 0.2 M ammonium acetate, pH 9.0, n-butanol (1:1) with the upper phase pumped from the tail to the head at a rate of 150 ml/hr. The peptide was retained and eluted after material at the solvent front. An example shown in [N-Ac,Ala^4] CCK-8

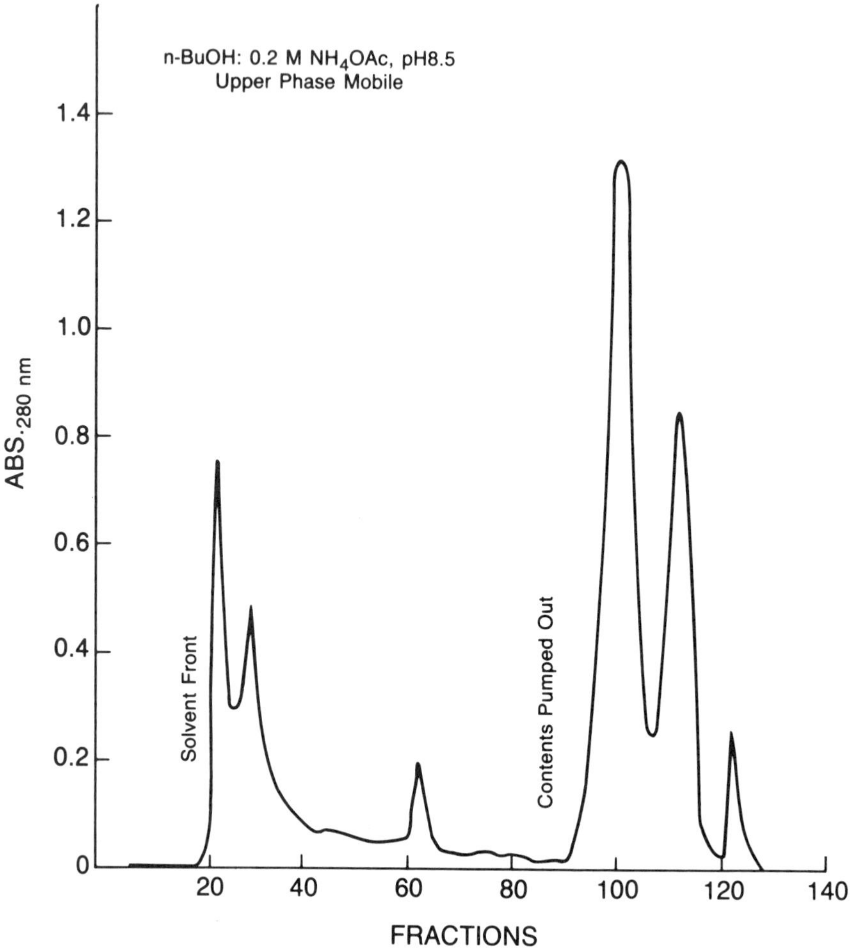

Fig. 14 Best conditions for chromatography of the CCK fragment peptide, Ac Asp Tyr Met Gly Trp Met Asp NH_2 on the horizontal flow-through instrument. A sample of 300 mg was run in the conditions given in the figure. After rotation was stopped at fraction 86, purified peptide was contained in the fractions 94–106 pumped out of the column. In these conditions pure peptide was better separated from impurities than when the lower phase was mobile in the same instrument. (Reprinted from Ref. 23.)

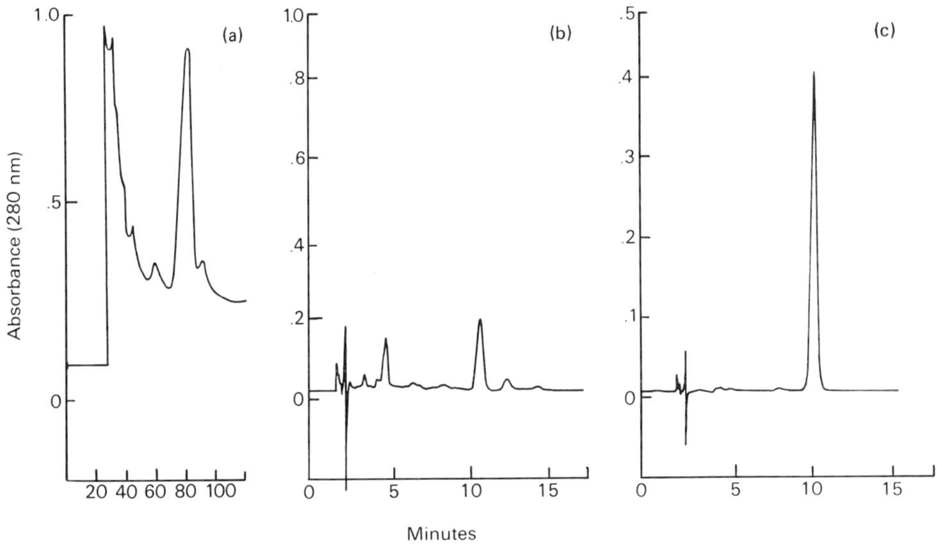

Fig. 15 (a) Purification of 100 mg of the synthetic cholecystokinin analog [N-Ac,Ala[4]] CCK-8 on the multilayer coil planet centrifuge in the n-butanol/0.2 M ammonium acetate, pH 9.0, solvent system. The stationary phase used was the lower phase and the mobile upper phase was pumped from the external tail end of the coil to the internal head end at 150 ml/hr. The rotational speed was 800 rpm. Fractions of 6 ml or 2 min/tube were collected. The solvent front emerged at tube 14, containing 22 mg of material and the purified peptide, 24 mg, in fractions 37–44, the contents of the peak at 80 min. (b) Analytical HPLC of 50 μg of the starting material on a μ-Bondapak C_{18} column (0.4 x 30 cm, Waters, Millipore) eluted isocratically in 0.1% phosphoric acid and 33% acetonitrile at a flow of 2 ml/min. Absorbance at 280 nm is shown. (c) Analytical chromatography of the purified peptide in the same conditions. (Reprinted from Refs. 22 and 24.)

(Fig. 15). The analytical HPLC of the peptide before and after countercurrent chromatography demonstrated the purification achieved. Other results of experiments are listed in Table 5.

The results demonstrate the great potential of this technology for rapid and high-capacity purification of peptides. Most of the examples of peptides chromatographed are smaller in molecular weight except for insulin. Recent work with hydrophobic small peptides on the horizontal flow-through planet centrifuge has shown successful results with the chloroform/acetic acid/dilute hydrochloric acid

Table 5 High-Speed Countercurrent Chromatography of Peptides

Compounds	Solvent system	Volume ratio	Conditions	Ref.
Dipeptides	n-Butanol/ acetic acid/ water	4:1:5	Lower phase mobile	[29]
Dipeptides	n-Butanol/ dichloro- acetic acid/ 0.1 M NH_4COOH	100:1:100--> 100:1:100→	Linear gradient Lower phase mobile	[18]
Gramicidins	Chloroform/ benzene/ methanol water	15:15:23:7	Upper phase mobile	[18]

Fig. 16 Compact horizontal flow-through (or dual-layer) coil planet centrifuge (Laboratory of Technical Development, NHLBI). The column holder (upper side) supports eight rods wrapped with two layers of tubing of 1.6 mm i.d. The coils are 8 in. wide. Other aspects of the construction are similar to the multilayer coil planet centrifuge. The heating pads are shown as well as the temperature controller (upper left). Urethane insulation is on the outside.

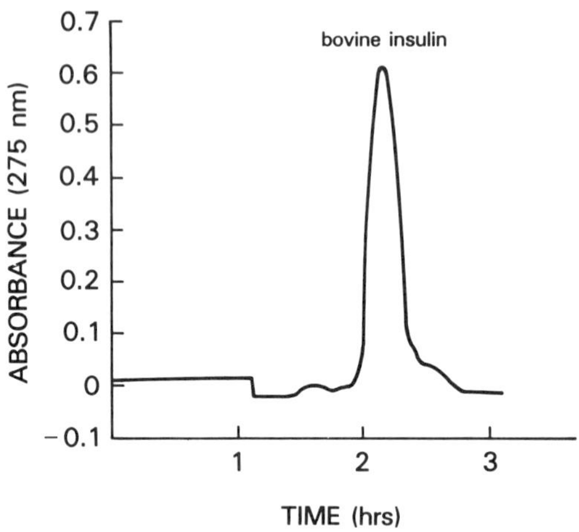

Fig 17 Countercurrent chromatogram of 100 mg bovine insulin on the compact horizontal flow-through coil planet centrifuge. Absorbance at 275 nm of the run is shown. Fractions of 6 ml were collected. The chromatography was conducted at 50°C with a flow of 50 ml/hr and a centrifugation rate of 600 rpm. The lower phase of 2% dichloroacetic acid/sec-butanol was pumped from head to tail. The solvent front emerged at 1 hr. The desamido species is separated and emerges later.

solvent system. This solvent system works well at room temperature in the multilayer coil; thus these types of peptides should be amenable to chromatography in this instrument in less time. With the coming availability of heated table top multilayer instruments from commercial sources, undoubtedly more experience with peptides, especially longer ones, will come about. Nevertheless, the multilayer coil planet centrifuge is a promising chromatograph that can serve the needs of research laboratories.

IV. COMPACT HORIZONTAL FLOW-THROUGH (OR DUAL-LAYER) COIL PLANET CENTRIFUGE

Aspects of the design features of the multilayer coil planet centrifuge enabling faster chromatography and the as yet not totally understood behavior of the solvent system affording very high resolution in a certain K range were added to the horizontal flow-through coil planet centrifuge resulting in a modified horizontal

Table 6 Summary of Operating Conditions of the Coil Planet Centrifuges for Peptide Separation

Conditions	Solvent System
Horizontal Flow-through Coil Planet Centrifuge	
Either phase mobile, room temperature, 24 ml/hr flow at 400 rpm	All butanol solvent systems Tables 1–3 Chloroform/benzene/methanol/water (15:15:23:7) [6]
Either phase mobile, room temperature, 120 ml/hr at 400 rpm	Chloroform/acetic acid/water (2:2:1) Dichloromethane/acetic acid/water (2:1:2)
Multilayer Coil Planet Centrifuge Using n-Butanol Solvent Systems	
Lower phase mobile, 40–50°C, 60–150 ml/hr at 800 rpm	0.5 to 1% trifluoroacetic acid/n-butanol (1:1) [22] 0.3 to 3% dichloroacetic acid/n-butanol (1:1) [21,22] n-butanol/acetic acid/water (9:1:10) [22]
Either phase mobile, 40–50°C, 60–150 ml/hr at 800 rpm	0.2 M ammonium acetate/n-butanol (1:1) [22,23]
Compact Horizontal Flow-through (or Dual-Layer) Coil Planet Centrifuge [22]	
At room temperature same conditions as horizontal flow-through CPC	All solvent systems
At 40–50°C flow increased to 100 ml/hr and centrifugation to 800 rpm	Same as for multilayer CPC

flow-through instrument. The tubing is wrapped around the columns in two layers, thus decreasing by half the length of the column coil or overall width of the apparatus. This allows the centrifugation to be increased and the flow rate can be faster. The instrument is heated as well for fast elution of the n-butanol solvent systems. This instrument has been constructed (Fig. 16) and chromatography of insulin and bombesin have been performed with results equivalent to that of the other machines. Insulin, 100 mg, was run at 50°C on the compact instrument in 2% dichloroacetic acid/sec-butanol (1:1) with the lower phase mobile at 50 ml/hr. The results are shown in Fig. 17 [22]. Since the diameter of the tubing used was 1.6 mm, the equivalent flow rate for 2.6-mm i.d. tubing would be 150 ml/hr. Thus a high flow rate is possible with this machine. The machine has the potential of being usable with all solvents at room temperature like the horizontal flow-through coil planet centrifuge and when heated for rapid chromatography.

V. CONCLUSION

As more is understood about the hydrodynamic process occurring in the countercurrent coil planet centrifuges, newer instruments are being designed by Yoichiro Ito. It is likely that an even better instrument for the various solvents suitable for the different kinds of peptides will soon be built. Nevertheless, the present instruments have served in the preparative purification of synthetic peptides rather well. Table 6 lists operating conditions for the different machines. From the experience to date particular conditions have been worked out for the different instruments. Most of the peptides studied with this technology are smaller peptides. Some of the hydrophobic ones can be chromatographed in chloroform or similar solvent systems and the rest are more suitably chromatographed in butanol systems.

The potential of this method has only begun to be realized. One important application would be in scaling up to process quantities. At present amounts of up to a few grams can be chromatographed on instruments of 2.6-mm i.d. tubing. Experiments are underway to increase the capacity. An impetus for this application would be in-line flow purification or extraction of microbial products from large-volume fermentation reactions such as genetically engineered hormones. Also, this technique is amenable to the isolation of biologically active peptides from tissues such as brain. Most neuropeptides are relatively hydrophobic. Therefore this technique can isolate and fractionate these substances or their metabolites into organic phases from which it would be easy to characterize and purify further.

ACKNOWLEDGMENT

Part of the results presented in this chapter was supported by a Small Business Innovation Research Grant, NS 22319, from the National Institutes of Health to Peptide Technologies Corporation.

REFERENCES

1. L. C. Craig, *J. Biol. Chem.*, *55*, 519 (1944).
2. Y. Ito and R. L. Bowman, *J. Chromatogr. Sci.*, *11*, 289 (1973).
3. Y. Ito and R. L. Bowman, *J. Chromatogr.*, *147*, 221 (1978).
4. Y. Ito and G. J. Putterman, *J. Chromatogr.*, *193*, 37 (1980).
5. Y. Ito, R. E. Hurst, R. L. Bowman, and E. A. Achter, *Sep. Purif. Met.*, *3*, 133 (1974).
6. Y. Ito, *Meth. Enzymol.*, *91* (I) p. 335; C. H. W. Hirs and S. N. Timasheff (eds.), Academic Press, New York, 1983.
7. I. A. Sutherland, J. S. Lee, and D. J. Gavreau, *Anal. Biochem.*, *89*, 213 (1978).
8. G. Johansson, R. Gysin, and S. D. Flanagan, *J. Biol. Chem.*, *256*, 9126 (1981).
9. B. Olde and G. Johansson, *Neuroscience*, *15*, 1247 (1985).
10. G. J. Putterman, F. Perini, E. L. White, and Y. Ito, *Peptides Structure and Biological Function* (E. Gross and J. Meienhofer, eds.), Pierce Chem. Co., Rockland, Ill., 1979, p. 113.
11. Y. Ito, *J. Chromatogr.*, *188*, 33 (1980).
12. J. M. Stewart and J. D. Young, *Solid-Phase Peptide Synthesis*, Pierce Chem. Co., Rockland, Ill. 1985.
13. M. Knight, T. R. Burke, and J. Ferkany, unpublished results.
14. M. Knight, A. M. Kask, and C. A. Tamminga, *J. Liq. Chromatogr.*, *7*, 351 (1984).
15. M. Knight, A. M. Kask, and C. A. Tamminga, *Peptides: Structure and Function* (V. J. Hruby and D. Rich, eds.), Pierce Chem. Co., Rockland, Ill., 1984, p. 75.
16. M. Knight and T. R. Burke, unpublished results.
17. W. D. Conway and Y. Ito, *J. Liq. Chromatogr.* *8*, 2195 (1985).
18. Y. Ito, J. Sandlin, and W. G. Bowers, *J. Chromatogr.*, *244*, 247 (1982).
19. Y. Ito, *J. Chromatogr.* *301*, 377 (1984).
20. Y. Ito and W. D. Conway, *J. Chromatogr* , *301*, 405 (1984).
21. M. Knight, Y. Ito, P. Peters, and C. diBello, *J. Liq. Chromatogr.* *8*, 2281 (1985).
22. M. Knight, Y. Ito, J. L. Sandlin, and A. M. Kask, *J. Liq. Chromatogr.*, *9*, 791 (1986).
23. M. Knight, Y. Ito, A. M. Kask, C. A. Tamminga, and T. N. Chase, *J. Liq. Chromatogr.*, *7*, 2525 (1984).

24. M. Knight, *LC, Liq. Chromatogr. HPLC Mag.*, *3*, 1062 (1985).
25. M. Knight, Y. Ito, J. D. Gardner, C. A. Tamminga, and T. N. Chase, *J. Chromatogr.* *301*, 277 (1984).
26. M. Knight, Y. Ito, and T. N. Chase, *J. Chromatogr.* *212*, 356 (1981).
27. M. Knight and W. A. Klee, *J. Biol. Chem.*, *254*, 10426 (1979).
28. M. Knight, T. R. Burke, and R. Felsted, unpublished results.
29. Y. Ito and J. Sandlin, unpublished results.
30. N. B. Mandava and Y. Ito, *J. Liq. Chromatogr.*, *7*, 303 (1984).
31. H. Bodenmuller and H. Chica Schaller, *Nature*, *293*, 379 (1981).

chapter eleven

Separation of Blood Cells, Particles, Organelles, Biopolymers, and Biomembranes by Countercurrent Chromatography

Ian A. Sutherland
National Institute for Medical Research
London, United Kingdom

I. INTRODUCTION

The application of countercurrent chromatography (CCC) to the separation of viable biological material depends on the availability of suitable aqueous phase systems. These must be able to act as a gentle host medium for the cells and subcellular substances they contain and have the appropriate hydrodynamic properties for CCC operation.

Albertsson first introduced the concept of partition between immiscible aqueous polymer phases for the separation of cells in the mid-1950s [1]. But it was at least another 20 years before the technique became widely used and accepted.

Countercurrent chromatography and its application to the separation of natural products using aqueous/organic phase systems has been developing in parallel. It is only recently that attempts have been made to merge the two technologies with the ultimate aim of producing a high-resolution chromatographic method for biological material.

This chapter describes the fundamental principles of aqueous phase partition (Sec. II); the established countercurrent distribution (CCD) methods of achieving high-resolution separations (Sec. III); the current state of development and application of countercurrent chromatography using aqueous phase partition (Sec. IV); and the direction of future research (Sec. V).

Readers who are familiar with aqueous phase partition can omit Sec. II and III, but those who have experienced difficulties in using polymer phase systems should read these sections as they outline some of the common pitfalls.

II. AQUEOUS PHASE SYSTEMS

There are a number of reviews on the use of two-phase aqueous systems for the separation of cells organelles and particles [2–5]. Understanding the way aqueous phases can be manipulated to achieve separation is an important prerequisite to applying the technology to countercurrent chromatography. The following is meant to act as a brief outline rather than a comprehensive guide. The reader requiring more exacting information should refer to the reviews.

A. Phase Diagram

When two water-soluble polymers are mixed, they separate into discrete aqueous phases, provided their respective concentrations are above critical levels. These aqueous phase systems have an extremely low interfacial tension so that delicate organelles or membranes at the interface are not subjected to disruptive stress. Low molecular weight substances, such as sucrose, sorbitol, and salts, distribute evenly between the phases to give a gentle host medium for a whole range of viable biological materials.

A typical phase diagram is shown in Fig. 1. The thick line is called the binodial and divides the single-phase solution at the lower concentrations (shaded) from the two-phase solutions at the higher concentrations. Temperature has a significant effect on the position of the binodial. Lowering the temperature from, say, 20°C to 4°C would move the binodial to the position of the dotted line. A phase system at point Q would therefore be a single phase at 20°C and two phases at 4°C, while a phase system at the point P would always be two phases.

At room temperature the concentrations of the phase system at point P are determined by U and L for upper and lower phase, respectively, and the volumes ratio is determined from the geometry of the tie line (UL):

$$\frac{\text{Volume of upper phase}}{\text{Volume of lower phase}} = \frac{PL}{UP} \qquad (1)$$

The length of the tie line is related to the interfacial tension between the phases [2] so that a phase system made up near the "critical point" (C) will have a very short tie line and hence a very

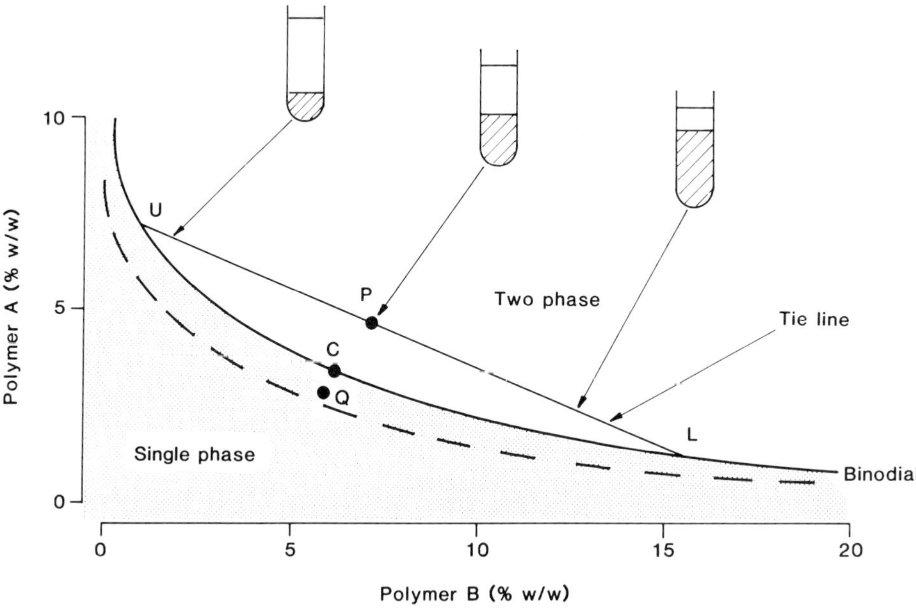

Fig. 1 Typical phase diagram for a polymer phase system (schematic).

low interfacial tension. Phase systems which operate near the critical point (C) are the most difficult to work with because temperature and polymer concentrations have to be closely controlled to maintain the correct position relative to the critical point.

While commercially available polymers are identified by their mean molecular weight, they are also given a batch number. Variability in molecular weight distribution of different batches can affect the operating point relative to the binodial. It is therefore strongly recommended to purchase relatively large batches of polymers and make stock solutions from the same batch if a large number of separations are anticipated. Alternatively, if stock solutions are made up from a new batch, adjust the polymer concentrations to compensate for any movement of the binodial.

B. Preparation of Phase Systems

The most commonly used polymer phase systems are dextran and polyethylene glycol (PEG). PEG stock solutions can be made up to concentrations of 40% (w/w). Sucrose, salts, and buffer solution should be made up 5–20 times stronger than required. The dextran solution requires special treatment as it can contain up to 10% water when in powder form. For a 20% (w/w) stock solution of dextran, layer 220 g on 780 g of water, heat indirectly via a water bath, stirring while hot, until all the dextran is dissolved. Dilute 5 g of the solution to 25 ml with water and measure the optical rotation. The concentration of the dextran (C_d) will be

$$C_d \text{ (w/w)} = \frac{\alpha^{25}}{W \times 199 \times L} \times 100$$

where α^{25} = optical rotation (degrees)

L = pathlength (dm)

W = weight of stock solution (g)

When the concentration (C_d) has been calculated, adjust the solution to 20% by adding water. Once stock solutions are made up, a point on the binodial is checked by mixing 1 g of PEG stock solution with 2 g of dextran stock solution in a measuring cylinder and shaking. A milky emulsion is formed. Add known increments of water, shake, and continue until the solution is clear. The phase system composition at the binodial is calculated from the amount of water added.

C. An Introduction to Phase Partition

The partitioning process involves preparing a phase system with a known amount of sample in a test tube or measuring cylinder, mixing the contents, and allowing them to settle out into two distinct phases again. The sample constituents will partition into the upper phase, interface, or lower phase.

Low molecular weight compounds and macromolecules will generally be soluble in the phase systems and their differential solubility can be described by their partition coefficient (K):

$$K = C_u/C_l \tag{3}$$

where C_u and C_l are the concentrations in the upper and lower phases, respectively.

Particulate material such as cells, organelles, and membrane vesicles will either be in suspension in the two phases or adsorbed at the interface between them. With the majority of phase systems, partition is between the upper phase and the interface. It is therefore convenient to consider the interface and lower phase as one and define the partition behaviour using the partition ratio (G):

$$G = \frac{A_u}{A_t - A_u} \tag{4}$$

where A_u is the amount of material in the upper phase and A_t the total amount of material added to the phase system. A typical 4-g phase system will take about 20 min to settle. Studies by Fisher [6] and Heywood-Waddington [7] have shown that the partition of cells and macromolecules is not necessarily an equilibrium process and that the partition ratio G can change with time. Other factors, therefore, such as the shape of the container, could become important as they may effect the settling time in the test tube.

While the procedure selected may affect the quantitative partition behavior, qualitative differences in partition will not be affected. Methods of improving resolution and selectivity will be dealt with in the next section.

D. Factors Affecting Phase Partition

Choice of Polymer

The choices of polymer used for a particular application depends to some extent on both scale and type of application. For example, the polymer pair most often quoted in the literature is PEG/dextran. This combination gives a reasonably fast settling time in unit gravity and can be applied to a wide range of biological material. However, for large-scale usage, dextran is expensive and other phase systems such as aquaphase PPT [8] are now becoming available as dextran substitutes. High concentrations of PEG can sometimes produce aggregation in separations of subcellular components [7]. Additives such as BSA and EDTA can be helpful; otherwise phase systems such as ficoll/dextran [9–11] have to be considered, but with their long settling times these are more difficult to use.

Countercurrent chromatography ideally requires phase systems with low viscosity, high density difference, and high interfacial tension. Such properties are not yet available for partitioning biological material and so apparatus development has so far been constrained to operate with the currently available phase systems. The remaining factors affecting phase partition in this section will concentrate on the dextran/PEG systems as they are currently the most frequently used.

Molecular Weight

Increasing the molecular weight of one polymer will force the sample to partition more toward the other. For example, in a phase system where the upper phase is PEG and the lower phase is dextran, the partition ratio increases with increasing molecular weight of dextran but decreases with increasing molecular weight of PEG.

Polymer Concentration

At polymer concentrations close to the critical points, particles partition mainly into the upper phase and have a high partition ratio (G). As the concentration of the polymers increases, the partition ratio decreases with particles partitioning more toward the interface and the lower phase. A simple selection strategy would be to make up a series of phase systems of increasing polymer concentration. For example in a PEG(% w/w)/dextran(% w/w) system a 4:4, 5:5, 6:6, and 7:7 should be sufficient to change the partition from the upper phase to the interface.

Salt Concentration

The partition ratio of a particular particle can be controlled by changing the salt concentration of the phase system. When no salt is present materials normally partition toward the interface. Adding salts such as potassium phosphate will create an ionic potential difference between the phases, resulting in a positively charged upper phase. Material with a net negative charge will therefore be redistributed into the upper phase in the presence of potassium phosphate buffer. Addition of potassium chloride will have the opposite effect [12] and bring material back down to the interface again. There are a number of salts which can be added to the phase system to manipulate the partition in this way [13]. A typical strategy demonstrating the effect of phosphate and chloride is given by Larsson [12].

Use of Charged Polymers

Using charged ligands covalently bound to PEG can have a much stronger effect on the partition ratio than salt. Positively charged PEG, trimethylamino-PEG, will repel positively charged material toward the lower phase, while negatively charged PEG, PEG-sulfonate, will have the opposite effect [14].

Hydrophobic Partition

Hydrophobic effects can be achieved by using polymers carrying uncharged ligands. For example, the addition of PEG-palmitate in quite low concentrations to a phase system can transfer cells or subcellular material to the upper phase on the basis of hydrophobic affinity [15,16].

A number of other fatty acids have been esterified with PEG and their effect on the partition of cells and organelles reported [17].

Affinity Partition

Attaching a ligand to one of the polymers which has a specific affinity for the component to be isolated can be a particularly powerful method of separation. For example, a ligand attached to PEG could partition the component into the upper phase while the bulk material remains at the interface. The effectiveness of an affinity ligand is characterized by the increase in logarithmic partition coefficient, $\Delta \log K$, of a solute ($\Delta \log G$ for a particle) caused by introducing the ligand polymer into the system [18].

$$\Delta \log K = \log K \text{ (with ligand)} - \log K \text{ (without ligand)} \tag{5}$$

Ideally a high $\Delta \log K$ would occur with effective purification achieved in one or two simple extractions. This is not always the case and frequently multistep procedures such as countercurrent distribution (Sec. III) or countercurrent chromatography (Sec. IV) are required.

III. COUNTERCURRENT DISTRIBUTION

A. Introduction

Countercurrent distribution (CCD) is a form of liquid-liquid extraction which can be used when the differences in partition are small and separation or enrichment cannot be achieved in a single step. The process can also be analytical, whereby cell–cell or protein–protein interactions can be measured.

The theoretical aspects of CCD will be described first (Sec. B) followed by a description of the different types of countercurrent distribution apparatus which operate at unit gravity (Sec. C) and at enhanced gravity (Sec. D). Finally (Sec. E) there will be a brief review of polymer phase system applications, highlighting the various historical milestones.

B. Theoretical Distribution

A simple example of CCD is illustrated in Fig. 2 for a sample containing three components of differing partition ratio (G). In the first step the sample is added to a suitable two-phase system. The

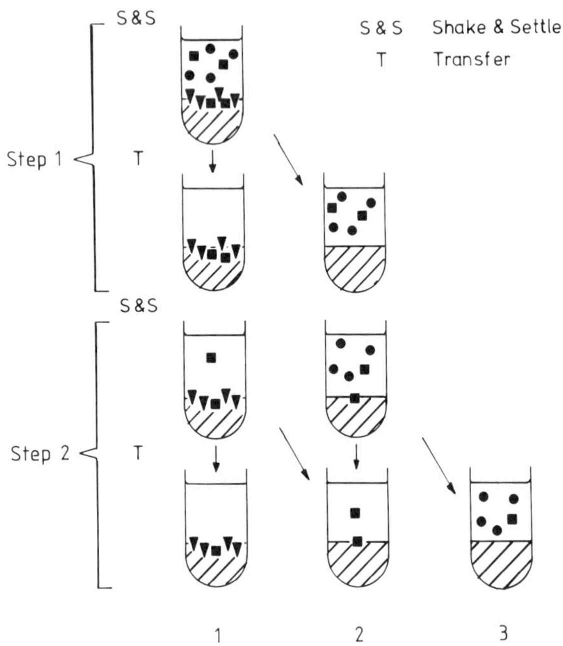

Fig. 2 A simple example of countercurrent distribution.

tube is then shaken and allowed to settle (S & S). After settling it can already be seen that the high K material (circles) favor the upper phase; the low K material (triangles) favors the interface and the K = 1 material (squares) distributes evenly between the two. The next step is the transfer stage (T) which involves transferring the top phase of tube 1 to fresh bottom phase in tube 2, and adding a new top phase to tube 1. The shake-and-settle step is then repeated and after a second transfer it can be seen that the high and low K material is completely separated and the K = 1 material is predominantly in the center tube. The distribution of material throughout the transfer train (Fig. 3) can be predicted mathematically [19] and the resolution will depend on the number of transfers.

C. Unit Gravity Apparatus

The successive dispensing and extracting involved in CCD is very laborious and automated methods have been designed to simplify the process. The original glass CCD devices developed by Craig and

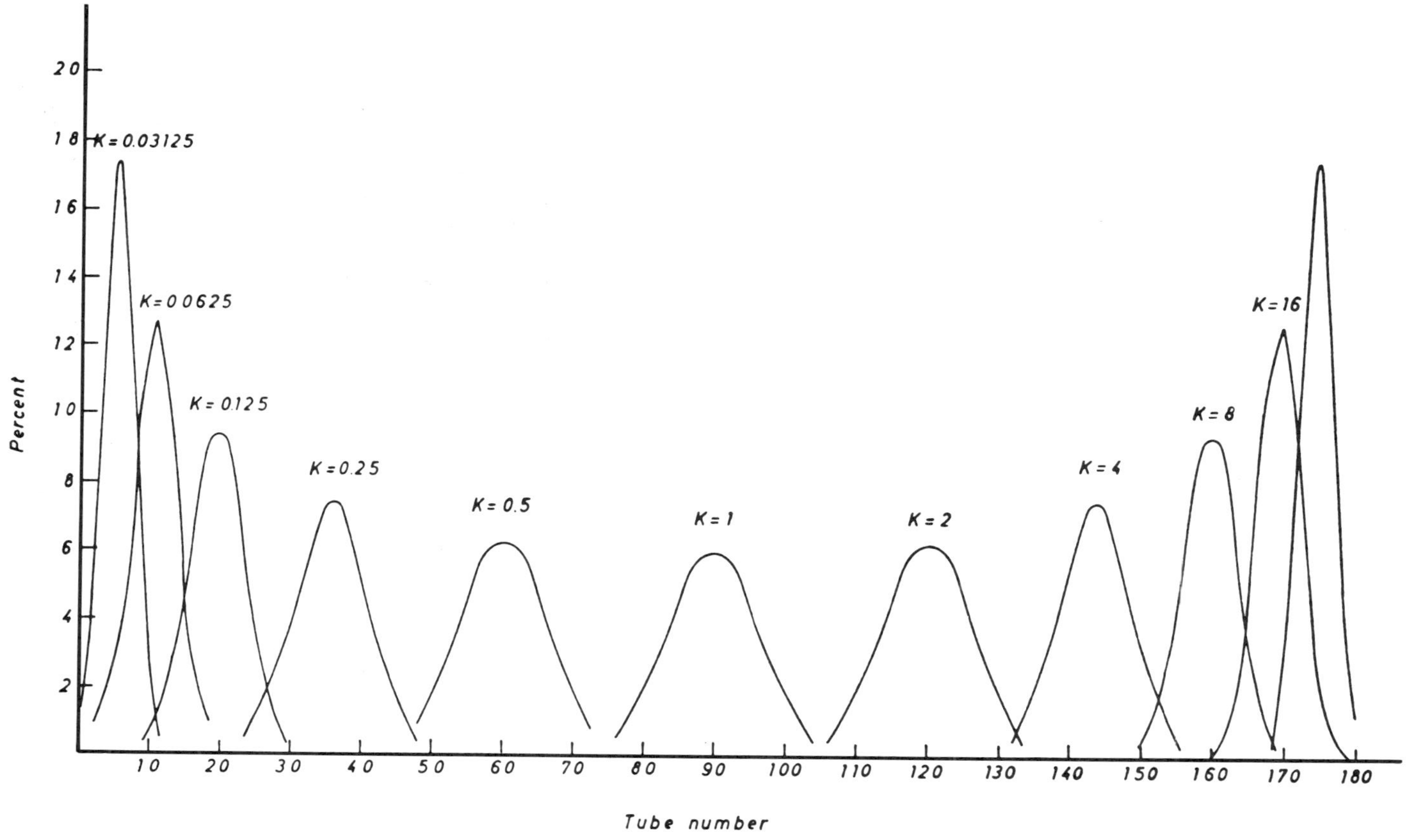

Fig. 3 Theoretical countercurrent distribution profiles for a range of partition coefficients [2].

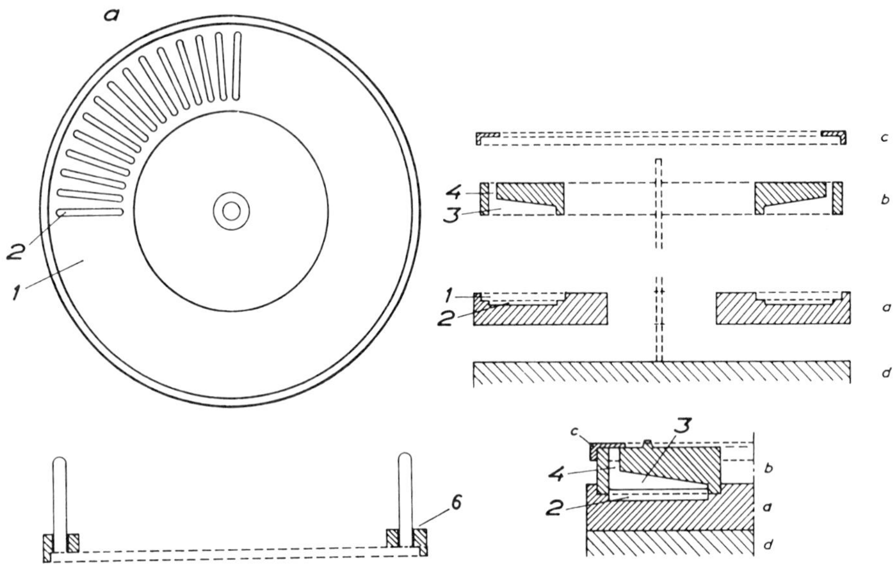

Fig. 4 The Albertsson thin-layer CCD apparatus [2].

Craig [20] for aqueous/organic phase systems are not suitable for the more viscous aqueous polymer phases due to their long settling times. Albertsson [21] therefore developed the thin-layer countercurrent distribution apparatus shown in Fig. 4. This consists essentially of two concentric Plexiglas disks or plates, the surfaces of which (1) are very accurately machined to avoid leakage. The lower plate (a) contains up to 120 radial cavities (2) arranged circumferentially in an annulus around the plate. The depth of these cavities is only 2 mm. The upper plate (b) has a similar geometry but the cavities are shaped (3) so that the plates can be inverted and the contents of the cavities easily drained into test tubes (4 and 6).

When the plates (a and b) are assembled with the lid (c), they are mounted onto a vortex-shaking table (d). The assembled section is shown in the bottom right-hand corner of Fig. 4 indicating the level of the liquids added to each chamber and the transfer plane (dotted).

Operation consists of filling all cavities except the first with the required quantity of each phase. The sample is made up with the stock solutions, in such a way that the phase system concentration is preserved, and added to the first cavity. Once assembled on the shaker, the process of mixing, settling, and upper phase transfer is automatically regulated by the control unit—tranfer being achieved by indexing the top plate (b) one cavity relative to the bottom plate (a). After completing the desired number of transfers, an annular fraction collector (e) containing test tubes can be attached to the top plate (b) and the whole assembly inverted to allow the chamber contents to flow into the test tubes.

The circular arrangement of the cavities makes the process extremely versatile. In a 100-transfer apparatus, for example, one sample can undergo up to 100 transfers or alternatively four samples can be run simultaneously for up to 25 transfers. The process can also be recycled with the top plate (b) indexing more than just one resolution.

An alternative thin-layer CCD apparatus (the Bioshef, Department of Biochemistry, University of Sheffield), based on the Albertsson design, recently became available. A powerful hydraulic transfer mechanism allows the rotor halves to be clamped together firmly. This combined with an improved rotor design which allows the contact surfaces to be lapped and polished helps prevent leakage between chambers. The whole apparatus is housed in a refrigerated cabinet and can be operated with one or two sets of 60-chamber rotors for 59 automatically controlled partition steps (Fig. 5).

D. Enhanced Gravity Apparatus

Pritchard et al. [22] describe a miniature version of the original stainless steel CCD machine developed by Craig and Post [23]. The device, illustrated in Fig. 6, consists of an upper and lower section each with 18 circumferentially arranged chambers. Each chamber is filled with the appropriate amount of phase system (top left of figure) and has a small ball added. The sample is added to chamber 1 and the sections are then assembled together with their appropriate seals. The principle of operation is quite simple. First the whole device is inverted a number of times (like a cocktail shaker) to allow the balls to thoroughly mix the phases. The whole device is then centrifuged in a swinging bucket rotor for 5 min at 300g. After these mixing and centrifugation steps, the upper section is indexed anticlockwise by one chamber and the whole process repeated 17 times. This method offers a cheap and simple way of performing a few countercurrent distribution steps but it is labor-intensive, with each run taking about 2 hr under continuous manual control.

Akerlund [24] developed a fully automated enhanced gravity version of Albertsson's thin-layer CCD apparatus. The major design difference between the machines is that the Akerlund version rotates at speeds of up to 3000 revolutions/min to separate the phases centrifugally. This significantly reduces the cycle time. The transfer plane is vertical as opposed to horizontal (Fig. 7), requiring the transfer step to take place while the rotor is still rotating. Mixing

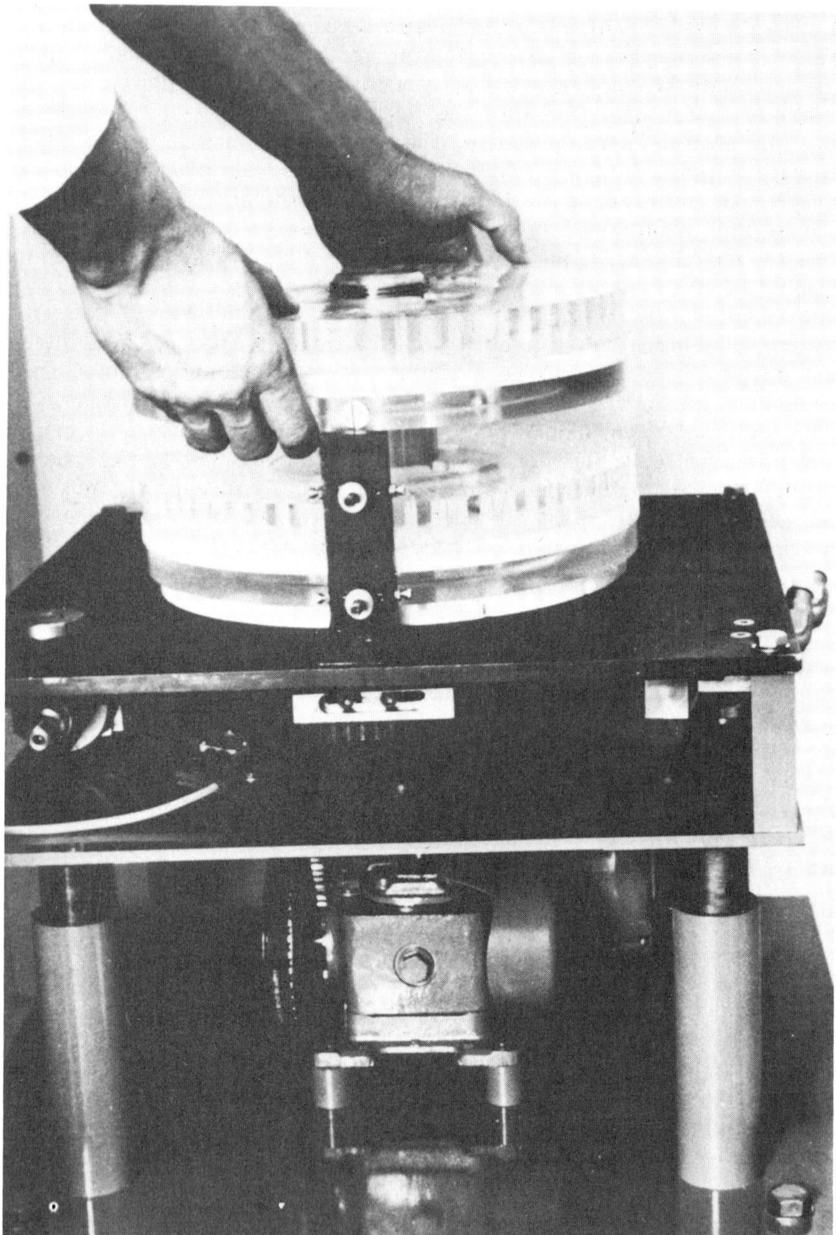

Fig. 5 Assembly of the second rotor to the Bioshef CCD apparatus.

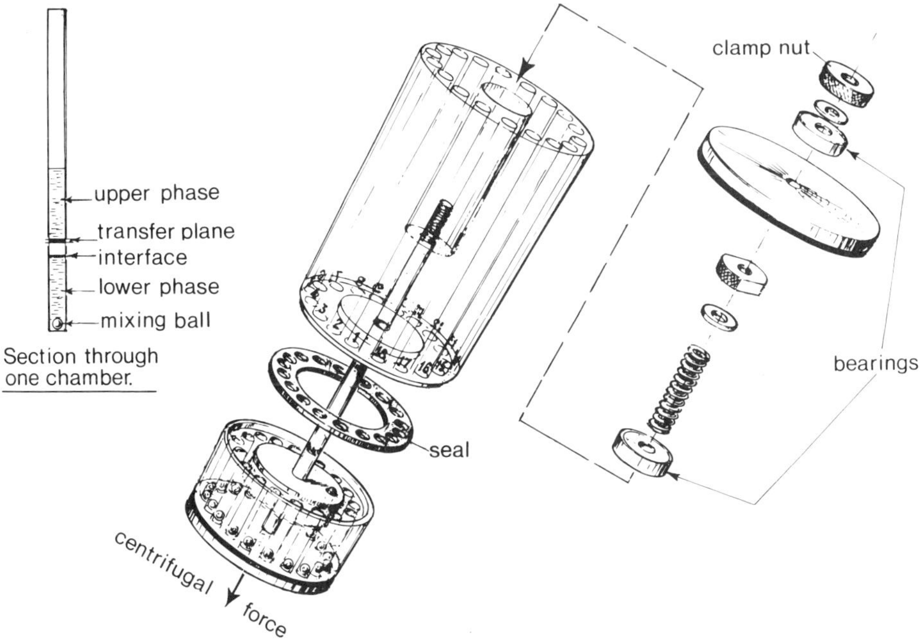

Fig. 6 The Pritchard "gun barrel" CCD device.

(Fig. 7a) is achieved by stopping or slowing down the rotor and using the conventional Albertsson vortex-mixing procedure. The advantage of Akerlund's system is that it is quick (10 sec mixing, 90 sec centrifugation including acceleration and deceleration), giving a total cycle time of 2 min. Also, the use of enhanced gravity allows separations with viscous phase systems, such as ficoll/dextran, that would not readily settle out in unit gravity. The disadvantage is that the engineering is complicated and the plates have to be extremely accurately machined to avoid leaks between the chambers when the apparatus is spinning.

D. Applications Using Countercurrent Distribution

The first use of countercurrent distribution (CCD) with aqueous phase systems was by Albertsson and Nyu [25] for the separation of proteins. This was followed by fractionations of bacteria [26] and other cells [27]. More applications followed; the countercurrent distribution of polio virus by Bergtsson et al. [28]; the relationship between size and distribution of DNA molecules by Frick and Lif [29];

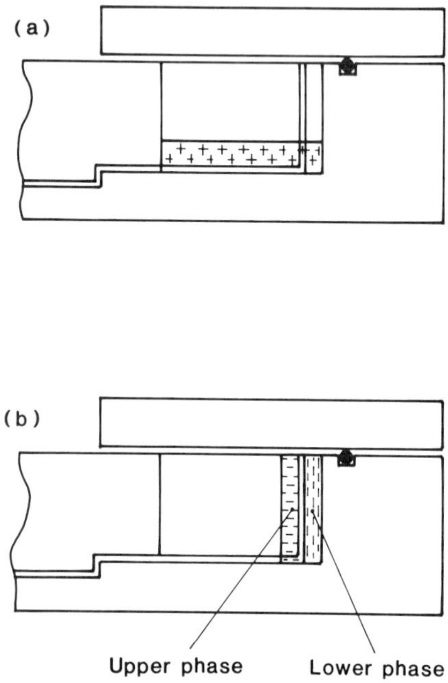

Fig. 7 Detail of the Akerlund enhanced gravity CCD apparatus showing (a) the mixing mode (unit gravity) and (b) the settling and transfer mode (enhanced gravity).

the first subcellular fraction of chloroplasts using CCD by Albertsson and Baltescheffsky [30]; and a separation of red blood cells on age differences by Walter [31–33].

Following the development of Albertsson's thin-layer countercurrent distribution apparatus in 1965 [21], there were numerous applications involving protein, bacteria, subcellular material, and cells. Some examples worthy of note were the separation of T- and B-lymphocytes by Walter and Nagaya [34]; the subfractionation of Golgi apparatus by Hino et al. [35]; and the characterization of subpopulations of cells by Malmström, Nelson, and others [36–39].

Countercurrent distribution was also used for studying cell–cell affinity by Walter [40] and protein–protein interactions by Backman [41–43].

More recently the countercurrent distribution principle has been used for direct chiral resolution of D- and L-tryptophan using proteins by Eckberg [44] and has been combined with affinity partitioning for enzyme purification by Johansson [45].

Most of these applications have used the Albertsson thin-film countercurrent distribution apparatus [21]. Notable exceptions have been the use of the Pritchard device for the fractionation of mucopolysaccharides [22] and rat liver homogenate [46,47], and the Akerlund enhanced gravity CCD for the fractionation of organelles [24] and membrane-based opiate receptors [48].

IV. COUNTERCURRENT CHROMATOGRAPHY

A. Introduction

Countercurrent chromatography (CCC) can be considered a continuous form of countercurrent distribution or alternatively, as liquid–liquid chromatography without a solid support.

The two essential requirements for successful CCC using aqueous polymer phase systems are the retention of one phase relative to the flow of the other and efficient mixing to facilitate mass transfer between the phases. Retention is achieved by passing the phases through helically wound tubes of PTFE (Fig. 8) which are held in an enhanced gravity field, the heavier phase being retained in the outer segments of each coil, which is equivalent to one stage of a CCD system. Mixing is achieved either by rotating the coils relative to the acceleration vector or by pumping one phase through the retained segments of the other.

The majority of countercurrent chromatography apparati, described in Chap. 3, cannot be used for the separation of viable biological material due to the high-viscosity, low-density difference and extremely low interfacial tension of the polymer phase systems. This is mainly due to the synchronous motion of most coil planet centrifuges which produces such high mixing rates that emulsification occurs.

One solution is to reduce the mixing rates to a level that can be tolerated by the phase systems. This can be done by using any of the following mixing schemes:

1. The coil is fixed relative to the centripetal acceleration vector (Fig. 9a). Segments of the nonpumped phase are retained in one half of the coil, while mixing between the phases is dictated by the flow rate of the pumped phase. This is the simplest configuration and forms the basis of the toroidal coil planet centrifuge described in Sec. C.
2. The coil is fixed relative to the mean acceleration vector but the acceleration field fluctuates about the mean in both amplitude and direction (Fig. 9b). The pump controlled mixing of scheme 1 is used, with additional mixing induced by the fluctuating acceleration field produced by the epicyclic coil planet centrifuge (Sec. D).

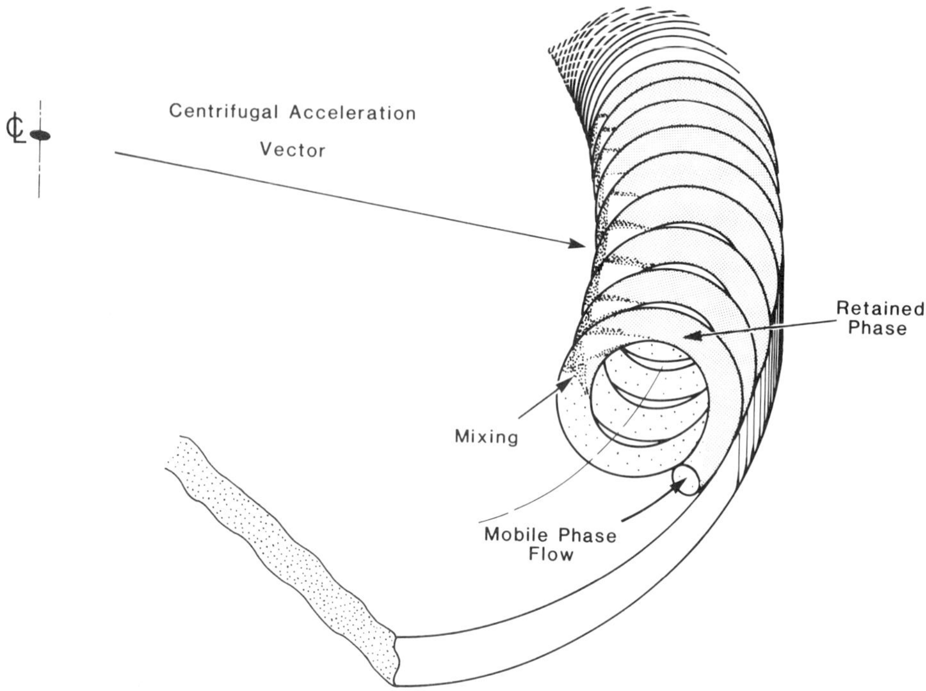

Fig. 8 Schematic diagram of the toroidal coil rotor.

3. The coil rotates slowly relative to a fixed acceleration field (Fig. 9c), inducing rotational and pumped mixing on one side of the coil and rotational mixing on the other. Coil rotation is produced by the nonsynchronous coil planet centrifuge (Sec. E).

Another solution is to develop new phase systems with physical properties more suited to countercurrent chromatography (Sec. V.C).

B. Phase Distribution Diagram

Countercurrent chromatography can only be successfully achieved if one phase can be retained in the coils relative to the other. Phase distribution diagrams relating retention of the stationary phase to the critical operating parameters, rotor speed and flow, are essential (Fig. 10). Retention is obtained by first filling the coils with the proposed stationary phase with the rotor at rest and then running the rotor at a set speed and flow rate, while collecting the

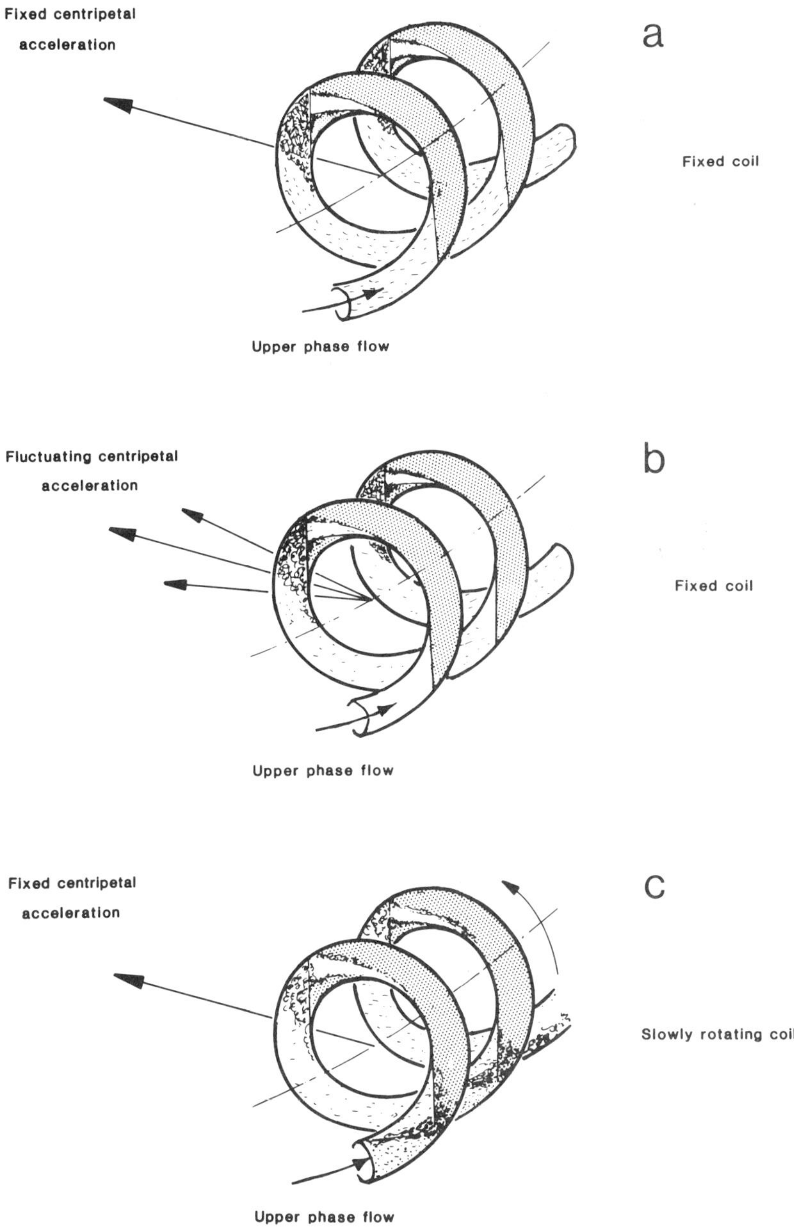

Fig. 9 The three mixing schemes used in countercurrent chromatography with polymer phase systems.

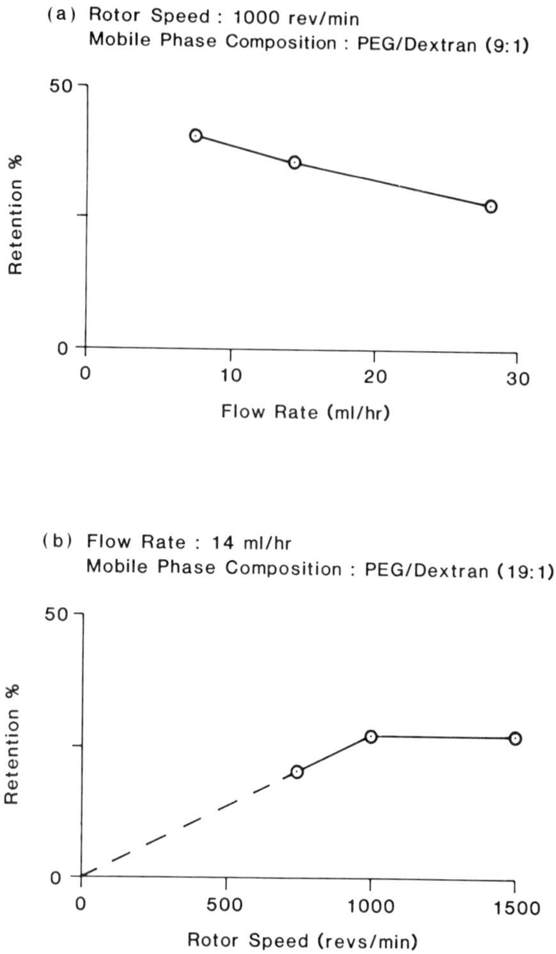

Fig. 10 Phase distribution diagrams for dextran (3.3% w/w / PEG (5.4% w/w) showing the (a) rotor speed and (b) flow.

Fig. 11 Toroidal coil rotor mounted in a Beckman J6 centrifuge.

eluent. If the volumes of the coil and inlet/outlet leads are known, it is possible to calculate the retention from the volume of stationary phase collected.

C. Toroidal Coil Centrifuge

The toroidal coil centrifuge (Sec. V in Chap. 3) uses mixing scheme 1 illustrated in Fig. 9a. It is ideal for fractionating samples with particle sizes less than 1 μm in diameter, but sedimentation occurs if the particles are larger. This apparatus can be used with phase systems that are particularly sensitive to emulsification, as it uses the most stable of the three mixing schemes.

A toroidal coil rotor is shown mounted in a Beckman J6 centrifuge in Fig. 11. The rotors generally have a system which avoids the use of rotating seals. This helps to give recoveries in the order of 70–100% [49]. However, the rotor is quite large (50 cm in diameter) and the planetary system required to avoid seals (Chap. 3) is quite complicated. Floor mounting with refrigeration is therefore required for optimum operation [47].

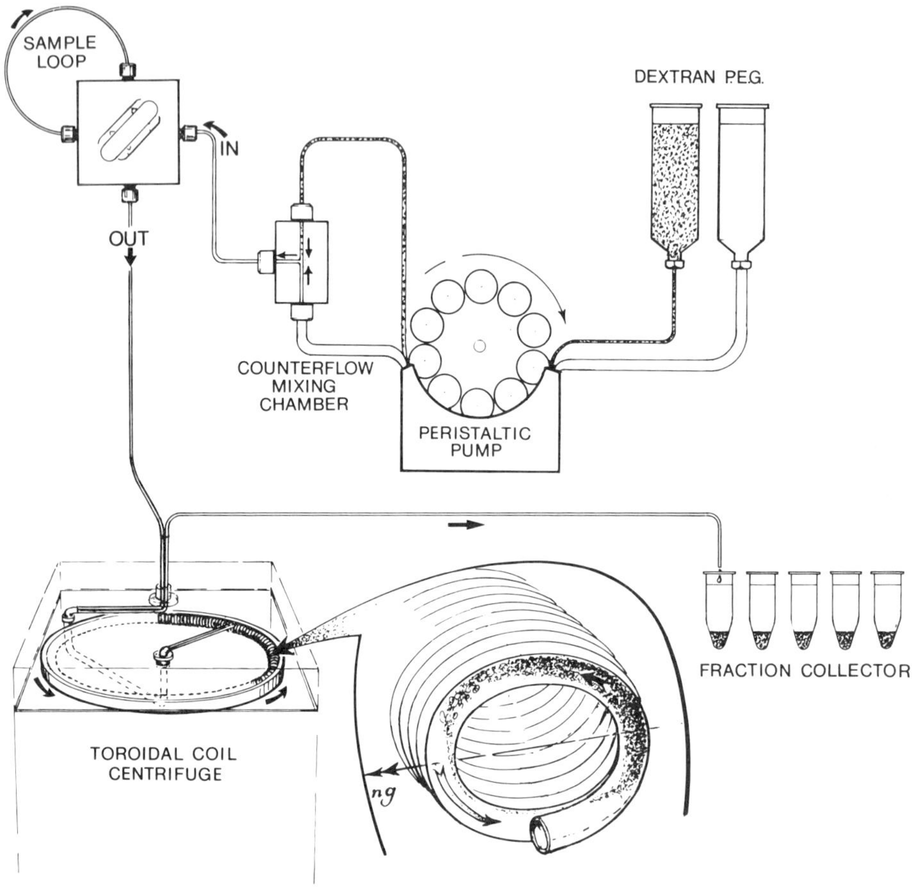

Fig. 12 Toroidal coil operating system (schematic).

The coil consists of a helically wound PTFE tubing mounted circumferentially on a rotating disk (Fig. 11). The coil is initially filled with one of the phases. The plate is rotated at 1000 revolutions/min while the other phase is pumped in. As the phases mix together, centrifugal force ensures that the lighter and heavier phases are respectively retained in the inner and outer halves of each coil unit. The pumped phase progressively displaces the other phase from the coils until an equilibrium is established. Continuous pumping of the mobile phase sets up a series of cascades (much like waterfalls) through the retained segments of the other phase in each coil unit.

The sample is injected with the mobile phase using a conventional liquid chromatography sample loop and undergoes a series of mixing and settling steps before it eventually elutes to the fraction collector (Fig. 12). Sample components partitioning toward the mobile phase will elute early while components favoring the stationary phase or interphase will be retained. As there is no solid support, either phase can be used as the mobile phase or even a mixture of the two. Adding a small proportion of the stationary phase in the above example would accelerate the elution of all the retained components and clear the coil system for another sample loading.

D. Epicyclic Coil Planet Centrifuge

The epicyclic coil planet (Chap. 3) uses scheme 2 mixing. Its major advantage is that the planetary mechanism (described in Chap. 3) is simple and can be mounted in a bench centrifuge. Like the toroidal coil centrifuge, sedimentation effects limit its use to sample sizes less than 1 μm.

The coil is mounted circumferentially around a drum attached to gear B (Fig. 13a) such that the locus of the point P on the periphery of the toroidal coil prescribes a cardioid. This produces an acceleration vector that varies in both magnitude and direction as shown in Fig. 13b [50]. The acceleration vectors for the conventional toroidal coil centrifuge are shown for comparison in Fig. 13c. Here the vector is fixed relative to the coil and does not change in magnitude or direction. The heavier phase is held in the outer segment of each coil unit by centrifugal action. Flow of the lighter phase results in a series of cascades of one phase through the other. The same fluid behavior mechanism applies to both processes except that the epicyclic coil planet centrifuge produces enhanced mixing of the phases that can be controlled by choice of the β-value (the ratio of the coil radius CP to the planet radius OC in Fig. 13a). A high β-value increases mixing, as the proportional variation of the direction and magnitude of the acceleration vector is greater at low β-values [50]. In the limit as β becomes very large the epicyclic coil planet centrifuge tends toward the toroidal coil centrifuge with its uniform acceleration field.

The epicyclic coil planet centrifuge is operated in exactly the same way as the toroidal coil centrifuge and gives qualitatively similar results.

E. Nonsynchronous Coil Planet Centrifuge

The two rotors described above cannot be used for fractionating cells greater than 1 μm in diameter due to sedimentation effects. Cell separations can be achieved using mixing scheme 3, where the coils

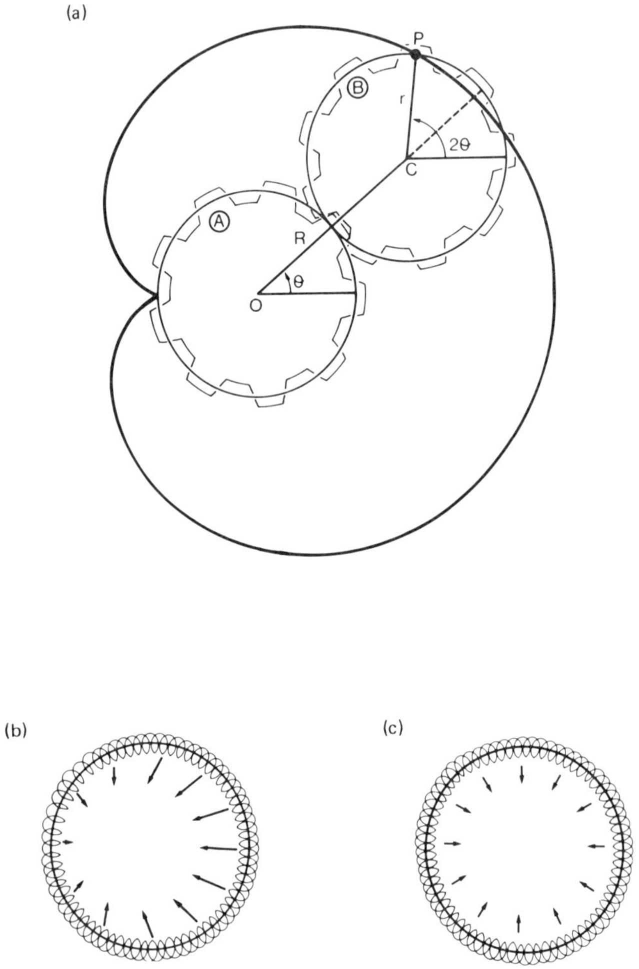

Fig. 13 Geometry of the epicyclic coil planet centrifuge showing (a) the locus of a point P and the centripetal acceleration field for (b) the epicyclic and (c) the conventional toroidal coil rotors.

slowly rotate relative to the acceleration vector. There are two non-synchronous coil planet centrifuges (See Sec. VIII in Chap. 3) that can achieve this type of motion, one with rotating seals and one without.

While the engineering solutions to produce such complex motion without rotating seals are both elegant and inventive, the complexity

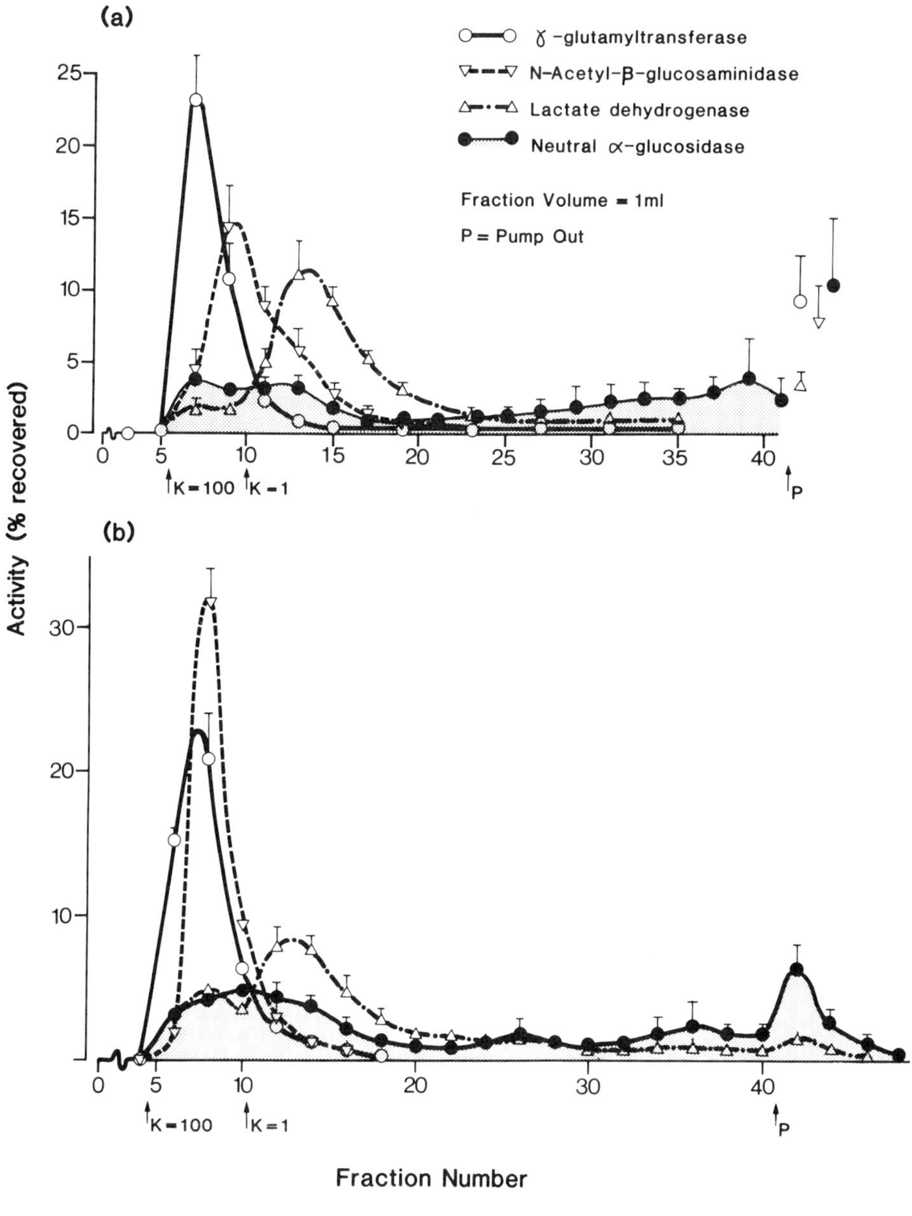

Fig. 14 A fractionation of rat liver homogenate in (a) the toroidal coil centrifuge and (b) the epicyclic coil planet centrifuge.

of the device itself has meant that it has not been extensively used. However, they do work. The slow coil rotation causes particulate material to move in small circles, thus avoiding sedimentation.

F. Applications Using CCC—Subcellular Particles

Rat Liver Homogenate

Both the toroidal coil and epicyclic coil planet centrifuges have been used for subcellular particle fractionations. For example, rat liver homogenate has been successfully fractionated on both machines (Fig. 14). The phase system used contained 3.3% (w/w) dextran T500, 5.4% PEG 6000, 10 mM sodium phosphate/phosphoric acid buffer (pH 7.4), 0.26 M sucrose, 0.05 mM Na_2-EDTA, and 1 mM ethanol. The phase mixture was shaken and allowed to separate overnight at 4°C into two phases. While this phase system was ideal for the toroidal coil fractionation (Fig. 14a), an additional 0.7% (w/w) dextran was used in the epicyclic CPC system. This prevents premature elution of the sample, perhaps induced by the extra mixing in the epicyclic system.

The coils were initially filled with the dextran-rich lower phase and rotated at 1000 revolutions/min. The mobile phase, comprising 94% PEG-rich upper phase, well mixed with 6% dextran-rich lower phase, was pumped into the coil at 14 ml/hr. This small proportion of lower phase in the mobile phase causes low-partition material that would otherwise be retained in the coils to elute slowly. When the system reached equilibrium (i.e., when the inflow and eluent both contained 6% dextran-rich lower phase), the sample was injected. After collecting 40 x 1 ml fractions, the centrifuge was stopped and the contents pumped out with water. The temperature throughout the fractionation was maintained at 4°C.

Sample preparation and enzyme assay procedures are outlined in detail by Heywood-Waddington et al. [49] and Sutherland et al. [47]. Fractionations in either machine are qualitatively similar with plasma membrane eluting early, lysosomes shortly after, and endoplasmic reticulum spread over three possible fractions. Heywood-Waddington [49] used rat liver homogenate as a standard fractionation for studying a number of operating parameters such as rotational speed, flow rate, coil geometry, and sample loading. The process was shown not to be critical, and small changes in these parameters did not significantly affect the order of elution or resolution of the process, provided certain boundary conditions were met.

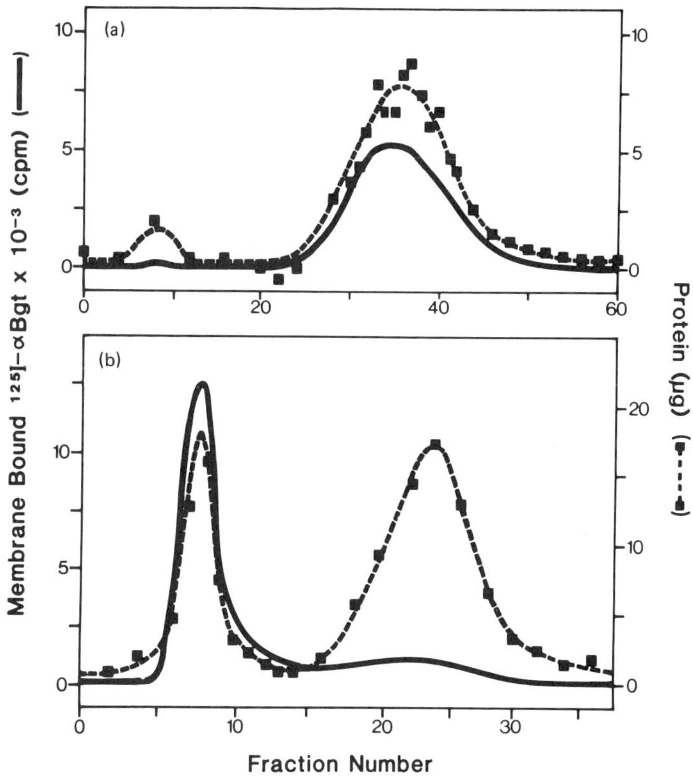

Fig. 15 A fractionation of torpedo membranes using toroidal coil centrifuge (a) without and (b) with an affinity ligand. The presence of the ligand produces a membrane fraction enriched in nicotinic cholinergic receptor sites.

Torpedo Membranes

Torpedo electroplax membranes enriched in nicotinic cholinergic receptor sites have been successfully purified by Flanagan et al. [51] using affinity partitioning techniques with phase systems operating near to the critical point. The phase system contained 4.6% (w/w) dextran T500, 3.8% (w/w) PEG 8000, 5 mM sodium phosphate pH 7.4, and 15 mM sodium chloride. The toroidal coil centrifuge was similar to that used for rat liver fractionation with similar operating conditions. The rotor speed was 1200 revolutions/min.

A fractionation without and with a ligand polymer is shown in Fig. 15a and b respectively. Without the ligand polymer, hexaethonium-polyethylene glycol, the labeled cholinergic receptor-rich membranes (membrane curve, solid line) closely follows the protein assay curve, suggesting that the cholinergic receptors are distributing with the bulk of the other membrane material in the bottom phase/interface and retarded in their progress through the toroidal coil. By replacing 0.05% of the PEG with the ligand polymer, the distribution of the cholinergic receptor-rich membranes is shifted to the mobile PEG-rich phase with elution earlier than the remaining material.

Flanagan [51] has elegantly demonstrated the power of linking affinity partition with countercurrent chromatography and supports his case by performing a complementary study with conventional countercurrent distribution techniques. He has also made major contributions by defining the procedures for operating the toroidal coil near to the critical point and examining the effects of sample loading. He found that it was essential to pump an emulsion of upper and lower phase (approximately in the ratio 10:1) to get an elution of material at all, unlike the phase system used for the rat liver fractionation where it was optional. He also concluded that sample loading was limited not by the process but by the ability to obtain a sufficiently concentrated sample.

G. Applications Using CCC—Whole Cells

Bacteria Cells

Bacteria cells are approximately 1 μm in diameter and present a borderline choice between the toroidal coil centrifuge or the nonsynchronous coil planet centrifuge. While these cells have been successfully separated using the toroidal coil planet centrifuge [52], the following examples on the purification of different strains of *E. coli* and the separation of *Salmonella typhimurium* cells use the nonsynchronous one.

The partition behavior for two different strains of *E. coli* in a PEG/dextran phase system is illustrated in Fig. 16 for various NaCl concentrations. The basic phase system used comprised 5% (w/w) dextran 500, 4% (w/w) PEG 6000, 0.01 M potassium phosphate pH 6.9. Both strains partition toward the upper phase at zero NaCl concentration. As the NaCl concentration increases, *E. coli* I starts to partition predominantly toward the interface at a concentration greater than 0.005 M, while *E. coli* II exhibits the same behavior above a NaCl concentration of 0.015 M.

A 200-coil column of 1-mm i.d. PTFE tubing was used. It was initially filled with the heavier phase and the centrifuge set to rotate at 750 rpm with a relative coil rotation of 5.25 rpm. The flow rate was 14 ml/hr.

The *E. coli* cells of Fig. 16a injected with the mobile phase were eluted at the solvent front when no NaCl was present but were

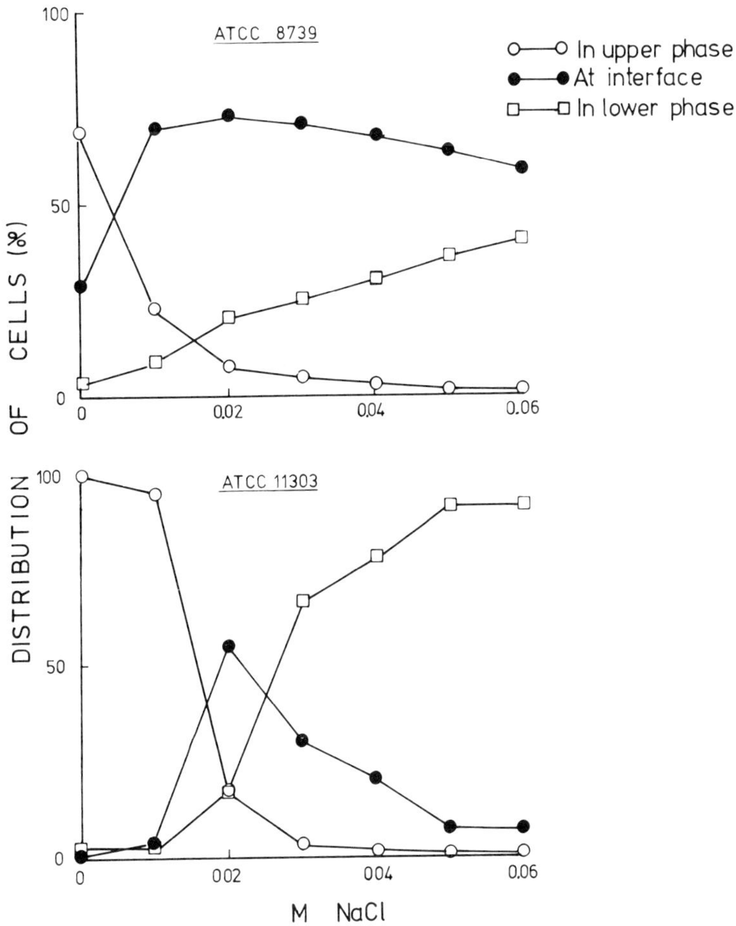

Fig. 16 The distribution of two different strains of *E. coli* in a two-phase polymer system.

retained in the column when a NaCl concentration of 0.06 M was used. This signified that a gradient separation was feasible. The column was then filled with the heavier phase at 0.02 M NaCl concentration. A mixture of the two *E. coli* strains (Fig. 16) was injected with the mobile phase. In order to produce an NaCl gradient (0.02–0.00 M), mobile upper phase was made up with zero NaCl concentration and fed via a 5-ml continuously stirred mixing chamber (initially filled with upper phase containing 0.02 M NaCl concentration) before being pumped into the column.

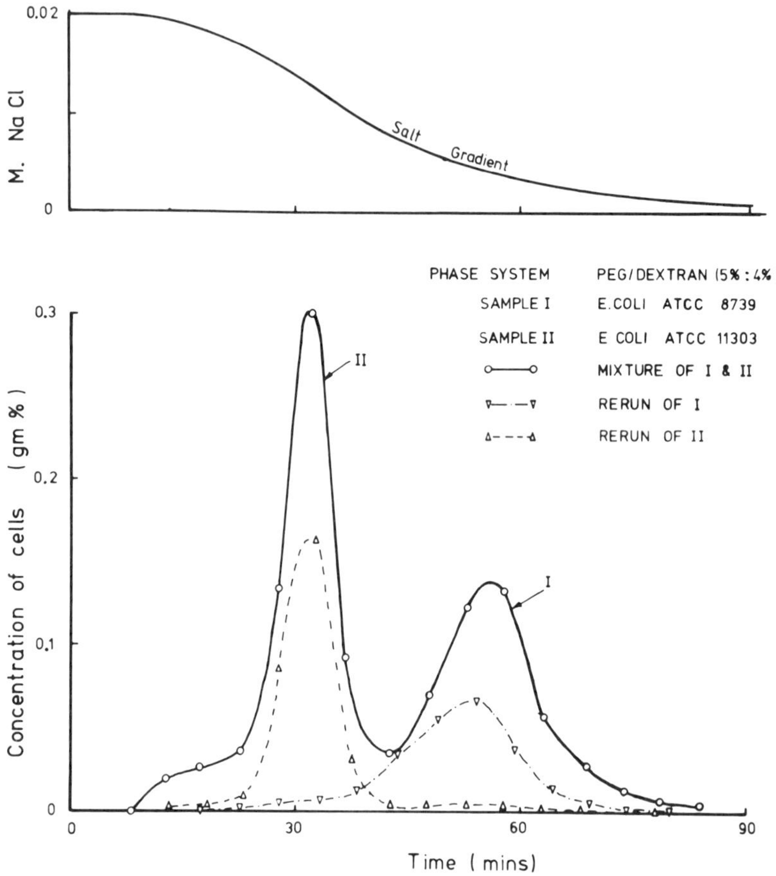

Fig. 17 A gradient separation of a mixture of the two strains *E. coli* cells shown in Fig. 14 using the nonsynchronous coil planet centrifuge.

The cells were eluted in two populations (Fig. 17). The most concentrated fractions from each peak were collected and rerun separately with up to 93% reproducibility in both cases. Earlier runs with the individual pure *E. coli* strains produced single peaks at the expected positions. Separations were repeated and found to be reproducible.

Countercurrent chromatography using the non-synchronous coil planet centrifuge without rotating seals [53] was used to separate cells of *Salmonella typhimurium* whose surfaces were identical except for the proportion of lipopolysaccharide (LPS) molecules with long vs. short polysaccharide chains [54].

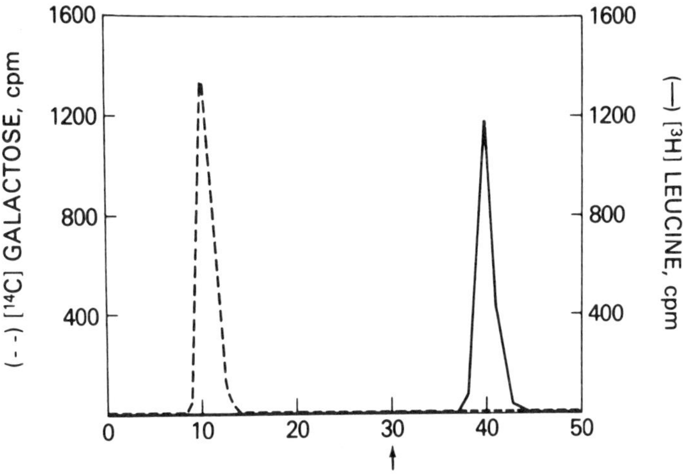

Fig. 18 Separation *Salmonella typhimurium* cells of differing surface properties. Those bearing long vs. short polysaccharide chains elute early.

The phase system used was composed of 6.2% (w/w) dextran 500, 4.4% (w/w) PEG 6000, 0.5 M Tris-chloride (pH 7.0), 10 mM potassium phosphate (pH 7.0), and 0.01% sodium azide. The apparatus was run at 600 rpm with a slow coil rotation of 5 rpm. The upper was the mobile phase with a flow rate of 8.5 ml/hr. Figure 18 demonstrates how cells bearing long-chain LPS elute early while those with short-chain LPS are retained in the column, to be pumped out at the end of the run. The authors conclude that CCC provides a tool for separating bacteria on the basis of outer membrane physical properties that are not discriminated by other means, and thus may be of help in analyzing the progress of membrane biogenesis.

Separation of Human Erythrocytes

Sutherland and Ito demonstrated that the nonsynchronous coil planet centrifuge could be used for larger cells by separating different species of red blood cells [52]. They used the version with rotating seals (Sec. VIII in Chap. 3). They demonstrated that cells of varying size could be separated on their partition behavior and were not affected by sedimentation. The partition behavior of dog, human, and sheep erythrocytes is shown in Fig. 19 for different salt concentrations and a typical fractionation of similar sized cells (dog and human) is shown in Fig. 20. Individual peak fractions (plotted as

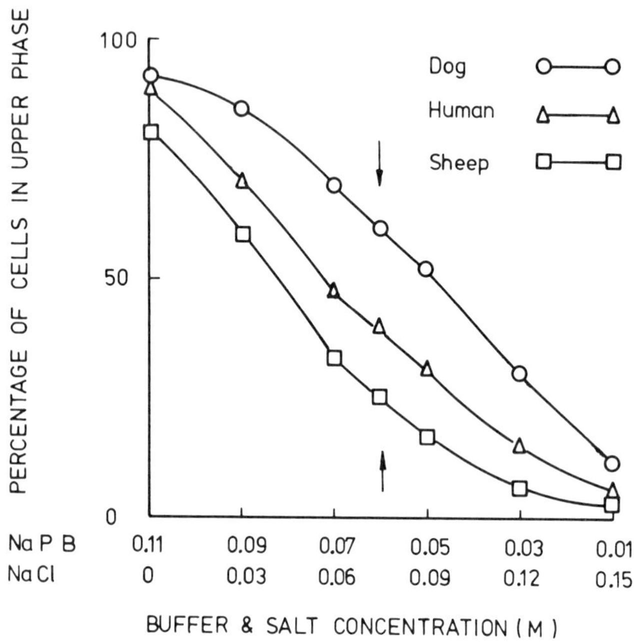

Fig. 19 The partition behavior of dog, human, and sheep erthrocytes for various salt concentrations.

shaded points in the figure) were bulked and rerun, in accordance with classical chromatographic methods of testing for reproducibility and purity. The results of rerunning the peak material under otherwise idential conditions are plotted for comparison on the same graph. Each peak occurs in the position expected, with a reproducibility of 99% for the dog cells and 74% for the human.

These cell separations were achieved with the similar operating conditions to the ones previously described except the process was stopped before elution and the coil contents pumped out. When the process was extended beyond elution, peaks would become spread possibly due to sedimentation effects in the outlet tubing.

Other workers [55] have reported difficulties in reproducing these results, partly due to discrepancies in phase system composition, and partly due to the fact that the process is a nonelution one. Clearly if these barriers were overcome a promising new cell separation process could emerge.

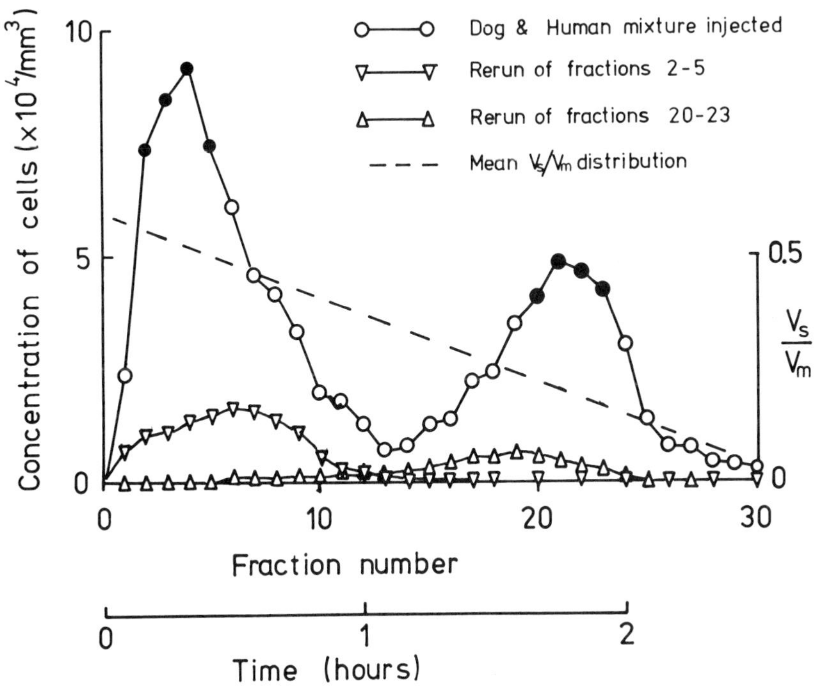

Fig. 20 Separation of a mixture of dog and human erythrocytes using the nonsynchronous coil planet centrifuge.

V. FUTURE ADVANCES IN COUNTERCURRENT CHROMATOGRAPHY

A. Summary

Countercurrent chromatography is easy to use, has a wide range of applications, and is suited to automation. But its major asset is its low cost and its high sample recovery rates. The lack of solid support minimizes adsorption problems and allows one set of coils to be used with a variety of different phase systems using off-the-shelf solvent systems.

These are powerful advantages in today's scientific community where both costs and performance have to be considered. There are two approaches in taking the technology forward. The first is to develop the hardware (Sec. B) to create gentler mixing and settling zones than can be achieved with existing phase systems. The second is to develop phase systems (Sec. C) with physical properties more suited for use with countercurrent chromatography.

B. Equipment Development

The problems of applying countercurrent technology to the separation of biological material using polymer phase systems have already been highlighted in Sec. IV (F and G). While separations of subcellular material and whole cells are possible, the mixing rates must be low to avoid emulsification, and there is a consequential loss in resolution. The rotor which may most easily be adapted to extend the range of application to include cells (Sec. E) is necessarily the most complicated to manufacture and is unlikely to find widespread commercial application.

The most practical rotor is the epicyclic coil planet centrifuge. A version is already marketed by P.C. Inc. (Potomac, MD) for applications with aqueous/organic phase systems and the device can be used, with specially wound toroidal coils, for membrane fractionations using aqueous polymer phase systems [47]. The full potential of this device has not yet been realized. The fluid dynamics are only just becoming understood [56] and the full significance of low β-coils with force field invertion [57] has not yet been explored. The recent development of the angle rotor version of this device by Ito [58], with its three-dimensional force field fluctuation, may provide the additional gentle mixing required in the toroidal coils, although the retention of hydrophilic phase systems in this system is not as good as the conventional epicyclic coil planet centrifuge.

C. Phase System Development

The increased popularity of aqueous two-phase partition is due in part to its versatility and breadth of application [5,12,59,60], but also to its sensitivity as a separation process. Furthermore techniques developed in the laboratory can be carried through to full production in industry. Polymer phases are being increasingly used on a large scale for biotechnological applications [61–63]. The availability of practical countercurrent chromatography techniques will enhance the progress of this emerging technique.

However, for the full potential of the process to be recognized, it is essential to develop new biocompatible phase systems which have physical properties more resembling the aqueous/organic phase systems that have been so successfully applied to countercurrent chromatography. These phase systems must be able to respond to the rapid mixing and settling cycles that are the essence of high-resolution countercurrent chromatography.

One approach suggested for direct enzyme purification from a fermentation broth is the extraction of the enzyme using the principle of reversed micelle formation in an oil/water phase system [64]. Enzymes do solubilize in an oil phase under certain conditions [65].

An alternative approach is to develop new, less viscous aqueous phase systems. Already Aquaphase PPT [8] and Reppal PES (Reppe Glykos AB, S-352 50 Vaxjo, Sweden) have been proposed as potential dextran substitutes. The motivation for the development of these products has come from the increased biotechnological application of PEG/dextran phase systems and the need to find a cheaper alternative to dextran. Along with basic material costs there is the need to develop improved emulsification characteristics to enhance mass transfer between the phases and faster settling times to speed up production runs.

These inevitable developments will lead to new phase systems that will most probably have characteristics that are more suited to countercurrent chromatography and could well form the basis for future advances in the technology.

ACKNOWLEDGMENTS

The author is indebted to his colleagues in the Engineering Department of the National Institute for Medical Research for their help and support, and in particular to Denis Rothwell and Deborah Heywood-Waddington for their critical evaluation of the manuscript and helpful suggestions.

REFERENCES

1. P.-Å. Albertsson, *Nature, 177,* 771 (1956).
2. P.-Å. Albertsson, *Partition of Cell Particles and Macromolecules*, 2nd ed., Wiley-Interscience, New York, 1971.
3. P.-Å. Albertsson, *Endeavour, 1,* 69 (1977).
4. P.-Å. Albertsson, B. Andersson, C. Larsson, and H.-E. Akerlund, in *Methods of Biochemical Analysis* (D. Glick, ed.), Vol. 28, Wiley-Interscience, New York, 1982, pp. 115–150.
5. H. Walter, D. E. Brooks, and D. Fisher (eds.), *Partitioning in Aqueous Two-Phase Systems: Theory, Methods, Uses and Applications to Biotechnology.* Academic, London, 1985.
6. F. D. Raymond and D. Fisher, *Biochem. Soc. Trans., 8,* 118 (1980).
7. D. Heywood-Waddington, T. J. Peters, and I. A. Sutherland, *Biochem. J. 235,* 245 (1986).
8. F. Tjerneld, S. Berner, A. Cajarville, and G. Johansson, *Enzyme and Microbial Technology*, *8*, 417 (1986).
9. B. Yu. Zaslavsky, L. G. Shchukina, and S. V. Rogozhin, *Mol. Biol.* (Moscow), *15,* 1315 (1981).

10. B. Yu. Zaslavsky, N. M. Mestechkina, and S. V. Rogozhin, *J. Chromatogr.*, *260*, 329 (1983).
11. G. Johansson, M. Joelsson, B. Olde, and V. P. Shanbhag, *J. Chromatogr.*, *331*, 11 (1985).
12. C. Larsson, in *Isolation of Membranes and Organelles from Plant Cells* (J. L. Hall and A. L. Moore, eds.), Academic, London, 1983, pp. 277–309.
13. G. Johansson, *Solvent Extr.*, *Proc. Int. Solvent Extr. Conf.*, Vol. 2, 1971, pp. 928–936.
14. G. Johansson, A. Hartman, and P.-Å. Albertsson, *Eur. J. Biochem.*, *33*, 379 (1973).
15. E. Eriksson, P.-Å. Albertsson, and G. Johansson, *Mol. Cell. Biochem.*, *10*, 123 (1976).
16. H. Westrin, P.-Å. Albertsson, and G. Johansson, *Biochim. Biophys. Acta*, *436*, 696 (1976).
17. G. Johansson and H. Westrin, *Plant Sci. Lett.*, *13*, 201 (1978).
18. G. Johansson, in *Methods in Enzymology* (W. B. Jakoby, ed.), Vol. 104, Academic, New York, 1984, pp. 356–364.
19. T. P. King and L. C. Craig, in *Methods of Biochemical Analysis* (D. Glick, ed.), Vol. 10, 1962, pp. 201–228.
20. L. C. Craig and D. Craig, *Technique of Organic Chemistry*, Vol. III, 2nd ed., Part 1 (A. Weissberger, ed.), Interscience, New York, 1956.
21. P.-Å. Albertsson, *Anal. Biochem.*, *11*, 121 (1965).
22. D. G. Pritchard, R. M. Halpern, J. A. Halpern, B. C. Halpern, and R. A. Smith, *Biochim. Biophys. Acta*, *404*, 289 (1975).
23. L. C. Craig and H. O. Post, *Anal. Chem.*, *21*, 500 (1949).
24. H.-E. Akerlund, *J. Biochem. Biophys. Meth.*, *9*, 133 (1984).
25. P.-Å. Albertsson and E. J. Nyns, *Nature*, *184*, 1465 (1959).
26. G. Baird, P.-Å. Albertsson, and B. Hofsten, *Nature*, *192*, 236 (1961).
27. P.-Å. Albertsson and G. D. Baird, *Exp. Cell Res.*, *28*, 296 (1962).
28. S. Bengtsson, L. Philipson, and P.-Å. Albertsson, *Biochem. Biophys. Res. Commun.*, *9*, 318 (1962).
29. G. Frick and T. Lif, *Arch. Biochem. Biophys.*, Suppl. 1, 271 (1962).
30. P.-Å. Albertsson and H. Baltescheffsky, *Biochem. Biophys. Res. Commun.*, *12*, 14 (1963).
31. H. Walter, F. W. Selby, and J. M. Brake, *Biochem. Biophys. Res. Commun.*, *15*, 497 (1964).
32. H. Walter, R. Winge, and F. W. Selby, *Biochim. Biophys. Acta*, *109*, 293 (1965).
33. H. Walter and F. W. Selby, *Biochim. Biophys. Acta*, *112*, 146 (1966).

34. H. Walter and H. Nagaya, *Cell. Immunol.*, *19,* 158 (1975).
35. Y. Hino, A. Asano, and R. Sato, *J. Biochem.*, *83,* 935 (1978).
36. P. Malmström, K. Nelson, A. Jönsson, H. O. Sjogren, H. Walter, and P.-Å. Albertsson, *Cell. Immunol.*, *37,* 409 (1978).
37. K. Nelson, P. Malmström, A. Jönsson, and H. O. Sjogren, *Cell. Immunol.*, *37,* 422 (1978).
38. P. Malmström, A. Jönsson, T. Hallberg, and H. O. Sjogren, *Cell. Immunol.*, *53,* 39 (1980).
39. P. Malmström, A. Jönsson, and H. O. Sjogren, *Cell. Immunol.*, *53,* 51 (1980).
40. H. Walter, E. J. Krob, and B. J. Moncla, *Exp. Cell Res.*, *115,* 379 (1978).
41. L. Backman and V. Shanbhag, *J. Chromatogr.*. *171,* 1 (1978).
42. L. Backman, *J. Chromatogr.*, *196,* 207 (1980).
43. L. Backman, *J. Chromatogr.*, *237,* 185 (1982).
44. B. Ekberg, B. Sellergren, and P.-Å. Albertsson, *J. Chromatogr.*, *333,* 211 (1985).
45. G. Johansson, M. Joelsson, and H.-E. Akerlund, *J. Biotechnol.*, *2,* 225 (1985).
46. W. B. Morris and T. J. Peters, *Eur. J. Biochem.*, *121,* 421 (1982).
47. I. A. Sutherland, D. Heywood-Waddington, and T. J. Peters, *J. Liqu. Chromatogr.*, *7,* 363 (1984).
48. B. Olde and G. Johansson, *Neuroscience*, *15,* No. 4, 1247 (1985).
49. D. Heywood-Waddington, I. A. Sutherland, W. B. Morris, and T. J. Peters, *Biochem. J.*, *217,* 751 (1984).
50. I. A. Sutherland, D. Heywood-Waddington, and T. J. Peters, *J. Liq. Chromatogr.*, *8,* No. 12, 2315 (1985).
51. S. D. Flanagan, G. Johansson, B. Yost, Y. Ito, and I. A. Sutherland, *J. Liq. Chromatogr.*, *7,* 385 (1984).
52. I. A. Sutherland and Y. Ito, *Anal. Biochem.*, *108,* 367 (1980).
53. Y. Ito, G. T. Bramblett, R. Bhatnagar, M. Huberman, L. Lieve, L. M. Cullinane, and W. Groves, *Sep. Sci. Technol.*, *18,* 33 (1983).
54. L. Lieve, L. M. Cullinane, Y. Ito, and G. Bramblett, *J. Liq. Chromatogr.*, *7,* 403 (1984).
55. J. M. Harris, M. G. Case, R. S. Snyder, and A. A. Chenault, *J. Liq. Chromatogr.*, *7,* 419 (1984).
56. I. A. Sutherland, S. Jones, and D. Heywood-Waddington, in Proc. of the Pittsburgh Conference, Atlantic City, March 10-15, 1986, p. 1045.
57. Y. Ito, *J. Chromatogr.*, *188,* 33 (1980).
58. Y. Ito, *J. Chromatogr.*, *358,* 313 (1986).

59. P.-Å. Albertsson, *Partition of Cell Particles and Macromolecules*, 3rd Ed., Wiley-Interscience, New York, 1986.
60. D. Fisher, *Biochem. J.*, *196*, 1 (1981).
61. B. Mathiasson, *Trends Biotechnol.*, *1*, 16 (1983).
62. B. Mathiasson, in *Methods of Enzymoloby*, Vol. 91, Pt. E, Immunochemical Techniques, Academic, New York, 1983, pp. 498–522.
63. H. Hustedt, K. H. Kroner, and M.-R. Kula, in *Partitioning in Aqueous Two-Phase Systems* (H. Walter, D. E. Brooks, and D. Fisher, eds.), Academic, London, 1985, pp. 529–587.
64. K. van 't Riet and M. Dekker, in *Proc. 3rd European Conference on Biotechnology*, Munchen, September 10-14, 1984, Vol. III, Verlag-Chemie, Weinheim, 1984, pp. 541–544.
65. P. L. Luisi, F. J. Bonner, A. Pellegrini, P. Wiget, and R. Wolp, *Helv. Chim. Acta*, *62*, 740 (1979).

chapter twelve

Countercurrent Chromatography/Mass Spectrometry

John M. Ruth
Agricultural Environmental Quality Institute
Beltsville Agricultural Research Center, ARS, USDA
Beltsville, Maryland

N. Bhushan Mandava*
Office of Pesticides and Toxic Substances
U.S. Environmental Protection Agency
Washington, D.C.

I. INTRODUCTION

The aim of this chapter is to outline the developments of gas chromatography/mass spectrometry (GC/MS) and of liquid chromatography/mass spectrometry (LC/MS), to compare the two, and to consider the prospects for countercurrent chromatography/mass spectrometry (CCC/MS). Two aspects of the comparison of GC/MS and LC/MS dominate the resulting view: (a) the basic histories of the two developments fall into the same general outline, showing the parallelism of the two methods of analysis, and (b) when one facet of the common outline is important in one development but relatively unimportant in the other, the fundamental differences between GC/MS and LC/MS are reemphasized.

As discussed later in Sec. III, the coupling of liquid chromatography and mass spectrometry was undertaken for two basic reasons: the desire to obtain the molecular structural information contained in mass spectra and the need for a more widely applicable detector for liquid chromatography. The same considerations suggest the investigation of CCC/MS.

The purpose of a mass spectrometer is to produce, usually in the very low-pressure gas phase of a high-vacuum system, an assortment of ionized molecular fragments derived from the sample, to sort the mixture of ions according to their apparent mass, and to

**Current affiliation*: Todhunter, Mandava, & Associates, Washington, D.C.

determine the abundances of the different ions in the mixture. The present discussion will be concerned only with the mass spectrometry of organic chemical compounds, including the subclass of biochemicals. The mass spectrum may be defined in familiar mathematical notation as a function of one independent variable. The independent variable is not strictly the mass m, but rather the ratio m/z of mass to the number of elementary units of electrical charge carried by the ion. For an ion written as A^-, z is 1, the charge is that of one negative electron, and the ratio m/z is equal to the mass m. If the ion is symbolized as B^{2+}, z has the value 2, the charge is the positive charge of two protons, and the ratio m/z is equal to one-half of the mass m. Under the usual operating conditions of organic chemical mass spectrometry, multiply charged ions are relatively rare, as compared with singly charged ions, and in the normal use of the very popular quadrupole instrument they are not observed at all. Another standard feature of mass spectral notation is the dot beside the plus sign in such an example as $C^{+\cdot}$. The dot is used to show that the ion contains an odd number of electrons, and the absence of the dot indicates an even number.

The dependent variable is the ionic intensity I. In the form of its experimentally measured value I_a, it may be observed as an ion current, proportional to the number of ions per unit time and measured over a relatively short time interval, as in the reading of a microammeter. It may also be obtained as the integral of such a current over a relatively longer fixed time interval, as in the optical blackening produced in a spectral line on a photographic plate. In the first case, when the time needed to record one scan of the mass spectrum is sufficiently short in comparison with the time required for a change of sample composition in the ion source, successive scans will present true pictures of the change of sample composition with time. In the second case, represented in the extreme by the photographic plate, all of the transient behavior of the ion population during one plate exposure is summed into a single spectrum which represents an average of the sample composition during that time interval. If the recording speed of the instrumental system is intermediate between the two extremes just described, the resulting data will also be intermediate in nature, representing successive averages of sample composition over a sequence of subintervals. Another source of complexity, the possible variation of sampling time of individual m/z values from the low end to the high end of the m/z scale within a single scan, may affect the data from different experimental systems differently. This effect of the "scan function" has been discussed in the literature [1].

As noted in the beginning, the mass spectral intensity I_a is conveniently written as a function $I_a(m/z)$ of the variable m/z:

$$I_a = I_a(m/z) \tag{1}$$

The intensity is usually tabulated in the form of a relative intensity, frequently designated as R.I. or R.A. (relative abundance), where the intensity at any particular m/z value is expressed as a percentage of the intensity $I_{a,max}$ of the strongest peak in the spectrum, called the "base peak." Then

$$R.I.(\%) = \frac{100}{I_{a,max}} I_a(m/z) \tag{2}$$

The intensity is sometime expressed as a percentage of the total intensity summed from an initial value m_i of m/z up to the m/z value M of the molecular ion:

$$\%\Sigma_m = \frac{100 \, I_a(m/z)}{\sum_{m/z=m_i}^{m/z=M} I_a(m/z)} \tag{3}$$

It is desired that the rate of production of ions during the time interval of the data acquisition be sufficient to yield an assortment of ions that is statistically reproducible to a degree considered satisfactory for the purpose in mind. The mass spectroscopist seeking to identify previously known compounds or to determine the molecular structures of previously unknown compounds has, of course, not much time to devote to formal statistical manipulations. Fortunately here, as in chemistry in general, it is often possible to produce a population large enough, in proportion to Avogadro's number of 6.023×10^{23} molecules/mole, to exceed the minimum statistical requirement by some obvious orders of magnitude, as far as the principal peaks of the spectrum are concerned. However, the mass spectroscopist is often concerned also with peaks of low intensity, most notably with the special case of the molecular ion. Then, he may be found examining single-ion profiles across his gas chromatographic peak, comparing them with the extensions of those profiles through background regions before and after the chromatographic peak, and thinking for a moment about signal-to-noise ratios, all to the mystification and frustration of his "management," which tends on such occasions to wait impatiently for the results.

II. MIXTURES BEFORE GAS CHROMATOGRAPHY/MASS SPECTROMETRY

The student of prehistoric civilizations soon learns that no stage in the growth of technology, such as the New Stone Age or the Bronze Age, is well described as beginning at a fixed point on a one-dimensional frame of reference labeled "time." The presence or absence of such a cultural phenomenon at various points within an area of finite size can only be described by a four-dimensional function in which the variables are time and the three coordinates of space. The development of metallurgy, for example, was perhaps not pursued as rapidly in northern as in central Europe, and possibly not as early there as in Anatolia and the Near East. In a parallel way, the mass spectroscopist's experiences in one institution might be recorded under the heading "Pre-GC/MS," even though many other localities had long been committed both to the continuing development and to the routine application of combined gas chromatographic/mass spectrometric analysis.

The mass spectra of samples introduced batchwise from a heated reservoir often suggest clearly the presence of a mixture. Introduction by means of a direct inlet probe permits a small degree of separation of compounds by fractional evaporation (not an efficient technique, especially under reduced pressure). Introduction from a small reservoir that is gradually warmed sometimes gives a similar degree of fractionation. Whenever an authentic spectrum of a suspected mixture component was available, it was the practice to subtract that spectrum from the spectrum of the mixture, hoping always that the resulting difference spectrum might be recognizable as that of another component or of a decipherable mixture. The examples to be described below will serve to illustrate what sort of things could be accomplished through the determined effort to imitate GC/MS without a GC.

A. Arithmetic Manipulation of Mixed Mass Spectra

The general nature of the subtraction process itself can be considered in terms of a mixture M of two compounds A and B, the mass spectrum of the mixture M, and the mass spectra of the components A and B. The spectrum of M is the linear combination of the spectra of A and B. We may take B as the suspected compound to be subtracted out and A as an unknown compound, whose spectrum will be obtained if it can be verified that B was properly identified. (The process is easily extended to the case in which A is still a mixture of two or more components.)

Let the term I_{Ai} represent the intensity I of the peak at m/z i in the spectrum of A. Let the sequence of such peaks

$$I_{A1}, I_{A2}, I_{A3}, \ldots, I_{Ai}, \ldots \tag{4}$$

constituting the entire mass spectrum of compound A, be denoted by

$$[I_{Ai}] \tag{5}$$

Similarly,

$$[I_{Bi}] \tag{6}$$

and

$$[I_{Mi}] \tag{7}$$

represent the spectra of B and M, respectively. Then

$$[I_{Mi}] = [r_A I_{Ai}] + [r_B I_{Bi}] \tag{8}$$

The coefficients r_A and r_B are constants whose values at this stage are not known.

If, as in the cases to be considered, a value n of m/z is observed at which the peak I_{An} should be of zero intensity, then the intensity at that m/z value must all belong to I_{Bn}. From Eq. (8), we see by setting I_{An} equal to zero that $r_B = I_{Mn}/I_{Bn}$. When the portion of the mixture spectrum belonging to B is subtracted out of the spectrum of the mixture M, the resulting difference spectrum may be given the symbol

$$[I_{Di}] \tag{9}$$

Since the subtraction is to be done in such a way that

$$I_{Dn} = 0 \tag{10}$$

we multiply expression (7), term by term, by I_{Bn}, then multiply expression (6) in the same way by I_{Mn}, and subtract the latter product from the former to set up Eq. (11):

$$[I_{Di}] = I_{Bn}[I_{Mi}] - I_{Mn}[I_{Bi}] = [I_{Bn}I_{Mi} - I_{Mn}I_{Bi}] \tag{11}$$

For any fixed value m of the subscript i, which specifies the m/z position of the spectral peak,

$$I_{Dm} = I_{Bn}I_{Mm} - I_{Mn}I_{Bm} \tag{12}$$

In particular, if m is set equal to the value n for which I_{An} is zero, it is immediately verified that $I_{Dn} = I_{Bn}I_{Mn} - I_{Mn}I_{Bn} = 0$. Because reference spectra are usually presented in the form of % intensity relative to the base peak, it is convenient to normalize the difference spectrum (11) by multiplying it through by the constant 100%/ I_{Dj}, in which j is the m/z value of the base peak:

$$[I_{Di}]\% = \frac{100[I_{Bn}I_{Mi} - I_{Mn}I_{Bi}]}{I_{Dj}} \tag{13}$$

The question is often asked whether the B spectrum should be in the % relative intensity form or in some other form. To answer that question, let us suppose that the B spectrum is multiplied through by an arbitrary constant k. That leads to the new forms of the difference spectrum:

$$[I'_{Di}] = kI_{Bn}[I_{Mi}] - I_{Mn}[kI_{bi}] = k(I_{Bn}[M_i] - I_{Mn}[I_{Bi}]) \tag{14}$$

If again we fix the index i, at the same value m used in Eq. (12), the result is

$$I'_{Dm} = k(I_{Bn}I_{Mm} - I_{Mn}I_{Bm}) \tag{15}$$

Dividing Eq. (12) by Eq. (16), member by member, and rearranging, we get

$$I'_{Dm} = kI_{Dm} \tag{16}$$

Here we see that the second version of the difference spectrum is obtainable from the first version simply by multiplying each peak intensity by the constant k. Conversion of Eq. (15) to % relative intensity gives

$$[I'_{Di}]\% = \frac{(100\%)k[I_{Bn}I_{Mi} - I_{Mn}I_{Bi}]}{kI_{Dj'}} \tag{17}$$

In Eq. (17), it is seen that the constant k cancels from numerator and denominator, and has no effect on the final relative intensities in the difference spectrum.

B. Applications to Codling Moth Pheromone Research

The codling moth project began several years before the writer had any acquaintance with it. At the Yakima, Washington laboratory of the USDA, the analytical chemistry investigations group engaged in insect research had developed a procedure for the isolation of a fraction that was believed to contain the female sex pheromone. They had obtained low-resolution mass spectra and other data. The opinion had become firmly established among their consultants and themselves that there were two oxygen atoms in the molecule, one of which was recognizable as a hydroxyl oxygen and the other of which was probably a cyclic ether. Roelofs [2], testing compounds by electroantennagram, had identified the pheromone as trans-8-trans-10-dodecadien-1-ol, but at Yakima it was evaluated as being biologically inactive.

Near the end of calendar year 1970, the investigations leader of analytical investigations at Beltsville asked the writer about the possibility of getting a high-resolution mass spectrum. The reply was that the sample, from the description supplied, seemed adequate for that purpose. The instrument was checked over and put into condition to do its best work, in anticipation of the arrival of the sample. The first fact that emerged from the high-resolution mass measurements was that the troublesome second oxygen did not exist. Other than the hydroxyl group, the rest of the molecule was a hydrocarbon group, which showed evidence of some branching and the equivalent of two double bonds. Other experiments were discussed by the writer with Yakima. The chemical reactions were carried out at Yakima and the products obtained were sent to Beltsville. Mass spectra of the hydrogenation product of the acetate established the branching. Mass spectra of the ozonolysis products, the NMR spectrum which was recorded at Beltsville, and syntheses done at Yakima contributed other information. The compound was identified [3] as (2Z, 6E)-7-methyl-3-propyl-2,6-decadien-1-ol.

Meanwhile, Yakima thought that it had discovered a multiplicity of pheromones scattered across the gas chromatogram, and trapped fractions for analysis. Many scans, high-resolution plate spectra, and mass measurements by peak matching were recorded. The spectra were obviously mixtures, not pure compounds. Only the investigations leader had funds to travel from Beltsville to Yakima, and the working members of the team never saw each other's laboratories at work. Whenever he had time, the mass spectroscopist returned to the pheromone mixture samples. The main compounds could have been of terpenoid nature, although he was inclined to look carefully at acetone polymerization products and other solvent impurities. Various subtractions of spectra were done, in the effort to "do GC/MS without a GC."

Ruth's memoranda covering work on pheromone 1B [4] and pheromone 2A [5], reproduced in the Appendix to this chapter, furnish examples of the manual subtraction treatment of mixture mass spectra. Sample 1B appeared enough like isophorone or some very similar compound, mixed with other material, that the writer sent some isophorone to Yakima for bioassay. It was found somewhat active, but not as strongly so as the pheromone. Number 2A also showed what looked like a molecular ion at m/z 138, plus other possibilities. Evidence for dehydration in the inlet line with possible aromatization (a known reaction) is discussed in the Appendix, Sec. B. The two examples, if worked through by the reader, will help to develop an appreciation of the place of the GC/MS data system in mass spectrometry.

Once the skeletal structure of the methylpropyldecadienol had been settled, it was necessary to make a choice among the four possible E,Z configurations about the two double bonds. The syntheses did not answer that question correctly. Several papers [6–10] dealt with that problem, and in the process it was established that none of the four was the pheromone. The bioassay at Yakima had failed, and Roelofs had been right. The multiplicity of pheromones [11], which had been introduced by the same bioassay failure, did not exist.

III. GAS CHROMATOGRAPHY/MASS SPECTROMETRY

Readers familiar with the classical methods of inorganic qualitative analysis will recall that the identification of a given ion, by a precipitation reaction or by a color test, depends altogether on the correct working of a separatory scheme. Without the separation of any particular ion at its own predetermined point in that scheme, its identification becomes uncertain. In general, all other chemistry is built on a foundation of analytical chemistry. Within that basic supporting structure, qualitative analytical chemistry is dependent on the coordinated performance of its two fundamental aspects, separation and identification, to which quantitative analysis adds the third: the methodology of precise measurements of amounts and concentrations. Although the problems, the apparatus, and the laboratory operations of up-to-date organic analytical chemistry look different from their early inorganic predecessors, the underlying principles remain the same.

After the development of mass spectrometry into the most powerful single tool for the identification of organic compounds in terms of their molecular structures, and the corresponding development of gas chromatography as, at that time at least, the most effective separatory technique, it was obvious that the coupling of the two together would be another great step forward. Although the refinement of methods for interfacing them required some effort, the combination was a logical expectation because (a) their sensitivities in terms of sample size were compatible; (b) samples or derivatives [12–17] sufficiently volatile and sufficiently stable for gas chromatography were adequate in those respects for mass spectrometry; (c) the scanning speed of the mass spectrometer, in some then-existing types and in others then under development, was adequate or more than adequate; and (d) the use of digital data processing technology would bring the recording speed within the limits set by the rates of gas chromatographic peak elution [1,18–20]. At that point, the task ahead was the selection of the best-matched existing apparatus for the basic stages in the system, the further improvement of those components, and the optimization of the interfaces between them.

A. The Beginning: Stream Splitting, with Direct Introduction

Holmes and Morrell of Phillip Morris, Inc. described the first achievement of a combined gas chromatographic/mass spectrometric (GC/MS) system, and presented the first data from it, at the Fourth Annual Meeting of ASTM Committee E-14 on Mass Spectrometry and Allied Topics in 1956 [21]. Their publication appeared the following year [22]. The gas chromatograph was coupled to a Consolidated Electrodynamics Corporation Model 21-103B mass spectrometer by sampling the gas chromatographic effluent stream through a capillary of appropriate dimensions. The capillary was selected to pass the sample to the ion source of the mass spectrometer in the shortest possible time and to maintain a suitable operating pressure in the ion source. This simple and effective stream splitter permitted the major portion of the effluent stream to bypass the mass spectrometer and flow into a trapping apparatus for recovery of the gas chromatographic fractions.

The mass spectrometer, of a type designed for the production of stable, closely reproducible spectral intensity patterns and at that time much used for the quantitative analysis of mixtures in the petroleum industry, was a slower scanning one than those routinely used in GC/MS today. It has been equipped with a photographic recording oscillograph that required a few minutes to make available a visible display of the spectrum. To permit the mass spectrum to be seen while the sample was in the ion source, the cirucitry was modified (Fig. 1), and an os-

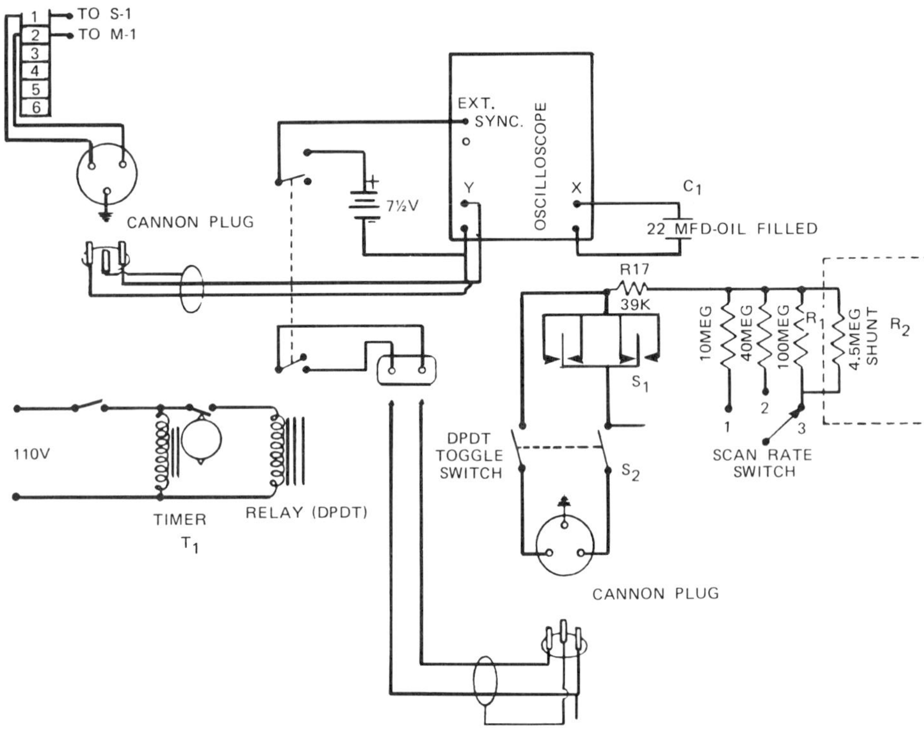

Fig. 1. Schematic diagram for connecting oscilloscope to mass spectrometer. (Reprinted with permission from J. C. Holmes and F. A. Morrell, *Applied Spectroscopy 11,* 86 (1957). Copyright 1957 Society for Applied Spectroscopy.)

cilloscope was connected. It was possible to present on the oscilloscope the m/z range from 28 to 44 in a 13-sec sweep. A longer segment could be displayed in the higher mass range, and a further increase could be obtained by operating at a higher magnet strength.

For a permanent record, an oscillographic recording could be made simultaneously with the oscilloscopic presentation. Holmes and Morrell pointed out the usefulness of the oscilloscope in following the course of a fast reaction, in studying the products of controlled pyrolysis, and in monitoring the effects of various adjustments during maintenance of the instrument. Three years later, the present writer found the similar oscilloscopic display on the Bendix Model 14 time-of-flight mass spectrometer so useful in batch sample operations that to work eventually without the oscilloscope on a different instrument gave for the first few days the feeling of groping in the dark.

Gohlke [23] reported a great improvement in instrumentation. He removed the oil diffusion pump and water-cooled baffle from an early Model 12-100 Bendix time-of-flight mass spectrometer and replaced them with a mercury diffusion pump, a Freon 22-refrigerated baffle having a baffle interior temperature of about -20°C, and a liquid air cold trap. The modifications reduced the instrument background to a very low, completely unobjectionable level. The sampling line to the mass spectrometer was connected from the GC effluent line between the end of the column and the thermal conductivity detector. A finely adjustable needle valve in the sampling line between ion source and the GC effluent line was used to control the sample flow rate, and thus to control the ion source pressure. A 10-ft x 1/4 in. column of Tide detergent, at a temperature of 138°C, with a helium inlet pressure of 10 psig, was used to demonstrate the separation of a mixture of acetone, benzene, toluene, ethylbenzene, and styrene.

The spectra were presented for viewing or photographing on a Tektronix 541 oscilloscope, where the mass range displayed was easily variable from 1 to 6000 mass units. The scanning speed of the time-of-flight mass spectrometer, typically operated at 2000 or 10,000 scans/sec, was much more than adequate for gas chromatography. Since the time width of a GC peak is several orders of magnitude longer than a spectral scan time of 500 or 100 μsecs the spectra as usually seen on the oscilloscope represent the integration of such superimposed scans over a time interval determined by the persistence of the phosphor on the oscilloscope screen. Gohlke used a technique which permitted successively photographing the full spectral range and four overlapping segments of it on the same film. The lag between the appearance of the GC peak on the chromatographic recorder and the appearance of the mass spectrum on the oscilloscopic screen was of the order of a second. It was found that a single GC peak containing two or three components could usually be resolved by carefully examining several spectra recorded at various times across the width of the GC peak (basically, the same technique still used in the GC/MS/computer systems).

A little later, the Bendix Corporation made available an analog unit to permit oscillographic recording of spectra. This circuit took a short time interval at one end of the scan and slowly swept that interval along the time axis of the superimposed repetitive scans toward the other end of the spectrum. That had the effect of (a) summing many scans at a given m/z value for a great improvement in ion statistics (as had been accomplished on the oscilloscope screen by the persistence of the phosphor) and (2) producing the improved spectrum at an apparent scan rate which was much slower than the electronic cycle time of the pulsed mass spectrometer. The apparent scan rate of the analog output was compatible both with the needs of most analytical operations and with the rate limitations of recording apparatus. For monitoring phenomena in the microsecond range, the photographic recording of individual sweeps of the oscilloscope was still possible.

Lindeman and Annis [24] used first a modified Consolidated Electrodynamics Corporation (CEC) Model 21-620 cycloidal path mass spectrometer and later a CEC Model 21-103B (a Dempster-type instrument with a resolution of 1 part in 600). The gas chromatographic packed columns were of 10- and 30-ft lengths. The carrier gas was helium, at a flow rate of 20 to 50 ml/min. The connections from the outlet of the thermal conductivity GC detector were made by capillaries; in the first case entering the mass spectrometer inlet system just above the gold pinhole leak; in the second case, between the leak and the inlet vacuum pump, with the ion source pressure regulated by adjusting one of the valves to the inlet vacuum pump. The scan rate of the 21-103B was increased to sweep the m/z range from 40 to 160 in 30 sec.

Lindeman and Annis plotted single-ion profiles (now among the indispensable tools in the software kit of the modern GC/MS/computer system) across multicomponent GC peaks and carefully verified that acceptable quantitative analyses could be obtained. Figure 2 presents the gas chromatogram of C_5-C_8 hydrocarbons in a California naphtha, from a 10-ft column. (The authors pointed out that much better resolution could be achieved by GC than that used for demonstrative purposes in the quantitation experiments.) Figure 3 shows plots of characteristic ion profiles for the nine components present in GC peak 6. It was reported that a number of components in the chromatogram, such as benzene, cyclohexane, 3-methylhexane, 2-methylhexane, and n-heptane, were determined readily from the mass spectra because of the presence of very little spectral interference. For trans-1,2-dimethylcyclohexane and the combined cis- and trans-dimethylcyclopentane, the investigators resorted to simultaneous equations using m/z 112 and 97.

Henneberg [25,26] recorded single-ion profiles across the whole chromatogram with the GC column connected to a modified CEC 21-620 mass spectrometer. In the case of packed columns, the effluent stream passed through the thermal conductivity detector and was then split, with a small portion going to the mass spectrometer. The single-ion chromatogram was compared with the thermal conductivity chromatogram. Repetition with the choice of a different ion yielded further information about the nature of the mixture. The single-ion plot was recommended for use with capillary columns.

Widmer and Gaumann [27] used the capillary stream sampler to connect a packed column to a Bendix Model 12-101 mass spectrometer. Sensitivity, GC resolution, speed of recording, and the difficulties encountered with very small concentrations were studied.

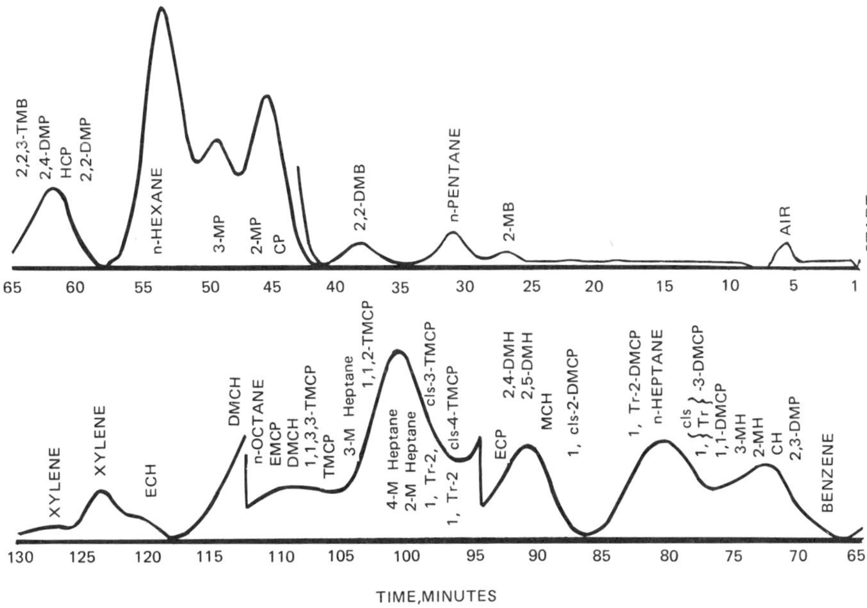

Fig. 2 Gas chromatogram of C_5-C_8 hydrocarbons in a California naphtha, (Reprinted with permission from L. P. Lindeman and J. L. Annis, *Anal. Chem. 32,* 1742 (1960). Copyright 1960 American Chemical Society.)

The problems with components of low concentration included memory effect, low abundance sensitivity, and gas noise found to have the same frequency distribution as the flicker noise of thermal cathodes.

B. Trapping Systems—Off-Line and Coupled

Before integrated GC/MS systems reached their present state of development, it had become the customary practice to recover GC fractions in small cold traps and to introduce the trapped fractions as batch samples into the mass spectrometer. The trapping technique required skill and concentration for successful application to a complex mixture. It became well developed and reliable in many laboratories, and continued in use as the standard method during the interim in which GC/MS systems were undergoing a parallel development. Only the most elementary problems in such fields as natural product identification can be solved by mass spectrometry alone. As a method for the isolation of compounds needed for other chemical and spectrometric work, the trapping technique has not been superceded.

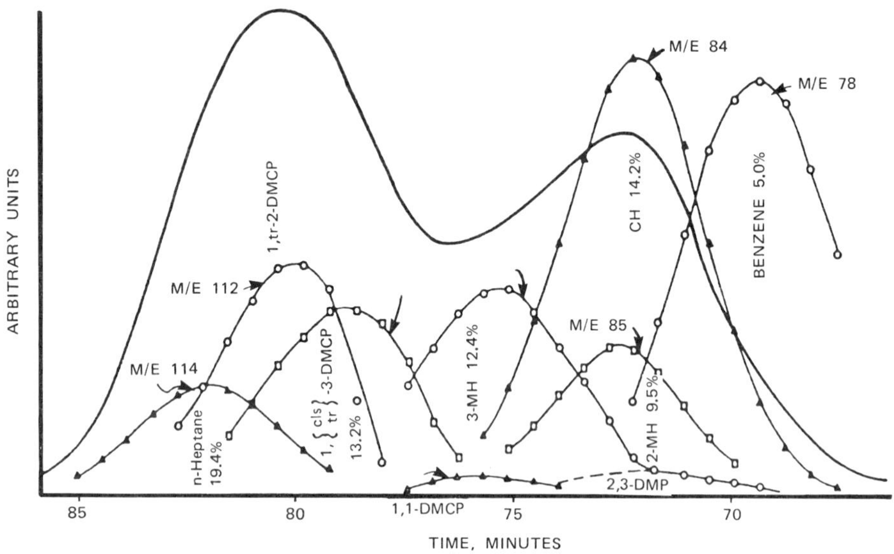

Fig. 3 Plot of intensities characteristic of compounds in peak 6. (Reprinted with permission from L. P. Lindeman and J. L. Annis, *Anal. Chem. 32*, 1742 (1960). Copyright 1960 American Chemical Society.)

Levy et al. [28] used columns of their own design and construction, operating at 300°C, to separate 67 compounds present in a refined paraffin wax. The starting material consisted of n-paraffins, 1-cyclopentyls, 1-cyclohexyls, 2- to 5-methyl-branched paraffins, 1-phenyls, and mono- and dimethyl cyclic compounds. Other, lower melting, branched paraffins had been removed in the course of wax-refining processes.

The substrate in the first column was a bottoms fraction (residue) from the molecular distillation of a microwax, which proved to be excellent for the separation of n-paraffins in the $C_{20}-C_{32}$ range, but left those peaks contaminated with 1-cyclopentyls, 1-cyclohexyls, and methyl-branched paraffins. In the second column, a substrate of an asphaltene residue further separated the n-paraffins from the isoparaffins and cycloparaffins. Fractions from the CEC X26-201 gas chromatograph were trapped and were subsequently analyzed in a CEC 21-103A mass spectrometer.

A very simple technique sufficed for the trapping of the high-boiling wax fractions. The collection vessel was a glass tube 22 mm long and 11 mm in diameter, closed at one end. In use, the open end was placed against the asbestos tape insulation covering the 1/8-in. exit tube. The carrier gas continued to flow out over the rim of the little glass cup. Cooling was in this case not necessary.

It was not desirable because it also cooled the exit line and caused condensation of wax in the exit tube. Collection efficiency with the tubes at room temperature was 65%. With the collection tubes kept at 75°C by a brass block and cartridge heater, the efficiency was raised to 85–90%.

In recalling briefly this paper from the Atlantic Refinining Co., it is not inappropriate to remind younger readers of the large part played by laboratories of the petroleum-refining industry in the founding stages of mass spectrometry and gas chromatography. In addition to their very broad and general contribution to chemistry, the instrumentation developed and the fundamental knowledge gained in response to the economic needs of a basic industry where applied directly to research in geochemistry and the fossil chemical evidence for the evolution of life [29–32]. Dean and Whitehead [33] and Bendoraitis et al. [34] found phytane (2,6,10,14-tetramethylhexadecane) and pristane (2,6,10,14-tetramethylpentadecane) in petroleum. Those two compounds were found by Oro et al. [35] in chert from the Gunflint iron formation (1.9 billion years old) of the north shore of Lake Superior. Robinson et al. [36] studied the variation of hydrocarbon analyses with depth in an oil shale core through 900 ft of the Green River Formation in Rio Blanco County, Colorado. The isoprenoid compounds, phytane, pristane, 2,6,10-trimethylpentadecane, 2,6,10-trimethyltridecane, and 2,6,10-trimethyldodecane were all found consistent with an origin in the phytol group of the chlorophyll molecular structure.

The next step beyond off-line trapping was the design of on-line trapping interfaces to hold up the second of two chromatographic fractions whose retention times were too close for convenient scanning in the slower mass spectrometers, or to allow concentrating the weaker fractions. Ebert [37] designed a valving and trapping system, installed between a Burrell gas chromatograph and a Bendix time-of-flight mass spectrometer. The leak valve was built by taking an on-or-off toggle valve, drilling out the passages to permit uninterrupted flow by way of the normal entrance and exit lines, drilling out the seat, and replacing some parts with those of a needle valve. The needle valve opening could be adjusted as desired and then left in a fixed position while the other valves were being operated. After the modifications, the chromatographic effluent flow through the original ports was continuous. The cutoff feature now opened or closed access to the leak, which led to a third line connected into the mass spectrometer. Other valves were installed before and after the leak valve, to permit trapping and holding a cut of the GC effluent.

Two of these systems were mounted in parallel, to allow using one as a reservoir while the GC effluent passed through the other. That provided a means of handling a second GC peak that followed quickly after the preceding one. The trapping technique made it possible to hold the GC fraction long enough for the analog unit to scan the mass spectrum at a rate suitable for oscillographic recording, which usually required less than a minute. The lines were wrapped with heating tape and asbestos to provide temperature control.

Miller [38] published an improved version of the Ebert manifold, again used to couple the gas chromatograph to a Bendix time-of-flight mass spectrometer. Because of corrosion problems encountered with the earlier system, copper and brass were eliminated, and only stainless steel and teflon were permitted to contact the sample vapor (Teflon only in valve seats and gaskets). A cold trap, capable of being heated rapidly by direct resistive heating, was substituted for the second direct-run trap and was used to concentrate minor constituents for greater sensitivity. All of the valves were of the quick-acting toggle type. Internal volume was minimized in all parts of the system, to avoid the degradation of chromatographic resolution by the inlet system. The manifold was enclosed in an oven to provide better temperature control during the analysis of higher boiling materials.

Levy et al. [39] continued the effort to solve various GC/MS interfacing problems. They used a gas chromatograph and a Bendix Model 14 time-of-flight mass spectrometer to study the catalytic cracking reactions of n-nonane in a laboratory reactor. The system of connecting lines included a sampling loop and valves between the reactor and gas chromatograph, and a vacuum manifold with leaks and valves between the gas chromatograph and mass spectrometer (Fig. 4). Three modifications of the mass spectrometer were made to aid in getting mass spectra from chromatographic peaks where the sample size varied rapidly over several orders of magnitude. This problem of the dynamic range of intensity values awaited ultimate solution by the autoranging circuits of the digital data processing interfaces of the present day integrated systems. The first of the three modifications was a high-voltage rotary selector switch to make possible rapid and reproducible changes of multiplier voltage. The second was a switch connecting to the pulsed gating circuitry of the multiplier to obtain a total ionization signal. That switch was used to investigate the time delay between the GC thermal conductivity detector's indication of a peak maximum and the appearance of the mass spectral intensity maximum for the same GC peak. The third modification used an electrical connection between the scanner and controller modules of the spectrometer to produce a second mass spectrum of a GC peak a short time interval after the preceding spectrum. Having more than one spectrum from different times during the passage of the same GC peak made it possible to detect composition changes within a composite GC peak.

C. Early Experiments with Capillary Columns

Gohlke [40], who had in the meantime moved from Dow's spectroscopy laboratory in Midland, Michigan to the company's eastern re-

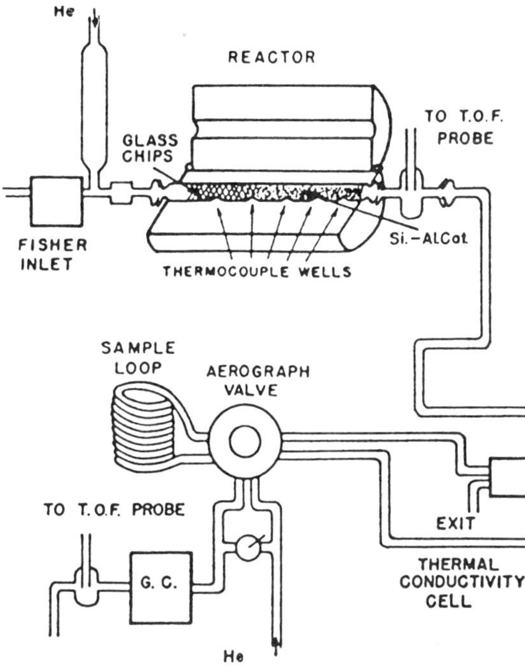

Fig. 4 Schematic diagram of catalytic cracking reactor connected to as chromatograph and mass spectrometer. (Reprinted with permission from E. J. Levy, E. D. Miller, and W. S. Beggs. *Anal. Chem. 35,* 946 (1963). Copyright 1963 American Chemical Society.)

search laboratory in Framingham, Massachusetts, connected a 100-ft capillary column directly into the ion source of a Bendix time-of-flight mass spectrometer. The internal diameter of the tube was 0.010 in. This column had an uncoated metal surface operating in the gas-solid adsorption chromatographic mode. The inlet end of the column was under a helium pressure of 1 psig, and the outlet end in the ion source discharged the effluent into a region of estimated pressure 10^{-6} torr. The mass spectrometer, operating at 10,000 scans/sec, was equipped with a bistable multivibrator which, on alternate scans, switched the spectral output between two collectors in the electron multiplier circuit. On odd-numbered cycles, 1, 3, 5 ..., (2 n − 1), ..., collector no. 1 was used to gather the total ion current, plotted as the vertical coordinate (abundance) in the gas chromatogram. On the even-numbered cycles, 2, 4, 6, ..., 2n, ..., collector no. 2 provided a normal mass spectrum. A block diagram of the circuit is shown in Fig. 5.

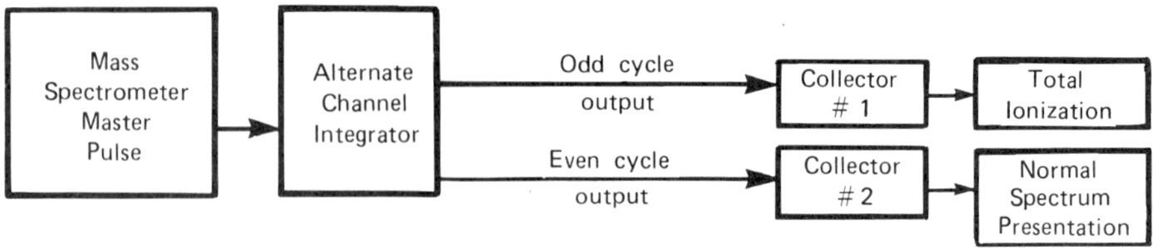

Fig. 5 Block diagram of alternate cycle integrator connections to mass spectrometer. (Reprinted with permission from R. S. Gohlke, *Anal. Chem. 34,* 1332 (1962). Copyright 1962 American Chemical Society.)

Noting that it was customary with stream-splitting apparatus to use about 1% of a packed-column GC effluent stream, and that the total flow through a capillary column was of the order of 1% of that from a packed column, McFadden et al. [41] introduced the entire effluent stream from a 250-ft (76.2-m) capillary column into the ion source of a Bendix time-of-flight mass spectrometer. The chromatographic equipment was operated in a conventional manner, except that the exit end of the column was under mass spectrometer vacuum. It was found that the reduction of the GC inlet pressure by a corresponding 15 lb/in.^{2} compensated for the effect of the vacuum, giving retention times close to those observed in normal operation where the column discharges at atmospheric pressure. Although the Dow 710 silicone oil coating in the capillary was not as well chosen for separatory properties as the neopentylglycol succinate used in a packed column, comparison of the packed-column chromatogram (Fig. 6) with the capillary chromatogram (Fig. 7) shows the superiority of the capillary system in the analysis of a sample of orange juice condensate oil. Data from other fruit and vegetable extracts were also presented.

Brunnée et al. [42] using an Atlas CH4 mass spectrometer and a capillary gas chromatograph investigated several experimental, theoretical, and statistical aspects of GC/MS. The connection to the ion source was made with a sort of stream-splitting device consisting of a length of smaller capillary than the column, which took a fraction of the effluent into the ion source while permitting the main stream to be discharged at atmospheric pressure. They also described [43] the performance of a double ion source, in which one part with 20-eV electrons produced the signal generating a gas chromatogram. The other part of the source employed 70-eV electrons in the production of mass spectra. Dorsey et al. [44] connected their capillary column to the inlet system of a CEC cycloidal-focusing mass spectrometer with a stream-splitting device of smaller

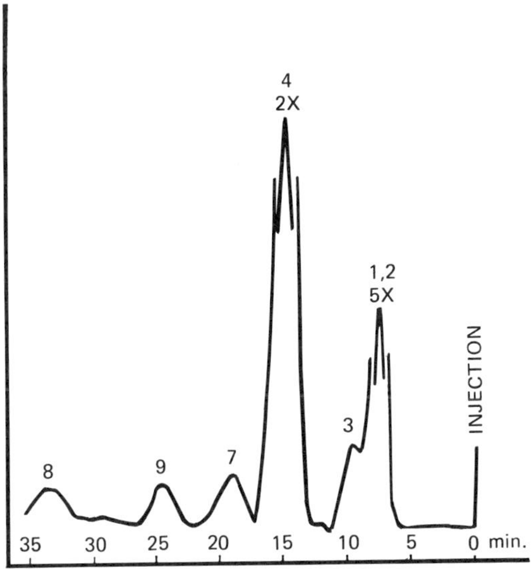

Fig. 6 Packed-column chromatogram of orange condensate oil on 7-ft neopentylglycol succinate column at 100°C (peaks identified in Fig. 7). (Reprinted with permission from W. H. McFadden, R. Teranishi, D. R. Black, and J. C. Day, *J. Food Sci. 28,* 316 (1963). Copyright 1963 Institute of Food Technologists.)

capillary tubing, similar to that used by Brunnée et al., to permit the carrier gas to leave the column at atmospheric pressure. Lipsky et al. [45] applied their Teflon helium separator to support-coated open tubular columns as well as to packed columns (Sec. III.D). Modzeleski et al. [31], in the analysis of alkanes from sedimentary rocks, used a Watson-Biemann separator to couple a capillary column to a Hitachi RMU-6E mass spectrometer. Other separators specifically designed for use with capillary or open tubular columns include the porous silver membrane microseparator of Grayson and Levy [46] (Sec. III.D) and the Llewellyn type designed by Black et al. [47].

Váradi and Ettre [48] studied the effect of evacuation of the exit end of a capillary column upon separatory power and retention time, using their QQ detector [49], a Machlett ML-494-A radiofrequency mass spectrometer tube. With the input pressure at 20 psig and the output at 10^{-4} torr, they reported that retention times were a little more than half those measured with the outlet at atmospheric pressure. The separatory capabilities of the column were said to be equal to or a little better than those demonstrated with the column discharge at atmospheric pressure (see also McFadden et al., above on the reduction of inlet pressure by one atmosphere.).

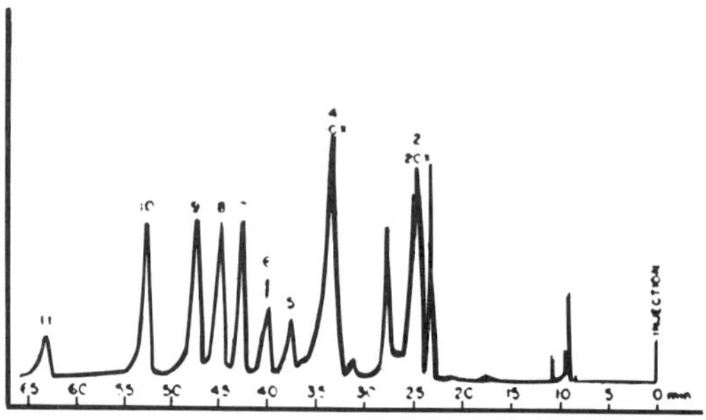

Fig. 7 Capillary chromatogram of orange condensate oil on 250-ft DOW 710 silicone oil column at 97°C. (Reprinted with permission from W. H. McFadden, R. Teranishi, D. R. Black, and J. C. Day, *J. Food Sci. 28*, 316 (1963). Copyright © 1963 Institute of Food Technologists.)

Banner et al. [50] coupled a capillary column to a rapid scanning high-resolution mass spectrometer. The stainless steel capillary was 100 ft (30.5 m) long by 0.010 in. (0.254 mm) in internal diameter, coated with Apiezon L. The sample splitting ratio was 50:1, with a column flow rate of 1 ml/min. The splitter consisted of a 30-cm length of 0.002-in. i.d. uncoated stainless steel capillary tubing connected into the column immediately before the flame ionization detector. This device maintained a constant flow at about 0.1 ml/min into the mass spectrometer ion source, independently of the flow rate in the column, resulting in an ion source pressure of about 5×10^{-4} torr. Helium was the only carrier gas that could be tolerated at that pressure in the ion source. The mass spectrometer was an AEI Model MS10 (12-in. radius, 90° sector magnetic analyzer), with a scanning speed of 0.5–300 sec in the repetitive exponential mode. Over the m/z range of 20–200 in 1 sec, the resolution (10% valley definition) was 200. Data from the analysis of hydrocarbons and fatty acid esters were presented. In additional comments at the end of the prepared paper, the authors described the replacement of the stainless steel sample splitter by a glass inlet system.

Studier and Hayatsu [51] built an inlet system in the form of a vacuum manifold to join their capillary column to a Bendix Model 12 time-of-flight mass spectrometer, permitting samples to be loaded by vacuum distillation without the presence of a solvent. Modifications of the source and the detection circuitry further increased the sensitivity. The apparatus was used to analyze complex mixtures of hydrocarbons formed in a Fischer-Tropsch synthesis.

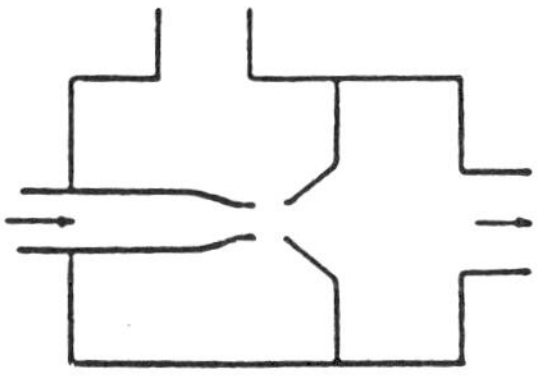

Fig. 8 The principle of the molecular jet separator. (From E. W. Becker, K. Bier, H. Burghoff, O. Hagena, P. Lohse, R. Schutte, P. Turowski and F. Zigan, in *Proceedings of the Second United Nations International Conference on the Peaceful Uses of Atomic Energy*, Volume 4, United Nations, Geneva (1958).)

D. Molecular Separators for GC Sample Enrichment

Another set of developments was concerned with the effort to make better use of a GC sample than was possible with the original stream-splitting connections, which wasted the major portion of the GC effluent from packed columns. It was obvious that the selective removal of carrier gas to yield a more concentrated input to the mass spectrometer ought to yield better results from the smaller peaks in a chromatogram. Becker et al. [52], while conducting basic research toward the preparation of intense molecular beams, noticed the potential utility of pressure diffusion in a free-expanding jet stream for the separation of isotopes. In the extensive studies which followed, the operating principles were described and tested. When the mean free path of the molecules in a gaseous mixture is small compared with the diamter of the jet (Fig. 8), the separatory effect is governed by diffusion theory. The application of Maxwell's equation [53,54] gives as a first approximation for b, the density of diffusion flow in a binary gas mixture, the equation

$$b = -\nu D\left(\text{grad } r_1 + r_1 r_2 \frac{M_2 - M_1}{r_1 M_1 + r_2 M_2}\frac{1}{p}\text{ grad } p + \frac{k_T}{T}\text{ grad } T\right) \tag{18}$$

In this expression, ν is the molecular density, D the diffusion constant, r the mole fraction, M the molecular mass, p the pressure, T the temperature, and k_T the thermal diffusion ratio. The three terms on the right side describe ordinary diffusion, pressure diffusion, and thermal diffusion. In a free, expanding jet, the pressure diffusion term dominates over the thermal diffusion term. The jet, or nozzle, process was applied to the separation of uranium isotopes in the form of UF_6 [55].

The separation of isotopes is based upon small differences in mass and other physical properties and requires many separatory units operating in a cascade arrangement. Because of the much larger differences in properties between organic molecules and the helium carrier gas, the necessary concentration of the gas chromatographic sample can be accomplished with a simpler apparatus. Ryhage [56] coupled two molecular separators of the Becker type in series and inserted them between a gas chromatograph and a mass spectrometer. The column was a 4-m by 2-mm glas tube packed with 1% SE-30 on 100- to 120-mesh acid washed and silanized Gas-Chrom P. The mass spectrometer was an Atlas CH4 equipped with a fast recording system, capable of scanning the m/z range from 12 to 500 in 1 to 2 sec. The first separator was connected to a forepump with a speed of 100 L/min at atmospheric pressure and the second to an oil diffusion pump having a speed of 80 L/sec at 10^{-6} torr. The flow rate of the helium carrier gas was kept at 2.5 ml/mm^2 of cross-sectional area per minute, for this and other columns of different diameters. In the example described, the flow rate was 8 ml/min, at an inlet pressure of 1.6 kg/cm^2. A valve between the low-pressure separator and the ion source was temporarily closed to keep the solvent peak from the GC injection out of the ion source. The accelerating potential was kept constant at 3000 v. Because the ionization potential of helium is 24.8 v, helium ions were almost eliminated by setting the energy of the ionizing electrons at 20 eV. The two jet separators increased the concentration of the organic chemical sample vapor in the gas stream entering the ion source by a factor of at least 100. Figure 9 shows a diagram of the system, which was the prototype for the LKB 9000.

Watson and Biemann [57–58] developed a separator in which a fritted-glass tube was enclosed in a vacuum jacket. Helium was pumped away through a line attached to the jacket, resulting in enrichment of the sample entering the ion source by a factor of 10 or more. In addition to the fritted tube and its evacuated jacket, the pressure reduction system included a small capillary between it and the gas chromatograph, another constriction at the other end, and provisions for heating the line between the separator and the ion source, as shown in Fig. 10. About 20% of the effluent from a 6-ft x 1/8-in. column entered the separator, which in the application described served as a combined enrichment device and stream splitter.

Another unique feature of the system was the use of a high-resolution mass spectrometer, a CEC Model 21-110. That instrument, of the Mattauch-Herzog type, recorded spectra on a photographic plate. As many as 30 spectra could be put on a single plate. When a mass spectrum is scanned, a large portion of the scanning time is wasted on the blank spaces between peaks. The simultaneous recording of all the lines in the spectrum makes much more efficient use of the time and thereby yields a great increase in sensitivity. It also avoids the effect of sample concentration change on the relative

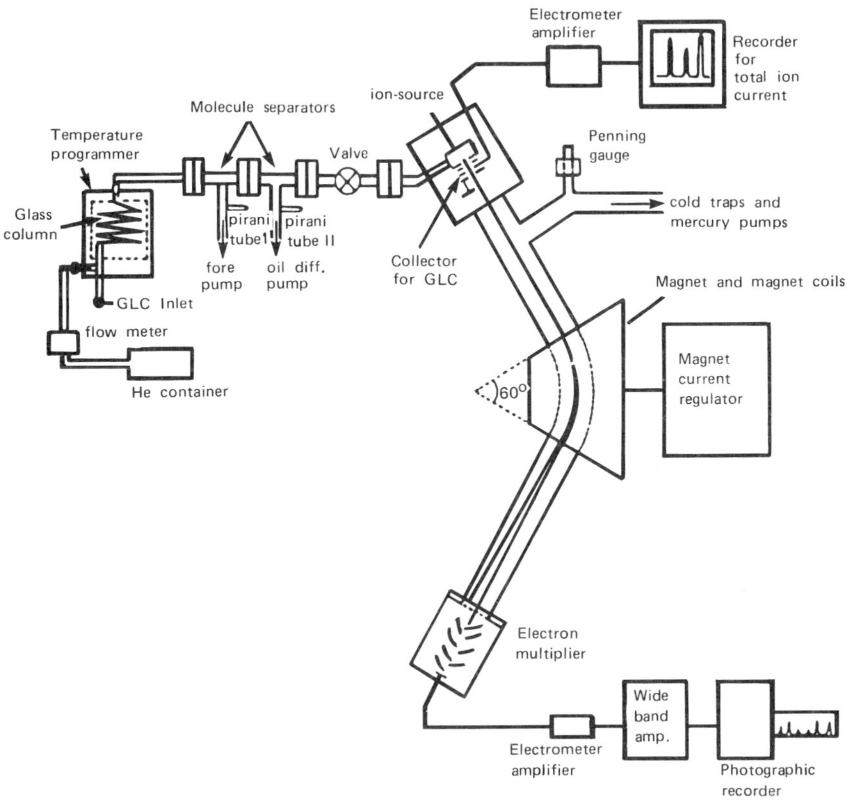

Fig. 9 Schematic diagram of gas chromatograph/mass spectrometer combination. (Reprinted with permission from R. Ryhage, *Anal. Chem. 36*, 759 (1964). Copyright 1964 American Chemical Society.)

intensities of mass peaks within a spectrum, an effect still familiar in the scanning of small, narrow chromatographic peaks. Another advantage is the recording of mass values with sufficient precision to determine the elemental compositions of the ions.

The glass-jacketed porous frit, the central element of the Watson-Biemann separator, became available in a version manufactured by the Corning Glass Works. The Watson-Biemann separator was very popular, and a number of modified versions were designed and tested. They include Markey's [59] retractable glass frit, the porous silver frit versions of Cree [60,61], Blumer [62], and Grayson and Levy [46], and the porous stainless steel separators of Krueger and McCloskey [63] and Copet and Evans [64]. MacLeod and Nagy [65] improved the performance of a fritted glass separator

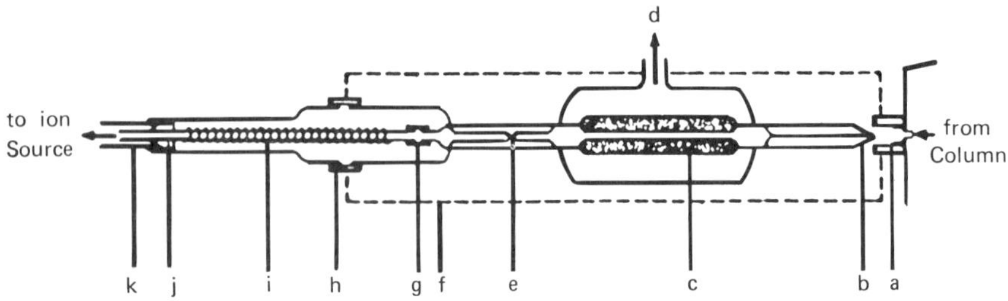

Fig. 10 All-glass pressure reduction system: (a) Teflon tubing; (b) capillary orifice; (c) fritted-glass tube; 5-μm pore size; (d) mechanical pump; (e) constriction, 0.12 mm i.d.; (f) heating mantle; (g) Teflon connection; (h) silicone rubber connection; (i) silver wire (for heat conduction); (j) Teflon plug; (k) heated glass connection to ion source. (Reprinted with permission from J. T. Watson and K. Biemann, *Anal. Chem. 36,* 1135 (1964). Copyright 1964 American Chemical Society.)

by silanization. Krueger and McCloskey described two interfaces, one for permanent mounting and one a separator-probe assembly, for intermittent use by way of the direct inlet probe vacuum lock on the CEC 21-110 high-resolution instrument. The catalysis of thermal reactions by the metal surface was essentially eliminated by silanization.

Grayson and Bellina [66] treated the requirements for the design of effusion separators and presented the solutions for the design parameters in nomographic form. They described apparatus and technique for determining the conductance of frits, tabulated data, and discussed the application of their results in terms of porous silver membranes. In the effusion separator of Brunnés et al. [67,68], the pores of the frit have been replaced by two concentric circular slits of adjustable width. This elegant simplification of the separator design implies some corresponding simplification in the theoretical treatment of its operation.

The most used separator recently has been the Gohlke single-stage glass jet [18], shown in Fig. 11. Some information on its operating parameters is provided by Bonelli et al. [69].

The membrane interface between gas chromatograph and mass spectrometer was described by Llewellyn and Littlejohn [70] at the Pittsburgh Conference in 1966. The membrane was usually of silicone polymer (polysiloxane). This separator is different from the others described in that it is the organic sample vapors that are transmitted through the membrane itself, by solution and diffusion, while the carrier gas is largely excluded by the membrane from entrance into the mass spectrometer. Llewellyn's patent [71] describes a dual-membrane separator with a throttle valve in the vacuum line from

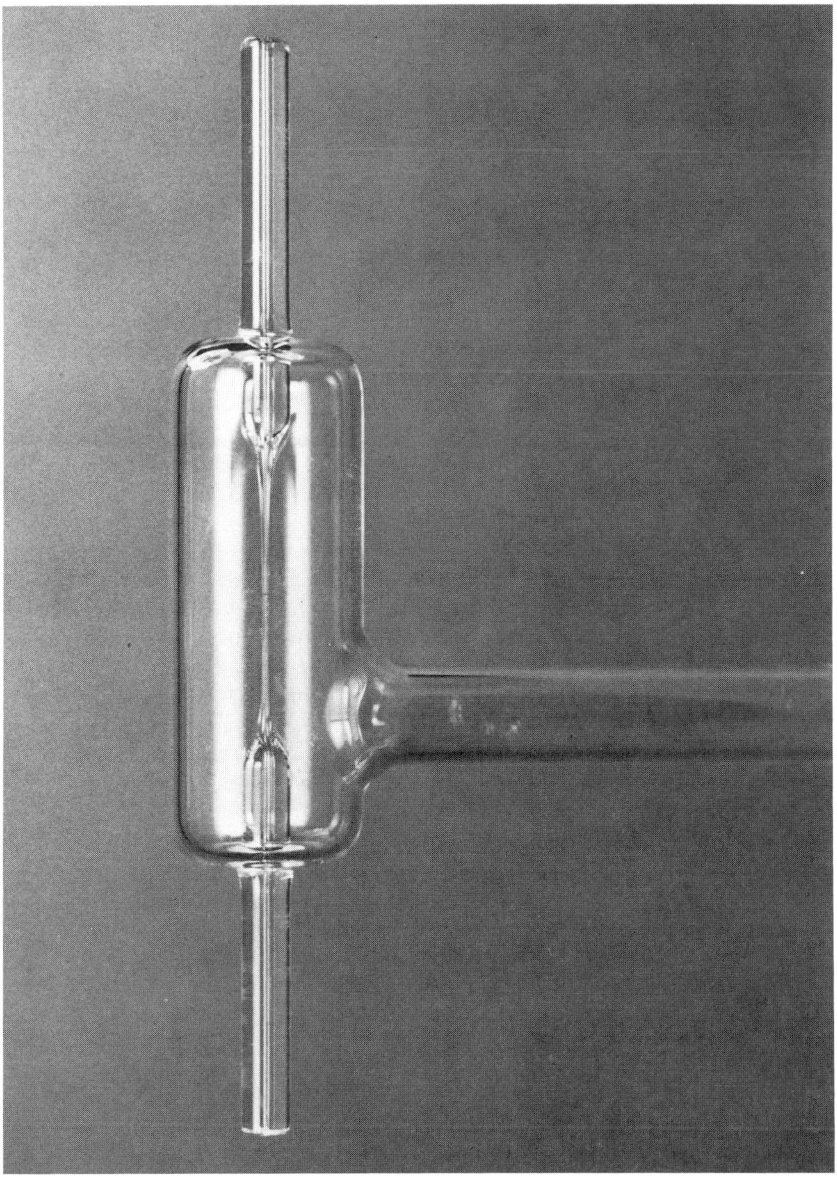

Fig. 11 Gohlke glass jet separator from Finnigan 4021 system.

the space between the membranes. The vacuum pump and valve provide for adjustment of the quantity of sample fed into the mass spectrometer as the size of the gas chromatographic peak varies. The range of transmission is given as being from 80 to 2%. Another patent [72] describes the use of a gas transfer switch and a dummy carrier gas stream between GC and MS. A German patent [73] discloses the splitting of the GC effluent stream between a GC detector and the duel-membrane separator, with provision for recycling material from the space between the membranes into the main (GC detector) stream and for control of the throttle valve by the GC detector or the MS detector.

A modification of the Llewellyn membrane separator was designed by Black et al. [47] to handle the output of a 0.030-in. (0.75-mm) i.d. column. Another variation, made by Hawes et al. [74], featured all-glass construction except for the membrane itself. Both were single-membrane separators.

The Teflon capillary separator described by Lipsky et al. [45] consisted of a 7-ft (2.13-m) length of thin walled (0.005-in., or 0.13-mm) Teflon capillary tubing (0.010-in., or 0.25-mm i.d.) with a surface area of 20 cm^2, enclosed in an evacuated jacket. The helium carrier gas passed through the wall of the Teflon tube and was removed by the vacuum pump attached to the jacket. The proper function of the apparatus depended on temperature control. It was reported that at 200°C above ambient not enough helium was removed but that at 300°C above ambient the loss of the sample was complete. Optimum performance was reported to have been achieved in the range of 250–260°C above ambient, where the relative permeability of the Teflon to helium and to the organic sample was most favorable. Grayson and Wolf [75] compared a similar Teflon separator with a fritted glass one of the Watson-Biemann type. The mechanism of operation of the Teflon separator is not well understood and it was not widely used.

Another very interesting membrane separator has a membrane of palladium-silver alloy. In this case, the carrier gas used is hydrogen, which can be totally removed by the separator. No other gas passes through it, and the only mechanisms suggested for sample loss are adsorption and catalytic reaction. Lucero and Haley [76] developed mathematical models describing the enrichment factor, time rate of response, and temperature response of semipermeable membrane separators in general and of palladium-silver alloy membranes in particular, in each case as functions of the mass transport properties and geometric characteristics of the membranes. The equations are applicable to the establishment of performance limits, the evaluation of the relative importance of design parameters, the design of experiments, and the interpretation of experimental results. Lovelock et al. [77] used the separator in gas chromatographic analyses. Further description of the GC/MS system built for a landing on Mars and the organic analysis of Martian soil was published by Simmonds et al. [78]. The separator was said to serve as its own indicator of adsorption because hydrogen removal requires a clean surface. Catalytic hydrogenation occurred only with certain specific types of unsaturated compounds. Temporary poisoning, for a time generally corresponding to no more than twice the GC peak width of the responsible compound, was found to be caused by SO_2, H_2S, mercaptans, and iodine compounds.

E. Atmospheric Pressure Ionization

Horning and coworkers [79] described an atmospheric pressure ion source in which ionization was accomplished by a 12.5-mCi nickel-63 radiation source. The source was separated by a 25-μm pinhole aperture from the high-vacuum system of a modified Finnegan Model 1015 quadrupole mass spectrometer. The modified system was originally designed by the Franklin GNC Corporation for its plasma chromatograph. Samples in the form of 1–2 μl of solution in benzene, methanol, ethanol, chloroform, or other solvents were injected into a stream of nitrogen carrier gas. Both positive and negative ion spectra were studied. The ion–molecule reactions and electron capture reactions involved were discussed. The sensitivity of the system was said to be adequate for a sample of 5–10 pg in single-ion monitoring and for as little as 25 pg for a scanned spectrum.

For an interesting example, the system was applied [80] to measurements of the nicotine content of urine from smokers and nonsmokers in the same laboratory areas, indicating that nonsmokers absorbed nicotine from the air to give about 5% of the urine nicotine concentration of the smokers. Ion source development continued, and studies of negative ion spectra were extended [81–83]. Continuing developments in instrumentation are described in more recent papers and reviews [84,85]. The system appears well fitted for certain applications in direct environmental sampling. One of its limitations is supplied by the fact that chemical ionization can hardly replace electron impact ionization in the general field of mass spectrometry.

F. Summary and the Current Trend: Flexible Silica Capillaries

Mass spectroscopists analyzing chromatographic fractions collected off-line could see the experimental logic and the practical potential of a combined GC/MS instrument. Initial experiments in which a portion of a GC effluent stream was split out and run through a mass spectrometer demonstrated the soundness of the idea but wasted much of the available sample, thus reducing sensitivity for the less abundant components. On-line trapping schemes served under certain circumstances but slowed the process, introduced complications for less volatile components, and sometimes blurred the resolution of close or partially resolved GC peaks. Several types of molecular separators for the selective removal of carrier gas contributed heavily to progress. The two parameters that describe the performance of a separator are the separation or enrichment factor N and the efficiency of yield Y [1,75]. The first, N, is the ratio (concentration in carrier gas entering mass spectrometer)/(concentration in carrier gas leaving chromatographic column). The second, Y, is the percentage of the injected sample that enters the mass spectrometer.

Early experiments with capillary columns pointed the way to further development, to a large extent consisting of better pumping systems. Capillary gas chromatography appealed to the imagination through the theoretical treatments of Golay [86–88], offered increased GC resolution, and could be handled without stream splitting or carrier gas separation. Atmospheric pressure ion sources proved useful in some environmental sampling situations but were not expected to compete with evacuated systems in operations on isolated samples or gas chromatographic effluents. The final advance in capillary column GC was the introduction of the fused silica capillary by Dandeneau and Zerenner in 1979 [89]. The subject has been reviewed by Jennings [90]. The success of fused silica capillaries has generated a popular movement to replace packed columns, in many uses, by capillaries.

Interest in GC/high-resolution mass spectrometry has continued, with improvements in magnets (Burns et al. [91]) and other hardware, making possible increased scanning speed. The resulting gas chromatographic/mass measurement capability is of especial value for work with heteroatomic unknown molecules, such as biochemicals or their derivatives. Where thermal volatilization is not possible, the high-resolution mass spectrometer gives good service in combination with the newer soft ionization techniques.

Except for these applications of high-resolution mass spectrometry, the most popular instrument today is the combination of a flexible silica capillary gas chromatograph with a quadrupole mass spectrometer, with the full GC effluent stream introduced directly into the ion source. For all types of mass spectrometry, the standard includes an up-to-date digital data-processing system. The earlier developments in the computer processing of high-resolution data were soon followed by the application to GC/MS [92], and very good systems have become available from commercial sources. The one with which the author is most familiar belongs to the Finnigan 4000 series [93]. The following example will serve to illustrate how the common numerical problems of mass spectrometry are simplified by the use of the GC/MS/computer system.

In the calibration of the instrument, the scale of m/z values is related to a scale of time values or of values of some other parameter by taking the known m/z values of a reference compound and finding their corresponding locations on the time scale or other parameter scale. Upon further study of the problem of dynamic range of intensity, which was introduced during the brief mention of ion statistics in Sec. I above, we find that in order to increase the reliability of measurement of very weak peaks, we must drive the stronger ones off the intensity scale to such an extent that their m/z values can no longer be determined accurately. The reason is that the stronger mass peaks in the spectrum have begun to saturate the detection system, which means that they are spreading laterally and becoming flatter on top. That often happens more or less accidentally when the mass spectroscopist has no ready access to an off-line gas chromatograph with which to check concentrations of samples brought to him. The situation is illustrated by the ion profiles across the large GC peak in Fig. 12.

In such a case, the m/z values of the more intense peaks often shift upward by a mass unit, while those of the less intense peaks still have correct values assigned. If we want a spectrum showing correct m/z values for the larger gas chromatographic peaks, we can often get it from the lower slopes of the oversize GC peaks. If we want a better look at the weaker mass peaks, we may find it in the central area of the large GC peaks. We cannot get both in the same scan, but we can plot a complete spectrum by intervening manually and combining parts of different scans, using ratios of large peaks and small peaks to peaks of intermediate size for peak size adjustment in the manual reconstruction of the spectrum from the fragmentary scans. That was what we did more or less routinely before we had computer assistance. If we had time, we could probably write a program to teach the computer how to do that for us.

The reason why the shift to an incorrect value is often in the direction of a larger m/z value may be that in organic compounds the more intense peaks are more likely to have strong peaks after them (C-13 isotopic peaks) than immediately before them. The data in Table 1 show the upward shift of the m/z 146 peak and the disappearance of what had been the m/z 147 peak as it is combined with the larger peak. The growth in intensity has probably filled in the valley between the peaks to such an extent that the system no longer recognizes separate peaks, and seeks a new centroid for the combined pair. In the example given, the erroneous value at m/z 147 might be accepted if the spectroscopist were under harassment to finish the job quickly. Peaks at m/z 147 are more common in the mass spectra of trimethylsilyl derivatives than peaks at 146, but the correct value in the example appears to be 146. It would be much more difficult to detect such malfunctions without the data system.

Table 1 The Shift of Computed m/z Value with Excessive Ion Intensity

	m/z = 146			m/z = 147		
Scan No.	Computed mass	% RA	I	Computed mass	% RA	I
750	146.12	80.78	6824	147.12	92.71	7832
751	146.12	88.28	7656	147.12	89.30	7744
752	146.12	97.96	10000	147.12	81.50	8320
753	146.12	100.00	15296	147.14	57.53	8800
754	146.12	100.00	26016	147.14	40.65	10576
755	146.12	100.00	43008	147.14	30.13	12960
756	146.12	100.00	68608	147.14	24.95	17120
757	146.12	100.00	93056	147.14	23.45	21824
758	146.12	100.00	103168	147.14	25.96	26784
759	146.22	100.00	136704	147.20	20.48	28000
760	146.44	100.00	193792	—	—	—
761	146.50	100.00	216064	—	—	—
762	146.58	100.00	239872	—	—	—
763	146.68	100.00	266240	—	—	—
764	146.86	100.00	310784	—	—	—
765	146.92	100.00	331264	—	—	—
766	146.94	100.00	340480	—	—	—
767	146.98	100.00	350208	—	—	—
768	147.04	100.00	367104	—	—	—
769	147.06	100.00	373248	—	—	—
770	147.10	100.00	382464	—	—	—
771	147.12	100.00	391168	—	—	—
772	147.14	100.00	397312	—	—	—
773	147.16	100.00	403968	—	—	—
774	147.20	99.63	413696	—	—	—
775	147.22	99.39	147792	—	—	—
776	147.24	98.56	421888	—	—	—
777	147.26	98.47	429568	—	—	—
778	147.26	98.47	429568	—	—	—
779	147.28	98.26	434688	—	—	—
780	147.26	97.81	433664	—	—	—
781	147.30	97.18	440320	—	—	—
782	147.30	97.30	442368	—	—	—
783	147.32	93.53	443904	—	—	—
784	147.32	96.68	446976	—	—	—
785	147.32	96.69	448512	—	—	—
786	147.34	92.47	452608	—	—	—
787	147.34	91.72	453632	—	—	—

Table 1 (continued)

Scan No.	m/z = 146 Computed mass	% RA	I	m/z = 147 Computed mass	% RA	I
788	147.34	92.02	454655	—	—	—
789	147.36	91.61	458240	—	—	—
790	147.36	95.63	459776	—	—	—
791	147.36	91.30	461824	—	—	—
792	147.38	89.98	464384	—	—	—
793	147.38	90.31	467456	—	—	—
794	147.40	88.59	468992	—	—	—
795	147.38	93.52	465408	—	—	—
796	146.70	63.79	279552	—	—	—
797	146.12	85.04	63680	147.14	20.94	15680
798	146.12	80.30	34432	147.14	25.52	10944
799	146.12	75.84	24416	147.14	29.52	9504
800	146.12	67.10	18144	147.14	31.30	8464
801	146.12	62.50	14000	147.14	34.71	7776
802	146.12	59.62	11008	147.14	39.25	7248
803	146.12	50.95	8576	147.12	42.63	7176
804	146.12	50.23	7128	147.12	48.82	6928
805	146.14	43.40	5896	147.12	49.47	6720
806	146.14	40.16	4960	147.12	52.91	6536
807	146.10	37.39	4272	147.12	57.07	6520
808	146.12	32.16	3556	147.12	56.37	6232
809	146.10	31.56	3444	147.12	57.84	6312
810	146.14	28.01	2940	147.10	58.99	6192

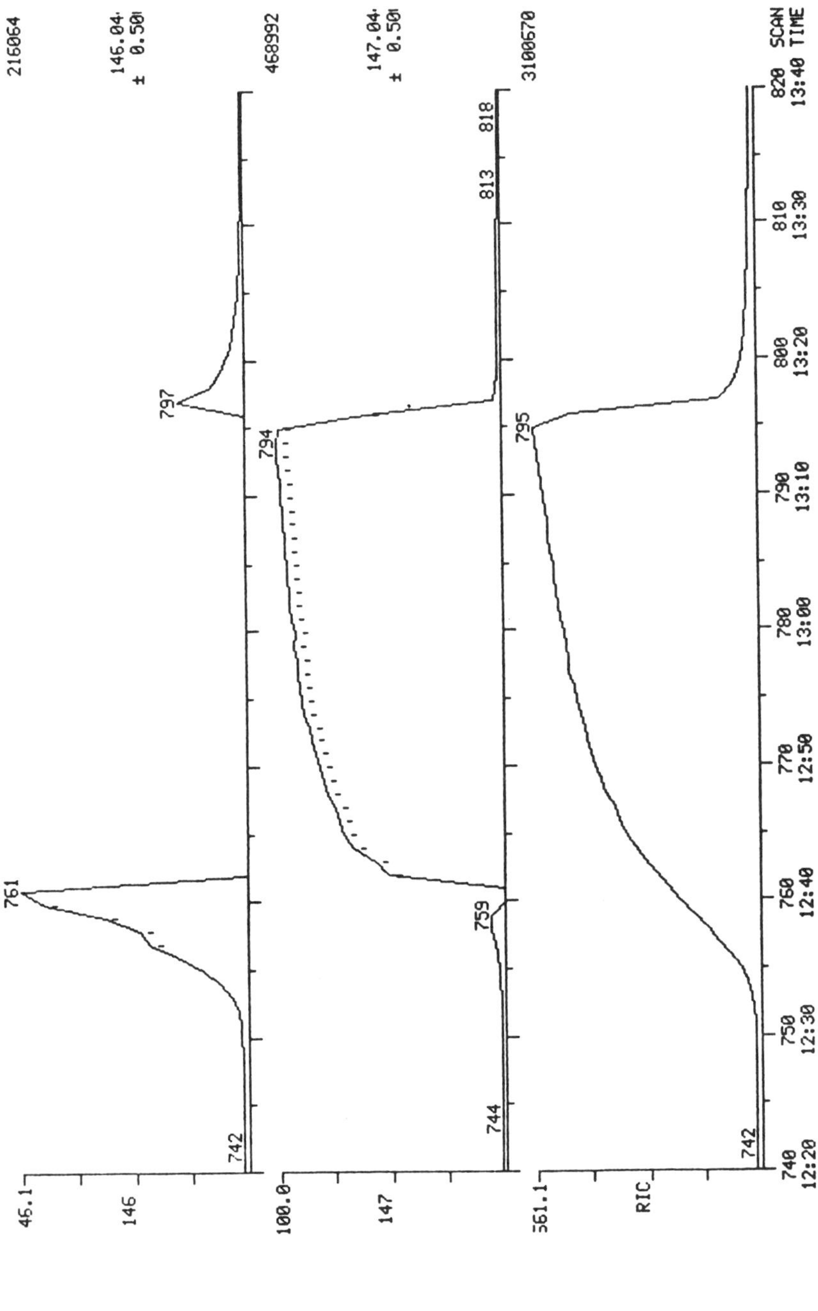

Fig. 12 The shift of computed m/z value with excessive sample pressure in the ion source.

IV. LIQUID CHROMATOGRAPHY/MASS SPECTROMETRY

The differences between LC/MS and GC/MS reflect the differences between liquid chromatography and gas chromatography. The mission of LC is to separate mixtures of compounds that are too labile, or too poorly volatile because of polarity or of high molecular weight, to be separated by GC. In the combination of LC with MS, the greatest difficulties arise from two sources, the greater flow of carrier fluid (perhaps 100–1000 times as much) in LC, and the more labile or less volatile nature of the sample components. Other problems associated with the carrier are the MS background contributed by any solvent that enters the ion source, the polar nature of some solvents, the use of programmed mixing of solvents, and the presence of nonvolatile components such as salts in the solvent.

Because of the great range of variability in the properties both of the sample and of the carrier, LC/MS has been slow to develop any system that is routinely applicable to a broad range of analytical problems. LC/MS has tended rather to focus on developments tailored to the needs of restricted areas of application, so that the larger area has been only incompletely covered by a mosaic of sometimes overlapping techniques. Their effectiveness has varied both between different techniques and within individual areas of applicability [94–105].

Yet the challenge of LC/MS development is of great pratical importance. It has been generally agreed that the reasons for pursuing the technology of LC/MS coupling are essentially two: (1) LC detectors have been, overall, a rather inadequate lot. Although the ultraviolet (UV) detector is sensitive where a sample component strongly absorbs the available wavelength, the obvious differences expected in the absorption spectra of sample components make up a very complex problem. More widely applicable detectors have not been sufficiently sensitive. (2) Beyond the problem of finding an LC detector, it is desirable to obtain the wealth of molecular structural information that LC/MS offers, provided that the sample or its ions can be volatilized without causing a degree of decomposition that renders the resulting spectrum useless.

A survey of LC/MS interfacing techniques discloses a certain parallelism between the approaches to LC/MS and GC/MS coupling. This parallelism begins with the pioneering split-stream direct introduction method and is carried along by the continuing dependence on trapping techniques, the expanding research involving capillary columns, the use of carrier removal techniques, and the applications of atmospheric pressure ionization. The fundamental differences are at the same time emphasized by the variations in those techniques from LC/MS to GC/MS and by the varying results obtained. Derivatization methods are present [106,107] but not necessarily for the same purpose. The more complex nature of the LC/MS problem is reflected in the growth of soft ionization techniques [108,109], especially in the ongoing development of the thermospray interface. Finally, the survey shows the progress that determined and powerful efforts have made in the pursuit of better and more widely applicable techniques.

A. Stream Splitting, with Direct Introduction

Tal'roze et al. [110,111] designed and constructed capillaries, including copper ones of slitlike cross-section, as stream-splitting devices for the direct admission of LC effluent into the ion source of a mass spectrometer. They employed various solvents, including mixtures, and noted the nature of the problems sometimes encountered. Dust particles, if not filtered out, blocked the small capillaries. Accumulations of less volatile sample components sometimes plugged them, leading to unstable flow as the plugs evaporated and reformed. Mixed solvents of limited miscibility underwent phase separations as solvent composition changed during evaporation, causing oscillations in spectral intensity [112]. Skurat [113] derived an equation for the capillary flow of a volatile liquid from a region of atmospheric or higher pressure into a vacuum. Tal'roze et al. [114] reviewed the LC/MS and other research.

Baldwin and McLafferty [115] approached the coupling problem by first admitting a liquid solution to a chemical ionization (CI) source from a glass capillary tube that had been drawn out to a very fine capillary outlet. The sample tube was inserted into the ion source by means of the existing probe inlet. Alternatively, they obtained spectra of relatively involatile compounds [116] by exposing the solid samples on the probe to the CI plasma. Arpino et al. [117] then reported a special inlet probe (Fig. 13) for continuous introduction of about 1% of an LC effluent stream into a CI source. Arpino et al. [118] added a cryogenic pump, modified the source block, and incorporated a data system, described by McLafferty et al. [119]. Sensitivities were generally in the nanogram range.

Henion [120,121] used a glass probe for interfacing a Waters LC to a Finnigan 1015 or 3300 CI mass spectrometer. The systems were applied to the separation of drugs and drug metabolites. Serum and Melera [122] and Melera [123] designed another direct inlet CI system. All of these CI projects had an advantage in sensitivity over the Russian group, whose instruments were fitted with electron impact ion sources. In the electron impact mode, the admissible quantity of solution is severely limited and stable operating conditions are correspondingly difficult to maintain. The CI technique, using the solvent as CI reagent, provides a better chance for detecting minor LC peaks.

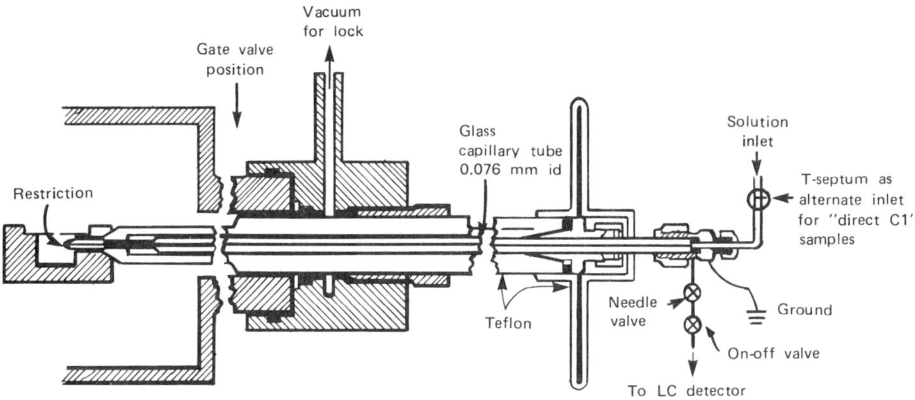

Fig. 13 Inlet probe for continuous introduction of solutions, which replaced the direct introduction sample probe of the RMH-2, utilizing the same vacuum lock system. (Reprinted with permission from P. Arpino, M. A. Baldwin, and F. W. McLafferty, *Biomedical Mass Spectrometry 1*, 80 (1974). Copyright 1974 John Wiley & Sons Ltd.)

Performance of the capillary as a stream-splitting device was in some respects improved by the insertion of a coaxial wire inside the capillary [104]. Melera's diaphragm pinhole orifice [122–124] injects into the ion source a liquid jet, which soon breaks into droplets. The clogging problems of the capillary are reduced by the use of the pinhole nebulizer. Development of the direct liquid introduction CI technique continues.

B. Trapping Systems—Off-Line and Coupled

Huber et al. [125] built an automatic fraction collector for micro-preparative high-speed LC and discussed the theory of the process. The LC/MS interface designed by Lovins et al. [126,127] is a probe with a removable tip, packed either with gold gauze or with absorbent material. Collection of an LC fraction occurs while the probe tip is in the vacuum lock, after which the probe is inserted into the ion source manually or by semiautomatic motor-driven travel. The analogy with the on-line GC/MS cold-trapping system is obviously close. The problems associated with closely spaced chromatographic peaks, including the effects of stopping the chromatographic flow, are of necessity similar.

Elbert et al. [128] returned to an off-line trapping technique and illustrated it with polymer analyses. Schmid et al. [129] studied transfer efficiences in off-line fraction collecting, which, for samples in the picrogram-microgram range, were better than 50% in all cases. With a little extra work, the transfer efficiency could be made to approach 100%.

Karger et al. [130] inserted a modified segmented flow extractor between a liquid chromatograph, operated in the reversed phase mode, and a mass spectrometer. Solutes were then extracted into a volatile organic solvent that could be put into the mass spectrometer with a moving belt interface. The apparatus was successfully applied to carbamate pesticide analysis. Salt removal was said to be essentially complete. Other aspects of the operation, such as the choice of solvent, are also discussed.

For the analysis of LC fractions by various soft ionization techniques, efficient off-line trapping and transfer will undoubtedly remain essential.

C. Microbore and Capillary Columns

A growing amount of work is being reported on the direct introduction of the whole effluent stream from small-diameter LC columns into a chemical ionization source. Henion and Maylin [131] injected the full stream from a packed microcolumn (5 cm x 0.5 mm) through a glass capillary probe into the CI source of a Finnigan 1015 mass spectrometer. Spectra were routinely obtained from drug samples ranging from 1 to 5 ng, and SIM results on 20–35 pg. Other results were reported by Henion [132]. Henion and Wachs [133] designed a new probe, terminated by a diaphragm with a 5-μm pinhole leak. With other conditions adjusted, the pinhole produced a steady, 1-in. long jet of the total column effluent. Inserted into the standard probe inlet of a Hewlett-Packard 5985B mass spectrometer, it gave a stable ion current baseline from m/z 100 to 500 over 4-hr periods of time. Lee and Henion [134] reviewed developments in micro-LC/MS.

Tiebach et al. [135] also used a probe of their own design to conduct the effluent from a microbore packed column (16 cm x 0.5 mm) into the ion source of a Finnigan MAT 4500 mass spectrometer. About 15 different trans-line capillaries were tested. Breakdowns of eluent flow occurred with most of them because of the formation of ice crystals in the desolvation zone of the transfer line. Plots of the positive ion and negative ion mass spectra of aflatoxin M_1 (500-ng sample) are shown, along with reconstructed ion current (RIC) plots and single-ion profiles for 30-ng samples of aflatoxins. Hirter et al. [136] built another probe, with which the entire output of a packed microcolumn (25 cm x 1 mm) was introduced into a Nermag R 10-10 C quadrupole mass spectrometer. Several applications were described. Henion [137] and Bruins [138] have recently reviewed the subject of microbore LC/MS.

Another small-diameter column type is the packed microcapillary column, a glass capillary of 50–200 μm internal diameter, described by Tsuda and Novotny [139]. After tests of various columns, they chose an internal diameter range of 50-80μm, packed with particles of about 30-μm size, as a good compromise between increased analytical performance and reduced operational difficulties. Column lengths from 14-40μm, depending on diameter, are given for some of their columns. One of the applications in mind was to LC/MS.

The use of open capillary columns in liquid-solid chromatography was reported by Nota et al. [140] in 1970. Niessen and Poppe [141] employed such columns in the coupling by direct introduction of the entire effluent stream into the CI source of a Finnigan 3100 mass spectrometer with a modified vacuum pumping system. The fused silica columns were 6.82 m x 25 μm, 2.03 m x 10 μm, and 1.81 m x 5 μm. Diaphragms of about 4-μm pinhole diameter were used, requiring a makeup liquid flow to stabilize a jet. The makeup liquid can be chosen for its CI properties, provided only that it be miscible with the LC liquid phase. Detection limits were considered very promising.

D. Molecular Separators for Sample Enrichment

Because carrier gas separators had already been developed for GC/MS and had proved their effectiveness, it was natural that they be tried in LC/MS. Takeuchi et al. [142] ran the output of a micro LC column (7 cm x 0.5 mm) through the single-stage glass jet separator of a JMS-Q10A quadrupole mass spectrometer. The solvent, serving as CI reagent, was 70% acetonitrile/30% water, 8 μl/min, for an aromatic hydrocarbon mixture and 100% acetonitrile, 8 μl/min, for a fatty acid methyl ester mixture. Tsuge et al. [143] stated that the jet alone had some problems with relatively less volatile compounds. They modified it into a nebulizing jet by running the small capillary from the LC through the entrance jet and surrounding it with a concentric jet of helium. The skimmer (the exit on the mass spectrometer side of the opposing pair of capillary tips) was independently heated to a temperature sufficient to keep the solute particles left by the fog droplets from plating out on its inner surface. The nebulizing jet (atomizer) was applied by Hirata et al. [144] to obtain LC/MS data in the form of CI mass spectra of a variety of materials, including free amino acids, drug components in a cold medication, dilauryl phthalate and its impurities, oligomers of styrene and of ethylene glycol, and steroids. A subsequent paper of Yoshida et al. [145] reported an improved jet nebulizer (Fig. 14) with a very short counter nozzle (skimmer), which they again used to interface their micro-LC column (total effluent) to the JMS-Q10a quadrupole mass spectrometer. With this modification, they were able to observe the $(M + 1)^+$ peak of glutamic acid (5% intensity relative to the base peak) by injecting 1 μl of water solution containing 5 g of glutamic acid into the micro-LC, which was running at 16 L/min of methanol.

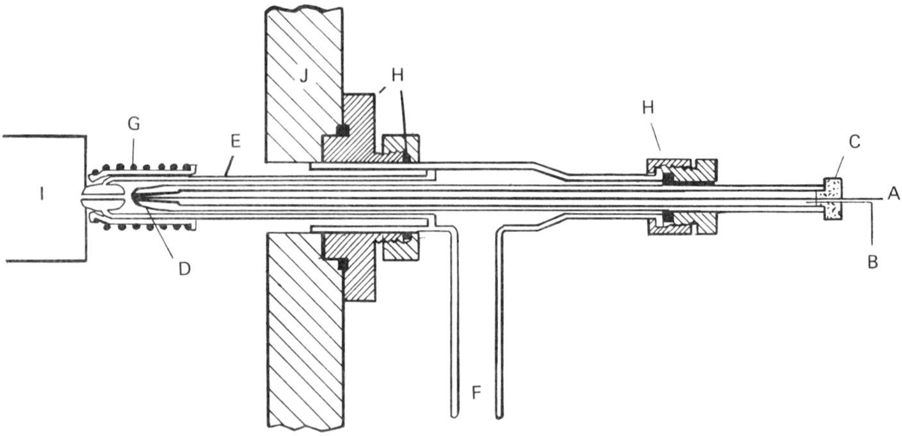

Fig. 14 Schematic diagram of vacuum nebulizing interface for LC/MS coupling. A: Coaxial capillary tube from LC. B: Needle for introduction of nebulizing gas (He). C: Silicone septum. D: Nebulizing tip (Pyrex). E: Nebulizing tube (Pyrex). F: To rotary pump. G: Heater. H: Silicone O-ring. I: Ion chamber. J: Body of MS. (Reprinted with permission from Y. Yoshida, H. Yoshida, S. Tsuge, T. Takeuchi, and K. Mochizuki, *Journal of HRC & CC*, 3, 16 (1980). Copyright 1980 Huething Publishing Ltd.)

Other compounds included steroids, aromatic amines, and chlorinated hydrocarbon insecticides, yielding mass spectral scans and MID analyses. The MID results were satisfactory, even with gradient elution.

Vestal's [97,146–147] laser vaporization apparatus (Fig. 15) employed a sequence of steps: (a) liquid jet formation from a nozzle; (b) quick evaporation of solvent from a CO_2 laser; (c) passage through a sampling aperture; (d) ionization of solvent (if CI was desired) by a crossed electron beam; (e) passage of beam through skimmer to remove much of solvent vapor; (f) passage through a quadrupole with RF only; (g) ionization by a crossed electron beam (if EI was desired); (h) passage through collimator and mass-analyzing quadrupole; (i) detection of positive and negative ions by multipliers. In a later development, the laser was replaced by a copper tube heated by four oxyhydrogen flames [148–151] for rapid solvent evaporation. Here, in either case, the vaporized portion of the solvent served as a nebulizing gas for the remaining liquid phase. Arpine [104] speaks of these interfaces of Tsuge et al., and Vestal as hybrid interfaces, in the sense that they employ the input of a jet of liquid droplets, as in the direct inlet nebulizers, simultaneously with the removal of some solvent by the skimmer.

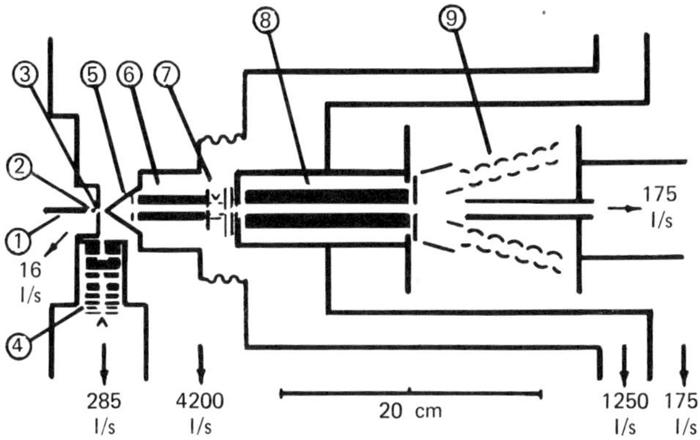

Fig. 15 Schematic diagram of the crossed-beam LC/MS interface: (1) liquid nozzle; (2) laser beam; (3) sampling aperture; (4) CI electron gun; (5) skimmer; (6) quadrupole with RF only; (7) EI ion source; (8) quadrupole mass analyzer; (9) multipliers. (From M. L. Vestal, in *Trace Organic Analysis: A New Frontier in Analytical Chemistry*, ed. by H. S. Hertz and S. N. Chesler, National Bureau of Standards Special Pulication 519, U.S. Government Printing Office, Washington, D.C., 1979, p. 647.)

Jones and Yang [152] used a three-stage silicone membrane interface to couple their LC to an EAI QUAD 300D quadrupole mass spectrometer. The experiments involved reversed phase chromatography and a water/methanol gradient (5%/min). The apparatus worked well for sufficiently volatile materials, but its limitation of temperature to about 250°C imposed a corresponding limitation on the involatility of the sample analyzed.

In the wire transport system of Scott et al. [94,153–155], the LC eluent was deposited on a moving wire, which passed through two evacuated chambers (Fig. 16) in entering and leaving the mass spectrometer. The solvent was evaporated and pumped away in the two vacuum lock chambers. The portion of the wire between an insulated pulley and a grounded pulley was heated by passing a direct current through it. The residual solute on the wire was then evaporated and the mass spectrum was obtained. System performance in the analyses of vitamin A acetate, peptides, and other materials was promising, although it was noted that sensitivity was somewhat limited by the aged condition of the mass spectrometer.

McFadden et al. [156] described a system in which the wire was replaced by a moving belt. Sample transfer yields were found to be in the range of 25–40% under the most favorable conditions. The detection limit for carbaryl was found to be less than 1 ng. The

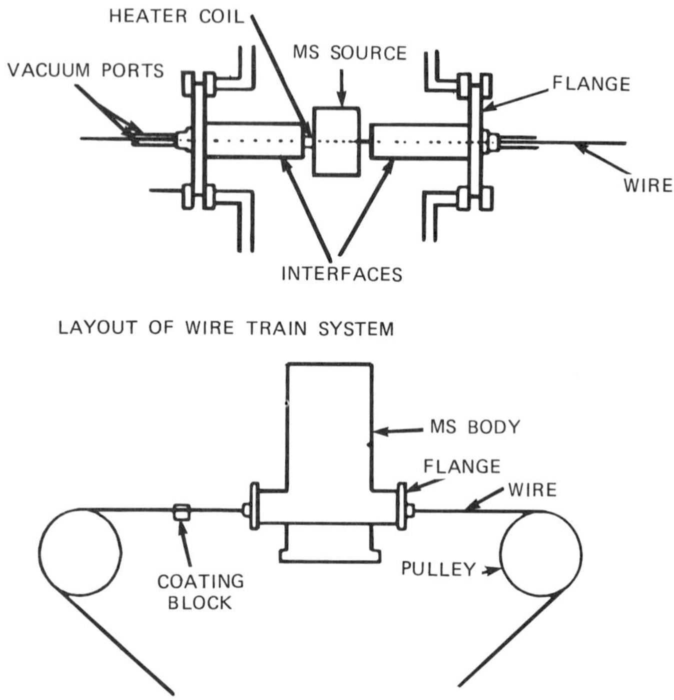

Fig. 16 Diagram of interfaces located in MS source. (From R. P. W. Scott, in *Trace Organic Analysis: A New Frontier in Analytical Chemistry*, ed. by H. S. Hertz and S. N. Chesler, National Bureau of Standards Special Publication 519, U.S. Government Printing Office, Washington, D.C., 1979, p. 637.)

apparatus of Erdahl and Privett [157] made use of a perforated belt that was said to improve its performance by entraining more solvent. Another belt interface was designed by Yorke et al. [158]. Van der Greef et al. [159] coupled a Finnigan MAT moving belt interface with a double-focusing Finnigan MAT 8230 mass spectrometer that was equipped with a combined EI-CI source. They found it beneficial to deactivate the metal belt with a 50 ppm solution of Carbowax 20M to reduce memory effects and improve desorption characteristics of the belt. They were able to do isotopic dilution analysis for progesterone in the ppb range. Desorption of cyclosporin A was improved, so that the $(M + H)^+$ ion at m/z 1202 became the base peak in its CI spectrum. For one more example of recent applications, Games, et al. [160] looked at a variety of materials and have discussed the advantages and limitations of the belt system. Finnegan

MAT has continued to improve the belt transport system [161]. Several authors have investigated the belt as a mechanism for coupling LC into soft ionization or other techniques [162–165].

E. Atmospheric Pressure Ionization

An atmospheric pressure ionization (API) system was applied by Horning et al. [166,167] to the mass spectrometric analysis of LC effluent. The entire stream was injected. Two types of ion sources [168] were used, in which the electrons were generated by a corona discharge or by a ^{63}Ni foil. A small flow of nitrogen (8–10 ml/min) was maintained during analysis of the continuous LC solvent flow. The principal positive ions observed with carrier gas alone are those derived from water, cluster ions of composition $H^+(H_2O)_n$. In an API source, water is always present.

The background spectrum in an API source depends on the solvent used, the carrier gas (nitrogen), the condition of the room atmosphere as regards water content and traces of impurities, the temperature of the ion source, and probably variations in other operating parameters. If no other LC/MS system were available, the API apparatus might be more used as a rather specialized research tool, but the background situation, including variations in the chemistry of interaction between background and sample components, would probably create difficulties in any attempt at a routine analytical application. In the soft ionization studies of large molecules, that sort of background would be relatively much less important. The study of API systems, sometimes for special uses, has continued [169,170].

F. Summary and State of the Art: Thermospray

The most exciting development in the LC/MS problem in recent years is the thermospray process. In one way or another its roots probably extend in various directions [171–175]. It evolved from Vestal's laser vaporization system [97,146–147] through the intermediate form of his oxyhydrogen flame apparatus [148–151], as described in Section IV.D, into that of the electrically heated copper block [176–181]. The first mention of the name "thermospray" itself is probably that in the abstract from the ASMS meeting in Minneapolis in 1981 [176]. In looking at the papers describing the evolution of the apparatus, for example [178], the reader gets an impression of an intensity of concentration on basic principles that has, in the past 20 years, unfortunately become impossible in some work environments.

Developments have continued. Vestal [182] designed and built an interface for magnetic mass spectrometers, in which the high electrical potential carried on the source requires a well-considered isolation. Vestal and Fergusson [183] replaced the heated copper block by the electrically directly heated input capillary. The latter appara-

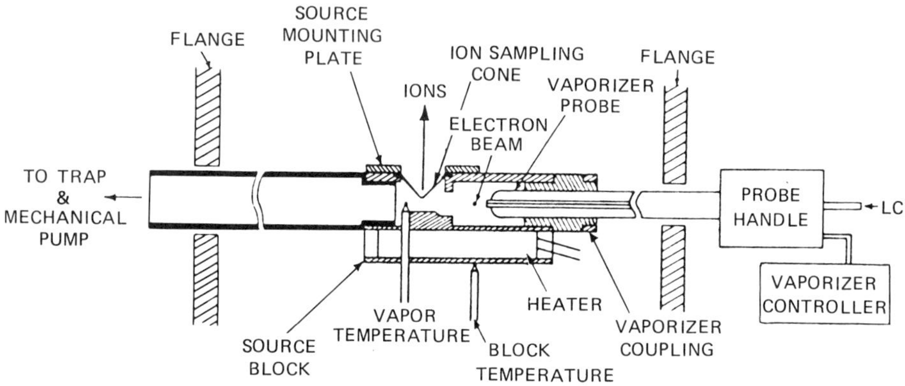

Fig. 17 Thermospray LC/MS interface using direct electrical heating of capillary. (Reprinted with permission from M. L. Vestal and G. J. Fergusson, *Anal. Chem.* 57, 2373 (1985). Copyright 1985 American Chemical Society.)

tus is shown schematically in Fig. 17. The most noticeable difference from the laser system of Fig. 15, aside from the substitution of electrical heating, is the redesign of the thermospray portion of the vacuum pumping system. That was done during the second developmental stage, that of the oxyhydrogen flame apparatus [150]. In Fig. 17, the jet of LC effluent is directed into the pumping aperture, so that the jet acts as its own ejector pump [178]. The large diffusion pump used in the laser system was removed, and its job is being done by a 300 L/min mechanical pump.

The success of the thermospray apparatus depends on the fact that it is both an effective sample transfer device and a soft ionization source. The authors noted [178] that its response was quite dependent on the properties of the sample molecule and that it is not yet possible to predict the relative response of different molecules, or to eliminate day-to-day variations in ionization efficiency and in fragmentation patterns. Ion intensities depended strongly on liquid flow. Not all large molecules yielded molecular ions. Since that paper [178] was published, the electrically heated capillary modification with its feedback-controlled triac power supply already shows improved stability [183], and the effects of operating parameters on performance are discussed at length. A study of thermospray solvent selectin by Liberato and Yergey [184] introduces more order into that phase of the subject.

There will no doubt always be room for other LC/MS techniques alongside thermospray. Some in which continuing progress is seen, and which may find their own special places, are the stream-splitting direct inlet; the microcolumn direct inlet; the jet nebulizer; the moving belt, on-line or off-line, partly as an interface for various soft ionization methods; and atmospheric pressure ionization. The reader who wants to expand on the limited scope of the present outline will find a good starting point in the reviews listed earlier and in the LC/MS bibliography of Edmonds et al. [185].

V. COUNTERCURRENT CHROMATOGRAPHY/ MASS SPECTROMETRY

In considerations of GC/MS v. LC/MS, it is usually said or implied that the sample inventory is allocated to the one or to the other according to some preconceived notions of suitability: Whatever can be done by GC/MS is done by GC/MS, after which the more exotic residuum goes to LC/MS. In the real world things do not necessarily work out that way. (a) The more utilitarian result is that where not all techniques are equally accessible, whatever is available will be used to the extent that its capabilities can be stretched. (b) Another consideration is that where it is desired to reduce environmental stress on a compound to the absolute minimum in order to reduce degradation to its smallest possible extent (i.e., to produce the best possible product), the gentlest treatment will be chosen. (c) Still another is that where alternatives are accessible and the user is impatient or harassed, or the maximized performance is not felt to be necessary, that route will be chosen which tends to minimize the personal effort. (d) In any complex mixture, there may well be components which by themselves would be treated by different procedures, but the mixture must go the one way or the other. Thus, the split between GC/MS and LC/MS depends in practice on an assortment of influences. When it comes to CCC/MS, the same reasoning of course applies.

A. The Monitoring of Preparative CCC

In an on-line coupled procedure, CCC will present the chromatographic stream to the mass spectrometer interface as a liquid phase. If it is being used as an analytical tool, the whole stream may be utilized where it is not too large. If the task at hand is the monitoring of a preparative operation, then by definition it is not permissible to consume all of the product, and the output will usually be too large to be accommodated by that approach. In preparative work, a split-stream or off-line procedure is implied. On the laboratory scale, the analysis of aliquots from collected fractions might be expected, but an industrial process might well find a use for a sufficiently stable and reliable on-line, split-stream technique.

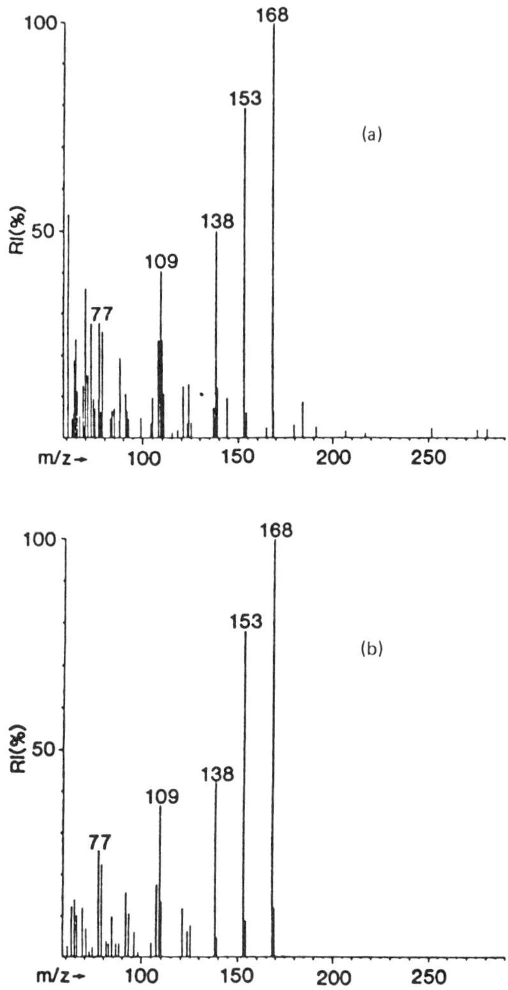

Fig. 18 Mass spectra of 3-methyl-1-4-(methylthio)anisole, with intensity relative to base peak (RI %) plotted against the mass/charge ratio m/z: (a) CCC fraction 47 (Fig. 14a of Chap. 8); (b) 3-methyl-4-(methylthio)anisole. (From J. M. Ruth, Y. Ito, N. B. Mandava, and V. P. Flanagan, *J. Liquid Chromatogr. 8*, 2239 (1985).)

B. Mass Spectra of Trapped Fractions

In the earlier discussions of GC/MS (Sec. III) and LC/MS (Sec. IV), the description began in each case with the first attempts at direct, on-line interfacing. Sections III.B and IV.B reviewed the experiments with on-line stopped-flow interfacing, which came a little later in each development, almost as if from a logical effort to improve on the first continuous flow processes. The same sections acknowledged the continuing usefulness of off-line fraction collecting as a process that (a) was applied before the advent of on-line experiments and (b) because it often has its own reasons for preference, is not likely to be superseded, even if all laboratories come to possess the full range of up-to-date equipment.

CCC/MS is now in the off-line trapping stage of its development, and examples of this phase will be quoted. Some GC/MS spectra of CCC fractions were done with Finnigan 3200 and Hitachi RMU-6 mass spectrometers interfaced to an Incos data system, and some other GC/MS spectra, on a Finnigan 4021 system [186–188]. The spectra of 3-methyl-4-(methylthio)anisole, the impurity obtained from a fenthion analytical standard and of an authentic sample are shown in Fig. 18. The corresponding pair for the fenthion CCC fraction and for an authentic sample are presented in Fig. 19, and the spectra of the CCC fractions containing the methyl and n-octyl esters of the herbicide 2,4,5-trichlorophenoxyacetic acid, together with standard samples, in Fig. 20. A GC/mass spectrum of the methyl ester of abscisic acid from a CCC fraction (Fig. 21) was published by Mandava and Ito [189]. CCC fractions containing four s-triazine herbicides were concentrated and run as probe samples, as reported by Mandava et al. [190] (Figs. 22,23). The countercurrent chromatographic treatment of those samples was described in Chap. 8.

Fales et al. [191] described the separation of methyl violet 2B by CCC and identification of the components by ^{252}Cf plasma desorption mass spectrometry. The dye methyl violet 2B has the structure 1, with R_1–R_5 = CH_3, and R_6 = H. When R_1–R_6 = CH_3, the product is called crystal violet. The ^{252}Cf plasma desorption mass

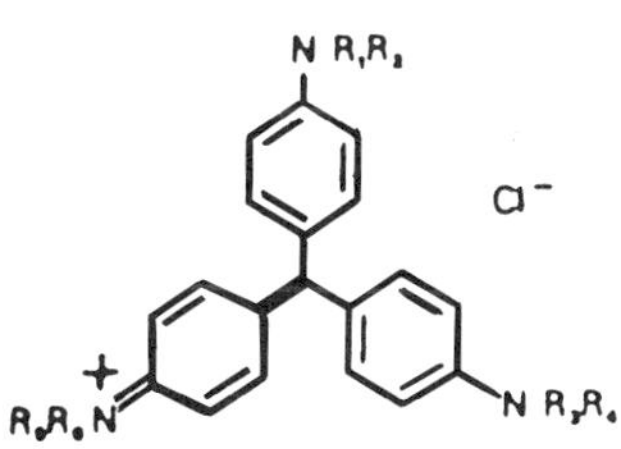

1

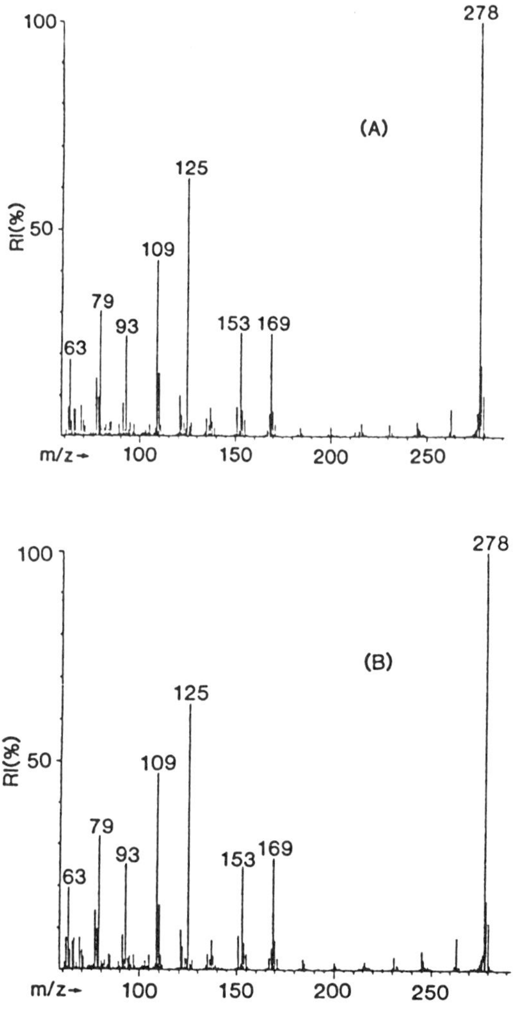

Fig. 19 Mass spectra of fenthion, with intensity relative to the base peak, (RI %), plotted against the mass/charge ratio m/z: (a) CCC fraction 63 (Fig. 14A of Chap. 8); (b) Fenthion. (From J. M. Ruth, Y. Ito N. B. Mandava, and V. P. Flanagan, *J. Liquid Chromatogr. 8,* 2239 (1985).)

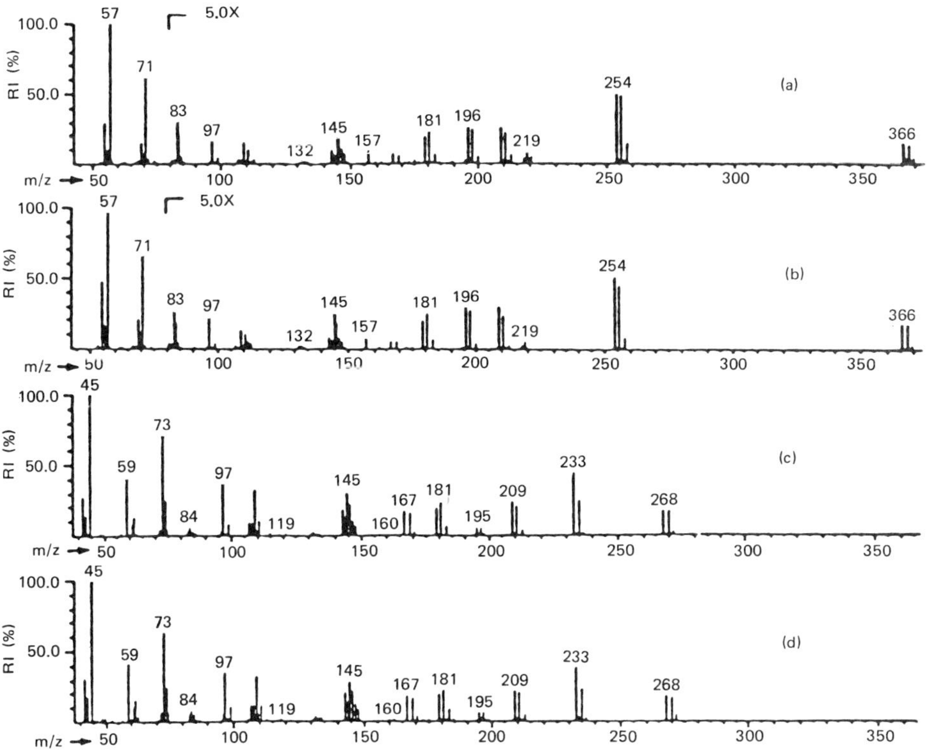

Fig. 20 Mass spectra of the 2,4,5-T esters, with intensity relative to the base peak, (RI %), plotted against the mass/charge ratio m/z: (a) the component in CCC fractions 103–123 (Fig. 10 of Chap. 8); (b) the n-octyl ester of 2,4,5-T; (c) the component in CCC fractions 160–190 (Fig. 10 of Chap. 8); (d) the methyl ester of 2,4,5-T. (From J. M. Ruth, Y. Ito, N. B. Mandava, and V. P. Flanagan, *J. Liquid Chromatogr. 8,* 2239 (1985).)

spectrum (PDMS) of the sample suggested that the sample was largely composed of crystal violet, but the occurrence of an ion 14 mass units lower was compatible with the presence of some of the pentamethyl homolog. Chromatography on silica gel using chloroform/acetic acid/0.1 N hydrochloric acid (2:2:1) yielded only one spot, R_F = 0.64.

The separation was done on a high-speed countercurrent chromatograph (PC, Inc., Potomac, MD) with a multilayer coil (revolutional radius, 4 in.; rotational radius range, 2–2.5 in.) of Teflon tubing (1.6 mm i.d.). Using the preequilibrated solvent system

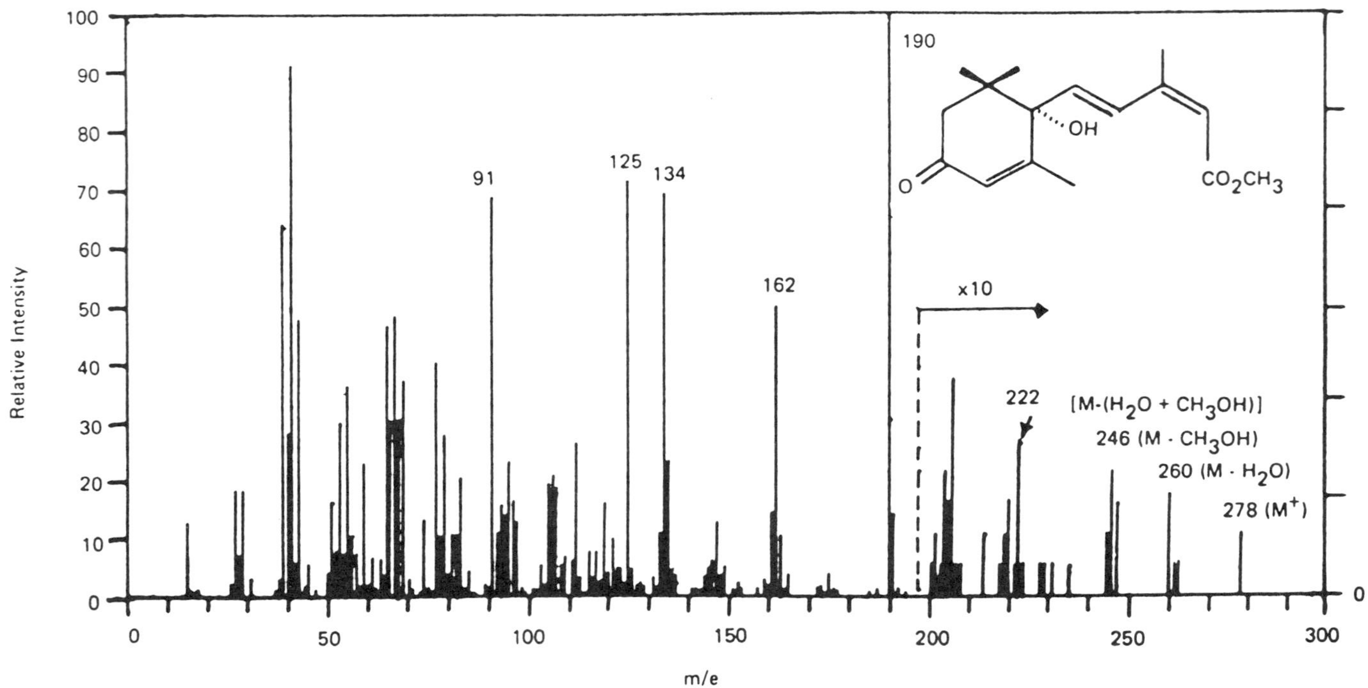

Fig. 21 Mass spectrum of the methyl ester of abscisic acid. (From N. B. Mandava and Y. Ito, *J. Liquid Chromatogr.* 7, 303 (1984).)

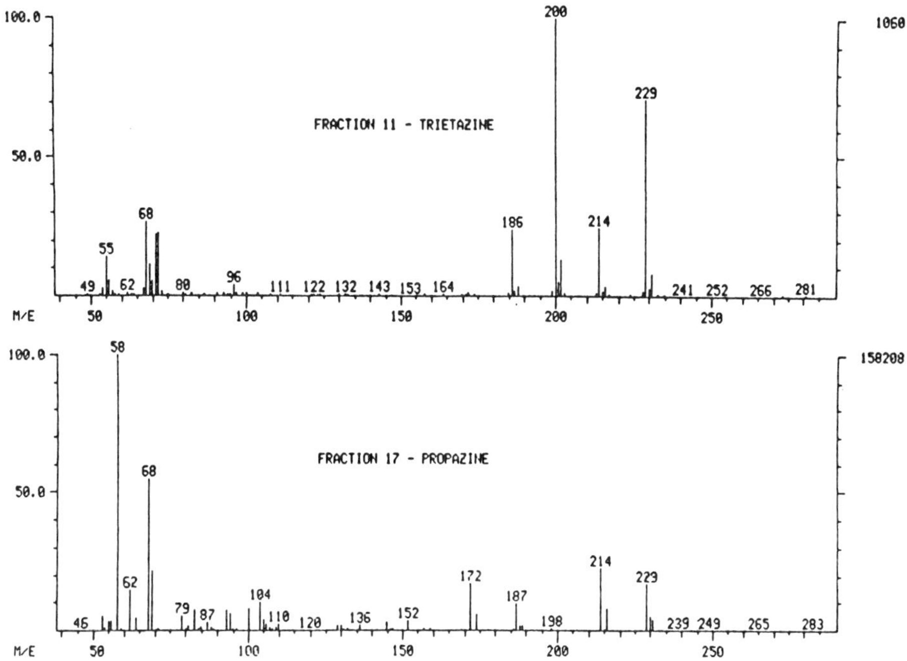

Fig. 22 Mass spectra of fractions from CCC separation (Fig. 11 of Chap. 8). Top: fraction 11 (trietazine). Bottom: fraction 17 (propazine). (From N. B. Mandava, Y. Ito, and J. M. Ruth, *J. Liquid Chromatogr.* *8,* 2221 (1985).)

chloroform/acetic acid/0.1 N hydrochloric acid (2:2:1), the coil was charged with the aqueous (lighter) phase as stationary phase. The sample consisted of 6 mg of the dye in 5 ml of the moving phase (the heavier, chloroform phase). The mobile phase was pumped through at a rate of 240 ml/hr, and 84 fractions of 8 ml each were collected, analyzed with a Perkin-Elmer Model 552 spectrophotometer and evaporated to dryness.

The separation (Fig. 24) showed the presence of at least seven dye species. The two major ones correspond to the hexamethyl and pentamethyl compounds. The triarylmethane dyes eluted in the expected order of increasing polarity. In addition to the cation peaks, the spectra of the first three compounds (Fig. 25) each show a peak 14 mass units lower, considered most likely caused by hydrogenation of the $C{=}N^+$ linkage during the condensed state ionization process, with subsequent loss of a methyl group. The minor peaks in the chromatogram are also discussed in the paper.

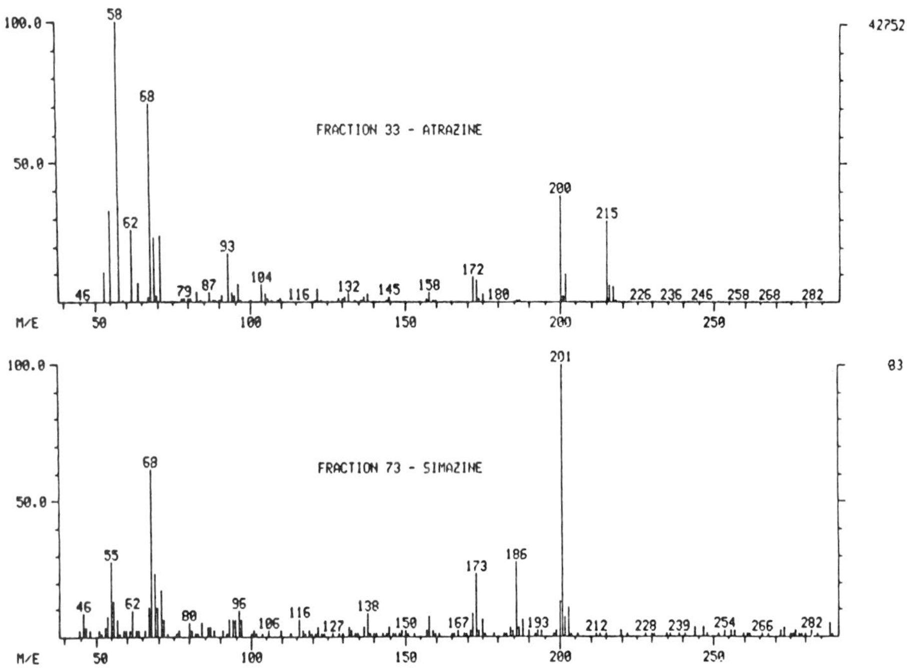

Fig. 23 Mass spectra of fractions from CCC separation (Fig. 11 of Chap. 8). Top: fraction 33 (atrazine). Bottom: fraction 73 (simazine). (From N. B. Mandava, Y. Ito, and J. M. Ruth, *J. Liquid Chromatogr. 8,* 2221 (1985).)

Yang et al. [192] reported the separation of the components of commercial digitonin by HPLC and CCC, followed by the identification of the products by ^{252}Cf plasma desorption mass spectrometry. The CCC apparatus used was an ITO multilayer separator/extractor (PC, Inc., Bethesda, MD), which has been described by Ito and Conway [193]. The solvent system was chloroform/methanol/acetic acid/water (5:3:1:3) of which the heavier (chloroform) phase was used as the mobile phase (4 ml/min). The rotor speed was 800 rpm. The elution coefficient $E_{H/L}$ of digitonin in this system was 0.487. A solution of 90 mg of digitonin was dissolved in 10 ml of a 1:1 mixture of the two phases and injected into the pumping line at the column head. Fractions (8 ml each) were collected, the solvents were evaporated under vacuum, the residues were weighed, and the weights were plotted to display the chromatogram (Fig. 26). Separation of digitonin (structure 2), the main component, with a molecular weight of 1229 (average, C = 12.01), from four other saponins,

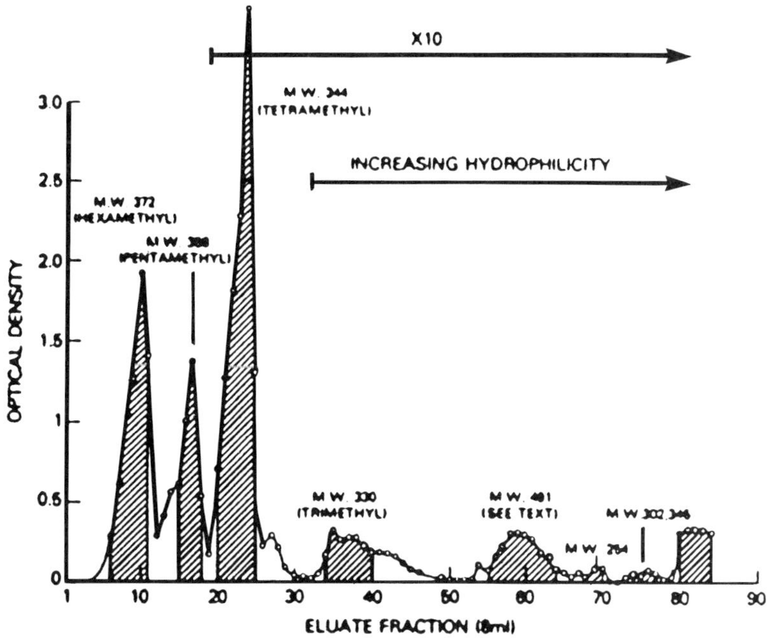

Fig. 24 Countercurrent chromatogram of methyl violet 2B certified (lot no. 011097, Aldrich). (Reprinted with permission from H. M. Fales, L. K. Pannell, E. A. Sokoloski, and P. Carmeci, *Anal. Chem. 57*, 376 (1985). Copyright 1985 American Chemical Society.)

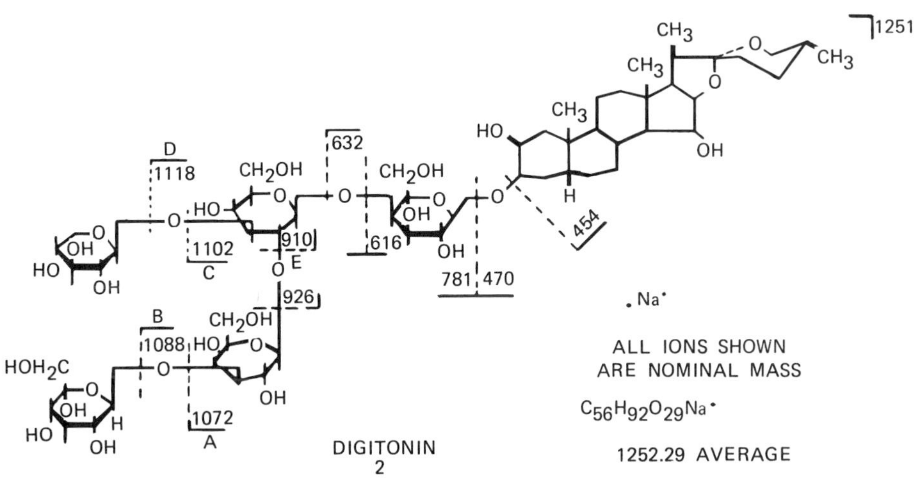

Structure 2

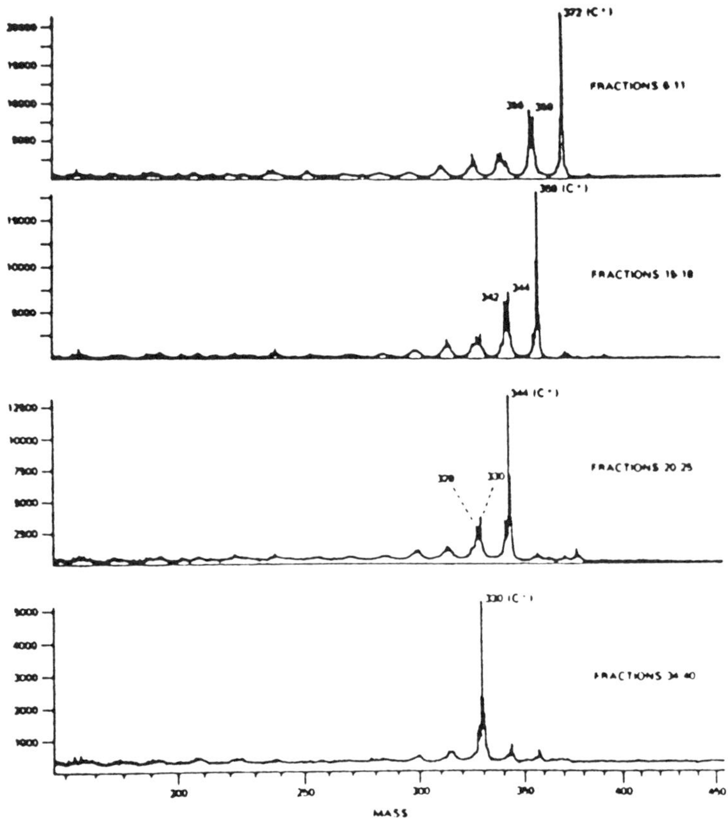

Fig. 25 Positive ion californium-252 plasma desorption mass spectra of fractions 6–40 from the countercurrent chromatogram of methyl violet 2B. (Reprinted with permission from H. M. Fales, L. K. Pannell, E. A. Sokoloski, and P. Carmeci, *Anal. Chem.* 57, 376 (1985). Copyright 1985 American Chemical Society.)

including one previously inknown, was quite good. The most intense ion in their plasma desorption mass spectra (Fig. 27) is $[M + Na]^+$.

C. The Range of Retention Times in CCC

Ito [194] has built several different kinds of CCC apparatus and has investigated many different sets and ranges of operating parameters. The recent devlopment of high-speed countercurrent chromatography has made it competitive on the time scale with HPLC. A good data system will have no difficulty in handling the on-line output from CCC/MS.

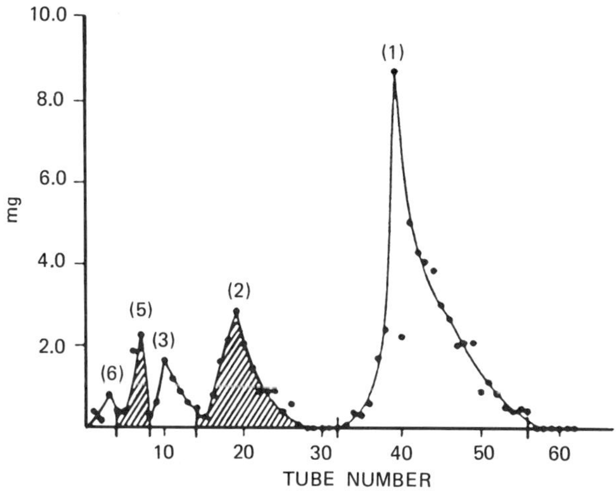

Fig. 26 High-speed countercurrent chromatogram of commercial digitonin. Solvent system: chloroform/methanol/acetic acid/water (5:3:1:3). (Figure 26 and Structure 2 (p. 733) reprinted with permission from Y. M. Yang, H. A. Lloyd, L. K. Pannell, H. M. Fales, R. D. Macfarlane, C. J. McNeal, and Y. Ito, *Biomedical and Environmental Mass Spectrometry 13*, 439 (1986). Copyright 1986 John Wiley & Sons Ltd.)

The 1-sec and 2-sec scans familar in capillary GC/MS are adjustable in several different ways for optimum performance under CCC conditions, as by lengthening the actual scan time, by lengthening the rest period between scans, and by changing MS detector parameters. Where a slower CCC instrument is used, the same adjustments are easily made.

D. Solvent Comparison, CCC Phase Interchange, and Notation

The general complex of solvent problems is in one sense simpler in CCC than in HPLC in the absence (so fa, at least) in CCC of solvent composition programming. In addition, CCC has other advantages over HPLC, which result from the absence of any solid phase in CCC and the consequent elimination of those problems (solute loss by adsorption, chemical reaction of solutes through surface interaction, and peak tailing caused by adsorption) associated with the presence in HPLC of a solid phase. CCC also has the advantage of a more facile manipulation of certain operating parameters, as in the reversal of elution order (and consequent change in peak dispersion) by the simple interchange of mobile and stationary phases [193], and in the greater freedom of change of the solvent system.

Because of the nature of packed-column chromatography, the countercurrent chromatographic phase interchange is not applicable

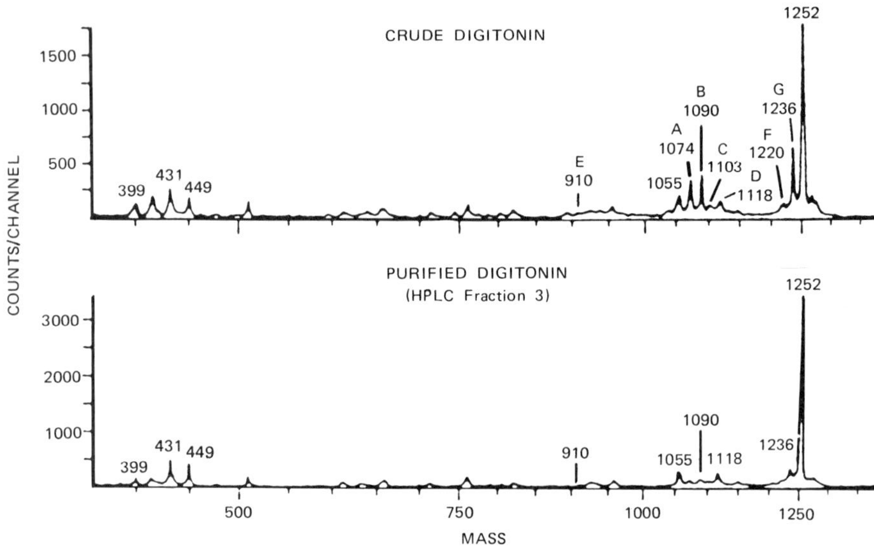

Fig. 27 (a) ^{252}Cf-plasma desorption mass spectrum of commercial digitonin; (b) ^{252}Cf-plasma desorption mass spectrum of purified digitonin. (Reprinted with permission from Y. M. Yang, H. A. Lloyd, L. K. Pannell, H. M. Fales, R. D. Macfarlane, C. J. McNeal, and Y. Ito, *Biomedical and Environmental Mass Spectrometry 13,* 439 (1986). Copyright 1986 John Wiley & Sons Ltd.)

in the same sense, and the partition coefficient K_C is usually defined as the ratio of the solute concentration C_s in the stationary phase to its concentration C_m in the mobile phase. In CCC, Ito has adopted a convention such that the order of elution of components has a constant relationship to the order of size of their partition coefficients. It has become customary to speak in terms of the coefficient C_m/C_s. The result is that components elute in the order of decreasing size of their partition coefficients, as defined in that way. The one with the largest partition coefficient emerges first and the one with the smallest coefficient comes out last.

The interchange of phases (the stationary phase becoming mobile and the mobile phase becoming stationary) in countercurrent chromatography then makes it necessary to invert the partition coefficient, so that again the components elute in the order of decreasing size of the new partition coefficients, which are the reciprocals of the other set (K_C and $1/K_c$, in column chromatographic language). In CCC, the ratio C_m/C_s follows the same definition as before, with the mobile phase in the numerator, but the subscript m now refers to the liquid which was formerly stationary. It is thus customary to tabulate two sets of partition coefficients such that each set contains the reciprocals of the corresponding coefficients in the other set. To avoid the confusion which may arise in the comparison of the double set of partition coefficients with the single set that has been conventionally defined in packed-column chromatography, it has been found desirable to adopt a modified terminology and a modified notation for use in countercurrent chromatography.*

Here the constant is called an "elution coefficient" K_E, which is defined by the ratio C_mC_s discussed above:

$$K_E = \frac{C_m}{C_s} \tag{19}$$

The order of elution is then always in the order of decreasing K_E. Since either phase may be mobile, there are two elution coefficients. Let C_L be the concentration of solute in the lighter phase (the upper phase in a static system with separated phases) and let C_H be the concentration in the heavier phase (the lower phase in a static situation). The two values of the elution coefficient are

$$\text{Lighter phase mobile:} \quad K_E = K_E(L/H) = \frac{C_L}{C_H} \tag{20}$$

$$\text{Heavier phase mobile:} \quad K_E = K_E(H/L) = \frac{C_H}{C_L} \tag{21}$$

For each component, the two elution coefficients are reciprocals of each other, so that

$$K_E(L/H) \bullet K_E(H/L) = 1 \tag{22}$$

and the convention that elution proceeds in the decreasing order of size of the elution coefficients is then maintained for both modes of operation. The name and notation adopted here avoid confusion with the single-valued partition coefficient of packed-column chromatography, and the mathematical symmetry of the reciprocal pair is made clear. Where a single term and a single symbol are needed for brevity in discussion, the general form of the elution coefficient K_E covers both modes of operation.

*See Section on preferred usage of terminology in CCC in Chap. 2.

The routine interchange of mobile and stationary phases in CCC makes it possible always to choose whichever of the two phases is the more desirable as the mobile phase for CCC/MS and to try each in turn. In many cases that may be a considerable convenience, depending on the sort of interface to the mass spectrometer that is available. Examples of the results of phase interchange are given in Chap. 8 and in Sec. V.B of this chapter.

E. Conclusion

In Sec. IV, many experiments in LC/MS interfacing were recalled from the literature, and in Sec. IV.F. some brief impressions of the current situation were noted. A variety of LC/MS interfaces are available, each of which is applicable in CCC/MS. Since LC/MS is now in a very active phase of its development, improved versions of existing interfaces and perhaps new ones based on different operating mechanisms may be expected. The choice for CCC/MS will be influenced, as in LC/MS, by the range of problem types forseen by the prospective user.

One aspect of CCC separation is its gentler and therefore potentially more effective treatment of macromolecular solute mixtures. It may for that reason be considered a natural adjunct of the rapidly developing soft ionization techniques of mass spectrometry. This feature is of obvious interest both in biomolecular chemistry and in synthetic polymer chemistry. It is of course equally evident that in any anticipation of CCC/MS interfacing, the developing situation in LC/MS interfacing will logically be monitored from month to month, with both similarities and differences considered.

ACKNOWLEDGMENTS

The authors are grateful to Wanda Wyczolkowska for translation from the Japanese language and to William H. Longenecker for his assistance in searches for some of the less easily accessible library materials.

APPENDIX: EXAMPLES OF THE MANUAL PROCESSING OF MIXTURE MASS SPECTRA—CODLING MOTH PHEROMONE RESEARCH

A. Codling Moth Sex Pheromone No. 1B

The following memorandum and its accompanying tables are reproduced in full as an example of the manual treatment of the mass spectra of mixtures. Tables I—VI of the original are renumbered here as Tables 2—7, respectively.

AGRICULTURAL ENVIRONMENTAL QUALITY INSTITUTE
ANALYTICAL CHEMISTRY LABORATORY
Beltsville, Maryland 20705

August 17, 1972

Subject: Codling Moth Sex Attractant No. 1B

To: L. M. McDonough
3706 Nob Hill Boulevard
Yakima, Washington 98902

I have gotten together the data on No. 1B that we have been discussing by telephone and will enclose it all here.

The sample was run from the small cold-trap inlet through the molecular leak, by-passing the large reservoir. Perfluorokerosene (Perfluoroalkane-225) was used as a standard for the mass measurements, supplemented by other known lines where desirable in the plate work. Scans 1 and 2 were background scans. Scan 3 was started as the sample began warming to room temperature, after pumping out solvent at CO_2- acetone bath temperature. All scans were run from the high mass end downward, as is usual on the 21-110. Because of the sample vapor pressure increase during the run, the peaks at the lower end are out of proportion to those at the other end of the spectrum. However, this scan is still interesting and useful for the comparison of peaks near each other in the spectrum, because it shows both the most volatile portion of the sample and the least probability of any thermal degradation. (The ion source was at 205°C, and the glass lines were at 180–200°C, at various points.) The sample had probably pretty well evaporated by the time scan 4 was recorded, so that any pressure variation effect is much smaller, if present at all. The base peak (m/e 82) is not readable in scan 4. The molecular ion peak (m/e 138) was assigned a value of 30% relative intensity, after examining the later scans, and the spectrum was tabulated on that basis for comparison with the others. The 30% value may possibly give a little too much weight to the apparent slow decrease of the relative intensity of m/e 138 with time (in succeeding scans). On the other hand, the decrease of relative intensity of some other m/e values with time suggests this effect at m/e 138 may be real.

After scan 4, the plate exposures were made rather quickly. After that, the molecular ion was measured by matching peaks on the oscilloscope. Then the perfluoroalkane was pumped out, after closing its valve, and scans 5 and 6 were made in succession. Comparison of those two will show that they are quite similar, and scan 4 is not much different either. After scan 6, some time was spent in peak-matching. At this time, the cold trap reservoir was warmed up, and was at 150°C during scan 7. The sample was getting weak by that time. That can be seen from the peak heights in Table 2. The relative intensities in Table 4 indicate that a small quantity of less volatile material remained at scan 7, after most of the sample had been used up. (Note that the intensity ratio for m/e 138 to 82 remains fairly constant, however.)

A few very weak peaks are omitted in Tables 2 and 3, which were done for comparison of the scans. Table 4 shows mass values obtained by peak-matching, which are the ones that I gave you by telephone the day of the run. Table 5 is the high resolution spectrum from manual measurement and processing of data from the glass plate. Table 6 is the complete spectrum from scan 5, with the ionic composition as obtained from Table 5.

Disregarding scan 3, and remembering that scan 7 shows the residue of less volatile minor contaminants, we may compare relative intensities at a given m/e value, in succeeding scans, to try to locate peaks that do not belong to the major component of the sample. Peaks 28(a) and 29(a) should be partly at least from CO_2 (see m/e 44). The one at m/e 107 likewise is an impurity peak, increasing in relative intensity as the volatile sample is depleted. That one would be under suspiciion anyway, if the sample is a ketone, since a ketone ought not to lose a CH_3O fragment. The one at m/e 105 is probably derived from the 107, and can then be thrown out. The peak at m/e 122 probably should be rejected, both because of relative intensity increase and because an M minus oxygen is not easily believable here. However, 123 appears to belong to the sample. A part of the intensity at m/e 53 and 55 may be extraneous, but 54 looks good. On the bais of composition alone, m/e 113 can be rejected. (Besides, M-25 never looks reasonable). Also 119 looks questionable, and the whole group from 126 to 134 can be rejected. There are some weak impurity peaks above 138, such as the $C_9H_{16}O$ component at 140, the 141, 142, 145, and the $C_{10}H_{16}O$ at 152. (Perhaps the 91 and 107 are from the 152.) Most of the ones rejected are very weak. Such things are usually expected in working with small samples. The elimination of 91 and 107 is a great convenience. The spectrum then looks very much like the isophorone spectrum in the Aldermaston 8-peak index, allowing for some extraneous contribution to the light hydrocarbon fragments and some probable difference in ion source temperature or other instrumental parameters. (See Table 7.) However, the possibility that one of the isomeric compounds may have a similar spectrum has to be kept in mind. The symmetrical open-chain phorone can be eliminated, on the basis of the 8-peak index. (The Dow spectrum of isophorone was obtained from very impure material. I used the other one.)

The most interesting features of the spectrum are the loss of the methyl group to give m/e 123, the retro-Diels-Alder reaction to give the base peak at m/e 82 (see Djerassi, p. 153, on piperitone), and the subsequent loss of CO from the 82 to give m/e 54, with this connection further indicated by the one metastable peak observed, at m/e 35.6. (Very few metastable transitions are seen in normal scanning with instruments of the Mattauch-Herzog design. I have had the necessary circuit modification installed for detecting metastables by accelerator decoupling, and will do some more work with that when I have time.)

I put some isophorone from the ENT chemical collection into the airmail yesterday afternoon. In addition to Tables 2–7, I will enclose copies of the applicable pages from the Aldermaston index and the Dow collection. You undoubtedly have the "Dictionary of Organic compounds," but I will include the data it has on the isomeric compounds, since I have copies here. I will look for the complete isophorone spectrum from the MCA collection or run one on the ENT material if it is fit. If there are not too many interruptions, I will try to work up the data on No. 2A next.

John M. Ruth
Research Chemist

Enclosures:
Tables 2–7
Copies of literature
Manuscript notes

cc:
S. A. Hall
K. R. Hill

ARS:AEQI:JMRuth:mr: 8-17-72

B. Codling Moth Sex Pheromone No. 2A

The following memorandum and its accompanying tables are reproduced in full as an example of the manual treatment of the mass spectra of mixtures. Tables I–XI of the original are renumbered here as Tables 8–18, respectively.

AGRICULTURAL ENVIRONMENTAL QUALITY INSTITUTE
Beltsville, Maryland 20705

August 28, 1972

Subject: Codling Moth Sex Pheromone No. 2A

To: L. M. McDonough
3706 Nob Hill Boulevard
Yakima, Washington 98902

The mass spectra of sample 2A may represent two components. If so, that has not necessarily handicapped us. It may be that we have obtained as much information from the mixture as we could have from the separated components. As was the case with sample 1B, empirical compositions have been measured, and sets of isomeric structures can be suggested. As before, if a spectrum from the literature is offered for comparison, that is done with the knowledge that isomeric compounds, or perhaps compounds that yield degradation products isomeric to the given one, must also be considered.

The sample was introduced in the same way described before for number 1B. Scans 8 and 9 are background scans. Scan 10 (high resolution) shows the mass spectrum of the material coming out of the cold inlet as the trap warmed to room temperature. Four spectra were then recorded on a glass plate (positions 17, 18, 19, and 20 on plate 85), which would have taken only a few minutes in this case. During this time, the cold trap inlet was warmed up, to offset some continuing decrease in intensity. (This intensity decrease could indicate some polymerization along with the dehydration to be noted later, or it may indicate the fractional distillation of an original mixture.) As soon as the plate exposures had been made, scan 11 (cold trap now at 123 degrees C) and scan 12 (cold trap at 142 degrees C) were recorded. For these two, the slits were opened to 10 mils each (lowest resolution), to get greater sensitivity and to permit easier counting in the assignment of mass numbers on the charts. The slits were then closed down, and a few of the more important peaks were measured by matching on the oscilloscope. (I always prefer to match a few peaks, because of the known possibility of losing a plate by breakage in the plate removal machinery). The peak-matching took a bit more time than the scans had. After that, scans 13 and 14 were recorded, with the slits again wide open, the trap now at 153 degrees C, and the anode current being raised to compensate for the decreasing partial pressure of sample.

Table 8 records peak heights for scans 10, 11, 12, 13, and 14. For Table 9, intensities were calculated relative to that of m/e 43(a), the $C_2H_3O^+$ ion. For Table 10, they were recalculated relative to that of m/e 105, the $C_8H_9^+$. Table 11 contains the mass values obtained from the plate spectrum, and Table 12 the mass measurements by peak-matching.

From Table 9, it may be seen that m/e 123 and 138 are important only in the first scan (scan 10). The remaining scans show small peaks there, but samples of this sort show small peaks at nearly every mass number, with the slits wide open. This difference shows better in the peak heights of Table 8 than in relative intensisities. The last four scans (11–14) were measured on the most sensitive galvanometer tracing, but the galvanometer for the first one (scan 10) is the intermediate one. To compare the signals for a given m/e value, that for scan 10 must be multiplied by a factor of 10 for the galvanometer and by another factor to account for some additional loss at high resolution. Scan 14 must be divided by 1.5, to allow for anode current increase. For example, a comparison of signal intensities at m/e 138 would give scans 10, 11, 12, 13, and 14 the adjusted peak heights of (120), 3, 2, 0.3, and 0.25, respectively. This makes it clear that the peak at m/e 138 is not really significant after scan 10. Scan 10 also has a strong peak at m/e 120 (18 mass units below the 138). Mass measurements confirm that the difference between them is one water molecule. Since the 120 remains strong in succeeding scans, while 138 has essentially disappeared, it is possible to conclude that the loss of water is a chemical reaction of the neutral molecule, not a loss from the molecular ion. The dehydration then may be a thermal effect, or may involve the glass or metal (gallium and gold) surfaces of the inlet line in some other way than just thermal energy transfer. If the dehydration is a chemical reaction of the unionized molecule, which has gone to completion, then the m/e 98, which is $C_6H_{10}O^+$, must come from a different starting material.

The alternative explanation would be that a single compound of composition $C_9H_{14}O$ is still present in the later scans, but that the intensity of its molecular ion has been reduced to almost nothing by the warming of the inlet cold trap. This explanation does not really make any sense to me, because the temperature of the ion source and the line leading to it has not changed. Only the cold trap at the extreme outer end of the line has changed in temperature. There are enough right angle bends, glass valves, and generally irregular curves in this hot line, both before it gets to the gold molecular leak and in between the leak and the ion source, to make it appear impossible for the trap temperature to have any effect upon the temperature of the gas entering the ion source. On purely physical grounds then, it seems necessary to conclude that in the later scans the compound of m/e 138 is no longer entering the ion source, but that two molecular species, the m/e 120 derived by thermal or catalytic dehydration of the 138, and another species which is either of molecular weight 98 or contributes a fragment of mass 98, are both entering the ion source.

At this point, we may then ask whether the two species could have originated from two different chemical reactions of a single pure starting material of molecular weight 138. That is of course possible, through some combination of rearrangement, polymerization, pyrolysis, and simple dehydration reactions.

If there was only one starting material, we may try to identify the 138, by-passing the more spectacular chemistry of the inlet line. If there were two starting materials, we would of course like to identify both, although we might be satisfied to identify only the active one, if the second is inactive. (We could then boast, justifiably or not, of our judgment in picking the right one.) If I had to pick one to start with, I would pick the 138, which is an isomer of 1B but is not identical to 1B. The second component entering the ion source, if not derived from the same starting material, might be another isomer of MW 138, if we can find one that loses C_3H_4 in a rearrangement to give the m/e 98. (Ring splitting? McLafferty rearrangement of noncyclic molecule?) Or, it might be something more different.

To get back to the hydrocarbon which appears to have been derived by dehydration of the $C_9H_{14}O$ to give C_9H_{12}, it appears from Table 10 (where m/e 120, 119, and 105 should belong to it alone, and 91 partly to it) that this hydrocarbon is a trimethylbenzene, not a methylethylbenzene. In Table 13, various hydrocarbon spectra are compared with the peaks from scan 11 of Table 10. The m/e 119 peak, presumably arising through ring expansion to give a dimethyltropylium structure, is important here. It might even appear that the data favor 1,2,4- or 1,3,5-substitution in preference to 1,2,3-. Perhaps biochemical arguments would support that conclusion.

At this point, we have to stop for a moment and look at the literature on methyl group migration. (See some of the papers included with the 1B report.)

In Table 14, a mesitylene spectrum from the literature (as a somewhat arbitrary model for the trimethylbenzene) has been subtracted from each of the spectra of Table 10, eliminating the 105 peak and, hopefully, most of the other contribution from the dehydration product. The results are renormalized to m/e 43(a). As pointed out above, it must be remembered that the relative intensities fail to show some valuable information.

There is a strong decrease in absolute signal intensities, from left to right across the table, with a consequent apparent increase in the relative importance of some sort of small aliphatic hydrocarbon residue at low mass values. Scans 13 and 14 are really very weak and not of much value except as indicators of trends with time in the experiment. However, it should be noted that certain peaks, such as m/e 83 and 98, show a decrease in relative intensity with time, while the aliphatic hydrocarbon peaks show an increase (Table 14). It is of some interest to compare such significant peaks as m/e 43(a), 55, 58, 83, and 98 (Table 14) with a hexenone spectrum or with a compound that produces a similar spectrum after losing water (hexenones, Table 15; diacetone alcohol, Table 16). Table 17 records a subtraction of scan 11 from scan 10, with the elimination of m/e 105 of the dehydration product. Table 18 contains scan 10, with compositions assigned.

The selection of specific structures for 1B and 2A from among the possible isomeric forms of $C_9H_{14}O$ would be greatly aided, of course, by infrared or NMR spectra. In the meantime, I will try to find what is available from the literature.

John M. Ruth
Research Chemist

Enclosures:
Tables 8–18

cc:
S. A. Hall
K. R. Hill

ARS:AEQI:JMRuth:MRadden:8-28-72

Table 2 Low-Resolution Scans of Codling Moth Sex Attractant No. 1B

	Peak height (mm)				
m/z	Scan 3	Scan 4	Scan 5	Scan 6	Scan 7
26	39.	34.	11.	10.	6.
27	390.	345.	93.	68.	31.
28(a)	630.	450.	50.	50.	40.
28(b)	670.	670.	350.	350.	560.
28(c)	100.	90.	40.	20.	12.
29(a)	35.	40.	18.	16.	22.
29(b)	250.	200.	67.	50.	23.
32	95.	90.	59.	58.	74.
37	11.	16.	4.	3.	2.
38	43.	29.	12.	10.	3.
39	620.	535.	150.	110.	33.
40(a)	9.	13.	6.	6.	12.
40(b)	73.	69.	22.	16.	8.
41	450.	470.	130.	102.	46.
42(a)	50.	54.	16.	11.	5.
42(b)	65.	85.	28.	23.	13.
43(a)	130.	140.	37.	33.	18.
43(b)	180.	255.	74.	55.	37.
44(a)	51.	70.	44.	43.	61.
44(b)	10.	31.	20.	15.	12.
45	16.	14.	6.	4.	2.
46	21.	11.	2.	2.	1.
50	34.	34.	17.	11.	5.
51	51.	100.	33.	23.	9.
52	37.	53.	16.	14.	3.
53	110.	180.	58.	47.	11.
54	220.	310.	72.	50.	10.
55	145.	300.	84.	63.	28.
56	92.	190.	52.	39.	20.
57	270.	290.	70.	67.	27.
65	24.	33.	18.	15.	6.
67(a)	24.	55.	43.*	32.	13.
67(b)	46.	110.			
69(a)	28.	80.	51.*	42.	20.
69(b)	25.	60.			
70	39.	63.	21.	16.	10.
71	95.	75.	25.	21.	9.
77	22.	60.	30.	23.	9.
79	23.	80.	38.	27.	13.

Table 2 (continued)

m/z	Peak height (mm)				
	Scan 3	Scan 4	Scan 5	Scan 6	Scan 7
81	16.	125.	40.	28.	10.
82	1080.	—**	650.	430.	57.
83	75.	280.	80.	61.	20.
85	33.	37.	13.	11.	5.
91	19.	41.	41.	37.	24.
93	7.	33.	17.	15.	8.
94	5.	15.	6.	4.	2.
95	31.	210.	55.	35.	8.
96	10.	90.	23.	16.	5.
98	12.	14.	5.	5.	2.
99	9.	9.	3.	3.	1.
105	1.	22.	18.	14.	10.
107	3.	32.	58.	58.	39.
110	2.	13.	8.	5.	2.
111	14.	4.	5.	3.	2.
112	6.	14.	5.	3.	2.
113	14.	7.	4.	2.	1.
115	4.	14.	5.	3.	1.
117	9.	44.	10.	7.	2.
118	5.	3.	2.	1.	1.
119	18.	80.	25.	18.	5.
120	7.	17.	6.	4.	1.
121	1.	12.	7.	6.	2.
122	6.	11.	18.	16.	10.
123	16.	140.	28.	18.	4.
124	3.	29.	7.	4.	3.
125	4.	6.	4.	3.	2.
126	2.	5.	2.	2.	1.
127	4.	4.	2.	2.	0.5
128	2.	1.	2.	1.	0.5
132	7.	12.	4.5	3.	0.5
133	4.	6.	5.	3.	1.
134	6.	44.	11.	8.	2.
138	72.	860.	180.	109.	13.
139	5.	80.	17.	13.	1.5
140	6.	9.	6.5	5.	4.
142	3.	0.	1.	0.3	0.5

*Not resolved.
**Off scale. Assigned mass 138 a value of 30% RI for calculations.

Table 3 Low-Resolution Scans of Codling Moth Sex Attractant No. 1B

	Relative intensity (%)				
m/z	Scan 3	Scan 4	Scan 5	Scan 6	Scan 7
26	4.	1.	2.	2.	11.
27	36.	12.	14.	16.	54.
28(a)	58.	16.	8.	12.	70.
28(b)	—*	—*	—*	—*	—*
28(c)	9.	3.	6.	5.	21.
29(a)	3.	1.	3.	4.	39.
29(b)	23.	7.	10.	12.	40.
32	—*	—*	—*	—*	—*
37	1.	0.6	0.6	0.7	4.
38	4.	1.	2.	2.	5.
39	57.	19.	23.	26.	58.
40(a)	—*	—*	—*	—*	—*
40(b)	7.	2.	3.	4.	14.
41	42.	16.	20.	24.	81.
42(a)	5.	2.	3.	3.	9.
42(b)	6.	3.	4.	5.	23.
43(a)	12.	5.	6.	8.	32.
43(b)	17.	9.	11.	13.	65.
44(a)	5.	2.	7.	10.	107.
44(b)	0.9	1.	3.	3.	21.
45	1.	0.5	0.9	0.9	4.
46	2.	0.4	0.3	0.5	2.
50	3.	1.	3.	3.	9.
51	5.	3.	5.	5.	16.
52	3.	2.	3.	3.	5.
53	10.	6.	9.	11.	19.
54	20.	1.	1.	12.	18.
55	13.	10.	13.	15.	49.
56	9.	7.	8.	9.	35.
57	25.	10.	13.	16.	47.
65	2.	1.	3.	3.	11.
67(a)	2.	2.	7.**	7.**	23.**
67(b)	3.	3.			
69(a)	3.	3.	8.**	10.**	35.**
69(b)	2.	2.			
70	4.	2.	3.	4.	18.
71	9.	3.	4.	5.	16.
77	2.	2.	5.	5.	16.
79	2.	3.	6.	6.	23.

Table 3 (continued)

m/z	Relative intensity (%)				
	Scan 3	Scan 4	Scan 5	Scan 6	Scan 7
81	1.	4.	6.	7.	18.
82	100.	—**	100.	100.	100.
83	7.	10.	12.	14.	35.
85	3.	1.	2.	3.	9.
91	2.	1.	6.	9.	42.
93	0.6	1.	3.	3.	14.
94	0.5	0.5	0.9	0.9	4.
95	3.	7.	9.	8.	14.
96	0.9	3.	4.	4.	9.
98	1.	0.5	0.8	1.	4.
99	0.8	0.3	0.5	0.7	2.
105	0.1	0.8	3.	3.	18.
107	0.3	1.	9.	13.	68.
110	0.2	0.5	1.	1.	4.
111	1.	0.1	0.8	0.7	4.
112	0.6	0.5	0.8	0.7	4.
113	1.	0.2	0.6	0.5	2.
115	0.4	0.5	0.8	0.7	2.
117	0.8	2.	2.	2.	4.
118	0.5	0.1	0.3	0.2	2.
119	2.	3.	4.	4.	9.
120	0.6	0.6	0.9	0.9	2.
121	0.1	0.4	1.	1.	4.
122	0.	0.4	3.	4.	18.
123	1.	5.	4.	4.	7.
124	0.3	1.	1.	0.9	5.
125	0.4	0.2	0.6	0.7	4.
126	0.2	0.2	0.3	0.5	2.
127	0.4	0.1	0.3	0.5	0.9
128	0.2	0.0	0.3	0.2	0.9
132	0.6	0.4	0.7	0.7	0.9
133	0.4	0.2	0.8	0.7	2.
134	0.6	2.	2.	2.	4.
138	7.	30.***	28.	25.	23.
139	0.5	3.	3.	3.	3.
140	0.6	0.3	1.	1.	7.
142	0.3	0.0	0.2	0.1	0.9

*The large air peaks were omitted in calculating relative intensities of the sample peaks.

**Not resolved.

***Because m/z 82 was off scale in this scan, the relative intensity of m/z was estimated as 30%, and this provided the basis of calculation for the remainder of the spectrum.

Table 4 Codling Moth Sex Pheromone No. 1B — Mass Measurements by Peak Matching

m/z (low resolution)	m/z found	m/z for assumed composition	Composition	h_a/h_b ratio within doublets	Probable origin
82	82.0415	82.0419	$C_5H_6O^{+\cdot}$		Main comp.
91	91.0551	91.0548	$C_7H_7^+$		Impurity
93	93.0702	93.0704	$C_7H_9^+$		Impurity
95	95.0865	95.0861	$C_7H_{11}^+$		Note 1
105	105.0702	105.0704	$C_8H_9^+$		Impurity
107	107.0885	107.0861	$C_8H_{11}^+$		Impurity
119	119.0864	119.0861	$C_9H_{11}^+$		Note 2
121(a)	121.0653	121.0653	$C_8H_9O^+$	1/4	Note 1
121(b)	121.1019	121.1017	$C_9H_{13}^+$	1/4	Note 1
123(a)	123.0811	123.0810	$C_8H_{11}O^+$	4/1	Main comp.
123(b)	123.1160	123.1174	$C_9H_{15}^+$	4/1	Impurity
138	138.1049	138.1045	$C_9H_{14}O^{+\cdot}$		Main comp.

[1]Origin uncertain, but not important.
[2]Origin undecided.
Abbreviation: comp. = component.

Table 5 High-Resolution Plate Mass Spectrum of Codling Moth Sex Pheromone No. 1B. Plate 85, Position 15

m/z (low resolution)	Line intensity	D (mm)	m/z found	m/z for assumed composition	Composition
16(a)	v.st.	-0.0131	—	15.9949	$O^{+\cdot}$
16(b)	w.	0.0437	16.0194	16.0187	NH_2^+
16(c)	v.st.	0.0712	16.0313	16.0313	$CH_4^{+\cdot}$
17(a)	v.st.	2.2864	17.0026	17.0027	OH^+
17(b)	w.	2.3417	17.027	17.0265	$NH_3^{+\cdot}$
18	v.st.	4.5218	18.0117	18.0106	$H_2O^{+\cdot}$
19(a)		6.6497	18.9992	18.9984	F^+
19(b)		6.6680	19.0079	19.0070	$^{18}OH^+$
19(c)		6.6874	19.0170	19.0168	$HDO^{+\cdot}$
19.5		7.7329	19.5121	19.5117	$C_3H_3^{2+\cdot}$
20(a)		8.7113	—	19.9812	Ar^{++}
20(b)	v.st.	8.7640	20.0066	20.0062	$HF^{+\cdot}$
20(c)		8.7795	20.0141	20.0148	$H_2{}^{18}O^{+\cdot}$
21	v.w.	10.8080	21.0052	21.0053	$C_2H_2O^{2+}$
22		12.7863	21.9949	21.9949	CO_2^{2+}
24		16.6545	—	24.0000	$C_2^{+\cdot}$
25(a)	v.w.	18.5248	24.9992	24.9984	CF_2^{2+}
25(b)		18.5416	25.0083	25.0078	C_2H^+
25.5		19.4707	25.5123	25.5117	$C_4H_3^{2+\cdot}$
26(a)	v.w.	20.3686	26.0042	26.0031	CN^+
26(b)		20.3912	26.0166	26.0156	$C_2H_2^{+\cdot}$
27(a)	w.	22.1822	27.0120	27.0109	$CHN^{+\cdot}$
27(b)	v.st	22.2049	27.0248	27.0235	$C_2H_3^+$
28(a)	v.st.	23.9208	27.9962	27.9949	$CO^{+\cdot}$
28(b)	v.st.	23.9382	—	28.0061	$N_2^{+\cdot}$
28(c)	v.st.	23.9807	28.0304	28.0313	$C_2H_4^{+\cdot}$
29(a)	sh.	25.6682	29.0025	29.0032	$N^{15}N^{+\cdot}$
29(b)		25.7316	29.0393	29.0391	$C_2H_5^+$
30(a)		27.3690	29.9989	29.9992	$C^{18}O^{+\cdot}$
30(b)		27.3884	30.0104	30.0106	$CH_2^{+\cdot}$
30(c)	v.w.	27.4300	30.0350	30.0344	CH_4N^+
30(d)		27.4443	30.0434	30.0425	$C^{13}CH_5^+$

Table 5 (continued)

m/z (low resolution)	Line intensity	D (mm)	m/z found	m/z for assumed composition	Composition
31(a)	st.; sh.b.	29.0468	—	30.9984	CF^+
31(b)		29.0789	31.0177	30.0184	CH_3O^+
31.5		29.8944	31.5096	31.5117	$C_5H_3^{2+\cdot}$
32(a)	st.	30.6825	31.9893	31.9898	$O_2^{+\cdot}$
32(b)		30.7031	32.0019	32.0018	$^{13}CF^+$
32(c)	v.w.	30.7249	32.0152	32.0156	$C_5H_4^{2+}$
32(d)		30.7411	32.0251	32.0262	$CH_4O^{+\cdot}$
32.5		31.5458	32.5186	32.5196	$C_5H_5^{2+\cdot}$
33		32.3474	33.0140		
34		33.9144	33.9931	33.9941	$O^{18}O^{+\cdot}$
35		35.4536	34.9688	34.9689	Cl^+
36(a)		37.0204	35.9762	35.9767	$HCl^{+\cdot}$
36(b)		37.0572	—	36.0000	$C_3^{+\cdot}$
37(a)	v.w.	38.5382	36.9657	36.9659	$^{37}Cl^+$
37(b)	st.	38.6025	37.0079	37.0078	C_3H^+
37½		39.3676	37.5121	37.5117	$C_6H_3^{2+\cdot}$
38(a)	w.	40.0635	37.9737	37.9737	$H^{37}Cl^{+\cdot}$
38(b)	st.	40.1268	38.0158	38.0156	$C_3H_2^{+\cdot}$
38½		40.8803	38.5190	38.5196	$C_6H_5^{2+\cdot}$
39	v.st.	41.6320	39.0244	39.0235	$C_3H_3^+$
39½		42.3753	39.5273	39.5274	$C_6H_7^{2+\cdot}$
40(a)		43.0146	—	39.9624	$Ar^{+\cdot}$
40(b)	w.	43.0627	39.9952	39.9949	$C_2O^{+\cdot}$
40(c)	shoulder on (d) (w)	43.0982	40.0195	40.0187	$C_2H_2N^+$
40(d)	v.st.	43.1138	40.0302	40.0313	$C_3H_4^{+\cdot}$
40½	v.w.	43.8477	40.5335	40.5352	$C_6H_9^{2+\cdot}$
41(a)	w.	44.5291	41.0036	41.0027	C_2HO^+
41(b)	w. [not complete-resolved from (c)]	44.5644	41.0281	41.0265	$C_2H_3N^{+\cdot}$
41(c)	v.st.	44.5816	41.0400	41.0391	$C_3H_5^+$

Table 5 (continued)

m/z (low resolution)	Line intensity	D (mm)	m/z found	m/z for assumed composition	Composition
42(b)	st.	45.9776	42.0120	42.0106	$C_2H_2O^{+\cdot}$
42(d)	v.st.[(a) and (c) v.w.]	46.0273	42.0468	42.0469	$C_3H_6^{+\cdot}$
43(a)	w.	47.3811	43.0008	42.9984	$C_2F^{+\cdot}$
43(b)	w.	47.3920	43.0085	43.0058	$CHNO^{+\cdot}$
43(c)	st.	47.4086	43.0202	43.0184	$C_2H_3O^+$
43(d)	st. (w shoulder in front)	47.4594	43.0563	43.0548	$C_3H_7^+$
43½	v.w.	48.0995	43.5113	43.5117	$C_7H_3^{2+\cdot}$
44(a)	st.	48.7690	—	43.9898	$CO_2^{+\cdot}$
44(c)	strongest	48.8167	44.0240	44.0262	$C_2H_4O^{+\cdot}$
44(d)	w.	48.8635	44.0576	44.0581	$C_2{}^{13}CH_7^+$
44½	v.w.	49.5020	44.5167	44.5196	$C_7H_5^{2+\cdot}$
45(b)	st.	50.1580	44.9909	44.9932	$^{13}CO_2^{+\cdot}$
45(d)	st.	50.1969	45.0191	45.0215	$CH_3NO^{+\cdot}$
45(e)	strongest	50.2085	45.0275	45.0296	$C^{13}CH_4O^{+\cdot}$
45½		50.8815	45.5169		
46	Several very weak ones				
50(a)	Strong	56.9307	—	49.9968	$CF_2^{+\cdot}$
50(b)	Strong	56.9543	50.0150	50.0156	$C_4H_2^{+\cdot}$
51(b)	v. strong	58.2335	—	51.0046	CHF_2^+
51(c)	Strong	58.2548	51.0212	51.0235	$C_4H_3^+$
51½		58.9020	51.5257	51.5274	$C_8H_7^{2+\cdot}$
51(a)	v. st.	58.2311	—	51.0046	CHF_2^+
51(b)	st.	58.2530	51.0216	51.0235	$C_4H_3^+$
51½		58.9000	51.5246	51.5274	$C_8H_7^{2+\cdot}$
52(b)	w.	59.5187	—	52.0080	$^{13}CHF_2^+$
52(c)	st.	59.5456	52.0291	52.0313	$C_4H_4^{+\cdot}$
52½(a)	v.w.	60.1658	52.5167	52.5170	$C_7H_5O^{2+\cdot}$
52½(b)	w.	60.1892	52.5350	52.5352	$C_8H_9^{2+\cdot}$

Table 5 (continued)

m/z (low resolution)	Line intensity	D (mm)	m/z found	m/z for assumed composition	Composition
53(a)	w.	60.7842	53.0029	53.0027	C_3HO^+
53(b)	w.	60.8073	53.0211	53.0209	$C_7H_6O^{2+}$
53(c)	st.	60.8298	53.0388	53.0391	$C_4H_5^+$
53½(a)	v.w.	61.4447	53.5247	53.5248	$C_7H_7O^{2+\cdot}$
53½(b)	v.w.	61.4681	53.5432	53.5430	$C_8H_{11}^{2+\cdot}$
54(d)		62.1026	—	54.0469	$C_4H_6^{+\cdot}$
55(a)	w.	63.2960	55.0002	54.9984	C_3F^+
55(b)	w.	63.3202	55.0196	55.0184	$C_3H_3O^+$
55(c)	st. (sh. in front)	63.3643	55.0550	55.0548	$C_4H_7^+$
56(a)	v.w.	64.5465	56.0079	56.0062	$C_3HF^{+\cdot}$
56(b)	w.	64.5698	56.0268	56.0262	$C_3H_4O^{+\cdot}$
56(c)	st.	64.6140	—	56.0626	$C_4H_8^{+\cdot}$
57(a)	v.w.	65.7878	57.0154	57.0141	$C_3H_2F^+$
57(b)	w.	65.8114	57.0346	57.0340	$C_3H_5O^+$
57(c)	st.	65.8551	57.0703	57.0704	$C_4H_9^+$
57½	w.	66.4158	57.5284	57.5274	$C_9H_7^{2+\cdot}$
58(a)	[shoulder on (b)]	67.0301	58.0325	58.0313	$C_9H_8^{2+}$
58(b)	st.	67.0400	58.0407	58.0419	$C_3H_6O^{+\cdot}$
58(c)	weaker than (b)	67.0805	58.0740	58.0738	$C_3{}^{13}CH_9^+$
58½		67.6401	58.5353	58.5352	$C_9H_9^{2+\cdot}$
59(c)		68.2600	59.0484	59.0497	$C_3H_7O^+$
59½(b)	w.	68.8335	59.5251	59.5248	$C_8H_7O^{2+\cdot}$
59½(c)	v.w.	68.8535	59.5417	59.5430	$C_9H_{11}^{2+\cdot}$
60(b)		69.4300	60.0229	60.0211	$C_2H_4O_2^{+\cdot}$
60(c)		69.4367	60.0285	60.0288	$C_8H_8O^{2+}$
60½		70.0376	60.5322	60.5327	$C_8H_9O^{2+\cdot}$
61(a)		70.6029	61.0079	61.0078	C_5H^+
61½		71.2330	61.5403	61.5405	$C_8H_{11}O^{2+\cdot}$
62(a)		71.7710	—	61.9968	$C_2F_2^{+\cdot}$
62(b)		71.7928	62.0153	62.0156	$C_5H_2^{+\cdot}$

Table 5 (Continued)

m/z (low resolution)	Line intensity	D (mm)	m/z found	m/z for assumed composition	Composition
63(a)		72.9513	63.0037	63.0046	$C_2HF_2^+$
63(b)		72.9730	63.0223	63.0235	$C_5H_3^+$
63½		73.5603	63.5264	63.5274	$C_{10}H_7^{2+\cdot}$
64(a)	v.w.	74.0642	63.9605	63.9619	$SO_2^{+\cdot}$
64(b)		74.1237	64.0119	64.0125	$C_2H_2F_2^{+\cdot}$
64(c)	strongest	74.1444	64.0298	64.0313	$C_5H_4^{+\cdot}$
64½		74.7265	64.5334		
65(b)		75.2861	65.0194	65.0203	$C_2H_3F_2^+$
65(c)	stronger	73.3081	65.0385	65.0391	$C_5H_5^+$
65½(a)		75.8307	65.4941	65.4960	$C_3F_8^{2+\cdot}$
65½(b)		75.8836	65.5403		
66(b)		76.3985	65.9909	65.9917	$COF_2^{+\cdot}$
66(c)		76.4210	66.0106	66.0106	$C_4H_2O^{+\cdot}$
66(e)	strongest	76.4603	66.0451	66.0469	$C_5H_6^{+\cdot}$
67(c)	strong	77.5661	67.0184	67.0184	$C_4H_3O^+$
67(e)	v. strong	77.6067	67.0542	67.0548	$C_5H_7^+$
68(d)	st.	78.6998	68.0236	68.0262	$C_4H_4O^{+\cdot}$
68(f)	stronger	78.7407	68.0600	68.0626	$C_5H_8^{+\cdot}$
69(a)	too strong	79.7887	68.9962	68.9952	CF_3^+
69(b)	(b) & (c) are about equal	79.8305	69.0336	69.0340	$C_4H_5O^+$
69(c)		79.8694	69.0685	69.0704	$C_5H_9^+$
70(a)		80.9030	—	69.9986	$^{13}CF_3^+$
70(a)		80.9010	—	69.9986	$^{13}CF_3^+$
70(b)		80.9467	70.0390	70.0418	$C_4H_6O^{+\cdot}$
70(c)	stronger than (b)	80.9879	70.0762	70.0782	$C_5H_{10}^{+\cdot}$
71(b)		82.0610	71.0488	71.0497	$C_4H_7O^+$
71(c)	strongest	82.1050	71.0888	71.0861	$C_5H_{11}^+$
72(a)	st.	83.1650	72.0564	72.0575	$C_4H_8O^{+\cdot}$
72(b)	st.	83.2023	72.0960	72.0894	$C_4{}^{13}CH_{11}^+$
73	four weak ones	—	—	—	—

Table 5 (continued)

m/z (low resolution)	Line intensity	D (mm)	m/z found	m/z for assumed composition	Composition
74(a)	strongest	84.2694	73.9967	73.9968	$C_3F_2^{+\cdot}$
74(b)		85.2897	74.0156	74.0156	$C_6H_2^{+\cdot}$
75	strongest	86.3518	75.0047	75.0046	$C_3HF_2^+$
75(b)		86.3715	75.0231	75.0235	$C_6H_3^+$
76(b)	weak	87.4460	76.0307	76.0313	$C_6H_4^{+\cdot}$
77(b)	weak	88.4942	77.0200	77.0203	$C_3H_3F_2^+$
77(c)	st.	88.5150	77.0397	77.0391	$C_6H_5^+$
78		89.5731	78.0451	78.0469	$C_6H_6^{+\cdot}$
79(a)	w.	90.5714	78.9995	78.9984	C_5F^+
79(c)	st.	90.6286	79.0544	79.0548	$C_6H_7^+$
80(c)	st.	91.6745	80.0610	80.0626	$C_6H_8^{+\cdot}$
81(a)	st.	92.6394	—	80.9952	$C_2F_3^+$
81(b)	w.	92.6787	81.0334	81.0340	$C_5H_5O^+$
81(c)	st.	92.7154	81.0690	81.0704	$C_6H_9^+$
82(a)		93.6737	82.0028	82.0030	$C_2HF_3^{+\cdot}$
82(b)	(too strong)	93.7155	82.0437	82.0419	$C_5H_6O^{+\cdot}$
82(c)		93.7472	82.0747	81.0738	$C_5{}^{13}CH_9^+$
83(a)	w.	94.7015	83.0103	83.0109	$C_2H_2F_3^+$
83(b)	st.	94.7390	83.0472	83.0452	$C_4{}^{13}CH_6O^{+\cdot}$ and $C_5H_7O^+$
83(c)		94.7761	83.0837	83.0861	$C_6H_{11}^+$
84(b)	not resolved	95.7541	84.0485	84.0486	$C_3{}^{13}C_2H_6O^{+\cdot}$
84(c)	v. st.	95.7615	84.0558	84.0575	$C_5H_8O^{+\cdot}$
84(d)		95.7991	84.0930	84.0939	$C_6H_{12}^{+\cdot}$
85(a)	v. st.	96.6794	84.9664	84.9657	$CClF_2^+$
85(e)	w.	96.7783	85.0648	85.0653	$C_5H_9O^+$
85(f)	st.	96.8140	85.1003	85.1017	$C_6H_{13}^+$
86(c)	w.	97.7120	85.9967	85.9968	$C_4F_2^{+\cdot}$
86(f)	w.	97.7873	86.0720	86.0732	$C_5H_{10}O^{+\cdot}$
86(g)	(weaker than (f))	97.8203	86.1051	86.1051	$C_5{}^{13}CH_{13}^+$

Table 5 (continued)

m/z (low resolution)	Line intensity	D (mm)	m/z found	m/z for assumed composition	Composition
87(a)		98.6747	86.9628	86.9627	$C^{37}ClF_2^+$
87(b)	v.w.	98.7158	87.0041	87.0046	$C_4HF_2^+$
87(c)		98.7353	87.0238	87.0235	$C_7H_3^+$
89		100.7251	89.0384	89.0391	$C_7H_5^+$
90(b)	v.w.	101.7117	90.0459	90.0469	$C_7H_6^{+\cdot}$
91	st.	102.6940	91.0546	91.0548	$C_7H_7^+$
92	st.	103.6671	92.0594	92.0581	$C_6{}^{13}CH_7^+$
93(a)	st.	104.5686	—	92.9952	$C_3F_3^+$
93(c)	st. [(b) is negligible]	104.6398	93.0693	93.0704	$C_7H_9^+$
- - -	- - -	- - -	- - -	- - -	- - -
93(a)		104.5696	—	92.9952	$C_3F_3^+$
94(a)		105.5318	93.9994	93.9986	$C_2{}^{13}CF_3^+$
94(b)	w.	105.5711	94.0405	94.0419	$C_6H_6O^{+\cdot}$
94(c)		105.6048	94.0758	94.0738	$C_6{}^{13}CH_9^+$
95(a)		106.4970	95.0121	95.0109	$C_3H_2F_3^+$
95(b)	w.	106.5334	95.0504	95.0497	$C_6H_7O^+$
95(c)		106.5689	95.0877	95.0861	$C_7H_{11}^+$
96(b)	w.	107.4883	96.0578	96.0575	$C_6H_8OO^{+\cdot}$
96(c)	st.	107.5212	96.0926	96.0939	$C_7H_{12}^{+\cdot}$
97(b)	w.	108.4712	97.1003	97.1017	$C_7H_{13}^+$
98(d)		109.3850	98.0746	98.0732	$C_6H_{10}O^{+\cdot}$
98(e)		109.4191	98.1111	98.1095	$C_7H_{14}^{+\cdot}$
99(b)		110.2545	99.0062	99.0046	$C_5HF_2^+$
99(d)		110.3254	99.0824	99.0810	$C_6H_{11}O^+$
99(e)		110.3593	99.1188	99.1174	$C_7H_{15}^+$
99.5		110.6999	99.4851	99.4841	$^{199}Hg^{2+}$
100(a)	v.st.	111.1737	99.9958	99.9936	$C_2F_4^{+\cdot}$
100(c)		111.2611	100.0902	100.0888	$C_6H_{12}O^{+\cdot}$
101(2)		111.6274	100.4861	100.4851	$^{201}Hg^{2+}$
100(a)		112.0435	100.9368	100.9361	CCl_2F^+
101(b)	(not fully resolved)	112.0896	100.9868	100.9853	$^{202}Hg^{2+}$

Table 5 (continued)

m/z (low resolution)	Line intensity	D (mm)	m/z found	m/z for assumed composition	Composition
101(c)	st.	112.1024	101.0006	101.0014	$C_2HF_4^+$
102(c)		113.0654	102.0479	102.0469	$C_8H_6^{+\cdot}$
103(b)		113.9871	103.0553	103.0548	$C_8H_7^+$
104		114.9033	104.0616	104.0626	$C_8H_8^{+\cdot}$
105(a)		115.7494	—	104.9952	$C_4F_3^+$
105(b)	w.	115.8837	105.0331	105.0340	$C_7H_5O^+$
105(c)		115.8168	105.0698	105.0704	$C_8H_9^+$
106(a)		116.6577	106.0024	106.0030	$C_4HF_3^{+\cdot}$
106(b)		116.6923	106.0409	106.0419	$C_7H_6O^{+\cdot}$
106(c)		116.7239	106.0760	106.0782	$C_8H_{10}^{+\cdot}$
107(b)	w. [(a) is v.w.]	117.5972	107.0493	107.0497	$C_7H_7O^+$
107(c)	st.	117.6293	107.0851	107.0861	$C_8H_{11}^+$
108(b)	[(a) is v.w.]	118.4972	108.0570	108.0575	$C_7H_8O^{+\cdot}$
108(c)		118.5273	108.0907	108.0894	$C_7{}^{13}CH_{11}^+$
109(d)		119.3936	109.0653	109.0653	$C_7H_9O^+$
109(e)		119.4258	109.1016	109.1017	$C_8H_{13}^+$
110(c)		120.2850	110.0727	110.0732	$C_7H_{10}O^{+\cdot}$
110(d)		120.3168	110.1087	110.1095	$C_8H_{14}^{+\cdot}$
111(d)		121.1718	111.0794	111.0810	$C_7H_{11}O^+$
111(e)		121.2048	111.1170	111.1174	$C_8H_{15}^+$
112(a)	st.	121.9736	—	111.9936	$C_3F_4^{+\cdot}$
112(a)	st.	121.9700	—	111.9936	$C_3F_4^{+\cdot}$
112(b)	v.w.	129.0527	112.0881	112.0888	$C_7H_{12}O^{+\cdot}$
112(c)	weak	122.0844	112.1243	112.1252	$C_8H_{16}^{+\cdot}$
113(a)	st.	122.8496	113.0007	113.0014	$C_3HF_4^+$
113(b)	v.w.	122.9324	113.0957	113.0966	$C_7H_{13}O^+$
113(c)	w.	122.9634	113.1313	113.1330	$C_8H_{17}^+$
114(a)		123.7231	114.0052	114.0048	$C_2{}^{13}CHF_4^+$
114(b)	v.w.	123.8080	114.1031	114.1045	$C_7H_{14}O^{+\cdot}$
114(c)	weaker than (b)	123.8362	114.1356	114.1364	$C_7{}^{13}CH_{17}^+$

Table 5 (continued)

m/z (low resolution)	Line intensity	D (mm)	m/z found	m/z for assumed composition	Composition
115(a)	v.w.	124.5980	115.0158	115.0171	$C_3H_3F_4^+$
115(b)	w.	124.6307	115.0536	115.0548	$C_9H_7^+$
116(a)		125.4149	115.9634	115.9641	$C_2ClF_3^{+\cdot}$
116(b)		125.4989	116.0610	116.0626	$C_9H_8^{+\cdot}$
117(a)		126.3005	116.9951	116.9952	$C_5F_3^+$
117(b)		126.3640	117.0692	117.0704	$C_9H_9^+$
118(a)	v.w.	127.1250	117.9597	117.9611	$C_2{}^{37}ClF_3^{+\cdot}$
118(b)	v.w.	127.1579	117.9983	117.9986	$C_4{}^{13}CF_3^+$
118(c)	v.w. [stronger than (b)]	127.2226	118.0742	118.0738	$C_8{}^{13}CH_9^+$
119(a)	v.st.	128.0036	—	118.9920	$C_2F_5^+$
119(b)	v.w.	128.0508	119.0476	119.0497	$C_8H_7O^+$
119(c)		128.0817	119.0840	119.0861	$C_9H_{11}^+$
120(a)		128.8524	119.9941	119.9954	$C{}^{13}CF_5^+$
120(b)	v.w., trace	128.9042	120.0553		
120(c)		128.9340	120.0906	120.0894	$C_8{}^{13}CH_{11}^+$
121(a)		129.7550	121.0642	121.0653	$C_8H_9O^+$
121(b)	stronger than (a)	129.7848	121.0996	121.1017	$C_9H_{13}^+$
122(a)	v.w.	130.6013	122.0719	122.0732	$C_8H_{10}O^{+\cdot}$
122(b)		130.6314	122.1079	122.1095	$C_9H_{14}^{+\cdot}$
123(a)		131.4448	123.0805	123.0810	$C_8H_{11}O^+$
123(b)	w. comp. to (a)	131.4733	123.1146	123.1129	$C_8{}^{13}CH_{14}^{+\cdot}$
124(a)	st.	132.2055	—	123.9936	$C_4F_4^{+\cdot}$
124(b)	st.	132.2823	124.0860	124.0888	$C_8H_{12}O^{+\cdot}$
124(c)	v.w.	132.3141	124.1242	124.1252	$C_9H_{16}^{+\cdot}$
125(a)		133.0406	124.9994	125.0014 124.9970	$C_4HF_4^+$ $C_3{}^{13}CF_4^{+\cdot}$
125(b)		133.1200	125.0952	125.0966	$C_8H_{13}O^+$
125(c)	weaker than (b)	133.1499	125.1313	125.1330	$C_9H_{17}^+$

Table 5 (continued)

m/z (low resolution)	Line intensity	D (mm)	m/z found	m/z for assumed composition	Composition
126(a)		133.8750	126.0084	126.0093	$C_4H_2F_4^{+\cdot}$
126(b)	v.w.	133.9526	126.1024	126.1045	$C_8H_{14}O^{+\cdot}$
126(c)		133.9828	126.1390	126.1408	$C_9H_{18}^{+\cdot}$
127(a)	v.w.	134.7049	127.0159	127.0171	$C_4H_3F_4^+$
127(b)	v.w.	134.7356	127.0533	127.0548	$C_{10}H_7^+$
127(c)	v.w.	134.7830	127.1109	127.1123	$C_8H_{15}O^+$
127(d)	w.	134.8126	127.1470	127.1487	$C_9H_{19}^+$
128(b)		135.5632	128.0622	128.0626	$C_{10}H_8^{+\cdot}$
129(a)		136.3256	128.9951	128.9952	$C_6F_3^+$
129(b)		136.3861	129.0693	129.0704	$C_{10}H_9^+$
130(a)	v.w.	137.1450	130.0016	130.0030	$C_6HF_3^{+\cdot}$
130(b)		137.2060	130.0767	130.0782	$C_{10}H_{10}^{+\cdot}$
131(a)	v.st.	137.9487	130.9926	130.9920	$C_3F_5^+$
131(b)		138.0234	131.0849	131.0861	$C_{10}H_{11}^+$
132(a)	v.w.	138.7084	131.9328	131.9345	$C_2Cl_2F_2^{+\cdot}$
132(b)	st.	138.7588	131.9953	131.9954	$C_2{}^{13}CF_{5+}$
132(c)		138.8362	132.0913	132.0939	$C_{10}H_{12}^{+\cdot}$
133(a)		139.5724	133.0062	133.0077	$C_3H_2F_5^+$
133(b)		139.6478	133.1001	133.1017	$C_{10}H_{13}^+$
134		140.4556	134.1079	134.1095	$C_{10}H_{14}^{+\cdot}$
135(a)		141.1374	134.9615	134.9625	$C_2ClF_4^+$
135(b)	v.w.	141.2325	135.0808	135.0810	$C_9H_{11}O^+$
136(a)	st. comp. to (b)	141.9589	—	135.9936	$C_5F_4^{+\cdot}$
136(b)		142.0336	136.0877	136.0888	$C_9H_{12}O^{+\cdot}$
136(c)	stronger than (b)	142.0626	136.1242	136.1252	$C_{10}H_{16}^{+\cdot}$
137(a)		142.7245	136.9589	136.9595	$C_2{}^{37}ClF_4^+$
137(b)		142.7569	136.9998	137.0014	$C_5HF_4^+$
137(c)	w.	142.8328	137.0957	137.0966	$C_9H_{13}O^+$
137(d)	v.w.	142.8600	137.1301	137.1330	$C_{10}H_{17}^+$

Table 5 (continued)

m/z (low resolution)	Line intensity	D (mm)	m/z found	m/z for assumed composition	Composition
138(a)	v.w.	143.5520	138.0061	138.0048	$C_4{}^{13}CHF_4{}^+$
138(b)	v.w.	143.6001	138.0671	138.0681	$C_8H_{10}O_2{}^{+\cdot}$
138(c)	v.st.	143.6294	138.1042	138.1045	$C_9H_{14}O^{+\cdot}$
138(d)	v.w.	143.6574	138.1397	138.1408	$C_{10}H_{18}{}^{+\cdot}$
139		144.4195	139.1082	139.1078	$C_8{}^{13}CH_{14}O^{+\cdot}$
140(a)	st.; not resolved	145.2066	140.1119	140.1112	$C_7{}^{13}C_2H_{14}O^{+\cdot}$
140(b)	st.; not resolved	145.2019	140.1174	140.1201	$C_9H_{16}O^{+\cdot}$
140(c)	v.w.	145.2404	140.1551	140.1565	$C_{10}H_{20}{}^{+\cdot}$
141(a)	w.	145.8962	140.9942	140.9952	$C_7F_3{}^+$
141(b)	w.	145.0275	141.1626	141.1643	$C_{10}H_{21}{}^+$
142(a)	v.w.	146.6780	141.9979	141.9986	$C_6{}^{13}CF_3{}^+$
142(b)	v.w.	146.8102	142.1680	142.1677	$C_9{}^{13}CH_{21}{}^+$
143		147.4496	—	142.9920	$C_4F_5{}^+$
144		148.2273	143.9975	143.9998	$C_4HF_5{}^{+\cdot}$
145(a)		149.0046	145.0061	145.0077	$C_4H_2F_5{}^+$
145(b)	v.w.	149.0768	145.0999	145.1017	$C_{11}H_{13}{}^+$
150		152.8071	—	149.9904	$C_3F_6{}^{+\cdot}$
152(a)		154.3235	152.0016	152.0016	$C_2{}^{13}CHF_6{}^+$
152(b)		154.4118	152.1191	152.1201	$C_{10}H_{16}O^{+\cdot}$
174		170.2957	—	173.9904	$C_5F_6{}^{+\cdot}$
178(a)	w.	173.0707	177.9623	177.9609	$C_4ClF_5{}^{+\cdot}$
178(b)	v.w.	173.1933	178.1388		

Abbreviations: st., strong; w., weak; v., very; sh., shoulder; ext., extremely.
Broken line across table: Indicates that some time elapsed between the two series of measurements above and below the line, with possible change of plate position on the comparator.
Calibration: By Reference [93], p. 347, Eqs. (32) and (33).
Calibration segments are described by the numerical format: SEGMENT NUMBER [FIRST STANDARD LINE, SECOND STANDARD LINE, FIRST CALCULATED LINE, LAST CALCULATED LINE]. Data given after Table 11. These notes also apply to Table 11.

Table 6 Codling Moth Sex Pheromone No. 1B. Scan 5 (3/26/72)

m/z	Peak height (mm)	Relative intensity (%)	Composition (from plate spectrum)
12	2.0	0.3	$C^{+\cdot}$
13	1.5	0.2	CH^{+}
14(a)	17.	2.6	N^{+}
14(b)	4.	0.6	$CH_2^{+\cdot}$
15	22.	3.4	CH_{3+} (based on $CH_4^{+\cdot}$ at 16)
16(a)	9.	1.4	$O^{+\cdot}$
16(b)	3.	0.5	$CH_4^{+\cdot}$
17	34.	5.2	OH^{+}
18	140.	22.	$H_2O^{+\cdot}$
19	0.5	0.08	$^{18}OH^{+}$ and $HDO^{+\cdot}$
20(a)	0.7	0.1	Ar^{2+}
20(b)	6.	0.9	$H_2{}^{18}O^{+\cdot}$
22	0.6	0.09	CO_2^{2+}
24	0.4	0.06	$C_2^{+\cdot}$
25	1.7	0.3	C_2H^{+}
26	11.	1.7	$C_2H_2^{+\cdot}$
27	93.	14.	$C_2H_3^{+}$
28(a)	50.	7.7	$CO^{+\cdot}$
28(b)	350.	54.	$N_2^{+\cdot}$ (air)
28(c)	40.	6.2	$C_2H_4^{+\cdot}$
29(a)	18.	2.8	$^{13}CO^{+\cdot}$ and $N^{15}N^{+\cdot}$, etc.

Table 6 (Continued)

m/z	Peak height (mm)	Relative intensity (%)	Composition (from plate spectrum)
29(b)	67.	10.	$C_2H_5^+$
30(a)	1.6	0.2	$C^{18}O^{+\cdot}$
30(b)	2.0	0.3	$CH_2O^{+\cdot}$
31	3.	0.5	CH_3O^+
32(a)	59.	9.1	$O_2^{+\cdot}$ (air)
32(b)	0.6	0.09	$CH_4O^{+\cdot}$
35.6	0.5	0.08	Metastable (82 → 54)
36	1.	0.2	$CH_3^{+\cdot}$
37	4.	0.6	C_3H^+
38	12.	1.8	$C_3H_2^{+\cdot}$
39	150.	23.	$C_3H_3^+$
39.5	1.	0.2	$C_6H_2^{2+}$
40(a)	6.	0.9	$Ar^{+\cdot}$ (air)
40(b)	22.	3.4	$C_3H_4^{+\cdot}$
41	130.	20.	$C_3H_5^+$
42(a)	16.	2.5	$C_2H_2O^{+\cdot}$
42(b)	28.	4.3	$C_3H_6^{+\cdot}$
43(a)	37.	5.7	$C_2H_3O^+$
43(b)	74.	11.	$C_3H_7^+$
44(a)	44.	6.8	$CO_2^{+\cdot}$
44(b)	20.	3.1	$C_2H_4O^{+\cdot}$

Table 6 (Continued)

m/z	Peak height (mm)	Relative intensity (%)	Composition (from plate spectrum)
45	6.	0.9	$^{13}CO_2^{+\cdot}$, $C^{13}CH_4O^{+\cdot}$, etc.
45.5	0.5	0.008	Very weak, not measured
46	2.	0.3	(Problem on plate in this area)
46.5	0.4	0.06	(Problem on plate in this area)
49	2.	0.3	Not measured
50	17.	2.6	$C_4H_2^{+\cdot}$
51	33.	5.1	$C_4H_3^{+}$
52	16.	2.5	$C_4H_4^{+\cdot}$
52.5	0.8	0.1	$C_7H_5O^{2+\cdot}$ and $C_8H_9^{2+\cdot}$
53	58.	8.9	$C_4H_5^{+}$ (trace of C_3HO^{+})
54	72.	11.	$C_4H_6^{+\cdot}$
55	84.	13.	$C_4H_7^{+}$ (weak $C_3H_3O^{+}$)
56	52.	8.	$C_4H_8^{+\cdot}$ (weak $C_3H_4O^{+\cdot}$)
57(a)	10.	1.5	$C_3H_5O^{+}$
57(b)	70.	11.	$C_4H_9^{+}$
57.5	0.5	0.08	$C_9H_7^{2+\cdot}$
58	13.	2.0	$C_3H_6O^{+\cdot}$
59	2.	0.3	$C_3H_7O^{+}$
59.5	0.3	0.05	$C_8H_7O^{2+\cdot}$ and $C_9H_{11}^{2+\cdot}$
60	1.8	0.3	$C_2H_4O_2^{+\cdot}$ and $C_8H_8O^{2+}$
60.5	0.4	0.06	$C_8H_9O^{2+\cdot}$

Table 6 (Continued)

m/z	Peak height (mm)	Relative intensity (%)	Composition (from plate spectrum)
61	2.2	0.3	C_5H^+
61.5	0.7	0.1	$C_8H_{11}O^{2+\cdot}$
62	3.	0.5	$C_5H_2^{+\cdot}$
62.5	0.1	0.02	
63	11.	1.7	$C_5H_3^+$
64	3.5	0.5	$C_5H_4^{+\cdot}$
65	18.	2.8	$C_5H_5^+$
66	10.	1.5	$C_5H_6^{+\cdot}$
67	43.	6.6	$C_5H_7^+$ and some $C_4H_3O^+$
68	17.	2.6	$C_5H_8^{+\cdot}$ and some $C_4H_4O^{+\cdot}$
69	51.	7.8	$C_5H_9^+$ and $C_4H_5O^+$ (about equal)
70	21.	3.2	$C_5H_{10}^{+\cdot}$ and some $C_4H_6O^{+\cdot}$
71	25.	3.8	$C_5H_{11}^+$ and some $C_4H_7O^+$
72	10.	1.5	$C_4{}^{13}CH_{11}^+$ and $C_4H_8O^{+\cdot}$
73	1.	0.2	
74	3.5	0.5	$C_6H_2^{+\cdot}$
75	2.	0.3	$C_6H_3^+$
76	1.	0.2	$C_6H_4^{+\cdot}$
77	30.	4.6	$C_6H_5^+$
78	10.	1.5	$C_6H_6^{+\cdot}$
79	38.	5.8	$C_6H_7^+$

Table 6 (Continued)

m/z	Peak height (mm)	Relative intensity (%)	Composition (from plate spectrum)
80	12.	1.8	$C_6H_8^{+\cdot}$
81	40.	6.2	$C_6H_9^+$ (and a little $C_5H_5O^+$)
82	650.	100.	$C_5H_6O^{+\cdot}$
83	80.	12.	$C_4{}^{13}CH_6O^{+\cdot}$ and $C_5H_7O^+$ (trace of $C_6H_{11}^+$)
84	18.	2.8	$C_3{}^{13}C_2H_6O^{+\cdot}$ and $C_5H_8O^{+\cdot}$ (trace of $C_6H_{12}^{+\cdot}$)
85	13.	2.0	$C_6H_{13}^+$ (weak $C_5H_9O^+$)
86	2.5	0.4	$C_5H_{10}O^{+\cdot}$
87	0.8	0.1	$C_7H_3^+$
89	1.3	0.2	$C_7H_5^+$
90	0.7	0.1	$C_7H_6^{+\cdot}$
91	41.	6.3	$C_7H_7^+$
92	6.	0.9	$C_6{}^{13}CH_7^+$
93	17.	2.6	$C_7H_9^+$
94	6.	0.9	$C_6{}^{13}CH_9^+$ (weak $C_6H_6O^{+\cdot}$)
95	55.	8.5	$C_7H_{11}^+$ (weak $C_6H_6O^+$)
96	23.	3.5	$C_7H_{12}^{+\cdot}$ (weak $C_6H_8O^{+\cdot}$)
97	6.	0.9	$C_7H_{13}^+$ (weak $C_6H_9O^+$)
98	5.	0.8	$C_6H_{10}O^{+\cdot}$ and $C_7H_{14}^{+\cdot}$
99	3.	0.5	$C_6H_{11}O^+$ and $C_7H_{15}^+$

Table 6 (Continued)

m/z	Peak height (mm)	Relative intensity (%)	Composition (from plate spectrum)
100	2.8	0.4	$C_6H_{12}O^{+\cdot}$ (impurity?)
101	0.6	0.09	$202Hg^{2+}$ (from manometer)
102	0.8	0.1	$C_8H_6^{+\cdot}$
103	3.5	0.5	$C_8H_7^+$
104	2.0	0.3	$C_8H_8^{+\cdot}$
105	18.	2.8	$C_8H_9^+$ (weak $C_7H_5O^+$)
106	5.	0.8	$C_7H_6O^{+\cdot}$ and $C_8H_{10}^{+\cdot}$
107	58.	8.9	$C_8H_{11}^+$ (probably impurity); weak $C_7H_7O^+$
108	10.	1.5	$C_7H_8O^{+\cdot}$ and $C_7{}^{13}CH_{11}^+$
109	13.	2.0	$C_7H_9O^+$ and $C_8H_{13}^+$
110	8.	1.2	$C_7H_{10}O^{+\cdot}$ and $C_8H_{14}^{+\cdot}$
111	5.	0.8	$C_7H_{11}O^+$ and $C_8H_{15}^+$
112	5.	0.8	$C_8H_{16}^{+\cdot}$ (trace of $C_7H_{12}O^{+\cdot}$)
113	4.	0.6	$C_8H_{17}^+$ (trace of $C_7H_{13}O^+$)
114	1.	0.2	$C_7H_{14}O^{+\cdot}$ (impurity)
115	5.	0.8	$C_9H_7^+$
116	2.	0.3	$C_9H_8^{+\cdot}$
117	10.	1.5	$C_9H_9^+$
118	2.	0.3	$C_8{}^{13}CH_9^+$
119	25.	3.8	$C_9H_{11}^+$

Table 6 (Continued)

m/z	Peak height (mm)	Relative intensity (%)	Composition (from plate spectrum)
120	6.	0.9	$C_8{}^{13}CH_{11}{}^+$
121	7.	1.1	$C_9H_{13}{}^+$ and $C_8H_9O^+$ (4:1)
122	18.	2.8	$C_9H_{14}{}^{+\cdot}$ (trace of $C_8H_{10}O^{+\cdot}$)
123	28.	4.3	$C_8H_{11}O^+$, and ($C_9H_{15}{}^+$ plus $C_8{}^{13}CH_{14}{}^{+\cdot}$), 4:1
124	7.	1.1	$C_8H_{12}O^{+\cdot}$ (v. weak $C_9H_{16}{}^{+\cdot}$)
125	4.	0.6	$C_8H_{13}O^+$ (trace of $C_9H_{17}{}^+$)
126	2.	0.3	$C_9H_{18}{}^{+\cdot}$ (trace of $C_8H_{14}O^{+\cdot}$)
127	2.	0.3	$C_9H_{19}{}^+$ + etc.
128	2.	0.3	$C_{10}H_8{}^{+\cdot}$
129	1.	0.2	$C_{10}H_9{}^+$
130	1.2	0.2	$C_{10}H_{10}{}^{+\cdot}$
131	5.5	0.8	$C_{10}H_{11}{}^+$
132	4.5	0.7	$C_{10}H_{12}{}^{+\cdot}$
133	5.0	0.8	$C_{10}H_{13}{}^+$
134	11.	1.7	$C_{10}H_{14}{}^{+\cdot}$ (impurity)
135	3.	0.5	$C_9{}^{13}CH_{14}{}^{+\cdot}$ (trace of $C_9H_{11}O^+$)
136	3.	0.5	$C_{10}H_{16}{}^{+\cdot}$ and some $C_9H_{12}O^{+\cdot}$
137	1.	0.2	$C_9H_{13}O^+$ (trace of $C_{10}H_{17}{}^+$)
138	180.	28.	$C_9H_{14}O^{+\cdot}$ (trace of $C_{10}H_{18}{}^{+\cdot}$ and $C_8H_{10}O_2{}^{+\cdot}$)

Table 6 (Continued)

m/z	Peak height (mm)	Relative intensity (%)	Composition (from plate spectrum)
139	17.	2.6	$C_8{}^{13}CH_{14}O^{+\cdot}$
140	6.5	1.0	$C_7{}^{13}C_2H_{14}O^{+\cdot}$ and $C_9H_{16}O^{+\cdot}$ (impurity)
141	2.	0.3	$C_{10}H_{21}{}^{+}$
142	1.	0.2	$C_9{}^{13}CH_{21}{}^{+}$
143	0.5	0.08	
144	0.2	0.03	
145	1.8	0.3	$C_{11}H_{13}{}^{+}$
152	2.5	0.4	$C_{10}H_{16}O^{+\cdot}$ (impurity)

Note: Plate spectrum begins at m/z 16.

Table 7 Comparison of Pheromone 1B[a] with Isophorone from Aldermaston 8-Peak Index

	Relative intensity (%)	
m/z	Pheromone 1B	Spectrum W0093
82	100.	100.
39	23.	28.
138	28.	17.
27	14.	17.
41	20.	13.
54	11.	13.
53	9.	9.
29	10.	7.

[a]After rejecting peaks which probably belong mostly to impurities.

Table 8 Peak Heights for Low-Resolution Scans of Codling Moth Sex Pheromone No. 2A

m/z	Peak height (mm)				
	Scan 10	Scan 11	Scan 12	Scan 13	Scan 14
15	105.	70.	48.	—*	—*
27	80.	58.	65.	19.	28.
29(b)	85.	70.	49.	16.	20.
39	70.	93.	70.	17.	21.
41(b)	105.	135.	103.	31.	41.
42(a)	26.	30.	19.	3.	6.
42(b)	53.	63.	41.	21.	20.
43(a)	480.	300.	200.	28.	35.
43(b)	100.	120.	85.	27.	36.
44(a)	34.	110.	110.	53.	73.
44(b)	14.	40.	35.	12.	15.
55	130.	105.	65.	14.	17.
56	65.	63.	48.	13.	15.
57	95.	110.	80.	20.	23.
58	100.	83.	54.	11.	14.
59	12.	7.	6.	1.	1.
69(b)	16.	17.	13.	7.	6.
70	16.	19.	13.	6.	7.
71	39.	33.	23.	6.	7.
72(b)	10.	13.	10.	3.	5.
77	10.	23.	16.	4.	6.
79	15.	15.	11.	3.	5.
82	2.	18.	13.	3.	6.
83(a)	145.	100.	50.	6.	5.
85	17.	20.	14.	4.	4.
91	18.	22.	15.	6.	7.
95	16.	7.	5.	2.	2.

Table 8 (continued)

	Peak height (mm)				
m/z	Scan 10	Scan 11	Scan 12	Scan 13	Scan 14
98	95.	57.	30.	2.5	2.
105	60.	147.	107.	13.	17.
119	11.	31.	20.	3.	3.
120	34.	105.	68.	9.	10.
123	74.	4.	2.	0.5	0.6
138	12.	3.	2.	0.3	0.5

*Not recorded.

Table 9 Relative Intensities (%) for Low-Resolution Scans of Codling Moth Sex Pheromone No. 2A (Base Peak at m/z 43(a))

	Intensity (%) relative to m/z 43(a)				
m/z	Scan 10	Scan 11	Scan 12	Scan 13	Scan 14
15	22.	23.	24.	—*	—*
27	17.	19.	33.	68.	80.
29(b)	18.	23.	25.	57.	57.
39	15.	31.	35.	61.	60.
41(b)	22.	45.	52.	111.	117.
42(a)	5.	10.	10.	11.	17.
42(b)	11.	21.	21.	75.	57.
43(a)	100.	100.	100.	100.	100.
43(b)	21.	40.	43.	96.	103.
44(a)	7.	37.	55.	189.	209.
44(b)	3.	13.	18.	43.	43.
55	27.	35.	33.	50.	49.
56	14.	21.	24.	46.	43.
57	20.	37.	40.	71.	66.

Table 9 (continued)

m/z	Intensity (%) relative to m/z 43(a)				
	Scan 10	Scan 11	Scan 12	Scan 13	Scan 14
58	21.	28.	27.	39.	40.
59	2.	2.	3.	4.	3.
69(b)	3.	6.	7.	25.	17.
70	3.	6.	7.	21.	20.
71	8.	11.	12.	21.	20.
72(b)	2.	4.	5.	11.	14.
77	2.	8.	8.	14.	17.
79	3.	5.	6.	11.	14.
82	0.4	6.	7.	11.	17.
83(a)	30.	33.	25.	21.	14.
85	4.	7.	7.	14.	11.
91	4.	7.	8.	21.	20.
95	3.	2.	3.	7.	6.
98	20.	19.	15.	9.	6.
105	12.	49.	54.	46.	49.
119	2.	10.	10.	11.	9.
120	7.	35.	34.	32.	29.
123	15.	1.	1.	2.	2.
138	2.	1.	1.	1.	1.

*Not recorded.

Table 10 Relative Intensities (%) for Low-Resolution Scans of Codling Moth Sex Pheromone No. 2A (Base Peak at m/z 105)

m/z	Intensity (%) relative to m/z 105				
	Scan 10	Scan 11	Scan 12	Scan 13	Scan 14
15	175.	48.	45.	—*	—*
27	133.	39.	61.	146.	165.
29(b)	142.	48.	46.	123.	118.

Table 10 (continued)

m/z	Intensity (%) relative to m/z 105				
	Scan 10	Scan 11	Scan 12	Scan 13	Scan 14
39	117.	63.	65.	131.	124.
41(b)	175.	92.	96.	238.	241.
42(a)	43.	20.	18.	23.	35.
42(b)	88.	43.	38.	162.	118.
43(a)	800.	204.	187.	215.	206.
43(b)	167.	82.	79.	208.	212.
44(a)	57.	75.	103.	408.	429.
44(b)	23.	27.	33.	92.	88.
55	217.	71.	61.	108.	100.
56	108.	43.	45.	100.	88.
57	158.	75.	75.	154.	135.
58	167.	56.	50.	85.	82.
59	20.	5.	6.	8.	6.
69(b)	27.	12.	12.	54.	35.
70	27.	13.	12.	46.	41.
71	65.	22.	21.	46.	41.
72(b)	17.	9.	9.	23.	29.
77	17.	16.	15.	31.	35.
79	25.	10.	10.	23.	29.
82	3.	12.	12.	23.	35.
83(a)	242.	68.	47.	46.	29.
85	28.	14.	13.	31.	24.
91	30.	15.	14.	46.	41.
95	27.	5.	5.	15.	12.
98	158.	39.	28.	19.	12.
105	100.	100.	100.	100.	100.
119	18.	21.	19.	23.	18.
120	57.	78.	64.	69.	59.
123	123.	3.	2.	4.	4.
138	20.	2.	2.	2.	3.

*Not recorded.

Table 11 High-Resolution Plate Mass Spectrum of Codling Moth Sex Pheromone No. 2A. Plate 85, Position 18

m/z (low resolution)	Line intensity	D (mm)	m/z found	m/z for assumed composition	Composition
16(a)	v.st.	3.2204	—	15.9949	$O^{+\cdot}$
16(b)		3.2769	16.0193	16.0187	NH_2^+
16(c)		3.3044	16.0312	16.0313	$CH_4^{+\cdot}$
17(a)	v.st.	5.5168	17.0023	17.0027	OH^+
17(b)		5.5695	17.0257	17.0265	$NH_3^{+\cdot}$
18	v.st.	7.7477	18.0104	18.0106	$H_2O^{+\cdot}$
20(a)		11.9357	—	19.9812	Ar^{2+}
22		16.0074	21.9954	21.9949	CO_2^{2+}
24		19.8776	—	24.0000	$C_2^{+\cdot}$
25		21.7625	25.0078	25.0078	C_2H^+
26		23.6100	26.0157	26.0156	$C_2H_2^{+\cdot}$
27		25.4222	27.0238	27.0235	$C_2H_3^+$
28(a)	st.	27.1374	27.9955	27.9949	$CO^{+\cdot}$
28(b)	v.st.	27.1561	—	28.0061	$N_2^{+\cdot}$
28(c)	st.	27.2004	28.0315	28.0313	$C_2H_4^{+\cdot}$
29(a)	st.(+ $^{13}CO^{+\cdot}$ sh.f.)	28.8836	29.0021	29.0032	$N^{15}N^{+\cdot}$
29(b)	st.	28.9471	29.0390	29.0391	$C_2H_5^+$
30(c)	w.	30.6561	30.0420	30.0425	$C^{13}CH_5^+$
31(a)	st.; sh.b.	32.2593	—	30.9984	CF^+
31(b)		32.2921	31.0181	30.0184	CH_3O^+

Table 11 (Continued)

m/z (low resolution)	Line intensity	D (mm)	m/z found	m/z for assumed composition	Composition
32(a)		33.8937	31.9891	31.9898	O_2^{+}
32(b)	w.; sh.b.	33.9141	32.0016	32.0018	$^{13}CF^{+}$
32(c)	w.	33.9530	32.0254	32.0262	$CH_4O^{+\cdot}$
33		35.5581	33.0141	33.0141	CH_2F^{+}
35	w.	38.6622	34.9688	34.9689	Cl^{+}
36(a)	w.	40.2282	35.9762	35.9767	$HCl^{+\cdot}$
36(b)	w.	40.2649	—	36.0000	$C_3^{+\cdot}$
37(b)		41.8083	37.0075	37.0078	C_3H^{+}
38(b)		43.3325	38.0161	38.0156	$C_3H_2^{+\cdot}$
39		44.8349	39.0236	39.0235	$C_3H_3^{+}$
40(a)		45.2175	—	39.9624	$Ar^{+\cdot}$
40(d)		46.3174	40.0306	40.0313	$C_4H_4^{+\cdot}$
41(a)	w.	47.7300	41.0033	41.0027	C_2HO^{+}
41(c)	st.	47.7821	41.0394	41.0391	$C_3H_5^{+}$
42(a)		49.1780	42.0123	42.0106	$C_2H_2O^{+\cdot}$
42(b)	st.	49.2291	42.0481	42.0469	$C_3H_6^{+\cdot}$
43(c)	v.st.	50.6068	43.0199	43.0184	$C_2H_3O^{+}$
43(d)	st.	50.6567	43.0554	43.0548	$C_3H_7^{+}$
44(a)	st.	51.9663	—	43.9898	$CO_2^{+\cdot}$
44(c)	v.st.	52.0133	44.0236	44.0262	$C_2H_4O^{+\cdot}$
44(d)		52.0611	44.0579	44.0581	$C_1{}^{13}CH_7^{+}$

Table 11 (Continued)

m/z (low resolution)	Line intensity	D (mm)	m/z found	m/z for assumed composition	Composition
50(a)		60.1195	—	49.9968	$CF_2^{+\cdot}$
50(b)		60.1435	50.0153	50.0156	$C_4H_2^{+\cdot}$
51(a)		61.4191	—	51.0046	CHF_2^{+}
51(b)		61.4431	51.0229	51.0235	$C_4H_3^{+}$
52(a)		62.7054	52.0069	52.0080	$^{13}CHF_2^{+}$
52(b)		62.7336	52.0290	52.0313	$C_4H_4^{+\cdot}$
53		64.0167	—	53.0391	$C_4H_5^{+}$
54		65.2880	54.0463	54.0469	$C_4H_6^{+\cdot}$
55(a)	w.	66.4790	—	54.9984	C_3F^{+}
55(b)	w.	66.5034	55.0180	55.0184	$C_3H_3O^{+}$
55(c)		66.5490	55.0546	55.0548	$C_4H_7^{+}$
56		67.7976	56.0607	56.0626	$C_4H_8^{+\cdot}$
57(c)	w.	68.9924	57.0321	57.0340	$C_3H_5O^{+}$
57(d)	st.	69.0357	57.0675	57.0704	$C_4H_9^{+}$
58(a)	v.w.	70.2094	58.0301	58.0293	$C_2H_4NO^{+}$
58(b)		70.2217	58.0403	58.0419	$C_3H_6O^{+\cdot}$
58(c)	v.w.	70.2614	58.0730	58.0738	$C_3{}^{13}CH_{9+}$
59(a)		71.4296	59.0394	59.0391	$C_9H_{10}^{2+}$
59(b)		71.4352	59.0441	59.0452	$C_2{}^{13}CH_6O^{+\cdot}$
62(a)		74.9460	—	61.9968	$C_2F_2^{+\cdot}$
62(b)		74.9683	62.0158	62.0156	$C_5H_2^{+\cdot}$

Table 11 (Continued)

m/z (low resolution)	Line intensity	D (mm)	m/z found	m/z for assumed composition	Composition
63(b)		76.1482	63.0230	63.0235	$C_5H_3^+$
64(a)	w.	77.2980	64.0124	64.0125	$C_2H_2F_2^{+\cdot}$
64(b)	v.w.	77.3199	64.0313	64.0313	$C_5H_4^{+\cdot}$
65(a)		78.4595	65.0197	65.0203	$C_2H_3F_2^+$
65(b)		78.4805	65.0380	65.0391	$C_5H_5^+$
66(b)		79.5700	65.9901	65.9917	$COF_2^{+\cdot}$
66(c)		79.6322	66.0447	66.0469	$C_5H_6^{+\cdot}$
67(b)		80.7770	67.0530	67.0548	$C_5H_7^+$
68(b)		81.9121	68.0603	68.0626	$C_5H_8^{+\cdot}$
69(a)	v.st.	82.9581	—	68.9952	CF_3^+
- - -	- - -	- - -	- - -	- - -	- - -
69(a)		82.9560	—	68.9952	CF_3^+
69(b)	v.w.	82.9977	69.0326	69.0340	$C_4H_5O^+$
69(c)	w.	83.0377	69.0686	69.0704	$C_5H_9^+$
70(a)		84.0698	—	69.9986	$^{13}CF_3^+$
70(c)	[(b) is v.w.]	84.1575	70.0779	70.0782	$C_5H_{10}^{+\cdot}$
71(b)		85.2692	71.0862	71.0861	$C_5H_{11}^+$
72(a)	v.w.	86.3320	72.0570	72.0575	$C_4H_8O^{+\cdot}$
72(b)	w.	86.3715	72.0932	72.0939	$C_5H_{12}^{+\cdot}$
74(a)		88.4344	—	73.9968	$C_3F_2^{+\cdot}$
74(b)		88.4530	74.0141	74.0156	$C_6H_2^{+\cdot}$

Table 11 (Continued)

m/z (low resolution)	Line intensity	D (mm)	m/z found	m/z for assumed composition	Composition
75(a)		89.5154	75.0044	75.0046	$C_3HF_2^+$
75(b)	v.w.	89.5353	75.0230	75.0235	$C_6H_3^+$
77(a)	v.w.	91.6561	77.0199	77.0203	$C_3H_3F_2^+$
77(b)		91.6757	77.0385	77.0391	$C_6H_5^+$
78		92.7346	78.0455	78.0469	$C_6H_6^{+\cdot}$
79(a)	v.w.	93.7311	78.9991	78.9984	C_5F^+
79(b)		93.7879	79.0536	79.0548	$C_6H_7^+$
80	v.w.	94.8332	80.0605	80.0626	$C_6H_8^{+\cdot}$
81(a)		95.7977	—	80.9952	$C_2F_3^+$
81(c)		95.8740	81.0694	81.0704	$C_6H_9^+$
82(a)		96.8306	82.0023	82.0030	$C_2HF_3^{+\cdot}$
82(b)		96.8706	82.0414	82.0419	$C_5H_6O^{+\cdot}$
82(c)		96.9070	82.0771	81.0782	$C_6H_{10}^{+\cdot}$
83(a)		97.8579	83.0101	83.0109	$C_2H_2F_3^+$
83(b)	st.	97.8975	83.0491	83.0497	$C_2H_7O^+$
83(c)	v.w.	97.9341	83.0851	83.0861	$C_6H_{11}^+$
84(a)	w,	98.9152	84.0538	84.0530	$C_4{}^{13}CH_7O^+$
84(b)	v.w.	98.9549	84.0931	84.0939	$C_6H_{12}^{+\cdot}$
85(a)		99.8338	84.9659	84.9657	$CClF_2^+$
85(b)	v.w.	99.9332	85.0649	85.0653	$C_5H_9O^+$
85(c)		99.9695	85.1010	85.1017	$C_6H_{13}^+$

Table 11 (Continued)

m/z (low resolution)	Line intensity	D (mm)	m/z found	m/z for assumed composition	Composition
86		100.8660	85.9966	85.9968	$C_4F_2^{+\cdot}$
87		101.8274	86.9622	86.9627	$C^{37}ClF_2^{+}$
89	v.w.	103.8774	89.0392	89.0391	$C_7H_5^{+}$
91	st.	105.8440	91.0547	91.0548	$C_7H_7^{+}$
92		106.8178	92.0611	92.0626	$C_7H_8^{+\cdot}$
93(a)		107.7169	—	92.9952	$C_3F_3^{+}$
93(b)	v.w.	107.7881	93.0694	93.0704	$C_7H_9^{+}$
94(a)		108.6780	93.9991	93.9986	$C_2{}^{13}CF_3^{+}$
94(b)	v.w.	108.7178	94.0408	94.0419	$C_6H_6O^{+\cdot}$
94(c)	ext.w.	108.7524	94.0770	94.0782	$C_7H_{10}^{+\cdot}$
95(a)		109.6403	95.0096	95.0109	$C_3H_2F_3^{+}$
95(b)	weaker than (c)	109.6770	95.0482	95.0497	$C_6H_7O^{+}$
95(c)	v.w.	109.7115	95.0846	95.0861	$C_7H_{11}^{+}$
96(b)	v.w.	110.6314	96.0560	96.0575	$C_6H_8O^{+\cdot}$
96(c)	v.w.	110.6661	96.0927	96.0939	$C_7H_{12}^{+\cdot}$
97(a)		111.5106	96.9890	96.9901	$C_2F_3O^{+}$
97(b)		111.6150	97.1001	97.1017	$C_7H_{13}^{+}$
98(a)		112.4549	97.9962	97.9968	$C_5F_2^{+\cdot}$
98(b)		112.5253	98.0715	98.0732	$C_6H_{10}O^{+\cdot}$
98(c)	v.w.	112.5591	98.1076	98.1095	$C_7H_{14}^{+\cdot}$
99(a)	v.w.	113.3945	99.0035	99.0046	$C_5HF_2^{+}$

Table 11 (Continued)

m/z (low resolution)	Line intensity	D (mm)	m/z found	m/z for assumed composition	Composition
99(b)	v.w.	113.4614	99.0754	99.0765	$C_5{}^{13}CH_{10}O^{+\cdot}$
99(c)	v.w.	113.4992	99.1161	99.1174	$C_7H_{15}{}^+$
100(a)	v.st.	114.3134	—	99.9936	$C_2F_4{}^{+\cdot}$
100(b)		114.4003	100.0875	100.0888	$C_6H_{12}O^{+\cdot}$
101(a)	v.w.	115.1820	100.9343	100.9361	CCl_2F^+
101(b)	st.	115.2419	100.9994	100.9970	$C^{13}CF_4{}^{+\cdot}$
				or 101.0014	$C_2HF_4{}^+$
102(a)	v.w.	116.1637	102.0029	102.0048	$C^{13}CHF_4{}^+$
102(b)	v.w.	116.2027	102.0454	102.0469	$C_8H_6{}^{+\cdot}$
103(a)	v.w.	117.0123	102.9311	102.9332	$CCl^{37}ClF^+$
103(b)		117.1241	103.0537	103.0548	$C_8H_7{}^+$
104		118.0404	104.0612	104.0626	$C_8H_8{}^{+\cdot}$
105(a)		118.8847	104.9940	104.9952	$C_4F_3{}^+$
105(b)	st.	118.9524	105.0689	105.0704	$C_8H_9{}^+$
106(a)		119.7932	106.0023	106.0030	$C_4HF_3{}^{+\cdot}$
106(b)		119.8567	106.0729	106.0738	$C_7{}^{13}CH_9{}^+$
107		120.7638	107.0848	107.0861	$C_8H_{11}{}^+$
109(a)	v.w.	122.4380	108.9651	108.9657	$C_3ClF_2{}^+$
109(b)	v.w.	122.4596	108.9894	108.9901	$C_3F_3O^+$
109(c)		122.5584	109.1009	109.1017	$C_8H_{13}{}^+$
111		124.3369	111.1174	111.1174	$C_8H_{15}{}^+$

Table 11 (Continued)

m/z (low resolution)	Line intensity	D (mm)	m/z found	m/z for assumed composition	Composition
112(a)		125.1046	—	111.9936	$C_3F_4^{+\cdot}$
112(b)		125.2195	112.1250	112.1252	$C_8H_{16}^{+\cdot}$
113(a)		125.9836	113.0011	113.0014	$C_3HF_4^{+}$
113(b)		126.0974	113.1318	113.1330	$C_8H_{17}^{+}$
114		126.8557	114.0051	114.0048	$C_2{}^{13}CHF_4^{+}$
115(a)	v.w.	127.7303	115.0164	115.0171	$C_3H_3F_4^{+}$
115(b)	w.	127.7623	115.0535	115.0548	$C_9H_7^{+}$
116		128.5467	115.9645	115.9641	$C_2ClF_3^{+\cdot}$
117(a)		129.4308	116.9956	116.9952	$C_5F_3^{+}$
117(b)		129.4942	117.0697	117.0704	$C_9H_9^{+}$
118(a)	v.w.	130.2541	117.9599	117.9611	$C_2{}^{37}ClF_3^{+\cdot}$
118(b)	v.w.	130.2866	117.9980	117.9986	$C_4{}^{13}CF_3^{+}$
118(c)	v.w.	130.3537	118.0768	118.0782	$C_9H_{10}^{+\cdot}$
119(a)		131.1321	118.9926	118.9920	$C_2F_5^{+}$
119(b)		131.2110	119.0856	119.0861	$C_9H_{11}^{+}$
120(a)		131.9802	119.9944	119.9954	$C^{13}CF_5^{+}$
120(b)		132.0630	120.0294	120.0939	$C_9H_{12}^{+\cdot}$
121		132.9090	121.0964	121.0973	$C_8{}^{13}CH_{12}^{+\cdot}$
123		134.5708	123.0806	123.0810	$C_8H_{11}O^{+}$
124		135.3310	—	123.9936	$C_4F_4^{+\cdot}$
126(a)		136.9980	126.0073	126.0093	$C_4H_2F_4^{+\cdot}$

Table 11 (Continued)

m/z (low resolution)	Line intensity	D (mm)	m/z found	m/z for assumed composition	Composition
126(b)		137.1068	126.1393	126.1408	$C_9H_{18}^{+\cdot}$
127(a)		137.8281	127.0161	127.0171	$C_4H_3F_4^+$
127(b)		137.9356	127.1471	127.1487	$C_9H_{19}^+$
131		141.0686	130.9928	130.9920	$C_3F_5^+$
132(a)		141.8285	131.9341	131.9345	$C_2Cl_2F_2^{+\cdot}$
132(b)		141.8779	131.9955	131.9954	$C_2{}^{13}CF_5^+$
133		142.6915	133.0073	133.0077	$C_3H_2F_5^+$
135		144.2550	134.9627	134.9625	$C_2ClF_4^+$
136(a)		145.0747	—	135.9936	$C_5F_4^{+\cdot}$
136(b)	v.w.	145.1510	136.0898	136.0888	$C_9H_{12}O^{+\cdot}$
137(a)		145.8396	136.9588	136.9595	$C_2{}^{37}ClF_4^+$
137(b)		145.8730	137.0011	137.0014	$C_5HF_4^+$
138		146.7445	138.1052	138.1045	$C_9H_{14}O^{+\cdot}$

The following list contains line measurements taken from a stronger exposure (plate 85, position 20) to locate weaker lines.

112		125.0926	—	111.9936	$C_3F_4^{+\cdot}$
116(a)		128.5359	115.9645	115.9641	$C_2ClF_3^{+\cdot}$
116(b)		128.6183	116.0604	116.0626	$C_9H_8^{+\cdot}$
117(a)		129,4200	—	116.9952	$C_5F_3^+$
135(a)		144.2444	—	134.9625	$C_2ClF_4^+$
136(a)		145.0647	135.9935	135.9936	$C_5F_4^{+\cdot}$

Table 11 (Continued)

m/z (low resolution)	Line intensity	D (mm)	m/z found	m/z for assumed composition	Composition
136(b)		145.1405	136.0890	136.0888	$C_9H_{12}O^{+\cdot}$
137(a)		145.8297	136.9586	136.9595	$C_2{}^{37}ClF_4^+$
137(b)		145.8624	136.9999	137.0014	$C_5HF_4^+$
138(a)		146.6575	138.0068	138.0092	$C_2H_2F_4^{+\cdot}$
138(b)		146.7345	138.1045	138.1045	$C_9H_{14}O^{+\cdot}$
141		149.0000	140.9945	140.9952	$C_7F_3^+$
143		150.5523	—	142.9920	$C_4F_5^+$

Calibration data for Table 5: 1[16(a), 20(a), 16(b), 22]; 2[24, 28(b), 25(a), 28(c)]; 3[28(b), 31(a), 29(a), 31.5]; 4[31(a), 36(b), 32(a), 36(a)]; 5[36(b), 40(a), 37(a), 39.5]; 6[40(a), 44(a), 40(b), 45.5]; 7[50(a), 51(b), 50(b), 51.5]; 8[51(a), 52(b), 51(b), 52(c)]; 9[52(c), 54(d), 52.5(a), 53.5(b)]; 10[54(d), 56(c), 55(a), 56(b)]; 11[56(c), 62(a), 57(a), 61.5]; 12[62(a), 70(a), 62(b), 69(c)]; 13[70(a), 81(a), 70(b), 81(c)]; 14[81(a), 93(a), 82(a), 93(c)]; 15[93(a), 105(a), 94(a), 105(c)]; 16[105(a), 112(a), 106(a), 111(e)]; 17[112(a), 119(a), 112(b), 119(c)]; 18[119(a), 124(a), 120(a), 124(c)]; 19[124(a), 136(a), 125(a), 136(c)]; 20[136(a), 143, 137(a), 142(b)]; 21[143,174,144,152(b)]; 22[150, 174, 178(a), 178(b)].

Calibration data for Table 11, Plate 85, Position 18: 1[16(a), 20(a), 16(b), 22]; 2[24, 28(b), 25, 28(c)]; 3[28(b), 31(a), 29(a), 31(b)]; 4[31(a), 36(b), 32(a), 36(a)]; 5[36(b), 40(a), 37(b), 40(d)]; 6[40(a), 44(a), 41(a), 44(d)]; 7[50(a), 51(a), 50(b), 51(b)];8[51(a), 53, 52(a), 52(b)]; 9[53, 55(a), 54, 55(c)]; 10[55(a), 62(a), 56, 62(b)]; 11[62 (a), 69(a), 63(b), 68(b)]; 12[69(a), 70(a), 69(b), 70(c)]; 13[70(a), 74(a), 71(b), 74(b)]; 14[74(a), 81(a), 75(a), 81(c)]; 15[81(a), 93(a), 82(a), 93(b)]; 16[93(a), 100(a), 94(a), 100(b)]; 17[100(a), 112(a), 101(a), 112(b)]; 18[112(a), 124, 113(a), 123]; 19[124, 136(a), 126(a), 136(b)]; 20[136(a), 143, 137(a), 138].

Calibration data for Table 11, Plate 85, Position 20: 1[112, 117(a), 116(a), 116(b)]; 2[135, 143, 136(a), 141].

Table 12 Codling Moth Sex Pheromone No. 2A—Mass Measurements by Peak Matching

Nominal m/z	m/z found	m/z for assumed composition	Composition	Intensity ratios in doublets in scan 10
83(a)	83.0488	83.0497	$C_5H_7O^+$	Intensity of $C_5H_7O^+$ is 30 times that of $C_6H_{11}^+$
83(b)*	83.0851	83.0861	$C_6H_{11}^+$	
91	91.0550	91.0548	$C_7H_7^+$	
98(a)	98.0732	98.0732	$C_6H_{10}O^{+\cdot}$	Intensity of $C_6H_{10}O^{+\cdot}$ is 15 times that of $C_7H_{14}^{+\cdot}$
98(b)*	98.1098	98.1095	$C_7H_{14}^{+\cdot}$	
105	105.0702	105.0704	$C_8H_9^+$	
119	119.0861	119.0861	$C_9H_{11}^+$	
120	120.0938	120.0939	$C_9H_{12}^{+\cdot}$	
138	138.1050	138.1045	$C_9H_{14}O^{+\cdot}$	

*Looks like an impurity.

Table 13 Relative Intensities in Scan 11 of Codling Moth Attractant #2A Compared with Alkylbenzene Spectra from the Literature

	Relative intensity (%)			
Compound	at m/z 91	at m/z 105	15 m/z 119	at m/z 120
CMA #2A, Scan 11	15	100	21	78
1,2,3-Trimethylbenzene	8	100	12	55
1,2,4-Trimethylbenzene	8	100	15	59
1,3,5-Trimethylbenzene	9	100	16	67
1-Methyl-2-ethylbenzene	10	100	2	31
1-Methyl-3-ethylbenzene	10	100	5	32
1-Methyl-4-ethylbenzene	8	100	3	31
n-Propylbenzene	100	86	1	47
Isopropylbenzene	5	100	0.8	25

Note: See also scans 10 and 12.

Table 14 Relative Intensities (%) for Low-Resolution Scans of Codling Moth Sex Pheromone No. 2A (Base Peak at m/z 43(a))

	Intensity (%) relative to m/z 43(a)*					
m/z	Mesitylene (API 263)	Scan 10 − API 263	Scan 11 − API 263	Scan 12 − API 263	Scan 13 − API 263	Scan 14 − API 263
15	Not given	[22]	[23]	[24]	**	**
27	10	15	14	27	63	75
29(b)	1	18	23	24	57	57
39	15	13	24	27	54	53
41	6	21	42	48	108	114
42(a)	—	5	10	10	11	17
42(b)	—	11	21	20	75	57
43(a)	—	100	100	100	100	100
43(b)	—	21	40	42	97	103
44(a)	—	7	37	55	190	208
44(b)	—	3	13	18	43	43
55	—	27	35	32	50	48
56	—	13	21	24	46	43
57	1	20	36	40	71	65
58	3	21	26	25	38	38
59	3	2	1	2	2	1
69(b)	—	3	6	6	25	17
70	—	3	6	6	21	20
71	—	8	11	11	21	20
72(b)	—	2	4	5	11	14
77	13	1	2	1	9	11
79	6	2	2	2	8	11
82	—	—	6	6	11	17
83(a)	—	30	33	25	21	14
85	—	3	7	7	14	12
91	9	3	3	3	17	15

Table 14 (continued)

m/z	Intensity (%) relative to m/z 43(a)*					
	Mesitylene (API 263)	Scan 10 – API 263	Scan 11 – API 263	Scan 12 – API 263	Scan 13 – API 263	Scan 14 – API 263
95	—	3	2	3	7	6
98	—	20	19	15	9	6
105	100	—	—	—	—	—
119	16	1	4	4	5	3
120	67	-1	6	-1	1	-4
123	—	15	1	1	2	2
138	3	1	1	1	1	1

*Note that this choice of base peak will assign values >100% at some m/z values.
**Not recorded.

Table 15 Hexenone Spectra from the Literature

m/z	Relative intensity (%)	
	5-Hexen-2-one Atlas: MOR0020	4-Methylpent-4-en-2-one (Index)
26	3.	—
27	14.	—
28	3.	18. (air?)
29	9.	14.
32	0.1	11. (air?)
39	11.	17.
41	6.	12.
43	100.	100.
55	19.	19.
58	0.7	—
83	4.	24.
98	4.	13.

Table 16 Diacetone Alcohol (ENT-45). Oven t = 156°C; Ion Source t = 165°C

m/z	Peak height (mm)	Relative intensity (%)	Notes
12	145.	0.5	
13	400.	1.5	
14	1450.	5.5	
15	7500.	28.	
16	130.	0.5	$0^{+\cdot}$ left out
24	95.	0.4	
25	410.	1.6	
26	1570.	5.9	
27	2500.	9.5	
28	500.	1.9	$N_2^{+\cdot}$ left out
29	1200.	4.5	
30	35.	0.1	
31	210.	0.8	
32	—	—	$0_2^{+\cdot}$ left out
36	240.	0.9	
36.5	7.	0.03	Metastable
37	820.	3.1	
38	950.	3.6	
39	2000.	7.6	
40	350.	1.3	
41	1140.	4.3	
42	2500.	9.5	
43	26400.	100.	
44	1000.	3.8	
45	110.	0.4	
49	25.	0.09	
50	118.	0.4	

Table 16 (continued)

m/z	Peak height (mm)	Relative intensity (%)	Notes
51	132.	0.5	
52	95.	0.4	
53	430.	1.6	
54	60.	0.2	
55	2000.	7.6	
56	120.	0.5	
57	420.	1.6	
58	11000.	42.	
59	490.	1.9	
60	40.	0.2	
61	16.	0.06	
62	20.	0.08	
63	24.	0.09	
65	24.	0.09	
67	48.	0.2	
69	50.	0.2	
70.3	7.	0.03	Metastable
77	24.	0.09	
79	67.	0.3	
82	37.	0.1	
83	2300.	8.7	
84	200.	0.8	
85	15.	0.06	
95	24.	0.09	
98	1600.	6.1	
99	100.	0.4	
105	6.	0.02	
116	0.	0.0	M^+

Table 16 (continued)

m/z	Peak height (mm)	Relative intensity (%)	Notes
123	140.	0.5	Impurity
124	15.	0.06	Impurity
138	30.	0.1	Impurity

Table 17 Difference of Scans 10 and 11 of Codling Moth Sex Pheromone No. 2A

m/z	I(10) − 60/147XI(11) (mm)	RI (%)
15	76	21
27	56	16
29(b)	56	16
39	32	9
41(b)	50	14
42(a)	14	4
42(b)	27	8
43(a)	358	100
43(b)	51	14
44(a)	-11	-3
44(b)	-2	-0.6
55	87	24
56	39	11
57	50	14
58	66	18
59	9	3
69(b)	9	3
70	8	2

Table 17 (continued)

m/z	I(10) − 60/147XI(11) (mm)	RI (%)
71	26	7
72(b)	5	1
77	0.6	0.2
79	9	3
82	-5	-1
83(a)	104	29
85	9	3
91	9	3
95	13	4
98	71	20
105	0	0
119	-2	-0.6
120	-13	-4
123	72	20
138	11	3

Table 18 Codling Moth Sex Pheromone No. 2A. Scan 10 (3/26/72)

m/z	Peak height (mm)	Relative intensity (%)	Composition (from plate spectrum)
12	3.	0.6	$C^{+\cdot}$ (Not on plate)
13	7.	1.5	CH^{+} (Not on plate)
14(a)	27.	5.6	N^{+} (air) (Not on plate)
14(b)	23.	4.8	$CH_2^{+\cdot}$ (Not on plate)
15	100.	21.	CH_3^{+} (Not on plate)

Table 18 (Continued)

m/z	Peak height (mm)	Relative intensity (%)	Composition (from plate spectrum)
16	75.	—	(From water)
17	3300.	—	(From water)
18	14800.	—	(From water)
19(a)	7.	1.5	
19(b)	4.	0.8	
20(a)	3.	0.6	Ar^{2+}
20(b)	8.	1.7	
20(c)	12.	2.5	
25	2.	0.4	C_2H^+
26	25.	5.2	$C_2H_2^{+\cdot}$
27	80.	17.	$C_2H_3^+$
28(a)	120.	25.	$CO^{+\cdot}$
28(b)	380.	79.	$N_2^{+\cdot}$ (air)
28(c)	17.	3.5	$C_2H_4^{+\cdot}$
29(a)	17.	3.5	$N^{15}N^{+\cdot}$ (air)
			$^{13}CO^{+\cdot}$, etc.
29(b)	80.	17.	$C_2H_5^+$
30(a)	3.	0.6	
30(b)	6.	1.3	$C^{13}CH_5^+$
31	1.	0.2	CH_3O^+
32(a)	85.	18.	$O_2^{+\cdot}$ (air)

Table 18 (Continued)

m/z	Peak height (mm)	Relative intensity (%)	Composition (from plate spectrum)
32(b)	2.	0.4	$CH_4O^{+\cdot}$
36	4.	0.8	$C_3^{+\cdot}$
37	14.	2.9	C_3H^+
38	13.	2.7	$C_3H_2^{+\cdot}$
39	70.	15.	$C_3H_3^+$
40(a)	6.	1.3	$Ar^{+\cdot}$ (air)
40(b)	16.	3.3	$C_3H_4^{+\cdot}$
41(a)	4.	0.8	C_2HO^+
41(b)	105.	22.	$C_3H_5^+$
42(a)	25.	5.2	$C_2H_2O^{+\cdot}$
42(b)	55.	11.	$C_3H_6^{+\cdot}$
43(a)	480.	100.	$C_2H_3O^{+\cdot}$
43(b)	100.	21.	$C_3H_7^+$
44(a)	34.	7.1	$CO_2^{+\cdot}$
44(b)	14.	2.9	$C_2H_4O^{+\cdot}$
44(c)	3.	0.6	$C_2{}^{13}CH_7^+$
45(a)	1.	0.2	
45(b)	3.	0.6	
47	3.	0.6	
49	5.	1.0	
50	10.	2.1	$C_4H_2^{+\cdot}$

Table 18 (Continued)

m/z	Peak height (mm)	Relative intensity (%)	Composition (from plate spectrum)
51	10.	2.1	$C_4H_3^+$
52	3.	0.6	$C_4H_4^{+\cdot}$
53	27.	5.6	$C_4H_5^+$
54	11.	2.3	$C_4H_6^{+\cdot}$
55	130.	27.	$C_4H_7^+$ (trace of $C_3H_3O^+$)
56	65.	14.	$C_4H_8^{+\cdot}$
57	95.	20.	$C_4H_9^+$ (weak $C_3H_5O^+$)
58	100.	21.	$C_3H_6O^{+\cdot}$
59	12.	2.5	$C_2{}^{13}CH_6O^{+\cdot}$ and $C_9H_{10}{}^{2+}$
61	0.5	0.1	
61.5	0.5	0.1	
62	1.	0.2	$C_2H_2^{+\cdot}$
63	2.	0.4	$C_5H_3^+$
64	1.	0.2	$C_5H_4^{+\cdot}$
65	8.	1.7	$C_5H_5^+$
66	1.	0.2	$C_5H_6^{+\cdot}$
67	11.	2.3	$C_5H_7^+$
68	5.	1.	$C_5H_8^{+\cdot}$
69(a)	1.	0.2	$C_4H_5O^+$
69(b)	16.	3.3	$C_5H_9^+$
70	16.	3.3	$C_5H_{10}^{+\cdot}$

Table 18 (Continued)

m/z	Peak height (mm)	Relative intensity (%)	Composition (from plate spectrum)
71	39.	8.1	$C_5H_{11}^{+}$
72(a)	2.	0.4	$C_4H_8O^{+\cdot}$
72(b)	10.	2.1	$C_5H_{12}^{+\cdot}$
73	1.	0.2	
74	2.	0.4	$C_6H_2^{+\cdot}$
75	1.	0.2	$C_6H_3^{+}$
77	10.	2.1	$C_6H_5^{+}$
78	5.	1.0	$C_6H_6^{+\cdot}$
79	15.	3.1	$C_6H_7^{+}$
80	11.	2.3	$C_6H_8^{+\cdot}$
81	6.	1.3	$C_6H_9^{+}$
82	3.	0.6	$C_5H_6O^{+\cdot}$ and $C_6H_{10}^{+\cdot}$
83(a)	145.	30.	$C_5H_7O^{+}$
83(b)	6.	1.3	$C_6H_{11}^{+}$
84(2)	5.	1.0	$C_4{}^{13}CH_7O^{+}$
84(b)	3.	0.6	$C_6H_{12}^{+\cdot}$
85	17.	3.5	$C_6H_{13}^{+}$ (trace of $C_5H_9O^{+}$)
86	2.	0.4	
89	1.	0.2	$C_7H_5^{+}$
91	18.	3.8	$C_7H_7^{+}$
92	3.	0.6	$C_7H_8^{+\cdot}$

Table 18 (Continued)

m/z	Peak height (mm)	Relative intensity (%)	Composition (from plate spectrum)
93	1.	0.2	$C_7H_9^+$
94	3.	0.6	$C_6H_6O^{+\cdot}$ (trace of $C_7H_{10}^{+\cdot}$)
95	16.	3.3	$C_6H_7O^+$ (trace of $C_7H_{11}^+$)
96	4.	0.8	$C_6H_8O^{+\cdot}$ and $C_7H_{12}^{+\cdot}$
97	5.	1.0	$C_7H_{13}^+$
98(a)	95.	20.	$C_6H_{10}O^{+\cdot}$
98(b)	6.	1.3	$C_7H_{14}^{+\cdot}$
99(a)	9.	1.9	$C_5{}^{13}CH_{10}O^{+\cdot}$
99(b)	3.	0.6	$C_7H_{15}^+$
100	1.	0.2	$C_6H_{12}O^{+\cdot}$
102	0.5	0.1	$C_8H_6^{+\cdot}$
103	3.	0.6	$C_8H_7^+$
104	3.	0.6	$C_8H_8^{+\cdot}$
105	60.	13.	$C_8H_9^+$
106	5.	1.0	$C_7{}^{13}CH_9^+$
107	5.	1.0	$C_8H_{11}^+$
109	2.	0.4	$C_8H_{13}^+$
111	3.	0.6	$C_8H_{15}^+$
112	5.	1.0	$C_8H_{16}^{+\cdot}$
113	6.	1.3	$C_8H_{17}^+$
117	4.	0.8	$C_9H_9^+$

Table 18 (Continued)

m/z	Peak height (mm)	Relative intensity (%)	Composition (from plate spectrum)
118	1.	0.2	$C_9H_{10}^{+\cdot}$
119	11.	2.3	$C_9H_{11}^{+}$
120	34.	7.1	$C_9H_{12}^{+\cdot}$
121	8.	1.7	$C_8{}^{13}CH_{12}^{+\cdot}$
123	74.	15.	$C_8H_{11}O^{+}$
124	9.	1.9	
126	2.	0.4	$C_9H_{18}^{+\cdot}$
127	2.5	0.5	$C_9H_{19}^{+}$
136	1.	0.2	$C_9H_{12}O^{+\cdot}$
138	12.	2.5	$C_9H_{14}O^{+\cdot}$
139	1.	0.2	
140	2.5	0.5	

REFERENCES

1. W. H. McFadden, *Techniques of Combined Gas Chromatography/Mass Spectrometry: Applications in Organic Analysis*, Wiley-Interscience, New York, 1973.
2. W. Roelofs, A. Comeau, A. Hill, and G. Milicevic, *Science, 174,* 297 (1971).
3. L. M. McDonough, D. A. George, B. A. Butt, J. M. Ruth, and K. R. Hill, *Science, 177,* 177 (1972).
4. J. M. Ruth, memorandum to L. M. McDonough, on the subject of "Codling Moth Sex Attractant No. 1B," August 17, 1972 (Appendix A of this chapter).
5. J. M. Ruth, memorandum to L. M. McDonough, on the subject of "Codling Moth Sex Pheromone No. 2A," August 28, 1972 (Appendix B of this chapter).

6. S. B. Bowlus and J. A. Katzenellenbogen, *Tetrahedr. Lett. 1973,* No. 15, 1277 (1973).
7. M. P. Cooke, Jr., *Tetrahedr. Lett. 1973,* No. 15, 1281 (1973).
8. M. P. Cooke, Jr., *Tetrahedr. Lett. 1973,* No. 22, 1983 (1973).
9. J.-P. Morizur, G. Muzard, J.-J. Basselier, and J. Kossanyi, Bulletin *de la Société Chimique de France 1975*, Nos. 1-2, 257 (1975).
10. C. Ouannes and Y. Langlois, *Tetrahedr. Lett. 1975,* No. 40, 3461 (1975).
11. D. A. George and L. M. McDonough, *Nature, 239,* 109 (1972).
12. A. E. Pierce, *Silylation of Organic Compounds,* Pierce Chemical Co., Rockford, Illinois, 1968.
13. K. Blau and G. S. King, *Handbook of Derivatives for Chromatrography,* Heyden, London, 1978.
14. J. A. Perry and C. A. Feit, in *GLC and HPLC Determination of Therapeutic Agents,* Part 1 (K. Tsuji and W. Morozowich, eds.), Marcel Dekker, New York, 1978, p. 137.
15. D. R. Knapp, *Handbook of Analytical Derivatization Reactions,* Wiley-Interscience, New York, 1979.
16. P. Vouros, in *Mass Spectrometry,* Part B (C. Merritt, Jr. and C. N. McEwen, eds), Marcel Dekker, New York, 1980, p. 129.
17. J. Drozd, *Chemical Derivatization in Gas Chromatography* (Journal of Chromatography Library, Vol. 19), Elsevier, Amsterdam, 1981.
18. B. J. Gudzinowicz, M. J. Gudzinowicz, and H. F. Martin, *The Integrated CG/MS Analytical System,* Part III of *Chromatograph Science, Vol. 7, Fundamentals of Integrated Gas Chromatography—Mass Spectrometry,* Marcel Dekker, New York, 1976.
19. G. A. Junk, *Int. J. Mass Spectrom. Ion Phys., 8,* 1 (1972).
20. W. L. Budde and J. W. Eichelberger, *Organic Analysis Using Gas Chromatography/Mass Spectrometry—A Techniques and Procedures Manual,* Ann Arbor Science, Ann Arbor, Michigan 1979.
21. J. C. Holmes and F. A. Morrell, Fourth Annual Conference on Mass Spectrometry, ASTM Committee E-14, Cincinnati, Ohio, 1956.
22. J. C. Holmes and F. A. Morrell, *Appl. Spectrosc., 11,* 86 1957.
23. R. S. Gohlke, *Anal. Chem., 31,* 535 (1959).
24. L. P. Lindeman and J. L. Annis, *Anal. Chem., 32,* 1742 (1960).
25. D. Henneberg, *Z. Anal. Chemie, 183,* 12 (1961).

26. D. Henneberg, in *Gas Chromatography 1960,* Proc. of the Third Symposium (R. P. W. Scott, ed.), Butterworths, London, 1960, pp. 129–130.
27. H. Widmer and T. Gäumann, *Helv. Chim. Acta, 45,* 2175 (1962).
28. E. J. Levy, R. R. Doyle, R. A. Brown, and F. W. Melpolder, *Anal. Chem., 33,* 698 (1961).
29. I. A. Breger (ed.), *Organic Geochemistry,* Macmillan, New York, 1963.
30. B. Nagy and U. Colombo, *Fundamental Aspects of Petroleum Geochemistry,* Elsevier, Amsterdam, 1967.
31. V. E. Modzeleski, W. D. MacLeod, Jr., and B. Nagy, *Anal. Chem., 40,* 987 (1968).
32. G. Eglinton and M. T. J. Murphy, *Organic Geochemistry: Methods and Results,* Springer-Verlag, New York, 1969.
33. R. A. Dean and E. V. Whitehead, *Tetrahedr. Lett.* No. 21, 768 (1961).
34. J. G. Bendoraitis, B. L. Brown, and L. S. Hepner, *Anal. Chem., 34,* 49, (1962).
35. J. Oró, D. W. Nooner, A. Zlatkis, S. A. Wikström, and E. S. Barghoorn, *Science, 148,* 77 (1965).
36. W. E. Robinson, J. J. Cummins, and G. U. Dinneen, *Geochim. Cosmochim. Acta, 29,* 249 (1965).
37. A. A. Ebert, Jr., *Anal. Chem., 33,* 1865 (1961).
38. D. O. Miller, *Anal. Chem., 35,* 2033 (1963).
39. E. J. Levy, E. D. Miller, and W. S. Beggs, *Anal. Chem., 35,* 946 (1963).
40. R. S. Gohlke, *Anal. Chem., 34, 1332* (1962).
41. W. H. McFadden, R. Teranishi, D. R. Black, and J. C. Day, *J. Food Sci., 28,* 316 (1963).
42. C. Brunnée, L. Jenckel, and E. Kronenberger, *Z. Anal. Chem., 189,* 50 (1962).
43. C. Brunnée, L. Jenckel, and E. Kronenberger, *Z. Anal. Chem., 197,* 42 (1963).
44. J. A. Dorsey, R. H. Hunt, and M. J. O'Neal, *Anal. Chem., 35,* 511 (1963).
45. S. R. Lipsky, C. G. Horvath, and W. J. McMurray, *Anal. Chem., 38,* 1585 (1966).
46. M. A. Grayson and R. L. Levy, *J. Chromatogr. Sci., 9,* 687 (1971).
47. D. R. Black, R. A. Flath, and R. Teranishi, *J. Chromatogr. Sci., 7,* 284 (1969).
48. P. F. Váradi and K. Ettre, *Anal. Chem., 35,* 410 (1963).
49. P. F. Váradi and K. Ettre, *Anal. Chem., 34,* 1417 (1962).
50. A. E. Banner, R. M. Elliott, and W. Kelly, in *Gas Chroma-*

tography 1964, Proceedings of the Fifth Symposium (A. Goldup, ed.), Institute of Petroleum, London, 1965, pp. 180–188.
51. M. H. Studier and R. Hayatsu, *Anal. Chem., 40,* 1011 (1968).
52. E. W. Becker, K. Bier, H. Burghoff, O. Hagena, P. Lohse, R. Schutte, P. Turowski, and F. Zigan, *Proceedings of the Second United Nations International Conference on the Peaceful Uses of Atomic Energy,* Held in Geneva, 1 September–13 September 1958, *Vol. 4, Production of Nuclear Materials and Isotopes,* United Nations, Geneva, 1958, pp. 455–457.
53. E. W. Becker, in *Separation of Isotopes* (H. London, ed.), George Newnes Ltd., London, 1961, pp. 360–367.
54. S. Chapman and T. G. Cowling, *The Mathematical Theory of Non-uniform Gases,* Cambridge University Press, Cambridge, 1953, pp. 244–258.
55. E. W. Becker, in *Uranium Enrichment* (S. Villani, ed.), Springer-Verlag, Berlin, 1979, pp. 245–268.
56. R. Ryhage, *Anal. Chem., 36,* 759 (1964).
57. J. T. Watson and K. Biemann, *Anal. Chem., 36,* 1135 (1964).
58. J. T. Watson and K. Biemann, *Anal. Chem., 37,* 844 (1965).
59. S. P. Markey, *Anal. Chem., 42,* 306 (1970).
60. R. F. Cree, *Pittsburgh Conference on Analytical Chemistry and Applied Spectroscopy,* Paper 188, Pittsburgh, Pennsylvania, 1967, p. 96.
61. R. F. Cree, U. S. Patent 3,430,417 (Cl. 55-16), 04 Mar 1969, Appl. 15 Jun 1966. (*CA, 70,* P 92965q.).
62. M. Blumer, *Anal. Chem., 40,* 1590 (1968).
63. P. M. Krueger and J. A. McCloskey, *Anal. Chem., 41,* 1930 (1969).
64. A. Copet and J. Evans, *Organic Mass Spectrom. 3,* 1457 (1970).
65. W. D. MacLeod, Jr., and B. Nagy, *Anal. Chem., 40,* 841 (1968).
66. M. A. Grayson and J. J. Bellina, Jr., *Anal. Chem., 45,* 487 (1973).
67. C. Brunnée, H. J. Bültemann, and G. Kappus, G., *Seventeenth Annual Conference on Mass Spectrometry and Allied Topics,* ASTM Committee E-14, Dallas, Texas, Paper 46, 1969, pp. 121–126.
68. C. Brunnée, Ger. Offen. 1,919,460 (Cl. B Old), 29 Oct 1970 (Appl. 17 apr 1969), as listed in *CA, 74:* P 27814h.
69. E. J. Bonelli, M. S. Story, and J. B. Knight, *Dynamic Mass Spectrom. 2,* 177 (1971).
70. P. M. Llewellyn and D. P. Littlejohn, *Pittsburgh Conference on Analytical Chemistry and Applied Spectroscopy,* Pittsburgh, Pennsylvania, Paper 146, 1966, p. 86.

71. P. M. Llewellyn, U. S. Patent 3,398,505,filed 27 March, 1967, granted 27 Aug 1968 [*CA 70,* 12884z (1969)].
72. P. M. Llewellyn and D. P. Littlejohn, U. S. Patent 3,471,692, filed 27 Mar 1967, granted 07 Oct 1969 [*CA 72,* 9023d (1970)].
73. P. M. Llewellyn, Ger. Offen. 1,934,378, filed 10 July, 1968, granted 15 Jan 1970 [As reported in *CA 72,* 85998r (1970)].
74. J. E. Hawes, R. Mallaby, and V. P. Williams, *J. Chromatogr. Sci., 7,* 690 (1969).
75. M. A. Grayson and C. J. Wolf, *Anal. Chem., 39,* 1438 (1967).
76. D. P. Lucero and F. C. Haley, *Gas Chromatogr. 6,* 477 (1968).
77. J. E. Lovelock, K. W. Charlton, and P. G. Simmonds, *Anal. Chem., 41,* 1048 (1969).
78. P. G. Simmonds, G. R. Shoemake, and J. E. Lovelock, *Anal. Chem., 42,* 881 (1970).
79. E. C. Horning, M. G. Horning, D. I. Carroll, I. Dzidic, and R. N. Stillwell, *Anal. Chem., 45,* 936 (1973)
80. E. C. Horning, M. G. Horning, D. I. Carroll, R. N. Stillwell, and I. Dzidic, *Life Sci. 13,* 1331 (1973).
81. D. I. Carroll, I. Dzidic, R. N. Stillwell, M. G. Horning, and E. C. Horning, *Anal. Chem., 46,* 706 (1974).
82. E. C. Horning, D. I. Carroll, I. Dzidic, S.-N. Lin, R. N. Stillwell, and J.-P. Thenot, *J. Chromatography, 142,* 481 (1977).
83. D. I. Carroll, I. Dzidic, R. N. Stillwell, and E. C. Horning, in *National Bureau of Standards Special Publication 519,* Trace Organic Analysis: A New Frontier in Analytical Chemistry, Proceedings of the 9th Materials Research Symposium, April 10-13, 1978, held at N. B. S., Gaithersburg, Md., issued April 1979.
84. H. Kambara, *Bunseki, 12,* 848 (1979).
85. R. K. Mitchum and W. A. Korfmacher, *Spectra* (Finnigan), *8,* 12 (1982).
86. M. J. E. Golay, *Gas Chromatography 1958,* Proceedings of the Second Symposium (D. H. Desty, Ed.), Butterworths, London, 1958, p. 36.
87. M. J. E. Golay, *Nature, 199,* 370 (1963).
88. M. J. E. Golay, *Anal. Chem., 40,* 382 (1968).
89. R. D. Dandeneau and E. H. Zerenner, *J. High Res. Chromatog. Chromatog. Commun., 2,* 351 (1979).
90. W. G. Jennings, *Comparisons of Fused Silica and Other Glass Columns in Gas Chromatography,* Dr. Alfred Huthig Verlag, Heidelberg, 1981.
91. P. Burns, B. N. Green, and D. S. Millington, *28th Annual Conference on Mass Spectrometry and Allied Topics,* American

Society for Mass Spectrometry, New York City, 1980, pp. 345–346.
92. R. A. Hites and K. Biemann, *Anal. Chem.*, *40*, 1217 (1968).
93. J. M. Ruth, Mass Spectrometry, in *Handbook of Natural Pesticides, Vol. 2,* (N. B. Mandava, ed.), CRC, Boca Raton, 1985, pp. 323–438.
94. R. P. W. Scott, *Contemp. Liq. Chromatography* (Techniques of Chemistry, Vol. XI), Wiley-Interscience, New York 1976, pp. 273–295.
95. V. L. Tal'rose, Chmn., *Advances in Mass Spectrometry, 7B* (N. R. Daly, ed.), Heyden, on behalf of The Institute of Petroleum, London, 1978, pp. 949–961.
96. B. G. Dawkins and F. W. McLafferty, in *GLC and HPLC Determination of Therapeutic Agents*, Part 1 (Chromatographic Science Series, Vol. 9, Part 1) (K. Tsuji and W. Morozowich, eds.), Marcel Dekker, New York, 1978, pp. 259–275.
97. M. L. Vestal, *National Bureau of Standards Special Publication 519, Trace Organic Analysis: A New Frontier in Analytical Chemistry*. Proceedings of the 9th Materials Research Symposium (H. S. Hertz and S. N. Chesler, eds.), 1979, pp. 647–654.
98. W. H. McFadden, *J. Chromatogr. Sci.*, *17*, 2 (1979).
99. P. J. Arpino and G. Guiochon, *Anal. Chem.*, *51*, 682A (1969).
100. D. E. Games, *Anal. Proc. Chem. Soc. London, 1980,* 110 (1980).
101. D. E. Games, *Anal. Proc. Chem. Soc. London, 1980,* 322 (1980).
102. W. H. McFadden, *J. Chromatogr. Sci.*, *18*, 97 (1980).
103. D. E. Games, in *Soft Ionization Biological Mass Spectrometry*, Proceedings of the Chemical Society Symposium on Advances in Mass Spectrometry Soft Ionization Methods, London, July 1980 (H. R. Morris, ed.), Heyden, London, 1981, pp. 54–68.
104. P. J. Arpino, *Liquid Chromatography Detectors* (Chromatographic Science, Vol. 23) (T. M. Vickrey, ed.), Marcel Dekker, New York, 1983, pp. 243–322.
105. P. J. Arpino, *J. Chromatography*, *323*, 3 (1985).
106. J. F. Lawrence and R. W. Frei, *Chemical Derivatization in Liquid Chromatography* (Vol. 7 of Journal of Chromatography Library), Elsevier, Amsterdam, 1976.
107. W. Morozowich and M. J. Cho, *GLC and HPLC Determination of Therapeutic Agents*, Part 1 (Chromatographic Science Series, Vol. 9, Part 1) (K. Tsuji and W. Morozowich, eds.), Marcel Dekker, New York, 1978, pp. 209–234.
108. H.-R. Schulten and H. D. Beckey, *J. Chromatography*, *83*, 315 (1973).

109. H. R. Morris (ed.), *Soft Ionization Biological Mass Spectrometry*, Proceedings of the Chemical Society Symposium on Advances in Mass Spectrometry Soft Ionization Methods, London, July 1980, Heyden, London, 1981.
110. V. L. Tal'rose, G. V. Karpov, I. G. Gorodetskii, and V. E. Skurat, *Russian J. Phys. Chem.*, *42*, 1658 (1968).
111. V. L. Tal'roze, G. V. Karpov, I. G. Gorodetskii, and V. E. Skurat, *Russian J. Phys. Chem.*, *42*, 1664 (1968).
112. V. L. Tal'roze, G. V. Karpov, I. G. Gorodetskii, and V. E. Skurat, *Russian J. Phys. Chem.* *43*, 198 (1969).
113. V. E. Skurat, *Russian J. Phys. Chem.* *46*, 570 (1972).
114. V. L. Tal'roze, V. D. Grishin, V. E. Skurat, and G. D. Tantsyrev, *Recent Developments in Mass Spectroscopy*, Proceedings of the International Conference on Mass Spectroscopy, Kyoto (K. Ogata and T. Hayakawa, eds.), University Park Press, Baltimore, pp. 1218–1226 (originally published by University of Tokyo Press in 1970).
115. M. A. Baldwin and F. W. McLafferty, *Organic Mass Spectrom.*, *7*, 1111 (1973).
116. M. A. Baldwin and F. W. McLafferty, *Organic Mass Spectrom.*, *7*, 1353 (1973).
117. P. Arpino, M. A. Baldwin, and F. W. McLafferty, *Biomedical Mass Spectrom.*, *1*, 80 (1974).
118. P. J. Arpino, B. G. Dawkins, and F. W. McLafferty, *J. Chromatogr. Sci.*, *12*, 574 (1974).
119. F. W. McLafferty, R. Knutti, R. Venkataraghavan, P. J. Arpino, and B. G. Dawkins, *Anal. Chem.*, *47*, 1503 (1975).
120. J. D. Henion, in *Advances in Mass Spectrometry*, Vol. 7B (N. R. Daly, ed.), Heyden, on behalf of The Institute of Petroleum, London, 1978, pp. 865–872.
121. J. D. Henion, *Anal. Chem.*, *50*, 1687 (1978).
122. J. W. Serum and A. Melera, 26th Annual Meeting on Mass Spectrometry and Allied Topics, ASMS, Saint Louis, Missouri, Paper No. RP2, 1978.
123. A. Melera, *Abstracts, 1979 Pittsburgh Conference on Analytical Chemistry and Applied Spectroscopy*, Cleveland, Ohio, Paper No. 85, 1979.
124. A. Melera, *Advances in Mass Spectrometry*, Vol. 8B (A. Quayle, ed.), Heyden, on behalf of The Institute of Petroleum, London, 1980, p. 1957.
125. J. F. K. Huber, A. M. van Urk-Schoen, and G. B. Sieswerda, *Z. Anal. Chem.*, *264*, 257 (1973).
126. R. E. Lovins, S. R. Ellis, G. D. Tolbert, and C. R. McKinney, *Anal. Chem.*, *45*, 1553 (1973).
127. R. E. Lovins, S. R. Ellis, G. D. Tolbert, and C. R. McKinney, in *Advances in Mass Spectrometry*, Vol. 6 (A. R.

West, ed.), Applied Science Publishers Ltd., Barking, Essex, on behalf of The Institute of Petroleum, London, 1974, Paper 51, pp. 457–462.
128. S. Elbert, B. Gruhn, E. Wipfelder, and H. Heusinger, *Anal. Chem.*, *48*, 1270 (1976).
129. E. R. Schmid, I. Fogy, F. Heresch, E. Kenndler, and J. F. K. Huber, *Mikrochimica Acta [Wien]*, *2*, 207 (1979).
130. B. L. Karger, D. P. Kirby, Paul Vouros, R. L. Foltz, and B. Hidy, *Anal. Chem.*, *51*, 2324 (1979).
131. J. D. Henion and G. A. Maylin, *Biomed. Mass Spectrom.*, *7*, 115 (1980).
132. J. D. Henion, *J. Chromatogr. Sci.*, *19*, 57 (1981).
133. J. D. Henion and T. Wachs, *Anal. Chem.*, *53*, 1963 (1981).
134. E. D. Lee and J. D. Henion, *J. Chromatogr. Sci.*, *23*, 253 (1985).
135. R. Tiebach, W. Blaas, and M. Kellert, *J. Chromatography*, *323*, 121 (1985).
136. P. Hirter, H. J. Walther, and P. Dätwyler, *J. Chromatography*, *323*, 89 (1985).
137. J. Henion, Micro LC/MS Coupling, *Microcolumn High-Performance Liquid Chromatography* (Journal of Chromatography Library, Vol. 28) (P. Kucera, ed.), Elsevier, Amsterdam, 1984, pp. 260–300.
138. A. P. Bruins, *J. Chromatography*, *323*, 99 (1985).
139. T. Tsuda and M. Novotny, *Anal. Chem.*, *50*, 271 (1978).
140. G. Nota, G. Marino, V. Buonocore, and A. Ballio, *J. Chromatography*, *46*, 103 (1970).
141. W. M. A. Niessen and H. Poppe, *J. Chromatography*, *323*, 37 (1985).
142. T. Takeuchi, Y. Hirata, and Y. Okumura, *Anal. Chem.*, *50*, 659 (1978).
143. S. Tsuge, Y. Hirata, and T. Takeuchi, *Anal. Chem.* *51*, 166 (1979).
144. Y. Hirata, T. Takeuchi, S. Tsuge, and Y. Yoshida, *Org. Mass Spectrom.*, *14*, 126 (1979).
145. Y. Yoshida, H. Yoshida, S. Tsuge, T. Takeuchi, and K. Mochizuki, *J. High Res. Chromatogr. Chromatogr. Commun.* *3*, 16 (1980).
146. C. R. Blakley, M. J. McAdams, and M. L. Vestal, *J. Chromatography*, *158*, 261 (1978).
147. C. R. Blakley, M. J. McAdams, and M. L. Vestal, in *Advances in Mass Spectrometry*, *Vol. 8B* (A Quayle, ed.), Heyden, on behalf of The Institute of Petroleum, London, 1980, pp. 1616–1623.
148. C. R. Blakley, M. J. McAdams, and M. L. Vestal, *27th Annual Conference on Mass Spectrometry and Allied Subjects*, ASMS, Seattle, Washington, 1979, pp. 622–623.

149. C. R. Blakley, J. J. Carmody, and M. L. Vestal, *J. Am. Chem. Soc.*, *102*, 5931 (1980).
150. C. R. Blakley, J. J. Carmody, and M. L. Vestal, *Anal. Chem.*, *52*, 1636 (1980).
151. C. R. Blakley, J. C. Carmody, and M. L. Vestal, *Clin. Chem.*, *26*, 1467 (1980).
152. P. R. Jones and S. K. Yang, *Anal. Chem.*, *47*, 1000 (1975).
153. R. P. W. Scott, C. G. Scott, M. Munroe, and J. Hess, Jr., in *Symposium on the Poisoned Patient: The Role of the Laboratory*, Proceedings of Ciba Foundation Symposium 26, new series, London, 1974 (R Porter, ed.), Elsevier, New York, 1974, pp. 155–169.
154. R. P. W. Scott, C. G. Scott, M. Munroe, and J. Hess, Jr., *J. Chromatography*, *99*, 395 (1974).
155. R. P. W. Scott, *National Bureau of Standards Special Publication 519, Trace Organic Analysis: A New Frontier in Analytical Chemistry*, Proceedings of the 9th Materials Research Symposium (H. S. Hertz and S. N. Chesler, eds.), 1979, pp. 637–645.
156. W. H. McFadden, H. L. Schwartz, and S. Evans, *J. Chromatography*, *122*, 389–396 (1976).
157. W. L. Erdahl and O. S. Privett, *Lipids*, *12*, 797 (1977).
158. D. A. Yorke, P. Burns, and D. S. Millington, *27th Annual Conference on Mass Spectrometry and Allied Topics*, ASMS, Seattle, Washington, Paper RPM0C11, 1979, p. 624.
159. J. van der Greef, A. C. Tas, M. C. ten Noever de Brauw, M. Höhn, G. Meijerhoff, and U. Rapp, *J. Chromatography*, *323*, 81 (1985).
160. D. E. Games, P. Hirter, W. Kuhnz, E. Lewis, N. C. A. Weerasinghe, and S. A. Westwood, *J. Chromatography*, *203*, 131 (1981).
161. N. J. Alcock, C. Eckers, D. E. Games, M. P. L. Games, M. S. Lant, M. A. McDowall, M. Rossiter, R. W. Smith, S. A. Westwood, and H.-Y. Wong, *J. Chromatography*, *251*, 165 (1982).
162. E. D. Hardin and M. L. Vestal, *28th Annual Conference on Mass Spectrometry and Allied Topics*, ASMS, New York City, 1980, pp. 616–617.
163. A. Benninghoven, A. Eicke, M. Junack, W. Sichtermann, J. Krizek, and H. Peters, *Org. Mass Spectrom.*, *15*, 459 (1980).
164. R. D. Smith and A. L. Johnson, *Anal. Chem.*, *53*, 1120 (1981).
165. P. Dobberstein, E. Korte, G. Meyerhoff, and R. Pesch, *Int. J. Mass Spectrom. Ion Phys.*, *46*, 185 (1983).

166. E. C. Horning, D. I. Carroll, I. Dzidic, K. D. Haegele, M. G. Horning, and R. N. Stillwell, *J. Chromatography, 99,* 13 (1974).
167. E. C. Horning, D. I. Carroll, I. Dzidic, K. D. Haegele, M. G. Horning, and R. N. Stillwell, *J. Chromatogr. Sci., 12,* 725 (1974).
168. D. I. Carroll, I. Dzidic, R. N. Stillwell, K. D. Haegele, and E. C. Horning, *Anal. Chem., 47,* 2369 (1975).
169. H. Kambara, Y. Mitsui, and I. Kanomata, *Anal. Chem., 51,* 1447 (1979).
170. M. Tsuchiya and T. Taira, *Int. J. Mass Spectrom. Ion Phys., 34,* 351 (1980).
171. M. Dole, L. L. Mack, R. L. Hines, R. C. Mobley, L. D. Ferguson, and M. B. Alice, *J. Chem. Phys., 49,* 2240 (1968).
172. R. J. Buehler, E. Flanigan, L. J. Greene, and L. Friedman, *J. Am. Chem. Soc., 96,* 3990 (1974).
173. D. S. Simons, B. N. Colby, and C. A. Evans, Jr., *Int. J. Mass Spectrom. Ion Phys., 15,* 291 (1974).
174. B. P. Stimpson and C. A. Evans, Jr., *Biomed. Mass Spectrom., 5,* 52 (1978).
175. M. A. Posthumus, P. G. Kistemaker, H. L. C. Meuzelaar, and M. C. ten Noever de Brauw, *Anal. Chem. 50,* 985 (1978).
176. J. J. Carmody, C. R. Blakley, and M. L. Vestal, *29th Annual Conference on Mass Spectrometry and Allied Topics,* ASMS, Minneapolis, Minnesota, 1981, p. 156.
177. C. R. Blakley and M. L. Vestal, *29th Annual Conference on Mass Spectrometry and Allied Topics,* ASMS, Minneapolis, Minnesota, 1981, p. 275.
178. C. R. Blakley and M. L. Vestal, *Anal. Chem., 55,* 750 (1983).
179. C. G. Edmonds, E. E. Jenkins, J. A. McCloskey, C. R. Blakley, M. L. Vestal, N. C. De, A. Farber, S. P. Dutta, and G. B. Chheda, *29th Annual Conference on Mass Spectrometry and Allied Topics,* ASMS, Minneapolis, Minnesota, 181, pp. 469-470.
180. C. G. Edmonds, H. Pang, J. A. McCloskey, C. R. Blakley, and M. L. Vestal, *30th Annual Conference on Mass Spectrometry and Allied Topics,* ASMS, Honolulu, 1982, pp. 606–607.
181. N. C. De, A. Mittelman, S. P. Dutta, C. G. Edmonds, E. E. Jenkins, J. A. McCloskey, C. R. Blakley, M. L. Vestal, and G. B. Chheda, *J. Carbohydr. Nucleosides Nucleotides, 8,* 363 (1981).
182. M. L. Vestal, *Anal. Chem., 56,* 2590 (1984).
183. M. L. Vestal and G. J. Fergusson, *Anal. Chem., 57,* 2373 (1985).

184. D. J. Liberato and A. L. Yergey, *Anal. Chem.*, *58*, 6 (1986).
185. C. G. Edmonds, J. A. McCloskey, and V. A. Edmonds, *Biomed. Mass Spectrom.*, *10*, 237 (1983).
186. J. M. Ruth, E. A. B. Brown, and N. B. Mandava, *Abstracts, 1981 Pittsburgh Conference on Analytical Chemistry and Applied Spectroscopy*, Atlantic City, Abstract No. 65, 1981.
187. J. M. Ruth and Y. Ito, *95th Annual Meeting of the Association of Official Analytical Chemists*, Washington, D.C., Kontes Workshop, Program, 1981, p. 24.
188. J. M. Ruth, Y. Ito, N. B. Mandava, and V. P. Flanagan, *J. Liq. Chromatogr.*, *8*, 2239 (1985).
189. N. B. Mandava and Y. Ito, *J. Liq. Chromatogr.*, *7*, 303 (1984).
190. N. B. Mandava, Y. Ito, and J. M. Ruth, *J. Liq. Chromatogr.*, *8*, 2221 (1985).
191. H. M. Fales, L. K. Pannell, E. A. Sokoloski, and P. Carmeci, *Anal. Chem.*, *57*, 376 (1985).
192. Y. M. Yang, H. A. Lloyd, L. K. Pannell, H. M. Fales, R. D. Macfarlane, C. J. McNeal, and Y. Ito, *Biomed. Environ. Mass Spectrom.*, *13*, 439 (1986).
193. Y. Ito and W. D. Conway, *Anal. Chem.*, *56*, 534a (1984).
194. Y. Ito, *Advances in Chromatography*, *24*, 181 (1984).

Part 3

APPENDIX

appendix one

Application of Countercurrent Chromatography

Table 1 Applications of High-Speed CCC

Substance*	Solvent system*	Volume ratio	Remarks	Reference
Insecticides:				
Demeton O and S	Pen/EtOH/H_2O	6:5:1	S=P-O & O=P-S isomers	1
Herbicides:				
S-Triazines	Hex/EtOAc/MeOH/H_2O	8:2:5:5	Mass spec analysis	2
Plant hormones:				
Auxins	Hex/EtOAc/MeOH/H_2O	3:7:5:5		3
IAA & ABA	$CHCl_3$/AcOH/H_2O	2:2:1		3
Antibiotics:				
Siderocheline A	$CHCl_3$/MeOH/H_2O	7:13:8		4
Efrotomycin	CCl_4/$CHCl_3$/MeOH/H_2O	5:5:6:4		4
Pentalenolactone	$CHCl_3$/MeOH/H_2O	1:1:1		4
Bu 2313 B	Hex/CH_2Cl_2/MeOH/H_2O	5:1:1:1		4
A 201 E	CCl_4/$CHCl_3$/MeOH/H_2O	2:5:5:5		4
Tirandamycin A and B	Hex/EtOAc/MeOH/H_2O	70:30:15:6		4
Valinomycin	Hex/MeOH/H_2O	10:9:1	Lactone-lactam	5
Piericidin A1	Hex/Et_2O/MeOH/H_2O	4:1:4:1	Alkenyl hydroxypyridine	5
Concanamycin C	Hex/EtOAc/MeOH/H_2O	1:1:1:1	Macrolide glycoside	5
Actinomycin C1, C2, & C3	Hex/Et_2O/MeOH/H_2O	1:5:4:5	Aminophenoxazinone cyclic peptides	5, 6

Quinomycin A & C	Hex/EtOAc/MeOH/H_2O	2:3:2:3	Quinoxaline cyclic peptides	5
Tomaymycin	Hex/EtOAc/MeOH/H_2O	1:3:1:3	Anthramycin	5
Acetoxycycloheximide E73	$CHCl_3$/toluene/MeOH/H_2O	5:4:5:4	Cycloheximide	5
Toyocamycin	n-BuOH/H_2O		Nucleoside	5
Pyridazine 3-carboxylic acid	n-BuOH/0.2N AcOH	1:1	Amino acid	5
Acivicin	$CHCl_3$/MeOH/0.2NAcOH	1:1:1		7
Macquarimicin	CCl_4/$CHCl_3$/MeOH/H_2O	1:1:1:1	New class of antibiotics	8
Arizonins	CCl_4/$CHCl_3$/MeOH/H_2O	5:5:5:2		
		5:5:10:4		
		3:2:3:2		
		3:2:5:2	Kalafungen Family	9
	Hex/EtOAc/MeOH/H_2O	2:3:3:2	NMR analysis	
		2:5:3:2		
		1:1:1:1		
Tiacumicins	CCl_4/$CHCl_3$/MeOH/H_2O	3:7:7:3	18-membered macrolide UV, IR, MS, NMR	10
Steroids:				
Hydrocortisone	CH_2Cl_2/Hex/MeOH/H_2O	40:10:17:8		11
1,2-Dihydrotriamcinolone	CH_2Cl_2/MeOH/H_2O	40:13:7		11
6-Methylprednisolone derivative	CH_2Cl_2/Bu_3N/HCOOH/H_2O	5000:24:5:1000		11

Table 1 (Continued)

Substance*	Solvent system*	Volume ratio	Remarks	Reference
Artemisinin & Arteannuin B	Isooctane/EtOAc/MeOH/H_2O	7:3:6:4	FAB-MS	12
Arteannuin B & Artemisitene	Isooctane/EtOAc/MeOH/H_2O	7:3:6:4		12
Rutin	EtOAc/H_2O		Mutagen of red wine	13
Digitonin (commercial)	$CHCl_3$/AcOH/MeOH/H_2O	5:1:3:3	Californium 252 PDMS	14
Acetophenone, o-Nitrophenol, p-Nitrophenol, & Phenol	$CHCl_3$/MeOH/H_2O	3:1:3	For HSCCC/FT-IR	15
DNP Amino Acids	$CHCl_3$/AcOH/0.1N HCl	2:2:1	Standard test system	16, 17, 18, 19, 20
Pigments:				
Methyl Violet 2B	$CHCl_3$/AcOH/H_2O	2:2:1	Californium 252 PDMS	21
Direct Dye 2	EtOAc/n-BuOH/H_2O	1:10:10		22
Direct Dye 4	EtOAc/n-BuOH/H_2O	1:5:5		22
Disperse Dyes 9 & 10	EtOAc/Ethylene glycol		NMR spectra	22
Rhodamine B & Evans Blue	N_2(80 psi)/1mM SDS		Foam CCC	23

Norepinephrine & Dopamine	n-BuOH/sat.$BaCl_2$ sol.	1:1	Isotope labelled	24
Purines & Pyrimidines	n-BuOH/1M PO_4 (K,pH 6.5)	1:1		17
Tannins	n-BuOH/0.1M NaCl	1:1	Poor resolution in HPLC	25
	n-BUOH/1mM Phytic Acid in 0.1M NaCl (pH 4.0)	1:1		25
Peptides and Proteins:				
Dipeptides	n-BuOH/AcOH/H_2O	4:1:5	Large-scale prep. CCC	26
	n-BuOH/DCA/0.1M NH_4CO_2H	100:1:100 → 100:0:100	Gradient elution	17
Gramicidins	$CHCl_3$/Ben/MeOH/H_2O	15:15:23:7		17
Cholecystokinin fragment	n-BuOH/0.2M NH_4OAc	1:1	at 50°C	27
Bombesin	n-BuOH/DCA/H_2O	100:1:100	at 45°-50°C	28
Bovine Insulin	sec.-BuOH/DCA/H_2O	200:1:200	at 50°C	29
BSA & sheep Hb	N_2(80psi)/0.2M Na_2HPO_4 (pH 8.9)		Foam CCC	23

*Abbreviations: ABA, abscisic acid; AcOH, acetic acid; Ben, benzene; BSA, bovine serum albumin; Bu_3N, tributylamine; BuOH, butanol; DCA, dichloroacetic acid; EtOAc, ethyl acetate; ETOH, ethanol; Hb, hemoglobin; Hex, n-hexane; IAA, indole-3-acetic acid; MeOH, methanol; NH_4OAc, ammonium acetate; Pen, n-pentane; SDS, Sodium dodecyl sulfate.

Table 2 Applications of Centrifugal CCC Other Than HSCCC

Substance[a]	Solvent system[b] (volume ratio)	Apparatus[c]
Prostaglandins	$CHCl_3/HOAc/H_2O$ (2:2:1)	CHCPC[30]
Catecholamine metabolites	EtOAc/10%HOAc, 5%NaCl (1:1)	ACPC[31]
Urinary VMA & HVA	EtOAc/10%HOAc, 5%NaCl (1:1)	ACPC[31], VCPC[32]
S-Triazines	$Hex/EtOAc/MeOH/H_2O$ (8:2:5:5)	HCPC[33]
Parathion-ethyl & p-Nitrophenol	CCl_4/H_2O	HCPC[34]
Fenthions	Hex/MeOH	CHCPC[35]
Plant hormones:		
indole auxins	$Hex/EtOAc/MeOH/H_2O$ (3:7:5:5)	TCPC[36,37]
indole auxins & ABA	$CHCl_3/HOAc/H_2O$ (2:2:1)	TCPC[36,37]RC[37]
gibberellines	$Et_2O/MeOH/0.5MPO_4$ (pH5.9) (3:1:2)	TCPC[36]
cytokinins	$EtOAC/MeOH/0.5PO_4$ (pH7) (3:1:3)	TCPC[36]
Antitumor drugs:		
daunorubicin derivatives	$CHCl_3/DCE/Hex/MeOH/H_2O$ (2:2:2:7:2)	CHCPC[38]
diaziquone	DCE/MeOH/10mM Maleate (Na, pH6.1) (3:3:1)	CHCPC[39]
	$DCE/MeOH/10mMPO_4$ (Na, pH6.1) (3:3:1)	CHCPC[39]
	$CHCl_3$/Ethylene glycol	CHCPC[39]
	$Hex/EtOAc/NO_2CH_3/MeOH$ (8:2:2:3)	CHCPC[39]
Antibiotics:		
candicidin, levorin, trichomycin & hemycin	$CHCl_3$/MeOH/Borate Buffer (pH8.2) (4:4:3)	VCPC[40]
echinomycin & triostin A	$TCE/MeOH/H_2O$ (7:3:1)	VCPC[41]

Table 2 (Continued)

Substance[a]	Solvent system[b] (volume ratio)	Apparatus[c]
Antibiotics: (continued)		
quinomycin C & echinomycin	Hep/EtOAc/$(CH_3)_2CO/H_2O$ (1:1:3:1)	VCPC[41]
aureomycin & terramycin	n-BuOH/0.01NHCl (1:1)	HCPC[34]
tetracycline	$CH_3NO_3/CHCl_3$/Pyridine/0.1MEDTA (pH7) (20:10:3:33)	HCPC[34]
trichomycin	$CHCl_3$/MeOH/Borate Buffer (4:4:3)	HCPC[34]
globoroseamycin	$CHCl_3$/MeOH/Borate Buffer (4:4:3)	HCPC[34]
nystatin	$CHCL_3$/MeOH/Borate Buffer (2:4:3)	HCPC[34]
erythromycin	Methyl-isobutyl-ketone/Acetone/0.2 M Phosphate-Citrate Buffer (pH6.5) (20:1:21)	HCPC[34]
kangdisu & colistin E	n-BuOH/2%DCA, 5%NaCl (6:7)	HCPC[34]
Medicinal Herbs:		
quercetin & rutin	EtOAc/EtOH/0.07MNaOH (2:1:2)	HCPC[34]
squalidine & platyphilline	$CHCl_3$/0.07M PO_4 (pH6.2) (1:1)	HCPC[34]
scopolamine, hyoscyamine, & nor-hyoscyamine	$CHCl_3$/0.07M PO_4 (pH6.5) (1:1)	HCPC[34]
scopolamine, anisodine, & 8-isopropyl-scopolamine	$CHCl_3$/0.07M PO_4 (pH6.3) (1:1)	HCPC[34]
Saponins	$CHCl_3/C_6H_6$/HOAc/MeOH/H_2O (45:2:3:60:40)	CDCCC[42]
Sugars:		
sucrose, glucose, & fucose	n-BuOH/EtOH/H_2O (4:1:4)	CDCCC[42]

Table 2 (Continued)

Substance[a]	Solvent system[b] (volume ratio)	Apparatus[c]
Fatty Acids	Hep/MeOH/HOAc (1:1:1)	CDCCC[42]
P-Nitrophenol conjugates	EtOAc/0.5–1.0MKH_2PO_4 (5:1) (EtOAc presaturated with water)	CHCPC[43]
Purines & Pyrimidines	n-BuOH/1M PO_4 (K, pH6.5) (1:1)	ACPC[31]
Heparin	1%HPC in n-BuOH/Aqueous NaCl gradient	RC[44]
Pigments	isoAmOH/EtOH/HOAc/H_2O (4:2:1:5)	CPC[45]
DNP Amino Acids	$CHCl_3$/HOAc/0.1NHCl (2:2:1)	RC[46,47,48,49,50]
		TCPC[51,52]
		CHCPC[46,53,54,55]
		HCPC[56,57]
		ACPC[31]
		EC[58]
		VCPC[59,60,61]
		TCC[59,62,63,64]
		CDCCC[42]
Dipeptides	n-BuOH/HOAc/H_2O (4:1:5)	VCPC[61], TCPC[51], ACPC[31], HCPC[57], RC[49], CHCPC[46,53,65,55]
	n-BuOH/DCA/NH_4COOH (100:1:100) → (100:0:100) DCA gradient	VPC[61], ACPC[31]
Gramicidins	$CHCl_3/C_6H_6$/MeOH/H_2O (15:15:23:7)	CHCPC[66,65]

Table 2 (Continued)

Substance[a]	Solvent system[b] (volume ratio)	Apparatus[c]
Synthetic Peptides:		
5,6 9 & 15 amino acid peptides	n-BuOH/HOAc/H_2O (4:1:5)	CHCPC[67,65,68]
des-enkephalin γ-endorphin (12 amino acids)	n-BuOH/0.1%TFA (1:1)	CHCPC[69]
cholecystokinin fragments (4–8 amino acids)	n-BuOH/HOAc/H_2O	CHCPC[70]
	n-BUOH/0.2–0.5M NH_4OAc (1:1)	CHCPC[70,27]
	n-BuOH/0.1%TFA (1:1)	CHCPC[70]
sulfated cholecystokinin fragments	n-BuOH/0.2–0.5M NH_4OAc (1:1)	CHCPC[70,71]
	n-BuOH/EtOH/Hex/HOAc/H_2O (3:1:2:1:5)	CHCPC[70]
	n-BuOH/HOAc/H_2O (4:1:5)	CHCPC[70,71]
Bovine Insulin	sec.-BuOH/DCA/H_2O	VCPC[72], ACPC[31]

Abbreviations:
[a]ABA, abscisic acid; DNP, dinitrophenyl; HVA, homovanillic acids; VMA, vanillylmandelic acid.
[b]BuOH, butanol; DCA, dichloroacetic acid; DCE, dichloroethane; EDTA, ethylenediaminetetraacetic acid; Et_2O, diethyl ether; EtOAc, ethyl acetate; EtOH, ethanol; Hep, n-heptane; Hex, n-hexane; HOAc, acetic acid; HPC, hexadecylpyridinium chloride; isoAmOH, isoamyl alcohol; MeOH, methanol; NH_4OAc, ammonium acetate; TFA, trifluoroacetic acid; TCE, trichloroethane.
[c]CPC, coil plant centrifuge; ACPC, angle rotor CPC; CDCCC, centrifugal droplet countercurrent chromatograph; CHCPC combined horizontal flow-through CPC; EC, elution centrifuge; HCPC, horizontal flow-through CPC; RC, rotating coil assembly; TCC, toroidal coil centrifuge; TCPC, toroidal CPC; VCPC, (vertical) flow-through CPC.

Table 3 Compounds Separated by DCCC and Solvent Systems

Separated compounds	Solvent systems
Saponins (spirostanol glycosides)	$CHCl_3$-MeOH-H_2O (7:13:8)
Saponins (triterpenoid glycosides)	$CHCl_3$-MeOH-H_2O (7:13:8)
	$CHCl_3$-MeOH-H_2O (5:6:4)
	$CHCl_3$-MeOH-n-PrOH-H_2O (9:12:1:8)
	$CHCl_3$-C_6H_6-EtOAc-MeOH-H_2O (45:2:3:60:40)
Saponins (ginsenosides)	$CHCl_3$-MeOH-n-PrOH-H_2O (45:60:6:40)
	$CHCl_3$-MeOH-iso-PrOH-H_2O (5:6:1:4)
	$CHCl_3$-MeOH-n-PrOH-H_2O (4:6:1:4)
Flavonoid glycosides	$CHCl_3$-MeOH-H_2O (7:13:8)
	$CHCl_3$-MeOH-H_2O (5:6:4)
	$CHCl_3$-MeOH-H_2O (4:4:3)
	$CHCl_3$-MeOH-n-PrOH-H_2O (5:6:1:4)
	$CHCl_3$-MeOH-n-BuOH-H_2O (10:10:1:6)
	n-BuOH-AcOH-H_2O (4:1:5)
Isoflavone glycosides	$CHCl_3$-MeOH-H_2O (7:13:8)
Anthraquinone glycosides	$CHCl_3$-MeOH-H_2O (5:5:3)
	$CHCl_3$-MeOH-n-PrOH-H_2O (9:12:1:8)
	n-BuOH-acetone-H_2O (33:10:50)
Chromenes, Lignans	$CHCl_3$-MeOH-H_2O (5:6:4)
	$CHCl_3$-MeOH-H_2O (4:4:3)
Xanthones	$CHCl_3$-MeOH-H_2O (13:7:4)
	$CHCl_3$-MeOH-n-PrOH-H_2O (9:12:1:8)
Cardenolides	$CHCl_3$-MeOH-H_2O (5:6:4)
	$CHCl_3$-MeOH-H_2O (5:10:6)
	$CHCl_3$-MeOH-H_2O (5:9:7)
Iridoid glycosides	$CHCl_3$-MeOH-H_2O (43:37:20)
	$CHCl_3$-MeOH-H_2O (5:5:3)

Table 3 (continued)

Separated compounds	Solvent systems
Secoiridoid glycosides	$CHCl_3$-MeOH-n-PrOH-H_2O (9:12:1:8)
Terpenoids (diterpenes)	$CHCl_3$-MeOH-H_2O (7:13:8)
Terpenoids (gibberellines)	$CHCl_3$-MeOH-H_2O (5:5:3)
Terpenoid glycosides	$CHCl_3$-MeOH-H_2O (7:13:8)
Alkaloids	$CHCl_3$-MeOH-H_2O (13:7:8)
	$CHCl_3$-MeOH-H_2O (5:5:3)
	$CHCl_3$-MeOH-5% HCl (5:5:3)
	C_6H_6-$CHCl_3$-MeOH-H_2O (5:5:7:2)
	$CHCl_3$-MeOH-n-PrOH-H_2O (45:60:2:40)

Source: Ref. 73.

Table 4 Non-Aqueous Solvent Systems for CCC

Solvent systems (by volume)[a]	Mobile phase	Apparatus[a]	Substance
HEX:EtOAc:NO_2CH_3:MeOH; 560:125:125:190	Upper	DCCC	Essential oils
55:9:28:9	Upper	DCCC	Essential oils
54:13:13:20	Both	HFTCPC	Diaziquone
HEPT:Acetone:MEOH; 48:376:576	Lower	DCCC	Oleanolic acid
			Hederagenin
HEPT:Acetone:MeOH; 50:10:40	Lower	DCCC	Vulpinic acid
HEPT:CH_2Cl_2:CH_3CN; 50:15:35	Lower	DCCC	Triterpenes
			Steroids
HEPT:HOAc:MeOH; 33:33:33	Upper	CDCCC	Saturated fatty acids
HEPT:HOAc:CH_3CN:MeOH; 57:14:14:14	Upper	CDCCC	Unsaturated fatty acids

Table 4 (Continued)

Solvent systems (by volume)[a]	Mobile phase	Apparatus[a]	Substance
Ethylene glycol: Et_2O	Upper	HFTCPC	Salicylates
Ethylene glycol: EtOAc	Upper	HFTCPC	Salicylates
Ethylene glycol: $CHCl_3$	Lower	HFTCPC	Diaziquone

[a]Abbreviations: HEX, hexane; EtOAc, ethyl acetate; MeOH, methanol; HEPT, heptane; HOAc, acetic acid; Et_2O, diethyl ether; DCCC, droplet countercurrent chromatography; HFTCPC, horizontal flow-through coil planet centrifuge; CDCCC, centrifugal DCCC.
Source: Ref. 74.

REFERENCES

1. Y. W. Lee and C. E. Cook, *J. Liq. Chromatogr.*, *8*, 2253 (1985).
2. N. B. Mandava, Y. Ito, and J. M. Ruth, *J. Liq. Chromatogr.*, *8*, 2221 (1985).
3. N. B. Mandava and Y. Ito, *J. Liq. Chromatogr.*, *7*, 303, (1984).
4. G. M. Brill, J. B. McAlpine, and J. E. Hochlowski, *J. Liq. Chromatogr.*, *8*, 2259 (1985).
5. D. G. Martin, C. Biles, and R. E. Peltonen, *Am. Lab.*, *18*, 21 (1986).
6. D. G. Martin, R. E. Peltonen, and J. W. Nielsen, *J. Antibiotics*, *39*, 721 (1986).
7. D. G. Martin, S. A. Mizsak, and J. W. Nielsen, *J. Antibiotics*, *39*, 603 (1986).
8. J. B. McAlpine, D. N. Whittern, J. E. Hochlowski, and L. M. Ranfranz, 26th Interscience Conference on Antimicrobial Agents and Chemotherapy, Poster Session, September 29, 1986.
9. J. E. Hochlowski, G. M. Brill, W. W. Andres, and J. B. McAlpine, 26th Interscience Conference on Antimicrobial Agents and Chemotherapy, Poster Session, September 29, 1986.
10. J. E. Hochlowski, S. J. Swanson, D. N. Whittern, A. M. Buko, and J. B. McAlpine, Interscience Conference on Antimicrobial Agents and Chemotherapy, Poster Session, September 29, 1986.
11. R. G. Williams, Presented at the 1985 Pittsburgh Conference & Exposition on Analytical Chemistry and Applied Spectroscopy, New Orleans, No. 300, 1985.

12. N. Acton, D. L. Klaymen, and I. J. Rollman, *J. Chromatogr.*, *355*, 448 (1986).
13. Chun-Li Yu, B. Swaminathan, L. G. Butler, and D. E. Pratt, *Mutation Res.*, *170*, 103 (1986).
14. Y. M. Yang, H. A. Lloyd, L. K. Rannel, H. M. Fales, R. D. Macfarlane, C. J. McNeal, and Y. Ito, *Biomed. Environ. Mass Spectrom.*, *13*, 439 (1986).
15. R. J. Romanach, J. A. de Haseth, and Y. Ito, *J. Liq. Chromatogr.*, *8*, 2209 (1985).
16. Y. Ito, *J. Chromatogr.*, *214*, 122 (1981).
17. Y. Ito, J. Sandlin, and W. G. Bowers, *J. Chromatogr.*, *244*, 247 (1982).
18. J. L. Sandlin and Y. Ito, *J. Liq. Chromatogr.*, *7*, 323 (1984).
19. Y. Ito, *J. Chromatogr.*, *358*, 325 (1986).
20. J. L. Sandlin, and Y. Ito, *J. Liq. Chromatogr.*, *8*, 2153 (1985).
21. H. M. Fales, L. K. Pannel, E. A. Sokoloski, and P. Carmeci, *Anal. Chem.*, *57*, 376 (1985).
22. H. S. Freeman and C. S. Williard, *Dyes and Pigments, Issue 6* (1986).
23. Y. Ito, *J. Liq. Chromatogr.*, *8*, 2131 (1985).
24. A. Weisz, S. P. Markey, and Y. Ito, *J. Chromatogr.*, *383*, 132 (1986).
25. L. J. Putman, and L. G. Butler, *J. Chromatogr.*, *318*, 85 (1985).
26. J. L. Sandlin, and Y. Ito, *J. Chromatogr.*, *348*, 131 (1985).
27. M. Knight, Y. Ito, A. M. Kask, C. A. Tamminga, and T. N. Chase, *J. Liq. Chromatogr.*, *7*, 2525 (1984).
28. M. Knight, Y. Ito, P. Peters, and C. DiBello, *J. Liq. Chromatogr.*, *8*, 2281 (1985).
29. M. Knight, Y. Ito, J. L. Sandlin, and A. M. Kask, *J. Liq. Chromatogr.*, *9*, 791 (1986).
30. E. A. B. Brown and Y. Ito, *J. Biochem. Biophys. Methods*, *3*, 77 (1980).
31. Y. Ito and R. L. Bowman, *Anal. Biochem.*, *65*, 310 (1975).
32. R. E. Hurst and Y. Ito, *Clin. Chem.*, *8*, 814 (1972).
33. N. B. Mandava, Y. Ito, and J. M. Ruth, *J. Liq. Chromatogr.*, *8*, 2221 (1985).
34. T. Y. Zhang, *J. Chromatogr.*, *315*, 287 (1984).
35. J. M. Ruth, Y. Ito, N. B. Mandava, and V. P. Flanagan, *J. Liq. Chromatogr.*, *8*, 2239 (1985).
36. N. B. Mandava and Y. Ito, *J. Chromatogr.*, *247*, 315 (1982).
37. N. B. Mandava and Y. Ito, *J. Liq. Chromatogr.*, *7*, 303 (1984).
38. H. Nakazawa, P. A. Andrews, N. R. Bachur, and Y. Ito, *J. Chromatogr.*, *205*, 482 (1981).

39. R. D. Freeman, H. Nakazawa, F. E. Chou, N. R. Bachur, and Y. Ito, *J. Chromatogr.*, *252*, 283 (1982).
40. J. W. Lightbown, P. Newland, I. A. Sutherland, and J. W. A. Dymond, *Proc. Analyst. Div. Chem. Soc.*, *14*, 34 (1977).
41. I. A. Sutherland, S. J. Lee, and D. J. Gauvreau, *Anal. Biochem.*, *89*, 213 (1978).
42. W. Murayama, T. Kobayashi, Y. Kosuge, H. Yano, Y. Nunogaki, and K. Nunogaki, *J. Chromatogr.*, *239*, 643 (1982).
43. W. D. Conway and Y. Ito, *J. Liq. Chromatogr.*, *7*, 275 (1984).
44. R. E. Hurst, Y. P. J. Sheng, and Y. Ito, *Anal. Biochem.*, *85*, 230 (1978).
45. Y. Ito, I. Aoki, and E. Kimura, *Anal. Chem.*, *41*, 1579 (1969).
46. W. D. Conway, Countercurrent Chromatography, in *Handbook of Naturally Occurring Pesticides, Vol. 1, Methods, Theory and Practice* (N. B. Mandava, ed.), CRC, Boca Raton, FL, 1983.
47. Y. Ito and R. Bhatnagar, *J. Liq. Chromatogr.*, *7*, 257 (1984).
48. Y. Ito and R. L. Bowman, *Anal. Biochem.*, *78*, 506 (1977).
49. Y. Ito and R. L. Bowman, *J. Chromatogr.*, *136*, 189 (1977).
50. Y. Ito, *J. Chromatogr.*, *196*, 295 (1980).
51. Y. Ito, *J. Chromatogr.*, *192*, 75 (1980).
52. Y. Ito, *Anal. Biochem.*, *102*, 150 (1980).
53. Y. Ito, *J. Chromatogr.*, *188*, 43 (1980).
54. Y. Ito, *Anal. Biochem.*, *100*, 271 (1980).
55. Y. Ito, *Methods in Enzymology, Vol. 91, Enzyme Structure*, Part I, 335 (C. H. W. Hirs and Serge N. Timasheff, eds.), Academic Press, New York, 1983.
56. Y. Ito and R. L. Bowman, *Anal. Biochem.*, *82* (1977).
57. Y. Ito and R. L. Bowman, *J. Chromatogr.*, *147*, 221 (1978).
58. Y. Ito, R. L. Bowman, and F. W. Noble, *Anal. Biochem.*, *49*, 1 (1972).
59. Y. Ito and R. L. Bowman, *Anal. Chem.*, *43*, 69A (1971).
60. Y. Ito and R. L. Bowman, *Science*, *173*, 420 (1971).
61. Y. Ito and R. L. Bowman, *J. Chromatogr. Sci.*, *11*, 284 (1973).
62. Y. Ito and R. L. Bowman, *J. Chromatogr. Sci.*, *8*, 315 (1970).
63. Y. Ito and R. L. Bowman, *Anal. Biochem.*, *85*, 614 (1978).
64. Y. Ito and R. L. Bowman, *Science*, *167*, 281 (1970).
65. G. J. Putterman, F. Perini, E. L. White, and Y. Ito, *Proc. Sixth American Peptide Symposium*, Pierce Chemical Co., p. 113, 1979.
66. Y. Ito, *Protein, Nucleic Acid and Enzyme*, *26*, 1020 (1981) (Japanese).
67. Y. Ito and G. J. Putterman, *J. Chromatogr.*, *193*, 37 (1980).
68. G. J. Putterman, M. B. Spare, and F. Perini, *J. Liq. Chromatogr.*, *7*, 341 (1984).
69. M. Knight, Y. Ito, and T. N. Chase, *J. Chromatogr.*, *212*, 456 (1981).

70. M. Knight, A. M. Kask, and C. A. Tamminga, *J. Liq. Chromatogr.*, *7*, 351 (1984).
71. M. Knight, C. A. Tamminga, Y. Ito, J. D. Gardner, and T. N. Chase, *J. Chromatogr.*, *301*, 277 (1984).
72. Y. Ito, R. E. Hurst, R. L. Bowman, and E. K. Achter, *Sep. Purif. Methods*, *3*, 133 (1974).
73. K. Hostettmann, in *Advances in Chromatography*, Marcel Dekker, New York, Vol. 21 (J. C. Giddings, E. Grushka, J. Cazes, and P. R. Brown, eds.), Chap. 5, p. 165, 1983.
74. W. D. Conway and Y. Ito, *LC Magazine*, *2*, 368 (1984).

appendix two

Glossary of Terms Used in Chromatography

GENERAL CHROMATOGRAPHY

ADSORBATE — A gas, liquid, or dissolved substance taken up by adsorption.

ADSORBENT — A substance that adsorbs or is capable of adsorbing. Examples: charcoal, silica, water, mercury, other metals, cellulose.

ADSORPTION — Adhesion of the molecules of a gas, liquid, or dissolved substance onto a surface.

ADSORPTION CHROMATOGRAPHY — A technique which deals with the separation of a chemical mixture (gas or liquid) by passing it through or over the adsorbent bed (column packing) that adsorbs different compounds at different rates.

ADSORPTION ISOTHERM — The relationship between the concentration C or gas pressure p and the amount w, in grams, of a solute, gas, or vapor taken up per gram of adsorbent at a constant temperature.

AFFINITY — The force that causes the atoms of certain elements to combine; close relationship or similarity of structure.

AFFINITY CHROMATOGRAPHY — A chromatographic technique that utilizes the ability of biological molecules to bond to certain ligands specifically and reversibly (e.g., used in protein biochemistry).

APPARATUS — The means or system by which something is kept in action or a desired result is obtained. Any device or machine or a specific use.

CASCADE — The arrangement of a process in a succession of stages, such that the output of one stage constitutes the input to the next stage.

CHROMATO- — (Greek: chroma or chromatos, color) color or pigmentation [chromatography].

CHROMATOGRAM — The arrangement of zones or bands resulting from a chromatographic separation.

CHROMATOGRAPH — (noun) the apparatus used to carry out a chromatographic procedure; (verb) to employ chromatography to separate substances.

CHROMATOGRAPHIC ADSORPTION — Preferential adsorption of chemical compounds (gases or liquids) in an ascending molecular weight sequence onto a solid adsorbent material, such as activated carbon, alumina, or silica gel; used for analysis and separation of chemical mixtures.

CHROMATOGRAPHY — A method of separating and analyzing mixtures of chemical substances by chromatographic adsorption, partition, or other principle.

CIRCULAR CHROMATOGRAPHY — See RADIAL CHROMATOGRAPHY.

CIRCULAR PAPER CHROMATOGRAPHY — See RADIAL CHROMATOGRAPHY.

COEFFICIENT — A number, constant for a given substance, used as a multiplier in measuring the change in some property of the substance under given conditions (example: Partition coefficient).

COLUMN — A straight or coiled, cylindrical vessel or tube of glass, metal, or plastic material for performing chromatographic operations (to effect separation).

COLUMN BLEED — The loss of carrier liquid during gas chromatography caused by evaporation into the gas under analysis.

COLUMN CHROMATOGRAPHY — Chromatographic technique of two general types: packed columns usually contain either a granular adsorbent or granular support material coated with a thin layer of high-boiling solvent (partition liquid); open tubular columns may contain a thin film of partitioning liquid on the column walls.

COLUMN DEVELOPMENT CHROMATOGRAPHY — technique using columnar apparatus for separating or concentrating one or more components from a physical mixture by means of an adsorbent. The various components are preferentially held at different rates, effecting a separation.

COMPONENT — a substance present in a mixture.

COMPOSITION — The elements or compounds making up a material or produced from it by analysis.

COMPOUND — A substance (chemical compound) whose molecules consist of unlike atoms and whose constituents cannot be separated by physical means.

CONCENTRATE — To increase the amount of a dissolved substance per unit volume or per unit mass by evaporation or other process.

CONCENTRATION — In solutions, the mass, the volume, or the number of moles of solute present per unit quantity of solvent or of solution.

CONCENTRATION GRADIENT — A difference in the concentration of a solute, varying continuously through the solution.

CONCENTRATION SCALE — Any numerical system defining the quantitative relationships of the components of a mixture; see CONCENTRATION.

COUNTERCURRENT DISTRIBUTION — A profile of a compound's concentration in the set of fractions obtained from a countercurrent extraction process.

COUNTERCURRENT EXTRACTION — A liquid-liquid extraction process in which the solvent and the process stream in contact with each other flow in opposite directions. Also known as countercurrent separation.

COUNTERCURRENT SEPARATION — See COUNTERCURRENT EXTRACTION.

DETECTOR — An apparatus or device for indicating the presence of something, as, for example, by the measurement of its ultraviolet absorption or some other property.

DEVICE — An invention or contrivance for some specific purpose.

DISPLACEMENT CHROMATOGRAPHY — A form of column development or elution chromatography in which the solvent is sorbed more strongly than the solute components; the freed sample migrates down the column, pushed by the solvent.

DISTRIBUTE — To divide or arrange according to a classification; to scatter or spread out.

DISTRIBUTION — The arrangement produced when something is distributed; the process by means of which something is distributed.

DISTRIBUTION COEFFICIENT — The ratio of the concentrations of solute dissolved in two immiscible phases at equilibrium.

DISTRIBUTION ISOTHERM — In a liquid-liquid extraction with two phases at equilibrium, a plot of partition coefficient against concentration, at a constant temperature.

DISTRIBUTION LAW — The law states that if a substance is dissolved in two immiscible phases at equilibrium, the distribution coefficient is a constant.

EFFICIENCY — The quality of being efficient; ability to produce a desired effect, product, etc., with a minimum of work. In an ion-exchange system, a measure of the effectiveness of the system expressed as the amount of regenerant required to remove a given unit of adsorbed material.

EFFLUENT — A liquid that flows out (e.g., of the column in chromatograph). Liquid that has passed through a processing operation.

ELECTRON CAPTURE DETECTOR — Extremely sensitive gas chromatography detector that is a modification of the argon ionization detector, with conditions adjusted to favor the formation of negative ions.

ELUANT — A liquid used to extract one material from another, as in chromatography.

ELUATE — The solution that results from an elution process.

ELUTE — To remove adsorbed materials (e.g., from a column) by the use of a solvent.

ELUTION — The removal of adsorbed species from a porous bed or chromatographic column by means of a stream of liquid or gas.

ELUTION PROFILE — A graph or diagram presenting the chromatographic separation of constituents from an elution process.

EMULSION — A stable dispersion of one liquid in a second immiscible liquid (example: oil dispersion in water).

EMULSION BREAKER — A material added to bring about the combined sedimentation and coalescence of the emulsified droplets of the dispersed phase, so that they settle out of the carrier liquid. The effect is sometimes accomplished mechanically, as in a centrifuge.

ENZYME — Any of a group of catalytic proteins produced by living cells, that mediate and promote the chemical processes of life without themselves being altered or destroyed.

EQUILIBRIUM CONSTANT — a constant at a given temperature, relating the chemical activities in the reversible reaction $aA + bB \rightleftharpoons xX + yY$.

EVAPORATOR — A device used to vaporize part or all of the solvent from a solution; the valuable product is usually either a solid or concentrated solution of the solute.

EXTRACT — Material separated from liquid or solid mixture by a solvent.

EXTRACTANT — The liquid used to extract a solute from a mixture or solution.

EXTRACTION — A method of separation in which a solid or solution is contacted with a liquid solvent (the two being essentially mutually insoluble) to transfer one or more components into the solvent.

EXTRACTION DISTILLATION — A distillation process to separate components from eutectic mixtures.

EXTRACTOR — An apparatus for carrying out solvent contact with liquids or solids for the removal of specific components.

FRACTION — One of the portions of material, of varying composition, obtained in the application of a separation process to a mixture.

FRACTIONATING COLUMN — An apparatus used widely for separation of fluid (gas or liquid) components by vapor-liquid fractionization, liquid-liquid extraction, or liquid-solid adsorption.

FRACTIONATION — Separation of a mixture in successive stages, each removing from the mixture some proportion of one of the substances, as by distillation or by differential solubility in water-solvent mixtures.

GAS CHROMATOGRAPH — The instrument used to carry out a gas chromatograph process.

GAS CHROMATOGRAPHY — A separation technique involving the passage of a moving gaseous phase through a column containing a fixed absorbent phase; it is used principally as a quantitative analytical technique for volatile compounds.

GAS-LIQUID CHROMATOGRAPHY — A form of gas chromatography in which the fixed phase (column packing) consists of an involatile liquid solvent distributed on an inert solid support. Abbreviated GLC. Also called gas-liquid partition chromatography, or just gas chromatography.

GAS-SOLID CHROMATOGRAPHY — A form of gas chromatography in which the moving phase is a gas and the stationary phase is a surface-active sorbent (charcoal, silica gel, or activated alumina). Abbreviation: GSC.

GEL — A two-phase colloidal system consisting of a solid and a liquid, with the external appearance of a solid.

GEL FILTRATION CHROMATOGRAPHY — A type of column chromatography which separates molecules on the basis of size; substances of higher molecular weight pass through the column first. Also known as molecular exclusion chromatography; molecular sieve chromatography; size exclusion chromatography.

GEL PERMEATION CHROMATOGRAPHY — See GEL FILTRATION CHROMATOGRAPHY.

GRADIENT ELUTION ANALYSIS — A form of liquid chromatography in which the eluting solvent is changed with time, either by gradually mixing a second solvent of greater eluting power with the first, less powerful solvent or by a gradual change in pH or other property.

INFLUENT — Anything that flows in (as in countercurrent distribution).

INSTRUMENT — A device for indicating or measuring conditions, performance, position, direction, or other physical or chemical property, or for controlling an operation.

ION EXCHANGE — A chemical reaction in which mobile hydrated ions of a solid are exchanged, equivalent for equivalent, for ions of like charge in solution: the solid has an open fishnet-like structure, and the mobile ions neutralize the charged, or potentially charged, groups attached to the solid matrix; the solid matrix is termed the ion exchanger.

ION-EXCHANGE CHROMATOGRAPHY — A chromatographic procedure in which the stationary phase consists of ion-exchange resin, which may be acidic or basic.

ION-EXCLUSION CHROMATOGRAPHY — Chromatography in which the adsorbent material is saturated with the same mobile ions (cationic or anionic) as are present in the sample-carrying eluent (solvent), thus repelling the similar sample ions.

ISOLATION — Separation of a pure chemical substance from a compound or mixture.

ISOTHERM — A line or curve (on a graph) representing the dependence of one quantity on another, at constant temperature.

LIQUID CHROMATOGRAPHY — A form of chromatography employing a liquid as the moving phase and a solid or a liquid on a solid support as the stationary phase; techniques include column chromatography, gel permeation chromatography, and partition chromatography.

MACHINE — A contrivance or device or apparatus.

MOBILE — (Latin: movable) moving or capable of moving or being moved.

MOLECULAR EXCLUSION CHROMATOGRAPHY — See GEL FILTRATION CHROMATOGRAPHY.

MOLECULAR SIEVE CHROMATOGRAPHY — See GEL FILTRATION CHROMATOGRAPHY.

MONITOR — In chromatography, any device recording the performance of a process, as, for example, by measuring ultraviolet absorbance or refractive index.

PAPER CHROMATOGRAPHY — Procedure for analysis of complex chemical mixtures by the separation of the compounds (in solution) on a special grade of paper.

PARTITION — The process of separating or dividing.

PHASE — A solid, liquid, or gaseous homogeneous form existing as a distinct part in contact with other forms; portion of a physical system (gas, liquid, solid) that is homogeneous throughout, has definable boundaries, and can be separated physically from other phases. Example: ice, liquid water, and water vapor are phases of water.

PHASE DIAGRAM — A graphical representation of the equilibrium relationship between phases (such as vapor-liquid, liquid-solid) of a chemical compound, mixture of compounds, or solution.

PHASE EQUILIBRIUM — The equilibrium relationship between phases (such as vapor, liquid, solid) of a chemical compound or mixture under various conditions of temperature, pressure, and composition.

POLAR — Having polarity.

POLARITY — The condition of being positive or negative with respect to some reference point or object.

PURITY — The state of a chemical compound when no impurity can be detected by any experimental method; absolute purity is never reached in practice.

RADIAL CHROMATOGRAPHY — A separation process using a circular disk of absorbent paper which has a strip (wick) cut from edge to center to dip into a solvent; the solvent climbs the wick, touches the sample, and resolves it into concentric rings (the chromatogram). Also known as circular chromatography; circular paper chromatography; radial paper chromatography.

RADIAL PAPER CHROMATOGRAPHY — See RADIAL CHROMATOGRAPHY.

RETENTION — The state of being retained; also, the power of or capacity for retaining.

RETENTION INDEX — In gas chromatography, the relationship of retention volume with arbitrarily assigned numbers to the compound being analyzed; used to indicate the volume retention behavior during analysis.

RETENTION TIME — In gas or liquid chromatography, the time at which the center, or maximum, of a symmetrical peak occurs on gas or liquid chromatogram.

RETENTION VOLUME — In gas chromatography, the product of retention time and flow rate.

REVERSE BONDED-PHASE CHROMATOGRAPHY — A technique of bonded-phase chromatography in which the stationary phase is nonpolar and the mobile phase is polar.

REVERSED-PHASE PARTITION CHROMATOGRAPHY — Paper chromatography in which the low-polarity phase (such as paraffin, paraffin jelly, or grease) is put onto the support (paper) and the high-polarity phase (such as water, acids, or organic solvents) is allowed to flow over it.

SEPARATORY FUNNEL — A funnel-shaped device used for the careful and accurate separation of two immiscible liquids; a stopcock on the funnel stem controls the rate and amount of outflow of the lower liquid.

SHELBEL EXTRACTOR — A liquid-liquid contact vessel used in liquid-liquid extraction processes, a vertical cylinder with interspersed open spaces and wire mesh packing along its height, with liquid agitators in the open spaces, or a vertical cylinder fully filled with a wire mesh packing. Also known as Shelbel column; Shelbel-York extractor; York-Shelbel column.

SOLUBILITY — The ability of a substance to form a solution in another substance.

SOLUBLE — Capable of being dissolved.

SOLUTE — The substance dissolved in a solvent.

SOLUTION — A single, homogeneous liquid, solid, or gas phase that is a mixture in which the components (liquid, gas, solid, or combinations thereof) are uniformly distributed throughout the mixture.

SOLVENT — The part of a solution that is present in the largest amount, or the compound that is normally liquid in the pure state (as for solutions of solids or gases in liquids).

SOLVENT EXTRACTION — The separation of materials of different chemical types and solubilities by selective solvent action; i.e., some materials are more soluble in one solvent than in another. Hence there is preferential extractive action; used to refine organic chemicals and other products (e.g., petroleum products and vegetable oils).

SORBENT — A substance that is capable of absorbing or adsorbing.

SORPTION — Absorption or adsorption.

STATIONARY — Not moving.

THEORETICAL PLATE — A distillation column plate or tray that produces perfect distillation (i.e., produces the same difference in composition as that existing between a liquid mixture and the vapor in equilibrium with it). The packed-column equivalent of a theoretical plate is HETP, or height (of packing) equivalent to a theoretical plate.

THIN-LAYER CHROMATOGRAPHY — Chromatography on thin layers of adsorbents rather than in columns; the adsorbent can be alumina, silica gel, silicates, charcoals, cellulose, etc.

TWO-DIMENSIONAL CHROMATOGRAPHY — A paper chromatography technique in which the sample is resolved by standard procedures (ascending, descending, or horizontal solvent movement) and then turned at right angles in a second solvent and rechromatographed. (the same procedure can also be applied to thin-layer chromatography.)

ZONE — Any area or region considered separate or distinct from others.

COUNTERCURRENT CHROMATOGRAPHY

ANGLE ROTOR COIL PLANET CENTRIFUGE (ACPC) — A rotary seal free flow-through coil planet centrifuge which produces type I-L (scheme II) or type J-L synchronous planetary motion. See text Fig. 17 and Appendix Figs. 1a and 6 of Chap. 3.

AQUEOUS PHASE — One phase of a two-phase solvent system mainly composed of water.

BINODIAL — A curve dividing the whole area of a phase diagram in such a way that the area below yields a single phase and the area above, immiscible two phases. See Chap. 11, Fig. 1.

CARRYOVER — Loss of stationary phase from the column during elution of the mobile phase, after emergence of the solvent front.

CENTRIFUGAL DROPLET COUNTERCURRENT CHROMATOGRAPHY (CDCCC) — A countercurrent chromatographic scheme utilizing a centrifugal force to move droplets of mobile phase through columns of stationary phase. See Chap. 7.

CENTRIFUGAL FORCE DISTRIBUTION DIAGRAM — A diagram which indicates distribution of centrifugal force vectors at various portions of the rotating column.

COAXIAL COIL — A coil coaxially wound around the column holder.

COIL PLANET CENTRIFUGE (CPC) — A centrifuge which subjects the coil holder to a planetary motion.

COIL UNIT — One helical turn of a coil.

COLUMN UNIT — A short column to become a unit of a long column.

COMBINED HORIZONTAL FLOW-THROUGH COIL PLANET CENTRIFUGE (CHCPC) — A rotary seal-free flow-through coil planet centrifuge which generates two different modes of planetary motion, i.e., scheme I (type I synchronous) for the pulley-driven holder and scheme IV (type J synchronous) for the gear-driven holder.

CONTINUOUS EXTRACTION — A method of extraction performed under continuous sample feeding.

COUNTERCURRENT CHROMATOGRAPHY (CCC) — Liquid partition chromatography without solid support.

CRITICAL POINT — See PLAIT POINT.

CROSS-AXIS SYNCHRONOUS FLOW-THROUGH COIL PLANET CENTRIFUGE (XCPC) — A rotary seal-free flow-through coil planet centrifuge which generates type X synchronous planetary motion. See Appendix Fig. 1a of Chap. 3.

DROPLET COUNTERCURRENT CHROMATOGRAPHY (DCCC) — A CCC scheme which utilizes a set of vertical tubular columns to move droplets of mobile phase in the columns of stationary phase. See Chap. 5.

DUAL COUNTERCURRENT CHROMATOGRAPHY (DUAL CCC) — A CCC scheme utilizing a true countercurrent process of two immiscible phases through the column.

ECCENTRIC PARALLEL COIL — A coiled column mounted in parallel and eccentric orientation to the axis of the coil holder.

ELUTION CENTRIFUGE (EC) — A rotary seal-free flow-through centrifuge which generates scheme III or type L synchronous planetary motion. See text Fig. 17 and Appendix Fig. 1a of Chap. 3.

EPICYCLIC COIL PLANET CENTRIFUGE — A scheme IV or type J synchronous coil planet centrifuge. A new descriptive name of the horizontal flow-through coil planet centrifuge.

FLAT HELIX COLUMN — A coiled column which is prepared by winding a tube around a flat core.

FOAM COUNTERCURRENT CHROMATOGRAPHY (FOAM CCC) — A dual CCC scheme utilizing foam and liquid phases.

GYRATION LOCULAR COUNTERCURRENT CHROMATOGRAPHY (GLCCC) — A CCC scheme utilizing gyrating motion of vertical locular columns around the column axis.

HEAD — One terminal of a rotating coil defined by the action of an Archimedean screw force. All objects in the coil tend to move toward the head.

HELIX COUNTERCURRENT CHROMATOGRAPHY (HCCC) — A CCC scheme utilizing a helical tube, particularly a toroidal coil in a centrifugal force field.

HIGH-PERFORMANCE COUNTERCURRENT CHROMATOGRAPHY (HPCCC) — Most advanced forms of countercurrent chromatography including continuous extraction, high-speed CCC, and dual CCC utilizing coaxial coils in type J (scheme IV) or related types of planetary motion.

HIGH-SPEED COUNTERCURRENT CHROMATOGRAPHY (HSCCC) — A CCC scheme capable of yielding rapid and efficient separations with a coaxial multilayer coil under type J (scheme IV) or related types of synchronous planetary motion.

HORIZONTAL FLOW-THROUGH COIL PLANET CENTRIFUGE (HCPC) — A rotary seal-free flow-through centrifuge with a horizontal axis of revolution to generate scheme IV or type J synchronous planetary motion.

HYDRODYNAMIC EQUILIBRIUM SYSTEM (HDES) — One of the two basic forms of CCC utilizing a hydrodynamic equilibrium of two solvent phases established in a rotating coil.

HYDROSTATIC EQUILIBRIUM SYSTEM (HSES) — One of the two basic forms of CCC utilizing a stationary coil relative to the acceleration field.

LIQUID-LIQUID DUAL COUNTERCURRENT CHROMATOGRAPHY (L-L DUAL CCC) — A typical dual CCC scheme utilizing a pair of immiscible solvent phases.

LOCULAR COLUMN — a column for locular CCC which is prepared by placing centrally perforated partitions in a tubular column at regular intervals, thus dividing the column space into multiple small partition units called locules.

LOCULAR COUNTERCURRENT CHROMATOGRAPHY (LCCC) — A CCC scheme utilizing a locular column including rotation LCCC and gyration LCCC.

LOCULE — a partition unit of the locular column.

LOWER PHASE — The heavier phase of a two-phase solvent system.

MIXING ZONE — Portion of a rotating column subjected to vigorous mixing of the two solvent phases observed under particular types of synchronous planetary motion such as type J (scheme IV), type J-L, type X, etc.

MOBILE PHASE — One phase of a two-phase solvent system used to elute the column.

MONOLITHIC INTEGRATED FLOW CIRCUIT (MIFC) — A molded column which incorporates multiple separation chambers (either tubular or locular) and interconnection channels in a compact planar form.

MULTILAYER COIL — A column for high-speed CCC consisting of multiple layers of coils coaxially mounted on a spool-shaped column holder.

NONAQUEOUS PHASE — One of a two-phase solvent system mainly composed of organic solvent(s).

NONHELIX COUNTERCURRENT CHROMATOGRAPHY (NHCCC) — A group of CCC schemes utilizing a separation column which lacks helical configuration, including DCCC, RLCCC, and GLCCC.

NONPLANETARY CENTRIFUGE — A rotary seal-free flow-through centrifuge based on a scheme V or type J nonplanetary design. See text Fig. 17 and Appendix Fig. 1b of Chap. 3.

NONSYNCHRONOUS FLOW-THROUGH COIL PLANET CENTRIFUGE (NSCPC) — A flow-through coil planet centrifuge which provides an adjustable rotation/revolution ratio.

NONSYNCHRONOUS PLANETARY MOTION — Planetary motion in which the ratio of rotation to revolution is variable.

ORGANIC PHASE — One phase of a two-phase solvent system mainly composed of organic solvent(s). Same as nonaqueous phase.

ORIGINAL COIL PLANET CENTRIFUGE (THE CPC) — The earliest model of the coil planet centrifuge which lacks flow-through mechanism.

PARTITION COEFFICIENT (K) — Ratio between solute concentrations in mutually equilibrated two solvent phases. K may be expressed in the following several ways: $K(C_u/C_l)$ or $K(C_l/C_u)$; $K(C_{aq}/C_{nonaq})$ or $K(C_{nonaq}/C_{aq})$; $K(C_m/C_s)$ or $K(C_s/C_m)$, etc., where C_u is solute concentration in the upper phase; C_l, that in the lower phase; C_{aq}, solute concentration in the aqueous phase; C_{nonaq}, that in the nonaqueous phase; C_m, solute concentration in the mobile phase; C_s, that in the stationary phase. Also see Chap. 2, Sec. 3 for preferred usage of terminology.

PARTITION RATIO (G) — Ratio of the number of particles partitioned into the upper phase to that of the remainder (at the interface and in the lower phase).

PHASE DIAGRAM — A diagram which illustrates two-phase compositions by the aid of a binodial, tie lines, plait (critical) point, etc. In CCC, the phase diagram is mainly used for characterization of polymer phase systems. See Chap. 11, Fig. 1.

PHASE DISTRIBUTION CURVE — a curve drawn in the phase distribution diagram to indicate distribution of two solvent phases in the column under various revolutional speeds either with or without introduction of flow.

PHASE DISTRIBUTION DIAGRAM — A diagram which indicates distribution of two solvent phases in the column under various revolutional speeds by the aid of either the phase distribution curve or the phase retention curve.

PHASE RETENTION CURVE — A curve drawn in a phase retention (or distribution) diagram to indicate the percentage retention of the stationary phase relative to the total column capacity as a function of the applied revolutional speed.

PHASE RETENTION DIAGRAM — A diagram which illustrates retention of the stationary phase with respect to the applied revolutional speeds by the use of the phase retention curve(s).

PLAIT POINT — In the phase diagram, the crossing between the binodial and the line drawn through the middle points of tie lines. It is also called the critical point.

PLANETARY MOTION — Simultaneous rotation (around its own axis) and revolution (around the central axis of the system).

PLUG FLOW — A two-phase flow pattern in which segments of one phase occupy the entire cross-sectional area of the tube forming a plug.

POLYMER PHASE SYSTEM — A two-phase or multiphase solvent system composed of one or more polymers as major components.

RETENTION OF STATIONARY PHASE — Amount of the stationary phase retained in the column during elution.

RETENTION TIME — In chromatogram, distance from the starting point to the peak maximum expressed in a time scale.

RETENTION VOLUME — In chromatogram, distance from the starting point to the peak maximum expressed in elution volume.

REVOLUTION — In planetary motion, the motion of the column holder turning around the main axis of the centrifuge system.

ROTATION — In planetary motion, the motion of the column holder turning around its own axis.

ROTARY COIL ASSEMBLY — A preparative CCC device consisting of a horizontal rotary frame which holds a set of coiled columns connected in series.

ROTATION LOCULAR COUNTERCURRENT CHROMATOGRAPHY (RLCCC) — A preparative CCC scheme utilizing slow rotation of an inclined locular column assembly in a gravitational field.

SCHEME I PLANETARY MOTION — Planetary motion characterized by revolution and synchronous counterrotation of the column holder which is held parallel to and at a given distance from the main axis of the centrifuge system. This is also called type I synchronous planetary motion. See text Fig. 17 and Appendix Figs. 1a and 6 of Chap. 3.

SCHEME II PLANETARY MOTION — Planetary motion chracterized by revolution and synchronous rotation of the column holder which maintains a fixed angle (between 0° and 90°) and distance relative to the main axis of the centrifuge system. Also called type I-L synchronous planetary motion. See text Fig. 17 and Appendix Figs. 1a and 6 of Chap. 3.

SCHEME III PLANETARY MOTION — Planetary motion characterized by synchronous counterrotation of the column holder which is held radially a given distance away from the main axis of the centrifuge system. It is also named type L synchronous planetary motion. See text Fig. 17 and appendix Figs. 1a and 6 of Chap. 3.

SCHEME IV PLANETARY MOTION — Planetary motion characterized by revolution and synchronous rotation of the column holder which is held parallel to and at a fixed distance from the main axis of the centrifuge system. It is also named type J synchronous planetary motion (See text Fig. 17 and appendix Figs. 1a and 6 of Chap. 3). It may also be called epicyclic planetary motion.

SCHEME V NONPLANETARY MOTION — Composite motion consisting of rotation of the central column holder at 2ω, rotation of the rotary frame (supporting the column holder) at ω, and revolution of the flow tubes at ω. This motion permits continuous elution through the rotating column without the use of rotary seals. It is also called type J nonplanetary motion. See text Fig. 17 and Appendix Fig. 1b of Chap. 3.

SCHEME VI PLANETARY MOTION — Nonsynchronous rotation and revolution of the column holder which is held parallel to and at a fixed distance from the main axis of the centrifuge system. Also called type I nonsynchronous planetary motion. See text Fig. 17 and Appendix Fig. 1b of Chap. 3.

SETTLING ZONE — Portion of a rotating column showing clear separation of two solvent layers. It has been observed in a spiral column (in the remote area from the central axis of the centrifuge) coaxially mounted in the type J (scheme IV) coil planet centrifuge.

SOLVENT FRONT — Mobile phase front emerging from the column.
STATIONARY PHASE — One phase of a two-phase solvent system retained in the column during elution.
SYNCHRONOUS PLANETARY MOTION — Planetary motion consisting of revolution and rotation (or counterrotation) at the same angular velocity.
TAIL — One terminal of a rotating coil opposite to the head. See HEAD.
THEORETICAL PLATE NUMBER (N) — Measure of partition efficiency given by $N = (4R/w)^2$ where R is retention volume or time and w, the peak width expressed by the same unit.
TIE LINE — A straight line drawn between a pair of nodes on the binodial in a phase diagram. Every point on the tie line yields the same two-phase composition in different volume ratios, where each node represents phase composition of the upper and the lower phases. See Chap. 11, Fig. 1.
TOROIDAL COIL — A coiled column circumferentially mounted in the periphery of the centrifuge bowl (toroidal coil centrifuge) or around cylindrical coil holder of a coil planet centrifuge (toroidal CPC).
TOROIDAL COIL CENTRIFUGE (TCC) — A flow-through centrifuge or a nonplanetary centrifuge equipped with a toroidal coil.
TOROIDAL COIL PLANET CENTRIFUGE (TCPC) — A type J (scheme IV) or similar coil planet centrifuge equipped with a toroidal coil.
TWISTED PAIR COLUMN — A column prepared by folding a tube in two and twisting it to produce a ropelike configuration. It has been substituted for the toroidal coil to reduce hydrostatic pressure.
TWO-PHASE SOLVENT SYSTEM — A solvent system providing a pair of mutually immiscible solvent phases.
TYPE I SYNCHRONOUS PLANETARY MOTION — A new descriptive term for scheme I planetary motion. See Appendix Figs. 1a and 6 of Chap. 3.
TYPE I-L SYNCHRONOUS PLANETARY MOTION — A new descriptive term for scheme II planetary motion. It is a hybrid between type I and type L synchronous planetary motions. See Appendix Figs. 1a and 6 of Chap. 3.
TYPE J SYNCHRONOUS PLANETARY MOTION — A new descriptive term for scheme IV planetary motion. See Appendix Figs. 1a and 6 of Chap. 3.
TYPE J-L SYNCHRONOUS PLANETARY MOTION — A hybrid between type J and type L synchronous planetary motions. See Appendix Fig. 1a of Chap. 3.
TYPE J NONPLANETARY MOTION — A new descriptive term of scheme V nonplanetary motion. See Appendix Fig. 1b of Chap. 3.

TYPE L SYNCHRONOUS PLANETARY MOTION — A new descriptive term for scheme III planetary motion. See Appendix Figs. 1a and 6 of Chap. 3.

TYPE X SYNCHRONOUS PLANETARY MOTION — A planetary motion characterized by revolution and synchronous rotation of the holder in such a way that the axis of the holder maintains its perpendicular orientation at a given distance from the main axis of the centrifuge system. Consequently, it forms a cross between the two axes as indicated by the name. See Appendix Fig. 1a of Chap. 3.

UNILATERAL HYDRODYNAMIC EQUILIBRIUM — A particular form of hydrodynamic equilibrium state where the two immiscible solvent phases are distributed along the length of the coil in such a way that one phase (head phase) entirely occupies the head side and the other phase (tail phase), the tail side. The unilateral hydrodynamic equilibrium has been observed in a coaxial coil subjected to particular modes of synchronous planetary motion such as type J (scheme IV), type J-L, type X, etc., as well as in a coil rotating about its horizontal axis in the gravity.

UNILATERAL HYDRODYNAMIC EQUILIBRIUM SYSTEM (UHDES) — A particular form of HDES which utilizes a unilateral hydrodynamic distribution of the two solvent phases. It provides the basis for high-performance CCC schemes.

UPPER PHASE — The lighter phase of a two-phase solvent system.

VERTICAL FLOW-THROUGH COIL PLANET CENTRIFUGE (VCPC) — A rotary-seal-free flow-through coil planet centrifuge which produces scheme I or type I synchronous planetary motion. See text Fig. 17 and Appendix Fig. 1a of Chap. 3.

β-VALUE — A dimensionless parameter given by the ratio of the radius of rotation (r) to the radius of revolution (R), or $\beta = r/R$, in various modes of planetary motion. It is a major factor to determine hydrodynamic behavior of the two solvent phases in a rotating coil in each planetary motion.

appendix three

Symbols

SYMBOLS USED IN CHAPTER 1

D_C	Capacity ratio
K_C	Partition coefficient (partition chromatography)
t_R	Total retention time of the solute mixture
t_m	Total retention of the nonsorbed substances
V_s	Volume of the stationary phase
V_m	Volume of the mobile phase (dead retention volume)
ϕ_s	Cross-section of the stationary phase (that portion of the cross-sectional area of the column occupied, on the average, by the stationary phase)
ϕ_m	Cross-section of the mobile phase (that portion of the cross-sectional area of the column occupied, on the average, by the mobile phase)
C_s	Equilibrium concentration (moles/unit volume) of the stationary phase
C_m	Equilibrium concentration (moles/unit volume) of the mobile phase
V_o	Total mobile phase volume
V	Volume coordinate in Fig. 5, Chap. 1
V_e	Volume of the effluent or eluent in chromatography (see tangent method)
V_a	Surface volume composed of the molecules of the adsorbed solvent and solute on the surface of the adsorbent (expressed as ml/g)

V_R	Overall retention volume
M_c	Solute mass per unit mass of the stationary phase
M_s	Solute mass retained in the stationary phase
M_m	Solute mass in the mobile phase
M_{sp}	Total mass of the stationary phase
n_m	Number of moles of solute in the mobile phase
n_s	Number of moles of solute in the stationary phase
K_D	Distribution coefficient (countercurrent distribution)
a	Adsorbtion coefficient (dimensions: reciprocal concentration)
K_A	Adsorption coefficient (dimensionless)
R	Retardation factor
R	Relative rate parameter
R_F	In planar systems, the ratio of the distances traveled from the origin by the solute zone and the solvent front
f	A factor whose value ranges from 0.8 to 0.9
t_e	Total elution time taken by a solute molecule to traverse the column
t_s	Time spent by solute molecule in the stationary phase
t_{R_1}	Retention time of component 1
t_{R_2}	Retention time of component 2
t	Time width of the peak in Fig. 5 of Chap. 1
Δt_1	Time width of the zone 1
Δt_2	Time width of the zone 2
R_s	Resolution
N	Number of theoretical plates
n	Total number of moles of solute
H	Height equivalent to a theoretical plate
L	Length of the column
σ	Standard deviation of the length of the chromatographic zone
Δz	Approximate width of the chromatographic zone as measured at the column outlet
W_t	Distance (width) between intersections of tangents with baseline of chromatographic peak (Chap. 1, Fig. 4)
F	Flow rate
U	Mobile phase velocity
z	Coordinate z (distance along column axis)
$\frac{\partial n}{\partial t}$	Change in the total number of moles in both phases in the volume element with time
$\frac{\partial n_m}{\partial t}$	Change in the number of moles of solute in the mobile phase in the volume element with time

$\frac{\partial n_m}{\partial z}$ — Instantaneous gradient of the number of moles of solute in the mobile phase within the volume element

SYMBOLS USED IN CHAPTER 2

K_D	Distribution constant, partition constant, or partition coefficient defined as the ratio of concentration of solute in upper phase to concentration of solute in lower phase
C_1, C_2	Concentrations of solute in first and second phases
K_C	Distribution constant or partition coefficient in chromatography, defined as a ratio of concentrations of solute between the stationary and mobile phases. (Most frequently, but not necessarily always, $K_D = 1/K_C$)
x and y	Fractions remaining in lower and upper phases
n	Number of transfers
T	Total amount of solute
T_n	Amount of solute in tube number n
D_D	Mass distribution ratio in countercurrent distribution
R_D	Phase volume ratio ($R_D = V_U/V_L$)
Q_U, Q_L	Quantities of solute in upper and lower phases
C_U, C_L	Concentration of solute in upper and lower phases
V_U, V_L	Volume in upper and lower phases ($R = V_U/V_L$)
D_C	Capacity factor in chromatography ($D_D = 1/D_C$, when $K_D = 1/K_C$)
$T_{n,r}$	Amount of solute in the rth tube for the nth transfer
$T_{n,r-1}$	Amount of solute in the (r-1)th tube for the nth transfer
T_o	Amount in stage zero
r_{max}	Maximum amount of solute in tube r
z	Number of stages by which the point under discussion is displaced from the maximum
$T_{r,z}$	Amount of solute in rth tube for stage z
σ	Standard deviation of the peak location
W	Peak width
$D_{D,A}$	Mass distribution ratio for solute A
$D_{D,B}$	Mass distribution ratio for solute B
R_D	Phase volume ratio
$K_{D,A}$	Partition coefficient for solute A

$K_{D,B}$	Partition coefficient for solute B
$T_{w,n}$	Amount of solute in the drawn series in a single withdrawal, in CCD
r^*	Number of the highest stage in the CCD train
δ	Solubility parameter

SYMBOLS USED IN CHAPTER 3

a	Radius of a particle
d	Helical diameter of a toroidal coil in helix CCC
F_g	Gravitational force
F_s	Stokes drag
g	Gravitational acceleration
K	Partition coefficient as specified
N	Partition efficiency expressed in terms of theoretical plates
n	Number of coils
P	Minimum pressure to sustain flow in helix CCC
R	Radius of rotation or revolution in a planetary motion
	Retention volume (time) of solute peak in elution
R_{SF}	Retention volume (time) of solvent front
r	Radius of rotation in planetary motion
r_1	Radius of revolution in elution centrifuge
r_2	Radius of rotation in elution centrifuge
t	Time
V_C	Total column capacity
V_e	Stokes Equilibrium velocity
V_F	Free space in flow tubes
w	Peak width
α	Magnitude of acceleration
α_1	Magnitude of acceleration component acting parallel to the holder axis in the elution centrifuge
α_2	Magnitude of acceleration component acting perpendicularly to the holder axis in the elution centrifuge
β	Ratio of the rotational radius to the revolutional radius or $\beta = r/R$
γ	Angle of acceleration
η	Viscosity of a liquid
θ	Angle to specify the location of a point
θ_e	Equilibrium angle of a particle in a rotating coil
θ_o	Angle to specify the original location of a point
ρ	Density of a particle
ρ_o	Density of a liquid
ρ_1	Density of a lower phase

ρ_u	Density of an upper phase
τ	Relaxation time of a particle
ψ	Angle formed between the holder axis and the central axis of the coil planet centrifuge (See Chap. 3 Appendix)
ω	Angular velocity
ω'	Angular velocity of rotation in nonsynchronous planetary motion (See Fig. 17 in Chap. 3)
ω_e	Critical angular velocity
ω_{rel}	Angular velocity of a particle relative to the rotating tube
$\bar{\omega}_{rel}$	Average angular velocity of a particle relative to the rotating tube

SYMBOLS USED IN CHAPTER 4

α	Separation factor, a ratio of partition coefficients or capacity factors, always expressed as a value greater than unity
f	Mobile phase flow rate
K	Partition coefficient, defined as the ratio of the concentration of solute in the stationary phase to the concentration of solute in the mobile phase. It is related to k' by the phase volume ratio, $K = k'(V_m/V_s)$
k'	Capacity factor, defined as the ratio of the quantity of solute in the stationary phase to the quantity of solute in the mobile phase
$K_1^{80\%}$	Value of K_1 required for the partition coefficient term in the exact resolution equation to reach 80% of its asymptotic value
N	Number of theoretical plates
R_s	Chromatographic resolution
S_F	Fractional volume of column occupied by stationary phase. $S_F = V_s/V_c$
t_c, V_c	Time required to elute a mobile phase volume equal to the column volume
t_c'	Corrected retention time for t_c. This corresponds to the time required to elute a volume of mobile phase equivalent to the volume of stationary phase contained in the column. See V_s, t_s
t_f, V_f	Time required to elute a volume corresponding to that of the feed lines; t_f equals t_i if feed line volume is negligible or zero. V_f includes line volume from injection port to start of column, plus volume

	from column end to the point of detection, which may be either the monitor cell center or the delivery tip at the collection vessel
t_i, V_i	Time of sample injection
t_m, V_m	Time required for an imaginary solvent front in the mobile phase to traverse the column, also equivalent to the time spent by every solute in the mobile phase as the solute passes through the column
t_r, t; V_R, V	Retention time, measured from t_f to peak maximum. See Fig. 2
T_R', t'; V_R', V'	Corrected retention time, measured from t_m
t_s	Time spent in the stationary phase by a solute molecule during passage of the solute through the column. This also corresponds to the corrected retention time t_R'. Note that t_s is a general term and does not correspond specifically to V_s except when $K = 1$. See t_c'
V_c	Column volume
V_m	Volume of mobile phase in a column. See t_m
V_R	Retention volume. See t_R
V_s	Volume of stationary phase retained in a column. See t_c'
V_f	Volume of feed lines. See t_f
V_i	Point corresponding to time of sample injection. See t_i
W_b, W	The 4σ base width of a chromatographic peak, usually measured as the distance on the baseline between the points of intersection of tangents to the sides of the peak

(*Note:* Parameters designated by V are more fully defined under the corresponding parameter designated by t.)

SYMBOLS USED IN CHAPTER 8

C_m	In countercurrent chromatography, the concentration of a solute in the mobile phase
C_s	In countercurrent chromatography, the concentration of a solute in the stationary phase
K_E	General symbol for the countercurrent chromatographic elution coefficient, equal to C_m/C_s

$K_E(H/L)$	In countercurrent chromatography, the elution coefficient for the case in which the heavier phase is mobile
$K_E(L/H)$	In countercurrent chromatography, the elution coefficient for the case in which the lighter phase is mobile

SYMBOLS USED IN CHAPTER 11

A_u	Amount of material in upper phase
A_t	Total amount of material added to phase system
C	Critical point
C_d	Concentration of dextran expressed in percent (w/w)
C_1	Concentration in lower phase
C_u	Concentration in upper phase
G	Partition ratio
K	Partition coefficient
L	Pathlength (dm)
W	Weight of stock solution (g)
α^{25}	Optical rotation of dextran measured in degrees at 25°C

SYMBOLS USED IN CHAPTER 12

m	Mass of an ion, in daltons
Z	Number of unit charges (1, 2, etc.) carried by an ion
M	Mass of the molecular ion in daltons (molecular weight)
m/Z	Ratio of mass in daltons to the number of charges carried by an ion; this ratio is the horizontal coordinate in the usual graph of a mass spectrum
I	Intensity of an ion, in general
I_a	Intensity of an ion, as measured in the units in which it is received from the detection circuitry
R.I.(%)	Intensity of an ion, expressed as a percentage of that of the most intense peak in the spectrum
$\%\Sigma m$	Intensity of an ion, expressed as a percentage of the sum of the intensities from m/z = m to m/z = M
I_{Xi}	Intensity I of an ion of m/z = i in the mass spectrum of compound X
r_X	Mole fraction of compound X present in a mixture
b	Density of diffusion flow in a binary gas mixture
ν	Molecular density of the gas

M_1, M_2	Molecular weights of gases 1 and 2 in Eq. (18)
r_1, r_2	Mole fractions of gases 1 and 2 in Eq. (18)
D	Diffusion constant
p	Pressure
T	Temperature
k_T	Thermal diffusion ratio in Eq. (18)
C_m	In countercurrent chromatography, the concentration of a solute in the mobile phase
C_s	In countercurrent chromatography, the concentration of a solute in the stationary phase
K_C	Partition coefficient as defined in liquid partition chromatography, equal to C_s/C_m
K_E	General symbol for the countercurrent chromatographic elution coefficient, equal to C_m/C_s
$K_E(H/L)$	In countercurrent chromatography, the elution coefficient for the case in which the heavier phase is mobile
$K_E(L/H)$	In countercurrent chromatography, the elution coefficient for the case in which the lighter phase is mobile

appendix four

Manufacturers of Countercurrent Chromatographic Instruments

Company	Product
Brinkman Instruments Company (U.W. Distributor for Büchi Lab.-Tech., Ltd., Switzerland) Cantiague Road Westbury, NY 11590 (516) 334-7500	Droplet countercurrent chromatograph (See Fig. 1)
Kontes Spruce Street P.O. Box 729 Vineland, NJ 08360-0709 (609) 692-8500	Combined horizontal flow-through coil planet centrifuge (See Fig. 2a) Rotating coil assembly (preparative countercurrent chromatograph) (See Fig. 2b)
P.C. Inc. 11805 Kim Place Potomac, MD 20854 (301) 299-9386	Ito multilayer coil separator-extractor (See Fig. 3)
Pharma-Tech Research Corp. 6807 York Road Baltimore, MD 21218 (301) 377-7018	High-speed countercurrent chromagraph (See Fig. 4)
Sanki Engineering, Ltd Imazato 2-16-17 Nagaoka-cho, Kyoto 617 Japan	Centrifugal droplet countercurrent chromatograph (See Fig. 5)
Tokyo Rikakikai Co., Ltd. No. 2 (Toei Bldg., 10F) Muromachi-4-chome Nihonbashi, Chuo-ku Tokyo 103 Japan (03) 245-0481	Droplet countercurrent chromagraph (See Fig. 6a) Rotation locular countercurrent chromatograph (See Fig. 6b)

Figure 1 Droplet countercurrent chromatograph.

Figure 2

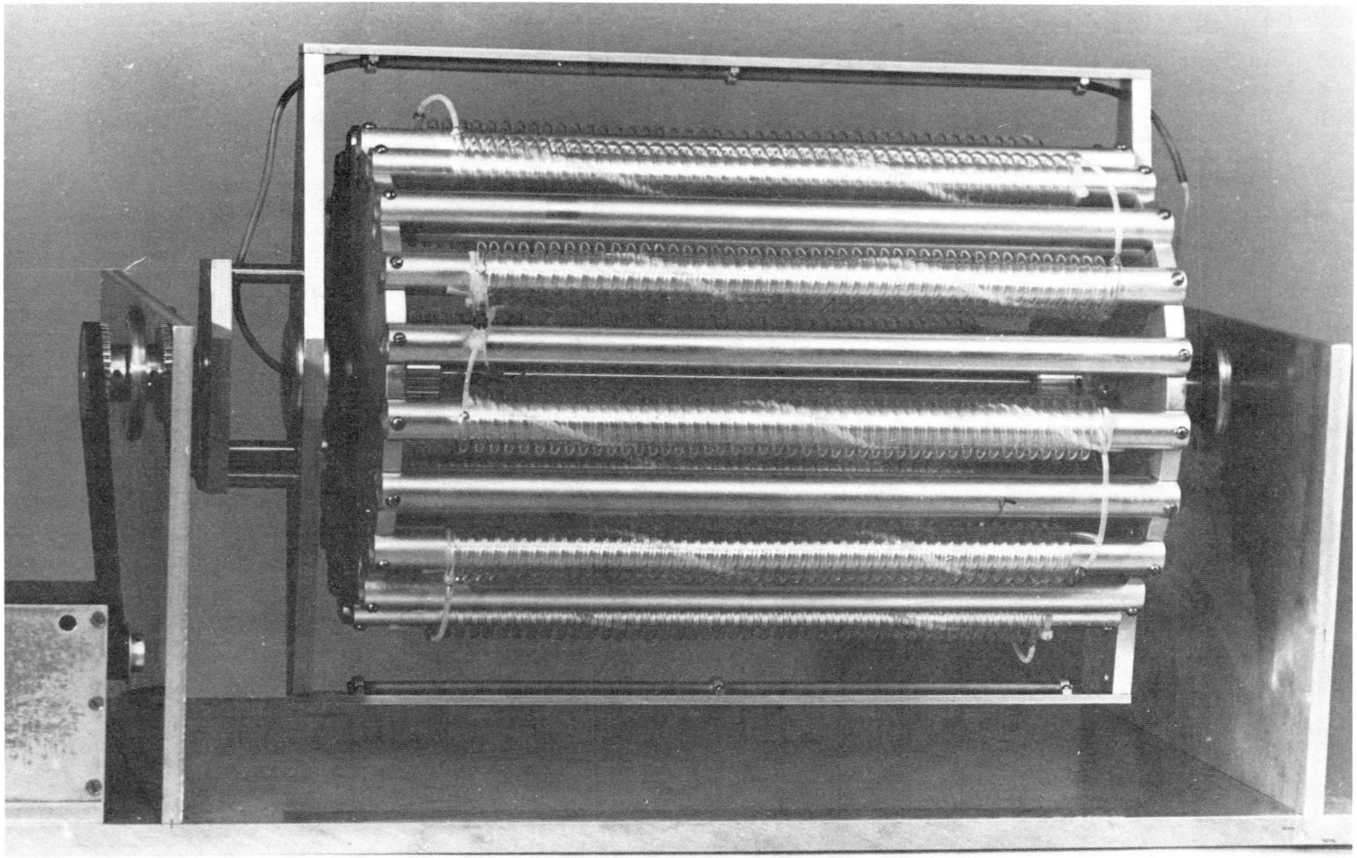

Figure 2 (a) Combined horizontal flow-through coil planet centrifuge and (b) Rotating coil assembly (preparative countercurrent chromatograph).

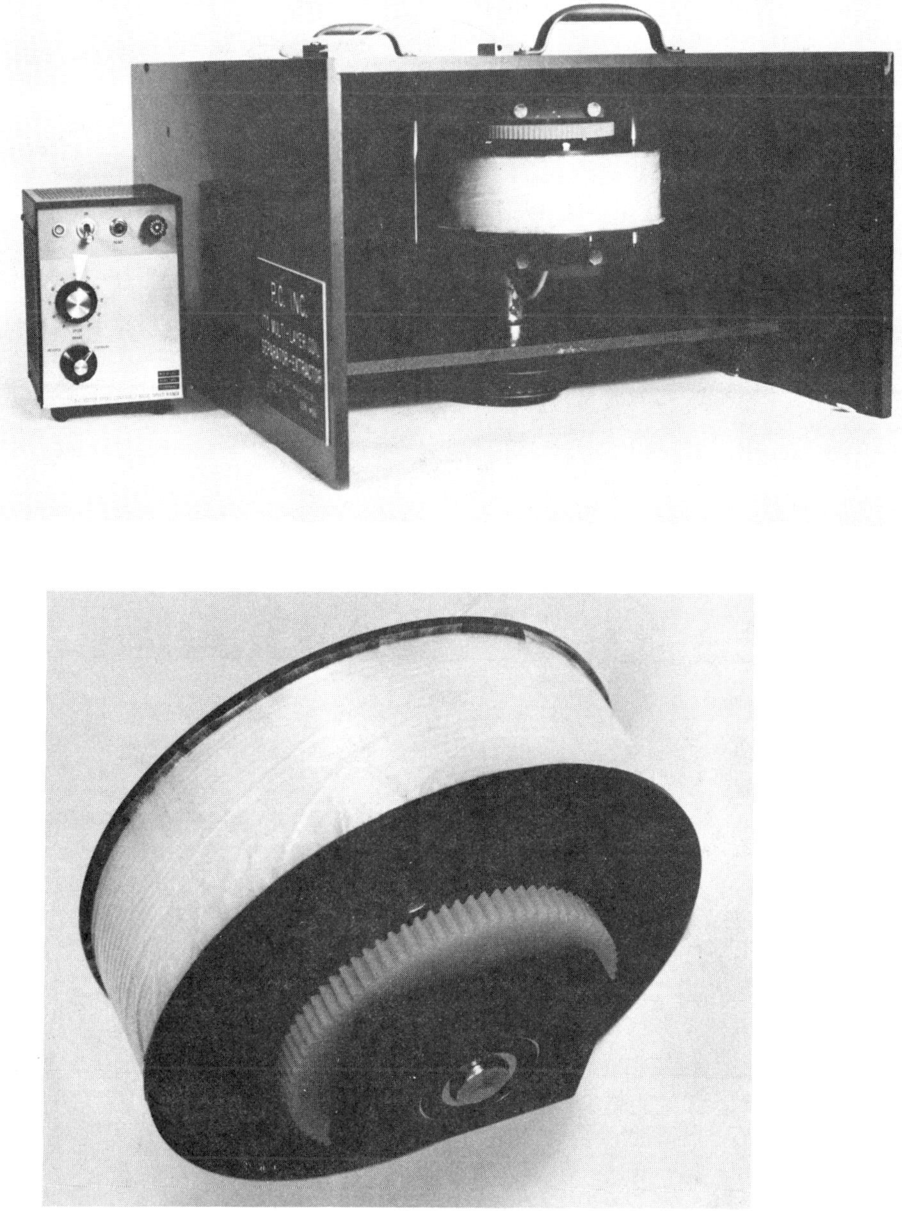

Figure 3 Ito multilayer coil separator-extractor.

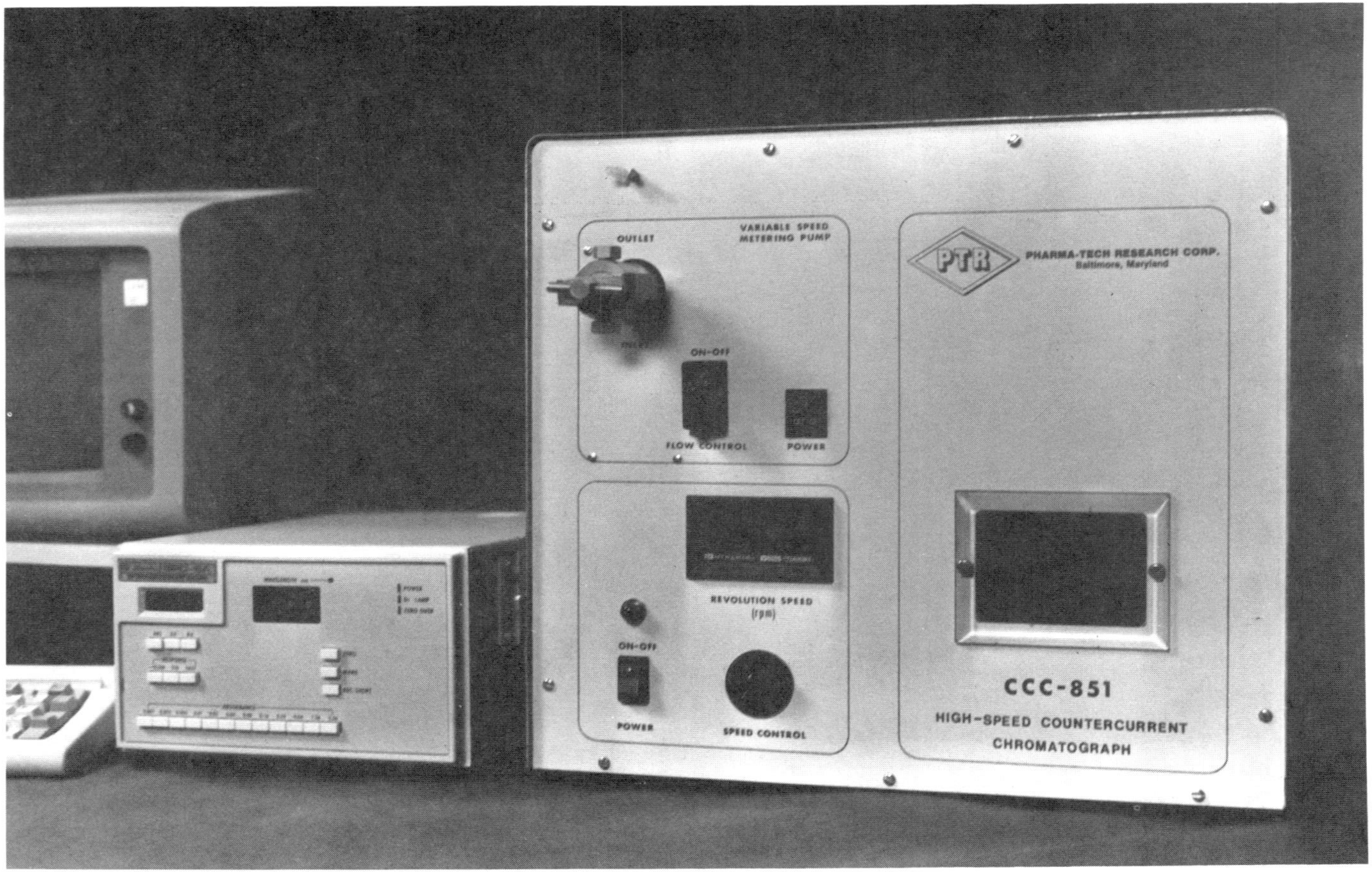

Figure 4 High-speed countercurrent chromatograph.

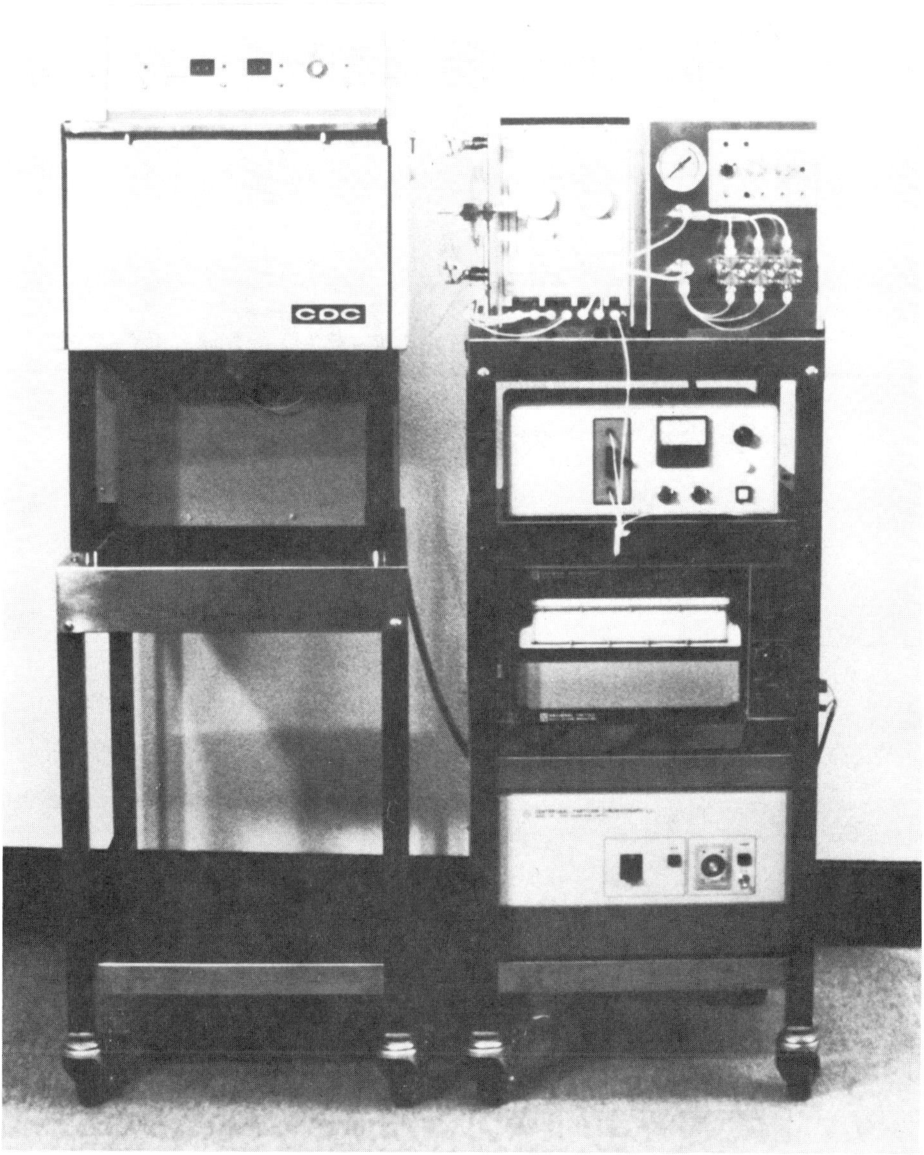

Figure 5 Centrifugal droplet countercurrent chromatograph.

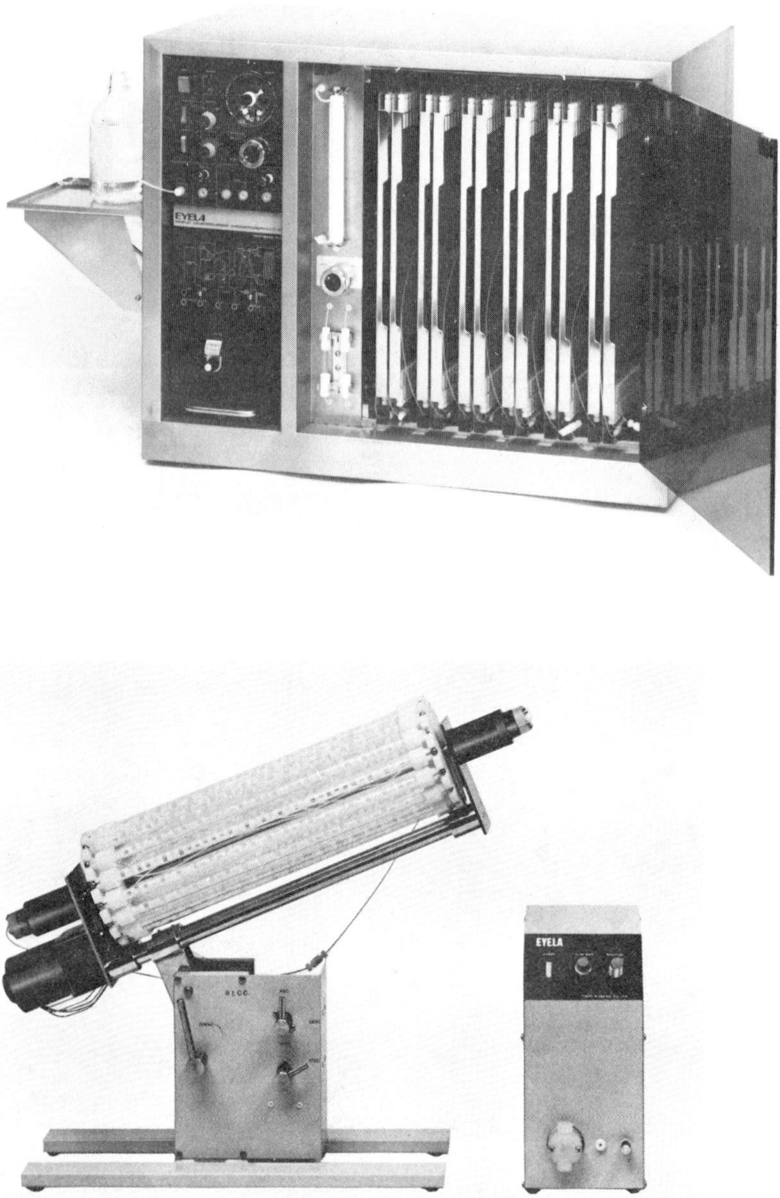

Figure 6 (a) Droplet countercurrent chromatograph and (b) Rotation locular countercurrent chromatograph.

Index

E

F

G

H

L

M

N

O

P

U

V